Mathematics and Molecular Modeling

Mathematics and Molecular Modeling offers an interdisciplinary forum for monographs and multi-author surveys related to the design, analysis, and simulation of molecular models at all scales. Topics covered range from electronic structure over molecular dynamics to deterministic and stochastic coarse-grained models as relevant for, e.g., conformation dynamics of biomolecules or evolution of defects in solids. Computational and mathematical aspects of electronic and atomic structure of materials, molecular optics, molecular scale imaging, control of molecular systems, crystal growth, supramolecular architectures and data-based modeling are also covered. Texts highlight the state of the art in a particular area, in a way accessible to graduate students and researchers in applied and computational mathematics as well as to interested researchers from other fields.

Eric Cancès • Gero Friesecke
Editors

Density Functional Theory

Modeling, Mathematical Analysis, Computational Methods, and Applications

Springer

Editors
Eric Cancès
CERMICS
Ecole des Ponts and Inria Paris
Marne-la-Vallée, France

Gero Friesecke
Department of Mathematics
Technische Universität München
Garching bei München, Germany

Mathematics and Molecular Modeling
ISBN 978-3-031-22339-6 ISBN 978-3-031-22340-2 (eBook)
https://doi.org/10.1007/978-3-031-22340-2

Mathematics Subject Classification: 81-10, 35Q40, 65K10, 65H17, 81V55, 81Q05, 81V70, 65N25

This Springer imprint is published by the registered company Springer Nature Switzerland AG
The registered company address is: Gewerbestrasse 11, 6330 Cham, Switzerland

Preface

Density functional theory (DFT) provides the most widely used models for simulating molecules and materials based on the fundamental laws of quantum mechanics. It earned its main inventor Walter Kohn a Nobel prize, and nowadays plays a central role in a huge spectrum of applications in chemistry, physics, and materials science. DFT has become an interdisciplinary field as testified by the list of authors of this book, coming from chemistry, physics, materials science, mathematics, and scientific computing.

Quantum mechanics describes a system of N interacting particles in the physical three-dimensional space by a partial differential equation in $3N$ spatial variables. Hence, standard numerical methods incur an exponential increase of computational effort with N, a phenomenon known as the curse of dimensionality; in practice these methods already fail beyond $N = 2$. DFT overcomes this problem by (1) reformulating the N-body problem involving functions of $3N$ variables in terms of the electronic density, a function of 3 variables, (2) approximating it by a pioneering hybrid approach which keeps important ab initio contributions and re-models the remainder in a data-driven way.

In fact many different types of data were used by different researchers to parameterize the remainder, ranging from numerical results for reference systems over experimental properties of atoms and small molecules to exact constraints in asymptotic regimes.

Kohn–Sham DFT models a molecular system with N electrons as follows. For simplicity we describe the common case of a spin-unpolarized system. The quantum state of the electrons is described by a set of $N/2$ doubly occupied spatial orbitals $\{\varphi_i\}_{i=1,\ldots,N/2}$, which are functions on the physical space $\mathbb{R}^3$ that are orthonormal with respect to the L^2 inner product. The total electronic energy is

$$E[\{\varphi_i\}] = \sum_{i=1}^{N/2} \int_{\mathbb{R}^3} |\nabla \varphi_i(\mathbf{r})|^2 \mathrm{d}\mathbf{r} + \int_{\mathbb{R}^3} v_{\mathrm{ne}}(\mathbf{r})\rho(\mathbf{r})\mathrm{d}\mathbf{r} + E_{\mathrm{Hxc}}[\rho]. \tag{1}$$

The first term models the kinetic energy of the electrons, v_{ne} is the electrostatic potential generated by the nuclei, and E_{Hxc} is the so-called Hartree-exchange-correlation functional which models the electron-electron interactions and interaction-induced corrections to the kinetic energy by some explicit functional of the density. The density ρ is expressed in terms of the orbitals as

$$\rho(\mathbf{r}) = 2\sum_{i=1}^{N/2} |\varphi_i(\mathbf{r})|^2 . \tag{2}$$

The Kohn–Sham ground state is obtained by minimizing the energy functional E over $\{\varphi_i\}$ subject to the orthonormality constraints $\int_{\mathbb{R}^3} \varphi_i^* \varphi_j = \delta_{ij}$. The Euler–Lagrange equations (stationarity conditions) are the *Kohn–Sham equations* [1]

$$\left(-\frac{1}{2}\nabla^2 + v_{\mathrm{ne}}(\mathbf{r}) + v_{\mathrm{Hxc}}(\mathbf{r})\right)\varphi_i(\mathbf{r}) = \varepsilon_i\varphi_i(\mathbf{r}), \tag{3}$$

where the ε_i are Lagrange multipliers and $v_{\mathrm{Hxc}}(\mathbf{r})$ is the Hartree-exchange-correlation potential defined as the functional derivative of $E_{\mathrm{Hxc}}[\rho]$ with respect to $\rho(\mathbf{r})$. The orbitals satisfying Eq. (3) are called the Kohn–Sham orbitals. Mathematically, the Kohn–Sham equations are a system of nonlinear elliptic eigenvalue problems, since the potential v_{Hxc} depends on the φ_i's through the density ρ.

There are a great many challenges associated with this system, touching upon different areas of physical and chemical modeling, mathematical analysis, and scientific computing, such as:

- Design of accurate yet computationally practical Hartree-exchange-correlation functionals, going beyond the local density approximation (LDA) introduced in Kohn and Sham's seminal 1965 paper [1] and the—currently widely used—semilocal and hybrid functionals and remaining applicable in strongly correlated regimes (stretched chemical bonds, transition metal oxides, Mott insulators).
- Design of effective numerical methods to solve the Kohn–Sham equations for large molecular systems, possibly coupled to an environment (linear scaling methods, coarse-graining, quantum-mechanics/molecular-mechanics (QM/MM), and other multilayer approaches).
- Theoretical understanding of the errors incurred both by the modeling and the numerical treatment (comparison to exact many-body quantum mechanics in scaling limits, choice of discretization or background basis, iterative algorithms with stopping criteria, adaptive floating-point arithmetic, massive parallelization for hybrid CPU/GPU, and future exascale architectures).
- Unearthing the intrinsic mathematical properties of different Kohn–Sham models in important basic situations such as bond dissociation, dispersion forces, or defects in crystals.

These challenges give rise to a variety of exciting and difficult problems for mathematicians and computational scientists, and bring one immediately to the edge of current knowledge in variational methods, nonlinear partial differential equations, large-scale numerical optimization and linear algebra, or model reduction. This book introduces and reviews the main models of DFT, covering their derivation and mathematical properties, numerical treatment, and selected applications. It intends to be an accessible yet state-of-art text on DFT for graduate students and researchers in applied and computational mathematics, physics, chemistry, and materials science. We hope that it helps to attract mathematicians and computer scientists willing to contribute to a very lively research topic and an essential tool to meet many scientific challenges of the twenty-first century, such as computer-aided drug design or computational materials engineering for green energy production and storage. We also hope that this volume will be useful to computational chemists and physicists seeking to better understand the mathematical foundations of DFT, and the state-of-the-art numerical methods used to solve the Kohn–Sham equations.

This book is organized as follows. This preface is followed by a prologue on the early days of DFT by Mel Levy, one of the pioneers of the field, who—among other things—introduced the celebrated concept of constrained search. The 11 chapters of the book can be classified into two groups:

- DFT models and their derivation, mathematical justification and analysis: Chaps. 1 (J. Toulouse), 2 (A. Savin), 3 (M. Lewin, E.H. Lieb, R. Seiringer), 4 (G. Friesecke, A. Gerolin, P. Gori-Giorgi), 5 (S. Kvaal), 6 (D. Gontier, J. Lu, C. Ortner);
- Numerical methods and applications: Chaps. 7 (E. Cancès, A. Levitt, Y. Maday, C. Yang), 8 (L. Lin, J. Lu, L. Ying), 9 (H. Chen, R. Schneider), 10 (X. Dai, A. Zhou), 11 (L. Genovese, T. Deutsch), 12 (K. Bhattacharya, V. Gavini, M. Ortiz, M. Ponga, P. Suryanarayana).

The chapters are self-contained and can be read independently, but we advise readers with mathematical or scientific computing backgrounds unfamiliar with DFT to get in shape with Sects. 1.1 and 1.3.1 in Chap. 1 (basics of DFT) and the first three sections of Chap. 7 (mathematical formulation and discretization methods), and consult the first three sections of Chap. 3 for a mathematical justification of the LDA.

Let us finally mention that this book focuses on Kohn–Sham DFT methods for electronic ground-state calculation within the Born–Oppenheimer (clamped nuclei) approximation. It does not cover time-dependent DFT (computation of excited states, linear response theory…), orbital-free DFT (simulation of warm dense matter...), nor classical DFT (classical statistical mechanics of the liquid state). Likewise, post-DFT methods (e.g., Green's function methods such as GW or Bethe–Salpeter), coupling of DFT with finer but computationally more expensive electronic structure methods for strongly correlated electrons (e.g., DFT+DMFT), coupling of DFT with molecular dynamics and implicit solvent models (multilayer models such as QM/MM), or machine learning methods trained on data generated by DFT calculations lie beyond the scope of this book. So are the other two most popular families of methods to compute ground-state electronic structures of molecules and materials, namely post-Hartree–Fock wave function methods (coupled-cluster, multiconfiguration methods, tensor network methods,...) and Monte Carlo methods. Each of these topics would be worth a book of their own in this series.

Paris, France
Munich, Germany
December 2021

Eric Cancès
Gero Friesecke

Reference

1. W. Kohn and L.J. Sham. Self-consistent equations including exchange and correlation. *Phys. Rev.* **140**, A1133 (1965).

Prologue: Early Days of Modern DFT (1964–1979)

Mel Levy

It is a pleasure to write about the early days of modern DFT from a personal perspective.

At that time, the local density approximation dominated in approximating exchange-correlation for calculations. In basic theory, there was intense interest in the definition of $F[\rho]$, the universal functional of the density for the kinetic energy plus the electron-electron repulsion energy, where, of course,

$$E_0 = \min_{\rho} \left\{ \int_{\mathbb{R}^3} \rho(\mathbf{r})\, v_{\mathrm{ne}}(\mathbf{r})\, \mathrm{d}\mathbf{r} + F[\rho] \right\}.$$

Along these lines, on a chalkboard at a special session of a theoretical chemistry conference in Boulder, Colorado, in the summer of 1975, Walter Kohn wrote out the proof, by contradiction, of the 1964 Hohenberg–Kohn theorem. As I recall, the audience applauded enthusiastically. Then there was a lively discussion about the $F[\rho]$ in this HK paper of 1964, because this functional, born by this proof by contradiction, was identified specifically for only each trial density that is a non-degenerate ground state of some extremal potential (non-degenerate v-representable). There were questions. What happens when the trial density is not some ground state? What happens when the trial density is a ground state, but is a degenerate one? These questions were in my thoughts when I began the serious study of DFT in 1978.

Then, in looking through a 1978 issue of IJQC, I noticed a long paper by Jeremy Percus. In it he derived a number of interesting Legendre transform-type bounds, but, almost as an aside, he also happened to identify a noninteracting kinetic energy functional that was expressed as the minimum kinetic energy of all ensembles of idempotent density matrices that yield a noninteracting v-representable density. Influenced by this paper, it suddenly occurred to me a couple of weeks later that the original HK functional could be identified as a constrained search. Moreover, it could also be generalized by using this constrained search. Namely,[1]

$$F[\rho] = \min_{\Psi \to \rho} \langle \Psi \,|\, \widehat{T} + \widehat{W}_{\mathrm{ee}} \,|\, \Psi \rangle \tag{1}$$

because

$$E_0 = \min_{\rho} \min_{\Psi \to \rho} \langle \Psi \,|\, \widehat{H} \,|\, \Psi \rangle \tag{2}$$

and because the same expectation value, involving v_{ne}, must be given by all wave functions that yield $\rho(\mathbf{r})$.

I remember finding expressions (1) and (2) a bit frightening as I wrote them down and stared at them, because they were so simply expressed and because I knew that they generalized DFT to include densities that are degenerate and to densities that are not necessarily v-representable. I also realized that they led to a deeper understanding of DFT. For instance, I soon realized that the philosophy behind Eqs. (1) and (2) could be applied to generalize Gilbert's functional of the electron-electron repulsion energy, as a functional of the one-matrix, to include a larger class of trial one matrices. All this appears in my 1979 paper.

Yes, the 15-year period from 1964 to 1979 was important indeed. Kohn–Sham theory and the thermal DFT of Mermin were formulated in 1965. Then von Barth and Hedin and Rajagopal and Calloway formulated spin-density functional theory in 1972 and 1973. In the mid-1970s, the adiabatic connection formula was derived by Langreth and Perdew and Gunnarsson and Lundqvist. This formula is, of course, important for the development of approximate exchange-correlation functionals. The concept of electronegativity was elucidated through DFT by Parr, Donnelly, Levy, and Palke in the 1978 paper that launched conceptual DFT. Then the time-independent ensemble theory for the density functional treatment of excited states was formulated by Theophilou in 1979.

Forty years ago, those of us immersed in DFT knew that we were gambling on a new field, especially since there were influential vocal skeptics. But we felt that there just might be good times ahead…

Mel Levy
Department of Chemistry and Quantum Theory Group
Tulane University
New Orleans, LA, USA

[1] See Sect. 1.1.1 for a precise definition of the notation.

Contents

Contributors

Kaushik Bhattacharya California Institute of Technology, Pasadena, CA, USA

Eric Cancès CERMICS, Ecole des Ponts and Inria Paris, Marne-la-Vallée, France

Huajie Chen School of Mathematical Sciences, Beijing Normal University, Beijing, China

Xiaoying Dai LSEC, Institute of Computational Mathematics and Scientific/Engineering Computing, Academy of Mathematics and Systems Science, Chinese Academy of Sciences, Beijing, China

Thierry Deutsch Univ. Grenoble Alpes, CEA, IRIG-MEM, Grenoble, France

Gero Friesecke Department of Mathematics, Technische Universität München, Garching bei München, Germany

Vikram Gavini University of Michigan, Ann Arbor, MI, USA

Luigi Genovese Univ. Grenoble Alpes, CEA, IRIG-MEM, Grenoble, France

Augusto Gerolin Department of Chemistry and Biomolecular Sciences, University of Ottawa, Ottawa, ON, Canada
Department of Mathematics and Statistics, University of Ottawa, Ottawa, ON, Canada

David Gontier CEREMADE, Université Paris-Dauphine, PSL University, Paris, France

Paola Gori-Giorgi Department of Theoretical Chemistry, Vrije Universiteit Amsterdam, Amsterdam, The Netherlands

AI4Science Microsoft Research, Amsterdam, The Netherlands

Simen Kvaal Hylleraas Centre for Quantum Molecular Sciences, Department of Chemistry, University of Oslo, Oslo, Norway

Antoine Levitt CERMICS, Ecole des Ponts and Inria Paris, Marne-la-Vallée, France

Mel Levy Department of Chemistry and Quantum Theory Group, Tulane University, New Orleans, LA, USA

Mathieu Lewin CNRS & CEREMADE, Université Paris-Dauphine, PSL University, Paris, France

Elliott H. Lieb Departments of Mathematics and Physics, Jadwin Hall, Princeton University, Princeton, NJ, USA

Lin Lin Department of Mathematics, University of California, Berkeley and Applied Mathematics and Computational Research Division, Lawrence Berkeley National Laboratory, Berkeley, CA, USA

Jianfeng Lu Departments of Mathematics, Physics, and Chemistry, Duke University, Durham, NC, USA

Yvon Maday Sorbonne Université, Université Paris-Cité, CNRS, Laboratoire Jacques-Louis Lions (LJLL), Paris, France
Institut Universitaire de France, Paris, France

Michael Ortiz California Institute of Technology, Pasadena, CA, USA

Christoph Ortner Department of Mathematics, University of British Columbia, Vancouver, BC, Canada

Mauricio Ponga University of British Columbia, Vancouver, BC, Canada

Andreas Savin Laboratoire de Chimie Théorique, CNRS and Sorbonne University, Paris, France

Reinhold Schneider Institut für Mathematik, Technische Universität Berlin, Berlin, Germany

Robert Seiringer IST Austria (Institute of Science and Technology Austria), Klosterneuburg, Austria

Phanish Suryanarayana Georgia Institute of Technology, Atlanta, GA, USA

Julien Toulouse Laboratoire de Chimie Théorique (LCT), Sorbonne Université and CNRS, Paris, France
Institut Universitaire de France, Paris, France

Chao Yang Computational Research Division, Lawrence Berkeley National Laboratory, Berkeley, CA, USA

Lexing Ying Department of Mathematics and Institute for Computational and Mathematical Engineering, Stanford University, Stanford, CA, USA

Aihui Zhou LSEC, Institute of Computational Mathematics and Scientific/Engineering Computing, Academy of Mathematics and Systems Science, Chinese Academy of Sciences, Beijing, China

List of Symbols

Parameters of the Electronic Problem

N	Number of electrons or electron pairs
N_{occ}	Number of electron occupied one-body electronic states
$M, N_{\text{nuc}}, N_{\text{n}}$	Number of nuclei
$Z_m, \mathbf{R}_m$	Nuclear charges, positions
β	Inverse temperature

Electron Coordinates

$\mathbf{r}$	Single-electron position coordinate vector
r_α	Cartesian components of single-electron position coordinate vector
$r_{ij} = \|\mathbf{r}_i - \mathbf{r}_j\|$	Euclidean distance between electrons i and j
$\mathbf{x}_i = (\mathbf{r}_i, s_i)$	Space-spin coordinates of ith electron

Wavefunctions, Density Matrices, Densities, Orbitals

$\Psi, \Psi(\mathbf{x}_1, \ldots, \mathbf{x}_N)$	Many-particle wavefunction
Γ	Many-particle density matrix
Γ_k	k-particle reduced density matrix
γ	Single-particle density matrix/1-particle reduced density matrix
$\rho, \rho(\mathbf{r})$	Single-particle density
$\rho_2, \rho_2(\mathbf{r}_1, \mathbf{r}_2)$	Pair density
Φ	Kohn–Sham wavefunction

φ_i, φ_a	Occupied/unoccupied Kohn–Sham orbitals or spin-orbitals
i, j, a, b	Summation indices: i, j for occupied orbitals, a, b for unoccupied orbitals
Φ_I	With $I = (i_1, \cdots, i_N)$, or $\lvert\varphi_{i_1} \cdots \varphi_{i_N}\rangle$ or $\varphi_{i_1} \wedge \cdots \wedge \varphi_{i_N}$: Slater determinant built from orbitals $\varphi_{i_1}, ..., \varphi_{i_N}$

Functionals

$F, F[\rho]$	General functional
E_{H}	Hartree energy functional
E_{Hxc}	Hartree-exchange-correlation energy functional
E_{xc}	Exchange-correlation energy functional
E_{x}	Exchange energy functional
E_{c}	Correlation energy functional
T	Kinetic energy functional
T_{S}	Non-interacting kinetic energy functional
F_{LL}	Levy–Lieb constrained-search functional
F_{HK}	Hohenberg–Kohn functional
F_{DM}	Constrained-search density matrix functional

Energy Levels, Eigenvalues, Energy Densities, Fermi–Dirac Function

E_0	Ground state energy
$\varepsilon_i, \varepsilon_a$	Kohn–Sham eigenvalues
μ	Chemical potential, Fermi level
e_{xc}	Exchange-correlation energy per unit volume
ϵ_{xc}	Exchange-correlation energy per particle
f_β	Fermi–Dirac function

Potentials and Kernels

v	General (single-particle) potential
v_{ext}	External potential
$v_{\mathrm{s}}, v_{\mathrm{eff}}, v_{\mathrm{KS}}$	Kohn–Sham effective potential
v_{H}	Hartree potential $v_{\mathrm{H}} = \rho \ast \lvert\cdot\rvert^{-1}$
$v_{\mathrm{x}}, v_{\mathrm{c}}, v_{\mathrm{xc}}, v_{\mathrm{Hxc}}$	Exchange, correlation, exchange-correlation, Hxc potentials
$f_{\mathrm{x}}, f_{\mathrm{c}}, f_{\mathrm{xc}}, f_{\mathrm{Hxc}}$	Exchange, correlation, exchange-correlation, Hxc kernels

Operators

A or $\widehat{A}$	Generic linear operators
A^* or $A^\dagger$	Adjoint of A
H	Many-body Hamiltonian operators
h	General single-particle Hamiltonian
h_ρ or h_γ	Kohn–Sham Hamiltonian
v_C	Coulomb operator
$\Delta = \nabla^2$	Laplace operator
χ	Reducible polarizability operator
χ_0	Irreducible polarizability operator
$\mathcal{F}$	Fourier transform
$\mathcal{F}_{KS}$	Kohn–Sham map

Matrices

$A^\top$	Transpose of A
A^*	Conjugate transpose of A

Discretization Parameters, Basis Functions, Discretized Models

N_b	Number of basis functions
N_g	Number of grid points
ξ_μ	Localized orbital (Gaussian, Slater, numerical,...)
Ξ	Set of localized orbitals
C, X	Coefficient matrices (of discretized KS orbitals in localized basis sets)
P	Discretized one-body density matrix
S	Overlap matrix
H	Hamiltonian matrices
V	Generic matrix
V_H	Hartree matrix
V_x	Exchange matrix
V_c	Correlation matrix
V_{xc}, V_{Hxc}	xc, Hxc matrices

Chapter 1
Review of Approximations for the Exchange-Correlation Energy in Density-Functional Theory

Julien Toulouse

Abstract In this chapter, we provide a review of the ground-state Kohn–Sham density-functional theory of electronic systems and some of its extensions, we present exact expressions and constraints for the exchange and correlation density functionals, and we discuss the main families of approximations for the exchange-correlation energy: semilocal approximations, single-determinant hybrid approximations, multideterminant hybrid approximations, dispersion-corrected approximations, as well as orbital-dependent exchange-correlation density functionals. The chapter aims at providing both a consistent bird's-eye view of the field and a detailed description of some of the most used approximations. It is intended to be readable by chemists/physicists and applied mathematicians.

1.1 Basics of Density-Functional Theory

1.1.1 The Many-Body Problem

We consider an N-electron system (atom or molecule) in the Born–Oppenheimer and non-relativistic approximations. The electronic Hamiltonian in the position representation is, in atomic units,

$$H = -\frac{1}{2}\sum_{i=1}^{N} \nabla_{\mathbf{r}_i}^2 + \frac{1}{2}\sum_{i=1}^{N}\sum_{\substack{j=1\\ i\neq j}}^{N} \frac{1}{|\mathbf{r}_i - \mathbf{r}_j|} + \sum_{i=1}^{N} v_{\text{ne}}(\mathbf{r}_i), \tag{1.1}$$

J. Toulouse (✉)
Laboratoire de Chimie Théorique (LCT), Sorbonne Université and CNRS, Paris, France

Institut Universitaire de France, Paris, France
e-mail: toulouse@lct.jussieu.fr

E. Cancès, G. Friesecke (eds.), *Density Functional Theory*, Mathematics and Molecular Modeling, https://doi.org/10.1007/978-3-031-22340-2_1

where $\nabla^2_{\mathbf{r}_i} = \Delta_{\mathbf{r}_i}$ is the Laplacian with respect to the electron coordinate $\mathbf{r}_i$ and $v_{\text{ne}}(\mathbf{r}_i) = -\sum_{\alpha=1}^{N_{\text{n}}} Z_\alpha/|\mathbf{r}_i - \mathbf{R}_\alpha|$ is the nuclei-electron interaction depending on the positions $\{\mathbf{R}_\alpha\}$ and charges $\{Z_\alpha\}$ of the N_{n} nuclei. The stationary electronic states are determined by the time-independent Schrödinger equation,

$$H\Psi(\mathbf{x}_1, \mathbf{x}_2, \ldots, \mathbf{x}_N) = E\Psi(\mathbf{x}_1, \mathbf{x}_2, \ldots, \mathbf{x}_N), \tag{1.2}$$

where $\Psi(\mathbf{x}_1, \mathbf{x}_2, \ldots, \mathbf{x}_N)$ is a wave function written with space-spin coordinates $\mathbf{x}_i = (\mathbf{r}_i, \sigma_i) \in \mathbb{R}^3 \times \{\uparrow, \downarrow\}$ (with $\{\uparrow, \downarrow\} \cong \mathbb{Z}_2$ being the set of spin coordinates) which is antisymmetric with respect to the exchange of two coordinates, and E is the associated energy.

Using Dirac notation, the Schrödinger equation (1.2) can be rewritten in a representation-independent formalism,

$$\hat{H} |\Psi\rangle = E |\Psi\rangle, \tag{1.3}$$

where the Hamiltonian is formally written as

$$\hat{H} = \hat{T} + \hat{W}_{\text{ee}} + \hat{V}_{\text{ne}},$$

with the kinetic-energy operator $\hat{T}$, the electron-electron interaction operator $\hat{W}_{\text{ee}}$, and the nuclei-electron interaction operator $\hat{V}_{\text{ne}}$.

The quantity of primary interest is the ground-state energy E_0. The variational theorem establishes that E_0 can be expressed as an infimum,

$$E_0 = \inf_{\Psi \in \mathcal{W}^N} \langle \Psi | \hat{H} | \Psi \rangle, \tag{1.4}$$

where the search is over the set of N-electron antisymmetric normalized wave functions Ψ having a finite kinetic energy,

$$\mathcal{W}^N = \left\{ \Psi \in \bigwedge^N L^2(\mathbb{R}^3 \times \{\uparrow, \downarrow\}; \mathbb{C}), \Psi \in H^1((\mathbb{R}^3 \times \{\uparrow, \downarrow\})^N; \mathbb{C}), \langle \Psi | \Psi \rangle = 1 \right\}, \tag{1.5}$$

where $\bigwedge^N$ is the N-fold antisymmetrized tensor product, L^2 and H^1 are the standard Lebesgue and Sobolev spaces (i.e., respectively, the space of functions that are square integrable and the space of functions that are square integrable together with their first-order derivatives), and $\langle \cdot | \cdot \rangle$ designates the L^2 inner product. Density-functional theory (DFT) is based on a reformulation of the variational theorem in terms of the one-electron density defined as[1]

$$\rho_\Psi(\mathbf{r}) = N \int_{\{\uparrow,\downarrow\}\times(\mathbb{R}^3\times\{\uparrow,\downarrow\})^{N-1}} |\Psi(\mathbf{x},\mathbf{x}_2,\ldots,\mathbf{x}_N)|^2 \, \mathrm{d}\sigma \mathrm{d}\mathbf{x}_2 \ldots \mathrm{d}\mathbf{x}_N, \tag{1.6}$$

which is normalized to the electron number, $\int_{\mathbb{R}^3} \rho_\Psi(\mathbf{r})\mathrm{d}\mathbf{r} = N$.

1.1.2 The Universal Density Functional

Building on the work of Hohenberg and Kohn [118], Levy [153] and Lieb [160] proposed to define the following universal density functional $F[\rho]$ using a *constrained-search* approach,

$$F[\rho] = \min_{\Psi\in\mathcal{W}^N_\rho} \langle\Psi|\,\hat{T} + \hat{W}_{\mathrm{ee}}\,|\Psi\rangle = \langle\Psi[\rho]|\,\hat{T} + \hat{W}_{\mathrm{ee}}\,|\Psi[\rho]\rangle\,, \tag{1.7}$$

where the minimization is done over the set of N-electron wave functions Ψ yielding the fixed density ρ [via Eq. (1.6)],

$$\mathcal{W}^N_\rho = \{\Psi \in \mathcal{W}^N, \rho_\Psi = \rho\}.$$

In Eq. (1.7), for a given density ρ, $\Psi[\rho]$ denotes a minimizing wave function, which is known to exist [160] but is possibly not unique. This so-called Levy–Lieb functional $F[\rho]$ is defined on the set of N-representable densities [160]:

$$\begin{aligned}\mathcal{D}^N &= \{\rho \mid \exists\Psi \in \mathcal{W}^N \text{ s.t. } \rho_\Psi = \rho\} \\ &= \{\rho \in L^1(\mathbb{R}^3) \mid \rho \geqslant 0, \int_{\mathbb{R}^3} \rho(\mathbf{r})\mathrm{d}\mathbf{r} = N, \sqrt{\rho} \in H^1(\mathbb{R}^3)\}.\end{aligned} \tag{1.8}$$

We note that an alternative universal density functional can be defined by a Legendre–Fenchel transformation, or equivalently by a constrained-search over N-electron ensemble density matrices [160]. This so-called Lieb functional has the advantage of being convex but in this chapter we will simply use the Levy–Lieb functional of Eq. (1.7).

The exact ground-state energy can then be expressed as

$$E_0 = \inf_{\rho\in\mathcal{D}^N} \left\{ F[\rho] + \int_{\mathbb{R}^3} v_{\mathrm{ne}}(\mathbf{r})\rho(\mathbf{r})\mathrm{d}\mathbf{r} \right\}, \tag{1.9}$$

[1] An integration over a spin coordinate σ just means a sum over the two values $\sigma \in \{\uparrow,\downarrow\}$, i.e. $\int_{\{\uparrow,\downarrow\}} \mathrm{d}\sigma = \sum_{\sigma\in\{\uparrow,\downarrow\}}$ and $\int_{\mathbb{R}^3\times\{\uparrow,\downarrow\}} \mathrm{d}\mathbf{x} = \sum_{\sigma\in\{\uparrow,\downarrow\}} \int_{\mathbb{R}^3} \mathrm{d}\mathbf{r}$.

and if a minimizer exists then it is a ground-state density $\rho_0(\mathbf{r})$ for the potential $v_{\text{ne}}(\mathbf{r})$. Hence, the ground-state energy can in principle be obtained by minimizing over the density ρ, i.e. a simple function of 3 real variables, which is a tremendous simplification compared to the minimization over a complicated many-body wave function Ψ. However, the explicit expression of $F[\rho]$ in terms of the density is not known, and the direct approximations for $F[\rho]$ that have been tried so far turn out not to be accurate enough.

If there is a unique wave function $\Psi[\rho]$ (up to a phase factor) in Eq. (1.7), we can define kinetic and potential contributions to $F[\rho]$,

$$F[\rho] = T[\rho] + W_{\text{ee}}[\rho],$$

where $T[\rho] = \langle\Psi[\rho]|\,\hat{T}\,|\Psi[\rho]\rangle$ and $W_{\text{ee}}[\rho] = \langle\Psi[\rho]|\,\hat{W}_{\text{ee}}\,|\Psi[\rho]\rangle$. The kinetic-energy functional $T[\rho]$ is the contribution which is particularly difficult to approximate as an explicit functional of the density.

1.1.3 The Kohn–Sham Scheme

1.1.3.1 Decomposition of the Universal Functional

Following the idea of Kohn and Sham (KS) [135], the difficulty of approximating $F[\rho]$ directly can be circumvented by decomposing $F[\rho]$ as

$$F[\rho] = T_{\text{s}}[\rho] + E_{\text{Hxc}}[\rho], \tag{1.10}$$

where $T_{\text{s}}[\rho]$ is the non-interacting kinetic-energy functional which can be defined with a constrained search,[2]

$$T_{\text{s}}[\rho] = \min_{\Phi\in\mathcal{S}^N_\rho} \langle\Phi|\,\hat{T}\,|\Phi\rangle = \langle\Phi[\rho]|\,\hat{T}\,|\Phi[\rho]\rangle\,, \tag{1.11}$$

where the minimization is over the set of N-electron single-determinant wave functions Φ yielding the fixed density ρ:

$$\mathcal{S}^N_\rho = \{\Phi \in \mathcal{S}^N, \rho_\Phi = \rho\}.$$

[2] It is also possible to define the non-interacting kinetic-energy functional analogously to the Levy–Lieb functional in Eq. (1.7) by minimizing over wave functions $\Psi \in \mathcal{W}^N_\rho$, i.e. $T_{\text{s,LL}}[\rho] = \min_{\Psi\in\mathcal{W}^N_\rho} \langle\Psi|\,\hat{T}\,|\Psi\rangle$ [160]. In this case, the corresponding minimizing KS wave function can generally be a linear combination of Slater determinants. However, we often have $T_{\text{s,LL}}[\rho] = T_{\text{s}}[\rho]$, in particular for densities ρ that come from a non-interacting ground-state wave function which is not degenerate. In this chapter, we will usually assume this nondegeneracy condition.

Here, $\mathcal{S}^N$ is the set of N-electron single-determinant wave functions built from orthonormal spin orbitals

$$\mathcal{S}^N = \{\Phi = \phi_1 \wedge \phi_2 \wedge \ldots \wedge \phi_N \mid \forall i \; \phi_i \in H^1(\mathbb{R}^3 \times \{\uparrow, \downarrow\}; \mathbb{C}), \; \forall i, j \; \langle \phi_i | \phi_j \rangle = \delta_{i,j}\},$$

where $\phi_1 \wedge \phi_2 \wedge \ldots \wedge \phi_N$ designates the normalized N-fold antisymmetrized tensor product of N spin orbitals. The functional $T_s[\rho]$ is defined over the entire set of N-representable densities $\mathcal{D}^N$ since any N-representable density can be obtained from a single-determinant wave function [80, 105, 160]. In Eq. (1.11), for a given density ρ, $\Phi[\rho]$ denotes a minimizing single-determinant wave function (again known to exist [160] but possibly not unique), also referred to as a *KS wave function*. The remaining functional $E_{\text{Hxc}}[\rho]$ that Eq. (1.10) defines is called the Hartree-exchange-correlation functional. The idea of the KS scheme is then to use the exact expression of $T_s[\rho]$ by reformulating the minimization over densities in Eq. (1.9) as a minimization over single-determinant wave functions Φ,

$$E_0 = \inf_{\Phi \in \mathcal{S}^N} \left\{ \langle \Phi | \hat{T} + \hat{V}_{\text{ne}} | \Phi \rangle + E_{\text{Hxc}}[\rho_\Phi] \right\}, \tag{1.12}$$

and if a minimum exists then any minimizing single-determinant wave function in Eq. (1.12) gives a ground-state density $\rho_0(\mathbf{r})$. Thus, the exact ground-state energy can in principle be obtained by minimizing over single-determinant wave functions only. Even though a wave function has been reintroduced compared to Eq. (1.9), it is only a single-determinant wave function Φ and therefore it still represents a tremendous simplification over the usual variational theorem involving a correlated (multideterminant) wave function Ψ. The advantage of Eq. (1.12) over Eq. (1.9) is that a major part of the kinetic energy can be treated exactly with the single-determinant wave function Φ, and only $E_{\text{Hxc}}[\rho]$ needs to be approximated as an explicit functional of the density.

In practice, $E_{\text{Hxc}}[\rho]$ is decomposed as

$$E_{\text{Hxc}}[\rho] = E_{\text{H}}[\rho] + E_{\text{xc}}[\rho], \tag{1.13}$$

where $E_{\text{H}}[\rho]$ is the Hartree energy functional,

$$E_{\text{H}}[\rho] = \frac{1}{2} \int_{\mathbb{R}^3 \times \mathbb{R}^3} \frac{\rho(\mathbf{r}_1)\rho(\mathbf{r}_2)}{|\mathbf{r}_1 - \mathbf{r}_2|} \mathrm{d}\mathbf{r}_1 \mathrm{d}\mathbf{r}_2, \tag{1.14}$$

representing the classical electrostatic repulsion energy for the charge distribution $\rho(\mathbf{r})$ and which is calculated exactly, and $E_{\text{xc}}[\rho]$ is the exchange-correlation energy functional that remains to be approximated. If there is a unique KS wave function $\Phi[\rho]$ (up to a phase factor), we can further decompose $E_{\text{xc}}[\rho]$ as

$$E_{\text{xc}}[\rho] = E_{\text{x}}[\rho] + E_{\text{c}}[\rho], \tag{1.15}$$

where $E_x[\rho]$ is the exchange energy functional,

$$E_x[\rho] = \langle\Phi[\rho]|\,\widehat{W}_{ee}\,|\Phi[\rho]\rangle - E_H[\rho], \tag{1.16}$$

and $E_c[\rho]$ is the correlation energy functional,

$$E_c[\rho] = \langle\Psi[\rho]|\,\widehat{T}+\widehat{W}_{ee}\,|\Psi[\rho]\rangle - \langle\Phi[\rho]|\,\widehat{T}+\widehat{W}_{ee}\,|\Phi[\rho]\rangle = T_c[\rho] + U_c[\rho],$$

which contains a kinetic contribution $T_c[\rho] = \langle\Psi[\rho]|\,\widehat{T}\,|\Psi[\rho]\rangle - \langle\Phi[\rho]|\,\widehat{T}\,|\Phi[\rho]\rangle$ and a potential contribution $U_c[\rho] = \langle\Psi[\rho]|\,\widehat{W}_{ee}\,|\Psi[\rho]\rangle - \langle\Phi[\rho]|\,\widehat{W}_{ee}\,|\Phi[\rho]\rangle$. Using the fact that $\Phi[\rho]$ is a single-determinant wave function, it can be shown that the exchange functional can be expressed as

$$E_x[\rho] = -\frac{1}{2}\sum_{\sigma\in\{\uparrow,\downarrow\}}\int_{\mathbb{R}^3\times\mathbb{R}^3}\frac{|\gamma_\sigma(\mathbf{r}_1,\mathbf{r}_2)|^2}{|\mathbf{r}_1-\mathbf{r}_2|}\mathrm{d}\mathbf{r}_1\mathrm{d}\mathbf{r}_2, \tag{1.17}$$

where γ_σ, for $\sigma\in\{\uparrow,\downarrow\}$, is the spin-dependent one-particle KS density matrix,

$$\gamma_\sigma(\mathbf{r},\mathbf{r}') \tag{1.18}$$

$$= N\int_{(\mathbb{R}^3\times\{\uparrow,\downarrow\})^{N-1}} \Phi[\rho](\mathbf{r}',\sigma,\mathbf{x}_2,\ldots,\mathbf{x}_N)^*\,\Phi[\rho](\mathbf{r},\sigma,\mathbf{x}_2,\ldots,\mathbf{x}_N)\mathrm{d}\mathbf{x}_2\ldots\mathrm{d}\mathbf{x}_N,$$

which shows that $E_x[\rho]\leqslant 0$. Moreover, from the variational definition of $F[\rho]$, we see that $E_c[\rho]\leqslant 0$.

1.1.3.2 The Kohn–Sham Equations

The single-determinant wave function Φ in Eq. (1.12) is constructed from a set of N orthonormal occupied spin-orbitals $\{\phi_i\}_{i=1,\ldots,N}$. To enforce S_z spin symmetry, each spin-orbital is factorized as $\phi_i(\mathbf{x}) = \varphi_i(\mathbf{r})\chi_{\sigma_i}(\sigma)$, where $\varphi_i\in H^1(\mathbb{R}^3,\mathbb{C})$ is a spatial orbital and χ_{σ_i} is a spin function from $\{\uparrow,\downarrow\}$ to $\{0,1\}$ such that $\forall\sigma_i,\sigma\in\{\uparrow,\downarrow\}$, $\chi_{\sigma_i}(\sigma) = \delta_{\sigma_i,\sigma}$ (σ_i is the spin of the spin-orbital i). Alternatively, when this is convenient, we will sometimes reindex the spatial orbitals, $\{\varphi_i\}\longrightarrow\{\varphi_{i\sigma}\}$, including explicitly the spin σ in the index. Writing the total electronic energy in Eq. (1.12) in terms of spin-orbitals and integrating over the spin variables, we obtain:

$$E[\{\varphi_i\}] = \frac{1}{2}\sum_{i=1}^{N}\int_{\mathbb{R}^3}|\nabla\varphi_i(\mathbf{r})|^2\mathrm{d}\mathbf{r} + \int_{\mathbb{R}^3}v_{ne}(\mathbf{r})\rho(\mathbf{r})\mathrm{d}\mathbf{r} + E_{Hxc}[\rho], \tag{1.19}$$

where the density is expressed in terms of the orbitals as

$$\rho(\mathbf{r}) = \sum_{i=1}^{N}|\varphi_i(\mathbf{r})|^2. \tag{1.20}$$

The minimization over Φ can then be recast into a minimization of $E[\{\varphi_i\}]$ with respect to the spatial orbitals $\{\varphi_i\}$ with the constraint of keeping the orbitals orthonormalized. The stationary condition with respect to variations of $\varphi_i(\mathbf{r})$ leads to the *KS equations* [135],

$$\left(-\frac{1}{2}\nabla^2 + v_{\text{ne}}(\mathbf{r}) + v_{\text{Hxc}}(\mathbf{r})\right)\varphi_i(\mathbf{r}) = \varepsilon_i\varphi_i(\mathbf{r}), \tag{1.21}$$

where ε_i is the Lagrange multiplier associated to the normalization condition of φ_i and $v_{\text{Hxc}}(\mathbf{r})$ is the Hartree-exchange-correlation potential defined as the functional derivative of $E_{\text{Hxc}}[\rho]$ with respect to $\rho(\mathbf{r})$,

$$v_{\text{Hxc}}(\mathbf{r}) = \frac{\delta E_{\text{Hxc}}[\rho]}{\delta\rho(\mathbf{r})}, \tag{1.22}$$

which is itself a functional of the density. The orbitals satisfying Eq. (1.21) are called the KS orbitals. They are the eigenfunctions of the KS one-electron Hamiltonian,

$$h_{\text{s}}(\mathbf{r}) = -\frac{1}{2}\nabla^2 + v_{\text{s}}(\mathbf{r}), \tag{1.23}$$

where

$$v_{\text{s}}(\mathbf{r}) = v_{\text{ne}}(\mathbf{r}) + v_{\text{Hxc}}(\mathbf{r}) \tag{1.24}$$

is the KS potential, and ε_i are then the KS orbital energies. Note that Eq. (1.21) constitutes a set of coupled self-consistent equations since the potential depends on all the occupied orbitals $\{\varphi_i\}_{i=1,\ldots,N}$ through the density [Eq. (1.20)]. The operator $h_{\text{s}}(\mathbf{r})$ defines the KS system which is a system of N non-interacting electrons in an effective external potential $v_{\text{s}}(\mathbf{r})$ ensuring that the density $\rho(\mathbf{r})$ in Eq. (1.20) is the same as the exact ground-state density $\rho_0(\mathbf{r})$ of the physical system of N interacting electrons. The exact ground-state energy E_0 is then obtained by injecting the KS orbitals in Eq. (1.19). The other (unoccupied) eigenfunctions in Eq. (1.21) define virtual KS orbitals $\{\varphi_a\}_{a\geqslant N+1}$.

Note that to define the potential $v_{\text{Hxc}}(\mathbf{r})$ in Eq. (1.22) a form of differentiability of the functional $E_{\text{Hxc}}[\rho]$, also referred to as v-representability of the density, has been assumed. Justifying this is in fact subtle and has been debated [56, 57, 109, 142–144, 164] (see also Chap. 5 by Kvaal in this volume). Here, we will simply assume that a form of differentiability of $E_{\text{Hxc}}[\rho]$ holds on at least a restricted set of densities that allows one to define the potential $v_{\text{Hxc}}(\mathbf{r})$ up to an additive constant. For a further restricted set of densities that should include ground-state densities of electronic Hamiltonians of molecular systems [Eq. (1.1)], it is expected that the KS potential $v_{\text{s}}(\mathbf{r})$ tends to a constant as $|\mathbf{r}| \to \infty$ and we choose this constant to be zero. Note also that the assumption of the existence of the KS potential $v_{\text{s}}(\mathbf{r})$ in Eq. (1.23), which does not depend on spin coordinates, implies that each spin-orbital must indeed have a definite S_z spin value.

Following the decomposition of $E_{\text{Hxc}}[\rho]$ in Eq. (1.13), the potential $v_{\text{Hxc}}(\mathbf{r})$ is written as

$$v_{\text{Hxc}}(\mathbf{r}) = v_{\text{H}}(\mathbf{r}) + v_{\text{xc}}(\mathbf{r}), \tag{1.25}$$

where $v_{\text{H}}(\mathbf{r}) = \delta E_{\text{H}}[\rho]/\delta\rho(\mathbf{r}) = \int_{\mathbb{R}^3} \rho(\mathbf{r}')/|\mathbf{r} - \mathbf{r}'|\text{d}\mathbf{r}'$ is the Hartree potential and $v_{\text{xc}}(\mathbf{r}) = \delta E_{\text{xc}}[\rho]/\delta\rho(\mathbf{r})$ is the exchange-correlation potential. Likewise, following the decomposition of $E_{\text{xc}}[\rho]$ in Eq. (1.15), and assuming that both $E_{\text{x}}[\rho]$ and $E_{\text{c}}[\rho]$ are differentiable with respect to ρ, the potential $v_{\text{xc}}(\mathbf{r})$ can be further decomposed as

$$v_{\text{xc}}(\mathbf{r}) = v_{\text{x}}(\mathbf{r}) + v_{\text{c}}(\mathbf{r}), \tag{1.26}$$

where $v_{\text{x}}(\mathbf{r}) = \delta E_{\text{x}}[\rho]/\delta\rho(\mathbf{r})$ is the exchange potential and $v_{\text{c}}(\mathbf{r}) = \delta E_{\text{c}}[\rho]/\delta\rho(\mathbf{r})$ is the correlation potential. Thus, the KS equations are similar to the Hartree–Fock (HF) equations, with the difference that they involve a local exchange potential $v_{\text{x}}(\mathbf{r})$ instead of the nonlocal HF exchange potential, and an additional correlation potential. At least for ground-state densities of finite molecular systems, the exchange potential has the long-range asymptotic behavior (see, e.g., Ref. [93]),

$$v_{\text{x}}(\mathbf{r}) \underset{|\mathbf{r}|\to\infty}{\sim} -\frac{1}{|\mathbf{r}|}, \tag{1.27}$$

whereas the correlation potential decays faster [4].

1.1.3.3 Extension to Spin Density-Functional Theory

To deal with an external magnetic field, DFT has been extended from the total density to spin-resolved densities [13, 203]. Without external magnetic fields, this spin density-functional theory is in principle not necessary, even for open-shell systems (see, e.g., Ref. [263]). However, the dependence on the spin densities allows one to construct approximate exchange-correlation functionals that are more accurate, and is therefore almost always used in practice for open-shell systems.

The spin density $\rho_{\sigma,\Psi}$ with $\sigma \in \{\uparrow, \downarrow\}$ associated to a wave function Ψ is defined as

$$\rho_{\sigma,\Psi}(\mathbf{r}) = N \int_{(\mathbb{R}^3\times\{\uparrow,\downarrow\})^{N-1}} |\Psi(\mathbf{x}, \mathbf{x}_2, \ldots, \mathbf{x}_N)|^2 \,\text{d}\mathbf{x}_2 \ldots \text{d}\mathbf{x}_N,$$

and integrates to the number of σ-spin electrons N_σ, i.e. $\int_{\mathbb{R}^3} \rho_{\sigma,\Psi}(\mathbf{r})\text{d}\mathbf{r} = N_\sigma$. For $\rho_\uparrow \in \mathcal{D}^{N_\uparrow}$ and $\rho_\downarrow \in \mathcal{D}^{N_\downarrow}$, the universal density functional is now defined as [200],

$$F[\rho_\uparrow, \rho_\downarrow] = \min_{\Psi \in \mathcal{W}^N_{\rho_\uparrow,\rho_\downarrow}} \langle\Psi|\hat{T} + \hat{W}_{\text{ee}}|\Psi\rangle = \langle\Psi[\rho_\uparrow, \rho_\downarrow]|\hat{T} + \hat{W}_{\text{ee}}|\Psi[\rho_\uparrow, \rho_\downarrow]\rangle, \tag{1.28}$$

where the search is over the set of normalized antisymmetric wave functions Ψ with $N = N_\uparrow + N_\downarrow$ electrons and yielding the fixed spin densities $\rho_\uparrow$ and $\rho_\downarrow$:

$$\mathcal{W}^N_{\rho_\uparrow,\rho_\downarrow} = \{\Psi \in \mathcal{W}^N, \ \rho_{\uparrow,\Psi} = \rho_\uparrow, \ \rho_{\downarrow,\Psi} = \rho_\downarrow\}.$$

In Eq. (1.28), $\Psi[\rho_\uparrow, \rho_\downarrow]$ designates a minimizing wave function.

A spin-dependent KS scheme is obtained by decomposing $F[\rho_\uparrow, \rho_\downarrow]$ as

$$F[\rho_\uparrow, \rho_\downarrow] = T_s[\rho_\uparrow, \rho_\downarrow] + E_H[\rho] + E_{xc}[\rho_\uparrow, \rho_\downarrow], \tag{1.29}$$

where $T_s[\rho_\uparrow, \rho_\downarrow]$ is defined as

$$T_s[\rho_\uparrow, \rho_\downarrow] = \min_{\Phi \in \mathcal{S}^N_{\rho_\uparrow,\rho_\downarrow}} \langle \Phi | \widehat{T} | \Phi \rangle = \langle \Phi[\rho_\uparrow, \rho_\downarrow] | \widehat{T} | \Phi[\rho_\uparrow, \rho_\downarrow] \rangle , \tag{1.30}$$

with a constrained search over the set of single-determinant wave functions Φ yielding the fixed spin densities $\rho_\uparrow$ and $\rho_\downarrow$:

$$\mathcal{S}^N_{\rho_\uparrow,\rho_\downarrow} = \{\Phi \in \mathcal{S}^N, \ \rho_{\uparrow,\Phi} = \rho_\uparrow, \ \rho_{\downarrow,\Phi} = \rho_\downarrow\}.$$

Here, $\Phi[\rho_\uparrow, \rho_\downarrow]$ denotes a minimizing KS single-determinant wave function, $E_H[\rho]$ is the Hartree energy which is a functional of the total density $\rho = \rho_\uparrow + \rho_\downarrow$ only [Eq. (1.14)], and $E_{xc}[\rho_\uparrow, \rho_\downarrow]$ is the spin-dependent exchange-correlation energy functional. The ground-state energy is then obtained as

$$E_0 = \inf_{\Phi \in \mathcal{S}^N} \left\{ \langle \Phi | \widehat{T} + \widehat{V}_{ne} | \Phi \rangle + E_H[\rho_\Phi] + E_{xc}[\rho_{\uparrow,\Phi}, \rho_{\downarrow,\Phi}] \right\}. \tag{1.31}$$

Writing the spatial orbitals of the spin-unrestricted determinant as $\{\varphi_{i\sigma}\}_{i=1,\ldots,N}$ (with the index explicitly including the spin σ now for clarity), we arrive at the spin-dependent KS equations,

$$\left(-\frac{1}{2}\nabla^2 + v_{ne}(\mathbf{r}) + v_H(\mathbf{r}) + v_{xc,\sigma}(\mathbf{r}) \right) \varphi_{i\sigma}(\mathbf{r}) = \varepsilon_{i\sigma}\varphi_{i\sigma}(\mathbf{r}), \tag{1.32}$$

with the spin-dependent exchange-correlation potential,

$$v_{xc,\sigma}(\mathbf{r}) = \frac{\delta E_{xc}[\rho_\uparrow, \rho_\downarrow]}{\delta \rho_\sigma(\mathbf{r})}, \tag{1.33}$$

and the spin density,

$$\rho_\sigma(\mathbf{r}) = \sum_{i=1}^{N_\sigma} |\varphi_{i\sigma}(\mathbf{r})|^2 . \tag{1.34}$$

As before, if there is a unique KS wave function $\Phi[\rho_\uparrow, \rho_\downarrow]$ (up to a phase factor), we can decompose $E_{xc}[\rho_\uparrow, \rho_\downarrow]$ into exchange and correlation contributions,

$$E_{xc}[\rho_\uparrow, \rho_\downarrow] = E_x[\rho_\uparrow, \rho_\downarrow] + E_c[\rho_\uparrow, \rho_\downarrow], \tag{1.35}$$

with $E_x[\rho_\uparrow, \rho_\downarrow] = \langle \Phi[\rho_\uparrow, \rho_\downarrow] | \widehat{W}_{ee} | \Phi[\rho_\uparrow, \rho_\downarrow] \rangle - E_H[\rho]$. It turns out that the spin-dependent exchange functional $E_x[\rho_\uparrow, \rho_\downarrow]$ can be exactly expressed in terms of the spin-independent exchange functional $E_x[\rho]$ [183],

$$E_x[\rho_\uparrow, \rho_\downarrow] = \frac{1}{2}\left(E_x[2\rho_\uparrow] + E_x[2\rho_\downarrow]\right), \tag{1.36}$$

which is known as the *spin-scaling relation* and stems directly from the fact the $\uparrow$- and $\downarrow$-spin electrons are uncoupled in the exchange energy [see Eq. (1.17)]. Therefore, any approximation for $E_x[\rho]$ can be easily extended to an approximation for $E_x[\rho_\uparrow, \rho_\downarrow]$. Unfortunately, there is no such relation for the spin-dependent correlation functional $E_c[\rho_\uparrow, \rho_\downarrow]$.

Obviously, in the spin-unpolarized case, i.e. $\rho_\uparrow = \rho_\downarrow = \rho/2$, this spin-dependent formalism reduces to the spin-independent one.

1.1.4 The Generalized Kohn–Sham Scheme

An important extension of the KS scheme is the so-called *generalized Kohn–Sham* (GKS) scheme [222], which recognizes that the universal density functional $F[\rho]$ of Eq. (1.7) can be decomposed in other ways than the KS decomposition of Eq. (1.10). In particular, we can decompose $F[\rho]$ as

$$F[\rho] = \min_{\Phi \in \mathcal{S}^N_\rho} \left\{ \langle \Phi | \widehat{T} | \Phi \rangle + E_H[\rho_\Phi] + S[\Phi] \right\} + \bar{S}[\rho], \tag{1.37}$$

where $S[\Phi]$ is any functional of a single-determinant wave function $\Phi \in \mathcal{S}^N$ leading to a minimum in Eq. (1.37), and $\bar{S}[\rho]$ is the corresponding complementary density functional that makes Eq. (1.37) exact. Defining the S-dependent GKS exchange-correlation functional as

$$E^S_{xc}[\Phi] = S[\Phi] + \bar{S}[\rho_\Phi], \tag{1.38}$$

we can express the exact ground-state energy as

$$E_0 = \inf_{\Phi \in \mathcal{S}^N} \left\{ \langle \Phi | \widehat{T} + \widehat{V}_{ne} | \Phi \rangle + E_H[\rho_\Phi] + E^S_{xc}[\Phi] \right\}, \tag{1.39}$$

and if a minimum exists then any minimizing single-determinant wave function in Eq. (1.39) gives a ground-state density $\rho_0(\mathbf{r})$. Similarly to the KS equations [Eq. (1.21)], Eq. (1.39) leads to the one-electron GKS equations,

$$\left(-\frac{1}{2}\nabla^2 + v_{\text{ne}}(\mathbf{r}) + v_{\text{H}}(\mathbf{r}) + v_{\bar{S}}(\mathbf{r})\right)\varphi_{i\sigma}(\mathbf{r}) + \frac{\delta S[\Phi]}{\delta\varphi^*_{i\sigma}(\mathbf{r})} = \varepsilon_{i\sigma}\varphi_{i\sigma}(\mathbf{r}), \tag{1.40}$$

where $v_{\bar{S}}(\mathbf{r}) = \delta\bar{S}[\rho]/\delta\rho(\mathbf{r})$ is a local potential and $\delta S[\Phi]/\delta\varphi^*_{i\sigma}(\mathbf{r})$ generates a one-electron (possibly nonlocal) operator.

In the special case $S[\Phi] = 0$, we recover the KS exchange-correlation density functional:

$$E^{S=0}_{\text{xc}}[\Phi] = E_{\text{xc}}[\rho_\Phi]. \tag{1.41}$$

Due to the freedom in the choice of $S[\Phi]$, there is an infinity of GKS exchange-correlation functionals $E^S_{\text{xc}}[\Phi]$ giving the exact ground-state energy via Eq. (1.39). This freedom and the fact that Φ carries more information than ρ_Φ gives the possibility to design more accurate approximations for the exchange-correlation energy.

Of course, by starting from the density functional $F[\rho_\uparrow, \rho_\downarrow]$ in Eq. (1.28), this GKS scheme can be extended to the spin-dependent case, leading to GKS exchange-correlation functionals of the form $E^S_{\text{xc}}[\Phi] = S[\Phi] + \bar{S}[\rho_{\uparrow,\Phi}, \rho_{\downarrow,\Phi}]$.

1.2 Exact Expressions and Constraints for the Kohn–Sham Exchange and Correlation Functionals

1.2.1 The Exchange and Correlation Holes

Let us consider the pair density associated with the wave function $\Psi[\rho]$ defined in Eq. (1.7),

$$\rho_2(\mathbf{r}_1, \mathbf{r}_2) \tag{1.42}$$
$$= N(N-1)\int_{\{\uparrow,\downarrow\}^2\times(\mathbb{R}^3\times\{\uparrow,\downarrow\})^{N-2}} |\Psi[\rho](\mathbf{x}_1, \mathbf{x}_2, \ldots, \mathbf{x}_N)|^2 \, \mathrm{d}\sigma_1\mathrm{d}\sigma_2\mathrm{d}\mathbf{x}_3\ldots\mathrm{d}\mathbf{x}_N,$$

which is a functional of the density, and normalized to the number of electron pairs, $\int_{\mathbb{R}^3\times\mathbb{R}^3}\rho_2(\mathbf{r}_1, \mathbf{r}_2)\mathrm{d}\mathbf{r}_1\mathrm{d}\mathbf{r}_2 = N(N-1)$. The pair density is proportional to the probability density of finding two electrons at positions $(\mathbf{r}_1, \mathbf{r}_2)$ with all the other electrons being anywhere. The pair density is useful to express the expectation value of the electron-electron interaction operator,

$$\langle\Psi[\rho]|\, \widehat{W}_{\text{ee}}\, |\Psi[\rho]\rangle = \frac{1}{2}\int_{\mathbb{R}^3\times\mathbb{R}^3} \frac{\rho_2(\mathbf{r}_1, \mathbf{r}_2)}{|\mathbf{r}_1 - \mathbf{r}_2|}\mathrm{d}\mathbf{r}_1\mathrm{d}\mathbf{r}_2. \tag{1.43}$$

Mirroring the decomposition of the Hartree-exchange-correlation energy performed in the KS scheme [Eq. (1.13)], the pair density can be decomposed as

$$\rho_2(\mathbf{r}_1, \mathbf{r}_2) = \rho(\mathbf{r}_1)\rho(\mathbf{r}_2) + \rho_{2,\text{xc}}(\mathbf{r}_1, \mathbf{r}_2). \tag{1.44}$$

The product of the densities $\rho(\mathbf{r}_1)\rho(\mathbf{r}_2)$ corresponds to the case of independent electrons [up to a change of normalization, i.e. $\int_{\mathbb{R}^3\times\mathbb{R}^3} \rho(\mathbf{r}_1)\rho(\mathbf{r}_2)\text{d}\mathbf{r}_1\text{d}\mathbf{r}_2 = N^2$ instead of $N(N-1)$] and the exchange-correlation pair density $\rho_{2,\text{xc}}(\mathbf{r}_1, \mathbf{r}_2)$ represents the modification of the pair density due to exchange and correlation effects between the electrons. It can be further written as

$$\rho_{2,\text{xc}}(\mathbf{r}_1, \mathbf{r}_2) = \rho(\mathbf{r}_1)h_{\text{xc}}(\mathbf{r}_1, \mathbf{r}_2), \tag{1.45}$$

where $h_{\text{xc}}(\mathbf{r}_1, \mathbf{r}_2)$ is the *exchange-correlation hole*. Introducing the conditional density $\rho^{\text{cond}}(\mathbf{r}_1, \mathbf{r}_2) = \rho_2(\mathbf{r}_1, \mathbf{r}_2)/\rho(\mathbf{r}_1)$ of the remaining $N-1$ electrons at $\mathbf{r}_2$ given that one electron has been found at $\mathbf{r}_1$, the exchange-correlation hole can be interpreted as the modification of $\rho^{\text{cond}}(\mathbf{r}_1, \mathbf{r}_2)$ due to exchange and correlation effects:

$$\rho^{\text{cond}}(\mathbf{r}_1, \mathbf{r}_2) = \rho(\mathbf{r}_2) + h_{\text{xc}}(\mathbf{r}_1, \mathbf{r}_2). \tag{1.46}$$

The positivity of $\rho_2(\mathbf{r}_1, \mathbf{r}_2)$ implies that

$$h_{\text{xc}}(\mathbf{r}_1, \mathbf{r}_2) \geq -\rho(\mathbf{r}_2).$$

Moreover, from Eq. (1.46), we have the following sum rule:

$$\forall \mathbf{r}_1 \in \mathbb{R}^3, \quad \int_{\mathbb{R}^3} h_{\text{xc}}(\mathbf{r}_1, \mathbf{r}_2)\text{d}\mathbf{r}_2 = -1. \tag{1.47}$$

We can separate the exchange and correlation contributions in the exchange-correlation hole. For this, consider the pair density $\rho_{2,\text{KS}}(\mathbf{r}_1, \mathbf{r}_2)$ associated with the KS single-determinant wave function $\Phi[\rho]$ defined in Eq. (1.11). It can be decomposed as

$$\rho_{2,\text{KS}}(\mathbf{r}_1, \mathbf{r}_2) = \rho(\mathbf{r}_1)\rho(\mathbf{r}_2) + \rho_{2,\text{x}}(\mathbf{r}_1, \mathbf{r}_2), \tag{1.48}$$

where $\rho_{2,\text{x}}(\mathbf{r}_1, \mathbf{r}_2)$ is the exchange pair density, which is further written as

$$\rho_{2,\text{x}}(\mathbf{r}_1, \mathbf{r}_2) = \rho(\mathbf{r}_1)h_{\text{x}}(\mathbf{r}_1, \mathbf{r}_2), \tag{1.49}$$

where $h_{\text{x}}(\mathbf{r}_1, \mathbf{r}_2)$ is the *exchange hole*. Just like the exchange-correlation hole, the exchange hole satisfies the conditions

$$h_{\text{x}}(\mathbf{r}_1, \mathbf{r}_2) \geq -\rho(\mathbf{r}_2),$$

and

$$\forall \mathbf{r}_1 \in \mathbb{R}^3, \quad \int_{\mathbb{R}^3} h_{\mathrm{x}}(\mathbf{r}_1, \mathbf{r}_2) \mathrm{d}\mathbf{r}_2 = -1. \tag{1.50}$$

Moreover, since the exchange hole can be written as [compare with Eq. (1.17)]

$$h_{\mathrm{x}}(\mathbf{r}_1, \mathbf{r}_2) = -\frac{1}{\rho(\mathbf{r}_1)} \sum_{\sigma \in \{\uparrow, \downarrow\}} |\gamma_\sigma(\mathbf{r}_1, \mathbf{r}_2)|^2, \tag{1.51}$$

where $\gamma_\sigma(\mathbf{r}_1, \mathbf{r}_2) = \sum_{i=1}^{N_\sigma} \varphi_{i\sigma}^*(\mathbf{r}_2)\varphi_{i\sigma}(\mathbf{r}_1)$ is the spin-dependent one-particle KS density matrix, it thus appears that the exchange hole is always non-positive,

$$h_{\mathrm{x}}(\mathbf{r}_1, \mathbf{r}_2) \leq 0. \tag{1.52}$$

From Eqs. (1.16), (1.43), (1.48), and (1.49), it can be seen that the exchange energy functional can be written in terms of the exchange hole,

$$E_{\mathrm{x}}[\rho] = \frac{1}{2} \int_{\mathbb{R}^3 \times \mathbb{R}^3} \frac{\rho(\mathbf{r}_1) h_{\mathrm{x}}(\mathbf{r}_1, \mathbf{r}_2)}{|\mathbf{r}_1 - \mathbf{r}_2|} \mathrm{d}\mathbf{r}_1 \mathrm{d}\mathbf{r}_2, \tag{1.53}$$

leading to the interpretation of E_{x} as the electrostatic interaction energy of an electron and its exchange hole. It is useful to write the exchange energy functional as

$$E_{\mathrm{x}}[\rho] = \int_{\mathbb{R}^3} \rho(\mathbf{r}_1) \varepsilon_{\mathrm{x}}[\rho](\mathbf{r}_1) \mathrm{d}\mathbf{r}_1, \tag{1.54}$$

where $\varepsilon_{\mathrm{x}}[\rho](\mathbf{r}_1)$ is the exchange energy density per particle,

$$\varepsilon_{\mathrm{x}}[\rho](\mathbf{r}_1) = \frac{1}{2} \int_{\mathbb{R}^3} \frac{h_{\mathrm{x}}(\mathbf{r}_1, \mathbf{r}_2)}{|\mathbf{r}_1 - \mathbf{r}_2|} \mathrm{d}\mathbf{r}_2, \tag{1.55}$$

which is itself a functional of the density. It is also convenient to define the exchange energy density $e_{\mathrm{x}}[\rho](\mathbf{r}) = \rho(\mathbf{r})\varepsilon_{\mathrm{x}}[\rho](\mathbf{r})$. For finite systems, we have the exact asymptotic behavior [18, 168]

$$\varepsilon_{\mathrm{x}}[\rho](\mathbf{r}) \underset{|\mathbf{r}| \to +\infty}{\sim} -\frac{1}{2|\mathbf{r}|}. \tag{1.56}$$

The *correlation hole* is defined as the difference

$$h_{\mathrm{c}}(\mathbf{r}_1, \mathbf{r}_2) = h_{\mathrm{xc}}(\mathbf{r}_1, \mathbf{r}_2) - h_{\mathrm{x}}(\mathbf{r}_1, \mathbf{r}_2),$$

and, from Eqs. (1.47) and (1.50), satisfies the sum rule

$$\forall \mathbf{r}_1 \in \mathbb{R}^3, \int_{\mathbb{R}^3} h_c(\mathbf{r}_1, \mathbf{r}_2)\mathrm{d}\mathbf{r}_2 = 0, \tag{1.57}$$

which implies that the correlation hole has negative and positive contributions.[3] In contrast with the exchange hole which is a smooth function of the interelectronic coordinate $\mathbf{r}_{12} = \mathbf{r}_2 - \mathbf{r}_1$, the correlation hole satisfies the electron-electron cusp condition (i.e., it has a derivative discontinuity in $\mathbf{r}_{12}$) [132, 252],

$$\forall \mathbf{r}_1 \in \mathbb{R}^3, \ h'_c(\mathbf{r}_1, \mathbf{r}_1) = h_c(\mathbf{r}_1, \mathbf{r}_1), \tag{1.58}$$

where $h'_c(\mathbf{r}_1, \mathbf{r}_1) = (\partial \widetilde{h}_c(\mathbf{r}_1, r_{12})/\partial r_{12})_{r_{12}=0}$ is the first-order derivative of the spherically averaged correlation hole $\widetilde{h}_c(\mathbf{r}_1, r_{12}) = (1/4\pi r_{12}^2) \int_{S(\mathbf{0},r_{12})} h_c(\mathbf{r}_1, \mathbf{r}_1 + \mathbf{r}_{12})\mathrm{d}\mathbf{r}_{12}$ and $S(\mathbf{0}, r_{12})$ designates the sphere centered at $\mathbf{0}$ and of radius $r_{12} = |\mathbf{r}_{12}|$. The potential contribution to the correlation energy can be written in terms of the correlation hole:

$$U_c[\rho] = \frac{1}{2} \int_{\mathbb{R}^3 \times \mathbb{R}^3} \frac{\rho(\mathbf{r}_1) h_c(\mathbf{r}_1, \mathbf{r}_2)}{|\mathbf{r}_1 - \mathbf{r}_2|} \mathrm{d}\mathbf{r}_1 \mathrm{d}\mathbf{r}_2. \tag{1.59}$$

In order to express the total correlation energy $E_c[\rho] = T_c[\rho] + U_c[\rho]$ in a form similar to Eq. (1.59), we need to introduce the adiabatic-connection formalism.

1.2.2 The Adiabatic Connection

The idea of the *adiabatic connection* [102, 146, 147] (see, also, Ref. [106]) is to have a continuous path between the non-interacting KS system and the physical system while keeping the ground-state density constant. This allows one to obtain a convenient expression for the correlation functional $E_c[\rho]$ as an integral over this path. An infinity of such paths are possible, but the one most often considered consists in switching on the electron-electron interaction linearly with a coupling constant λ. The Hamiltonian along this adiabatic connection is

$$\widehat{H}^\lambda = \widehat{T} + \lambda \widehat{W}_{ee} + \widehat{V}^\lambda, \tag{1.60}$$

where $\widehat{V}^\lambda$ is the external local potential operator imposing that the ground-state density is the same as the ground-state density of the physical system for all $\lambda \in \mathbb{R}$. Of course, Eq. (1.60) relies on a v-representability assumption, i.e. the external

[3] Therefore, the correlation hole is really a "hole" only in some region of space, and a "bump" in other regions.

potential is assumed to exist for all λ. The Hamiltonian (1.60) reduces to the KS non-interacting Hamiltonian for $\lambda = 0$ and to the physical Hamiltonian for $\lambda = 1$.

Just as for the physical system, it is possible to define a universal functional associated with the system of Eq. (1.60) for each value of the parameter λ,

$$F^{\lambda}[\rho] = \min_{\Psi \in \mathcal{W}_{\rho}^{N}} \langle \Psi | \hat{T} + \lambda \hat{W}_{ee} | \Psi \rangle = \langle \Psi^{\lambda}[\rho] | \hat{T} + \lambda \hat{W}_{ee} | \Psi^{\lambda}[\rho] \rangle , \quad (1.61)$$

where $\Psi^{\lambda}[\rho]$ denotes a minimizing wave function. This functional can be decomposed as

$$F^{\lambda}[\rho] = T_{s}[\rho] + E_{Hxc}^{\lambda}[\rho], \quad (1.62)$$

where $E_{Hxc}^{\lambda}[\rho]$ is the Hartree-exchange-correlation functional associated with the interaction $\lambda \hat{W}_{ee}$. One can write this functional as $E_{Hxc}^{\lambda}[\rho] = E_{H}^{\lambda}[\rho] + E_{x}^{\lambda}[\rho] + E_{c}^{\lambda}[\rho]$, where the Hartree and exchange contributions are simply linear in λ,

$$E_{H}^{\lambda}[\rho] = \frac{1}{2} \int_{\mathbb{R}^3 \times \mathbb{R}^3} \rho(\mathbf{r}_1) \rho(\mathbf{r}_2) \frac{\lambda}{|\mathbf{r}_1 - \mathbf{r}_2|} d\mathbf{r}_1 d\mathbf{r}_2 = \lambda E_{H}[\rho],$$

and

$$E_{x}^{\lambda}[\rho] = \langle \Phi[\rho] | \lambda \hat{W}_{ee} | \Phi[\rho] \rangle - E_{H}^{\lambda}[\rho] = \lambda E_{x}[\rho].$$

The correlation contribution is nonlinear in λ,

$$E_{c}^{\lambda}[\rho] = \langle \Psi^{\lambda}[\rho] | \hat{T} + \lambda \hat{W}_{ee} | \Psi^{\lambda}[\rho] \rangle - \langle \Phi[\rho] | \hat{T} + \lambda \hat{W}_{ee} | \Phi[\rho] \rangle . \quad (1.63)$$

We will assume that $F^{\lambda}[\rho]$ is of class C^1 as a function of λ for $\lambda \in [0, 1]$ and that $F^{\lambda=0}[\rho] = T_{s}[\rho]$, the latter condition being guaranteed for nondegenerate KS systems [see footnote on the definition of $T_{s}[\rho]$ just before Eq. (1.11)]. Taking the derivative of Eq. (1.63) with respect to λ and using the Hellmann–Feynman theorem for the wave function $\Psi^{\lambda}[\rho]$,[4] we obtain

$$\frac{\partial E_{c}^{\lambda}[\rho]}{\partial \lambda} = \langle \Psi^{\lambda}[\rho] | \hat{W}_{ee} | \Psi^{\lambda}[\rho] \rangle - \langle \Phi[\rho] | \hat{W}_{ee} | \Phi[\rho] \rangle . \quad (1.64)$$

---

[4] In this context, the Hellmann–Feynman theorem states that in the derivative

$$\frac{\partial F^{\lambda}[\rho]}{\partial \lambda} = \langle \frac{\partial \Psi^{\lambda}[\rho]}{\partial \lambda} | \hat{T} + \lambda \hat{W}_{ee} | \Psi^{\lambda}[\rho] \rangle + \langle \Psi^{\lambda}[\rho] | \hat{W}_{ee} | \Psi^{\lambda}[\rho] \rangle + \langle \Psi^{\lambda}[\rho] | \hat{T} + \lambda \hat{W}_{ee} | \frac{\partial \Psi^{\lambda}[\rho]}{\partial \lambda} \rangle$$

the first and third terms involving the derivative of $\Psi^{\lambda}[\rho]$ vanish. This is due to the fact that $\Psi^{\lambda}[\rho]$ is obtained via the minimization of Eq. (1.61) and thus any variation of $\Psi^{\lambda}[\rho]$ which keeps the density constant (which is the case for a variation with respect to λ) gives a vanishing variation of $F^{\lambda}[\rho]$.

Integrating over λ from 0 to 1, and using $E_c^{\lambda=1}[\rho] = E_c[\rho]$ and $E_c^{\lambda=0}[\rho] = 0$, we arrive at the *adiabatic-connection formula* for the correlation energy functional of the physical system

$$E_c[\rho] = \int_0^1 d\lambda \ \langle \Psi^\lambda[\rho] | \hat{W}_{ee} | \Psi^\lambda[\rho] \rangle - \langle \Phi[\rho] | \hat{W}_{ee} | \Phi[\rho] \rangle . \tag{1.65}$$

By introducing the correlation hole $h_c^\lambda(\mathbf{r}_1, \mathbf{r}_2)$ associated to the wave function $\Psi^\lambda[\rho]$, the adiabatic-connection formula for the correlation energy can also be written as

$$E_c[\rho] = \frac{1}{2} \int_0^1 d\lambda \int_{\mathbb{R}^3 \times \mathbb{R}^3} \frac{\rho(\mathbf{r}_1) h_c^\lambda(\mathbf{r}_1, \mathbf{r}_2)}{|\mathbf{r}_1 - \mathbf{r}_2|} d\mathbf{r}_1 d\mathbf{r}_2, \tag{1.66}$$

or, noting that $h_c^\lambda(\mathbf{r}_1, \mathbf{r}_2)$ is the only quantity that depends on λ in Eq. (1.66), in a more compact way,

$$E_c[\rho] = \frac{1}{2} \int_{\mathbb{R}^3 \times \mathbb{R}^3} \frac{\rho(\mathbf{r}_1) \bar{h}_c(\mathbf{r}_1, \mathbf{r}_2)}{|\mathbf{r}_1 - \mathbf{r}_2|} d\mathbf{r}_1 d\mathbf{r}_2, \tag{1.67}$$

where $\bar{h}_c(\mathbf{r}_1, \mathbf{r}_2) = \int_0^1 d\lambda \ h_c^\lambda(\mathbf{r}_1, \mathbf{r}_2)$ is the coupling-constant-integrated correlation hole. This leads to the interpretation of E_c as the electrostatic interaction energy of an electron with its coupling-constant-integrated correlation hole. As for the exchange energy, the correlation energy functional can be written as

$$E_c[\rho] = \int_{\mathbb{R}^3} \rho(\mathbf{r}_1) \varepsilon_c[\rho](\mathbf{r}_1) d\mathbf{r}_1, \tag{1.68}$$

where $\varepsilon_c[\rho](\mathbf{r}_1)$ is the correlation energy density per particle

$$\varepsilon_c[\rho](\mathbf{r}_1) = \frac{1}{2} \int_{\mathbb{R}^3} \frac{\bar{h}_c(\mathbf{r}_1, \mathbf{r}_2)}{|\mathbf{r}_1 - \mathbf{r}_2|} d\mathbf{r}_2, \tag{1.69}$$

which is a functional of the density. We can also define the correlation energy density $e_c[\rho](\mathbf{r}) = \rho(\mathbf{r}) \varepsilon_c[\rho](\mathbf{r})$.

Finally, note that the sum-rule and cusp conditions of Eqs. (1.57) and (1.58) apply to the λ-dependent correlation hole in the form

$$\forall \mathbf{r}_1 \in \mathbb{R}^3, \ \int_{\mathbb{R}^3} h_c^\lambda(\mathbf{r}_1, \mathbf{r}_2) d\mathbf{r}_2 = 0, \tag{1.70}$$

and

$$\forall \mathbf{r}_1 \in \mathbb{R}^3, \ h_c^{\lambda\,\prime}(\mathbf{r}_1, \mathbf{r}_1) = \lambda \ h_c^\lambda(\mathbf{r}_1, \mathbf{r}_1). \tag{1.71}$$

1.2.3 One-Orbital and One-Electron Spatial Regions

For systems composed of only one spin-↑ (or, symmetrically, one spin-↓) electron (e.g., the hydrogen atom) with ground-state density $\rho_{1e}(\mathbf{r}) = |\varphi_{1\uparrow}(\mathbf{r})|^2$ where $\varphi_{1\uparrow}(\mathbf{r})$ is the unique occupied KS orbital, the exchange hole in Eq. (1.51) simplifies to $h_x(\mathbf{r}_1, \mathbf{r}_2) = -\rho(\mathbf{r}_2)$, and consequently the exchange energy cancels out the Hartree energy:

$$E_x[\rho_{1e}] = -E_H[\rho_{1e}]. \tag{1.72}$$

Furthermore, the correlation energy vanishes:

$$E_c[\rho_{1e}] = 0. \tag{1.73}$$

This must of course also be true for the spin-dependent version of the functionals introduced in Sect. 1.1.3.3, i.e.

$$E_x[\rho_{1e}, 0] = -E_H[\rho_{1e}] \tag{1.74}$$

and

$$E_c[\rho_{1e}, 0] = 0. \tag{1.75}$$

For systems composed of two opposite-spin electrons (e.g., the helium atom or the dihydrogen molecule) in a unique doubly occupied KS orbital $\varphi_1(\mathbf{r}) = \varphi_{1\uparrow}(\mathbf{r}) = \varphi_{1\downarrow}(\mathbf{r})$ with ground-state density $\rho_{2e}^{\uparrow\downarrow}(\mathbf{r}) = 2|\varphi_1(\mathbf{r})|^2$, the exchange hole simplifies to $h_x(\mathbf{r}_1, \mathbf{r}_2) = -\rho(\mathbf{r}_2)/2$, and consequently the exchange energy is equal to half the opposite of the Hartree energy:

$$E_x[\rho_{2e}^{\uparrow\downarrow}] = -\frac{1}{2}E_H[\rho_{2e}^{\uparrow\downarrow}]. \tag{1.76}$$

These are constraints for the exchange and correlation density functionals in the special cases $N = 1$ and $N = 2$.

These special cases can be extended to more general systems. For systems with $N \geqslant 1$ electrons containing a spatial region $\Omega_{1o}^{\uparrow}$ in which, among the occupied KS orbitals, only one spin-↑ (or, symmetrically, one spin-↓) orbital is not zero (or, more generally, takes non-negligible values), we have again in this region

$$\forall \mathbf{r}_1, \mathbf{r}_2 \in \Omega_{1o}^{\uparrow}, \; h_x(\mathbf{r}_1, \mathbf{r}_2) = -\rho(\mathbf{r}_2),$$

and therefore the contribution to the exchange energy density per particle coming from this region must locally cancel out the contribution to the Hartree energy density per particle coming from the same region,

$$\forall \mathbf{r}_1 \in \Omega_{1o}^{\uparrow},\ \varepsilon_{\mathrm{x}}^{\Omega_{1o}^{\uparrow}}(\mathbf{r}_1) = -\varepsilon_{\mathrm{H}}^{\Omega_{1o}^{\uparrow}}(\mathbf{r}_1), \tag{1.77}$$

where

$$\varepsilon_{\mathrm{H}}^{\Omega}(\mathbf{r}_1) = (1/2)\int_{\Omega} \rho(\mathbf{r}_2)/|\mathbf{r}_1-\mathbf{r}_2| d\mathbf{r}_2 \text{ and } \varepsilon_{\mathrm{x}}^{\Omega}(\mathbf{r}_1) = (1/2)\int_{\Omega} h_{\mathrm{x}}(\mathbf{r}_1,\mathbf{r}_2)/|\mathbf{r}_1-\mathbf{r}_2| d\mathbf{r}_2.$$

Similarly, for systems with $N \geqslant 2$ electrons containing a spatial region $\Omega_{1o}^{\uparrow\downarrow}$ in which, among the occupied KS orbitals, only one doubly occupied orbital is not zero, we have in this region

$$\forall \mathbf{r}_1, \mathbf{r}_2 \in \Omega_{1o}^{\uparrow\downarrow},\ h_{\mathrm{x}}(\mathbf{r}_1,\mathbf{r}_2) = -\frac{1}{2}\rho(\mathbf{r}_2),$$

and therefore the contribution to the exchange energy density per particle coming from this region must locally be equal to half the opposite of the contribution to the Hartree energy density per particle coming from the same region,

$$\forall \mathbf{r}_1 \in \Omega_{1o}^{\uparrow\downarrow},\ \varepsilon_{\mathrm{x}}^{\Omega_{1o}^{\uparrow\downarrow}}(\mathbf{r}_1) = -\frac{1}{2}\varepsilon_{\mathrm{H}}^{\Omega_{1o}^{\uparrow\downarrow}}(\mathbf{r}_1). \tag{1.78}$$

Thus, we see, particularly clearly for these $\Omega_{1o}^{\uparrow}$ or $\Omega_{1o}^{\uparrow\downarrow}$ regions, that the Hartree functional introduces a spurious *self-interaction* contribution which must be eliminated by the exchange functional. Even though the concepts of $\Omega_{1o}^{\uparrow}$ and $\Omega_{1o}^{\uparrow\downarrow}$ regions are formal, in practice they can be approximately realized in chemical systems. For example, the unpaired electron in a radical approximately corresponds to a $\Omega_{1o}^{\uparrow}$, and an electron pair in a single covalent bond, in a lone pair, or in a core orbital approximately corresponds to a $\Omega_{1o}^{\uparrow\downarrow}$ region.

We can also consider one-electron regions Ω_{1e} that we define as[5]

$$\forall \mathbf{r}_1, \mathbf{r}_2 \in \Omega_{1e},\ \forall \lambda \in (0,1],\ \rho_2^{\lambda}(\mathbf{r}_1,\mathbf{r}_2) = 0, \tag{1.79}$$

where $\rho_2^{\lambda}(\mathbf{r}_1,\mathbf{r}_2)$ is the pair density associated to the wave function $\Psi^{\lambda}[\rho]$ along the adiabatic connection. This implies

$$\forall \mathbf{r}_1, \mathbf{r}_2 \in \Omega_{1e},\ \bar{h}_{\mathrm{xc}}(\mathbf{r}_1,\mathbf{r}_2) = -\rho(\mathbf{r}_2),$$

[5] In the definition of Eq. (1.79) we exclude the point $\lambda = 0$ in order to allow for the possibility of a discontinuity in λ there due to a degeneracy.

where $\bar{h}_{\text{xc}}(\mathbf{r}_1, \mathbf{r}_2) = h_{\text{x}}(\mathbf{r}_1, \mathbf{r}_2) + \bar{h}_{\text{c}}(\mathbf{r}_1, \mathbf{r}_2)$ and, consequently, the contribution to the exchange-correlation energy density per particle coming from this region must locally cancel out the contribution to the Hartree energy density per particle coming from the same region,

$$\forall \mathbf{r}_1 \in \Omega_{1\text{e}}, \; \varepsilon_{\text{xc}}^{\Omega_{1\text{e}}}(\mathbf{r}_1) = -\varepsilon_{\text{H}}^{\Omega_{1\text{e}}}(\mathbf{r}_1), \tag{1.80}$$

where $\varepsilon_{\text{xc}}^{\Omega}(\mathbf{r}_1) = (1/2) \int_{\Omega} \bar{h}_{\text{xc}}(\mathbf{r}_1, \mathbf{r}_2)/|\mathbf{r}_1 - \mathbf{r}_2| d\mathbf{r}_2$. For regions that are simultaneously one-electron and one-orbital regions, this simply implies that the contribution to the correlation energy must vanish,

$$\forall \mathbf{r}_1 \in \Omega_{1\text{e}} \cap \Omega_{1\text{o}}^{\uparrow}, \; \varepsilon_{\text{c}}^{\Omega_{1\text{e}} \cap \Omega_{1\text{o}}^{\uparrow}}(\mathbf{r}_1) = 0, \tag{1.81}$$

where $\varepsilon_{\text{c}}^{\Omega}(\mathbf{r}_1) = (1/2) \int_{\Omega} \bar{h}_{\text{c}}(\mathbf{r}_1, \mathbf{r}_2)/|\mathbf{r}_1 - \mathbf{r}_2| d\mathbf{r}_2$, and we say that the correlation functional must not introduce a self-interaction error. However, the definition of $\Omega_{1\text{e}}$ regions also includes the case of an electron entangled in several orbitals, such as the region around one hydrogen atom in the dissociated dihydrogen molecule. In this latter case, the Hartree functional introduces an additional spurious contribution (beyond the spurious self-interaction) which must be compensated by a *static correlation* (or *strong correlation*) contribution in the exchange-correlation functional.

1.2.4 Coordinate Scaling

1.2.4.1 Uniform Coordinate Scaling

We consider a norm-preserving uniform scaling of the spatial coordinates in the N-electron wave function along the adiabatic connection $\Psi^{\lambda}[\rho]$ [introduced in Eq. (1.61)] while leaving untouched the spin coordinates [154, 155, 157],

$$\Psi_{\gamma}^{\lambda}[\rho](\mathbf{r}_1, \sigma_1, \ldots, \mathbf{r}_N, \sigma_N) = \gamma^{3N/2} \Psi^{\lambda}[\rho](\gamma \mathbf{r}_1, \sigma_1, , \ldots, \gamma \mathbf{r}_N, \sigma_N),$$

where $\gamma \in (0, +\infty)$ is a scaling factor. The scaled wave function $\Psi_{\gamma}^{\lambda}[\rho]$ yields the scaled density

$$\rho_{\gamma}(\mathbf{r}) = \gamma^3 \rho(\gamma \mathbf{r}),$$

with $\int_{\mathbb{R}^3} \rho_{\gamma}(\mathbf{r}) d\mathbf{r} = \int_{\mathbb{R}^3} \rho(\mathbf{r}) d\mathbf{r} = N$, and minimizes $\langle \Psi | \hat{T} + \lambda \gamma \hat{W}_{\text{ee}} | \Psi \rangle$ since

$$\langle \Psi_{\gamma}^{\lambda}[\rho] | \hat{T} + \lambda \gamma \hat{W}_{\text{ee}} | \Psi_{\gamma}^{\lambda}[\rho] \rangle = \gamma^2 \langle \Psi^{\lambda}[\rho] | \hat{T} + \lambda \hat{W}_{\text{ee}} | \Psi^{\lambda}[\rho] \rangle \, .$$

We thus conclude that the scaled wave function at the density ρ and coupling constant λ corresponds to the wave function at the scaled density ρ_γ and coupling constant $\lambda\gamma$,

$$\Psi_\gamma^\lambda[\rho] = \Psi^{\lambda\gamma}[\rho_\gamma],$$

or, equivalently,

$$\Psi_\gamma^{\lambda/\gamma}[\rho] = \Psi^\lambda[\rho_\gamma],$$

and that the universal density functional satisfies the scaling relation

$$F^{\lambda\gamma}[\rho_\gamma] = \gamma^2 F^\lambda[\rho],$$

or, equivalently,

$$F^\lambda[\rho_\gamma] = \gamma^2 F^{\lambda/\gamma}[\rho]. \tag{1.82}$$

At $\lambda = 0$, we find the scaling relation of the KS wave function $\Phi[\rho]$ introduced in Sect. 1.1.3.1:

$$\Phi[\rho_\gamma] = \Phi_\gamma[\rho].$$

This directly leads to the scaling relation for the non-interacting kinetic density functional [see Eq. (1.11)],

$$T_\mathrm{s}[\rho_\gamma] = \gamma^2 T_\mathrm{s}[\rho],$$

for the Hartree density functional [see Eq. (1.14)],

$$E_\mathrm{H}[\rho_\gamma] = \gamma E_\mathrm{H}[\rho],$$

and for the exchange density functional [see Eq. (1.16)],

$$E_\mathrm{x}[\rho_\gamma] = \gamma E_\mathrm{x}[\rho]. \tag{1.83}$$

However, the correlation density functional $E_\mathrm{c}[\rho]$ has the more complicated scaling (as $F[\rho]$),

$$E_\mathrm{c}^\lambda[\rho_\gamma] = \gamma^2 E_\mathrm{c}^{\lambda/\gamma}[\rho],$$

and, in particular for $\lambda = 1$,

$$E_\mathrm{c}[\rho_\gamma] = \gamma^2 E_\mathrm{c}^{1/\gamma}[\rho]. \tag{1.84}$$

These scaling relations allow one to find the behavior of the density functionals in the high- and low-density limits. In the *high-density limit* ($\gamma \to \infty$), it can be shown from Eq. (1.84) that, for nondegenerate KS systems, the correlation functional $E_c[\rho]$ goes to a constant,

$$\lim_{\gamma \to \infty} E_c[\rho_\gamma] = E_c^{\text{GL2}}[\rho], \tag{1.85}$$

where $E_c^{\text{GL2}}[\rho]$ is the second-order Görling–Levy (GL2) correlation energy [90, 91] (see Sect. 1.7.2). This is also called the *weak-correlation limit* since in this limit the correlation energy is negligible with respect to the exchange energy which is itself negligible with respect to the non-interacting kinetic energy: $|E_c[\rho_\gamma]| = O(\gamma^0) \ll |E_x[\rho_\gamma]| = O(\gamma) \ll T_s[\rho_\gamma] = O(\gamma^2)$. Equation (1.85) is an important constraint since atomic and molecular correlation energies are often close to the high-density limit. For example, for the ground-state density of the helium atom, we have $E_c[\rho] = -0.0421$ hartree and $\lim_{\gamma\to\infty} E_c[\rho_\gamma] = -0.0467$ hartree [119].

In the *low-density limit* ($\gamma \to 0$), it can be shown from Eq. (1.82) that the Hartree-exchange-correlation energy $E_{\text{Hxc}}[\rho]$ goes to zero linearly in γ,

$$E_{\text{Hxc}}[\rho_\gamma] \underset{\gamma\to 0}{\sim} \gamma\, W_{\text{ee}}^{\text{SCE}}[\rho], \tag{1.86}$$

where $W_{\text{ee}}^{\text{SCE}}[\rho] = \inf_{\Psi \in \mathcal{W}_\rho^N} \langle \Psi | \hat{W}_{\text{ee}} | \Psi \rangle$ is the strictly-correlated-electron (SCE) functional [86, 220, 221, 223]. This is also called the *strong-interaction limit* since in this limit the Hartree-exchange-correlation energy dominates over the non-interacting kinetic energy: $E_{\text{Hxc}}[\rho_\gamma] = O(\gamma) \gg T_s[\rho_\gamma] = O(\gamma^2)$. In this limit, the electrons strictly localize relatively to each other. In particular, for the uniform-electron gas, this corresponds to the Wigner crystallization. Thus, in this limit, each electron is within a one-electron region $\Omega_{1\text{e}}$ [as defined in Eq. (1.79)]. For more information on the SCE functional, see Chap. 4 by Friesecke et al. in this volume.

1.2.4.2 Non-uniform Coordinate Scaling

We can also consider non-uniform one-dimensional or two-dimensional coordinate scalings of the density [156, 184],

$$\rho_\gamma^{(1)}(x, y, z) = \gamma \rho(\gamma x, y, z) \tag{1.87}$$

and

$$\rho_\gamma^{(2)}(x, y, z) = \gamma^2 \rho(\gamma x, \gamma y, z), \tag{1.88}$$

which also preserve the number of the electrons. These non-uniform density scalings provide constraints for the exchange and correlation functionals. In particular, in the non-uniform one-dimensional high-density limit, the exchange functional remains finite and the correlation functional vanishes [89, 154]:

$$\lim_{\gamma\to\infty} E_{\mathrm{x}}[\rho_\gamma^{(1)}] > -\infty \tag{1.89}$$

and

$$\lim_{\gamma\to\infty} E_{\mathrm{c}}[\rho_\gamma^{(1)}] = 0. \tag{1.90}$$

Also, in the non-uniform two-dimensional low-density limit, we have [89, 154]:

$$\lim_{\gamma\to 0} \frac{1}{\gamma} E_{\mathrm{x}}[\rho_\gamma^{(2)}] > -\infty \tag{1.91}$$

and

$$\lim_{\gamma\to 0} \frac{1}{\gamma} E_{\mathrm{c}}[\rho_\gamma^{(2)}] = 0. \tag{1.92}$$

The conditions of Eqs. (1.89)–(1.92) are particularly useful because they also correspond to the limit of rapidly varying densities [158].

1.2.5 Atoms in the Limit of Large Nuclear Charge

A practical realization of the uniform high-density limit is provided by atomic ions in the limit of large nuclear charge, $Z \to \infty$, at fixed electron number N (see Refs. [65, 66, 123, 233]). In this limit, the exact ground-state atomic density $\rho_{N,Z}(\mathbf{r})$ becomes the density of the isoelectronic hydrogenic (i.e., without electron-electron interaction) atom $\rho_{N,Z}^{\mathrm{H}}(\mathbf{r})$, which obeys a simple scaling with Z:

$$\rho_{N,Z}(\mathbf{r}) \underset{Z\to\infty}{\sim} \rho_{N,Z}^{\mathrm{H}}(\mathbf{r}) = Z^3 \rho_{N,Z=1}^{\mathrm{H}}(Z\mathbf{r}).$$

One can thus apply Eqs. (1.83) and (1.85) with $\gamma = Z$, which reveals that in an isoelectronic series the exchange functional scales linearly with Z,

$$E_{\mathrm{x}}[\rho_{N,Z}] \underset{Z\to\infty}{\sim} E_{\mathrm{x}}[\rho_{N,Z=1}^{\mathrm{H}}]Z, \tag{1.93}$$

and, for nondegenerate KS systems, the correlation functional saturates to a constant,

$$\lim_{Z\to\infty} E_{\mathrm{c}}[\rho_{N,Z}] = E_{\mathrm{c}}^{\mathrm{GL2}}[\rho_{N,Z=1}^{\mathrm{H}}]. \tag{1.94}$$

Equations (1.93) and (1.94) are constraints for the exchange and correlation functionals, particularly relevant for highly ionized atoms but also for the core-electron regions of heavy atoms in neutral systems.

Another very interesting limit is the one of large nuclear charge of neutral atoms, $N = Z \to \infty$ (see, e.g., Ref. [129]). In this semiclassical limit, the exact ground-state atomic density $\rho_{N,Z}(\mathbf{r})$ tends to the Thomas–Fermi (TF) density of a neutral atom $\rho_Z^{\text{TF}}(\mathbf{r})$ which has a known scaling with Z [162, 163]:

$$\rho_{Z,Z}(\mathbf{r}) \underset{Z\to\infty}{\sim} \rho_Z^{\text{TF}}(\mathbf{r}) = Z^2 \rho_{Z=1}^{\text{TF}}(Z^{1/3}\mathbf{r}). \tag{1.95}$$

In this limit, it was suggested that the exact exchange and correlation energies have the approximate large-Z asymptotic expansions [28, 30, 51]

$$E_{\text{x}}[\rho_{Z,Z}] \underset{Z\to\infty}{\sim} -A_{\text{x}} Z^{5/3} + B_{\text{x}} Z + \cdots \tag{1.96}$$

and

$$E_{\text{c}}[\rho_{Z,Z}] \underset{Z\to\infty}{\sim} -A_{\text{c}} Z \ln Z + B_{\text{c}} Z + \cdots, \tag{1.97}$$

with the coefficients $A_{\text{x}} = 0.220827$, $A_{\text{c}} = 0.020727$, $B_{\text{x}} \approx 0.224$, $B_{\text{c}} \approx 0.0372$. Recently, it was argued that there is in fact a missing term in $Z \ln Z$ in the expansion of the exchange energy in Eq. (1.96) [10, 42].

1.2.6 Lieb–Oxford Lower Bound

Lieb and Oxford derived a lower bound for the indirect Coulomb energy (i.e., the two-particle Coulomb potential energy beyond the Hartree energy) [161], which, when expressed in terms of the exchange or exchange-correlation functional, takes the form [187]

$$E_{\text{x}}[\rho] \geqslant E_{\text{xc}}[\rho] \geqslant -C_{\text{LO}} \int_{\mathbb{R}^3} \rho(\mathbf{r})^{4/3} \text{d}\mathbf{r}, \tag{1.98}$$

where the optimal (i.e., smallest) constant C_{LO} (independent of the electron number N) was originally shown to be in the range $1.23 \leqslant C_{\text{LO}} \leqslant 1.68$ [161]. The range was later successively narrowed to $1.4442 \leqslant C_{\text{LO}} \leqslant 1.5765$ [36, 41, 159, 187]. This bound is approached only in the low-density limit where the correlation energy becomes comparable to the exchange energy. Numerical results suggest that for densities of most physical systems the Lieb–Oxford lower bound on the exchange-correlation energy is far from being reached [182].

For two-electron densities, there is a specific tighter bound,

$$E_x[\rho_{2e}] \geqslant E_{xc}[\rho_{2e}] \geqslant -C_2 \int_{\mathbb{R}^3} \rho_{2e}(\mathbf{r})^{4/3} d\mathbf{r}, \tag{1.99}$$

with the best known constant $C_2 = 1.234$ [161]. For one-electron densities, an even tighter bound is known for the exchange functional [74, 161],

$$E_x[\rho_{1e}] \geqslant -C_1 \int_{\mathbb{R}^3} \rho_{1e}(\mathbf{r})^{4/3} d\mathbf{r}, \tag{1.100}$$

with the optimal constant $C_1 = 1.092$. For two-electron spin-unpolarized densities, we have $E_x[\rho_{2e}^{\uparrow\downarrow}] = 2E_x[\rho_{1e}]$ with $\rho_{1e} = \rho_{2e}^{\uparrow\downarrow}/2$, and Eq. (1.100) implies [194]

$$E_x[\rho_{2e}^{\uparrow\downarrow}] \geqslant -\frac{C_1}{2^{1/3}} \int_{\mathbb{R}^3} \rho_{2e}^{\uparrow\downarrow}(\mathbf{r})^{4/3} d\mathbf{r}, \tag{1.101}$$

which is a much tighter bound than the bounds of Eqs. (1.98) and (1.99).

1.3 Semilocal Approximations for the Exchange-Correlation Energy

We review here the different classes of *semilocal approximations* for the exchange-correlation energy.

1.3.1 The Local-Density Approximation

In the *local-density approximation* (LDA), introduced by Kohn and Sham [135], the exchange-correlation functional is approximated as

$$E_{xc}^{\text{LDA}}[\rho] = \int_{\mathbb{R}^3} e_{xc}^{\text{UEG}}(\rho(\mathbf{r})) d\mathbf{r},$$

where $e_{xc}^{\text{UEG}}(\rho)$ is the exchange-correlation energy density of the infinite *uniform electron gas* (UEG) with the density ρ. The UEG represents a family of systems of interacting electrons with an arbitrary spatially constant density $\rho \in [0, +\infty)$ that acts as a parameter. Thus, in the LDA, the exchange-correlation energy density of an inhomogeneous system at a spatial point of density $\rho(\mathbf{r})$ is approximated as the exchange-correlation energy density of the UEG of the same density.

In the spin-dependent version of LDA, sometimes specifically referred to as the local-spin-density approximation (LSDA), the exchange-correlation functional is approximated as [13]

$$E_{\mathrm{xc}}^{\mathrm{LSDA}}[\rho_\uparrow, \rho_\downarrow] = \int_{\mathbb{R}^3} e_{\mathrm{xc}}^{\mathrm{UEG}}(\rho_\uparrow(\mathbf{r}), \rho_\downarrow(\mathbf{r}))\mathrm{d}\mathbf{r},$$

where $e_{\mathrm{xc}}^{\mathrm{UEG}}(\rho_\uparrow, \rho_\downarrow)$ is the exchange-correlation energy density of the UEG with spin densities $\rho_\uparrow$ and $\rho_\downarrow$. For spin-unpolarized systems, we recover the spin-independent LDA as $E_{\mathrm{xc}}^{\mathrm{LDA}}[\rho] = E_{\mathrm{xc}}^{\mathrm{LSDA}}[\rho/2, \rho/2]$.

The function $e_{\mathrm{xc}}^{\mathrm{UEG}}$ is a sum of exchange and correlation contributions, $e_{\mathrm{xc}}^{\mathrm{UEG}} = e_{\mathrm{x}}^{\mathrm{UEG}} + e_{\mathrm{c}}^{\mathrm{UEG}}$, and it is convenient to introduce exchange and correlation energies per particle, $\varepsilon_{\mathrm{x}}^{\mathrm{UEG}}$ and $\varepsilon_{\mathrm{c}}^{\mathrm{UEG}}$, such that $e_{\mathrm{x}}^{\mathrm{UEG}} = \rho\ \varepsilon_{\mathrm{x}}^{\mathrm{UEG}}$ and $e_{\mathrm{c}}^{\mathrm{UEG}} = \rho\ \varepsilon_{\mathrm{c}}^{\mathrm{UEG}}$. The expression of the exchange energy per particle of the spin-unpolarized UEG is

$$\varepsilon_{\mathrm{x}}^{\mathrm{UEG}}(\rho) = C_{\mathrm{x}}\ \rho^{1/3}, \tag{1.102}$$

where $C_{\mathrm{x}} = -(3/4)(3/\pi)^{1/3}$, and the spin-polarized version is simply obtained from the spin-scaling relation [Eq. (1.36)], leading to

$$\varepsilon_{\mathrm{x}}^{\mathrm{UEG}}(\rho_\uparrow, \rho_\downarrow) = \varepsilon_{\mathrm{x}}^{\mathrm{UEG}}(\rho)\phi_4(\zeta),$$

where $\zeta = (\rho_\uparrow - \rho_\downarrow)/\rho$ is the spin polarization and $\phi_4(\zeta)$ is defined by the general spin-scaling function

$$\phi_n(\zeta) = \frac{(1+\zeta)^{n/3} + (1-\zeta)^{n/3}}{2}. \tag{1.103}$$

The LDA exchange functional is associated with the names of Dirac [44] and Slater [228]. For a rigorous mathematical derivation of Eq. (1.102), see Ref. [64].

The correlation energy per particle $\varepsilon_{\mathrm{c}}^{\mathrm{UEG}}(\rho_\uparrow, \rho_\downarrow)$ of the UEG cannot be calculated analytically. This quantity has been obtained numerically for a sample of densities and fitted to a parametrized function satisfying the known high- and low-density expansions. Expressed in terms of the Wigner–Seitz radius $r_{\mathrm{s}} = (3/(4\pi\rho))^{1/3}$, the first terms of the high-density expansion ($r_{\mathrm{s}} \to 0$) have the form

$$\varepsilon_{\mathrm{c}}^{\mathrm{UEG}}(\rho_\uparrow, \rho_\downarrow) = A(\zeta)\ln r_{\mathrm{s}} + B(\zeta) + C(\zeta) r_{\mathrm{s}} \ln r_{\mathrm{s}} + O(r_{\mathrm{s}}), \tag{1.104}$$

with spin-unpolarized coefficients $A(0) = (1 - \ln 2)/\pi^2$, $B(0) = -0.046921$, $C(0) = 0.009229$, and fully spin-polarized coefficients $A(1) = A(0)/2$, $B(1) = -0.025738$, $C(1) = 0.004792$. The first terms of the low-density expansion ($r_{\mathrm{s}} \to +\infty$) have the form

$$\varepsilon_{\mathrm{c}}^{\mathrm{UEG}}(\rho_\uparrow, \rho_\downarrow) = \frac{a}{r_{\mathrm{s}}} + \frac{b}{r_{\mathrm{s}}^{3/2}} + \frac{c}{r_{\mathrm{s}}^2} + O\left(\frac{1}{r_{\mathrm{s}}^{5/2}}\right), \tag{1.105}$$

where the coefficients $a = -0.895930$, $b = 1.325$, and $c = -0.365$ are assumed to be independent of ζ. The low-density limit of the UEG corresponds to the Wigner crystallization. For a recent review of results on the UEG, see Ref. [165].

The two most used parametrizations are the one of Vosko, Wilk, and Nusair (VWN) [265] and the more recent one of Perdew and Wang (PW92) [199] which we give here. In this parametrization, the UEG correlation energy per particle is estimated using the approximate spin-interpolation formula

$$\varepsilon_{\mathrm{c}}^{\mathrm{PW92}}(\rho_\uparrow, \rho_\downarrow) = \varepsilon_{\mathrm{c}}(r_{\mathrm{s}}, 0) + \alpha_{\mathrm{c}}(r_{\mathrm{s}}) \frac{f(\zeta)}{f''(0)}(1-\zeta^4) + [\varepsilon_{\mathrm{c}}(r_{\mathrm{s}}, 1) - \varepsilon_{\mathrm{c}}(r_{\mathrm{s}}, 0)] f(\zeta)\zeta^4, \tag{1.106}$$

where $\varepsilon_{\mathrm{c}}(r_{\mathrm{s}}, \zeta)$ is the UEG correlation energy per particle as a function of r_{s} and ζ, $f(\zeta) = [(1+\zeta)^{4/3} + (1-\zeta)^{4/3} - 2]/(2^{4/3} - 2)$ is a spin-scaling function borrowed from the exchange energy, and $\alpha_{\mathrm{c}}(r_{\mathrm{s}}) = (\partial^2 \varepsilon_{\mathrm{c}}(r_{\mathrm{s}}, \zeta)/\partial \zeta^2)_{\zeta=0}$ is the spin stiffness. This spin-interpolation formula was first proposed in the VWN parametrization based on a study of the ζ dependence of the UEG correlation energy per particle at the random-phase approximation (RPA) level. A unique parametrization function

$$\begin{aligned} &G(r_{\mathrm{s}}, A, \alpha_1, \beta_1, \beta_2, \beta_3, \beta_4) \\ &= -2(1+\alpha_1 r_{\mathrm{s}}) A \ln \left[1 + \frac{1}{2A\left(\beta_1 r_{\mathrm{s}}^{1/2} + \beta_2 r_{\mathrm{s}} + \beta_3 r_{\mathrm{s}}^{3/2} + \beta_4 r_{\mathrm{s}}^2\right)} \right], \end{aligned}$$

is then used for approximating $\varepsilon_{\mathrm{c}}(r_{\mathrm{s}}, 0)$, $\varepsilon_{\mathrm{c}}(r_{\mathrm{s}}, 1)$, and $-\alpha_{\mathrm{c}}(r_{\mathrm{s}})$, where

$$\begin{aligned} \varepsilon_{\mathrm{c}}(r_{\mathrm{s}}, 0) &= G(r_{\mathrm{s}}, A_0, \alpha_{1,0}, \beta_{1,0}, \beta_{2,0}, \beta_{3,0}, \beta_{4,0}), \\ \varepsilon_{\mathrm{c}}(r_{\mathrm{s}}, 1) &= G(r_{\mathrm{s}}, A_1, \alpha_{1,1}, \beta_{1,1}, \beta_{2,1}, \beta_{3,1}, \beta_{4,1}), \\ -\alpha_{\mathrm{c}}(r_{\mathrm{s}}) &= G(r_{\mathrm{s}}, A_2, \alpha_{1,2}, \beta_{1,2}, \beta_{2,2}, \beta_{3,2}, \beta_{4,2}). \end{aligned}$$

The form of G was chosen to reproduce the form of the high- and low-density expansions. The parameters A_i, $\beta_{1,i}$, and $\beta_{2,i}$ (with $i \in \{0, 1, 2\}$) are fixed by the first two terms of the high-density expansion, while the parameters $\alpha_{1,i}$, $\beta_{3,i}$, and $\beta_{4,i}$ are fitted to quantum Monte Carlo (QMC) data [32] for $\varepsilon_{\mathrm{c}}(r_{\mathrm{s}}, 0)$ and $\varepsilon_{\mathrm{c}}(r_{\mathrm{s}}, 1)$, and to an estimation of $-\alpha_{\mathrm{c}}(r_{\mathrm{s}})$ extrapolated from RPA data. The parameters are given in Table I of Ref. [199].

We now discuss the merits and deficiencies of the LDA. By construction, the LDA is of course exact in the limit of uniform densities. More relevant to atomic and molecular systems is that the LDA exchange and correlation energies are asymptotically exact in the limit of large nuclear charge of neutral atoms $N = Z \to \infty$. Indeed, in this semiclassical Thomas–Fermi limit, the LDA gives the exact coefficients A_{x} and A_{c} of the leading terms in the asymptotic expansions of

Eqs. (1.96) and (1.97) [190]. However, the coefficients of the next terms are very different: $B_x^{\text{LDA}} \approx 0$ instead of $B_x \approx 0.224$ and $B_c^{\text{LDA}} \approx -0.00451$ instead of $B_c \approx 0.0372$ [28].

Due to the scaling of the UEG exchange energy per particle,

$$\varepsilon_x^{\text{UEG}}(\gamma^3\rho_\uparrow, \gamma^3\rho_\downarrow) = \gamma\varepsilon_x^{\text{UEG}}(\rho_\uparrow, \rho_\downarrow),$$

the LDA exchange functional correctly scales linearly under uniform coordinate scaling of the density [Eq. (1.83)]. Similarly, due the scaling of the UEG correlation energy per particle in the low-density limit [Eq. (1.105)],

$$\varepsilon_c^{\text{UEG}}(\gamma^3\rho_\uparrow, \gamma^3\rho_\downarrow) \underset{\gamma\to 0}{\sim} \gamma\,\frac{a}{r_s}, \tag{1.108}$$

the LDA correlation functional correctly scales linearly under uniform coordinate scaling to the low-density limit [Eq. (1.86)]. However, from the behavior of $\varepsilon_c^{\text{UEG}}$ in the high-density limit [Eq. (1.104)],

$$\varepsilon_c^{\text{UEG}}(\gamma^3\rho_\uparrow, \gamma^3\rho_\downarrow) \underset{\gamma\to\infty}{\sim} -A(\zeta)\,\ln\gamma, \tag{1.109}$$

we see that the LDA correlation functional diverges logarithmically under uniform coordinate scaling to the high-density limit whereas the exact correlation functional goes to a constant for nondegenerate KS systems [Eq. (1.85)]. Consequently, in the limit of large nuclear charge, $Z \to \infty$, at fixed electron number N, the LDA exchange energy correctly scales linearly with Z [Eq. (1.93)], albeit with an incorrect coefficient, and the LDA correlation energy does not reproduce the exact saturation behavior [Eq. (1.94)] for a nondegenerate isoelectronic series but incorrectly diverges [193]. Also, the LDA exchange and correlation functionals do not satisfy the non-uniform scaling conditions of Eqs. (1.89)–(1.92), but instead both diverge in these limits.

The LDA can also be thought of as approximating the exchange and the (coupling-constant-integrated) correlation holes of an inhomogeneous system in Eqs. (1.55) and (1.69) by the corresponding exchange and correlation holes of the UEG. Namely, considering the spin-independent version for simplicity, the LDA exchange hole is

$$h_x^{\text{LDA}}(\mathbf{r}_1, \mathbf{r}_2) = h_x^{\text{UEG}}(\rho(\mathbf{r}_1), r_{12}), \tag{1.110}$$

with

$$h_x^{\text{UEG}}(\rho, r_{12}) = -\rho\,\frac{9}{2}\left(\frac{j_1(k_F r_{12})}{k_F r_{12}}\right)^2, \tag{1.111}$$

where $r_{12} = |\mathbf{r}_2 - \mathbf{r}_1|$ is the interelectronic distance, $k_F = (3\pi^2\rho)^{1/3}$ is the Fermi wave vector, and j_1 is the spherical Bessel function of the first kind. Similarly, the LDA correlation hole is

$$\bar{h}_c^{\text{LDA}}(\mathbf{r}_1, \mathbf{r}_2) = \bar{h}_c^{\text{UEG}}(\rho(\mathbf{r}_1), r_{12}) = \int_0^1 d\lambda\, h_c^{\lambda,\text{UEG}}(\rho(\mathbf{r}_1), r_{12}).$$

Since the UEG is a physical system, the LDA exchange hole correctly fulfills the negativity and sum-rule condition [Eqs. (1.50) and (1.52)] and the LDA correlation hole correctly fulfills the sum-rule and electron-electron cusp condition [Eqs. (1.70) and (1.71)]. This constitutes a significant merit of the LDA. However, because the LDA exchange hole $h_x^{\text{LDA}}(\mathbf{r}_1, \mathbf{r}_2)$ only depends on $\rho(\mathbf{r}_1)$ and not on $\rho(\mathbf{r}_2)$, the LDA exchange functional does not entirely eliminate the self-interaction contribution of the Hartree functional, in particular in one and two-electron systems [Eqs. (1.72) or (1.74), and (1.76)], or in one-orbital spatial regions of many-electron systems [Eqs. (1.77) and (1.78)]. Similarly, the LDA correlation functional does not vanish in one-electron systems [Eqs. (1.73) or (1.75)], or more generally in one-orbital one-electron regions [Eq. (1.81)]. Thus, the LDA introduces a self-interaction error. Moreover, the LDA exchange-correlation functional does not entirely cancel out the Hartree energy in entangled one-electron spatial regions [Eq. (1.80], i.e. it introduces a static-correlation error.

Another deficiency of the LDA is that the (spin-independent) LDA exchange potential

$$v_x^{\text{LDA}}(\mathbf{r}) = \frac{\delta E_x^{\text{LDA}}[\rho]}{\delta \rho(\mathbf{r})} = \frac{4}{3} C_x\, \rho(\mathbf{r})^{1/3},$$

decays exponentially at infinity for finite molecular systems (since the density $\rho(\mathbf{r})$ decays exponentially), i.e. much too fast in comparison to the $-1/|\mathbf{r}|$ asymptotic behavior of the exact exchange potential [Eq. (1.27)]. Since asymptotic spatial regions are dominated by the highest occupied molecular orbital (HOMO) and are thus one-orbital regions (assuming the HOMO is not degenerate), this is another signature of the incorrectness of the LDA exchange functional in these one-orbital regions.

For a review of mathematical results on the LDA, see Chap. 3 by Lewin et al. in this volume.

1.3.2 The Gradient-Expansion Approximation

The next logical step beyond the LDA is the *gradient-expansion approximation* (GEA) [135], in which the exchange-correlation functional is systematically expanded in the gradient and higher-order derivatives of the density. One way of deriving the GEA is to start from the UEG, introduce a weak and slowly-varying external potential $\delta v(\mathbf{r})$, and expand the exchange-correlation energy in terms of the gradients of the density (see, e.g., Refs. [55, 134, 166, 246]). Alternatively, one can perform a semiclassical expansion (i.e., an expansion in powers of the reduced Planck constant $\hbar$) of the exact $E_{\mathrm{xc}}[\rho]$ in terms of the gradients of the external potential and use the mapping between the potential and the density to express it in terms of the gradients of the density (see, e.g., Ref. [49]).

The spin-independent gradient expansion of the exchange functional is known up to fourth order (GEA4) [246],

$$E_{\mathrm{x}}^{\mathrm{GEA4}}[\rho] = E_{\mathrm{x}}^{\mathrm{LDA}}[\rho] + C_{\mathrm{x}}^{(2)} \int_{\mathbb{R}^3} \frac{|\nabla\rho(\mathbf{r})|^2}{\rho(\mathbf{r})^{4/3}} \mathrm{d}\mathbf{r}$$
$$+ C_{\mathrm{x},1}^{(4)} \int_{\mathbb{R}^3} \frac{|\nabla^2\rho(\mathbf{r})|^2}{\rho(\mathbf{r})^2} \mathrm{d}\mathbf{r} + C_{\mathrm{x},2}^{(4)} \int_{\mathbb{R}^3} \frac{|\nabla\rho(\mathbf{r})|^2\, \nabla^2\rho(\mathbf{r})}{\rho(\mathbf{r})^3} \mathrm{d}\mathbf{r}, \quad (1.112)$$

involving the density gradient $\nabla\rho(\mathbf{r})$ and Laplacian $\nabla^2\rho(\mathbf{r})$. Sham [224] obtained the second-order coefficient $C_{\mathrm{x,S}}^{(2)} = -7/(432\pi(3\pi^2)^{1/3}) \approx -0.001667$. The calculation was done by starting with the screened Yukawa interaction $\mathrm{e}^{-\kappa r_{12}}/r_{12}$ and taking the limit $\kappa \to 0$ at the end of the calculation. It was later shown that this calculation contains an order-of-limit problem and that the correct Coulombic second-order coefficient is $C_{\mathrm{x}}^{(2)} = -5/(216\pi(3\pi^2)^{1/3}) \approx -0.002382$ [55, 134]. The fourth-order coefficients are $C_{\mathrm{x},1}^{(4)} = -73/(64800\pi^3) \approx -0.000036$, and $C_{\mathrm{x},2}^{(4)} \approx 0.00009$, where the last one has been numerically estimated [246]. Note that each term in Eq. (1.112) correctly fulfills the scaling relation of Eq. (1.83). The spin-dependent gradient exchange expansion is simply obtained from the spin-scaling relation [Eq. (1.36)].

Similarly, Ma and Brueckner [166] obtained the spin-independent second-order gradient expansion (GEA2) of the correlation functional,

$$E_{\mathrm{c}}^{\mathrm{GEA2}}[\rho] = E_{\mathrm{c}}^{\mathrm{LDA}}[\rho] + \int_{\mathbb{R}^3} C_{\mathrm{c}}^{(2)}(r_{\mathrm{s}}(\mathbf{r})) \frac{|\nabla\rho(\mathbf{r})|^2}{\rho(\mathbf{r})^{4/3}} \mathrm{d}\mathbf{r}, \quad (1.113)$$

with a second-order coefficient in the high-density limit $C_{\mathrm{c,MB}}^{(2)}(r_{\mathrm{s}} \to 0) = 0.004235$. It is believed [149] that this calculation contains a similar order-of-limit problem as in Sham's coefficient $C_{\mathrm{x,S}}^{(2)}$, in such a way that these two coefficients must be combined to obtain the correct second-order exchange-correlation coefficient in the high-density limit $C_{\mathrm{xc}}^{(2)}(r_{\mathrm{s}} \to 0) = C_{\mathrm{x,S}}^{(2)} + C_{\mathrm{c,MB}}^{(2)}(r_{\mathrm{s}} \to 0)$. The correct second-order correlation coefficient in the high-density limit is then $C_{\mathrm{c}}^{(2)}(r_{\mathrm{s}} \to 0) = C_{\mathrm{xc}}^{(2)}(r_{\mathrm{s}} \to 0) - C_{\mathrm{x}}^{(2)} = 0.004950$. Similarly, the second-order correlation coefficient as a function of r_{s} can be obtained by $C_{\mathrm{c}}^{(2)}(r_{\mathrm{s}}) = C_{\mathrm{xc}}^{(2)}(r_{\mathrm{s}}) - C_{\mathrm{x}}^{(2)}$, where

$C_{\rm xc}^{(2)}(r_{\rm s})$ has been parametrized in Ref. [206]. The spin-dependent generalization has the form [204, 271]

$$E_{\rm c}^{\rm GEA2}[\rho_\uparrow,\rho_\downarrow] = E_{\rm c}^{\rm LSDA}[\rho_\uparrow,\rho_\downarrow] + \sum_{\sigma,\sigma'\in\{\uparrow,\downarrow\}} \int_{\mathbb{R}^3} C_{\rm c}^{\sigma,\sigma',(2)}(r_{\rm s}(\mathbf{r}),\zeta(\mathbf{r}))\frac{\nabla\rho_\sigma(\mathbf{r})}{\rho_\sigma(\mathbf{r})^{2/3}}\cdot\frac{\nabla\rho_{\sigma'}(\mathbf{r})}{\rho_{\sigma'}(\mathbf{r})^{2/3}}\mathbf{dr}, \tag{1.114}$$

where the functions $C_{\rm c}^{\sigma,\sigma',(2)}(r_{\rm s},\zeta)$ have been numerically calculated in the high-density limit [204, 205].

The GEA should improve over the LDA for sufficiently slowly varying densities. Since the spin-independent GEA2 exchange energy per particle has the form

$$\varepsilon_{\rm x}^{\rm GEA2}(\rho,\nabla\rho) = \rho^{1/3}(C_{\rm x} + C_{\rm x}^{(2)}x^2),$$

where $x = |\nabla\rho|/\rho^{4/3}$ is a dimensionless reduced density gradient, the precise condition for exchange is $x \ll 1$. Unfortunately, for real systems like atoms and molecules, the reduced density gradient x can be large in some regions of space. In particular, in the exponential density tail, $\rho(\mathbf{r}) \underset{|\mathbf{r}|\to\infty}{\propto} {\rm e}^{-\alpha|\mathbf{r}|}$, the reduced density gradient diverges $x(\mathbf{r}) \xrightarrow[|\mathbf{r}|\to\infty]{} \infty$. But this is not as bad as it seems since $\rho\,\varepsilon_{\rm x}^{\rm GEA2}$ goes to zero anyway in this limit. The situation is more catastrophic for correlation. Indeed, in the high-density limit, the spin-independent GEA2 correlation energy per particle behaves as

$$\varepsilon_{\rm c}^{\rm GEA2}(\gamma^3\rho,\gamma^4\nabla\rho) \underset{\gamma\to\infty}{\sim} -A(0)\ln\gamma + \gamma^{1/2}C_{\rm c}^{(2)}(r_{\rm s}\to 0)y^2,$$

where $y = |\nabla\rho|/\rho^{7/6}$ is another reduced density gradient adapted to correlation. Therefore, in this limit, the GEA2 correlation correction diverges to $+\infty$ even faster than the LDA diverges to $-\infty$.

Another aspect of the deficiency of the GEA is that the corresponding GEA exchange and correlation holes have unphysical long-range parts which break the negativity [Eq. (1.52)] and sum-rule conditions [Eqs. (1.50) and (1.57)].

In practice, the GEA tends to deteriorate the results obtained at the LDA level. Truncated gradient expansions should not be directly used but need to be resummed.

1.3.3 Generalized-Gradient Approximations

The failures of the GEA led to the development, which really started in the 1980s, of *generalized-gradient approximations* (GGAs) with the generic form

$$E_{\rm xc}^{\rm GGA}[\rho_\uparrow,\rho_\downarrow] = \int_{\mathbb{R}^3} e_{\rm xc}^{\rm GGA}(\rho_\uparrow(\mathbf{r}),\rho_\downarrow(\mathbf{r}),\nabla\rho_\uparrow(\mathbf{r}),\nabla\rho_\downarrow(\mathbf{r}))\mathbf{dr}, \tag{1.115}$$

where $e_{\text{xc}}^{\text{GGA}}$ is some function. The GGAs are often called *semilocal* approximations in the sense that $e_{\text{xc}}^{\text{GGA}}$ does not only use the local value of the spin densities $\rho_{\uparrow}(\mathbf{r})$ and $\rho_{\downarrow}(\mathbf{r})$ but also "semilocal information" through its gradients[6] $\nabla\rho_{\uparrow}(\mathbf{r})$ and $\nabla\rho_{\downarrow}(\mathbf{r})$.

Many GGA functionals have been proposed. They generally provide a big improvement over LDA for molecular systems. However, their accuracy is still limited, in particular by self-interaction and static-correlation errors. We review here some of the most widely used GGA functionals.

B88 Exchange Functional

In the Becke 88 (B88 or B) exchange functional [18], the exchange energy density is written as

$$e_{\text{x}}^{\text{B88}}(\rho_{\uparrow}, \rho_{\downarrow}, \nabla\rho_{\uparrow}, \nabla\rho_{\downarrow}) = e_{\text{x}}^{\text{UEG}}(\rho_{\uparrow}, \rho_{\downarrow}) - \sum_{\sigma\in\{\uparrow,\downarrow\}} \rho_{\sigma}^{4/3} \frac{\beta x_{\sigma}^2}{1 + 6\beta x_{\sigma} \sinh^{-1}(x_{\sigma})}, \tag{1.116}$$

where $x_{\sigma} = |\nabla\rho_{\sigma}|/\rho_{\sigma}^{4/3}$. The fact that $e_{\text{x}}^{\text{B88}}$ depends linearly on $\rho_{\sigma}^{4/3}$ and nonlinearly only on the dimensionless reduced density gradient $x_{\sigma}(\mathbf{r})$ guarantees the scaling relation of Eq. (1.83). Using the exponential decay of the ground-state spin densities of Coulombic systems, $\rho_{\sigma}(\mathbf{r}) \underset{|\mathbf{r}|\to\infty}{\propto} \mathrm{e}^{-\alpha_{\sigma}|\mathbf{r}|}$, it can be verified that the chosen form for $e_{\text{x}}^{\text{B88}}$ satisfies the exact asymptotic behavior of the exchange energy density per particle [Eq. (1.56)], although the corresponding exchange potential does not satisfy the exact $-1/r$ asymptotic behavior [Eq. (1.27)] [54]. For small x_{σ}, $e_{\text{x}}^{\text{B88}}$ is correctly quadratic in x_{σ}. The parameter $\beta = 0.0042$ was found by fitting to HF exchange energies of rare-gas atoms. A very similar value of β can also be found by imposing the coefficient B_{x} of the approximate large-Z asymptotic expansion of the exchange energy of neutral atoms [Eq. (1.96)] [51]. It turns out that imposing the coefficient of the second-order gradient expansion [Eq. (1.112)] would lead to a value of β about two times smaller and would greatly deteriorate the accuracy of the functional for atoms and molecules.

LYP Correlation Functional

The Lee–Yang–Parr (LYP) [150] correlation functional is one of the rare functionals which have not been constructed starting from LDA. It originates from the Colle–Salvetti [38] correlation-energy approximation depending on the curvature of the HF hole. By using a gradient-expansion approximation of the curvature of the HF hole, LYP turned the Colle–Salvetti expression into a density functional depending on the density, the density gradient, and the Laplacian of the density. The dependence on the Laplacian of the density can be exactly eliminated by an integration by

[6] For generality and simplicity, we consider here that the GGAs depend on the spin density gradients $\nabla\rho_{\uparrow}$ and $\nabla\rho_{\downarrow}$, but due to rotational invariance GGAs actually depend only on the scalar quantities $(\nabla\rho_{\uparrow})^2$, $(\nabla\rho_{\downarrow})^2$, and $\nabla\rho_{\uparrow}\cdot\nabla\rho_{\downarrow}$.

parts [174], giving the following correlation energy density

$$
\begin{aligned}
e_{\mathrm{c}}^{\mathrm{LYP}}(\rho_\uparrow, \rho_\downarrow, \nabla\rho_\uparrow, \nabla\rho_\downarrow) = & -a\frac{4}{1+d\rho^{-1/3}}\frac{\rho_\uparrow\rho_\downarrow}{\rho} - a\, b\, \omega(\rho)\Biggl\{\rho_\uparrow\rho_\downarrow \\
& \times\Biggl[\sum_{\sigma\in\{\uparrow,\downarrow\}}\left(2^{11/3}C_{\mathrm{F}}\rho_\sigma^{8/3} - \left(\frac{5}{2} - \frac{\delta(\rho)}{18}\right)|\nabla\rho_\sigma|^2\right. \\
& \left. - \frac{\delta(\rho)-11}{9}\frac{\rho_\sigma}{\rho}|\nabla\rho_\sigma|^2\right) \\
& + \left(\frac{47}{18} - \frac{7\delta(\rho)}{18}\right)|\nabla\rho|^2\Biggr] - \frac{2}{3}\rho^2|\nabla\rho|^2 \\
& + \left(\frac{2}{3}\rho^2 - \rho_\uparrow^2\right)|\nabla\rho_\downarrow|^2 + \left(\frac{2}{3}\rho^2 - \rho_\downarrow^2\right)|\nabla\rho_\uparrow|^2\Biggr\},
\end{aligned}
$$

where $\omega(\rho) = \rho^{-11/3}\exp(-c\rho^{-1/3})/(1+d\rho^{-1/3})$, $\delta(\rho) = c\rho^{-1/3} + d\rho^{-1/3}/(1+d\rho^{-1/3})$, and $C_{\mathrm{F}} = (3/10)(3\pi^2)^{2/3}$. The parameters $a = 0.04918$, $b = 0.132$, $c = 0.2533$, and $d = 0.349$ were obtained in the original Colle–Salvetti expression by a fit to Helium data. Note that the LYP correlation energy vanishes for fully spin-polarized densities ($\rho_\uparrow = 0$ or $\rho_\downarrow = 0$) and therefore correctly vanishes for one-electron systems [Eq. (1.75)].

PW91 Exchange-Correlation Functional

The Perdew–Wang 91 (PW91) (see Refs. [29, 187, 189]) exchange-correlation functional is based on a model of the exchange hole $h_{\mathrm{x}}(\mathbf{r}_1, \mathbf{r}_2)$ in Eq. (1.55) and of the coupling-constant-integrated correlation hole $\bar{h}_{\mathrm{c}}(\mathbf{r}_1, \mathbf{r}_2)$ in Eq. (1.69). The idea is to start from the GEA model of these holes given as gradient expansions and remove the unrealistic long-range parts of these holes to restore important constraints satisfied by the LDA. Specifically, the spurious positive parts of the GEA exchange hole are removed to enforce the negativity condition of Eq. (1.52) and a cutoff in $|\mathbf{r}_1 - \mathbf{r}_2|$ is applied to enforce the normalization condition of Eq. (1.50). Similarly, a cutoff is applied on the GEA correlation hole to enforce the condition that the hole integrates to zero [Eq. (1.70)]. The exchange and correlation energies per particle calculated from these numerical holes are then fitted to functions of the density and density gradient chosen to satisfy a number of exact constraints.

The spin-independent PW91 exchange energy density is written as

$$
e_{\mathrm{x}}^{\mathrm{PW91}}(\rho, \nabla\rho) = e_{\mathrm{x}}^{\mathrm{UEG}}(\rho) F_{\mathrm{x}}^{\mathrm{PW91}}(s), \tag{1.117}
$$

where the so-called enhancement factor is

$$
F_{\mathrm{x}}^{\mathrm{PW91}}(s) = \frac{1 + 0.19645 s \sinh^{-1}(7.7956 s) + [0.2743 - 0.1508\exp(-100 s^2)] s^2}{1 + 0.19645 s \sinh^{-1}(7.7956 s) + 0.004 s^4}, \tag{1.118}
$$

with the reduced density gradient $s = |\nabla\rho|/(2k_F\rho) = x/[2(3\pi^2)^{1/3}]$ where $k_F = (3\pi^2\rho)^{1/3}$ is the Fermi wave vector. The spin-dependent PW91 exchange energy density is simply obtained from the spin-scaling relation [Eq. (1.36)]: $e_x^{PW91}(\rho_\uparrow, \rho_\downarrow, \nabla\rho_\uparrow, \nabla\rho_\downarrow) = [e_x^{PW91}(2\rho_\uparrow, 2\nabla\rho_\uparrow) + e_x^{PW91}(2\rho_\downarrow, 2\nabla\rho_\downarrow)]/2$. The enhancement factor $F_x^{PW91}(s)$ satisfies the second-order gradient expansion [Eq. (1.112)], $F_x^{PW91}(s) = 1 + \mu\ s^2 + O(s^4)$ with $\mu = -16\pi(\pi/3)^{2/3}C_x^{(2)} = 10/81 \approx 0.1235$, the local Lieb–Oxford bound, $F_x^{PW91}(s) \leqslant -C_{LO}/(C_x 2^{1/3}) \approx 1.804$, which is a sufficient and necessary condition for a spin-dependent GGA exchange functional to satisfy the Lieb–Oxford lower bound [Eq. (1.98)] for all densities [194] (note however that 1.804 is not an optimal bound), and the condition $\lim_{s\to\infty} s^{1/2}F_x^{PW91}(s) < \infty$ which guarantees the non-uniform scaling finiteness conditions of Eqs. (1.89) and (1.91) [158, 194].

The PW91 correlation energy density is written as

$$e_c^{PW91}(\rho_\uparrow, \rho_\downarrow, \nabla\rho_\uparrow, \nabla\rho_\downarrow) = \rho\left[\varepsilon_c^{UEG}(\rho_\uparrow, \rho_\downarrow) + H^{PW91}(\rho_\uparrow, \rho_\downarrow, t)\right], \quad (1.119)$$

where the gradient correction $H^{PW91}(\rho_\uparrow, \rho_\downarrow, t) = H_0(\rho_\uparrow, \rho_\downarrow, t) + H_1(\rho_\uparrow, \rho_\downarrow, t)$ depends on another reduced density gradient (adapted to correlation) $t = |\nabla\rho|/(2\phi_2(\zeta)k_s\rho) = y/[4\phi_2(\zeta)(3/\pi)^{1/6}]$ where $k_s = \sqrt{4k_F/\pi}$ is the Thomas–Fermi screening wave vector and the spin-scaling function $\phi_2(\zeta)$ is defined by Eq. (1.103), with

$$H_0(\rho_\uparrow, \rho_\downarrow, t) = \phi_2(\zeta)^3\frac{\beta^2}{2\alpha}\ln\left[1 + \frac{2\alpha}{\beta}t^2\frac{1 + \mathcal{A}t^2}{1 + \mathcal{A}t^2 + \mathcal{A}^2t^4}\right],$$

$$\mathcal{A} = \frac{2\alpha}{\beta}\left[\exp(-2\alpha\varepsilon_c^{UEG}(\rho_\uparrow, \rho_\downarrow)/(\phi_2(\zeta)^3\beta^2)) - 1\right]^{-1},$$

$$H_1(\rho_\uparrow, \rho_\downarrow, t)$$
$$= 16\left(\frac{3}{\pi}\right)^{1/3}[C_{xc}(r_s) - C_{c,MB}^{(2)}(r_s \to 0) - C_x^{(2)}]\phi_2(\zeta)^3t^2e^{-100\phi_2(\zeta)^4k_s^2t^2/k_F^2},$$

and $C_{xc}(r_s)$ is taken from Ref. [206]. The function $H_0(\rho_\uparrow, \rho_\downarrow, t)$ was chosen so that it fulfills the second-order gradient expansion [Eq. (1.114)], $H_0(\rho_\uparrow, \rho_\downarrow, t) = \beta\phi_2(\zeta)^3t^2 + O(t^4)$, using an approximate ζ dependence [271] and the Ma–Brueckner high-density-limit second-order coefficient [166] $\beta = 16(3/\pi)^{1/3}C_{c,MB}^{(2)}(r_s \to 0) \approx 0.06673$, and so that it cancels the LDA correlation in the large-t limit, $\lim_{t\to\infty} H_0(\rho_\uparrow, \rho_\downarrow, t) = -\varepsilon_c^{UEG}(\rho_\uparrow, \rho_\downarrow)$. The only fitted parameter is $\alpha = 0.09$. The function $H_1(\rho_\uparrow, \rho_\downarrow, t)$ only serves to restore the correct second-order gradient expansion, such that $H^{PW91}(\rho_\uparrow, \rho_\downarrow, t) = 16(3/\pi)^{1/3}C_{xc}(r_s)\phi_2(\zeta)^3t^2 + O(t^4)$, while keeping the large-t limit unchanged.

PBE Exchange-Correlation Functional

The Perdew–Burke–Ernzerhof (PBE) [188] exchange-correlation functional is a simplification of the PW91 functional with no fitted parameters which gives almost the same energies. The spin-independent PBE exchange energy density is written as

$$e_{\mathrm{x}}^{\mathrm{PBE}}(\rho, \nabla\rho) = e_{\mathrm{x}}^{\mathrm{UEG}}(\rho) F_{\mathrm{x}}^{\mathrm{PBE}}(s), \tag{1.120}$$

where the enhancement factor is

$$F_{\mathrm{x}}^{\mathrm{PBE}}(s) = 1 + \kappa - \frac{\kappa}{1 + \mu s^2/\kappa}. \tag{1.121}$$

The function $F_{\mathrm{x}}^{\mathrm{PBE}}(s)$ has the second-order gradient expansion $F_{\mathrm{x}}^{\mathrm{PBE}}(s) = 1 + \mu\ s^2 + O(s^4)$, and the parameter is chosen as $\mu = 16\pi(\pi/3)^{2/3} C_{\mathrm{c,MB}}^{(2)}(r_{\mathrm{s}} \to 0) \approx 0.21951$ so as to cancel the correlation second-order gradient expansion. The second parameter κ is chosen so as to saturate the local Lieb–Oxford bound, i.e. $\lim_{s\to\infty} F_{\mathrm{x}}^{\mathrm{PBE}}(s) = 1 + \kappa = -C_{\mathrm{LO}}/(C_{\mathrm{x}} 2^{1/3}) \approx 1.804$, leading to $\kappa = 0.804$. The same exchange functional form was in fact proposed earlier in the Becke 86 (B86) functional [17] with empirical parameters ($\mu = 0.235$, $\kappa = 0.967$).

A revised version of the PBE exchange functional, called revPBE, was proposed where the local Lieb–Oxford bound constraint is relaxed and the parameter $\kappa = 1.245$ is found instead by fitting to exchange-only total atomic energies for He and Ar, resulting in more accurate atomic total energies and molecular atomization energies [276]. Another revised version of the PBE exchange functional, called RPBE, was also proposed to achieve a similar improvement, while still enforcing the local Lieb–Oxford bound, by changing the form of the enhancement factor to $F_{\mathrm{x}}^{\mathrm{RPBE}}(s) = 1 + \kappa(1 - \exp(-\mu s^2/\kappa))$ with the same parameters as in the original PBE [103].

The PBE correlation energy density is written as

$$e_{\mathrm{c}}^{\mathrm{PBE}}(\rho_\uparrow, \rho_\downarrow, \nabla\rho_\uparrow, \nabla\rho_\downarrow) = \rho \left[\varepsilon_{\mathrm{c}}^{\mathrm{UEG}}(\rho_\uparrow, \rho_\downarrow) + H^{\mathrm{PBE}}(\rho_\uparrow, \rho_\downarrow, t) \right], \tag{1.122}$$

with the gradient correction

$$H^{\mathrm{PBE}}(\rho_\uparrow, \rho_\downarrow, t) = A(0)\phi_2(\zeta)^3 \ln\left[1 + \frac{\beta}{A(0)} t^2 \frac{1 + \mathcal{A}t^2}{1 + \mathcal{A}t^2 + \mathcal{A}^2 t^4}\right]$$

and

$$\mathcal{A} = \frac{\beta}{A(0)} \left[\exp(-\varepsilon_{\mathrm{c}}^{\mathrm{UEG}}(\rho_\uparrow, \rho_\downarrow)/(A(0)\phi_2(\zeta)^3)) - 1 \right]^{-1}.$$

As in the PW91 correlation functional, the function $H^{\mathrm{PBE}}(\rho_\uparrow, \rho_\downarrow, t)$ has the second-order gradient expansion $H^{\mathrm{PBE}}(\rho_\uparrow, \rho_\downarrow, t) = \beta\phi_2(\zeta)^3 t^2 + O(t^4)$ where

$\beta = 16(3/\pi)^{1/3}C^{(2)}_{\text{c,MB}}(r_{\text{s}} \to 0) \approx 0.06673$, and it cancels the LDA correlation in the large-t limit, $\lim_{t\to\infty} H^{\text{PBE}}(\rho_\uparrow, \rho_\downarrow, t) = -\varepsilon_{\text{c}}^{\text{UEG}}(\rho_\uparrow, \rho_\downarrow)$. In contrast with the PW91 correlation functional, under uniform coordinate scaling to the high-density limit, the PBE correlation functional correctly cancels out the logarithm divergence of the LDA correlation functional [Eq. (1.109)], i.e. $H^{\text{PBE}}(\gamma^3\rho_\uparrow, \gamma^3\rho_\downarrow, \gamma^{1/2}t) \underset{\gamma\to\infty}{\sim} A(0)\phi_2(\zeta)^3 \ \ln\gamma$, where $A(0)\phi_2(\zeta)^3$ is a good approximation to the coefficient $A(\zeta)$ [271].

A variant of the PBE exchange-correlation functional, called PBEsol [196], targeted for solid-state systems, was proposed where the correct second-order exchange gradient-expansion coefficient is restored, i.e. $\mu_{\text{PBEsol}} = 16\pi(\pi/3)^{2/3}C_{\text{x}}^{(2)} = 10/81 \approx 0.1235$, and the second-order correlation gradient-expansion coefficient $\beta_{\text{PBEsol}} = 0.046$ is found by fitting to jellium surface exchange-correlation energies.

B97-GGA Exchange-Correlation Functional

The Becke 97 GGA (B97-GGA) exchange-correlation functional is the GGA part of the B97 hybrid functional [22] (see Sect. 1.4.1). The B97-GGA exchange energy density is

$$e_{\text{x}}^{\text{B97-GGA}}(\rho_\uparrow, \rho_\downarrow, \nabla\rho_\uparrow, \nabla\rho_\downarrow) = \sum_{\sigma\in\{\uparrow,\downarrow\}} e_{\text{x},\sigma}^{\text{UEG}}(\rho_\sigma)\ g_{\text{x}}(x_\sigma), \tag{1.123}$$

where $e_{\text{x},\sigma}^{\text{UEG}}(\rho_\sigma) = e_{\text{x}}^{\text{UEG}}(\rho_\sigma, 0)$ is the spin-σ contribution to the UEG exchange energy density and the gradient correction $g_{\text{x}}(x_\sigma)$ is a function of $x_\sigma = |\nabla\rho_\sigma|/\rho_\sigma^{4/3}$,

$$g_{\text{x}}(x_\sigma) = \sum_{i=0}^{m} c_{\text{x},i}\ u_{\text{x}}(x_\sigma)^i, \tag{1.124}$$

with $u_{\text{x}}(x_\sigma) = \gamma_{\text{x}}x_\sigma^2/(1+\gamma_{\text{x}}x_\sigma^2)$. The B97-GGA correlation energy density is written as the sum of opposite- and same-spin contributions

$$\begin{aligned} e_{\text{c}}^{\text{B97-GGA}}(\rho_\uparrow, \rho_\downarrow, \nabla\rho_\uparrow, \nabla\rho_\downarrow) &= e_{\text{c},\uparrow\downarrow}^{\text{B97-GGA}}(\rho_\uparrow, \rho_\downarrow, \nabla\rho_\uparrow, \nabla\rho_\downarrow) \\ &+ \sum_{\sigma\in\{\uparrow,\downarrow\}} e_{\text{c},\sigma\sigma}^{\text{B97-GGA}}(\rho_\sigma, \nabla\rho_\sigma), \end{aligned} \tag{1.125}$$

where

$$e_{\text{c},\uparrow\downarrow}^{\text{B97-GGA}}(\rho_\uparrow, \rho_\downarrow, \nabla\rho_\uparrow, \nabla\rho_\downarrow) = e_{\text{c},\uparrow\downarrow}^{\text{UEG}}(\rho_\uparrow, \rho_\downarrow)\ g_{\text{c},\uparrow\downarrow}(x_{\uparrow\downarrow}) \tag{1.126}$$

and

$$e_{\text{c},\sigma\sigma}^{\text{B97-GGA}}(\rho_\sigma, \nabla\rho_\sigma) = e_{\text{c},\sigma\sigma}^{\text{UEG}}(\rho_\sigma)\ g_{\text{c},\sigma\sigma}(x_\sigma). \tag{1.127}$$

In these expressions, $e_{\mathrm{c},\uparrow\downarrow}^{\mathrm{UEG}}(\rho_\uparrow,\rho_\downarrow) = e_{\mathrm{c}}^{\mathrm{UEG}}(\rho_\uparrow,\rho_\downarrow) - e_{\mathrm{c}}^{\mathrm{UEG}}(\rho_\uparrow,0) - e_{\mathrm{c}}^{\mathrm{UEG}}(\rho_\downarrow,0)$ and $e_{\mathrm{c},\sigma\sigma}^{\mathrm{UEG}}(\rho_\alpha) = e_{\mathrm{c}}^{\mathrm{UEG}}(\rho_\sigma,0)$ are estimations of the opposite- and same-spin contributions to the UEG correlation energy density [238, 239]. The opposite-spin gradient correction is taken as a function of $x_{\uparrow\downarrow} = \sqrt{(x_\uparrow^2 + x_\downarrow^2)/2}$,

$$g_{\mathrm{c},\uparrow\downarrow}(x_{\uparrow\downarrow}) = \sum_{i=0}^{m} c_{\mathrm{c},i}^{\uparrow\downarrow}\, u_{\mathrm{c}}^{\uparrow\downarrow}(x_{\uparrow\downarrow})^i, \tag{1.128}$$

with $u_{\mathrm{c}}^{\uparrow\downarrow}(x_{\uparrow\downarrow}) = \gamma_{\mathrm{c}}^{\uparrow\downarrow} x_{\uparrow\downarrow}^2/(1+\gamma_{\mathrm{c}}^{\uparrow\downarrow} x_{\uparrow\downarrow}^2)$, and the same-spin gradient correction is

$$g_{\mathrm{c},\sigma\sigma}(x_\sigma) = \sum_{i=0}^{m} c_{\mathrm{c},i}^{\sigma\sigma}\, u_{\mathrm{c}}^{\sigma\sigma}(x_\sigma)^i, \tag{1.129}$$

with $u_{\mathrm{c}}^{\sigma\sigma}(x_\sigma) = \gamma_{\mathrm{c}}^{\sigma\sigma} x_\sigma^2/(1+\gamma_{\mathrm{c}}^{\sigma\sigma} x_\sigma^2)$. The parameters $\gamma_{\mathrm{x}} = 0.004$, $\gamma_{\mathrm{c}}^{\uparrow\downarrow} = 0.006$, and $\gamma_{\mathrm{c}}^{\sigma\sigma} = 0.2$, were roughly optimized on atomic exchange and correlation energies. The other parameters $c_{\mathrm{x},i}$, $c_{\mathrm{c},i}^{\uparrow\downarrow}$, $c_{\mathrm{c},i}^{\sigma\sigma}$ for a polynomial degree $m = 2$ in Eqs. (1.124), (1.128), and (1.129) were optimized in the B97 hybrid functional in the presence of a fraction of HF exchange energy (see Sect. 1.4.1).

The Hamprecht–Cohen–Tozer–Handy (HCTC) [104] exchange-correlation functional uses the same form as the B97-GGA exchange-correlation functional but with a polynomial degree $m = 4$ and the parameters $c_{\mathrm{x},i}$, $c_{\mathrm{c},i}^{\uparrow\downarrow}$, $c_{\mathrm{c},i}^{\sigma\sigma}$ were optimized without HF exchange a set of energetic properties (atomic total energies, ionization energies, atomization energies), nuclear gradients, and accurate exchange-correlation potentials.

1.3.4 Meta-Generalized-Gradient Approximations

The *meta-generalized-gradient approximations* (meta-GGAs or mGGAs) are of the generic form, in their spin-independent version,

$$E_{\mathrm{xc}}^{\mathrm{mGGA}}[\rho,\tau] = \int_{\mathbb{R}^3} e_{\mathrm{xc}}^{\mathrm{mGGA}}(\rho(\mathbf{r}), \nabla\rho(\mathbf{r}), \nabla^2\rho(\mathbf{r}), \tau(\mathbf{r}))\mathrm{d}\mathbf{r}, \tag{1.130}$$

i.e., they use more ingredients than the GGAs, namely the Laplacian of the density $\nabla^2\rho(\mathbf{r})$ and/or the non-interacting positive kinetic energy density $\tau(\mathbf{r})$ associated with a single-determinant wave function Φ,

$$\begin{aligned}\tau(\mathbf{r}) = \tau_\Phi(\mathbf{r}) &= \frac{N}{2}\int_{\{\uparrow,\downarrow\}\times(\mathbb{R}^3\times\{\uparrow,\downarrow\})^{N-1}} |\nabla_{\mathbf{r}}\Phi(\mathbf{x},\mathbf{x}_2,\ldots,\mathbf{x}_N)|^2\,\mathrm{d}\sigma\mathrm{d}\mathbf{x}_2\ldots\mathrm{d}\mathbf{x}_N \\ &= \frac{1}{2}\sum_{i=1}^{N}|\nabla\varphi_i(\mathbf{r})|^2,\end{aligned} \tag{1.131}$$

where $\{\varphi_i\}_{i=1,\ldots,N}$ are the orbitals occupied in Φ. The meta-GGAs are considered as part of the family of semilocal approximations, in the sense that $\tau(\mathbf{r})$ contains semilocal information with respect to the orbitals.

Meta-GGAs can be viewed as implicit functionals of the density only, i.e. $E_{\text{xc}}^{\text{mGGA}}[\rho, \tau_{\Phi[\rho]}]$, since $\tau(\mathbf{r})$ can be considered itself as an implicit functional of the density via the KS single-determinant wave function $\Phi[\rho]$. This view in which $E_{\text{xc}}^{\text{mGGA}}[\rho]$ is a proper approximation to the exchange-correlation density functional $E_{\text{xc}}[\rho]$ of the KS scheme is normally adopted when constructing meta-GGAs approximations. However, the calculation of the functional derivative of $E_{\text{xc}}^{\text{mGGA}}[\rho]$ with respect to the density then requires the use of the complicated optimized-effective-potential method (see Sect. 1.7). Therefore, in practical calculations, meta-GGAs are usually reinterpreted as explicit functionals of a single-determinant wave function Φ, i.e. $E_{\text{xc}}^{\text{mGGA}}[\rho_\Phi, \tau_\Phi]$, [2, 9, 73, 180, 181, 232, 242, 275] or, in other words, approximations to an exact GKS exchange-correlation functional (see Sect. 1.1.4).

In the latter approach, which we will here refer to as the meta-Kohn–Sham (mKS) scheme, we introduce a functional $E_{\text{xc}}^{\text{mKS}}[\rho, \tau]$ (to which meta-GGAs are approximations) defined for ρ and τ simultaneously representable by a single-determinant wave function $\Phi \in \mathcal{S}^N$ and which defines the GKS functional $E_{\text{xc}}^{\mathcal{S}}[\Phi] = E_{\text{xc}}^{\text{mKS}}[\rho_\Phi, \tau_\Phi]$ [see Eq. (1.38)] giving the exact ground-state energy via Eq. (1.39),

$$E_0 = \inf_{\Phi \in \mathcal{S}^N} \left\{ \langle \Phi | \hat{T} + \hat{V}_{\text{ne}} | \Phi \rangle + E_{\text{H}}[\rho_\Phi] + E_{\text{xc}}^{\text{mKS}}[\rho_\Phi, \tau_\Phi] \right\}, \tag{1.132}$$

which, by taking variations with respect to the orbitals, gives the mKS equations:

$$\left(-\frac{1}{2}\nabla^2 + v_{\text{ne}}(\mathbf{r}) + v_{\text{H}}(\mathbf{r}) + v_{\text{xc}}^{\text{mKS}}(\mathbf{r}) \right) \varphi_i(\mathbf{r}) = \varepsilon_i \varphi_i(\mathbf{r}). \tag{1.133}$$

Here, $v_{\text{xc}}^{\text{mKS}}(\mathbf{r}) = v_{\text{xc},1}^{\text{mKS}}(\mathbf{r}) + v_{\text{xc},2}^{\text{mKS}}(\mathbf{r})$ contains a usual local potential

$$v_{\text{xc},1}^{\text{mKS}}(\mathbf{r}) = \frac{\delta E_{\text{xc}}^{\text{mKS}}[\rho, \tau]}{\delta \rho(\mathbf{r})},$$

and a non-multiplicative operator [2, 9, 73, 242, 275]

$$v_{\text{xc},2}^{\text{mKS}}(\mathbf{r}) = -\frac{1}{2}\nabla \cdot \left(\frac{\delta E_{\text{xc}}^{\text{mKS}}[\rho, \tau]}{\delta \tau(\mathbf{r})} \nabla \right), \tag{1.134}$$

evaluated with $\rho(\mathbf{r}) = \sum_{i=1}^N |\varphi_i(\mathbf{r})|^2$ and $\tau(\mathbf{r}) = (1/2) \sum_{i=1}^N |\nabla \varphi_i(\mathbf{r})|^2$. Interestingly, the mKS equations can be rewritten as a Schrödinger-like equation with a position-dependent mass $m(\mathbf{r})$ [50],

$$\left(-\frac{1}{2}\nabla \cdot \frac{1}{m(\mathbf{r})} \nabla + v_{\text{ne}}(\mathbf{r}) + v_{\text{H}}(\mathbf{r}) + v_{\text{xc},1}^{\text{mKS}}(\mathbf{r}) \right) \varphi_i(\mathbf{r}) = \varepsilon_i \varphi_i(\mathbf{r}), \tag{1.135}$$

where $m(\mathbf{r}) = \left(1 + \delta E_{\text{xc}}^{\text{mKS}}[\rho, \tau]/\delta\tau(\mathbf{r})\right)^{-1}$. As in the KS scheme, the functional $E_{\text{xc}}^{\text{mKS}}[\rho, \tau]$ is decomposed into exchange and correlation contributions: $E_{\text{xc}}^{\text{mKS}}[\rho, \tau] = E_{\text{x}}^{\text{mKS}}[\rho, \tau] + E_{\text{c}}^{\text{mKS}}[\rho, \tau]$. In the spin-dependent version of the mKS scheme, we consider a similar functional of the spin-resolved densities and non-interacting positive kinetic energy densities $E_{\text{xc}}^{\text{mKS}}[\rho_\uparrow, \rho_\downarrow, \tau_\uparrow, \tau_\downarrow]$ and the spin-scaling relation of Eq. (1.36) is generalized to

$$E_{\text{x}}^{\text{mKS}}[\rho_\uparrow, \rho_\downarrow, \tau_\uparrow, \tau_\downarrow] = \frac{1}{2}\left(E_{\text{x}}^{\text{mKS}}[2\rho_\uparrow, 2\tau_\uparrow] + E_{\text{x}}^{\text{mKS}}[2\rho_\downarrow, 2\tau_\downarrow]\right). \tag{1.136}$$

Correspondingly, the spin-dependent versions of the meta-GGAs are formulated in terms of the spin-resolved quantities $\rho_\uparrow$, $\rho_\downarrow$, $\nabla\rho_\uparrow$, $\nabla\rho_\downarrow$, $\nabla^2\rho_\uparrow$, $\nabla^2\rho_\downarrow$, $\tau_\uparrow$, and $\tau_\downarrow$.

One motivation for the introduction of the variable $\tau(\mathbf{r})$ is that it appears in the expansion of the spherically averaged exchange hole [entering in Eq. (1.55)] for small interelectronic distances r_{12} [16], which for the case of a closed-shell system is

$$\frac{1}{4\pi r_{12}^2}\int_{S(\mathbf{0},r_{12})} h_{\text{x}}(\mathbf{r}_1, \mathbf{r}_1 + \mathbf{r}_{12})\mathrm{d}\mathbf{r}_{12} = -\frac{\rho(\mathbf{r}_1)}{2} \tag{1.137}$$

$$-\frac{1}{3}\left(\frac{1}{4}\nabla^2\rho(\mathbf{r}_1) - \tau(\mathbf{r}_1) + \frac{|\nabla\rho(\mathbf{r}_1)|^2}{8\rho(\mathbf{r}_1)}\right) r_{12}^2 + O(r_{12}^4),$$

where $S(\mathbf{0}, r_{12})$ designates the sphere centered at $\mathbf{0}$ and of radius $r_{12} = |\mathbf{r}_{12}|$. Thus $\tau(\mathbf{r})$ is needed to describe the curvature of the exchange hole.

Another important motivation is that $\tau(\mathbf{r})$ is useful for identifying different types of spatial regions of electronic systems [245]. This is done by comparing $\tau(\mathbf{r})$ with the von Weizsäcker kinetic energy density,

$$\tau^{\text{W}}(\mathbf{r}) = \frac{|\nabla\rho(\mathbf{r})|^2}{8\rho(\mathbf{r})}, \tag{1.138}$$

which is the exact non-interacting kinetic energy density for one-electron systems and two-electron spin-unpolarized systems, and, more generally, for one-orbital regions as introduced in Sect. 1.2.3. For example, the indicator

$$z(\mathbf{r}) = \frac{\tau^{\text{W}}(\mathbf{r})}{\tau(\mathbf{r})}, \tag{1.139}$$

which takes its values in the range [0, 1] [141], identifies one-orbital regions ($z = 1$). A better indicator is

$$\alpha(\mathbf{r}) = \frac{\tau(\mathbf{r}) - \tau^{\text{W}}(\mathbf{r})}{\tau^{\text{UEG}}(\mathbf{r})}, \tag{1.140}$$

where $\tau^{\text{UEG}}(\mathbf{r}) = (3/10)(3\pi^2)^{2/3}\rho(\mathbf{r})^{5/3}$ is the non-interacting kinetic energy density of the UEG. This indicator $\alpha(\mathbf{r})$ distinguishes one-orbital regions ($\alpha = 0$), slowly varying density regions ($\alpha \approx 1$), and regions of density overlap between closed shells that characterize noncovalent bonds ($\alpha \gg 1$).

Nowadays, $\nabla^2\rho(\mathbf{r})$ is rarely used to construct meta-GGAs because it contains similar information to $\tau(\mathbf{r})$, which can be seen by the second-order gradient expansion of $\tau(\mathbf{r})$ [27]:

$$\tau^{\text{GEA2}}(\mathbf{r}) = \tau^{\text{UEG}}(\mathbf{r}) + \frac{1}{72}\frac{|\nabla\rho(\mathbf{r})|^2}{\rho(\mathbf{r})} + \frac{1}{6}\nabla^2\rho(\mathbf{r}). \tag{1.141}$$

In comparison to GGAs, meta-GGAs are more versatile and generally constitute an improvement. Significantly, thanks to the use of τ, self-interaction errors in the correlation functional can be essentially eliminated with meta-GGAs. They still suffer however from self-interaction errors in the exchange functional. We now describe some of the most used meta-GGA functionals.

TPSS Exchange-Correlation Functional

In the Tao–Perdew–Staroverov–Scuseria (TPSS) [198, 249] functional, the exchange energy density is written as

$$e_{\text{x}}^{\text{TPSS}}(\rho, \nabla\rho, \tau) = e_{\text{x}}^{\text{UEG}}(\rho)F_{\text{x}}^{\text{TPSS}}(s, z), \tag{1.142}$$

where the enhancement factor is a function of $s = |\nabla\rho|/(2k_{\text{F}}\rho)$ and $z = \tau^{\text{W}}/\tau$,

$$F_{\text{x}}^{\text{TPSS}}(s, z) = 1 + \kappa - \frac{\kappa}{1 + x^{\text{TPSS}}(s, z)/\kappa}, \tag{1.143}$$

with $\kappa = 0.804$ so as to saturate the local Lieb–Oxford bound (just like in the PBE exchange functional) and

$$x^{\text{TPSS}}(s, z) = \left[\left(\frac{10}{81} + c\frac{z^2}{(1+z^2)^2}\right)s^2 + \frac{146}{2025}\tilde{q}_b^2 - \frac{73}{405}\tilde{q}_b\sqrt{\frac{1}{2}\left(\frac{3}{5}z\right)^2 + \frac{1}{2}s^4}\right.$$
$$\left. + \frac{1}{\kappa}\left(\frac{10}{81}\right)^2 s^4 + 2\sqrt{e}\frac{10}{81}\left(\frac{3}{5}z\right)^2 + e\mu s^6\right] / \left(1 + \sqrt{e}s^2\right)^2, \tag{1.144}$$

and $\tilde{q}_b = (9/20)(\alpha - 1)/[1 + b\alpha(\alpha - 1)]^{1/2} + 2s^2/3$ (where $\alpha = (\tau - \tau^{\text{W}})/\tau^{\text{UEG}} = (5s^2/3)(z^{-1} - 1)$) is a quantity that tends to the reduced density Laplacian $q = \nabla^2\rho/(4k_{\text{F}}^2\rho)$ in the slowly varying density limit [using Eq. (1.141)]. The function $x^{\text{TPSS}}(s, z)$ is chosen so as to satisfy the fourth-order gradient expansion [Eq. (1.112)] which can be written in the form of the enhancement factor $F_{\text{x}}^{\text{GEA4}}(s, z) = 1 + (10/81)s^2 + (146/2025)q^2 - (73/405)s^2q$. The constant $\mu = 0.21951$ is chosen to retain the same large-s behavior of the PBE exchange functional, i.e. $F_{\text{x}}^{\text{TPSS}}(s, z) \underset{s\to\infty}{\sim} F_{\text{x}}^{\text{PBE}}(s)$. The constants $c = 1.59096$ and $e = 1.537$ are chosen so as to eliminate the divergence of the potential at the nucleus for a two-electron exponential density and to yield the correct exchange energy (−0.3125 hartree) for the exact ground-state density of the hydrogen atom. Finally, the constant $b = 0.40$ is chosen, quite arbitrarily, as the smallest value that makes $F_x^{\text{TPSS}}(s, z)$ a monotonically increasing function of s.

The TPSS correlation functional is constructed by making minor refinements to the previously developed Perdew–Kurth–Zupan–Blaha (PKZB) [192] meta-GGA correlation functional,

$$\begin{aligned} e_c^{\text{TPSS}}(\rho_\uparrow, \rho_\downarrow, \nabla\rho_\uparrow, \nabla\rho_\downarrow, \tau_\uparrow, \tau_\downarrow) &= \rho\, \varepsilon_c^{\text{revPKZB}}(\rho_\uparrow, \rho_\downarrow, \nabla\rho_\uparrow, \nabla\rho_\downarrow, \tau_\uparrow, \tau_\downarrow) \\ &\quad \times \Big[1 + d\, \varepsilon_c^{\text{revPKZB}} \\ &\quad \times (\rho_\uparrow, \rho_\downarrow, \nabla\rho_\uparrow, \nabla\rho_\downarrow, \tau_\uparrow, \tau_\downarrow) z^3\Big], \end{aligned} \tag{1.145}$$

where the revised PKZB correlation energy per particle is

$$\begin{aligned} \varepsilon_c^{\text{revPKZB}}(\rho_\uparrow, \rho_\downarrow, \nabla\rho_\uparrow, \nabla\rho_\downarrow, \tau_\uparrow, \tau_\downarrow) &= \varepsilon_c^{\text{PBE}}(\rho_\uparrow, \rho_\downarrow, \nabla\rho_\uparrow, \nabla\rho_\downarrow) \left[1 + C(\zeta, \xi) z^2\right] \\ &- [1 + C(\zeta, \xi)]\, z^2 \sum_{\sigma \in \{\uparrow, \downarrow\}} \frac{\rho_\sigma}{\rho} \tilde{\varepsilon}_{c,\sigma}^{\text{PBE}}(\rho_\uparrow, \rho_\downarrow, \nabla\rho_\uparrow, \nabla\rho_\downarrow), \end{aligned} \tag{1.146}$$

with $\tilde{\varepsilon}_{c,\sigma}^{\text{PBE}}(\rho_\uparrow, \rho_\downarrow, \nabla\rho_\uparrow, \nabla\rho_\downarrow) = \max[\varepsilon_c^{\text{PBE}}(\rho_\sigma, 0, \nabla\rho_\sigma, 0), \varepsilon_c^{\text{PBE}}(\rho_\uparrow, \rho_\downarrow, \nabla\rho_\uparrow, \nabla\rho_\downarrow)]$
where $\varepsilon_c^{\text{PBE}}(\rho_\uparrow, \rho_\downarrow, \nabla\rho_\uparrow, \nabla\rho_\downarrow)$ is the PBE correlation energy per particle. Equation (1.146) constitutes a one-electron self-interaction correction on the PBE correlation functional. Indeed, for one-electron densities we have $z = 1$ and $\zeta = \pm 1$, and the TPSS correlation energy correctly vanishes [Eqs. (1.75)]. The TPSS correlation functional preserves many properties of the PBE correlation functional: it has correct uniform coordinate scaling in the high- and low-density limits, vanishing correlation energy in the large density-gradient limit, and the same second-order gradient expansion (since the additional terms beyond PBE are at least in z^2 and thus only change the fourth-order terms of the gradient expansion). The parameters $d = 2.8$ hartree^{-1} and $C(0, 0) = 0.53$ are chosen so as to recover the PBE surface correlation energy of jellium [145] over the range of valence-electron bulk densities. The rest of the function is taken as

$$C(\zeta, \xi) = \frac{0.53 + 0.87\zeta^2 + 0.50\zeta^4 + 2.26\zeta^6}{(1 + \xi^2[(1+\zeta)^{-4/3} + (1-\zeta)^{-4/3}]/2)^4},$$

where $\xi = |\nabla\zeta|/(2k_F)$ is a reduced spin-polarization gradient. The function $C(\zeta, \xi)$ is chosen so as to make the exchange-correlation energy independent of the spin polarization ζ in the low-density limit [Eq. (1.86)] and to avoid that the self-interaction correction introduces additional correlation energy density in the core-valence overlap region of monovalent atoms such as Li.

M06-L Exchange-Correlation Functional

In the Minnesota 06 local (M06-L) exchange-correlation functional [278], the exchange energy density is written as

$$
\begin{aligned}
&e_{\mathrm{x}}^{\text{M06-L}}(\rho_\uparrow, \rho_\downarrow, \nabla\rho_\uparrow, \nabla\rho_\downarrow, \tau_\uparrow, \tau_\downarrow) \\
&= \sum_{\sigma\in\{\uparrow,\downarrow\}} e_{\mathrm{x},\sigma}^{\text{PBE}}(\rho_\sigma, \nabla\rho_\sigma) f(w_\sigma) + e_{\mathrm{x},\sigma}^{\text{UEG}}(\rho_\sigma) h_{\mathrm{x}}(x_\sigma, Z_\sigma).
\end{aligned}
\tag{1.147}
$$

The first term in Eq. (1.147), which has the same form as in the previously developed M05 exchange functional [277], contains the spin-σ PBE exchange energy density $e_{\mathrm{x},\sigma}^{\text{PBE}}(\rho_\sigma, \nabla\rho_\sigma) = e_{\mathrm{x}}^{\text{PBE}}(\rho_\sigma, 0, \nabla\rho_\sigma, 0)$ and the kinetic-energy density correction factor

$$
f(w_\sigma) = \sum_{i=0}^{11} a_i w_\sigma^i, \tag{1.148}
$$

where $w_\sigma = (\tau_\sigma^{\text{UEG}}/\tau_\sigma - 1)/(\tau_\sigma^{\text{UEG}}/\tau_\sigma + 1)$ with $\tau_\sigma^{\text{UEG}} = (3/10)(6\pi^2)^{2/3}\rho_\sigma^{5/3}$ is an indicator of the delocalization of the exchange hole [24]. The second term in Eq. (1.147), which has the same form as in the VS98 exchange functional [264], contains the spin-σ UEG exchange energy density $e_{\mathrm{x},\sigma}^{\text{UEG}}(\rho_\sigma) = e_{\mathrm{x}}^{\text{UEG}}(\rho_\sigma, 0)$ and the correction factor

$$
h_{\mathrm{x}}(x_\sigma, Z_\sigma) = h(x_\sigma, Z_\sigma, d_{\mathrm{x},0}, d_{\mathrm{x},1}, d_{\mathrm{x},2}, d_{\mathrm{x},3}, d_{\mathrm{x},4}, \alpha_{\mathrm{x}}), \tag{1.149}
$$

where $x_\sigma = |\nabla\rho_\sigma|/\rho_\sigma^{4/3}$ and $Z_\sigma = 2(\tau_\sigma - \tau_\sigma^{\text{UEG}})/\rho_\sigma^{5/3}$ and h is the parametrized function

$$
h(x, Z, d_0, d_1, d_2, d_3, d_4, \alpha) = \frac{d_0}{\gamma(x, Z, \alpha)} + \frac{d_1 x^2 + d_2 Z}{\gamma(x, Z, \alpha)^2} + \frac{d_3 x^4 + d_4 x^2 Z}{\gamma(x, Z, \alpha)^3},
$$

with $\gamma(x, Z, \alpha) = 1 + \alpha(x^2 + Z)$.

The M06-L correlation energy is written as the sum of opposite- and same-spin contributions, similarly to the B97-GGA correlation functional [Eq. (1.125)],

$$
\begin{aligned}
&e_{\mathrm{c}}^{\text{M06-L}}(\rho_\uparrow, \rho_\downarrow, \nabla\rho_\uparrow, \nabla\rho_\downarrow, \tau_\uparrow, \tau_\downarrow) \\
&= e_{\mathrm{c},\uparrow\downarrow}^{\text{M06-L}}(\rho_\uparrow, \rho_\downarrow, \nabla\rho_\uparrow, \nabla\rho_\downarrow, \tau_\uparrow, \tau_\downarrow) + \sum_{\sigma\in\{\uparrow,\downarrow\}} e_{\mathrm{c},\sigma\sigma}^{\text{M06-L}}(\rho_\sigma, \nabla\rho_\sigma, \tau_\sigma),
\end{aligned}
$$

where

$$e_{\mathrm{c},\uparrow\downarrow}^{\mathrm{M06}}(\rho_\uparrow,\rho_\downarrow,\nabla\rho_\uparrow,\nabla\rho_\downarrow,\tau_\uparrow,\tau_\downarrow) = e_{\mathrm{c},\uparrow\downarrow}^{\mathrm{UEG}}(\rho_\uparrow,\rho_\downarrow)\ \left[g_{\mathrm{c},\uparrow\downarrow}(x_{\uparrow\downarrow}) + h_{\mathrm{c},\uparrow\downarrow}(x_{\uparrow\downarrow}, Z_{\uparrow\downarrow})\right],$$

and

$$e_{\mathrm{c},\sigma\sigma}^{\mathrm{M06\text{-}L}}(\rho_\sigma,\nabla\rho_\sigma) = e_{\mathrm{c},\sigma\sigma}^{\mathrm{UEG}}(\rho_\sigma)\ \left[g_{\mathrm{c},\sigma\sigma}(x_\sigma) + h_{\mathrm{c},\sigma\sigma}(x_\sigma, Z_\sigma)\right] D_\sigma(z_\sigma),$$

where the spin-decomposed UEG correlation energies $e_{\mathrm{c},\uparrow\downarrow}^{\mathrm{UEG}}(\rho_\uparrow,\rho_\downarrow)$ and $e_{\mathrm{c},\sigma\sigma}^{\mathrm{UEG}}(\rho_\sigma)$ were already defined after Eq. (1.127), and the gradient corrections $g_{\mathrm{c},\uparrow\downarrow}(x_{\uparrow\downarrow})$ and $g_{\mathrm{c},\sigma\sigma}(x_\sigma)$ are given in Eqs. (1.128) and (1.129). The additional correction factors are

$$h_{\mathrm{c}}^{\uparrow\downarrow}(x_{\uparrow\downarrow}, Z_{\uparrow\downarrow}) = h(x_{\uparrow\downarrow}, Z_{\uparrow\downarrow}, d_{\mathrm{c},0}^{\uparrow\downarrow}, d_{\mathrm{c},1}^{\uparrow\downarrow}, d_{\mathrm{c},2}^{\uparrow\downarrow}, d_{\mathrm{c},3}^{\uparrow\downarrow}, d_{\mathrm{c},4}^{\uparrow\downarrow}, \alpha_{\mathrm{c}}^{\uparrow\downarrow}), \quad (1.150)$$

where $x_{\uparrow\downarrow} = \sqrt{(x_\uparrow^2 + x_\downarrow^2)/2}$, $Z_{\uparrow\downarrow} = Z_\uparrow + Z_\downarrow$, and

$$h_{\mathrm{c}}^{\sigma\sigma}(x_\sigma, Z_\sigma) = h(x_\sigma, Z_\sigma, d_{\mathrm{c},0}^{\sigma\sigma}, d_{\mathrm{c},1}^{\sigma\sigma}, d_{\mathrm{c},2}^{\sigma\sigma}, d_{\mathrm{c},3}^{\sigma\sigma}, d_{\mathrm{c},4}^{\sigma\sigma}, \alpha_{\mathrm{c}}^{\sigma\sigma}). \quad (1.151)$$

The factor $D_\sigma(z_\sigma) = 1 - z_\sigma$, where $z_\sigma = \tau_\sigma^{\mathrm{W}}/\tau_\sigma$ and $\tau_\sigma^{\mathrm{W}} = |\nabla\rho_\sigma|^2/(8\rho_\sigma)$, ensures that the correlation energy correctly vanishes for one-electron systems [23].

The parameters $\gamma_{\mathrm{c}}^{\uparrow\downarrow} = 0.0031$, and $\gamma_{\mathrm{c}}^{\sigma\sigma} = 0.06$, were optimized on the correlation energies of He and Ne. The parameters $\alpha_{\mathrm{x}} = 0.001867$, $\alpha_{\mathrm{c}}^{\uparrow\downarrow} = 0.003050$, and $\alpha_{\mathrm{c}}^{\sigma\sigma} = 0.005151$ were taken from Ref. [264]. The constraints $a_0 + d_{\mathrm{x},0} = 1$, $c_{\mathrm{c},0}^{\uparrow\downarrow} + d_{\mathrm{c},0}^{\uparrow\downarrow} = 1$, and $c_{\mathrm{c},0}^{\sigma\sigma} + d_{\mathrm{c},0}^{\sigma\sigma} = 1$ are enforced to obtain the correct UEG limit. The remaining 34 free parameters a_i, $c_{\mathrm{c},i}^{\uparrow\downarrow}$, $c_{\mathrm{c},i}^{\sigma\sigma}$ for a polynomial degree $m = 4$ in Eqs. (1.148), (1.128), and (1.129), and $d_{\mathrm{x},i}$, $d_{\mathrm{c},i}^{\uparrow\downarrow}$, $d_{\mathrm{c},i}^{\sigma\sigma}$ in Eqs. (1.149), (1.150), and (1.151) were optimized on a large set of diverse physicochemical properties concerning main-group thermochemistry, reaction barrier heights, noncovalent interactions, electronic spectroscopy, and transition metal bonding.

SCAN Exchange-Correlation Functional

In the SCAN (strongly constrained and appropriately normed) [244] exchange-correlation functional, the exchange energy density is written as

$$e_{\mathrm{x}}^{\mathrm{SCAN}}(\rho,\nabla\rho,\tau) = e_{\mathrm{x}}^{\mathrm{UEG}}(\rho) F_{\mathrm{x}}^{\mathrm{SCAN}}(s,\alpha), \quad (1.152)$$

where the enhancement factor is a function of $s = |\nabla\rho|/(2k_{\mathrm{F}}\rho)$ and $\alpha = (\tau - \tau^{\mathrm{W}})/\tau^{\mathrm{UEG}}$,

$$F_{\mathrm{x}}^{\mathrm{SCAN}}(s,\alpha) = [h_{\mathrm{x}}^1(s,\alpha) + f_{\mathrm{x}}(\alpha)(h_{\mathrm{x}}^0 - h_{\mathrm{x}}^1(s,\alpha))]g_{\mathrm{x}}(s), \quad (1.153)$$

which interpolates between $\alpha = 0$ and $\alpha \approx 1$, and extrapolates to $\alpha \to \infty$ using the function

$$f_{\mathrm{x}}(\alpha) = \exp[-c_{1\mathrm{x}}\alpha/(1-\alpha)]\theta(1-\alpha) - d_{\mathrm{x}} \exp[c_{2\mathrm{x}}/(1-\alpha)]\theta(\alpha-1),$$

where θ is the Heaviside step function. The function $g_{\mathrm{x}}(s) = 1 - \exp(-a_1 s^{-1/2})$ is chosen to make $F_{\mathrm{x}}^{\mathrm{SCAN}}(s, \alpha)$ vanish like $s^{-1/2}$ as $s \to \infty$, which guarantees the non-uniform scaling finiteness conditions [Eqs. (1.89) and (1.91)] [158, 194], and $a_1 = 4.9479$ is taken to recover the exact exchange energy of the hydrogen atom. For $\alpha \approx 1$ (slowly varying density regions), $F_{\mathrm{x}}^{\mathrm{SCAN}}(s, \alpha) \approx h_{\mathrm{x}}^1(s, \alpha) g_{\mathrm{x}}(s)$, where $h_{\mathrm{x}}^1(s, \alpha)$ is a PBE-like resummation of the fourth-order gradient expansion [Eq. (1.112)],

$$h_{\mathrm{x}}^1(s, \alpha) = 1 + k_1 - \frac{k_1}{1 + x^{\mathrm{SCAN}}(s, \alpha)/k_1},$$

where

$$x^{\mathrm{SCAN}}(s, \alpha) = \mu s^2[1 + (b_4 s^2/\mu)\mathrm{e}^{-|b_4| s^2/\mu}] + [b_1 s^2 + b_2(1-\alpha)\mathrm{e}^{-b_3(1-\alpha)^2}]^2,$$

with $\mu = 10/81$, $b_2 = (5913/405000)^{1/2}$, $b_1 = (511/13500)/(2b_2)$, $b_3 = 0.5$, and $b_4 = \mu^2/k_1 - 1606/18225 - b_1^2$. For $\alpha = 0$ (one-orbital regions), $F_{\mathrm{x}}^{\mathrm{SCAN}}(s, \alpha = 0) = h_{\mathrm{x}}^0 g_{\mathrm{x}}(s)$ where $h_{\mathrm{x}}^0 = 1.174$ is chosen to saturate the local two-electron tight bound $F_{\mathrm{x}}^{\mathrm{SCAN}}(s, \alpha = 0) \leqslant 1.174$, which is a sufficient and necessary condition for a meta-GGA exchange functional to satisfy the global tight bound of Eq. (1.101) for all two-electron spin-unpolarized densities [194].

The SCAN correlation energy density is written as

$$\begin{aligned} e_{\mathrm{c}}^{\mathrm{SCAN}}&(\rho_\uparrow, \rho_\downarrow, \nabla\rho_\uparrow, \nabla\rho_\downarrow, \tau_\uparrow, \tau_\downarrow) \\ &= \rho\,[\varepsilon_{\mathrm{c}}^1(\rho_\uparrow, \rho_\downarrow, t) + f_{\mathrm{c}}(\alpha)(\varepsilon_{\mathrm{c}}^0(\rho_\uparrow, \rho_\downarrow, s) - \varepsilon_{\mathrm{c}}^1(\rho_\uparrow, \rho_\downarrow, t))], \end{aligned} \tag{1.154}$$

which is again an interpolation between $\alpha = 0$ and $\alpha = 1$, and an extrapolation to $\alpha \to \infty$ using the function

$$f_{\mathrm{c}}(\alpha) = \exp[-c_{1\mathrm{c}}\alpha/(1-\alpha)]\theta(1-\alpha) - d_{\mathrm{c}} \exp[c_{2\mathrm{c}}/(1-\alpha)]\theta(\alpha-1).$$

For $\alpha = 1$, the correlation energy par particle is taken as a revised version of the PBE correlation energy per particle,

$$\varepsilon_{\mathrm{c}}^1(\rho_\uparrow, \rho_\downarrow, t) = \varepsilon_{\mathrm{c}}^{\mathrm{UEG}}(\rho_\uparrow, \rho_\downarrow) + H_1^{\mathrm{SCAN}}(\rho_\uparrow, \rho_\downarrow, t), \tag{1.155}$$

where

$$H_1^{\mathrm{SCAN}}(\rho_\uparrow, \rho_\downarrow, t) = A(0)\phi_2(\zeta)^3 \ln\left[1 + w_1(1 - g(\mathcal{A}t^2))\right], \tag{1.156}$$

with $t = |\nabla\rho|/(2\phi_2(\zeta)k_s\rho)$, $w_1 = \exp[-\varepsilon_c^{\text{UEG}}(\rho_\uparrow,\rho_\downarrow)/(A(0)\phi_2(\zeta)^3)] - 1$, $\mathcal{A} = \beta(r_s)/(A(0)w_1)$, and $g(\mathcal{A}t^2) = 1/(1+4\mathcal{A}t^2)^{1/4}$. The function has a second-order gradient expansion $H_1^{\text{SCAN}}(\rho_\uparrow,\rho_\downarrow,t) = \beta(r_s)\phi_2(\zeta)t^2 + O(t^4)$, where the coefficient $\beta(r_s) = 0.066725(1+0.1r_s)/(1+0.1778r_s)$ is a rough fit of the density dependence of the second-order gradient expansion correlation coefficient beyond the Ma–Brueckner high-density-limit value and designed so that for $r_s \to \infty$ the second-order gradient expansion terms for exchange and correlation cancel each other [195]. For $\alpha = 0$, the correlation energy par particle is constructed to be accurate for one- and two-electron systems and is written as

$$\varepsilon_c^0(\rho_\uparrow,\rho_\downarrow,s) = [\varepsilon_c^{\text{LDA0}}(\rho) + H_0^{\text{SCAN}}(\rho,s)]G_c(\zeta). \tag{1.157}$$

The spin function $G_c(\zeta) = [1 - 2.3631(\phi_4(\zeta) - 1)](1-\zeta^{12})$ is designed to make the correlation energy vanish for one-electron densities ($\alpha = 0$ and $\zeta = \pm 1$) and to make the exchange-correlation energy independent of ζ in the low-density limit [Eq. (1.86)]. Equation (1.157) includes a LDA-type term [243]

$$\varepsilon_c^{\text{LDA0}}(\rho) = -\frac{b_{1c}}{1 + b_{2c}r_s^{1/2} + b_{3c}r_s},$$

and a gradient correction

$$H_0^{\text{SCAN}}(\rho,s) = b_{1c}\ln\left[1 + w_0(1 - g_\infty(\zeta = 0, s))\right],$$

with $w_0 = \exp(-\varepsilon_c^{\text{LDA0}}(\rho)/b_{1c}) - 1$ and $g_\infty(\zeta = 0, s) = \lim_{\zeta\to 0}\lim_{r_s\to\infty} g(\mathcal{A}t) = 1/(1+0.512104s^2)^{1/4}$. The parameter $b_{1c} = 0.0285764$ is determined so that the high-density limit of $\varepsilon_c^0(\rho_\uparrow,\rho_\downarrow,s)$ reproduces the exact correlation energy of the Helium isoelectronic series in the large-nuclear charge limit, i.e. $\lim_{Z\to\infty} E_c[\rho_{N=2,Z}] = E_c^{\text{GL2}}[\rho_{N=2,Z=1}^{\text{H}}] = -0.0467$ hartree [Eq. (1.94)]. The parameter $b_{3c} = 0.125541$ is determined to saturate the lower bound on the exchange-correlation energies of two-electron densities [Eq. (1.99)]. The parameter $b_{2c} = 0.0889$ is determined to reproduce the exact exchange-correlation energy of the He atom.

The remaining seven parameters ($k_1 = 0.065$, $c_{1x} = 0.667$, $c_{2x} = 0.8$, $d_x = 1.24$, $c_{1c} = 0.64$, $c_{2c} = 1.5$, and $d_c = 0.7$) are determined by fitting to the approximate asymptotic expansions of the exchange and correlation energies of neutral atoms in large nuclear charge limit [Eqs. (1.96) and (1.97)], the binding energy curve of compressed Ar_2, and jellium surface exchange-correlation energies.

1.4 Single-Determinant Hybrid Approximations

1.4.1 Hybrid Approximations

Based on arguments relying on the adiabatic-connection formalism, in 1993 Becke [19] proposed to mix a fraction of the exact or Hartree–Fock (HF) exchange energy E_x^{HF} with GGA functionals. In particular, he proposed a *three-parameter hybrid (3H) approximation* [20] of the form, written here in its spin-independent version,

$$E_{xc}^{3H}[\Phi] = a\ E_x^{HF}[\Phi] + b\ E_x^{GGA}[\rho_\Phi] + (1-a-b)\ E_x^{LDA}[\rho_\Phi] \\ + c\ E_c^{GGA}[\rho_\Phi] + (1-c)\ E_c^{LDA}[\rho_\Phi], \tag{1.158}$$

with empirical parameters a, b, and c. The functional $E_{xc}^{3H}[\Phi]$ is thought of as a functional of a single-determinant wave function $\Phi \in \mathcal{S}^N$ since $E_x^{HF}[\Phi]$ is itself a functional of Φ,

$$E_x^{HF}[\Phi] = \langle\Phi|\,\hat{W}_{ee}\,|\Phi\rangle - E_H[\rho_\Phi] \tag{1.159}$$
$$= -\frac{1}{2}\sum_{\sigma\in\{\uparrow,\downarrow\}}\sum_{i=1}^{N_\sigma}\sum_{j=1}^{N_\sigma}\int_{\mathbb{R}^3\times\mathbb{R}^3}\frac{\varphi_{i\sigma}^*(\mathbf{r}_1)\varphi_{j\sigma}(\mathbf{r}_1)\varphi_{j\sigma}^*(\mathbf{r}_2)\varphi_{i\sigma}(\mathbf{r}_2)}{|\mathbf{r}_1-\mathbf{r}_2|}\mathrm{d}\mathbf{r}_1\mathrm{d}\mathbf{r}_2,$$

where $\{\varphi_{i\sigma}\}_{i=1,\ldots,N_\sigma}$ are the orbitals occupied in Φ. In 1996, Becke proposed a simpler *one-parameter hybrid (1H) approximation* [21],

$$E_{xc}^{1H}[\Phi] = a\ E_x^{HF}[\Phi] + (1-a)\ E_x^{GGA}[\rho_\Phi] + E_c^{GGA}[\rho_\Phi], \tag{1.160}$$

where the fraction a of HF exchange has to be determined. For simplicity, we considered GGA functionals $E_x^{GGA}[\rho_\Phi]$ and $E_c^{GGA}[\rho_\Phi]$ in Eq. (1.160) but we can more generally use meta-GGA functionals $E_x^{mGGA}[\rho_\Phi, \tau_\Phi]$ and $E_c^{mGGA}[\rho_\Phi, \tau_\Phi]$.

These hybrid approximations should be considered as approximations of the GKS exchange-correlation functional $E_{xc}^S[\Phi]$ in Eq. (1.38) with $S[\Phi] = a\ E_x^{HF}[\Phi]$. The corresponding GKS equations [Eq. (1.1.4)] then include the term

$$\frac{\delta S[\Phi]}{\delta\varphi_{i\sigma}^*(\mathbf{r})} = a\int_{\mathbb{R}^3} v_{x,\sigma}^{HF}(\mathbf{r},\mathbf{r}')\varphi_{i\sigma}(\mathbf{r}')\mathrm{d}\mathbf{r}', \tag{1.161}$$

where $v_{\mathrm{x},\sigma}^{\mathrm{HF}}(\mathbf{r}, \mathbf{r}')$ is the nonlocal HF exchange potential[7]

$$v_{\mathrm{x},\sigma}^{\mathrm{HF}}(\mathbf{r}, \mathbf{r}') = -\sum_{j=1}^{N_\sigma} \frac{\varphi_{j\sigma}(\mathbf{r})\varphi_{j\sigma}^*(\mathbf{r}')}{|\mathbf{r} - \mathbf{r}'|}.$$

The main benefit of adding a fraction of HF exchange is to decrease the self-interaction error (see Sect. 1.2.3) introduced by semilocal exchange functionals which tends to favor too much delocalized electron densities over localized electron densities. The fraction of HF exchange should however be small enough to keep the compensation of errors usually occurring between the approximate semilocal exchange and correlation functionals. First, Becke used the value $a = 0.5$ in the so-called Becke Half-and-Half functional [19], but then fits to various experimental data often repeatedly gave an optimal parameter a around 0.20–0.25. A rationale has been proposed in favor of the value 0.25 [191]. By decreasing self-interaction errors in the exchange energy, hybrid approximations are often a big improvement over semilocal approximations for molecular systems with sufficiently large electronic gaps. However, for systems with small HOMO-LUMO gaps, such as systems with stretched chemical bonds or with transition metal elements, they tend to increase static-correlation errors.

An interesting extension of the hybrid approximations are the so-called local hybrids, which use a position-dependent fraction $a(\mathbf{r})$ of a (non-uniquely defined) HF exchange energy density $e_{\mathrm{x}}^{\mathrm{HF}}(\mathbf{r})$ [125] (see, Ref. [167] for a recent review), and which belong to the wider family of hyper-GGA functionals in which the correlation energy can also be expressed as a function of $e_{\mathrm{x}}^{\mathrm{HF}}(\mathbf{r})$ [197]. The local-hybrid approximations are much more flexible than the global hybrid approach exposed in this section but require more complicated and computationally expensive implementations. For this reason, they have not often been used and we will not consider them any further here.

We now describe some of the most used hybrid approximations.

B3LYP Exchange-Correlation Functional

The B3LYP exchange-correlation functional [237] is the most famous and widely used three-parameter hybrid approximation [Eq. (1.158)]. It uses the B88 exchange functional and the LYP correlation functional,

$$E_{\mathrm{xc}}^{\mathrm{B3LYP}}[\Phi] = a\ E_{\mathrm{x}}^{\mathrm{HF}}[\Phi] + b\ E_{\mathrm{x}}^{\mathrm{B88}}[\rho_{\uparrow,\Phi}, \rho_{\downarrow,\Phi}] + (1-a-b)\ E_{\mathrm{x}}^{\mathrm{LSDA}}[\rho_{\uparrow,\Phi}, \rho_{\downarrow,\Phi}]$$
$$+ c\ E_{\mathrm{c}}^{\mathrm{LYP}}[\rho_{\uparrow,\Phi}, \rho_{\downarrow,\Phi}] + (1-c)\ E_{\mathrm{c}}^{\mathrm{LSDA}}[\rho_{\uparrow,\Phi}, \rho_{\downarrow,\Phi}], \quad (1.162)$$

and the parameters $a = 0.20$, $b = 0.72$, and $c = 0.81$ were found by optimizing on a set of atomization energies, ionization energies, proton affinities of small molecules

[7] The possibility of combining a nonlocal HF potential with a local correlation potential was mentioned already in 1965 in the paper by Kohn and Sham [135].

and first-row total atomic energies [20]. A caveat is that the VWN parametrization of the RPA correlation energy (sometimes referred to as VWN3) of the UEG was actually used for $E_c^{\text{LSDA}}[\rho_\uparrow, \rho_\downarrow]$ instead of the VWN parametrization of the accurate correlation energy (sometimes referred to as VWN5) of the UEG [265].

B97 Exchange-Correlation Functional
The Becke 97 (B97) exchange-correlation functional [22] is a GGA hybrid of the form

$$E_{\text{xc}}^{\text{B97}}[\Phi] = a\ E_{\text{x}}^{\text{HF}}[\Phi] + (1-a)\ E_{\text{x}}^{\text{B97-GGA}}[\rho_{\uparrow,\Phi}, \rho_{\downarrow,\Phi}] + E_{\text{c}}^{\text{B97-GGA}}[\rho_{\uparrow,\Phi}, \rho_{\downarrow,\Phi}], \tag{1.163}$$

where the form of the B97-GGA exchange and correlation functionals were given in Eqs. (1.123) and (1.125). The fraction of HF exchange $a = 0.1943$ and the remaining parameters $c_{\text{x},0} = 1.00459$, $c_{\text{x},1} = 0.629639$, $c_{\text{x},2} = 0.928509$, $c_{\text{c},0}^{\uparrow\downarrow} = 0.9454$, $c_{\text{c},1}^{\uparrow\downarrow} = 0.7471$, $c_{\text{c},2}^{\uparrow\downarrow} = -4.5961$, $c_{\text{c},0}^{\sigma\sigma} = 0.1737$, $c_{\text{c},1}^{\sigma\sigma} = 2.3487$, and $c_{\text{c},2}^{\sigma\sigma} = -2.4868$ for a polynomial degree $m = 2$ in Eqs. (1.124), (1.128), and (1.129) were optimized on a set of total energies, atomization energies, ionization energies, and proton affinities. Note that, for $x_\sigma = 0$, the UEG limit is not imposed, which would require the parameters $c_{\text{x},0}$, $c_{\text{c},0}^{\uparrow\downarrow}$, and $c_{\text{c},0}^{\sigma\sigma}$ to be all strictly equal to 1. With the above optimized parameters, we see that it is nearly satisfied for the exchange energy and the opposite-spin correlation energy, but very far from it for the same-spin correlation energy, which is drastically reduced compared to the LDA.

PBE0 Exchange-Correlation Functional
The PBE0 exchange-correlation functional [1, 60] is a GGA hybrid using the PBE exchange and correlation functionals,

$$E_{\text{xc}}^{\text{PBE0}}[\Phi] = a\ E_{\text{x}}^{\text{HF}}[\Phi] + (1-a)\ E_{\text{x}}^{\text{PBE}}[\rho_{\uparrow,\Phi}, \rho_{\downarrow,\Phi}] + E_{\text{c}}^{\text{PBE}}[\rho_{\uparrow,\Phi}, \rho_{\downarrow,\Phi}], \tag{1.164}$$

and the fraction of the HF exchange is fixed at $a = 0.25$ according to the rationale of Ref. [191]. This functional is also known under the name PBE1PBE. The "1" in the latter name emphasizes that there is one parameter, a, while the "0" in the more common name PBE0 emphasizes that this parameter is not found by fitting.

TPSSh Exchange-Correlation Functional
The TPSSh exchange-correlation functional [234] is a meta-GGA hybrid using the TPSS exchange and correlation functionals,

$$E_{\text{xc}}^{\text{TPSSh}}[\Phi] = a\ E_{\text{x}}^{\text{HF}}[\Phi] + (1-a)\ E_{\text{x}}^{\text{TPSS}}[\rho_{\uparrow,\Phi}, \rho_{\downarrow,\Phi}, \tau_{\uparrow,\Phi}, \tau_{\downarrow,\Phi}] + E_{\text{c}}^{\text{TPSS}}[\rho_{\uparrow,\Phi}, \rho_{\downarrow,\Phi}, \tau_{\uparrow,\Phi}, \tau_{\downarrow,\Phi}], \tag{1.165}$$

and the fraction of the HF exchange $a = 0.10$ was determined by optimizing on a large set of atomization energies.

M06 and M06-2X Exchange-Correlation Functionals
The M06 exchange-correlation functional [279] is a meta-GGA hybrid using the M06-L exchange and correlation functionals,

$$E_{\mathrm{xc}}^{\mathrm{M06}}[\Phi] = a\, E_{\mathrm{x}}^{\mathrm{HF}}[\Phi] + (1-a)\, E_{\mathrm{x}}^{\mathrm{M06\text{-}L}}[\rho_{\uparrow,\Phi}, \rho_{\downarrow,\Phi}, \tau_{\uparrow,\Phi}, \tau_{\downarrow,\Phi}] + E_{\mathrm{c}}^{\mathrm{M06\text{-}L}}[\rho_{\uparrow,\Phi}, \rho_{\downarrow,\Phi}, \tau_{\uparrow,\Phi}, \tau_{\downarrow,\Phi}], \tag{1.166}$$

and the parameters in the M06-L exchange and correlation functionals were reoptimized together with the fraction of HF exchange $a = 0.27$ on the same large set of diverse physicochemical properties used for the M06-L functional. In the M06-2X exchange-correlation functional the fraction of HF exchange is doubled, i.e. $a = 0.54$, and the parameters were reoptimized with the function $h_{\mathrm{x}}(x_\sigma, Z_\sigma)$ in Eq. (1.149) set to zero and excluding transition metal properties in the training set. With this large fraction of HF exchange, the M06-2X functional is designed for systems without transition metal elements.

1.4.2 Range-Separated Hybrid Approximations

Based on earlier ideas of Savin [216] (exposed in detail in Sect. 1.5.2), in 2001, Iikura et al. [121] proposed a *long-range correction (LC) scheme* in which the exchange-correlation energy is written as, in its spin-independent version,

$$E_{\mathrm{xc}}^{\mathrm{LC}}[\Phi] = E_{\mathrm{x}}^{\mathrm{lr},\mu,\mathrm{HF}}[\Phi] + E_{\mathrm{x}}^{\mathrm{sr},\mu,\mathrm{GGA}}[\rho_\Phi] + E_{\mathrm{c}}^{\mathrm{GGA}}[\rho_\Phi]. \tag{1.167}$$

This scheme has also been referred to as the range-separated hybrid exchange (RSHX) scheme [77]. In Eq. (1.167), $E_{\mathrm{x}}^{\mathrm{lr},\mu,\mathrm{HF}}[\Phi]$ is the HF exchange energy for a long-range electron-electron interaction $w_{\mathrm{ee}}^{\mathrm{lr},\mu}(r_{12}) = \mathrm{erf}(\mu r_{12})/r_{12}$ (where erf is the error function and the parameter $\mu \in [0, +\infty)$ controls the range of the interaction),

$$E_{\mathrm{x}}^{\mathrm{lr},\mu,\mathrm{HF}}[\Phi] = -\frac{1}{2} \sum_{\sigma\in\{\uparrow,\downarrow\}} \sum_{i=1}^{N_\sigma} \sum_{j=1}^{N_\sigma} \int_{\mathbb{R}^3\times\mathbb{R}^3} \varphi_{i\sigma}^*(\mathbf{r}_1)\varphi_{j\sigma}(\mathbf{r}_1)\varphi_{j\sigma}^*(\mathbf{r}_2)\varphi_{i\sigma}(\mathbf{r}_2) w_{\mathrm{ee}}^{\mathrm{lr},\mu}(r_{12})\mathrm{d}\mathbf{r}_1\mathrm{d}\mathbf{r}_2, \tag{1.168}$$

and $E_{\rm x}^{{\rm sr},\mu,{\rm GGA}}[\rho]$ is a GGA exchange energy functional for the complementary short-range interaction $w_{\rm ee}^{{\rm sr},\mu}(r_{12}) = 1/r_{12} - w_{\rm ee}^{\rm lr}(r_{12})$. This latter functional can be thought of as an approximation to the short-range exchange functional

$$E_{\rm x}^{{\rm sr},\mu}[\rho] = \frac{1}{2}\int_{\mathbb{R}^3\times\mathbb{R}^3} \rho(\mathbf{r}_1) h_{\rm x}(\mathbf{r}_1,\mathbf{r}_2) w_{\rm ee}^{{\rm sr},\mu}(r_{12}) {\rm d}\mathbf{r}_1 {\rm d}\mathbf{r}_2, \tag{1.169}$$

where $h_{\rm x}(\mathbf{r}_1,\mathbf{r}_2)$ is the KS exchange hole of Sect. 1.2.1. For $\mu = 0$, the long-range HF exchange energy vanishes, i.e. $E_{\rm x}^{{\rm lr},\mu=0,{\rm HF}}[\Phi] = 0$, and the short-range exchange functional reduces to the standard KS exchange functional, i.e. $E_{\rm x}^{{\rm sr},\mu=0}[\rho] = E_{\rm x}[\rho]$. Reversely, for $\mu \to \infty$, the long-range HF exchange energy reduces to the full-range HF exchange energy, i.e. $E_{\rm x}^{{\rm lr},\mu\to\infty,{\rm HF}}[\Phi] = E_{\rm x}^{\rm HF}[\Phi]$, and the short-range exchange functional vanishes, i.e. $E_{\rm x}^{{\rm sr},\mu\to\infty}[\rho] = 0$. Significantly, for large μ, the short-range exchange functional becomes a local functional of the density [81, 255]:

$$E_{\rm x}^{{\rm sr},\mu}[\rho] \underset{\mu\to\infty}{\sim} -\frac{\pi}{4\mu^2}\int_{\mathbb{R}^3} \rho(\mathbf{r})^2 {\rm d}\mathbf{r}. \tag{1.170}$$

Like the hybrid approximations of Sect. 1.4.1, Eq. (1.167) should be considered as an approximation of the GKS exchange-correlation functional $E_{\rm xc}^S[\Phi]$ in Eq. (1.38) with $S[\Phi] = E_{\rm x}^{{\rm lr},\mu,{\rm HF}}[\Phi]$, and the corresponding GKS equations [Eq. (1.1.4)] then includes a long-range nonlocal HF exchange potential $v_{{\rm x},\sigma}^{{\rm lr},\mu,{\rm HF}}(\mathbf{r}_1,\mathbf{r}_2) = -\sum_{j=1}^{N_\sigma} \varphi_{j\sigma}(\mathbf{r}_1)\varphi_{j\sigma}^*(\mathbf{r}_2) w_{\rm ee}^{{\rm lr},\mu}(r_{12})$. Similarly to the hybrid approximations, the introduction of a fraction of long-range HF exchange reduces the self-interaction error (see, e.g., Ref. [179]). In addition, the short-range exchange part is easier to approximate with semilocal density-functional approximations, as Eq. (1.170) strongly suggests. In particular, the $-1/r$ asymptotic behavior of the exchange potential [Eq. (1.27)], which is difficult to satisfy with semilocal approximations, does not apply anymore to the short-range exchange potential.

In 2004, Yanai et al. [273], introduced a more flexible scheme called the Coulomb-attenuating method (CAM) [273] in which fractions of HF exchange are added at both short range and long range,

$$\begin{aligned} E_{\rm xc}^{\rm CAM}[\Phi] = a\, E_{\rm x}^{{\rm sr},\mu,{\rm HF}}[\Phi] + b\, E_{\rm x}^{{\rm lr},\mu,{\rm HF}}[\Phi] + (1-a)\, E_{\rm x}^{{\rm sr},\mu,{\rm GGA}}[\rho_\Phi] \\ + (1-b)\, E_{\rm x}^{{\rm lr},\mu,{\rm GGA}}[\rho_\Phi] + E_{\rm c}^{\rm GGA}[\rho_\Phi], \end{aligned} \tag{1.171}$$

where $E_{\rm x}^{{\rm sr},\mu,{\rm HF}}[\Phi] = E_{\rm x}^{\rm HF}[\Phi] - E_{\rm x}^{{\rm lr},\mu,{\rm HF}}[\Phi]$ is the short-range HF exchange energy and $E_{\rm x}^{{\rm lr},\mu,{\rm GGA}} = E_{\rm x}^{\rm GGA} - E_{\rm x}^{{\rm sr},\mu,{\rm GGA}}$ is a long-range GGA exchange energy. The reintroduction of HF exchange at short range further reduces the self-interaction error and improves thermodynamic properties such as atomization energies. Again, Eq. (1.171) should be considered as an approximation of the GKS exchange-correlation functional $E_{\rm xc}^S[\Phi]$ in Eq. (1.38) with $S[\Phi] = a\, E_{\rm x}^{{\rm sr},\mu,{\rm HF}}[\Phi] +$

$b\ E_{\text{x}}^{\text{lr},\mu,\text{HF}}[\Phi]$. Other forms of modified electron-electron interactions are also possible (see, e.g., Refs. [113, 217, 255]).

The approximations in Eqs. (1.167) and (1.171) are usually collectively referred to as *range-separated hybrid approximations*. Range-separated hybrids in the form of Eq. (1.171) are more flexible than the hybrid approximations of Sect. 1.4.1, and consequently are potentially more accurate, in particular for long-range electronic excitations. However, like the hybrid approximations, the presence of HF exchange tends to induce static-correlation errors for systems with small HOMO-LUMO gaps.

The range-separation parameter μ (also sometimes denoted as ω) is generally chosen empirically, e.g. by fitting to experimental data. In practice, a value around $\mu \approx 0.3 - 0.5$ bohr^{-1}, fixed for all systems, is often found to be optimal. It has also been proposed to adjust the value of μ in each system, e.g. by requiring that the opposite of the HOMO energy be equal to the ionization energy calculated by total energy differences [12, 235, 236]. These so-called optimally tuned range-separated hybrids are well suited for the calculation of charge-transfer electronic excitations but have the disadvantage of not being size consistent [130].

A natural idea is to use a position-dependent range-separation parameter $\mu(\mathbf{r})$ which allows the range of the modified interaction to adapt to the local average electron-electron distance in the diverse spatial regions of the system. These locally range-separated hybrids [11, 133, 139] are promising but they induced computational complications and are still in the early stages of development. We will thus not consider them any further here.

We now describe some of the most used approximations in the context of the range-separated hybrids.

Short-Range LDA Exchange Functional

The short-range LDA exchange functional [81, 216] can be obtained by using in Eq. (1.169) the LDA exchange hole [Eq. (1.110)], which leads to

$$E_{\text{x}}^{\text{sr},\mu,\text{LDA}}[\rho] = \int_{\mathbb{R}^3} e_{\text{x}}^{\text{sr},\mu,\text{UEG}}(\rho(\mathbf{r}))\text{d}\mathbf{r}, \tag{1.172}$$

with the short-range UEG exchange energy density

$$\begin{aligned} & e_{\text{x}}^{\text{sr},\mu,\text{UEG}}(\rho) \qquad (1.173) \\ &= e_{\text{x}}^{\text{UEG}}(\rho)\left[1 - \frac{8\widetilde{\mu}}{3}\left(\sqrt{\pi}\,\text{erf}\left(\frac{1}{2\widetilde{\mu}}\right) + (2\widetilde{\mu} - 4\widetilde{\mu}^3)\text{e}^{-1/(4\widetilde{\mu}^2)} - 3\widetilde{\mu} + 4\widetilde{\mu}^3\right)\right], \end{aligned}$$

where $\widetilde{\mu} = \mu/(2k_{\text{F}})$ is a dimensionless range-separation parameter. The spin-dependent version is obtained from the same spin-scaling relation as in the standard case [Eq. (1.36)]. The short-range LDA exchange functional becomes exact for large μ [Eq. (1.170)] and is the first building block for constructing short-range exchange GGA functionals.

CAM-B3LYP Exchange-Correlation Functional
The CAM-B3LYP exchange-correlation functional [273] uses Eq. (1.171) with short- and long-range versions of the B88 exchange functional and the same correlation functional used in B3LYP (i.e., $0.81\ E_{\text{c}}^{\text{LYP}} + 0.19\ E_{\text{c}}^{\text{LSDA}}$),

$$E_{\text{xc}}^{\text{CAM-B3LYP}}[\Phi] = a\ E_{\text{x}}^{\text{sr},\mu,\text{HF}}[\Phi] + b\ E_{\text{x}}^{\text{lr},\mu,\text{HF}}[\Phi] + (1-a)\ E_{\text{x}}^{\text{sr},\mu,\text{B88}}[\rho_{\uparrow,\Phi}, \rho_{\downarrow,\Phi}] + (1-b)\ E_{\text{x}}^{\text{lr},\mu,\text{B88}}[\rho_{\uparrow,\Phi}, \rho_{\downarrow,\Phi}] + 0.81 E_{\text{c}}^{\text{LYP}}[\rho_{\uparrow,\Phi}, \rho_{\downarrow,\Phi}] + 0.19 E_{\text{c}}^{\text{LSDA}}[\rho_{\uparrow,\Phi}, \rho_{\downarrow,\Phi}], \tag{1.174}$$

where the parameters $a = 0.19$ and $b = 0.65$ were optimized on atomization energies and the range-separation parameter $\mu = 0.33$ bohr^{-1} was taken from Ref. [250], where it was optimized on equilibrium distances of diatomic molecules. In this expression, the short-range B88 exchange functional $E_{\text{x}}^{\text{sr},\mu,\text{B88}}$ is defined by using in Eq. (1.169) the following generic GGA model for the exchange hole [121] (given here in its spin-independent version)

$$h_{\text{x}}^{\text{GGA}}(\rho, \nabla\rho, r_{12}) = -\rho\, \frac{9}{2} \left(\frac{j_1(k_{\text{GGA}} r_{12})}{k_{\text{GGA}} r_{12}} \right)^2, \tag{1.175}$$

with $k_{\text{GGA}} = k_{\text{F}}/\sqrt{e_{\text{x}}^{\text{GGA}}(\rho, \nabla\rho)/e_{\text{x}}^{\text{UEG}}(\rho)}$. The exchange-hole model of Eq. (1.175) properly yields the GGA exchange energy density $e_{\text{x}}^{\text{GGA}}(\rho, \nabla\rho)$ for $\mu = 0$ and thus allows one to extend any standard GGA exchange functional to a short-range GGA exchange functional. Note however that it does not fulfill the sum rule [Eq. (1.50)]. The long-range B88 exchange functional is then simply $E_{\text{x}}^{\text{lr},\mu,\text{B88}} = E_{\text{x}}^{\text{B88}} - E_{\text{x}}^{\text{sr},\mu,\text{B88}}$.

LC-ωPBE Exchange-Correlation Functional
The LC-ωPBE exchange-correlation functional [266, 267] uses a short-range version of the PBE exchange functional as well as the standard PBE correlation functional,

$$E_{\text{xc}}^{\text{LC-}\omega\text{PBE}}[\Phi] = E_{\text{x}}^{\text{lr},\mu,\text{HF}}[\Phi] + E_{\text{x}}^{\text{sr},\mu,\text{PBE}}[\rho_{\uparrow,\Phi}, \rho_{\downarrow,\Phi}] + E_{\text{c}}^{\text{PBE}}[\rho_{\uparrow,\Phi}, \rho_{\downarrow,\Phi}]. \tag{1.176}$$

The short-range PBE exchange functional is obtained by using in Eq. (1.169) the following GGA exchange hole model constructed to yield the PBE exchange energy [59],

$$h_{\text{x}}^{\text{PBE}}(\rho, \nabla\rho, r_{12}) = \rho J^{\text{PBE}}(s, k_{\text{F}} r_{12}), \tag{1.177}$$

where $s = |\nabla\rho|/(2k_F\rho)$ and

$$J^{\text{PBE}}(s,u) = \left[-\frac{\mathcal{A}}{u^2}\frac{1}{1+(4/9)\mathcal{A}u^2} + \left(\frac{\mathcal{A}}{u^2} + \mathcal{B} + \mathcal{C}[1+s^2\mathcal{F}(s)]u^2 + \mathcal{E}[1+s^2\mathcal{G}(s)]u^4\right) e^{-\mathcal{D}u^2} \right] e^{-s^2\mathcal{H}(s)u^2}. \tag{1.178}$$

Here, $\mathcal{A}$, $\mathcal{B}$, $\mathcal{C}$, $\mathcal{D}$, and $\mathcal{E}$ are constants chosen to obtain an oscillation-averaged UEG exchange hole for $s = 0$, and $\mathcal{F}(s)$, $\mathcal{G}(s)$ and $\mathcal{H}(s)$ are functions determined so that the hole yields the PBE exchange density for $\mu = 0$, and satisfies the sum rule [Eq. (1.50)] and the small-r_{12} expansion [Eq. (1.137)] using the gradient expansion of τ of Eq. (1.141). The range-separation parameter is fixed at $\mu = \omega = 0.4$ bohr^{-1} which has been found to be close to optimal for atomization energies, reaction barrier heights, and ionization energies [266].

ωB97X Exchange-Correlation Functional

The ωB97X exchange-correlation functional [34] has the form of Eq. (1.171) with $b = 1$:

$$E_{\text{xc}}^{\omega\text{B97X}}[\Phi] = a\ E_{\text{x}}^{\text{sr},\mu,\text{HF}}[\Phi] + E_{\text{x}}^{\text{lr},\mu,\text{HF}}[\Phi] + (1-a)\ E_{\text{x}}^{\text{sr},\mu,\text{B97-GGA}}[\rho_{\uparrow,\Phi},\rho_{\downarrow,\Phi}] + E_{\text{c}}^{\text{B97-GGA}}[\rho_{\uparrow,\Phi},\rho_{\downarrow,\Phi}]. \tag{1.179}$$

The short-range B97-GGA exchange density is defined as

$$e_{\text{x}}^{\text{sr},\mu,\text{B97-GGA}}(\rho_\uparrow,\rho_\downarrow,\nabla\rho_\uparrow,\nabla\rho_\downarrow) = \sum_{\sigma\in\{\uparrow,\downarrow\}} e_{\text{x},\sigma}^{\text{sr},\mu,\text{UEG}}(\rho_\sigma)\ g_{\text{x}}(x_\sigma),$$

where $e_{\text{x},\sigma}^{\text{sr},\mu,\text{UEG}}(\rho_\sigma) = e_{\text{x}}^{\text{sr},\mu,\text{UEG}}(\rho_\sigma, 0)$ is the spin-σ contribution to the short-range UEG exchange energy density [Eq. (1.173)] and the gradient correction $g_{\text{x}}(x_\sigma)$ where $x_\sigma = |\nabla\rho_\sigma|/\rho_\sigma^{4/3}$ has the same form as in Eq. (1.124) with polynomial degree $m = 4$. In Eq. (1.179), the correlation functional has the same form as the B97-GGA correlation functional but again with polynomial degree $m = 4$ in Eqs. (1.128) and (1.129). The fraction of short-range HF exchange $a \approx 0.16$, the range-separation parameter $\mu = \omega = 0.3$ bohr^{-1}, and the linear coefficients in Eqs. (1.124), (1.128), and (1.129) were optimized on sets of atomic energies, atomization energies, ionization energies, electron and proton affinities, reaction barrier heights, and noncovalent interactions, with the constraints $a + c_{\text{x},0} = 1$, $c_{\text{c},0}^{\uparrow\downarrow} = 1$, and $c_{\text{c},0}^{\sigma\sigma} = 1$ to enforce the correct UEG limit.

HSE Exchange-Correlation Functional
The Heyd–Scuseria–Ernzerhof (HSE) exchange-correlation functional [117] is of the form of Eq. (1.171) with $b = 0$ (i.e., no long-range HF exchange),

$$E_{\mathrm{xc}}^{\mathrm{HSE}}[\Phi] = a E_{\mathrm{x}}^{\mathrm{sr},\mu,\mathrm{HF}}[\Phi] + (1-a) E_{\mathrm{x}}^{\mathrm{sr},\mu,\mathrm{PBE}}[\rho_{\uparrow,\Phi}, \rho_{\downarrow,\Phi}] + E_{\mathrm{x}}^{\mathrm{lr},\mu,\mathrm{PBE}}[\rho_{\uparrow,\Phi}, \rho_{\downarrow,\Phi}] + E_{\mathrm{c}}^{\mathrm{PBE}}[\rho_{\uparrow,\Phi}, \rho_{\downarrow,\Phi}], \tag{1.180}$$

and involves the long-range PBE exchange functional $E_{\mathrm{x}}^{\mathrm{lr},\mu,\mathrm{PBE}} = E_{\mathrm{x}}^{\mathrm{PBE}} - E_{\mathrm{x}}^{\mathrm{sr},\mu,\mathrm{PBE}}$ complementary to the short-range PBE exchange functional constructed from the PBE exchange hole model [Eqs. (1.177) and (1.178)]. In order to reproduce reliable values for the band gap in semiconducting solids, the range-separation parameter is fixed at $\mu = 0.15$ bohr^{-1}, which is a very small value compared to the other range-separated hybrids. It means that the range of electron-electron distances covered by HF exchange is large, and the HSE functional could be thought of as a regular hybrid approximation but with the very long-range contribution of the HF exchange removed. This is particularly appropriate for solids since in these systems the very long-range HF exchange is effectively balanced by the correlation effects (a phenomenon known as screening). The fraction of (short-range) HF exchange is fixed at $a = 0.25$ like in the PBE0 hybrid functional.

1.5 Multideterminant Hybrid Approximations

1.5.1 Double-Hybrid Approximations

In 2006, Grimme [98] introduced a two-parameter *double-hybrid* (2DH) approximation, written here in its spin-independent version,

$$E_{\mathrm{xc}}^{\mathrm{2DH}} = a_{\mathrm{x}}\, E_{\mathrm{x}}^{\mathrm{HF}}[\Phi] + (1-a_{\mathrm{x}})\, E_{\mathrm{x}}^{\mathrm{GGA}}[\rho_\Phi] + (1-a_{\mathrm{c}}) E_{\mathrm{c}}^{\mathrm{GGA}}[\rho_\Phi] + a_{\mathrm{c}} E_{\mathrm{c}}^{\mathrm{MP2}}, \tag{1.181}$$

mixing a fraction a_{x} of the HF exchange energy with a GGA exchange functional, and a fraction a_{c} of the second-order Møller–Plesset (MP2) correlation energy $E_{\mathrm{c}}^{\mathrm{MP2}}$ with a GGA correlation functional. In Eq. (1.181), the first three terms are first calculated in a self-consistent manner, and then the last term $E_{\mathrm{c}}^{\mathrm{MP2}}$ is added perturbatively using the orbitals determined in the first step. The expression of $E_{\mathrm{c}}^{\mathrm{MP2}}$ is [247]

$$E_{\mathrm{c}}^{\mathrm{MP2}} = -\frac{1}{4} \sum_{i=1}^{N} \sum_{j=1}^{N} \sum_{a \geqslant N+1} \sum_{b \geqslant N+1} \frac{|\langle \phi_i \phi_j || \phi_a \phi_b \rangle|^2}{\varepsilon_a + \varepsilon_b - \varepsilon_i - \varepsilon_j}, \tag{1.182}$$

where i, j and a, b run over occupied and virtual spin orbitals, respectively, ε_k are spin orbital energies, and $\langle \phi_i \phi_j || \phi_a \phi_b \rangle = \langle \phi_i \phi_j | \phi_a \phi_b \rangle - \langle \phi_i \phi_j | \phi_b \phi_a \rangle$ are antisymmetrized two-electron integrals with (in physicists' notation)

$$\langle \phi_p \phi_q | \phi_r \phi_s \rangle = \int_{(\mathbb{R}^3 \times \{\uparrow,\downarrow\})^2} \frac{\phi_p^*(\mathbf{x}_1)\phi_q^*(\mathbf{x}_2)\phi_r(\mathbf{x}_1)\phi_s(\mathbf{x}_2)}{|\mathbf{r}_1 - \mathbf{r}_2|} \mathrm{d}\mathbf{x}_1 \mathrm{d}\mathbf{x}_2. \quad (1.183)$$

Note that the notation in Eq. (1.182) assumes that the one-electron wave-function space is spanned by a discrete set of spin orbitals. In the exact theory, the continuum limit of the set of virtual spin orbitals is implied.

The rigorous framework underlying these double-hybrid approximations was established by Sharkas et al. [226]. The idea is to decompose the universal density functional of Eq. (1.7) as

$$F[\rho] = \min_{\Psi \in \mathcal{W}_\rho^N} \langle \Psi | \widehat{T} + \lambda \widehat{W}_{\text{ee}} | \Psi \rangle + \bar{E}_{\text{Hxc}}^{\lambda}[\rho], \quad (1.184)$$

where $\lambda \in [0, 1]$ is a coupling constant and $\bar{E}_{\text{Hxc}}^{\lambda}[\rho]$ is a complementary density functional defined to make Eq. (1.184) exact. From Eqs. (1.10) and (1.62), we see that $\bar{E}_{\text{Hxc}}^{\lambda}[\rho] = E_{\text{Hxc}}[\rho] - E_{\text{Hxc}}^{\lambda}[\rho]$, where $E_{\text{Hxc}}[\rho]$ is the standard Hartree-exchange-correlation functional of the KS scheme and $E_{\text{Hxc}}^{\lambda}[\rho]$ is the Hartree-exchange-correlation functional along the adiabatic connection. The Hartree and exchange contributions are simply linear in λ,

$$\bar{E}_{\text{H}}^{\lambda}[\rho] = (1 - \lambda) E_{\text{H}}[\rho], \quad (1.185)$$

$$\bar{E}_{\text{x}}^{\lambda}[\rho] = (1 - \lambda) E_{\text{x}}[\rho], \quad (1.186)$$

where $E_{\text{H}}[\rho]$ and $E_{\text{x}}[\rho]$ are the standard Hartree and exchange functionals of the KS scheme. Moreover, from the uniform coordinate scaling relation of Eq. (1.84), we have

$$\bar{E}_{\text{c}}^{\lambda}[\rho] = E_{\text{c}}[\rho] - \lambda^2 E_{\text{c}}[\rho_{1/\lambda}], \quad (1.187)$$

where $E_{\text{c}}[\rho]$ is the standard correlation functional of the KS scheme and $\rho_{1/\lambda}(\mathbf{r}) = (1/\lambda)^3 \rho(\mathbf{r}/\lambda)$ is the scaled density. The decomposition in Eq. (1.184) leads to the following expression of the exact ground-state energy

$$E_0 = \inf_{\Psi \in \mathcal{W}^N} \left\{ \langle \Psi | \widehat{T} + \widehat{V}_{\text{ne}} + \lambda \widehat{W}_{\text{ee}} | \Psi \rangle + \bar{E}_{\text{Hxc}}^{\lambda}[\rho_\Psi] \right\}, \quad (1.188)$$

where the infimum is over general multideterminant wave functions $\Psi \in \mathcal{W}^N$. This constitutes a *multideterminant extension of the KS scheme*. Note that this multideterminant KS scheme can trivially be extended to spin-dependent density functionals and functionals depending on the kinetic-energy density [232].

The double-hybrid ansatz can be seen as a particular approximation within this multideterminant KS scheme [226]. To see this, we define a density-scaled one-parameter hybrid (DS1H) approximation by restricting the minimization in Eq. (1.188) to single-determinant wave functions $\Phi \in \mathcal{S}^N$,

$$E_0^{\text{DS1H},\lambda} = \inf_{\Phi \in \mathcal{S}^N} \left\{ \langle \Phi | \hat{T} + \hat{V}_{\text{ne}} + \lambda \hat{W}_{\text{ee}} | \Phi \rangle + \bar{E}_{\text{Hxc}}^{\lambda}[\rho_{\Phi}] \right\}, \tag{1.189}$$

obtaining an energy which necessarily depends on λ. A minimizing single-determinant wave function Φ^λ must satisfy the self-consistent eigenvalue equation

$$\left(\hat{T} + \hat{V}_{\text{ne}} + \lambda \hat{V}_{\text{Hx}}^{\text{HF}}[\Phi^\lambda] + \hat{\bar{V}}_{\text{Hxc}}^{\lambda}[\rho_{\Phi^\lambda}] \right) |\Phi^\lambda\rangle = \mathcal{E}_0^\lambda \, |\Phi^\lambda\rangle \,, \tag{1.190}$$

where $\hat{V}_{\text{Hx}}^{\text{HF}}[\Phi^\lambda]$ is the nonlocal HF potential operator evaluated with the DS1H wave function Φ^λ and $\hat{\bar{V}}_{\text{Hxc}}^{\lambda}[\rho_{\Phi^\lambda}]$ is the local Hartree-exchange-correlation potential operator generated by the energy functional $\bar{E}_{\text{Hxc}}^{\lambda}[\rho]$ and evaluated at the DS1H density ρ_{Φ^λ}. If written explicitly in terms of spin orbitals, Eq. (1.190) would have the form of the GKS equations [Eq. (1.40)]. The DS1H ground-state energy can be finally written as

$$\begin{aligned} E_0^{\text{DS1H},\lambda} &= \langle \Phi^\lambda | \hat{T} + \hat{V}_{\text{ne}} | \Phi^\lambda \rangle + E_{\text{H}}[\rho_{\Phi^\lambda}] + \lambda E_{\text{x}}^{\text{HF}}[\Phi^\lambda] \\ &\quad + (1-\lambda) E_{\text{x}}[\rho_{\Phi^\lambda}] + \bar{E}_{\text{c}}^{\lambda}[\rho_{\Phi^\lambda}], \end{aligned} \tag{1.191}$$

where the full Hartree energy $E_{\text{H}}[\rho]$ has been recomposed. The exchange-correlation energy in Eq. (1.191) is of the form of a hybrid approximation [Eq. (1.160)].

All that is missing in Eq. (1.191) is the correlation energy associated with the scaled interaction $\lambda \hat{W}_{\text{ee}}$. It can be calculated by a nonlinear Rayleigh–Schrödinger perturbation theory [5, 6, 69] starting from the DS1H reference. Consider the following energy expression with the perturbation parameter $\alpha \in [0, 1]$,

$$E_0^{\lambda,\alpha} = \inf_{\Psi \in \mathcal{W}^N} \left\{ \langle \Psi | \hat{T} + \hat{V}_{\text{ne}} + \lambda \hat{V}_{\text{Hx}}^{\text{HF}}[\Phi^\lambda] + \alpha\lambda \hat{W} | \Psi \rangle + \bar{E}_{\text{Hxc}}^{\lambda}[\rho_\Psi] \right\}, \tag{1.192}$$

where $\lambda \hat{W} = \lambda \left(\hat{W}_{\text{ee}} - \hat{V}_{\text{Hx}}^{\text{HF}}[\Phi^\lambda] \right)$ is the scaled Møller–Plesset perturbation operator. For $\alpha = 0$, the stationary equation associated with Eq. (1.192) reduces to the DS1H eigenvalue equation [Eq. (1.190)]. For $\alpha = 1$, Eq. (1.192) reduces to Eq. (1.188), so $E_0^{\lambda,\alpha=1}$ is the exact energy, independently of λ. The sum of the zeroth-order energy and first-order energy correction gives simply the DS1H energy, $E_0^{\text{DS1H},\lambda} = E_0^{\lambda,(0)} + E_0^{\lambda,(1)}$. Thanks to the existence of a Brillouin theorem just like in standard Møller–Plesset perturbation theory (see Refs. [5, 6, 69]), only double excitations contribute to the first-order wave-function correction $\Psi^{\lambda,(1)}$ and the second-order energy correction has a standard MP2 form

$$E_0^{\lambda,(2)} = \lambda^2 \langle \Phi^\lambda | \widehat{W} | \Psi^{\lambda,(1)} \rangle = \lambda^2 E_\mathrm{c}^\mathrm{MP2},$$

where $E_\mathrm{c}^\mathrm{MP2}$ has the expression in Eq. (1.182) with DS1H spin orbitals and associated orbital eigenvalues (which implicitly depend on λ). This second-order perturbation theory defines a density-scaled one-parameter double-hybrid (DS1DH) approximation

$$E_0^{\mathrm{DS1DH},\lambda} = E_0^{\mathrm{DS1H},\lambda} + E_0^{\lambda,(2)}, \tag{1.193}$$

which contains the exchange-correlation energy contribution

$$E_\mathrm{xc}^{\mathrm{DS1DH},\lambda} = \lambda E_\mathrm{x}^\mathrm{HF}[\Phi^\lambda] + (1-\lambda) E_\mathrm{x}[\rho_{\Phi^\lambda}] + \bar{E}_\mathrm{c}^\lambda[\rho_{\Phi^\lambda}] + \lambda^2 E_\mathrm{c}^\mathrm{MP2}. \tag{1.194}$$

To make connection with the double-hybrid ansatz of Eq. (1.181), we can also define a one-parameter double-hybrid (1DH) approximation, obtained by neglecting the density scaling in the correlation functional, i.e. $E_\mathrm{c}[\rho_{1/\lambda}] \approx E_\mathrm{c}[\rho]$ in Eq. (1.187),

$$E_\mathrm{xc}^{\mathrm{1DH},\lambda} = \lambda E_\mathrm{x}^\mathrm{HF}[\Phi^\lambda] + (1-\lambda) E_\mathrm{x}[\rho_{\Phi^\lambda}] + (1-\lambda^2) E_\mathrm{c}[\rho_{\Phi^\lambda}] + \lambda^2 E_\mathrm{c}^\mathrm{MP2}, \tag{1.195}$$

which, after using semilocal approximations for $E_\mathrm{x}[\rho]$ and $E_\mathrm{c}[\rho]$, has the form of Eq. (1.181) with parameters $a_\mathrm{x} = \lambda$ and $a_\mathrm{c} = \lambda^2$. In this rigorous formulation of the double-hybrid approximations, the fraction of HF exchange is thus connected to the fraction of MP2 correlation. Taking into account approximately the scaling of the density in $E_\mathrm{c}[\rho_{1/\lambda}]$, it has also been proposed to use $a_\mathrm{c} = \lambda^3$ [260]. Fromager [67] also proposed an extension of this rigorous formulation in order to justify the use of double-hybrid approximations with two parameters such that $a_\mathrm{c} \leqslant a_\mathrm{x}^2 = \lambda^2$.

An essential advantage of double-hybrid approximations is that the presence of nonlocal MP2 correlation allows one to use a larger fraction of nonlocal HF exchange, which helps decreasing the self-interaction error. This usually provides an improvement over hybrid approximations for molecular systems with sufficiently large electronic gaps. However, a large fraction of HF exchange and a fraction of MP2 correlation also generally means large static-correlation errors in systems with small HOMO-LUMO gaps.

The first and still best known double-hybrid approximation is B2PLYP [98], which is based on the B88 exchange functional and the LYP correlation functional,

$$E_\mathrm{xc}^\mathrm{B2PLYP} = a_\mathrm{x}\, E_\mathrm{x}^\mathrm{HF}[\Phi] + (1-a_\mathrm{x})\, E_\mathrm{x}^\mathrm{B88}[\rho_{\uparrow,\Phi}, \rho_{\downarrow,\Phi}] \\ + (1-a_\mathrm{c}) E_\mathrm{c}^\mathrm{LYP}[\rho_{\uparrow,\Phi}, \rho_{\downarrow,\Phi}] + a_\mathrm{c} E_\mathrm{c}^\mathrm{MP2},$$

and the parameters $a_\mathrm{x} = 0.53$ and $a_\mathrm{c} = 0.27$ have been optimized on a set of atomization energies. Interestingly, even though the two parameters have been optimized without any constraint, we have $a_\mathrm{c} \approx a_\mathrm{x}^2 = 0.28$ as predicted by Eq. (1.195).

It has also been proposed to use the spin-component-scaled (SCS) version of MP2 [95] to construct spin-component-scaled double-hybrid approximations of the form [136, 137]

$$E_{\mathrm{xc}}^{\text{SCS-DH}} = a_\mathrm{x}\, E_\mathrm{x}^{\mathrm{HF}}[\Phi] + (1 - a_\mathrm{x})\, E_\mathrm{x}^{\mathrm{GGA}}[\rho_\Phi] + (1 - a_\mathrm{c}) E_\mathrm{c}^{\mathrm{GGA}}[\rho_\Phi] + c_{\mathrm{OS}} E_{\mathrm{c,OS}}^{\mathrm{MP2}} + c_{\mathrm{SS}} E_{\mathrm{c,SS}}^{\mathrm{MP2}}, \tag{1.196}$$

which contains four empirical parameters a_x, a_c, c_{OS}, and c_{SS}. In this expression, $E_{\mathrm{c,OS}}^{\mathrm{MP2}}$ and $E_{\mathrm{c,SS}}^{\mathrm{MP2}}$ are the opposite-spin (OS) and same-spin (SS) contributions to the MP2 correlation energy obtained by restricting the sums over i and j in Eq. (1.182) to spin orbitals of opposite and same spins, respectively. Since in MP2 the same-spin component is usually overestimated relative to the opposite-spin component, this SCS variant is a simple way to achieve higher accuracy without increasing computational cost.

For reviews on different flavors of double hybrids and their assessments, the reader may consult Refs. [82, 171, 212, 241]. It has also been proposed to construct double-hybrid approximations where the MP2 correlation term is extended to a higher-order correlation method such as RPA [3, 100, 172, 173, 211] or coupled-cluster [35, 76]. More generally, the multideterminant extension of the KS scheme of Eq. (1.188) allows one to define hybrids combining any wave-function method with density functionals. For example, a multiconfiguration hybrid approximation based on Eq. (1.188) which combines a multiconfiguration self-consistent-field (MCSCF) wave function with density functionals has been proposed in the goal of tackling strongly correlated systems [225]. This approach has also been used to combine valence-bond (VB) theory [274] or variational two-electron reduced-density-matrix theory [176] with DFT.

1.5.2 *Range-Separated Double-Hybrid Approximations*

1.5.2.1 Range-Separated One-Parameter Double-Hybrid Approximations

In 2005, Ángyán et al. [6] introduced what could be called the first *range-separated one-parameter double-hybrid approximation*, i.e. combining HF exchange and MP2 correlation with density functionals using a one-parameter decomposition of the electron-electron interaction. This is based on the *range-separated multideterminant extension of the KS scheme* introduced earlier by Savin [216] (see, also, Refs. [215, 217, 255]) and which actually predates and inspired the multideterminant extension of the KS scheme of Eq. (1.188).

The idea is to decompose the universal density functional of Eq. (1.7) as

$$F[\rho] = \min_{\Psi \in \mathcal{W}_\rho^N} \langle \Psi | \widehat{T} + \widehat{W}_{\text{ee}}^{\text{lr},\mu} | \Psi \rangle + \bar{E}_{\text{Hxc}}^{\text{sr},\mu}[\rho], \tag{1.197}$$

where $\widehat{W}_{\text{ee}}^{\text{lr},\mu}$ is the long-range electron-electron interaction operator (associated with the pair potential $w_{\text{ee}}^{\text{lr},\mu}(r_{12}) = \text{erf}(\mu r_{12})/r_{12}$ as already used in the range-separated hybrids of Sect. 1.4.2) and $\bar{E}_{\text{Hxc}}^{\text{sr},\mu}[\rho]$ is the complementary short-range density functional defined to make Eq. (1.197) exact. As before, the parameter $\mu \in [0, +\infty)$ controls the range of the separation. The complementary short-range functional can be written as $\bar{E}_{\text{Hxc}}^{\text{sr},\mu}[\rho] = E_{\text{Hxc}}[\rho] - E_{\text{Hxc}}^{\text{lr},\mu}[\rho]$, where $E_{\text{Hxc}}[\rho]$ is the standard Hartree-exchange-correlation functional of the KS scheme and $E_{\text{Hxc}}^{\text{lr},\mu}[\rho]$ is the Hartree-exchange correlation functional associated with the long-range interaction $w_{\text{ee}}^{\text{lr},\mu}(r_{12})$. It is often convenient to decompose the short-range functional as (see Refs. [240, 254, 258] for an alternative decomposition)

$$\bar{E}_{\text{Hxc}}^{\text{sr},\mu}[\rho] = E_{\text{H}}^{\text{sr},\mu}[\rho] + E_{\text{x}}^{\text{sr},\mu}[\rho] + \bar{E}_{\text{c}}^{\text{sr},\mu}[\rho],$$

where $E_{\text{H}}^{\text{sr},\mu}[\rho]$ is the short-range Hartree functional,

$$E_{\text{H}}^{\text{sr},\mu}[\rho] = \frac{1}{2} \int_{\mathbb{R}^3 \times \mathbb{R}^3} \rho(\mathbf{r}_1)\rho(\mathbf{r}_2) w_{\text{ee}}^{\text{sr},\mu}(r_{12}) \mathbf{d}\mathbf{r}_1 \mathbf{d}\mathbf{r}_2,$$

with the short-range interaction $w_{\text{ee}}^{\text{sr},\mu}(r_{12}) = 1/r_{12} - w_{\text{ee}}^{\text{lr},\mu}(r_{12})$, $E_{\text{x}}^{\text{sr},\mu}[\rho]$ is the short-range exchange functional [Eq. (1.169)] which can also be written as

$$E_{\text{x}}^{\text{sr}}[\rho] = \langle \Phi[\rho] | \widehat{W}_{\text{ee}}^{\text{sr},\mu} | \Phi[\rho] \rangle - E_{\text{H}}^{\text{sr},\mu}[\rho],$$

with the KS single-determinant wave function $\Phi[\rho]$, and $\bar{E}_{\text{c}}^{\text{sr},\mu}[\rho]$ is the complementary short-range correlation functional. Just like for Eq. (1.188), the decomposition in Eq. (1.197) leads to the following expression of the exact ground-state energy

$$E_0 = \inf_{\Psi \in \mathcal{W}^N} \left\{ \langle \Psi | \widehat{T} + \widehat{V}_{\text{ne}} + \widehat{W}_{\text{ee}}^{\text{lr},\mu} | \Psi \rangle + \bar{E}_{\text{Hxc}}^{\text{sr},\mu}[\rho_\Psi] \right\}, \tag{1.198}$$

where the infimum is over general multideterminant wave functions $\Psi \in \mathcal{W}^N$.

To obtain an MP2/DFT hybrid scheme, we proceed analogously to Sect. 1.5.1. First, we define the following range-separated hybrid (RSH) approximation by restricting the minimization in Eq. (1.198) to single-determinant wave functions $\Phi \in \mathcal{S}^N$,

$$E_0^{\text{RSH},\mu} = \inf_{\Phi \in \mathcal{S}^N} \left\{ \langle \Phi | \widehat{T} + \widehat{V}_{\text{ne}} + \widehat{W}_{\text{ee}}^{\text{lr},\mu} | \Phi \rangle + \bar{E}_{\text{Hxc}}^{\text{sr},\mu}[\rho_\Phi] \right\}, \tag{1.199}$$

obtaining an energy which necessarily depends on μ. A minimizing single-determinant wave function Φ^μ must satisfy the self-consistent eigenvalue equation

$$\left(\hat{T} + \hat{V}_{\text{ne}} + \hat{V}_{\text{Hx}}^{\text{lr},\mu,\text{HF}}[\Phi^\mu] + \hat{\bar{V}}_{\text{Hxc}}^{\text{sr},\mu}[\rho_{\Phi^\mu}]\right) |\Phi^\mu\rangle = \mathcal{E}_0^\mu |\Phi^\mu\rangle, \quad (1.200)$$

where $\hat{V}_{\text{Hx}}^{\text{lr},\mu,\text{HF}}[\Phi^\mu]$ is the nonlocal long-range HF potential operator evaluated with the RSH wave function Φ^μ and $\hat{\bar{V}}_{\text{Hxc}}^{\text{sr},\mu}[\rho_{\Phi^\mu}]$ is the local short-range Hartree-exchange-correlation potential operator generated by the energy functional $\bar{E}_{\text{Hxc}}^{\text{sr},\mu}[\rho]$ and evaluated at the RSH density ρ_{Φ^μ}. The RSH ground-state energy can be finally written as

$$\begin{aligned} E_0^{\text{RSH},\mu} &= \langle\Phi^\mu|\, \hat{T} + \hat{V}_{\text{ne}}\, |\Phi^\mu\rangle + E_{\text{H}}[\rho_{\Phi^\mu}] + E_{\text{x}}^{\text{lr},\mu,\text{HF}}[\Phi^\mu] \\ &\quad + E_{\text{x}}^{\text{sr},\mu}[\rho_{\Phi^\mu}] + \bar{E}_{\text{c}}^{\text{sr},\mu}[\rho_{\Phi^\mu}], \end{aligned} \quad (1.201)$$

where the full Hartree energy $E_{\text{H}}[\rho]$ has been recomposed. The exchange-correlation energy in Eq. (1.201) has a similar form as in the LC scheme of Eq. (1.167).

To calculate the missing long-range correlation energy in Eq. (1.201), we can define a nonlinear Rayleigh–Schrödinger perturbation theory [5, 6, 69] starting from the RSH reference. We start from the following energy expression with the perturbation parameter $\alpha \in [0, 1]$,

$$E_0^{\mu,\alpha} = \inf_{\Psi\in\mathcal{W}^N} \left\{\langle\Psi|\, \hat{T}+\hat{V}_{\text{ne}}+\hat{V}_{\text{Hx}}^{\text{lr},\mu,\text{HF}}[\Phi^\mu]+\alpha\hat{W}^{\text{lr},\mu}\, |\Psi\rangle+\bar{E}_{\text{Hxc}}^{\text{lr},\mu}[\rho_\Psi]\right\}, \quad (1.202)$$

where $\hat{W}^{\text{lr},\mu} = \left(\hat{W}_{\text{ee}}^{\text{lr},\mu} - \hat{V}_{\text{Hx}}^{\text{lr},\mu,\text{HF}}[\Phi^\mu]\right)$ is the long-range Møller–Plesset perturbation operator. For $\alpha = 0$, the stationary equation associated with Eq. (1.202) reduces to the RSH eigenvalue equation [Eq. (1.200)]. For $\alpha = 1$, Eq. (1.202) reduces to Eq. (1.198), so $E_0^{\mu,\alpha=1}$ is the exact energy, independently of μ. The sum of the zeroth-order energy and first-order energy correction gives simply the RSH energy, $E_0^{\text{RSH},\mu} = E_0^{\mu,(0)} + E_0^{\mu,(1)}$. As in Sect. 1.5.1, only double excitations contribute to the first-order wave-function correction $\Psi^{\mu,(1)}$ and the second-order energy correction has a standard MP2 form

$$E_0^{\mu,(2)} = \langle\Phi^\mu|\, \hat{W}^{\text{lr},\mu}\, |\Psi^{\mu,(1)}\rangle = E_{\text{c}}^{\text{lr},\mu,\text{MP2}},$$

where $E_{\text{c}}^{\text{lr},\mu,\text{MP2}}$ has the same expression as in Eq. (1.182) with RSH spin orbitals and associated orbital eigenvalues (which implicitly depend on μ) but using the long-range two-electron integrals

$$\langle\phi_p\phi_q|\phi_r\phi_s\rangle^{\text{lr},\mu} = \int_{(\mathbb{R}^3\times\{\uparrow,\downarrow\})^2} \phi_p^*(\mathbf{x}_1)\phi_q^*(\mathbf{x}_2)\phi_r(\mathbf{x}_1)\phi_s(\mathbf{x}_2)w_{\text{ee}}^{\text{lr},\mu}(r_{12})\text{d}\mathbf{x}_1\text{d}\mathbf{x}_2, \quad (1.203)$$

instead of the standard two-electron integrals of Eq. (1.183). This second-order perturbation theory defines a RSH+MP2 approximation,

$$E_0^{\text{RSH+MP2},\mu} = E_0^{\text{RSH},\mu} + E_c^{\text{lr},\mu,\text{MP2}}, \tag{1.204}$$

which contains the exchange-correlation energy contribution

$$E_{\text{xc}}^{\text{RSH+MP2},\mu} = E_x^{\text{lr},\mu,\text{HF}}[\Phi^\mu] + E_x^{\text{sr},\mu}[\rho_{\Phi^\mu}] + \bar{E}_c^{\text{sr},\mu}[\rho_{\Phi^\mu}] + E_c^{\text{lr},\mu,\text{MP2}}. \tag{1.205}$$

When using semilocal density-functional approximations for the short-range functionals $E_x^{\text{sr},\mu}[\rho]$ and $\bar{E}_c^{\text{sr},\mu}[\rho]$, the RSH+MP2 exchange-correlation energy expression of Eq. (1.205) thus constitutes range-separated double-hybrid approximations similar to the double hybrids of Sect. 1.5.1. The optimal value for the range-separation parameter is often around $\mu \approx 0.5$ bohr^{-1} [77, 177]. This scheme has the advantage of dropping the long-range part of both the exchange and correlation density functionals, which are usually not well described by semilocal density-functional approximations. Moreover, using a long-range MP2 correlation energy has the advantage of leading to a correct qualitative description of London dispersion interaction energies [6, 78, 79, 251], while displaying a fast convergence with the one-electron basis size [63]. Similar to the SCS double hybrids [Eq. (1.196)], a SCS variant of the RSH+MP2 scheme has also been proposed [213].

The range-separated multideterminant extension of the KS scheme of Eq. (1.198) allows one to define various hybrid schemes combining any wave-function method with density functionals. For example, one can go beyond second order by using long-range coupled-cluster [75, 83, 84, 262] or random-phase approximations [124, 185, 257, 261, 262]. To describe strongly correlated systems, one can also use for the long-range part wave-function methods such as configuration interaction (CI) [31, 62, 152, 202], MCSCF [70, 71, 108], density-matrix renormalization group (DMRG) [107], or multireference perturbation theory [68]. Density-matrix functional theory (DMFT) [201, 209, 210] and Green-function methods [128, 207] have also been used for the long-range part.

We now consider the approximations used for $E_x^{\text{sr},\mu}[\rho]$ and $\bar{E}_c^{\text{sr},\mu}[\rho]$. In Sect. 1.4.2, we have already described the short-range exchange LDA [Eq. (1.172)] and some short-range exchange GGAs for $E_x^{\text{sr},\mu}[\rho]$. Here, we describe the short-range LDA correlation functional and another short-range GGA exchange-correlation functional.

Short-Range LDA Correlation Functional

The complementary short-range LDA (or LSDA) correlation functional is

$$\bar{E}_c^{\text{sr},\mu,\text{LSDA}}[\rho_\uparrow, \rho_\downarrow] = \int_{\mathbb{R}^3} \bar{e}_c^{\text{sr},\mu,\text{UEG}}(\rho_\uparrow(\mathbf{r}), \rho_\downarrow(\mathbf{r}))\text{d}\mathbf{r}, \tag{1.206}$$

where $\bar{e}_c^{\text{sr},\mu,\text{UEG}}(\rho_\uparrow, \rho_\downarrow) = \rho\ \bar{\varepsilon}_c^{\text{sr},\mu,\text{UEG}}(\rho_\uparrow, \rho_\downarrow)$ is the complementary short-range UEG correlation energy density. In this expression, $\bar{\varepsilon}_c^{\text{sr},\mu,\text{UEG}}(\rho_\uparrow, \rho_\downarrow)$ is defined by

$$\bar{\varepsilon}_c^{\text{sr},\mu,\text{UEG}}(\rho_\uparrow, \rho_\downarrow) = \varepsilon_c^{\text{UEG}}(\rho_\uparrow, \rho_\downarrow) - \varepsilon_c^{\text{lr},\mu,\text{UEG}}(\rho_\uparrow, \rho_\downarrow), \tag{1.207}$$

where $\varepsilon_c^{\text{UEG}}(\rho_\uparrow, \rho_\downarrow)$ and $\varepsilon_c^{\text{lr},\mu,\text{UEG}}(\rho_\uparrow, \rho_\downarrow)$ are the correlation energies per particle of the UEG with the standard Coulomb and long-range electron-electron interactions, respectively. A simple spin-independent parametrization of $\bar{\varepsilon}_c^{\text{sr},\mu,\text{UEG}}$ was given in Ref. [259]. A better spin-dependent parametrization was constructed in Ref. [186] which uses the PW92 parametrization for $\varepsilon_c^{\text{UEG}}(\rho_\uparrow, \rho_\downarrow)$ [Eq. (1.106)] and the following parametrization for $\varepsilon_c^{\text{lr},\mu,\text{UEG}}(\rho_\uparrow, \rho_\downarrow)$ in terms of $r_s = (3/(4\pi\rho))^{1/3}$ and $\zeta = (\rho_\uparrow - \rho_\downarrow)/\rho$:

$$\varepsilon_c^{\text{lr},\mu,\text{UEG}}(\rho_\uparrow, \rho_\downarrow) =$$

$$\frac{\left[\begin{array}{l}\phi_2(\zeta)^3 Q\left(\frac{\mu\sqrt{r_s}}{\phi_2(\zeta)}\right) + a_1(r_s,\zeta)\mu^3 + a_2(r_s,\zeta)\mu^4 + a_3(r_s,\zeta)\mu^5 + a_4(r_s,\zeta)\mu^6 \\ \qquad + a_5(r_s,\zeta)\mu^8\end{array}\right]}{(1 + b_0(r_s)^2\mu^2)^4}.$$

In this expression, $\phi_2(\zeta)$ is a spin-scaling function defined by Eq. (1.103), $Q(x)$ is a function determined from the small-μ and/or small-r_s limit,

$$Q(x) = \frac{2\ln(2) - 2}{\pi^2} \ln\left(\frac{1 + ax + bx^2 + cx^3}{1 + ax + dx^2}\right),$$

with $a = 5.84605$, $c = 3.91744$, $d = 3.44851$, $b = d - 3\pi\alpha/[4\ln(2) - 4]$, $\alpha = 4/(9\pi)^{1/3}$, and the functions $a_i(r_s, \zeta)$ are

$$a_1(r_s,\zeta) = 4b_0(r_s)^6 C_3(r_s,\zeta) + b_0(r_s)^8 C_5(r_s,\zeta),$$

$$a_2(r_s,\zeta) = 4b_0(r_s)^6 C_2(r_s,\zeta) + b_0(r_s)^8 C_4(r_s,\zeta) + 6b_0(r_s)^4 \varepsilon_c^{\text{PW92}}(r_s,\zeta),$$

$$a_3(r_s,\zeta) = b_0(r_s)^8 C_3(r_s,\zeta),$$

$$a_4(r_s,\zeta) = b_0(r_s)^8 C_2(r_s,\zeta) + 4b_0(r_s)^6 \varepsilon_c^{\text{PW92}}(r_s,\zeta),$$

$$a_5(r_s,\zeta) = b_0(r_s)^8 \varepsilon_c^{\text{PW92}}(r_s,\zeta),$$

where $\varepsilon_c^{\text{PW92}}(r_s, \zeta)$ is the PW92 parametrization of the UEG correlation energy per particle. The functions $C_i(r_s, \zeta)$ are determined from the large-μ limit,

$$C_2(r_s, \zeta) = -\frac{3g_c(0, r_s, \zeta)}{8r_s^3},$$

$$C_3(r_s, \zeta) = -\frac{g(0, r_s, \zeta)}{\sqrt{2\pi} r_s^3},$$

$$C_4(r_s, \zeta) = -\frac{9[g_c''(0, r_s, \zeta) + (1 - \zeta^2)D_2(r_s)]}{64r_s^3},$$

$$C_5(r_s, \zeta) = -\frac{9[g''(0, r_s, \zeta) + (1 - \zeta^2)D_3(r_s)]}{40\sqrt{2\pi} r_s^3},$$

where $g(0, r_s, \zeta)$ is the on-top pair-distribution function[8] of the Coulombic UEG and $g''(0, r_s, \zeta)$ is its second-order derivative with respect to r_{12} at $r_{12} = 0$, and similarly for their correlation parts $g_c(0, r_s, \zeta) = g(0, r_s, \zeta) - (1 - \zeta^2)/2$ and $g_c''(0, r_s, \zeta) = g''(0, r_s, \zeta) - \phi_8(\zeta)/(5\alpha^2 r_s^2)$ with $\phi_8(\zeta)$ defined by Eq. (1.103). The ζ-dependence of the latter quantities is assumed to be exchange-like, i.e. $g(0, r_s, \zeta) \approx (1 - \zeta^2)g(0, r_s, \zeta = 0)$ and $g''(0, r_s, \zeta) \approx \zeta_+^2 g''(0, r_s/\zeta_+^{1/3}, \zeta = 1) + \zeta_-^2 g''(0, r_s/\zeta_-^{1/3}, \zeta = 1)$ where $\zeta_\pm = (1 \pm \zeta)/2$. The on-top pair-distribution function has been parametrized in Ref. [85] as

$$g(0, r_s, \zeta = 0) = (1 - Br_s + Cr_s^2 + Dr_s^3 + Er_s^4)\mathrm{e}^{-Fr_s},$$

with $B = 0.7317 - d$, $C = 0.08193$, $D = -0.01277$, $E = 0.001859$, and $F = 0.7524$. The remaining functions were determined by fitting to QMC data:

$$b_0(r_s) = 0.784949 r_s,$$

$$g''(0, r_s, \zeta = 1) = \frac{2^{5/3}}{5\alpha^2 r_s^2} \frac{1 - 0.02267 r_s}{1 + 0.4319 r_s + 0.04 r_s^2},$$

$$D_2(r_s) = \frac{\mathrm{e}^{-0.547 r_s}}{r_s^2}(-0.388 r_s + 0.676 r_s^2),$$

$$D_3(r_s) = \frac{\mathrm{e}^{-0.31 r_s}}{r_s^3}(-4.95 r_s + r_s^2).$$

[8] For a general system, the pair-distribution function $g(\mathbf{r}_1, \mathbf{r}_2)$ is defined from the pair density $\rho_2(\mathbf{r}_1, \mathbf{r}_2)$ [Eq. (1.42)] as $\rho_2(\mathbf{r}_1, \mathbf{r}_2) = \rho(\mathbf{r}_1)\rho(\mathbf{r}_2)g(\mathbf{r}_1, \mathbf{r}_2)$. The on-top value is the value at electron coalescence, i.e. for $\mathbf{r}_1 = \mathbf{r}_2$.

Short-Range PBE(GWS) Exchange-Correlation Functional
The Goll–Werner–Stoll (GWS) variant of the short-range PBE exchange-correlation functional [83, 84] is a slight modification of the short-range PBE functional developed in Ref. [256]. The exchange energy density is

$$e_{\mathrm{x}}^{\mathrm{sr},\mu,\mathrm{PBE(GWS)}}(\rho,\nabla\rho) = e_{\mathrm{x}}^{\mathrm{sr},\mu,\mathrm{UEG}}(\rho)F_{\mathrm{x}}(s,\widetilde{\mu}), \tag{1.211}$$

with an enhancement factor of the same form as in the standard PBE exchange functional,

$$F_{\mathrm{x}}(s,\widetilde{\mu}) = 1+\kappa - \frac{\kappa}{1+b(\widetilde{\mu})s^2/\kappa}, \tag{1.212}$$

with $s = |\nabla\rho|/(2k_{\mathrm{F}}\rho)$ and $\widetilde{\mu} = \mu/(2k_{\mathrm{F}})$. In this expression, $\kappa = 0.840$, as in the standard PBE exchange functional, to saturate the local Lieb–Oxford bound (for $\mu = 0$) and $b(\widetilde{\mu}) = b^{\mathrm{PBE}}[b^{\mathrm{T}}(\widetilde{\mu})/b^{\mathrm{T}}(0)]\mathrm{e}^{-\alpha_{\mathrm{x}}\widetilde{\mu}^2}$ where $b^{\mathrm{PBE}} = 0.21951$ is the second-order gradient-expansion coefficient of the standard PBE exchange functional, and $b^{\mathrm{T}}(\widetilde{\mu})$ is a function coming from the second-order GEA of the short-range exchange energy [254, 256],

$$b^{\mathrm{T}}(\widetilde{\mu}) = \frac{-c_1(\widetilde{\mu})+c_2(\widetilde{\mu})\mathrm{e}^{1/(4\widetilde{\mu}^2)}}{c_3(\widetilde{\mu})+54c_4(\widetilde{\mu})\mathrm{e}^{1/(4\widetilde{\mu}^2)}}, \tag{1.213}$$

with $c_1(\widetilde{\mu}) = 1+22\widetilde{\mu}^2+144\widetilde{\mu}^4$, $c_2(\widetilde{\mu}) = 2\widetilde{\mu}^2(-7+72\widetilde{\mu}^2)$, $c_3(\widetilde{\mu}) = -864\widetilde{\mu}^4(-1+2\widetilde{\mu}^2)$, and $c_4(\widetilde{\mu}) = \widetilde{\mu}^2[-3-24\widetilde{\mu}^2+32\widetilde{\mu}^4+8\widetilde{\mu}\sqrt{\pi}\,\mathrm{erf}(1/(2\widetilde{\mu}))]$. Finally, $\alpha_{\mathrm{x}} = 19.0$ is a damping parameter optimized for the He atom.

Similarly, the correlation energy density has the same form as the standard PBE correlation functional,

$$\bar{e}_{\mathrm{c}}^{\mathrm{sr},\mu,\mathrm{PBE(GWS)}}(\rho_\uparrow,\rho_\downarrow,\nabla\rho_\uparrow,\nabla\rho_\downarrow) = \rho\left[\bar{\varepsilon}_{\mathrm{c}}^{\mathrm{sr},\mu,\mathrm{UEG}}(\rho_\uparrow,\rho_\downarrow) + H^\mu(\rho_\uparrow,\rho_\downarrow,t)\right],$$

with $t = |\nabla\rho|/(2\phi_2(\zeta)k_{\mathrm{s}}\rho)$ and the gradient correction

$$H^\mu(\rho_\uparrow,\rho_\downarrow,t) = A(0)\phi_2(\zeta)^3\ln\left[1+\frac{\beta(\mu)}{A(0)}t^2\frac{1+\mathcal{A}(\mu)t^2}{1+\mathcal{A}(\mu)t^2+\mathcal{A}(\mu)^2t^4}\right],$$

where

$$\mathcal{A}(\mu) = \frac{\beta(\mu)}{A(0)}\left[\exp(-\bar{\varepsilon}_{\mathrm{c}}^{\mathrm{sr},\mu,\mathrm{UEG}}(\rho_\uparrow,\rho_\downarrow)/(A(0)\phi_2(\zeta)^3)) - 1\right]^{-1},$$

and

$$\beta(\mu) = \beta^{\mathrm{PBE}}\left(\frac{\bar{\varepsilon}_{\mathrm{c}}^{\mathrm{sr},\mu,\mathrm{UEG}}(\rho_\uparrow,\rho_\downarrow)}{\bar{\varepsilon}_{\mathrm{c}}^{\mathrm{sr},\mu=0,\mathrm{UEG}}(\rho_\uparrow,\rho_\downarrow)}\right)^{\alpha_{\mathrm{c}}}, \tag{1.214}$$

and the value of $A(0)$ is given after Eq. (1.104). In Eq. (1.214), $\beta = 0.066725$ is the second-order gradient coefficient of the standard PBE correlation functional and $\alpha_c = 2.78$ is a damping parameter optimized for the He atom.

For $\mu = 0$, this short-range PBE exchange-correlation functional reduces to the standard PBE exchange-correlation functional and for large μ it reduces to the short-range LDA exchange-correlation functional.

1.5.2.2 Range-Separated Two-Parameter Double-Hybrid Approximations

In 2018, Kalai and Toulouse [127] introduced what we will call *range-separated two-parameter double-hybrid approximations*, combining HF exchange and MP2 correlation with density functionals using a two-parameter decomposition of the electron-electron in a way reminiscent of the CAM decomposition [Eq. (1.171)] (see, also, Refs. [40, 75]). This is based on a multideterminant extension of the KS scheme which generalizes the schemes of Sects. 1.5.1 and 1.5.2.1.

We first decompose the universal density functional of Eq. (1.7) as

$$F[\rho] = \min_{\Psi \in \mathcal{W}_\rho^N} \langle \Psi | \hat{T} + \hat{W}_{\text{ee}}^{\text{lr},\mu} + \lambda \hat{W}_{\text{ee}}^{\text{sr},\mu} | \Psi \rangle + \bar{E}_{\text{Hxc}}^{\text{sr},\mu,\lambda}[\rho], \tag{1.215}$$

where the parameter $\mu \in [0, +\infty)$ controls the range of the separation as always, the parameter $\lambda \in [0, 1]$ corresponds to the fraction of the short-range electron-electron interaction in the wave-function part, and $\bar{E}_{\text{Hxc}}^{\text{sr},\mu,\lambda}[\rho]$ is the complementary short-range density functional defined to make this decomposition exact. As before, the latter functional can be decomposed as

$$\bar{E}_{\text{Hxc}}^{\text{sr},\mu,\lambda}[\rho] = E_{\text{H}}^{\text{sr},\mu,\lambda}[\rho] + E_{\text{x}}^{\text{sr},\mu,\lambda}[\rho] + \bar{E}_{\text{c}}^{\text{sr},\mu,\lambda}[\rho].$$

The Hartree and exchange contributions are linear in λ,

$$E_{\text{H}}^{\text{sr},\mu,\lambda}[\rho] = (1-\lambda) E_{\text{H}}^{\text{sr},\mu}[\rho],$$

$$E_{\text{x}}^{\text{sr},\mu,\lambda}[\rho] = (1-\lambda) E_{\text{x}}^{\text{sr},\mu}[\rho], \tag{1.216}$$

where $E_{\text{H}}^{\text{sr},\mu}[\rho]$ and $E_{\text{x}}^{\text{sr},\mu}[\rho]$ are the short-range Hartree and exchange functionals introduced in Sect. 1.5.2.1, and the correlation contribution can be written as

$$\bar{E}_{\text{c}}^{\text{sr},\mu,\lambda}[\rho] = E_{\text{c}}[\rho] - E_{\text{c}}^{\mu,\lambda}[\rho],$$

where $E_c[\rho]$ is the standard KS correlation functional and $E_c^{\mu,\lambda}[\rho]$ is the correlation functional associated with the interaction $w_{ee}^{lr,\mu}(r_{12}) + \lambda w_{ee}^{sr,\mu}(r_{12})$. The exact ground-state energy can then be expressed as

$$E_0 = \inf_{\Psi \in \mathcal{W}^N} \left\{ \langle \Psi | \hat{T} + \hat{V}_{ne} + \hat{W}_{ee}^{lr,\mu} + \lambda \hat{W}_{ee}^{sr,\mu} | \Psi \rangle + \bar{E}_{Hxc}^{sr,\mu,\lambda}[\rho_\Psi] \right\}, \quad (1.217)$$

which constitutes a generalization of Eqs. (1.188) and (1.198).

To obtain a MP2/DFT hybrid scheme, we proceed in full analogy to Sects. 1.5.1 and 1.5.2.1. First, we define the following single-determinant range-separated two-parameter hybrid (RS2H) approximation,

$$E_0^{RS2H,\mu,\lambda} = \inf_{\Phi \in \mathcal{S}^N} \left\{ \langle \Phi | \hat{T} + \hat{V}_{ne} + \hat{W}_{ee}^{lr,\mu} + \lambda \hat{W}_{ee}^{sr,\mu} | \Phi \rangle + \bar{E}_{Hxc}^{sr,\mu,\lambda}[\rho_\Phi] \right\}, \quad (1.218)$$

and use it as a reference for defining a perturbation theory similarly to Eqs. (1.192) and (1.202). At second order, we obtain

$$E_0^{RS2H+MP2,\mu,\lambda} = E_0^{RS2H,\mu,\lambda} + E_c^{\mu,\lambda,MP2}, \quad (1.219)$$

where $E_c^{\mu,\lambda,MP2}$ is the MP2 correlation energy expression evaluated with RS2H spin orbitals and orbital eigenvalues, and the two-electron integrals associated with the interaction $w_{ee}^{lr,\mu}(r_{12}) + \lambda w_{ee}^{sr,\mu}(r_{12})$. This RS2H+MP2 scheme thus contains the exchange-correlation energy contribution

$$E_{xc}^{RS2H+MP2,\mu,\lambda} = E_x^{lr,\mu,HF}[\Phi^{\mu,\lambda}] + \lambda E_x^{sr,\mu,HF}[\Phi^{\mu,\lambda}] + (1-\lambda) E_x^{sr,\mu}[\rho_{\Phi^{\mu,\lambda}}] + \bar{E}_c^{sr,\mu,\lambda}[\rho_{\Phi^{\mu,\lambda}}] + E_c^{\mu,\lambda,MP2}, \quad (1.220)$$

where $\Phi^{\mu,\lambda}$ is a minimizing single-determinant wave function in Eq. (1.218).

A good approximation for the λ-dependence of the complementary correlation functional $\bar{E}_c^{sr,\mu,\lambda}[\rho]$ is [127]

$$\bar{E}_c^{sr,\mu,\lambda}[\rho] \approx \bar{E}_c^{sr,\mu}[\rho] - \lambda^2 \bar{E}_c^{sr,\mu\sqrt{\lambda}}[\rho], \quad (1.221)$$

where $\bar{E}_c^{sr,\mu}[\rho]$ is the short-range correlation functional introduced in Sect. 1.5.2.1. In particular, the λ-dependence in Eq. (1.221) is correct both in the high-density limit, for a non-degenerate KS system, and in the low-density limit. Thanks to Eqs. (1.216) and (1.221), the semilocal density-functional approximations for $E_x^{sr,\mu}[\rho]$ and $\bar{E}_c^{sr,\mu}[\rho]$ of Sect. 1.5.2.1 can be reused here without developing new ones. In Ref. [127], the short-range PBE(GWS) exchange and correlation functionals were used, and the optimal parameters $\mu = 0.46$ bohr^{-1} and $\lambda = 0.58$ were found on small sets of atomization energies and reaction barrier heights, i.e. values similar to the ones usually used separately in range-separated hybrids and double hybrids.

The RS2H+MP2 scheme improves a bit over the RSH+MP2 scheme of Sect. 1.5.2.1, particularly for interaction energies of hydrogen-bonded systems. Even if the presence of short-range MP2 correlation deteriorates in principle the convergence rate with the one-electron basis size, in practice the fraction of pure short-range MP2 correlation ($\lambda^2 \approx 0.34$) is small enough to keep a fast basis convergence. Accuracy can be improved, particularly for dispersion interactions, by supplanting the MP2 term by coupled-cluster or random-phase approximations [126]. Like for the approach of Sect. 1.5.2.1, many wave-function methods could be used in the general scheme of Eq. (1.217).

1.6 Semiempirical Dispersion Corrections and Nonlocal van der Waals Density Functionals

Among the previously considered exchange-correlation approximations, only the range-separated double hybrids of Sect. 1.5.2, thanks to their long-range nonlocal correlation component, are capable of fully describing London dispersion interactions, crucial for describing weakly bonded systems. To improve the other approximations (semilocal functionals, single-determinant hybrids, double hybrids without range separation) for weakly bonded systems, it has been proposed to add to them a semiempirical dispersion correction or a nonlocal van der Waals density functional. We now describe these approaches.

1.6.1 Semiempirical Dispersion Corrections

To explicitly account for London dispersion interactions, it has been proposed in the 2000s to add to the standard approximate functionals a *semiempirical dispersion correction* of the form [52, 96, 97, 272]

$$E_{\text{disp}} = -s \sum_{\alpha=1}^{N_{\text{n}}} \sum_{\substack{\beta=1 \\ \beta>\alpha}}^{N_{\text{n}}} f(R_{\alpha\beta}) \frac{C_6^{\alpha\beta}}{R_{\alpha\beta}^6}, \tag{1.222}$$

where $R_{\alpha\beta}$ is the distance between each pair of atoms and $C_6^{\alpha\beta}$ is the London dispersion coefficient between these atoms. Here, $f(R_{\alpha\beta})$ is a damping function which tends to 1 at large $R_{\alpha\beta}$ and tends to zero at small $R_{\alpha\beta}$, e.g.

$$f(R_{\alpha\beta}) = \frac{1}{1 + \mathrm{e}^{-d(R_{\alpha\beta}/R_{\alpha\beta}^{\text{vdW}} - 1)}}, \tag{1.223}$$

with the sum of tabulated atomic van der Waals radii $R_{\alpha\beta}^{\text{vdW}} = R_{\alpha}^{\text{vdW}} + R_{\beta}^{\text{vdW}}$ and a constant d, and s is a scaling parameter that can be adjusted for each approximate functional. The dispersion coefficient $C_6^{\alpha\beta}$ for any pair of atoms is empirically calculated from tabulated same-atom dispersion coefficients $C_6^{\alpha\alpha}$ and/or atomic polarizabilities. This approach was named "DFT-D" by Grimme [96].

The last version of DFT-D (referred to as DFT-D3) also includes $C_8^{\alpha\beta}$ two-body terms and $C_9^{\alpha\beta\gamma}$ three-body terms [99]. There have also been various proposals to make the determination of dispersion coefficients less empirical, such as the scheme of Becke and Johnson [25] based on the exchange-hole dipole moment, the scheme of Tkatchenko and Scheffler [253] based on a Hirshfeld atomic partitioning, or the scheme of Sato and Nakai [214] based on the local-response approximation [45].

The "DFT-D" approach provides a big and inexpensive improvement for the description of weakly bonded systems. One limitation is that the semiempirical dispersion correction, being just a force field in its simplest variant, affects only the molecular geometry of the system but not directly its electronic structure. Some of the most used DFT-D functionals are:

- The PBE-D exchange-correlation functional [97], based on the PBE functional with a scaling parameter $s = 0.75$;
- The B97-D exchange-correlation functional [97], based on the B97-GGA functional with a scaling parameter $s = 1.25$ and reoptimized linear coefficients in Eqs. (1.124), (1.128), and (1.129) in the presence of the semiempirical dispersion correction;
- The B3LYP-D exchange-correlation functional [97], based on the B3LYP hybrid functional with a scaling parameter $s = 1.05$;
- The ωB97X-D exchange-correlation functional [33], based on the ωB97X range-separated hybrid functional with a scaling parameter $s = 1$, a modified damping function, and reoptimized parameters in ωB97X in the presence of the semiempirical dispersion correction.

The semiempirical dispersion correction can also be added to double-hybrid approximations. For example, B2PLYP-D [218] is based on the B2PLYP double hybrid with a scaling parameter $s = 0.55$. The scaling parameter is small since the fraction of MP2 correlation in B2PLYP already partially takes into account dispersion interactions. It has also been proposed to add a semiempirical dispersion correction to the SCS version of the double hybrids [Eq. (1.196)], resulting in a family of dispersion-corrected spin-component-scaled double-hybrid (DSD) approximations [136–138]. An example of double hybrid is this latter family is DSD-BLYP [136], which uses the B88 exchange functional and the LYP correlation functional.

1.6.2 Nonlocal van der Waals Density Functionals

Another approach to describe dispersion interactions is to add to the standard approximate functionals a so-called *nonlocal van der Waals density functional* of the form [43, 151, 268–270]

$$E_c^{nl}[\rho] = \frac{1}{2} \int_{\mathbb{R}^3 \times \mathbb{R}^3} \rho(\mathbf{r}_1)\rho(\mathbf{r}_2)\phi(\mathbf{r}_1, \mathbf{r}_2) d\mathbf{r}_1 d\mathbf{r}_2, \tag{1.224}$$

where $\phi(\mathbf{r}_1, \mathbf{r}_2)$ is a correlation kernel. Two main families of such nonlocal correlation functionals exist: the "van der Waals density functionals" (vdW-DF) [43, 151] and the Vydrov–Van Voorhis (VV) functionals [268–270].

We will only describe the last version of the VV functionals, i.e. the VV10 nonlocal correlation functional [270]. In this functional, the correlation kernel is taken as

$$\phi^{VV10}(\mathbf{r}_1, \mathbf{r}_2) = -\frac{3}{2g(\mathbf{r}_1, r_{12})g(\mathbf{r}_2, r_{12})(g(\mathbf{r}_1, r_{12}) + g(\mathbf{r}_2, r_{12}))} + \beta\delta(\mathbf{r}_1 - \mathbf{r}_2),$$

where $r_{12} = |\mathbf{r}_2 - \mathbf{r}_1|$ is the interelectronic distance, β is a constant determining the local (delta-distribution) part of the kernel, and the function g is defined as

$$g(\mathbf{r}, r_{12}) = \omega_0(\mathbf{r})r_{12}^2 + \kappa(\mathbf{r}). \tag{1.225}$$

In Eq. (1.225), $\omega_0(\mathbf{r}) = \sqrt{\omega_g(\mathbf{r})^2 + \frac{\omega_p(\mathbf{r})^2}{3}}$ involves the square of the local plasma frequency $\omega_p(\mathbf{r})^2 = 4\pi\rho(\mathbf{r})$ and the square of the local band gap $\omega_g(\mathbf{r})^2 = C|\nabla\rho(\mathbf{r})|^4/\rho(\mathbf{r})^4$, where C is an adjustable parameter controlling the large-r_{12} asymptotic dispersion coefficients, and $\kappa(\mathbf{r}) = b\ k_F(\mathbf{r})^2/\omega_p(\mathbf{r})$, where $k_F(\mathbf{r}) = (3\pi^2\rho(\mathbf{r}))^{1/3}$ is the local Fermi wave vector and b is an adjustable parameter controlling the short-range damping of the large-r_{12} asymptote. As expected for dispersion interactions, in the large-r_{12} limit, $\phi^{VV10}(\mathbf{r}_1, \mathbf{r}_2)$ behaves as $1/r_{12}^6$:

$$\phi^{VV10}(\mathbf{r}_1, \mathbf{r}_2) \underset{r_{12}\to\infty}{\sim} -\frac{3}{2\omega_0(\mathbf{r}_1)\omega_0(\mathbf{r}_2)(\omega_0(\mathbf{r}_1) + \omega_0(\mathbf{r}_2))r_{12}^6}.$$

The constant $\beta = (3/b^2)^{3/4}/16$ is chosen to make $E_c^{nl}[\rho]$ vanish in the uniform density limit, thus leaving this limit unchanged when $E_c^{nl}[\rho]$ is added to another density functional. The adjustable parameters $C \approx 0.009$ and $b \approx 6$ are found by optimization of C_6 dispersion coefficients and of weak intermolecular interaction energies, respectively, the precise values depending on which exchange-correlation functional the VV10 correction is added to.

Nonlocal van der Waals density functionals are necessarily more computationally expensive than semiempirical dispersion corrections. However, they have the advantage of being less empirical and, since they are functionals of the density, of impacting directly on the electronic structure of the system. The VV10 nonlocal functional has been incorporated in a number of recently developed exchange-correlation functionals, for example:

- The ωB97X-V exchange-correlation functional [169], based on the ωB97X range-separated hybrid [Eq. (1.179)] with reoptimized linear coefficients in Eq. (1.124) with polynomial degree $m = 2$ and in Eqs. (1.128) and (1.129) with polynomial degree $m = 1$, as well as reoptimized VV10 parameters $C = 0.01$ and $b = 6.0$;
- The ωB97M-V exchange-correlation functional [170], based on the ωB97X range-separated hybrid [Eq. (1.179)] but with more general and combinatorially optimized meta-GGA exchange and correlation enhancement factors and the same VV10 parameters $C = 0.01$ and $b = 6.0$ as in ωB97X-V.

1.7 Orbital-Dependent Exchange-Correlation Density Functionals

We discuss here some exchange-correlation density functionals explicitly depending on the KS orbitals (for a review, see Ref. [140]). Since the KS orbitals are themselves functionals of the density, these exchange-correlation expressions are thus *implicit* functionals of the density (for notational simplicity, this dependence on the density of the orbitals and other intermediate quantities will not be explicitly indicated). In fact, the single-determinant and multideterminant hybrid approximations of Sects. 1.4 and 1.5 already belong to this family, with the caveat that the orbitals are obtained with a *nonlocal* potential. In this section, we are concerned with orbital-dependent exchange-correlation energy functionals with orbitals obtained with a local potential, i.e. staying within the KS scheme.[9] These approximations tend to be more computationally involved than the approximations previously seen and have thus been much less used so far.

[9] The boundary between the various single-determinant and multideterminant hybrids of Sects. 1.4 and 1.5 and the orbital-dependent functionals of the present section is however thin. For example, it is possible to optimize the orbitals using a local potential in hybrids or range-separated hybrids [8, 131, 229], and in double hybrids or range-separated double hybrids [230, 231].

1.7.1 Exact Exchange

The *exact exchange* (EXX) energy functional [Eq. (1.16)] can be expressed in terms of the KS orbitals,

$$E_x[\rho] = -\frac{1}{2} \sum_{\sigma\in\{\uparrow,\downarrow\}} \sum_{i=1}^{N_\sigma} \sum_{j=1}^{N_\sigma} \int_{\mathbb{R}^3\times\mathbb{R}^3} \frac{\varphi_{i\sigma}^*(\mathbf{r}_1)\varphi_{j\sigma}(\mathbf{r}_1)\varphi_{j\sigma}^*(\mathbf{r}_2)\varphi_{i\sigma}(\mathbf{r}_2)}{|\mathbf{r}_1 - \mathbf{r}_2|} \mathrm{d}\mathbf{r}_1 \mathrm{d}\mathbf{r}_2, \tag{1.226}$$

and has exactly the same form as the HF exchange [Eq. (1.159)], but the orbitals used in both expressions are in general different.

Since the exact exchange energy in Eq. (1.226) is not an explicit functional of the density, the corresponding exchange potential $v_x(\mathbf{r}) = \delta E_x[\rho]/\delta\rho(\mathbf{r})$ cannot be calculated directly. We can however find an workable equation for $v_x(\mathbf{r})$ by first considering the functional derivative of $E_x[\rho]$ with respect to the KS potential $v_s(\mathbf{r})$ and then applying the chain rule:

$$\frac{\delta E_x[\rho]}{\delta v_s(\mathbf{r})} = \int_{\mathbb{R}^3} \frac{\delta E_x[\rho]}{\delta\rho(\mathbf{r}')} \frac{\delta\rho(\mathbf{r}')}{\delta v_s(\mathbf{r})} \mathrm{d}\mathbf{r}'. \tag{1.227}$$

Introducing the non-interacting KS static linear-response function $\chi_0(\mathbf{r}', \mathbf{r}) = \delta\rho(\mathbf{r}')/\delta v_s(\mathbf{r})$, we can rewrite Eq. (1.227) as

$$\int_{\mathbb{R}^3} v_x(\mathbf{r}')\chi_0(\mathbf{r}', \mathbf{r})\mathrm{d}\mathbf{r}' = \frac{\delta E_x[\rho]}{\delta v_s(\mathbf{r})},$$

which is known as the *optimized-effective-potential* (OEP) equation for the exact-exchange potential [91, 92, 248].

Using first-order perturbation theory on the KS system, explicit expressions in terms of the orbitals can be derived for $\chi_0(\mathbf{r}', \mathbf{r})$ and $\delta E_x[\rho]/\delta v_s(\mathbf{r})$. The expression of $\chi_0(\mathbf{r}', \mathbf{r})$ is

$$\chi_0(\mathbf{r}', \mathbf{r}) = - \sum_{\sigma\in\{\uparrow,\downarrow\}} \sum_{i=1}^{N_\sigma} \sum_{a\geqslant N_\sigma+1} \frac{\varphi_{i\sigma}^*(\mathbf{r}')\varphi_{a\sigma}^*(\mathbf{r})\varphi_{i\sigma}(\mathbf{r})\varphi_{a\sigma}(\mathbf{r}')}{\varepsilon_{a\sigma} - \varepsilon_{i\sigma}} + \text{c.c.}\,,$$

where i and a run over occupied and virtual spatial orbitals, respectively, and c.c. stands for the complex conjugate. The expression of $\delta E_x[\rho]/\delta v_s(\mathbf{r})$ is

$$\frac{\delta E_x[\rho]}{\delta v_s(\mathbf{r})} = \sum_{\sigma\in\{\uparrow,\downarrow\}} \sum_{i=1}^{N_\sigma} \sum_{j=1}^{N_\sigma} \sum_{a\geqslant N_\sigma+1} \langle\varphi_{a\sigma}\varphi_{j\sigma}|\varphi_{j\sigma}\varphi_{i\sigma}\rangle \frac{\varphi_{a\sigma}(\mathbf{r})\varphi_{i\sigma}^*(\mathbf{r})}{\varepsilon_{a\sigma} - \varepsilon_{i\sigma}} + \text{c.c.}\,,$$

where $\langle \varphi_{a\sigma}\varphi_{j\sigma}|\varphi_{j\sigma}\varphi_{i\sigma}\rangle$ are two-electron integrals over KS spatial orbitals:

$$\langle \varphi_{a\sigma}\varphi_{j\sigma}|\varphi_{j\sigma}\varphi_{i\sigma}\rangle = \int_{\mathbb{R}^3\times\mathbb{R}^3} \frac{\varphi^*_{a\sigma}(\mathbf{r}_1)\varphi^*_{j\sigma}(\mathbf{r}_2)\varphi_{j\sigma}(\mathbf{r}_1)\varphi_{i\sigma}(\mathbf{r}_2)}{|\mathbf{r}_1-\mathbf{r}_2|}\mathrm{d}\mathbf{r}_1\mathrm{d}\mathbf{r}_2. \tag{1.228}$$

Applying this OEP method with the EXX energy (and no correlation energy functional) is an old idea [227, 248], but reasonably efficient calculations for molecules have been possible only relatively recently [87, 122]. The EXX occupied orbitals turn out to be very similar to the HF occupied orbitals, and thus the EXX ground-state properties are also similar to the HF ones. However, the EXX virtual orbitals (which see a $-1/r$ asymptotic potential for a neutral system) tend to be much less diffuse than the HF virtual orbitals (which see an exponentially decaying potential for a neutral system), and may be more adapted for calculating excited-state properties.

1.7.2 Second-Order Görling–Levy Perturbation Theory

In 1993, Görling and Levy [90, 91] developed a perturbation theory in terms of the coupling constant λ of the adiabatic connection (Sect. 1.2.2) which provides an explicit orbital-dependent second-order approximation for the correlation energy functional. The Hamiltonian along the adiabatic connection [Eq. (1.60)] can be written as

$$\begin{aligned} \widehat{H}^{\lambda} &= \widehat{T} + \lambda\widehat{W}_{\mathrm{ee}} + \widehat{V}^{\lambda} \\ &= \widehat{H}_{\mathrm{s}} + \lambda(\widehat{W}_{\mathrm{ee}} - \widehat{V}_{\mathrm{Hx}}) - \widehat{V}_{\mathrm{c}}^{\lambda}, \end{aligned} \tag{1.229}$$

where $\widehat{H}_{\mathrm{s}} = \widehat{H}^{\lambda=0} = \widehat{T} + \widehat{V}_{\mathrm{s}}$ is the KS non-interacting reference Hamiltonian (which will be assumed to have a nondegenerate ground state). Equation (1.229) was obtained by decomposing the potential operator keeping the density constant as $\widehat{V}^{\lambda} = \widehat{V}_{\mathrm{s}} - \lambda\widehat{V}_{\mathrm{Hx}} - \widehat{V}_{\mathrm{c}}^{\lambda}$, where $\widehat{V}_{\mathrm{s}} = \widehat{V}^{\lambda=0}$ is the KS potential operator, $\lambda\widehat{V}_{\mathrm{Hx}}$ is the Hartree-exchange potential operator which is linear in λ, and $\widehat{V}_{\mathrm{c}}^{\lambda}$ is the correlation potential which starts at second order in λ, i.e. $\widehat{V}_{\mathrm{c}}^{\lambda} = \lambda^2\widehat{V}_{\mathrm{c}}^{(2)} + \cdots$. Using a complete set of orthonormal eigenfunctions Φ_n and eigenvalues $\mathcal{E}_n$ of the KS Hamiltonian, $\widehat{H}_{\mathrm{s}}|\Phi_n\rangle = \mathcal{E}_n|\Phi_n\rangle$, the normalized ground-state wave function of the Hamiltonian $\widehat{H}^{\lambda}$ can be expanded as $\Psi^{\lambda} = \Phi + \lambda\Psi^{(1)} + \cdots$ where $\Phi = \Phi_0$ is the ground-state KS single-determinant wave function and $\Psi^{(1)}$ is its first-order correction given by

$$|\Psi^{(1)}\rangle = -\sum_{n\neq 0} \frac{\langle\Phi_n|\,\widehat{W}_{\mathrm{ee}} - \widehat{V}_{\mathrm{Hx}}\,|\Phi\rangle}{\mathcal{E}_n - \mathcal{E}_0}\,|\Phi_n\rangle\,.$$

Using the expression in Eq. (1.63), the correlation energy functional can also be expanded in powers of λ:

$$E_c^\lambda[\rho] = \langle\Psi^\lambda|\,\hat{T} + \lambda\hat{W}_{ee}\,|\Psi^\lambda\rangle - \langle\Phi|\,\hat{T} + \lambda\hat{W}_{ee}\,|\Phi\rangle\,. \\ = E_c^{(0)} + \lambda E_c^{(1)} + \lambda^2 E_c^{(2)} + \cdots\,. \tag{1.230}$$

Since $\Psi^{\lambda=0} = \Phi$, the zeroth-order term vanishes: $E_c^{(0)} = 0$. Using the expression of the first-order derivative of E_c^λ with respect to λ in Eq. (1.64), i.e. $\partial E_c^\lambda/\partial\lambda = \langle\Psi^\lambda|\,\hat{W}_{ee}\,|\Psi^\lambda\rangle - \langle\Phi|\,\hat{W}_{ee}\,|\Phi\rangle$, we find that the first-order term vanishes as well: $E_c^{(1)} = 0$. The second-order term corresponds to the *second-order Görling–Levy (GL2) correlation energy* and is given by

$$E_c^{\text{GL2}}[\rho] = E_c^{(2)} = \langle\Phi|\,\hat{W}_{ee}\,|\Psi^{(1)}\rangle = \langle\Phi|\,\hat{W}_{ee} - V_{\text{Hx}}\,|\Psi^{(1)}\rangle\,, \tag{1.231}$$

where the second equality comes from the fact that $\langle\Phi|\,\hat{V}_{\text{Hx}}\,|\Psi^{(1)}\rangle = 0$ since it is the derivative with respect to λ at $\lambda = 0$ of $\langle\Psi^\lambda|\,\hat{V}_{\text{Hx}}\,|\Psi^\lambda\rangle = \int_{\mathbb{R}^3} v_{\text{Hx}}(\mathbf{r})\rho(\mathbf{r})\mathrm{d}\mathbf{r}$, which does not depend on λ by virtue of the fact that the density $\rho(\mathbf{r})$ is constant along the adiabatic connection. Using the last expression in Eq. (1.231) allows one to express the GL2 correlation energy as

$$E_c^{\text{GL2}}[\rho] = -\sum_{n\neq 0} \frac{|\,\langle\Phi|\,\hat{W}_{ee} - \hat{V}_{\text{Hx}}\,|\Phi_n\rangle\,|^2}{\mathcal{E}_n - \mathcal{E}_0}. \tag{1.232}$$

It is instructive to decompose the GL2 correlation energy as

$$E_c^{\text{GL2}}[\rho] = E_c^{\text{MP2}} + E_c^{\text{S}}, \tag{1.233}$$

where E_c^{MP2} is a MP2-like correlation energy evaluated with KS spin orbitals,

$$E_c^{\text{MP2}} = -\frac{1}{4}\sum_{i=1}^{N}\sum_{j=1}^{N}\sum_{a\geqslant N+1}\sum_{b\geqslant N+1} \frac{|\,\langle\phi_i\phi_j|\,|\phi_a\phi_b\rangle\,|^2}{\varepsilon_a + \varepsilon_b - \varepsilon_i - \varepsilon_j}, \tag{1.234}$$

and E_c^{S} is the contribution coming from the single excitations (which does not vanish here, contrary to HF-based MP2 perturbation theory),

$$E_c^{\text{S}} = -\sum_{i=1}^{N}\sum_{a\geqslant N+1} \frac{|\,\langle\phi_i|\,\hat{V}_x^{\text{HF}} - \hat{V}_x\,|\phi_a\rangle\,|^2}{\varepsilon_a - \varepsilon_i}, \tag{1.235}$$

involving the difference between the integrals over the nonlocal HF exchange potential $\langle\phi_i|\,\hat{V}_x^{\text{HF}}\,|\phi_a\rangle = -\sum_{j=1}^{N}\langle\phi_i\phi_j|\phi_j\phi_a\rangle$ and over the local KS exchange potential $\langle\phi_i|\,\hat{V}_x\,|\phi_a\rangle = \int_{\mathbb{R}^3\times\{\uparrow,\downarrow\}} \phi_i^*(\mathbf{x})v_x(\mathbf{r})\phi_a(\mathbf{x})\mathrm{d}\mathbf{x}$.

Calculations of the GL2 correlation energy using either a non-self-consistent post-EXX implementation or a more complicated OEP self-consistent procedure have been tested (see, e.g., Refs. [53, 94, 175]) but the results are often disappointing. It is preferable to go beyond second order with random-phase approximations in the adiabatic-connection fluctuation-dissipation approach.

1.7.3 Adiabatic-Connection Fluctuation-Dissipation Approach

1.7.3.1 Exact Adiabatic-Connection Fluctuation-Dissipation Expression

Using the adiabatic-connection formula of Eq. (1.65), the correlation energy functional can be written as

$$\begin{aligned} E_c[\rho] &= \int_0^1 \mathrm{d}\lambda \, \langle\Psi^\lambda|\, \widehat{W}_{ee} \,|\Psi^\lambda\rangle - \langle\Phi|\, \widehat{W}_{ee} \,|\Phi\rangle \\ &= \frac{1}{2}\int_0^1 \mathrm{d}\lambda \int_{\mathbb{R}^3\times\mathbb{R}^3} \frac{\rho_{2,c}^\lambda(\mathbf{r}_1,\mathbf{r}_2)}{|\mathbf{r}_1-\mathbf{r}_2|}\mathrm{d}\mathbf{r}_1\mathrm{d}\mathbf{r}_2, \end{aligned} \tag{1.236}$$

where $\rho_{2,c}^\lambda(\mathbf{r}_1,\mathbf{r}_2) = \rho_2^\lambda(\mathbf{r}_1,\mathbf{r}_2) - \rho_{2,\mathrm{KS}}(\mathbf{r}_1,\mathbf{r}_2)$ is the correlation part of the pair density along the adiabatic connection. The pair density $\rho_2^\lambda(\mathbf{r}_1,\mathbf{r}_2)$ can be expressed with the pair-density operator $\widehat{\rho}_2(\mathbf{r}_1,\mathbf{r}_2) = \widehat{\rho}(\mathbf{r}_1)\widehat{\rho}(\mathbf{r}_2) - \delta(\mathbf{r}_1-\mathbf{r}_2)\widehat{\rho}(\mathbf{r}_1)$ where $\widehat{\rho}(\mathbf{r})$ is the density operator,

$$\begin{aligned} \rho_2^\lambda(\mathbf{r}_1,\mathbf{r}_2) &= \langle\Psi^\lambda|\, \widehat{\rho}_2(\mathbf{r}_1,\mathbf{r}_2) \,|\Psi^\lambda\rangle \\ &= \langle\Psi^\lambda|\, \widehat{\rho}(\mathbf{r}_1)\widehat{\rho}(\mathbf{r}_2) \,|\Psi^\lambda\rangle - \delta(\mathbf{r}_1-\mathbf{r}_2)\, \langle\Psi^\lambda|\, \widehat{\rho}(\mathbf{r}_1) \,|\Psi^\lambda\rangle, \end{aligned}$$

and the KS pair density $\rho_{2,\mathrm{KS}}(\mathbf{r}_1,\mathbf{r}_2)$ simply corresponds to the case $\lambda = 0$,

$$\begin{aligned} \rho_{2,\mathrm{KS}}(\mathbf{r}_1,\mathbf{r}_2) &= \rho_2^{\lambda=0}(\mathbf{r}_1,\mathbf{r}_2) \\ &= \langle\Phi|\, \widehat{\rho}(\mathbf{r}_1)\widehat{\rho}(\mathbf{r}_2) \,|\Phi\rangle - \delta(\mathbf{r}_1-\mathbf{r}_2)\, \langle\Phi|\, \widehat{\rho}(\mathbf{r}_1) \,|\Phi\rangle. \end{aligned}$$

Since the density does not change with λ, i.e. $\langle\Psi^\lambda|\, \widehat{\rho}(\mathbf{r}) \,|\Psi^\lambda\rangle = \langle\Phi|\, \widehat{\rho}(\mathbf{r}) \,|\Phi\rangle = \rho(\mathbf{r})$, the correlation pair density needed in Eq. (1.236) can thus be expressed as

$$\rho_{2,c}^\lambda(\mathbf{r}_1,\mathbf{r}_2) = \langle\Psi^\lambda|\, \widehat{\rho}(\mathbf{r}_1)\widehat{\rho}(\mathbf{r}_2) \,|\Psi^\lambda\rangle - \langle\Phi|\, \widehat{\rho}(\mathbf{r}_1)\widehat{\rho}(\mathbf{r}_2) \,|\Phi\rangle. \tag{1.237}$$

We would like to calculate $\rho^{\lambda}_{2,c}(\mathbf{r}_1, \mathbf{r}_2)$ without having to calculate the complicated many-body wave function Ψ^{λ}. For this, we consider the retarded dynamic linear-response function along the adiabatic connection in frequency space (or the so-called Lehmann representation)

$$\chi_{\lambda}(\mathbf{r}_1, \mathbf{r}_2; \omega) \tag{1.238}$$

$$= \sum_{n\neq 0} \frac{\langle\Psi^{\lambda}|\,\widehat{\rho}(\mathbf{r}_1)\,|\Psi^{\lambda}_n\rangle\,\langle\Psi^{\lambda}_n|\,\widehat{\rho}(\mathbf{r}_2)\,|\Psi^{\lambda}\rangle}{\omega - \omega^{\lambda}_n + i0^+} - \frac{\langle\Psi^{\lambda}|\,\widehat{\rho}(\mathbf{r}_2)\,|\Psi^{\lambda}_n\rangle\,\langle\Psi^{\lambda}_n|\,\widehat{\rho}(\mathbf{r}_1)\,|\Psi^{\lambda}\rangle}{\omega + \omega^{\lambda}_n + i0^+},$$

where the sums are over all eigenstates Ψ^{λ}_n of the Hamiltonian $\widehat{H}^{\lambda}$, i.e. $\widehat{H}^{\lambda}\,|\Psi^{\lambda}_n\rangle = E^{\lambda}_n\,|\Psi^{\lambda}_n\rangle$, except the ground state $\Psi^{\lambda} = \Psi^{\lambda}_0$, and $\omega^{\lambda}_n = E^{\lambda}_n - E^{\lambda}_0$ are the corresponding excitation energies. By contour integrating $\chi_{\lambda}(\mathbf{r}_1, \mathbf{r}_2, \omega)$ around the right half ω-complex plane, we arrive at the (zero-temperature) *fluctuation-dissipation theorem*,

$$n^{\lambda}_{2,c}(\mathbf{r}_1, \mathbf{r}_2) = -\int_{-\infty}^{+\infty} \frac{\mathrm{d}\omega}{2\pi}\,[\chi_{\lambda}(\mathbf{r}_1, \mathbf{r}_2, i\omega) - \chi_0(\mathbf{r}_1, \mathbf{r}_2, i\omega)], \tag{1.239}$$

which relates ground-state correlations in the time-independent system $\rho^{\lambda}_{2,c}(\mathbf{r}_1, \mathbf{r}_2)$ to the linear response of the system due to a time-dependent external perturbation $\chi_{\lambda}(\mathbf{r}_1, \mathbf{r}_2, \omega)$.

Combining Eqs. (1.236) and (1.239), we finally obtain the exact *adiabatic-connection fluctuation-dissipation* (ACFD) formula for the correlation energy [146, 147] (see, also, Ref. [106]):

$$E_c[\rho] = -\frac{1}{2}\int_0^1 \mathrm{d}\lambda \int_{-\infty}^{+\infty} \frac{\mathrm{d}\omega}{2\pi} \int_{\mathbb{R}^3\times\mathbb{R}^3} \frac{\chi_{\lambda}(\mathbf{r}_1, \mathbf{r}_2; i\omega) - \chi_0(\mathbf{r}_1, \mathbf{r}_2; i\omega)}{|\mathbf{r}_1 - \mathbf{r}_2|}\mathrm{d}\mathbf{r}_1\mathrm{d}\mathbf{r}_2. \tag{1.240}$$

The usefulness of the ACFD formula is due to the fact that there are practical ways of directly calculating $\chi_{\lambda}(\mathbf{r}_1, \mathbf{r}_2; \omega)$ without having to calculate the many-body wave function Ψ^{λ}. In linear-response time-dependent density-functional theory (TDDFT), one can find a Dyson-like equation for $\chi_{\lambda}(\mathbf{r}_1, \mathbf{r}_2; \omega)$,

$$\chi_{\lambda}(\mathbf{r}_1, \mathbf{r}_2; \omega) = \chi_0(\mathbf{r}_1, \mathbf{r}_2; \omega) \tag{1.241}$$

$$+ \int_{\mathbb{R}^3\times\mathbb{R}^3} \chi_0(\mathbf{r}_1, \mathbf{r}_3; \omega) f^{\lambda}_{\mathrm{Hxc}}(\mathbf{r}_3, \mathbf{r}_4; \omega) \chi_{\lambda}(\mathbf{r}_4, \mathbf{r}_2; \omega)\mathrm{d}\mathbf{r}_3\mathrm{d}\mathbf{r}_4,$$

where $f^{\lambda}_{\mathrm{Hxc}}(\mathbf{r}_3, \mathbf{r}_4; \omega)$ is the *Hartree-exchange-correlation kernel* associated to the Hamiltonian H^{λ}. Here, Eq. (1.241) will be considered as the definition for $f^{\lambda}_{\mathrm{Hxc}}$. In principle, the exact correlation energy can be obtained with Eqs. (1.240) and (1.241). In practice, however, we need to use an approximation for $f^{\lambda}_{\mathrm{Hxc}}$.

1.7.3.2 Random-Phase Approximations

In the *direct random-phase approximation* (dRPA, also just referred to as RPA, or sometimes as time-dependent Hartree), only the Hartree part of the kernel, which is linear in λ and independent from ω, is retained [146, 148],

$$f_{\text{Hxc}}^{\text{dRPA},\lambda}(\mathbf{r}_1, \mathbf{r}_2; \omega) = f_{\text{H}}^{\lambda}(\mathbf{r}_1, \mathbf{r}_2) = \lambda w_{\text{ee}}(\mathbf{r}_1, \mathbf{r}_2), \tag{1.242}$$

where $w_{\text{ee}}(\mathbf{r}_1, \mathbf{r}_2) = 1/|\mathbf{r}_1 - \mathbf{r}_2|$ is the Coulomb interaction, and the corresponding dRPA linear-response function then satisfies the equation

$$\begin{aligned}\chi_{\lambda}^{\text{dRPA}}(\mathbf{r}_1, \mathbf{r}_2; \omega) &= \chi_0(\mathbf{r}_1, \mathbf{r}_2; \omega) \\ &+ \lambda \int_{\mathbb{R}^3\times\mathbb{R}^3} \chi_0(\mathbf{r}_1, \mathbf{r}_3; \omega) w_{\text{ee}}(\mathbf{r}_3, \mathbf{r}_4) \chi_{\lambda}^{\text{dRPA}}(\mathbf{r}_4, \mathbf{r}_2; \omega) \mathrm{d}\mathbf{r}_3 \mathrm{d}\mathbf{r}_4.\end{aligned} \tag{1.243}$$

The physical contents of this approximation can be seen by iterating Eq. (1.243), which generates an infinite series,

$$\begin{aligned}\chi_{\lambda}^{\text{dRPA}}(\mathbf{r}_1, \mathbf{r}_2; \omega) &= \chi_0(\mathbf{r}_1, \mathbf{r}_2; \omega) \\ &+ \lambda \int_{\mathbb{R}^3\times\mathbb{R}^3} \chi_0(\mathbf{r}_1, \mathbf{r}_3; \omega) w_{\text{ee}}(\mathbf{r}_3, \mathbf{r}_4) \chi_0(\mathbf{r}_4, \mathbf{r}_2; \omega) \mathrm{d}\mathbf{r}_3 \mathrm{d}\mathbf{r}_4 \\ &+ \lambda^2 \int_{\mathbb{R}^3\times\mathbb{R}^3\times\mathbb{R}^3\times\mathbb{R}^3} \chi_0(\mathbf{r}_1, \mathbf{r}_3; \omega) w_{\text{ee}}(\mathbf{r}_3, \mathbf{r}_4) \chi_0(\mathbf{r}_4, \mathbf{r}_5; \omega) w_{\text{ee}}(\mathbf{r}_5, \mathbf{r}_6) \\ &\qquad \chi_0(\mathbf{r}_6, \mathbf{r}_2; \omega) \mathrm{d}\mathbf{r}_3 \mathrm{d}\mathbf{r}_4 \mathrm{d}\mathbf{r}_5 \mathrm{d}\mathbf{r}_6 \\ &+ \cdots,\end{aligned}$$

which, after plugging it into Eq. (1.240), leads to the dRPA correlation energy as the following perturbation expansion[10]

$$\begin{aligned}E_{\text{c}}^{\text{dRPA}}[\rho] &= -\frac{1}{2}\int_0^1 \mathrm{d}\lambda \int_{-\infty}^{+\infty} \frac{\mathrm{d}\omega}{2\pi} \Bigg[\lambda \int_{\mathbb{R}^3\times\mathbb{R}^3\times\mathbb{R}^3\times\mathbb{R}^3} \frac{\chi_0(\mathbf{r}_1, \mathbf{r}_3; i\omega)\chi_0(\mathbf{r}_4, \mathbf{r}_2; i\omega)}{|\mathbf{r}_1 - \mathbf{r}_2|\,|\mathbf{r}_3 - \mathbf{r}_4|} \mathrm{d}\mathbf{r}_1 \mathrm{d}\mathbf{r}_2 \mathrm{d}\mathbf{r}_3 \mathrm{d}\mathbf{r}_4 \\ &+ \lambda^2 \int_{\mathbb{R}^3\times\mathbb{R}^3\times\mathbb{R}^3\times\mathbb{R}^3\times\mathbb{R}^3\times\mathbb{R}^3} \frac{\chi_0(\mathbf{r}_1, \mathbf{r}_3; i\omega)\chi_0(\mathbf{r}_4, \mathbf{r}_5; i\omega)\chi_0(\mathbf{r}_6, \mathbf{r}_2; i\omega)}{|\mathbf{r}_1 - \mathbf{r}_2|\,|\mathbf{r}_3 - \mathbf{r}_4|\,|\mathbf{r}_5 - \mathbf{r}_6|} \\ &\qquad \mathrm{d}\mathbf{r}_1 \mathrm{d}\mathbf{r}_2 \mathrm{d}\mathbf{r}_3 \mathrm{d}\mathbf{r}_4 \mathrm{d}\mathbf{r}_5 \mathrm{d}\mathbf{r}_6 + \cdots \Bigg].\end{aligned} \tag{1.244}$$

---

[10] Using the operator viewpoint, the series in Eq. (1.244) can be formally summed in the form $E_{\text{c}}^{\text{dRPA}}[\rho] = 1/(4\pi) \int_{-\infty}^{+\infty} \mathrm{d}\omega\, \mathrm{Tr}[\ln(1 - \chi_0(i\omega) w_{\text{ee}}) + \chi_0(i\omega) w_{\text{ee}}]$ (see, e.g., Ref. [178]).

Using now the Lehmann representation [Eq. (1.238)] of the KS dynamic linear-response function in terms of the KS orbitals and their energies,

$$\chi_0(\mathbf{r}_1, \mathbf{r}_2; \omega) = \sum_{\sigma\in\{\uparrow,\downarrow\}} \sum_{i=1}^{N_\sigma} \sum_{a\geqslant N_\sigma+1} \left[\frac{\varphi_{i\sigma}^*(\mathbf{r}_1)\varphi_{a\sigma}(\mathbf{r}_1)\varphi_{a\sigma}^*(\mathbf{r}_2)\varphi_{i\sigma}(\mathbf{r}_2)}{\omega - (\varepsilon_{a\sigma} - \varepsilon_{i\sigma}) + i0^+} - \frac{\varphi_{i\sigma}^*(\mathbf{r}_2)\varphi_{a\sigma}(\mathbf{r}_2)\varphi_{a\sigma}^*(\mathbf{r}_1)\varphi_{i\sigma}(\mathbf{r}_1)}{\omega + (\varepsilon_{a\sigma} - \varepsilon_{i\sigma}) + i0^+} \right], \tag{1.245}$$

one can obtain, after quite some work,

$$E_{\mathrm{c}}^{\mathrm{dRPA}}[\rho] = -\frac{1}{2}\sum_{i=1}^{N}\sum_{j=1}^{N}\sum_{a\geqslant N+1}\sum_{b\geqslant N+1} \frac{|\langle\phi_i\phi_j|\phi_a\phi_b\rangle|^2}{\varepsilon_a+\varepsilon_b-\varepsilon_i-\varepsilon_j} + \sum_{i=1}^{N}\sum_{j=1}^{N}\sum_{k=1}^{N}\sum_{a\geqslant N+1}\sum_{b\geqslant N+1}\sum_{c\geqslant N+1} \frac{\langle\phi_i\phi_j|\phi_a\phi_b\rangle\langle\phi_j\phi_k|\phi_b\phi_c\rangle\langle\phi_k\phi_i|\phi_c\phi_a\rangle}{(\varepsilon_a+\varepsilon_b-\varepsilon_i-\varepsilon_j)(\varepsilon_a+\varepsilon_c-\varepsilon_i-\varepsilon_k)} + \cdots. \tag{1.246}$$

The dRPA correlation energy is the sum of all the direct terms (i.e., no exchange terms) of the perturbation expansion up to infinite order. In the language of diagrammatic perturbation theory, we say that the dRPA correlation energy is the sum of all direct ring diagrams. Of course, Eq. (1.246) is not the way to calculate the dRPA correlation energy in practice. This is done by solving the Dyson equation [Eq. (1.243)] without explicitly expanding in powers of λ, e.g. using matrix equations from linear-response TDDFT [72, 261] or coupled-cluster theory [219, 262].

Most dRPA correlation energy (combined with the EXX energy) calculations are done in a non-self-consistent way, but self-consistent OEP dRPA calculations have also been performed [26, 112]. One of the main advantage of dRPA is that it accounts for long-range dispersion interactions [46–48]. However, it shows large self-interaction errors. To overcome the latter drawback and improve the general accuracy, one can add exchange and beyond terms in various ways (see, e.g., Refs. [7, 14, 15, 39, 58, 88, 101, 110, 111, 114–116, 120, 178, 257, 261]). This remains an active area of research. For reviews on random-phase approximations, the reader may consult Refs. [37, 61, 208].

Acknowledgments I thank Éric Cancès and Gero Friesecke for discussions and comments on the manuscript. This review chapter grew out of my lecture notes for DFT courses given in several summer schools (ISTPC summer schools in June 2015 and 2017 in France, ICS summer school in July 2015 in France, and ESQC summer school in September 2019 in Italy). This work has received funding from the European Research Council (ERC) under the European Union's Horizon 2020 research and innovation programme (grant agreement EMC2 No 810367).

Suggested Further Reading

Books

R. M. Dreizler and E. K. U. Gross, *Density Functional Theory*. Springer-Verlag, Berlin (1990).

E. Engel and R. M. Dreizler. *Density Functional Theory: An Advanced Course, Theoretical and Mathematical Physics*, Springer-Verlag, Berlin Heidelberg (2011).

C. Fiolhais, F. Nogueira and M. A. L. Marques, eds., *A Primer in Density Functional Theory*, Vol. 620 of Lecture Notes in Physics, Springer, Berlin (2003).

R. G. Parr and W. Yang, *Density-Functional Theory of Atoms and Molecules*, Oxford University Press, New York (1989).

T. Tsuneda, *Density Functional Theory in Quantum Chemistry*, Springer, Tokyo (2014).

Review Articles

A. D. Becke. Perspective: Fifty years of density-functional theory in chemical physics. *J. Chem. Phys.* **140**, 18A301 (2014).

K. Burke. Perspective on density functional theory. *J. Chem. Phys.* **136**, 150901 (2012).

K. Capelle. A bird's-eye view of density-functional theory. *Braz. J. Phys.* **36**, 1318 (2006).

A. J. Cohen. P. Mori-Sánchez and W. Yang. Challenges for density functional theory. *Chem. Rev.* **112**, 289–320 (2012).

R. O. Jones. Density functional theory: Its origins, rise to prominence, and future. *Rev. Mod. Phys.* **87**, 897 (2015).

N. Mardirossian and M. Head-Gordon. Thirty years of density functional theory in computational chemistry: an overview and extensive assessment of 200 density functionals. *Mol. Phys.* **115**, 2315–2372 (2017).

J. P. Perdew and K. Schmidt, Jacob's ladder of density functional approximations for the exchange-correlation energy. *AIP Conf. Proc.* **577**, 1 (2001).

G.E. Scuseria and V.N. Staroverov. Development of approximate exchange-correlation functionals. In: *Theory and Applications of Computational Chemistry: The First 40 years*, edited by C.E. Dykstra, G. Frenking, K.S. Kim and G.E. Scuseria, pp. 669–724, Elsevier (2005).

V.N. Staroverov. Density-functional approximations for exchange and correlation. In: *A Matter of Density: Exploring the Electron Density Concept in the Chemical, Biological, and Materials Sciences*, edited by N. Sukumar, pp. 125–156, John Wiley & Sons, Hoboken, NJ (2013).

H. S. Yu, S. L. Li and D. G. Truhlar. Perspective: Kohn-Sham density functional theory descending a staircase. *J. Chem. Phys.* 145, 130901 (2016).

References

1. C. Adamo and V. Barone. Toward Reliable Density Functional Methods without Adjustable Parameters: The PBE0 Model. *J. Chem. Phys.* **110**, 6158 (1999).
2. C. Adamo, M. Ernzerhof and G.E. Scuseria. The meta-GGA functional: Thermochemistry with a kinetic energy density dependent exchange-correlation functional. *J. Chem. Phys.* **112**, 2643 (2000).
3. S. Ahnen, A.-S. Hehn, K.D. Vogiatzis, M.A. Rachsel, S. Leutwyler and W. Klopper. Accurate computations of the structures and binding energies of the imidazole...benzene and pyrrole...benzene complexes. *Chem. Phys.* **441**, 17–22 (2014).
4. C.O. Almbladh and U. von Barth. Exact results for the charge and spin densities, exchange-correlation potentials, and density-functional eigenvalues. *Phys. Rev. B* **31**, 3231 (1985).
5. J.G. Ángyán. Rayleigh–Schrödinger many-body perturbation theory for density functionals: A unified treatment of one- and two-electron perturbations. *Phys. Rev. A* **78**, 022510 (2008).
6. J.G. Ángyán, I.C. Gerber, A. Savin and J. Toulouse. Van der Waals forces in density functional theory: Perturbational long-range electron-interaction corrections. *Phys. Rev. A* **72**, 012510 (2005).
7. J.G. Ángyán, R.-F. Liu, J. Toulouse and G. Jansen. Correlation energy expressions from the adiabatic-connection fluctuation dissipation theorem approach. *J. Chem. Theory Comput.* **7**, 3116–3130 (2011).
8. A.V. Arbuznikov. Hybrid exchange correlation functional and potentials: Concept elaboration. *J. Struct. Chem.* **48**, S1–S31 (2007).
9. A.V. Arbuznikov, M. Kaupp, V.G. Malkin, R. Reviakine and O.L. Malkina. Validation study of meta-GGA functionals and of a model exchange-correlation potential in density functional calculations of EPR parameters. *Phys. Chem. Chem. Phys.* **4**(22), 5467–5474 (2002).
10. N. Argaman, J. Redd, A.C. Cancio and K. Burke. Leading correction to the local density approximation for exchange in large-Z atoms. *Phys. Rev. Lett.* **129**, 153001 (2022).
11. T. Aschebrock and S. Kümmel. Exploring local range separation: The role of spin scaling and one-electron self-interaction. *J. Chem. Phys.* **151**, 154108 (2019).
12. R. Baer, E. Livshits and U. Salzner. Tuned range-separated hybrids in density functional theory. *Annu. Rev. Phys. Chem.* **61**, 85–109 (2010).
13. U. von Barth and L. Hedin. A local exchange-correlation potential for the spin polarized case. I. *J. Phys. C* **5**, 1629–1642 (1972).
14. J.E. Bates and F. Furche. Communication: Random phase approximation renormalized many-body perturbation theory. *J. Chem. Phys.* **139**, 171103 (2013).
15. J.E. Bates, N. Sengupta, J. Sensenig and A. Ruzsinszky. Adiabatic connection without coupling constant integration. *J. Chem. Theory Comput.* **14**, 2979 (2018).
16. A.D. Becke. Hartree–Fock exchange energy of an inhomogeneous electron gas. *Int. J. Quantum Chem.* **23**, 1915–1922 (1983).
17. A.D. Becke. Density functional calculations of molecular bond energies. *J. Chem. Phys.* **84**, 4524 (1986).
18. A.D. Becke. Density-functional exchange-energy approximation with correct asymptotic behavior. *Phys. Rev. A* **38**, 3098 (1988).
19. A.D. Becke. A new mixing of Hartree–Fock and local density-functional theories. *J. Chem. Phys.* **98**, 1372 (1993).
20. A.D. Becke. Density-functional thermochemistry. III. The role of exact exchange. *J. Chem. Phys.* **98**, 5648 (1993).
21. A.D. Becke. Density-functional thermochemistry. IV. A new dynamical correlation functional and implications for exact-exchange mixing. *J. Chem. Phys.* **104**, 1040 (1996).
22. A.D. Becke. Density-functional thermochemistry. V. Systematic optimization of exchange-correlation functionals. *J. Chem. Phys.* **107**, 8554 (1997).
23. A.D. Becke. A new inhomogeneity parameter in density-functional theory. *J. Chem. Phys.* **109**, 2092 (1998).

24. A.D. Becke. Simulation of delocalized exchange by local density functionals. *J. Chem. Phys.* **112**, 4020 (2000).
25. A. D. Becke and E.R. Johnson. Exchange-hole dipole moment and the dispersion interaction revisited. *J. Chem. Phys.* **127**, 154108 (2007).
26. P. Bleiziffer, A. Heßelmann and A. Görling. Efficient self-consistent treatment of electron correlation within the random phase approximation. *J. Chem. Phys.* **139**, 084113 (2013).
27. M. Brack, B.K. Jennings and Y.H. Chu. On the extended Thomas–Fermi approximation to the kinetic energy density. *Phys. Lett. B* **65**(1), 1–4 (1976).
28. K. Burke, A. Cancio, T. Gould and S. Pittalis. Locality of correlation in density functional theory. *J. Chem. Phys.* **145**, 054112 (2016).
29. K. Burke, J.P. Perdew and Y. Wang. Derivation of a generalized gradient approximation: The PW91 density functional. In: *Electronic Density Functional Theory*, edited by J.F. Dobson, G. Vignale, M.P. Das, pp. 81–111, Plenum, NY (1997).
30. A. Cancio, G.P. Chen, B.T. Krull and K. Burke. Fitting a round peg into a round hole: Asymptotically correcting the generalized gradient approximation for correlation. *J. Chem. Phys.* **149**, 084116 (2018).
31. D. Casanova. Short-range density functional correlation within the restricted active space CI method. *J. Chem. Phys.* **148**, 124118 (2018).
32. D.M. Ceperley and B.J. Alder. Ground state of the electron gas by a stochastic method. *Phys. Rev. Lett.* **45**, 566 (1980).
33. J.-D. Chai and M. Head-Gordon. Long-range corrected hybrid density functionals with damped atom-atom dispersion corrections. *Phys. Chem. Chem. Phys.* **10**, 6615–6620 (2008).
34. J.-D. Chai and M. Head-Gordon. Systematic optimization of long-range corrected hybrid density functionals. *J. Chem. Phys.* **128**, 084106 (2008).
35. B. Chan, L. Goerigk and L. Radom. On the inclusion of post-MP2 contributions to double-hybrid density functionals. *J. Comput. Chem.* **37**, 183–193 (2016).
36. G.K.-L. Chan and N.C. Handy. Optimized Lieb–Oxford bound for the exchange-correlation energy. *Phys. Rev. A* **59**, 3075 (1999).
37. G.P. Chen, V.K. Voora, M.M. Agee, S.G. Balasubramani and F. Furche. Random-phase approximation methods. *Annu. Rev. Phys. Chem.* **68**, 421–445 (2017).
38. R. Colle and O. Salvetti. Approximate calculation of the correlation energy for the closed shells. *Theor. Chim. Acta* **37**, 329–334 (1975).
39. N. Colonna, M. Hellgren and S. de Gironcoli. Correlation energy with exact-exchange adiabatic connection fluctuation-dissipation theory: Systematic development and simple approximations. *Phys. Rev. B* **90**, 125150 (2014).
40. Y. Cornaton and E. Fromager. Double hybrid density-functional theory using the Coulomb-attenuating method. *Int. J. Quantum Chem.* **114**, 1199 (2014).
41. C. Cotar and M. Petrache. Equality of the jellium and uniform electron gas next-order asymptotic terms for Coulomb and Riesz potentials. *arXiv:* 1707.07664 (2019).
42. T.J. Daas, D.P. Kooi, A.J.A.F. Grooteman, M. Seidl and P. Gori-Giorgi. Gradient expansions for the large-coupling strength limit of the Møller–Plesset adiabatic connection. *J. Chem. Theory Comput.* **18**, 1584–1594 (2022).
43. M. Dion, H. Rydberg, E. Schröder, D.C. Langreth and B.I. Lundqvist. Van der Waals density functional for general geometries. *Phys. Rev. Lett.* **92**, 246401 (2004).
44. P.A.M. Dirac. Note on exchange phenomena in the Thomas atom. *Proc. Cambridge Phil. Soc.* **26**(3), 376 (1930).
45. J.F. Dobson and B.P. Dinte. Constraint satisfaction in local and gradient susceptibility approximations: Application to a van der Waals density functional. *Phys. Rev. Lett.* **76**, 1780 (1996).
46. J.F. Dobson and T. Gould. Calculation of dispersion energies. *J. Phys. Condens. Matter* **24**(7), 073201 (2012).
47. J.F. Dobson, K. McLennan, A. Rubio, J. Wang, T. Gould, H.M. Le and B.P. Dinte. Prediction of dispersion forces: is there a problem? *Aust. J. Chem.* **54**, 513–527 (2001).

48. J.F. Dobson, J. Wang, B.P. Dinte, K. McLennan and H.M. Le. Soft cohesive forces. *Int. J. Quantum Chem.* **101**, 579 (2005).
49. R.M. Dreizler and E.K.U. Gross. *Density Functional Theory*. Springer-Verlag, Berlin (1990).
50. F.G. Eich and M. Hellgren. Derivative discontinuity and exchange-correlation potential of meta-GGAs in density-functional theory. *J. Chem. Phys.* **141**, 224107 (2014).
51. P. Elliott and K. Burke. Non-empirical derivation of the parameter in the B88 exchange functional. *Can. J. Chem.* **87**, 1485 (2009).
52. M. Elstner, P. Hobza, T. Frauenheim, S. Suhai and E. Kaxiras. Hydrogen bonding and stacking interactions of nucleic acid base pairs: A density-functional-theory based treatment. *J. Chem. Phys.* **114**, 5149 (2001).
53. E. Engel. Orbital-dependent functionals for the exchange-correlation energy: A third generation of density functionals. In: *A Primer in Density Functional Theory*, edited by C. Fiolhais, F. Nogueira and M.A.L. Marques, pp. 56–122, Vol. 620 of Lecture Notes in Physics, Springer, Berlin (2003).
54. E. Engel, J.A. Chevary, L.D. Macdonald and S.H. Vosko. Asymptotic properties of the exchange energy density and the exchange potential of finite systems: Relevance for generalized gradient approximations. *Z. Phys. D* **23**, 7 (1992).
55. E. Engel and S.H. Vosko. Wave-vector dependence of the exchange contribution to the electron-gas response functions: An analytic derivation. *Phys. Rev. B* **42**, 4940 (1990).
56. H. Englisch and R. Englisch. Exact density functionals for ground-state energies. I. General results. *Phys. Stat. Sol.* **123**(2), 711–721 (1984).
57. H. Englisch and R. Englisch. Exact density functionals for ground-state energies. II. Details and remarks. *Phys. Stat. Sol.* **124**(1), 373–379 (1984).
58. J. Erhard, P. Bleiziffer and A. Görling. Power series approximation for the correlation kernel leading to Kohn–Sham methods combining accuracy, computational efficiency, and general applicability. *Phys. Rev. Lett.* **117**, 143002 (2016).
59. M. Ernzerhof and J.P. Perdew. Generalized gradient approximation to the angle- and system-averaged exchange hole. *J. Chem. Phys.* **109**, 3313 (1998).
60. M. Ernzerhof and G.E. Scuseria. Assessment of the Perdew–Burke–Ernzerhof exchange-correlation functional. *J. Chem. Phys.* **110**, 5029 (1999).
61. H. Eshuis, J. Bates and F. Furche. Electron correlation methods based on the random phase approximation. *Theor. Chem. Acc.* **131**, 1084 (2012).
62. A. Ferté, E. Giner and J. Toulouse. Range-separated multideterminant density-functional theory with a short-range correlation functional of the on-top pair density. *J. Chem. Phys.* **150**, 084103 (2019).
63. O. Franck, B. Mussard, E. Luppi and J. Toulouse. Basis convergence of range-separated density-functional theory. *J. Chem. Phys.* **142**, 074107 (2015).
64. G. Friesecke. Pair correlations and exchange phenomena in the free electron gas. *Commun. Math. Phys.* **184**, 143–171 (1997).
65. G. Friesecke and B.D. Goddard. Explicit large nuclear charge limit of electronic ground states for Li, Be, B, C, N, O, F, Ne and basic aspects of the periodic table. *SIAM J. Math. Analysis* **41**, 631 (2009).
66. G. Friesecke and B.D. Goddard. Atomic structure via highly charged ions and their exact quantum states. *Phys. Rev. A* **81**, 032516 (2010).
67. E. Fromager. Rigorous formulation of two-parameter double-hybrid density-functionals. *J. Chem. Phys.* **135**, 244106 (2011).
68. E. Fromager, R. Cimiraglia and H.J.A. Jensen. Merging multireference perturbation and density-functional theories by means of range separation: Potential curves for Be_2, Mg_2, and Ca_2. *Phys. Rev. A* **81**, 024502 (2010).
69. E. Fromager and H.J.A. Jensen. Self-consistent many-body perturbation theory in range-separated density-functional theory: A one-electron reduced-density-matrix-based formulation. *Phys. Rev. A* **78**, 022504 (2008).

70. E. Fromager, F. Réal, P. Wåhlin, U. Wahlgren and H.J.A. Jensen. On the universality of the long-/short-range separation in multiconfigurational density-functional theory. II. Investigating f^0 actinide species. *J. Chem. Phys.* **131**, 054107 (2009).
71. E. Fromager, J. Toulouse and H.J.A. Jensen. On the universality of the long-/short-range separation in multiconfigurational density-functional theory. *J. Chem. Phys.* **126**, 074111 (2007).
72. F. Furche. Molecular tests of the random phase approximation to the exchange-correlation energy functional. *Phys. Rev. B* **64**, 195120 (2001).
73. F. Furche and J.P. Perdew. The performance of semilocal and hybrid density functionals in 3*d* transition-metal chemistry. *J. Chem. Phys.* **124**, 044103 (2006).
74. S.R. Gadre, L.J. Bartolotti and N.C. Handy. Bounds for Coulomb energies. *J. Chem. Phys.* **72**, 1034–1038 (1980).
75. A.J. Garza, I.W. Bulik, T.M. Henderson and G.E. Scuseria. Range separated hybrids of pair coupled cluster doubles and density functionals. *Phys. Chem. Chem. Phys.* **17**, 22412–22422 (2015).
76. A.J. Garza, I.W. Bulik, T.M. Henderson and G.E. Scuseria. Synergy between pair coupled cluster doubles and pair density functional theory. *J. Chem. Phys.* **142**, 044109 (2015).
77. I.C. Gerber and J.G. Ángyán. Hybrid functional with separated range. *Chem. Phys. Lett.* **415**, 100–105 (2005).
78. I.C. Gerber and J.G. Ángyán. Potential curves for alkaline-earth dimers by density functional theory with long-range correlation corrections. *Chem. Phys. Lett.* **416**, 370–375 (2005).
79. I.C. Gerber and J.G. Ángyán. London dispersion forces by range-separated hybrid density functional with second order perturbational corrections: The case of rare gas complexes. *J. Chem. Phys.* **126**, 044103 (2007).
80. T.L. Gilbert. Hohenberg–Kohn theorem for nonlocal external potentials. *Phys. Rev. B* **12**, 2111 (1975).
81. P.M.W. Gill, R.D. Adamson and J.A. Pople. Coulomb-attenuated exchange energy density functionals. *Mol. Phys.* **88**, 1005 (1996).
82. L. Goerigk and S. Grimme. Double-hybrid density functionals. *WIREs Comput. Mol. Sci.* **4**, 576–600 (2014).
83. E. Goll, H.-J. Werner and H. Stoll. A short-range gradient-corrected density functional in long-range coupled-cluster calculations for rare gas dimers. *Phys. Chem. Chem. Phys.* **7**, 3917–23 (2005).
84. E. Goll, H.-J. Werner, H. Stoll, T. Leininger, P. Gori-Giorgi and A. Savin. A short-range gradient-corrected spin density functional in combination with long-range coupled-cluster methods: Application to alkali-metal rare-gas dimers. *Chem. Phys.* **329**, 276–282 (2006).
85. P. Gori-Giorgi and J.P. Perdew. Short-range correlation in the uniform electron gas: Extended Overhauser model. *Phys. Rev. B* **64**, 155102 (2001).
86. P. Gori-Giorgi and M. Seidl. Density functional theory for strongly-interacting electrons: perspectives for physics and chemistry. *Phys. Chem. Chem. Phys.* **12**, 14405–14419 (2010).
87. A. Görling. New KS method for molecules based on an exchange charge density generating the exact local KS exchange potential. *Phys. Rev. Lett.* **83**, 5459 (1999).
88. A. Görling. Hierarchies of methods towards the exact Kohn–Sham correlation energy based on the adiabatic-connection fluctuation-dissipation theorem. *Phys. Rev. B* **99**, 235120 (2019).
89. A. Görling and M. Levy. Requirements for correlation energy density functionals from coordinate transformations. *Phys. Rev. A* **45**, 1509 (1992).
90. A. Görling and M. Levy. Correlation-energy functional and its high-density limit obtained from a coupling-constant perturbation expansion. *Phys. Rev. B* **47**, 13105 (1993).
91. A. Görling and M. Levy. Exact Kohn–Sham scheme based on perturbation theory. *Phys. Rev. A* **50**, 196 (1994).
92. A. Görling and M. Levy. DFT ionization formulas and a DFT perturbation theory for exchange and correlation, through adiabatic connection. *Int. J. Quantum Chem. Symp.* **56**, 93–108 (1995).

93. T. Grabo, T. Kreibich, S. Kurth and E.K.U. Gross. Orbital functionals in density functional theory: The optimized effective potential method. In: *Strong Coulomb Correlation in Electronic Structure: Beyond the Local Density Approximation*, edited by V. Anisimov, Gordon & Breach, Tokyo (2000).
94. I. Grabowski, S. Hirata, S. Ivanov and R.J. Bartlett. *Ab initio* density functional theory: OEP-MBPT(2). A new orbital-dependent correlation functional. *J. Chem. Phys.* **116**, 4415 (2002).
95. S. Grimme. Improved second-order Møller–Plesset perturbation theory by separate scaling of parallel- and antiparallel-spin pair correlation energies. *J. Chem. Phys.* **118**, 9095–9102 (2003).
96. S. Grimme. Accurate description of van der Waals complexes by density functional theory including empirical corrections. *J. Comput. Chem.* **25**, 1463–1473 (2004).
97. S. Grimme. Semiempirical GGA-type density functional constructed with a long-range dispersion correction. *J. Comput. Chem.* **27**, 1787–1799 (2006).
98. S. Grimme. Semiempirical hybrid density functional with perturbative second-order correlation. *J. Chem. Phys.* **124**, 034108 (2006).
99. S. Grimme, J. Antony, S. Ehrlich and H. Krieg. A consistent and accurate *ab initio* parametrization of density functional dispersion correction (DFT-D) for the 94 elements H-Pu. *J. Chem. Phys.* **132**, 154104 (2010).
100. S. Grimme and M. Steinmetz. A computationally efficient double hybrid density functional based on the random phase approximation. *Phys. Chem. Chem. Phys.* **18**(31), 20926–20937 (2016).
101. A. Grüneis, M. Marsman, J. Harl, L. Schimka and G. Kresse. Making the random phase approximation to electronic correlation accurate. *J. Chem. Phys.* **131**, 154115 (2009).
102. O. Gunnarsson and B.I. Lundqvist. Exchange and correlation in atoms, molecules, and solids by the spin-density-functional formalism. *Phys. Rev. B* **13**, 4274 (1976).
103. B. Hammer, L.B. Hansen and J.K. Nørskov. Improved adsorption energetics within density-functional theory using revised Perdew–Burke–Ernzerhof functionals. *Phys. Rev. B* **59**, 7413 (1999).
104. F.A. Hamprecht, A.J. Cohen, D.J. Tozer and N.C. Handy. Development and assessment of new exchange-correlation functionals. *J. Chem. Phys.* **109**, 6264 (1998).
105. J.E. Harriman. Orthonormal orbitals for the representation of an arbitrary density. *Phys. Rev. A* **24**, 680 (1981).
106. J. Harris and R. O. Jones. The surface energy of a bounded electron gas. *J. Phys. F* **4**, 1170 (1974).
107. E.D. Hedegård, S. Knecht, J.S. Kielberg, H.J.A. Jensen and M. Reiher. Density matrix renormalization group with efficient dynamical electron correlation through range separation. *J. Chem. Phys.* **142**, 224108 (2015).
108. E.D. Hedegård, J. Toulouse and H.J.A. Jensen. Multiconfigurational short-range density-functional theory for open-shell systems. *J. Chem. Phys.* **148**, 214103 (2018).
109. T. Helgaker. *Principles of Density-Functional Theory*. Lecture at the GdR CORREL Mini-School on Mathematics in Electronic Structure Theory, Université Pierre et Marie Curie, Paris (2017) http://folk.uio.no/helgaker/talks/Paris_2017.pdf.
110. M. Hellgren and U. von Barth. Correlation energy functional and potential from time-dependent exact-exchange theory. *J. Chem. Phys.* **132**, 044101 (2010).
111. M. Hellgren, N. Colonna and S. de Gironcoli. Beyond the random phase approximation with a local exchange vertex. *Phys. Rev. B* **98**, 045117 (2018).
112. M. Hellgren, D.R. Rohr and E.K.U. Gross. Correlation potentials for molecular bond dissociation within the self-consistent random phase approximation. *J. Chem. Phys.* **136**, 034106 (2012).
113. T.M. Henderson, A.F. Izmaylov, G.E. Scuseria and A. Savin. The importance of middle-range Hartree–Fock-type exchange for hybrid density functionals. *J. Chem. Phys.* **127**, 221103 (2007).
114. A. Heßelmann. Random-phase-approximation correlation method including exchange interactions. *Phys. Rev. A* **85**, 012517 (2012).

115. A. Heßelmann and A. Görling. Random phase approximation correlation energies with exact Kohn–Sham exchange. *Mol. Phys.* **108**, 359–372 (2010).
116. A. Heßelmann and A. Görling. Correct description of the bond dissociation limit without breaking spin symmetry by a random-phase-approximation correlation functional. *Phys. Rev. Lett.* **106**, 093001 (2011).
117. J. Heyd, G.E. Scuseria and M. Ernzerhof. Hybrid functionals based on a screened Coulomb potential. *J. Chem. Phys.* **118**, 8207 (2003).
118. P. Hohenberg and W. Kohn. Inhomogeneous electron gas. *Phys. Rev.* **136**, B 864 (1964).
119. C.-J. Huang and C.J. Umrigar. Local correlation energies of two-electron atoms and model systems. *Phys. Rev. A* **56**, 290 (1997).
120. F. Hummel, A. Grüneis, G. Kresse and P. Ziesche. Screened exchange corrections to the random phase approximation from many-body perturbation theory. *J. Chem. Theory Comput.* **15**, 3223–3236 (2019).
121. H. Iikura, T. Tsuneda, T. Yanai and K. Hirao. Long-range correction scheme for generalized-gradient-approximation exchange functionals. *J. Chem. Phys.* **115**, 3540–44 (2001).
122. S. Ivanov, S. Hirata and R.J. Bartlett. Exact exchange treatment for molecules in finite-basis-set Kohn–Sham theory. *Phys. Rev. Lett.* **83**, 5455 (1999).
123. S. Ivanov and M. Levy. Connections between high-density scaling limits of DFT correlation energies and second-order Z^1 quantum chemistry correlation energy. *J. Phys. Chem. A* **102**(18), 3151–3156 (1998).
124. B.G. Janesko, T.M. Henderson and G.E. Scuseria. Screened hybrid density functionals for solid-state chemistry and physics. *J. Chem. Phys.* **130**, 081105 (2009).
125. J. Jaramillo, G.E. Scuseria and M. Ernzerhof. Local hybrid functionals. *J. Chem. Phys.* **118**, 1068 (2003).
126. C. Kalai, B. Mussard and J. Toulouse. Range-separated double-hybrid density-functional theory with coupled-cluster and random-phase approximations. *J. Chem. Phys.* **151**, 074102 (2019).
127. C. Kalai and J. Toulouse. A general range-separated double-hybrid density-functional theory. *J. Chem. Phys.* **148**, 164105 (2018).
128. A.A. Kananenka and D. Zgid. Combining density functional theory and Green's function theory: Range-separated, nonlocal, dynamic, and orbital-dependent hybrid functional. *J. Chem. Theory Comput.* **13**(11), 5317–5331 (2017).
129. A.D. Kaplan, B. Santra, P. Bhattarai, K. Wagle, S.T.u.R. Chowdhury, P. Bhetwal, J. Yu, H. Tang, K. Burke, M. Levy and J.P. Perdew. Simple hydrogenic estimates for the exchange and correlation energies of atoms and atomic ions, with implications for density functional theory. *J. Chem. Phys.* **153**, 074114 (2020).
130. A. Karolewski, L. Kronik and S. Kümmel. Using optimally tuned range separated hybrid functionals in ground-state calculations: Consequences and caveats. *J. Chem. Phys.* **138**(20), 204115 (2013).
131. J. Kim, K. Hong, S.-Y. Hwang, S. Ryu, S. Choi and W.Y. Kim. Effects of the locality of a potential derived from hybrid density functionals on Kohn–Sham orbitals and excited states. *Phys. Chem. Chem. Phys.* **19**, 10177–10186 (2017).
132. J.C. Kimball. Short-range correlations and electron-gas response functions. *Phys. Rev. A* **7**, 1648 (1973).
133. S. Klawohn and H. Bahmann. Self-consistent implementation of hybrid functionals with local range separation. *J. Chem. Theory Comput.* **16**(2), 953–963 (2020).
134. L. Kleinman and S. Lee. Gradient expansion of the exchange-energy density functional: Effect of taking limits in the wrong order. *Phys. Rev. B* **37**, 4634 (1988).
135. W. Kohn and L.J. Sham. Self-consistent equations including exchange and correlation. *Phys. Rev.* **140**, A1133 (1965).
136. S. Kozuch, D. Gruzman and J.M.L. Martin. DSD-BLYP: A general purpose double hybrid density functional including spin component scaling and dispersion correction. *J. Phys. Chem. C* **114**, 20801–20808 (2010).

137. S. Kozuch and J.M.L. Martin. DSD-PBEP86: in search of the best double-hybrid DFT with spin-component scaled MP2 and dispersion corrections. *Phys. Chem. Chem. Phys.* **13**, 20104–20107 (2011).
138. S. Kozuch and J.M.L. Martin. Spin-component-scaled double hybrids: An extensive search for the best fifth-rung functionals blending DFT and perturbation theory. *J. Comput. Chem.* **34**(27), 2327–44 (2013).
139. A.V. Krukau, G.E. Scuseria, J.P. Perdew and A. Savin. Hybrid functionals with local range separation. *J. Chem. Phys.* **129**, 124103 (2008).
140. S. Kümmel and L. Kronik. Orbital-dependent density functionals: Theory and applications. *Rev. Mod. Phys.* **80**, 3 (2008).
141. S. Kurth, J.P. Perdew and P. Blaha. Molecular and Solid-State Tests of Density Functional Approximations: LSD, GGAs, and Meta-GGAs. *Int. J. Quantum Chem.* **75**, 889–909 (1999).
142. S. Kvaal, U. Ekström, A.M. Teale and T. Helgaker. Differentiable but exact formulation of density-functional theory. *J. Chem. Phys.* **140**, 18A518 (2014).
143. A. Laestadius, M. Penz, E.I. Tellgren, M. Ruggenthaler, S. Kvaal and T. Helgaker. Generalized Kohn–Sham iteration on Banach spaces. *J. Chem. Phys.* **149**, 164103 (2018).
144. P.E. Lammert. Differentiability of Lieb functional in electronic density functional theory. *Int. J. Quantum Chem.* **107**, 1943–1953 (2007).
145. N.D. Lang and W. Kohn. Theory of metal surfaces: Charge density and surface energy. *Phys. Rev. B* **1**, 4555 (1970).
146. D.C. Langreth and J.P. Perdew. The exchange-correlation energy of a metallic surface. *Solid State Commun.* **17**(11), 1425–1429 (1975).
147. D.C. Langreth and J.P. Perdew. Exchange-correlation energy of a metallic surface: Wave-vector analysis. *Phys. Rev. B* **15**, 2884 (1977).
148. D.C. Langreth and J.P. Perdew. Theory of nonuniform electronic systems. I. Analysis of the gradient approximation and a generalization that works. *Phys. Rev. B* **21**, 5469 (1980).
149. D.C. Langreth and S.H. Vosko. Exact electron-gas response functions at high density. *Phys. Rev. Lett.* **59**, 497 (1987).
150. C. Lee, W. Yang and R.G. Parr. Development of the Colle–Salvetti correlation-energy formula into a functional of the electron density. *Phys. Rev. B* **37**, 785 (1988).
151. K. Lee, É.D. Murray, L. Kong, B.I. Lundqvist and D.C. Langreth. Higher-accuracy van der Waals density functional. *Phys. Rev. B* **82**, 081101 (2010).
152. T. Leininger, H. Stoll, H.-J. Werner and A. Savin. Combining long-range configuration interaction with short-range density functionals. *Chem. Phys. Lett.* **275**(3), 151–160 (1997).
153. M. Levy. Universal variational functionals of electron densities, first-order density matrices, and natural spin-orbitals and solution of the v-representability problem *Proc. Natl. Acad. Sci. U.S.A.* **76**(12), 6062–5 (1979).
154. M. Levy. Density-functional exchange correlation through coordinate scaling in adiabatic connection and correlation hole. *Phys. Rev. A* **43**, 4637 (1991).
155. M. Levy. Coordinate scaling requirements for approximating exchange and correlation. In: *Density Functional Theory*, edited by E. Gross and R. Dreizler, pp. 11–31, Plenum Press, New York (1995).
156. M. Levy and H. Ou-Yang. Nonuniform coordinate scaling requirements for exchange-correlation energy. *Phys. Rev. A* **42**, 651 (1990).
157. M. Levy and J.P. Perdew. Hellmann–Feynman, virial, and scaling requisites for the exact universal density functionals. Shape of the correlation potential and diamagnetic susceptibility for atoms. *Phys. Rev. A* **32**, 2010 (1985).
158. M. Levy and J.P. Perdew. Tight bound and convexity constraint on the exchange-correlation-energy functional in the low-density limit, and other formal tests of generalized-gradient approximations. *Phys. Rev. B* **48**, 11638 (1993).
159. M. Lewin, E.H. Lieb and R. Seiringer. Improved Lieb–Oxford bound on the indirect and exchange energies. *Lett. Math. Phys.* **112**, 92 (2022).
160. E.H. Lieb. Density Functionals for Coulomb Systems. International journal of quantum chemistry. *Int. J. Quantum Chem.* **24**, 243–277 (1983).

161. E.H. Lieb and S. Oxford. Improved lower bound on the indirect Coulomb energy. *Int. J. Quantum Chem.* **19**(3), 427 (1981).
162. E.H. Lieb and B. Simon. Thomas–Fermi theory revisited. *Phys. Rev. Lett.* **31**, 681 (1973).
163. E.H. Lieb and B. Simon. The Thomas–Fermi theory of atoms, molecules, and solids. *Adv. Math.* **23**, 22–116 (1977).
164. I. Lindgren and S. Salomonson. Differentiability in density-functional theory. *Adv. Quantum Chem.* **43**, 95–117 (2003).
165. P.-F. Loos and P.M.W. Gill. The uniform electron gas. *WIREs Comput. Mol. Sci.* **6**, 410–429 (2016).
166. S.-K. Ma and K.A. Brueckner. Correlation energy of an electron gas with a slowly varying high density. *Phys. Rev.* **165**, 165 (1968).
167. T.M. Maier, A.V. Arbuznikov and M. Kaupp. Local hybrid functionals: Theory, implementation, and performance of an emerging new tool in quantum chemistry and beyond. *WIREs Comput. Mol. Sci.* **9**, e1378 (2019).
168. N.H. March. Asymptotic formula far from nucleus for exchange energy density in Hartree–Fock theory of closed-shell atoms. *Phys. Rev. A* **36**, 5077 (1987).
169. N. Mardirossian and M. Head-Gordon. ωB97X-V: A 10-parameter, range-separated hybrid, generalized gradient approximation density functional with nonlocal correlation, designed by a survival-of-the-fittest strategy. *Phys. Chem. Chem. Phys.* **16**, 9904–9924 (2014).
170. N. Mardirossian and M. Head-Gordon. ωB97M-V: A combinatorially optimized, range-separated hybrid, meta-GGA density functional with VV10 nonlocal correlation. *J. Chem. Phys.* **144**(21), 214110 (2016).
171. N. Mehta, M. Casanova-Páez and L. Goerigk. Semi-empirical or non-empirical double-hybrid density functionals: which are more robust? *Phys. Chem. Chem. Phys.* **20**, 23175–23194 (2018).
172. P.D. Mezei, G.I. Csonka, A. Ruzsinszky and M. Kállay. Construction and application of a new dual-hybrid random phase approximation. *J. Chem. Theory Comput.* **11**(10), 4615–4626 (2015).
173. P.D. Mezei, G.I. Csonka, A. Ruzsinszky and M. Kállay. Construction of a spin-component scaled dual-hybrid random phase approximation. *J. Chem. Theory Comput.* **13**(2), 796–803 (2017).
174. B. Miehlich, A. Savin, H. Stoll and H. Preuss. Results obtained with the correlation energy density functionals of Becke and Lee, Yang and Parr. *Chem. Phys. Lett.* **157**(3), 200–206 (1989).
175. P. Mori-Sánchez, Q. Wu and W. Yang. Orbital-dependent correlation energy in density-functional theory based on a second-order perturbation approach: Success and failure. *J. Chem. Phys.* **123**(6), 062204 (2005).
176. M. Mostafanejad, M.D. Liebenthal and A.E. DePrince III. Global hybrid multiconfiguration pair-density functional theory. *J. Chem. Theory Comput.* **16**(4), 2274–2283 (2020).
177. B. Mussard, P. Reinhardt, J.G. Ángyán and J. Toulouse. Spin-unrestricted random-phase approximation with range separation: Benchmark on atomization energies and reaction barrier heights. *J. Chem. Phys.* **142**, 154123 (2015), Erratum: *J. Chem. Phys.* **142**, 219901 (2015).
178. B. Mussard, D. Rocca, G. Jansen and J.G. Ángyán. Dialectric matrix formulation of correlation energies in the random phase approximation: Inclusion of exchange effects. *J. Chem. Theory Comput.* **12**, 2191–2202 (2016).
179. B. Mussard and J. Toulouse. Fractional-charge and fractional-spin errors in range-separated density-functional theory. *Mol. Phys.* **115**, 161–173 (2017).
180. R. Neumann and N.C. Handy. Investigations using the Becke95 correlation functional. *Chem. Phys. Lett.* **252**(1–2), 19–22 (1996).
181. R. Neumann, R.H. Nobes and N.C. Handy. Exchange functionals and potentials. *Mol. Phys.* **87**, 1–36 (1996).
182. M.M. Odashima and K. Capelle. How tight is the Lieb-Oxford bound? *J. Chem. Phys.* **127**, 054106 (2007).

183. G.L. Oliver and J.P. Perdew. Spin-density gradient expansion for the kinetic energy. *Phys. Rev. A* **20**, 397 (1979).
184. H. Ou-Yang and M. Levy. Nonuniform coordinate scaling requirements in density-functional theory. *Phys. Rev. A* **42**, 155 (1990).
185. J. Paier, B.G. Janesko, T.M. Henderson, G.E. Scuseria, A. Grüneis and G. Kresse. Hybrid functionals including random phase approximation correlation and second-order screened exchange. *J. Chem. Phys.* **132**, 094103 (2010).
186. S. Paziani, S. Moroni, P. Gori-Giorgi and G.B. Bachelet. Local-spin-density functional for multideterminant density functional theory. *Phys. Rev. B* **73**, 155111 (2006).
187. J.P. Perdew. Unified theory of exchange and correlation beyond the local density approximation. In: *Electronic Structure of Solids '91*, edited by P. Ziesche and H. Eschrig, pp. 11–20, Akademie Verlag, Berlin (1991).
188. J.P. Perdew, K. Burke and M. Ernzerhof. Generalized gradient approximation made simple. *Phys. Rev. Lett.* **77**, 3865 (1996).
189. J.P. Perdew, J.A. Chevary, S.H. Vosko, K.A. Jackson, M.R. Pederson, D.J. Singh and C. Fiolhais. Atoms, molecules, solids, and surfaces: Applications of the generalized gradient approximation for exchange and correlation. *Phys. Rev. B* **46**, 6671 (1992).
190. J.P. Perdew, L.A. Constantin, E. Sagvolden and K. Burke. Relevance of the slowly varying electron gas to atoms, molecules, and solids. *Phys. Rev. Lett.* **97**, 223002 (2006).
191. J.P. Perdew, M. Ernzerhof and K. Burke. Rationale for mixing exact exchange with density functional approximations. *J. Chem. Phys.* **105**, 9982–9985 (1996).
192. J.P. Perdew, S. Kurth, A. Zupan and P. Blaha. Accurate density functional with correct formal properties: A step beyond the generalized gradient approximation. *Phys. Rev. Lett.* **82**, 2544 (1999).
193. J.P. Perdew, E.R. McMullen and A. Zunger. Density-functional theory of the correlation energy in atoms and ions: A simple analytic model and a challenge. *Phys. Rev. A* **23**, 2785 (1981).
194. J.P. Perdew, A. Ruzsinszky, J. Sun and K. Burke. Gedanken densities and exact constraints in density functional theory. *J. Chem. Phys.* **140**, 18A533 (2014).
195. J.P. Perdew, A. Ruzsinszky, G.I. Csonka, L.A. Constantin and J. Sun. Workhorse semilocal density functional for condensed matter physics and quantum chemistry. *Phys. Rev. Lett.* **103**, 026403 (2009).
196. J.P. Perdew, A. Ruzsinszky, G.I. Csonka, O.A. Vydrov, G.E. Scuseria, L.A. Constantin, X. Zhou and K. Burke. Restoring the density-gradient expansion for exchange in solids and surfaces. *Phys. Rev. Lett.* **100**, 136406 (2008).
197. J.P. Perdew and K. Schmidt. Jacob's ladder of density functional approximations for the exchange-correlation energy. *AIP Conf. Proc.* **577**, 1–20 (2001).
198. J.P. Perdew, J. Tao, V.N. Staroverov and G.E. Scuseria. Meta-generalized gradient approximation: explanation of a realistic nonempirical density functional. J. Chem. Phys. **120**(15), 6898–911 (2004).
199. J.P. Perdew and Y. Wang. Accurate and simple analytic representation of the electron-gas correlation energy. *Phys. Rev. B* **45**, 13244 (1992).
200. J.P. Perdew and A. Zunger. Self-interaction correction to density-functional approximations for many-electron systems. *Phys. Rev. B* **23**, 5048 (1981).
201. K. Pernal. Long-range density-matrix-functional theory: Application to a modified homogeneous electron gas. *Phys. Rev. A* **81**, 052511 (2010).
202. R. Pollet, A. Savin, T. Leininger and H. Stoll. Combining multideterminantal wave functions with density functionals to handle near-degeneracy in atoms and molecules. *J. Chem. Phys.* **116**, 1250 (2002).
203. A.K. Rajagopal and J. Callaway. Inhomogeneous electron gas. *Phys. Rev. B* **7**, 1912 (1973).
204. M. Rasolt. Inhomogeneity corrections to the ground-state properties of itinerant ferromagnets. *Phys. Rev. B* **16**, 3234 (1977).
205. M. Rasolt and H.L. Davis. Exchange splitting of ferromagnetic nickel within the local potential approximation. *Phys. Lett. A* **86**(1), 45–47 (1981).

206. M. Rasolt and D.J.W. Geldart. Exchange and correlation energy in a nonuniform fermion fluid. *Phys. Rev. B* **34**, 1325 (1986).
207. E. Rebolini and J. Toulouse. Range-separated time-dependent density-functional theory with a frequency-dependent second-order Bethe–Salpeter correlation kernel. *J. Chem. Phys.* **144**, 094107 (2016).
208. X. Ren, P. Rinke, C. Joas and M. Scheffler. Random-phase approximation and its applications in computational chemistry and materials science. *J. Mater. Sci.* **47**(21), 7447–7471 (2012).
209. D.R. Rohr and K. Pernal. Open-shell reduced density matrix functional theory. *J. Chem. Phys.* **135**, 074104 (2011).
210. D.R. Rohr, J. Toulouse and K. Pernal. Combining density-functional theory and density-matrix-functional theory. *Phys. Rev. A* **82**, 052502 (2010).
211. A. Ruzsinszky, J.P. Perdew and G.I. Csonka. The RPA atomization energy puzzle. *J. Chem. Theory Comput.* **6**, 127–134 (2010).
212. J.C. Sancho-García and C. Adamo. Double-hybrid density functionals: merging wavefunction and density approaches to get the best of both worlds. *Phys. Chem. Chem. Phys.* **15**, 14581–14594 (2013).
213. G. Sansone, B. Civalleri, D. Usvyat, J. Toulouse, K. Sharkas and L. Maschio. Range-separated double-hybrid density-functional theory applied to periodic systems. *J. Chem. Phys.* **143**, 102811 (2015).
214. T. Sato and H. Nakai. Local response dispersion method. II. Generalized multicenter interactions. *J. Chem. Phys.* **133**, 194101 (2010).
215. A. Savin. Beyond the Kohn–Sham determinant. In: *Recent Advances in Density Functional Methods*, edited by D.P. Chong, pp. 129–153, World Scientific (1996).
216. A. Savin. On degeneracy, near-degeneracy and density functional theory. In: *Recent Developments of Modern Density Functional Theory*, edited by J.M. Seminario, pp. 327–357, Elsevier, Amsterdam (1996).
217. A. Savin and H.-J. Flad. Density functionals for the Yukawa electron-electron interaction. *Int. J. Quantum. Chem.* **56**, 327–332 (1995).
218. T. Schwabe and S. Grimme. Double-hybrid density functionals with long-range dispersion corrections: higher accuracy and extended applicability. *Phys. Chem. Chem. Phys.* **9**, 3397–3406 (2007).
219. G.E. Scuseria, T.M. Henderson and D.C. Sorensen. The ground state correlation energy of the random phase approximation from a ring coupled cluster doubles approach. *J. Chem. Phys.* **129**, 231101 (2008).
220. M. Seidl. Strong-interaction limit of density-functional theory. *Phys. Rev. A* **60**, 4387 (1999).
221. M. Seidl, P. Gori-Giorgi and A. Savin. Strictly correlated electrons in density-functional theory: A general formulation with applications to spherical densities. *Phys. Rev. A* **75**, 042511 (2007).
222. A. Seidl, A. Görling, P. Vogl, J.A. Majewski and M. Levy. Generalized Kohn–Sham schemes and the band-gap problem. *Phys. Rev. B* **53**, 3764 (1996).
223. M. Seidl, J.P. Perdew and M. Levy. Strictly correlated electrons in density-functional theory *Phys. Rev. A* **59**, 51 (1999).
224. L.J. Sham. Approximations of the exchange and correlation potentials. In: *Computational Methods in Band Theory*, edited by P. Marcus, J.F. Janak and A.R. Williams, pp. 458–468, Plenum, New York (1971).
225. K. Sharkas, A. Savin, H.J.A. Jensen and J. Toulouse. A multiconfigurational hybrid density-functional theory. *J. Chem. Phys.* **137**, 044104 (2012).
226. K. Sharkas, J. Toulouse and A. Savin. Double-hybrid density-functional theory made rigorous. *J. Chem. Phys.* **134**, 064113 (2011).
227. R.T. Sharp and G.K. Horton. A variational approach to the unipotential many-electron problem. *Phys. Rev.* **90**, 317 (1953).
228. J.C. Slater. A Simplification of the Hartree–Fock Method. *Phys. Rev.* **81**, 385 (1951).
229. S. Śmiga and L.A. Constantin. Unveiling the physics behind hybrid functionals. *J. Phys. Chem. A* **124**(27), 5606–5614 (2020).

230. S. Śmiga, O. Franck, B. Mussard, A. Buksztel, I. Grabowski, E. Luppi and J. Toulouse. Self-consistent double-hybrid density-functional theory using the optimized-effective-potential method. *J. Chem. Phys.* **145**, 144102 (2016).
231. S. Śmiga, I. Grabowski, M. Witkowski, B. Mussard and J. Toulouse. Self-consistent range-separated density-functional theory with second-order perturbative correction via the optimized-effective-potential method. *J. Chem. Theory Comput.* **16**(1), 211–223 (2020).
232. S.M.O. Souvi, K. Sharkas and J. Toulouse. Double-hybrid density-functional theory with meta-generalized-gradient approximations. *J. Chem. Phys.* **140**, 084107 (2014).
233. V.N. Staroverov, G.E. Scuseria, J.P. Perdew, J. Tao and E.R. Davidson. Energies of isoelectronic atomic ions from a successful meta-generalized gradient approximation and other density functionals. *Phys. Rev. A* **70**, 012502 (2004).
234. V.N. Staroverov, G.E. Scuseria, J. Tao and J.P. Perdew. Comparative assessment of a new nonempirical density functional: Molecules and hydrogen-bonded complexes. *J. Chem. Phys.* **119**, 12129 (2003).
235. T. Stein, L. Kronik and R. Baer. Prediction of charge-transfer excitations in coumarin-based dyes using a range-separated functional tuned from first principles. *J. Chem. Phys.* **131**, 244119 (2009).
236. T. Stein, L. Kronik and R. Baer. Reliable prediction of charge transfer excitations in molecular complexes using time-dependent density functional theory. *J. Am. Chem. Soc.* **131**, 2818–20 (2009).
237. P.J. Stephens, F.J. Devlin, C.F. Chabalowski and M.J. Frisch. *Ab initio* calculation of vibrational absorption and circular dichroism spectra using density functional force fields. *J. Phys. Chem.* **98**(45), 11623–11627 (1994).
238. H. Stoll, E. Golka and H. Preuss. Correlation energies in the spin-density functional formalism. *Theor. Chim. Acta* **55**, 29–41 (1980).
239. H. Stoll, C. Pavlidou and H. Preuss. On the calculation of correlation energies in the spin-density functional formalism. *Theor. Chim. Acta* **49**, 143–149 (1978).
240. A. Stoyanova, A.M. Teale, J. Toulouse, T. Helgaker and E. Fromager. Alternative separation of exchange and correlation energies in multi-configuration range-separated density-functional theory. *J. Chem. Phys.* **139**, 134113 (2013).
241. N.Q. Su and X. Xu. The XYG3 type of doubly hybrid density functionals. *WIREs Comput. Mol. Sci.* **6**, 721–747 (2016).
242. J. Sun, M. Marsman, G.I. Csonka, A. Ruzsinszky, P. Hao, Y.-S. Kim, G. Kresse and J.P. Perdew. Self-consistent meta-generalized gradient approximation within the projector-augmented-wave method. *Phys. Rev. B* **84**(3), 035117 (2011).
243. J. Sun, J.P. Perdew, Z. Yang and H. Peng. Communication: Near-locality of exchange and correlation density functionals for 1- and 2-electron systems. *J. Chem. Phys.* **144**, 191101 (2016).
244. J. Sun, A. Ruzsinszky and J.P. Perdew. Strongly constrained and appropriately normed semilocal density functional. *Phys. Rev. Lett.* **115**, 036402 (2015).
245. J. Sun, B. Xiao, Y. Fang, R. Haunschild, P. Hao, A. Ruzsinszky, G.I. Csonka, G.E. Scuseria and J.P. Perdew. Density functionals that recognize covalent, metallic, and weak bonds. *Phys. Rev. Lett.* **111**, 106401 (2013).
246. P.S. Svendsen and U. von Barth. Gradient expansion of the exchange energy from second-order density response theory. *Phys. Rev. B* **54**, 17402 (1996).
247. A. Szabo and N.S. Ostlund. *Modern Quantum Chemistry: Introduction to Advanced Electronic Structure Theory*. Dover, New York (1996).
248. J.D. Talman and W.F. Shadwick. Optimized effective atomic central potential. *Phys. Rev. A* **14**, 36 (1976).
249. J. Tao, J.P. Perdew, V.N. Staroverov and G.E. Scuseria. Generalized gradient approximation designed for molecules and solids. *Phys. Rev. Lett.* **91**, 146401 (2003).
250. Y. Tawada, T. Tsuneda, S. Yanagisawa, T. Yanai and K. Hirao. A long-range-corrected time-dependent density functional theory. *J. Chem. Phys.* **120**(18), 8425–33 (2004).

251. D.E. Taylor, J.G. Ángyán, G. Galli, C. Zhang, F. Gygi, K. Hirao, J.W. Song, K. Rahul, O.A. von Lilienfeld, R. Podeszwa, I.W. Bulik, T.M. Henderson, G.E. Scuseria, J. Toulouse, R. Peverati, D.G. Truhlar and K. Szalewicz. Blind test of density-functional-based methods on intermolecular interaction energies. *J. Chem. Phys.* **145**, 124105–20 (2016).
252. D.P. Tew. Second order coalescence conditions of molecular wave functions. *J. Chem. Phys.* **129**, 014104 (2008).
253. A. Tkatchenko and M. Scheffler. Accurate molecular van der Waals interactions from ground-state electron density and free-atom reference data. *Phys. Rev. Lett.* **102**, 073005 (2009).
254. J. Toulouse. *Extension multidéterminantale de la méthode de Kohn–Sham en théorie de la fonctionnelle de la densité par décomposition de l'interaction électronique en contributions de longue portée et de courte portée.* Ph.D. thesis, Université Pierre et Marie Curie (Paris 6) (2005), tel.archives-ouvertes.fr/tel-00550772.
255. J. Toulouse, F. Colonna and A. Savin. Long-range–short-range separation of the electron-electron interaction in density-functional theory. *Phys. Rev. A* **70**, 062505 (2004).
256. J. Toulouse, F. Colonna and A. Savin. Short-range exchange and correlation energy density functionals: Beyond the local-density approximation. *J. Chem. Phys.* **122**, 014110 (2005).
257. J. Toulouse, I.C. Gerber, G. Jansen, A. Savin and J.G. Ángyán. Adiabatic-connection fluctuation-dissipation density-functional theory based on range separation. *Phys. Rev. Lett.* **102**, 096404 (2009).
258. J. Toulouse, P. Gori-Giorgi and A. Savin. A short-range correlation energy density functional with multi-determinantal reference. *Theor. Chem. Acc.* **114**, 305–308 (2005).
259. J. Toulouse, A. Savin and H.-J. Flad. Short-range exchange-correlation energy of a uniform electron gas with modified electron-electron interaction. *Int. J. Quantum Chem.* **100**, 1047–1056 (2004).
260. J. Toulouse, K. Sharkas, E. Brémond and C. Adamo. Communication: Rationale for a new class of double-hybrid approximations in density-functional theory. *J. Chem. Phys.* **135**, 101102 (2011).
261. J. Toulouse, W. Zhu, J.G. Ángyán and A. Savin. Range-separated density-functional theory with the random-phase approximation: Detailed formalism and illustrative applications. *Phys. Rev. A* **82**, 032502 (2010).
262. J. Toulouse, W. Zhu, A. Savin, G. Jansen and J.G. Ángyán. Closed-shell ring coupled cluster doubles theory with range separation applied on weak intermolecular interactions. *J. Chem. Phys.* **135**, 084119 (2011).
263. C. A. Ullrich. (Spin-)density-functional theory for open-shell systems: Exact magnetization density functional for the half-filled Hubbard trimer. *Phys. Rev. A* **100**, 012516 (2019).
264. T. Van Voorhis and G. E. Scuseria. A novel form for the exchange-correlation energy functional. *J. Chem. Phys.* **109**, 400 (1998).
265. S.J. Vosko, L. Wilk and M. Nusair. Accurate spin-dependent electron liquid correlation energies for local spin density calculations: a critical analysis. *Can. J. Phys.* **58**, 1200 (1980).
266. O.A. Vydrov, J. Heyd, A.V. Krukau and G.E. Scuseria. Importance of short-range versus long-range Hartree–Fock exchange for the performance of hybrid density functionals. *J. Chem. Phys.* **125**, 074106 (2006).
267. O.A. Vydrov and G.E. Scuseria. Assessment of a long-range corrected hybrid functional. *J. Chem. Phys.* **125**, 234109 (2006).
268. O.A. Vydrov and T. Van Voorhis. Improving the accuracy of the nonlocal van der Waals density functional with minimal empiricism. *J. Chem. Phys.* **130**, 104105 (2009).
269. O.A. Vydrov and T. Van Voorhis. Nonlocal van der Waals density functional made simple. *Phys. Rev. Lett.* **103**, 063004 (2009).
270. O.A. Vydrov and T. Van Voorhis. Nonlocal van der Waals density functional: The simpler the better. *J. Chem. Phys.* **133**, 244103 (2010).
271. Y. Wang and J. Perdew. Correlation hole of the spin-polarized electron gas, with exact small-wave-vector and high-density scaling. *Phys. Rev. B* **44**, 13298 (1991).
272. Q. Wu and W. Yang. Empirical correction to density functional theory for van der Waals interactions. *J. Chem. Phys.* **116**, 515 (2002).

273. T. Yanai, D.P. Tew and N.C. Handy. A new hybrid exchange-correlation functional using the Coulomb-attenuating method (CAM-B3LYP). *Chem. Phys. Lett.* **393**, 51–57 (2004).
274. F. Ying, C. Zhou, P. Zheng, J. Luan, P. Su and W. Wu. λ-density functional valence bond: A valence bond-based multiconfiguration density functional theory with a single variable hybrid parameter. *Front. Chem.* **7**, 225 (2019).
275. F. Zahariev, S.S. Leang and M.S. Gordon. Functional derivatives of meta-generalized gradient approximation (meta-GGA) type exchange-correlation density functionals. *J. Chem. Phys.* **138**, 244108 (2013).
276. Y. Zhang and W. Yang. Comment on "Generalized gradient approximation made simple". *Phys. Rev. Lett.* **80**, 890 (1998).
277. Y. Zhao, N.E. Schultz and D.G. Truhlar. Exchange-correlation functional with broad accuracy for metallic and nonmetallic compounds, kinetics, and noncovalent interactions. *J. Chem. Phys.* **123**, 161103 (2005).
278. Y. Zhao and D.G. Truhlar. A new local density functional for main-group thermochemistry, transition metal bonding, thermochemical kinetics, and noncovalent interactions. *J. Chem. Phys.* **125**, 194101 (2006).
279. Y. Zhao and D.G. Truhlar. The M06 suite of density functionals for main group thermochemistry, thermochemical kinetics, noncovalent interactions, excited states, and transition elements: two new functionals and systematic testing of four M06-class functionals and 12 other functionals. *Theor. Chem. Acc.* **120**, 215–241 (2008).

Chapter 2
On Connecting Density Functional Approximations to Theory

Andreas Savin

Dedicated to Jean-Paul Malrieu on his 80th birthday

Abstract Usually, density functional models are considered approximations to density functional theory, However, there is no systematic connection between the two, and this can make us doubt about a linkage. This attitude can be further enforced by the vagueness of the argumentation for using spin densities. Questioning the foundations of density functional models leads to a search for alternative explanations. Seeing them as using models for pair densities is one of them. Another is considering density functional approximations as a way to extrapolate results obtained in a model system to those of a corresponding physical one.

2.1 Introduction

2.1.1 On approximations in DFT

Density functional theory is here. It has changed the way the computation of electronic systems is seen by the scientific community. It has a sound theoretical foundation. However, following exact theory is more complicated than solving the Schrödinger equation. Furthermore, it does not tell us how to systematically produce approximations. Usual approximations are convenient and (to a large degree) successful, but how to improve them?

A. Savin (✉)
Laboratoire de Chimie Théorique, CNRS and Sorbonne University, Paris, France
e-mail: andreas.savin@lct.jussieu.fr

E. Cancès, G. Friesecke (eds.), *Density Functional Theory*, Mathematics and Molecular Modeling, https://doi.org/10.1007/978-3-031-22340-2_2

2.1.2 *Excuses*

This is not a review. References are erratic and biased. My own publications dominate, not because they are more important, but because they are only given to complement argumentation.

2.1.3 *Summary*

After giving the notations, and recalling the Hohenberg–Kohn theorem, we review some practical solutions, such as the decomposition of the universal functional, in order to comply with different physical requirements. It is argued that this does not necessarily solve the problem. Refinements, such as using the spin density as a supplementary variable, are discussed. It is argued that the need for these refinements may hide a different foundation for the approximations. In order to introduce a "systematic" way to get closer to the physical Hamiltonian, model Hamiltonians are defined that via an adjustable parameter. It can bring the model Hamiltonian arbitrarily close to the physical Hamiltonian. Finally, examples show that simple mathematical recipes provide a quality similar to that of density functional approximations.

2.2 Schrödinger Equation and Notations

We start with a Schrödinger equation:

$$H\Psi = E\Psi. \tag{2.1}$$

The wave function Ψ depends on the coordinates of the electrons $\mathbf{r}_1, \mathbf{r}_2, \ldots, \mathbf{r}_N$ and their spins. We will be mainly concerned with ground state eigenvalues, $E = E_0$. We consider Hamiltonians of the form

$$H = T + V + W \tag{2.2}$$

where T is the operator for the kinetic energy,

$$T = -\frac{1}{2}\sum_{i=1}^{N} \nabla_i^2, \tag{2.3}$$

and V is a local one-particle potential

$$V = \sum_{i=1}^{N} v(\mathbf{r}_i)$$
$$= \int_{\mathbb{R}^3} v(\mathbf{r})\widehat{\rho}(\mathbf{r})\,d\mathbf{r}. \tag{2.4}$$

The density operator, $\widehat{\rho}(\mathbf{r})$, can be written using Dirac's δ function,

$$\widehat{\rho}(\mathbf{r}) = \sum_{i=1}^{N} \delta(\mathbf{r} - \mathbf{r}_i). \tag{2.5}$$

Its expectation value is the density

$$\rho(\mathbf{r}) = \langle\Psi|\,\widehat{\rho}(\mathbf{r})\,|\Psi\rangle\,. \tag{2.6}$$

Notice that it integrates to N,

$$N = \int_{\mathbb{R}} \rho(\mathbf{r})\,d\mathbf{r}. \tag{2.7}$$

W is a two-particle local potential,

$$W = \sum_{i<j}^{N} w(|\mathbf{r}_i - \mathbf{r}_j|) \tag{2.8}$$
$$= \frac{1}{2}\int_{\mathbb{R}^3}\int_{\mathbb{R}^3} w(|\mathbf{r}_i - \mathbf{r}_j|)\widehat{P}(\mathbf{r},\mathbf{r}')\,d\mathbf{r}\,d\mathbf{r}', \tag{2.9}$$

where $\widehat{P}$ is the pair density operator,

$$\widehat{P}(\mathbf{r},\mathbf{r}') = \sum_{i\neq j}^{N} \delta(\mathbf{r} - \mathbf{r}_i)\delta(\mathbf{r}' - \mathbf{r}_j). \tag{2.10}$$

The pair density is

$$P(\mathbf{r},\mathbf{r}') = \langle\Psi|\,\widehat{P}(\mathbf{r},\mathbf{r}')\,|\Psi\rangle\,. \tag{2.11}$$

As the interaction depends only on the distance between particles, often the dependence of P on $\mathbf{r}$ is reduced to that on $u = |\mathbf{r} - \mathbf{r}'|$, using instead of $\widehat{P}$ the spherically averaged operator

$$\widehat{P}_{\text{sph}}(\mathbf{r}, u) = \sum_{i \neq j}^{N} \delta(\mathbf{r} - \mathbf{r}_i)\delta(|\mathbf{r} - \mathbf{r}_j| - u), \tag{2.12}$$

yielding

$$P_{\text{sph}}(\mathbf{r}, u) = \langle\Psi|\, \widehat{P}_{\text{sph}}(\mathbf{r}, u)\, |\Psi\rangle\,. \tag{2.13}$$

Going one step further, one can also integrate over $\mathbf{r}$, to obtain the system-average

$$P_{\text{sys}}(u) = \int_{\mathbb{R}^3} \langle\Psi|\, \widehat{P}_{\text{sph}}(\mathbf{r}, u)\, |\Psi\rangle\, \mathbf{dr}. \tag{2.14}$$

For the electronic systems, $V = V_{\text{ne}}$, or $v = v_{\text{ne}}$, describes the Coulomb interaction between the nuclei and the electrons,

$$v_{\text{ne}}(\mathbf{r}) = -\sum_A \frac{Z_A}{|\mathbf{R}_A - \mathbf{r}|}. \tag{2.15}$$

A is an index for the nuclei, Z_A their nuclear charge, and $\mathbf{R}_A$ their position. Also, $W = V_{\text{ee}}$, or $w = v_{\text{ee}}$, describes the Coulomb interaction between electrons

$$v_{\text{ee}}(|\mathbf{r}_i - \mathbf{r}_j|) = \frac{1}{|\mathbf{r}_i - \mathbf{r}_j|}. \tag{2.16}$$

To characterize a given electronic system, one has only to specify N and v_{ne}.

Model systems are considered below, where $v \neq v_{\text{ne}}$ and $w \neq v_{\text{ee}}$. Of course, in this case the energies and wave functions depend also on the choice of v and w. No change of the non-local one-particle operator T is considered in this chapter, but such modifications can be found in the literature (see, e.g., [10] for a density functional context). The ground state energy can also be obtained using the variational principle,

$$E[v, w, N] = \min_{\Psi} \langle\Psi|\, H\, |\Psi\rangle = \min_{\Psi} \langle\Psi|\, T + V + W\, |\Psi\rangle\,. \tag{2.17}$$

2.3 The Density Functional Viewpoint

2.3.1 The Hohenberg–Kohn Theorem

In order to see how density functional theory can be useful, one generally argues using the Hohenberg–Kohn theorem [12] (cf. Chap. 3).

One of its formulations: "ρ yields v_{ne} and N, and thus everything" is useless, as we do not need to know the density to know the potential of the system under study. This formulation of the theorem is never used in practice.

However, the variational formulation of the Hohenberg–Kohn theorem is consequential. It states that

$$E[v, w, N] = \min_{\rho} \left(F[\rho, w] + \int_{\mathbb{R}^3} \rho(\mathbf{r}) v(\mathbf{r})\, \mathrm{d}\mathbf{r} \right), \tag{2.18}$$

where for F one uses either the Legendre transform form [18],

$$F[\rho, w] = \sup_{v} \left(E[v, w, N] - \int_{\mathbb{R}^3} v(\mathbf{r})\, \rho(\mathbf{r})\, \mathrm{d}\mathbf{r} \right) \tag{2.19}$$

or, equivalently, a constrained search for ensembles. For the sake of simplicity, in this chapter its pure state form [16, 18, 20]

$$F[\rho, w] = \min_{\Psi \to \rho} \langle \Psi | \, T + W \, | \Psi \rangle \tag{2.20}$$

is used. As F does not depend on v (which specifies the system) the functional is called *universal*. The dependence on N appears through that of ρ (Eq. (2.7)). As above for the energy functional, the dependence on the operator T is not included in the notation for F. For the physical system, $w = v_{\mathrm{ee}}$ is implicitly assumed; we write:

$$F[\rho] = F[\rho, w = v_{\mathrm{ee}}]. \tag{2.21}$$

The hope raised by Eq. (2.18) is that it can be used with some simple approximation for $F[\rho]$.

2.3.2 *Difficulty of Producing* $F[\rho]$

Obtaining F for a given $\rho(\mathbf{r})$ is possible, but still difficult: a constrained minimization, as required by Eq. (2.20), is more demanding than a minimization with the single constraint of normalizing the wave function, Eq. (2.17). The Legendre transformed form of F, Eq. (2.19), first requires us to compute E for all v, but then no $F[\rho]$ is needed.

Up to now, there has been no systematic way to construct approximations for $F[\rho]$. In practice, F is replaced by some model, $\widetilde{F}$: one speaks about a density functional approximation (DFA).

Note that when using Eq. (2.19), due to the variational principle, the errors will be of second order in F for first-order errors in v. Stated differently: there are many v that give values of F that are close. For example, adding to the potential a very rapidly oscillating function yields essentially the same value F. (For this, and other examples, see, e.g., [31, 32].) Again, it appears to be of little practical importance to follow the line *ρ gives v and thus everything*. However, obtaining E_0 from Eq. (2.18) is not necessarily affected by this problem once F is known or can be approximated. One can even wonder if the existence of many density functional approximations with a similar quality may be due to the indifference of F to changes in the approximation of an optimizing v in Eq. (2.19).

2.4 Practical Solutions for Density Functional Approximations

In order to create models, two main lines have emerged within density functional theory.

1. Using a simple ansatz for the density functional.
2. Considering DFT as an inspiration to develop other methods that do not require an explicit construction of a density functional.

The first approach is a *cutting the Gordian knot* type of solution. The second approach is in line with methods developed for wave functions, sometimes nothing but such a method.

2.4.1 Ansatz

2.4.1.1 Choice of the Ansatz

Most DFAs start with the so-called local density approximation (LDA). Within this model, a general functional $G[\rho]$ is replaced by the ansatz

$$\widetilde{G}[\rho] = \int_{\mathbb{R}^3} g(\rho(\mathbf{r}))\, d\mathbf{r}. \tag{2.22}$$

The function g has to be defined in some way. Traditionally, it is fixed in the uniform electron gas, a system with an infinite number of particles, and where ρ does not depend on the position (cf. Chap. 3). Typically g is either obtained analytically as a function of ρ, or computed for a series of values of ρ, and fitted to them satisfying asymptotic conditions.

LDA has the important advantage of being (to a certain extent) size-consistent, i.e., satisfying

$$E_{A...B} = E_A + E_B \tag{2.23}$$

where $E_{A...B}$ is the system composed of two parts, A and B, at infinite separation, while E_A and E_B are the energies of these parts computed independently. For the violations of size-consistency by LDA, see, e.g., [21, 27]. Another, major, advantage is its computational simplicity (just a numerical integration to obtain $\widetilde{G}$), and its linear scaling with system size. Both result from the local character of g: if ρ can be decomposed into contributions from two spatial parts,

$$\rho(\mathbf{r}) = \begin{cases} \rho_A(\mathbf{r}) \text{ for } \mathbf{r} \in \Omega_A(\mathbf{r}) \\ \rho_B(\mathbf{r}) \text{ for } \mathbf{r} \in \Omega_B(\mathbf{r}) \end{cases} \tag{2.24}$$

then so can g; $\widetilde{G}$ becomes the sum of the two contributions.

LDA can be extended by making g depend on other local quantities such as derivatives of the density, giving generalized gradient approximations (GGAs), etc. (cf. Chap. 1).

2.4.1.2 Finding the Right Functional to Approximate by Partitioning

Applying the LDA, Eq. (2.22), to $F[\rho]$, Eq. (2.21), does not provide the accuracy needed in most electronic structure calculations. The strategy chosen is to define some density functional $F_d[\rho]$, and approximate only the remaining part, $\bar{F}_d[\rho] = F[\rho] - F_d[\rho]$.

In the following, some choices for the partitioning of F will be presented.

2.4.1.3 Satisfying Electrostatics

In the classical limit, the electrostatic interaction is given by the nuclear repulsion,

$$V_{\text{nn}} = \sum_{A,B(>A)} \frac{Z_A Z_B}{|\mathbf{R}_A - \mathbf{R}_B|},$$

the interaction between the electron cloud and the positive charges of the nuclei,

$$\int v_{\text{ne}}(\mathbf{r})\, \rho(\mathbf{r})\, d\mathbf{r},$$

and the repulsion inside the electron clouds, the Hartree energy,

$$E_H[\rho] = \frac{1}{2} \int_{\mathbb{R}^3} \int_{\mathbb{R}^3} \frac{\rho(\mathbf{r})\rho(\mathbf{r}')}{|\mathbf{r} - \mathbf{r}'|} \mathbf{d}\mathbf{r}\mathbf{d}\mathbf{r}'. \tag{2.25}$$

There is a balance between these contributions. For example, between distant neutral atoms these compensate (there is no $|\mathbf{R}_A - \mathbf{R}_B|^{-1}$ term in the limit $|\mathbf{R}_A - \mathbf{R}_B|^{-1} \to \infty$). This balance is destroyed if E_H is approximated, for example, by using LDA, Eq. (2.22). An excess or deficit of repulsion produces an unphysical repulsion, or attraction of neutral atoms. Furthermore, even if this balance is enforced by parametrization for a given system, it is not kept for another, even closely related system (see, e.g., [29]). The solution to this problem was already proposed in the original Hohenberg–Kohn paper [12]: E_H is treated exactly, and only the remaining part approximated.

Finding good models for $F[\rho] - E_H[\rho]$ is still an active field of research; there are already approximations that work well for classes of systems, but we do not have yet a universally applicable model.

2.4.1.4 The Kohn–Sham Method: Imposing the Pauli Principle

The Pauli principle is hidden in the wave function used to define $F[\rho]$, Eq. (2.21). A way to impose it is to use a model system, with $F[\rho, w \neq v_{\text{ee}}]$, where the Pauli principle is imposed, and use approximations for the remaining part,

$$E_0 = \min_{\Psi} \left(\langle \Psi | T + V_{\text{ne}} + W | \Psi \rangle + \bar{E}_{\text{Hxc}}[\rho_\Psi, w] \right), \tag{2.26}$$

where the subscript Ψ indicates that ρ is obtained from this wave function, and

$$\bar{E}_{\text{Hxc}}[\rho, w] = F[\rho, v_{\text{ee}}] - F[\rho, w]. \tag{2.27}$$

This expression is derived using Eqs. (2.17), (2.18), and (2.20). In general, one takes into account the remark made above about $E_H[\rho]$, and defines

$$\bar{E}_H[\rho, w] = \frac{1}{2} \int_{\mathbb{R}^3} \int_{\mathbb{R}^3} \rho(\mathbf{r})\rho(\mathbf{r}') \left(\frac{1}{|\mathbf{r} - \mathbf{r}'|} - w(\mathbf{r}, \mathbf{r}') \right) \mathbf{d}\mathbf{r}\mathbf{d}\mathbf{r}'. \tag{2.28}$$

The remaining part,

$$\bar{E}_{\text{xc}}[\rho, w] = \bar{E}_{\text{Hxc}}[\rho, w] - \bar{E}_H[\rho, w], \tag{2.29}$$

is called the *exchange-correlation* energy.

With Eq. (2.26) one is back to an unconstrained variation of a wave function that is chosen to be anti-symmetric, thus satisfying the Pauli principle.

The problem is made simpler by a proper choice of w. For the Kohn–Sham model, one chooses the simplest one, namely $w = 0$.

The Kohn–Sham model is usually presented as a modified Schrödinger equation that is obtained by the variation of Ψ in Eq. (2.26),

$$H(w)\Psi(w) = \mathcal{E}(w)\Psi(w), \tag{2.30}$$

where

$$H(w) = T + V_{\text{ne}} + V_{\text{Hxc}}[\rho, w] + W, \tag{2.31}$$

$$V_{\text{Hxc}} = \sum_{i=1}^{N} v_{\text{Hxc}}(\mathbf{r}_i), \tag{2.32}$$

$$v_{\text{Hxc}}(\mathbf{r}, w) = \frac{\delta \bar{E}_{\text{Hxc}}[\rho, w]}{\delta \rho(r)}. \tag{2.33}$$

Observe that this step (with the extra problem of the existence of the functional derivative) is not needed to obtain E_0. Furthermore, $\mathcal{E}(w) = E[v_{\text{ne}} + v_{\text{Hxc}}, w, N]$, so that

$$E_0 = \mathcal{E}(w) + \bar{E}_{\text{Hxc}}[\rho_0, w] + \int_{\mathbb{R}^3} \rho_0 \left(v_{\text{ne}}(\mathbf{r}) - v_{\text{Hxc}}(\mathbf{r}, w)\right) \mathrm{d}\mathbf{r}, \tag{2.34}$$

where ρ_0 is a minimizing density.[1]

2.4.1.5 Using the Model Wave Function

One can also use the minimizing model wave function, $\Psi(w)$, and choose to approximate the *correlation* density functional

$$\bar{E}_{\text{c}}[\rho, w] = F[\rho, w] - \langle \Psi(w) | T + W | \Psi(w) \rangle . \tag{2.35}$$

2.4.1.6 Problems that Remain After Splitting *F*

Separating F into a defined part, F_d, and a remainder to be approximated, $\bar{F}_d$, does not necessarily guarantee that an approximation, like that given in Eq. (2.22), works better.

Separating the Hartree part, E_H, analogously to what was done in Eq. (2.29), removes a problem, but introduces a new one. Take the limiting case of one-

[1] For the standard Kohn–Sham model, $\mathcal{E}(w = 0)$ is a sum of orbital energies.

electron systems. There is no contribution of the interaction between electrons: $E_{\mathrm{Hxc}} = 0$. Thus, calculating the Hartree part exactly means that the remaining part has to cancel E_H exactly. But obtaining approximations for $-E_H$ is as difficult as obtaining them for E_H, and this was considered not to be reachable with approximations of LDA-type. This is known as the *self-interaction* problem.

Another (not unrelated) problem is due to degeneracy. For example, this appears when we consider two parts of the system far apart, and this even in the simplest molecules like H_2, or H_2^+ when they are stretched (the internuclear distance goes to infinity). Then, something related to the Einstein–Podolsky–Rosen effect shows up: an infinitesimal perturbation can produce a drastic change in the wave function, the density, etc., but not in the energy. Unfortunately, this conflicts with the general philosophy of constructing DFAs that are aimed to produce significant changes in the energies for small changes in the density.

One could imagine detecting degeneracy. However, the model systems, in particular the mean-field models (such as Kohn–Sham), do not necessarily have ground states presenting the same degeneracy as the physical system: while one can present some degeneracy, the other may not. While the physical wave functions have the symmetry of the Hamiltonian, the model wave function often breaks symmetry to reduce the energy. (Well-known is the breaking of spin symmetry which shows up when bonds are stretched.) The opposite can occur, too: the Kohn–Sham system can produce degeneracy, but not when Coulomb interaction is present (see, e.g., Fig. 11 in [32]).

Even more difficult is the case of near-degeneracy, i.e., when a small change in the parameters characterizing H can produce degeneracy. In this case, detecting degeneracy is not a trivial problem, and it exists in many-body calculations, too. The standard approach in such situations is to stop using a single Slater determinant as a reference.

2.4.1.7 Problems of the Model Systems

By construction, the minimizing model $\Psi(w)$ gives an exact ground state density. Some other properties can be reproduced, too. Trivially, all the expectation values of local one-particle operators can be found, as they need only the density to compute them. Surprisingly at first, the exact ionization potential can also be obtained. However, this can be easily understood, as it can be related to the asymptotic decay of the density (see, e.g., [6, 17]).

Often, quantities that are not proven to be reproduced exactly by the model system are nevertheless expected to be good approximations. However, there is a danger of over-stretching this analogy. For example, it is fashionable to judge DFAs by their ability to reproduce fundamental gaps (differences between the ionization potentials and the electron affinities) from differences between orbital energies (of the lowest unoccupied and highest occupied ones). However, this is wrong [22, 34]. Let us consider, for example, a system with zero electron affinity. For a neutral system the Kohn–Sham potential, $v_{\mathrm{ne}} + v_{\mathrm{Hxc}}$, Eq. (2.31), decays at large distances as $-1/r$ (see, e.g.,[17]), and we know that it supports Rydberg series. Thus, its gap (ionization potential) is necessarily larger than its first excitation energy. In fact, accurate Kohn–Sham orbital energy differences give good approximations to excitation energies. Let us take the He atom as an example [33]. An extremely accurate Kohn–Sham potential can be obtained from an extremely accurate density. The Kohn–Sham one-particle Hamiltonian lowest eigenvalue corresponds to the doubly occupied state (1s). However, higher eigenvalues exist. The next eigenvalue (2s) is $\approx$ 0.75 hartree above the lowest one. It can be compared to the excitation energies of the triplet and singlet ($\approx$ 0.73, and 0.76 hartree, respectively). However, the fundamental gap of the He atom is of $\approx$ 0.90 hartree. (This comparison should not to be confused with potentials produced by DFAs, as LDA for E_{xc} that generates a potential that does not support excited states, and has a ionization potential of $\approx$ 0.55 hartree.) Thus, in general, a DFA that produces an orbital energy difference that reproduces the exact fundamental gap can be expected not to be a good approximation to the exact Kohn–Sham system.

The preceding discussion leads to slippery ground. Could it be that the Kohn–Sham approximations are used because they produce convenient mean-field models? Could it be that (for specific purposes) they may be better than the exact Kohn–Sham system would be?

2.4.2 *Refining the Approximations*

2.4.2.1 Spin Densities

The quality of approximations improves considerably when spin densities are introduced, i.e., when the functional $\widetilde{G}$ is made to depend not only on ρ, but on its components, the spin-up, $\rho_\uparrow(\mathbf{r})$, and the spin-down $\rho_\downarrow(\mathbf{r})$ densities,

$$\rho(\mathbf{r}) = \rho_\uparrow(\mathbf{r}) + \rho_\downarrow(\mathbf{r}).$$

Equivalently, one may add to the dependence on ρ that on the spin polarization

$$\zeta(\mathbf{r}) = \frac{\rho_\uparrow(\mathbf{r}) - \rho_\downarrow(\mathbf{r})}{\rho(\mathbf{r})}. \tag{2.36}$$

A justification is brought by the fact that the exchange term acts only for electrons of the same spin, and that correlation is not the same for a pair of electrons of different spins as for that between two electrons of the same spin (which are kept apart by the Pauli principle).

An example of the importance of making the functional depend on $\rho_\uparrow$ and $\rho_\downarrow$ is shown in Fig. 2.1. According to the Hohenberg–Kohn theorem, neither the energy nor the value of F for the hydrogen atom should depend on ζ. However, for LDA where a dependence on ζ is introduced by adjusting the exchange-correlation of the

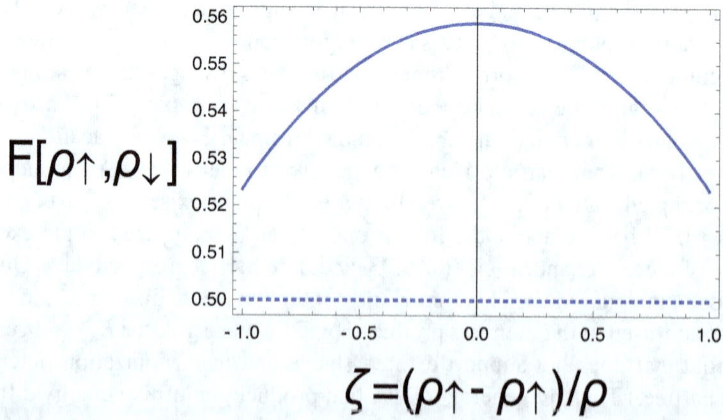

Fig. 2.1 Dependence of the local density approximation of the functional F on the spin polarization ζ, Eq. (2.36) for the exact density of the hydrogen atom; the exact value of F is 0.5 hartree (dotted line)

spin-polarized uniform electron gas, there is a clear dependence on ζ, the best value being obtained when $\zeta = \pm 1$, i.e., for maximal spin polarization.

In spite of contributing in a decisive way to the success of DFAs (most achievements in thermo-chemistry would be non-existent without using spin-densities), there is a problem: the theoretical foundation of this approach has never been established. There are several arguments in support of this affirmation. Here we give a few. One hears that the spin-density shows up in a weak magnetic field, and wrongly assumes that

1. a weak magnetic field should not affect the result,
2. a linear magnetic field should be sufficient, because the field is weak,
3. it is sufficient to take into account the interaction between the magnetic field and the spins (i.e., only a term $B_z S_z$),
4. the magnetic fields used for spin-polarized systems are weak.

The first point is wrong, because lifting degeneracy by a magnetic field can produce a different ground state. For example, putting a stretched H_2 molecule in a weak uniform magnetic field produces a triplet ground state, while in the absence of a magnetic field, it is a singlet. The second point is wrong, because it ignores a general problem: "a small perturbation parameter does not mean a small perturbation" [24]. For the specific case we are considering, we notice that even a weak linear magnetic field stabilizes states with high angular momentum below the ground state in the absence of the magnetic field. The variational principle cannot be applied, and the Hohenberg–Kohn theorem cannot be proven [28]. The third point is wrong, as we know from the elementary treatment of the Zeeman effect: the orbital momentum is as important as the spin, but if we introduce a dependence on it, we have a dependence on the external potential, and this is not allowed for a universal density

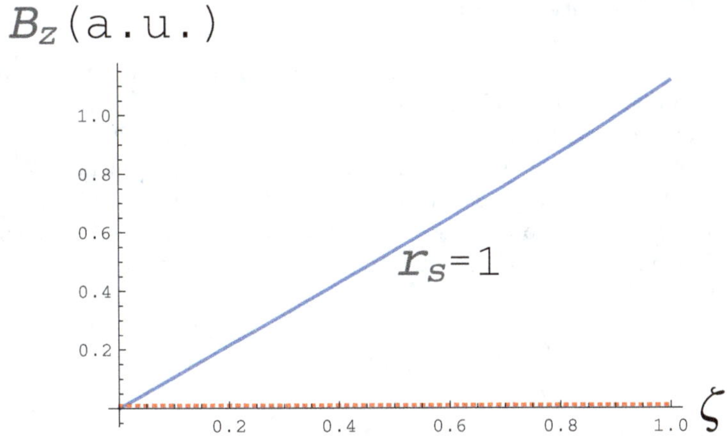

Fig. 2.2 Strength of the magnetic field, B, needed to stabilize the uniform electron gas with polarization ζ with respect to the unpolarized electron gas with density $\rho = 3/4\pi$, i.e., $r_s = 1$. The strongest magnetic field ever produced on earth is indicated by a horizontal dotted line

functional. Finally, the fourth point is wrong, because in order to produce a spin-polarized electron gas (for densities of chemical interest, $\rho \approx 3/4\pi$, i.e., $r_s = 1$) a strong electronic excitation is needed, and this can be produced only by a huge magnetic field (see Fig. 2.2).

There is, however, a different viewpoint: the spin-density is a proxy for another quantity that can be related to the spin-density. It was noticed long ago, for unrestricted Hartree–Fock calculations [38], that spin-densities can be connected to on-top pair density, $P(\mathbf{r}, \mathbf{r})$, cf. Eq. (2.11). Starting from

$$\rho(\mathbf{r}) = \rho_\uparrow(\mathbf{r}) + \rho_\downarrow(\mathbf{r})$$

$$P(\mathbf{r}, \mathbf{r}) = \left(\rho_\uparrow(\mathbf{r}) + \rho_\downarrow(\mathbf{r})\right)^2 - \left(\rho_\uparrow(\mathbf{r})^2 + \rho_\downarrow(\mathbf{r})^2\right) \quad \text{(single determinant)}$$

an alternative interpretation of the spin-density in DFT is obtained [2, 23]:

$$|\rho_\uparrow - \rho_\downarrow| = \sqrt{\rho(\mathbf{r})^2 - 2P(\mathbf{r}, \mathbf{r})} \quad \text{(single determinant)}. \tag{2.37}$$

Could it be that the theory behind DFAs is not DFT?

Let us mention that a relationship can also be found between spin-densities and first-order reduced density matrices ([35, 36]).

2.4.2.2 The Adiabatic Connection

The adiabatic connection was invoked in order to understand what a density approximation should do [9, 11, 15, 39]. The basic idea is that one constructs a model Hamiltonian depending continuously on a parameter, $H(\mu)$. The corresponding Schrödinger equation has an eigenvalue $E(\mu)$ and an eigenfunction $\Psi(\mu)$. We require that for a certain value of this parameter the model Hamiltonian becomes the physical one. Let us now choose $\mu = \infty$ for it. Furthermore, we assume that the Hellmann–Feynman theorem (or first-order perturbation theory) can be applied to this model:

$$\frac{\mathrm{d}}{\mathrm{d}\mu} E(\mu) = \langle \Psi(\mu) | \, \partial_\mu H(\mu) \, | \Psi(\mu) \rangle \,. \tag{2.38}$$

Suppose that the model system, say at μ_0, is accessible (for example, it is a Kohn–Sham calculation). We want to know how to correct the model energy, $E(\mu_0)$, to obtain $E = E(\mu = \infty)$. For the missing part, we use the notation, $\bar{E}(\mu_0)$:

$$E = E(\mu_0) + \bar{E}(\mu_0). \tag{2.39}$$

By integrating Eq. (2.38) we get what is also called the integrated Hellmann–Feynman formula [4]

$$\bar{E}(\mu_0) = E - E(\mu_0) = \int_{\mu_0}^{\infty} \langle \Psi(\mu) | \, \partial_\mu H(\mu) \, | \Psi(\mu) \rangle \, \mathrm{d}\mu. \tag{2.40}$$

If we consider (as above) that the model only changes V and W, we also write

$$E - E(\mu_0) = \int_{\mu_0}^{\infty} \langle \Psi(\mu) | \, \partial_\mu \left(V(\mu) + W(\mu) \right) | \Psi(\mu) \rangle \, \mathrm{d}\mu. \tag{2.41}$$

In density functional theory, one furthermore assumes that one can choose $V(\mu)$ such that the density does not change with μ. Using Eqs. (2.4) and (2.6), and the convention used here that $v \to v_{\mathrm{ne}}$ when $\mu \to \infty$, we can write

$$E - \langle \Psi(\mu_0) | \, T + W(\mu_0) + V_{\mathrm{ne}} \, | \Psi(\mu_0) \rangle = \int_{\mu_0}^{\infty} \langle \Psi(\mu) | \, \partial_\mu W(\mu) \, | \Psi(\mu) \rangle \, \mathrm{d}\mu. \tag{2.42}$$

Using a relationship analogous to Eqs. (2.9), and (2.11),

$$\langle \Psi(\mu) | \, \partial_\mu W(\mu) \, | \Psi(\mu) \rangle = \frac{1}{2} \int_{\mathbb{R}^3} \int_{\mathbb{R}^3} P_\mu(\mathbf{r}_1, \mathbf{r}_2, \mu) \partial_\mu w(|\mathbf{r}_1 - \mathbf{r}_2|, \mu) \, \mathrm{d}\mathbf{r}_1 \mathrm{d}\mathbf{r}_2. \tag{2.43}$$

Notice that as Ψ depends on μ, so does P. A comparison with Eq. (2.26) (where the dependence on w is replaced by that on μ) gives the correction to $E(\mu_0)$:

$$\bar{E}_{\text{Hxc}}(\mu_0) = \int_{\mathbb{R}^3} \mathrm{d}\mathbf{r}_1 \underbrace{\int_{\mu_0}^{\infty} \mathrm{d}\mu \int_{\mathbb{R}^3} d\mathbf{r}_2 \, P_\mu(\mathbf{r}_1, \mathbf{r}_2, \mu) \partial_\mu w(|\mathbf{r}_1 - \mathbf{r}_2|, \mu)}_{e(\mathbf{r}_1)} . \tag{2.44}$$

The integrand $e(\mathbf{r}_1)$ shows a superficial similarity with the function g appearing in LDA, Eq. (2.22). However, unlike LDA, the connection with ρ is not evident.

One can eliminate a known term from $\bar{E}_{\text{Hxc}}$, and correct correspondingly the r.h.s. For example, if we would like to have $\bar{E}_{\text{xc}}$, Eq. (2.29), we eliminate the contribution of $\bar{E}_H$, by taking the derivative w.r.t. μ in Eq. (2.28), i.e., by subtracting $\rho(\mathbf{r})\rho(\mathbf{r}')$ from P on the r.h.s. of Eq. (2.44).

2.4.2.3 Density or Pair-Density Functional Theory?

Starting from Eq. (2.44), we may ask whether we should construct functionals of the pair density, $P(\mathbf{r}, \mathbf{r}')$, instead of one that depends on $\rho(\mathbf{r})$. We first notice that the pair density, $P(\mathbf{r}, \mathbf{r}')$, yields, by integration over $\mathbf{r}'$, the density $\rho(\mathbf{r})$, up to a factor $N - 1$. The already mentioned relationship between spin-densities and the on-top pair density, Eq. (2.37), presents itself as a further argument. However, the conditions to be imposed on P such that it is fermionic are difficult, while those to be imposed on ρ are simple (ρ should be non-negative, and integrate to N).

In fact, LDA can be seen as replacing, in each point of space $\mathbf{r}_1$, $P(\mathbf{r}_1, \mathbf{r}_2)$ in Eq. (2.44) by that obtained in the uniform electron gas with density $\rho(\mathbf{r}_1)$ (see, e.g., [9]). This idea can be extended beyond LDA: many successful functionals have been constructed starting from this perspective (among them those developed by A.D. Becke, and by J.P. Perdew and co-workers, see, e.g., [1]).

Some people consider the random phase approximation (RPA) as a density functional model. It can also be seen as constructing a simplified form of P to be used in Eq. (2.44) (see, e.g., [5]).

Recently, new approximations using the pair density showed up (see, e.g., [37]).

2.4.3 *Approaching the Exact Result*

2.4.3.1 Limitations of the Mean Field Model

Even if by miracle we had the exact Kohn–Sham determinant (and potential), we would still miss information about the exact system (with Coulomb interaction). For example, we still would not have the exact energy. Unfortunately, the task of obtaining simple functionals capable of dealing with cases when a single Slater determinant is not a good approximation is not solved.

Sometimes ensembles of Kohn–Sham states are discussed. A formula expressing the correlation energy in terms of weighted Kohn–Sham orbital energies exists [25]. However, we do not know a simple expression for obtaining the weights, and it does not seem that they follow a Boltzmann distribution [30].

Long experience in quantum chemistry shows that a single Slater determinant is often a bad starting point for obtaining many properties such as the energy. There, it seems natural to consider multi-reference methods, i.e., wave functions where more than one determinant deserves a preferential treatment. The selection of determinants is an art, unless selective methods are used, such as CIPSI (configuration interaction by perturbation with multiconfigurational zeroth-order wave function selected by iterative process) [13]. In the following, a special way of generating a multi-determinant wave function will be discussed, namely using some (ideally) weak interaction operator W. Degenerate (and near-degenerate) states are detected by such operators, and this automatically introduces more than one Slater determinant if needed. Using more complicated wave functions is a price to pay for getting forms that make existing DFAs closer to a theoretically justifiable form.

2.4.3.2 Choosing *w*

Equation (2.44) suggests that it may be more easy to obtain approximations for E_{Hxc} when $w \neq 0$, i.e., $\mu > 0$. Indeed, if $\partial_\mu w$ is short-ranged, we can use some approximation of $P(\mathbf{r}, \mathbf{r}')$ that is valid only when $\mathbf{r}'$ is close to $\mathbf{r}$, and use an expansion around $\mathbf{r}$. In the limit of zero-range (δ-function) we obtain the on-top pair density $P(\mathbf{r}, \mathbf{r})$ that for a single Slater determinant produces a connection to the spin-density (Eq. (2.37)), i.e., a form that resembles LDA with spin-dependence. Furthermore, expanding P in $\mathbf{r}'$ around $\mathbf{r}$ produces *semi-local* terms such as density derivatives [7]. Finally, we can expect a better transferability between systems when electrons are close, justifying the transferability from other systems like the uniform electron gas, in other words, expecting “universality”.

Also, it seems advantageous to avoid using w that posses a singularity (like the Coulomb interaction), because this induces a strong dependence on the basis set used, and hence a very slow convergence to the exact results (cf. the difficulty of converging $\langle\Psi|\,\delta(r_{12})\,|\Psi\rangle$ with a finite basis set [3]).

A simple and computationally convenient form for w satisfying the requirements above is given by

$$w(\mathbf{r}, \mu) = \frac{\mathrm{erf}(\mu|\mathbf{r}|)}{|\mathbf{r}|}. \tag{2.45}$$

Its derivative is short-ranged,

$$\partial_\mu w(\mathbf{r}, \mu) = \frac{2}{\sqrt{\pi}} \mathrm{e}^{-\mu^2|\mathbf{r}|^2}. \tag{2.46}$$

and, when μ is very large

$$\partial_\mu w(\mathbf{r}, \mu) \to \frac{2\pi}{\mu^3}\delta(\mathbf{r}), \qquad \text{for } \mu \to \infty. \tag{2.47}$$

The interaction in Eq. (2.45) also has the properties:

- $w \to v_{ee}$ when $\mu \to \infty$,
- $w = 0$ when $\mu = 0$

i.e., by changing μ it is possible to switch between the Kohn–Sham and the physical system. This allows us to consider this method to be systematically improvable, in the sense that increasing μ brings the model closer to the physical Hamiltonian.

However, we do not know how far we have to get away from $w = 0$ to get reliable approximations. This can be explored numerically.

2.4.3.3 Errors of DFAs for $w > 0$

Below are results obtained with w given by Eq. (2.45) and the dependence on μ is analyzed.

First, to construct a density functional approximations to $\bar{E}_{xc}$, Eq. (2.29), uniform electron gas calculations are used [19, 26]. Now, the LDA, Eq. (2.22), can be applied to $\widetilde{G} = \bar{E}_{xc}$ for any value of μ.

The numerical results given below are for the 2-electron harmonium, a system with the Hamiltonian

$$H = T + \sum_{i=1}^{2} \omega^2 r_i^2 + \mathrm{erf}(\mu|\mathbf{r}_1 - \mathbf{r}_2|)/|\mathbf{r}_1 - \mathbf{r}_2|. \tag{2.48}$$

The variables can be separated, and the solutions can be found for real values of μ and ω by solving numerically a one-dimensional differential equation (see, e.g., [14]). For $\omega = 1/2$, which is chosen below, analytical solutions are known for the non-interacting ($\mu = 0$) and the fully interacting ($\mu = \infty$) system.

Figure 2.3 shows the errors made for the harmonium as a function of the choice of the parameter μ. At $\mu = 0$, the error is that given by the usual LDA. It decreases steadily, and around $\mu = 0.5 \ldots 1$ a change of behavior occurs, quickly reaching chemical accuracy (1 kcal/mol $\approx$ 2 mhartree).

As having $w \neq 0$ requires having more than one Slater determinant, the time required for computing the wave function rapidly increases with μ. However, as the convergence with the basis set is faster if w has no singularity, less computational effort is needed to obtain the wave function. Figure 2.4 shows the error that can be achieved in a given time. Calculations were done first for spherically symmetric basis functions to saturation (s-limit). Next a new value was obtained for the p-limit ($l = 1$), next for the d-limit, ($l = 2$), etc. For such a small system, there is no gain in

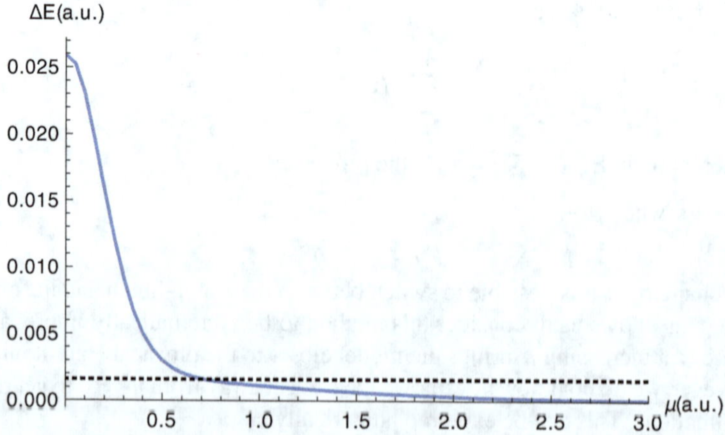

Fig. 2.3 Errors made by the local density approximation for the ground state energy of harmonium (Eq. (2.48)) as a function of the range separation parameter of the model, μ (Eq. (2.45))

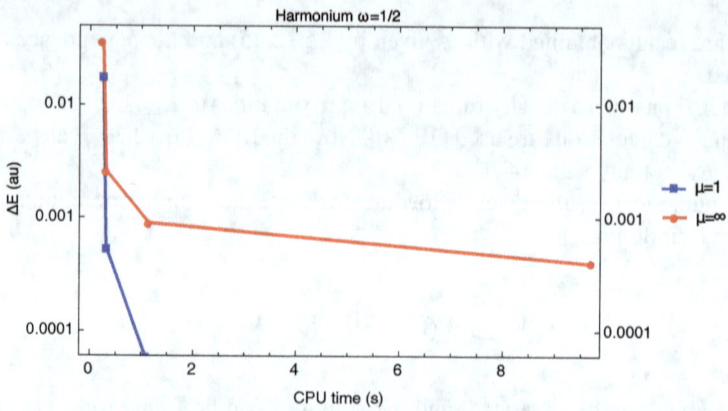

Fig. 2.4 Harmonium energy errors obtained by saturating the basis set with $l = 0, 1, 2, 3$, in a calculation with $\mu = 1$, blue, and for the Coulomb interaction, red

computing the integrals. However, one can see that a high accuracy is reached much faster when $\mu = 1$ than when $\mu = \infty$.

For this system, choosing a value of μ between 0.5 and 1 seems to provide a good compromise between the supplementary effort needed to have $w \neq 0$, and having a good density functional approximation.

2.4.3.4 Approaching the Exact Result Analytically

Instead of using *universal* models for P in Eq. (2.44), one can construct corrections for *a given* model Hamiltonian whose energies are determined by some v and

w, $\bar{E} = E[v_{\rm ne}, v_{\rm ee}, N] - E[v, w, N]$, Eq. (2.39). The role of the approximation is to correct for the difference between the energy of the exact system and that of the model system. We explore whether standard techniques from numerical analysis could compete with density functional approximations in estimating these corrections.

Notice that as $v \to v_{\rm ne}$ and $w \to v_{\rm ee}$, the correction vanishes: $\bar{E} \to 0$. One can also try to improve the result by using a set of model Hamiltonians for which obtaining the model energy is simpler than finding $E[v_{\rm ne}, v_{\rm ee}, N]$.

In a density functional context it is tempting to use v as given by some density functional approximation, or even to use the potential that yields the exact density ρ (to show the principle of the procedure). Below the simplest expression for the external potential is chosen, $v = v_{\rm ne}$. Of course, this brings the model system very far from the physical system when the interaction w is weak: the errors of the model at $w = 0$ are a very important part of the total energy. For example, for the harmonium studied above, at $\mu = 0$ the error is of 0.5 hartree, as shown in Fig. 2.5.

First, we analyze how the energy of the model system, $E(\mu)$, approaches that of the Coulomb system, i.e., how $E(\mu)$ approaches $E(\mu = \infty)$. From the large μ behavior of the interaction w of Eqs. (2.45) and (2.47), we derive

$$E = E(\mu) + a_{-k}\mu^{-k} + a_{-k-1}\mu^{-k-1} + \cdots \tag{2.49}$$

The coefficient a_{-k} in the equation above is proportional to

$$\int_{\mathbb{R}^3} P(\mathbf{r}, \mathbf{r}, \mu = \infty)\, d\mathbf{r}.$$

The coefficient a_{-k-1} is proportional to a_{-k}, and given by the cusp condition, as $\Psi(\mu)$ has to approach $\Psi = \Psi(\mu = \infty)$ when μ gets large [8], and k is equal to $2l + 2$, $2l$ being the power of the expansion of $P(\mathbf{r}, \mathbf{r} + \mathbf{u})$ in $|\mathbf{u}|$ around zero. In particular, for a pair of singlet coupled electrons (of anti-parallel spin), we have $k = 2$ and

$$P(\mathbf{r}, \mathbf{r}, \mu) = P(\mathbf{r}, \mathbf{r}, \mu = \infty)(1 + \frac{2}{\sqrt{\pi}}\mu^{-1} + \cdots), \tag{2.50}$$

yielding

$$\frac{a_{-3}}{a_{-2}} = \frac{4\sqrt{2}}{3\sqrt{\pi}}.$$

2.4.4 Taylor Series Truncation Error

We consider a Taylor series for large μ. First, we make a change of variable to $x(\mu)$ such that x monotonously approaches 0 as $\mu \to \infty$,

$$E(x=0) = E(x) - xE'(x) + \frac{1}{2}x^2 E''(x) + \cdots \tag{2.51}$$

Using the chain rule, we go back to the μ variable,

$$\begin{aligned} E(\mu=\infty) = E(\mu) - x(\mu)\frac{E'(\mu)}{x'(\mu)} \\ + \frac{1}{2}x(\mu)^2\left[E''(\mu) - E'(\mu)x''(\mu)/x'(\mu)\right]\left(x'(\mu)\right)^2 + \cdots \end{aligned} \tag{2.52}$$

Obtaining the first derivative of E with respect to the energy is not expensive, because it does not require the computation of a new wave function. Of course, the cost increases with higher derivatives.

A choice for a change of variable that makes the expansion correct at large μ, is $x = \mu^{-2}$, cf. Eq. (2.49). As we know the next term in this expansion, we can also choose $x = \mu^{-2} + \kappa\mu^{-3}$. Not surprisingly, the latter choice is more reliable than the first (cf. Fig. 2.5). However, it is a surprise that the approximation works very well up to $\mu \approx 1$, while the expansion was derived in the limit $\mu \to \infty$.

Comparing these results with Fig. 2.3, one notices that the range of models for which the density functional approximation works well is comparable to that for which the Taylor series works well.

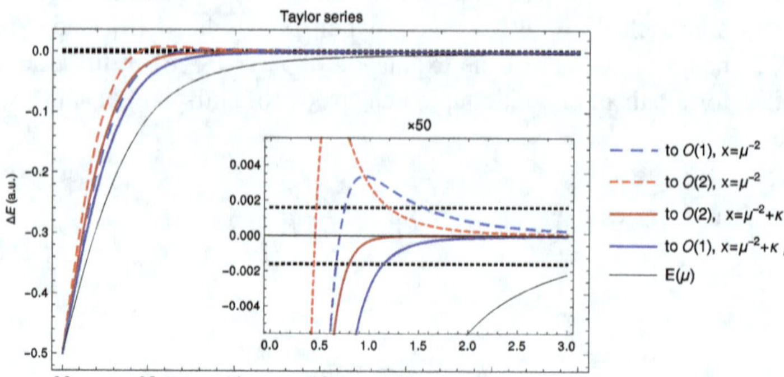

Fig. 2.5 Errors of different approximations for the ground state energy of harmonium: $E(\mu)$ (thin, black), and the Taylor series around μ to order 1 (blue), and to order 2 (red); the dashed curves correspond to a transformation to $x(\mu) = \mu^{-2}$, the others to $x(\mu) = \mu^{-2} + \kappa mu^{-3}$. The horizontal dotted lines indicate chemical accuracy (±1 kcal/mol). The inset shows a zoom on the same curves

2.5 Outlook

Most applications of density functional theory rely on the simplicity of using a single Slater determinant. This chapter does not intent to discourage the traditional search of density functionals. They are successful in practice, and there still is room for improvement. However, using the simple mathematical techniques discussed in the preceding section lead to good quality approximations, and this is encouraging. There are many paths that could be followed. One, of course, is to improve the mathematical techniques. Another is to change the interaction w to a form for which the extrapolations considered here would work better. Finally, one can envisage using density functional models in combination with the extrapolation approach presented here.

Acknowledgments Eric Cancès' comments on the first draft on the manuscript are gratefully acknowledged.

References

1. A.D. Becke. Hartree–Fock exchange energy of an inhomogeneous electron gas. *Int. J. Quantum. Chem.* **23**(6), 1915–1922 (1983).
2. A.D. Becke, A. Savin, and H. Stoll. Extension of the local-spin-density exchange-correlation approximation to multiplet states. *Theoret. Chim. Acta* **91**, 147–156 (1995).
3. E.R. Davidson. Natural expansions of exact wavefuncions. III. The Helium-atom ground state. *J. Chem. Phys.* **39**, 875 (1963).
4. S.T. Epstein, A.C. Hurley, R.E. Wyatt and R.G. Parr. Integrated and integral Hellmann–Feynman formulas. *J. Chem. Phys.* **47**, 1275 (1967).
5. H. Eshuis, J. Bates, and F. Furche. Electron correlation methods based on the random phase approximation. *Theor. Chem. Acc.* **131**, 1084 (2012).
6. S. Fournais, M. Hoffmann-Ostenof, T. Hoffmann-Ostenhof, and T.Ø. Sørensen. Analytic structure of many-body coulombic wave functions. *Commun. Math. Phys.* **289**, 291–310 (2009).
7. P.M.W. Gill, R.D. Adamson and J.A. Pople. Coulomb-attenuated exchange energy density functionals. *Mol. Phys.* **88**, 1005–1009 (1996).
8. P. Gori-Giorgi and A. Savin. Properties of short-range and long-range correlation energy density functionals from electron-electron coalescence. *Phys. Rev. A* 73, 032506 (2006).
9. O. Gunnarsson and B.I. Lundqvist. Exchange and correlation in atoms, molecules, and solids by the spin-density-functional formalism. *Phys. Rev. B* **13**, 4274 (1976).
10. C. Gutlé and A. Savin. Orbital spaces and density-functional theory. *Phys. Rev. A* 75, 032519, 2007.
11. J. Harris and R.O. Jones. The surface energy of a bounded electron gas-solid. *J. Phys. F* **4**, 1170–1186 (1974).
12. P. Hohenberg and W. Kohn. Inhomogeneous electron gas. *Phys. Rev. B* **136**, 864 (1964).
13. B. Huron, J.-P. Malrieu, and P. Rancurel. Iterative perturbation calculations of ground and excited state energies from multiconfigurational zeroth-order wavefunctions. *J. Chem. Phys.* **58**, 5745 (1973).
14. J. Karwowski and L. Cyrnek. Harmonium. *Ann. Phys. (Leipzig)* **13**(4), 181–193 (2004).
15. D.C. Langreth and J.P. Perdew. The exchange-correlation energy of a metallic surface. *Solid State Commun.* **17**(11), 1425–1429 (1975).

16. M. Levy. Universal variational functionals of electron densities, first-order density matrices, and natural spin-orbitals and solution of the v-representability problem. *Proc. Natl. Acad. Sci. U.S.A.* **76**(12), 6062–6065 (1979).
17. M. Levy, J.P. Perdew and V. Sahni. Exact differential equation for the density and ionization energy of a many-particle system. *Phys. Rev. A* **30**, 2745 (1984).
18. E. H. Lieb. Density functionals for coulomb systems. *Int. J. Quantum. Chem.* **24**(3), 243–277 (1983).
19. Simone Paziani, Saverio Moroni, Paola Gori-Giorgi, and Giovanni B. Bachelet. Local-spin-density functional for multideterminant density functional theory. *Phys. Rev. B* **73**, 155111–15519 (2006).
20. J. Percus. The role of model systems in the few-body reduction of the N-fermion problem. *Int. J. Quantum Chem.* **13**, 89–124 (1978).
21. J.P. Perdew. What do the Kohn–Sham orbital energies mean? How do atoms dissociate? In: *Density Functional Methods in Physics*, edited by R.M. Dreizler and J. da Providencia, pp. 265–308, Plenum, New York (1985).
22. J.P. Perdew and M. Levy. Physical Content of the Exact Kohn Sham Orbital Energies: Band Gaps and Derivative Discontinuities. *Phys. Rev. Lett.* **51**, 1884–1887 (1983).
23. J.P. Perdew, A. Savin and K. Burke. Escaping the symmetry dilemma through a pair-density interpretation of spin-density functional theory. *Phys. Rev. A* **51**, 4531 (1995).
24. F. Rellich. *Perturbation Theory of Eigenvalue Problems*. Gordon and Breach, New York (1969).
25. A. Savin. Expression of the exact electron-correlation-energy density functional in terms of first-order density matrices. *Phys. Rev. A* **52**, R1805–R1807 (1995).
26. A. Savin. On degeneracy, near degeneracy and density functional theory. In: *Recent Developments of Modern Density Functional Theory*, edited by J.M. Seminario, pp. 327–357, Elsevier, Amsterdam (1996).
27. A. Savin. Is size-consistency possible with density functional approximations? *Chem. Phys.* **356**, 91 (2009).
28. A. Savin. Absence of proof for the Hohenberg–Kohn theorem for a Hamiltonian linear in the magnetic field. *Mol. Phys.* **115**, 13 (2017).
29. A. Savin and F. Colonna. Local exchange-correlation energy density functional for monotonically decaying densities. *J. Mol. Struct. (Theochem)* **39**, 501–502 (2000).
30. A. Savin and F. Colonna. A spectral analysis of the correlation energy. *J. Mol. Struct. (Theochem)* **527**, 121 (2000).
31. A. Savin, F. Colonna, and M. Allavena. Analysis of the linear response function along the adiabatic connection from the Kohn–Sham to the correlated system. *J. Chem. Phys.* **115**, 6827 (2001).
32. A. Savin, F. Colonna, and R. Pollet. Adiabatic connection approach to density functional theory of electronic systems. *Int. J. Quantum. Chem.* **93**, 166–190 (2003).
33. A. Savin, C.J. Umrigar, and X. Gonze. Relationship of Kohn–Sham eigenvalues to excitation energies. *Chem. Phys. Lett.* **288**, 391 (1998).
34. L.J. Sham and M. Schlüter. Density-functional theory of the energy gap. *Phys. Rev. Letters* **51**, 1888 (1983).
35. V.N. Staroverov and E.R. Davidson. Distribution of effectively unpaired electrons. *Chem. Phys. Lett.* **330**, 161 (2000).
36. K. Takatsuka, T. Fueno, and K. Yamaguchi. Distribution of odd electrons in ground-state molecules. *Theor. Chim. Acta* **48**, 175–183 (1978).
37. L. Wilbraham, P. Verma, D.G. Truhlar, L. Gagliardi, and I. Ciofini. Multiconfiguration pair-density functional theory predicts spin-state ordering in iron complexes with the same accuracy as complete active space second-order perturbation theory at a significantly reduced computational cost. *J. Phys. Chem. Lett.* **8**, 2026–2030 (2017).

38. K. Yamaguchi and T. Fueno. Correlation effects in singlet biradical species. *Chem. Phys.* **19**, 35–42 (1977).
39. W. Yang. Generalized adiabatic connection in density functional theory. *J. Chem. Phys.* **109**, 10107 (1998).

Chapter 3
Universal Functionals in Density Functional Theory

Mathieu Lewin, Elliott H. Lieb, and Robert Seiringer

Abstract In this chapter we first review the Levy–Lieb functional, which gives the lowest kinetic and interaction energy that can be reached with all possible quantum states having a given density. We discuss two possible convex generalizations of this functional, corresponding to using mixed canonical and grand-canonical states, respectively. We present some recent works about the local density approximation, in which the functionals get replaced by purely local functionals constructed using the uniform electron gas energy per unit volume. We then review the known upper and lower bounds on the Levy–Lieb functionals. We start with the kinetic energy alone, then turn to the classical interaction alone, before we are able to put everything together. A later section is devoted to the Hohenberg–Kohn theorem and the role of many-body unique continuation in its proof.

3.1 Introduction

Density Functional Theory (DFT) attempts to describe all the relevant information about a many-body quantum system at or near its ground state in terms of its one-body density $\rho(\mathbf{r})$. In its orbital-free variational formulation, due to Levy [99] and Lieb [116], DFT relies completely on a universal functional $\rho \mapsto F_{\rm LL}[\rho]$ which gives the lowest (kinetic plus interaction) energy that can be reached with all possible quantum states having a given density function $\mathbf{r} \mapsto \rho(\mathbf{r})$. This functional is exact and it is able to describe interacting quantum Coulomb systems in their

M. Lewin (✉)
CNRS & CEREMADE, Université Paris-Dauphine, PSL University, Paris, France
e-mail: mathieu.lewin@math.cnrs.fr

E. H. Lieb
Departments of Mathematics and Physics, Jadwin Hall, Princeton University, Princeton, NJ, USA
e-mail: lieb@princeton.edu

R. Seiringer
IST Austria (Institute of Science and Technology Austria), Klosterneuburg, Austria
e-mail: robert.seiringer@ist.ac.at

E. Cancès, G. Friesecke (eds.), *Density Functional Theory*, Mathematics and Molecular Modeling, https://doi.org/10.1007/978-3-031-22340-2_3

ground states. It is of course not known explicitly and one of the main purpose of DFT is to find suitable approximations. In this chapter we review known upper and lower bounds on this functional and discuss some regimes in which it simplifies. In particular we focus on the *Local Density Approximation (LDA)* which becomes exact in the regime where the density is very flat on sufficiently large regions of space, as was recently proved in [106, 108]. A simpler older approximation is the Thomas–Fermi functional [53, 169], which is surprisingly accurate for heavy atoms [122, 123] and is reviewed in [114].

It turns out that there are several possible Levy–Lieb-type functionals. Instead of considering N-particle wavefunctions one can also work with mixed states [116], or even with grand-canonical states for which only the average number of particles is fixed. The latter has not been thoroughly discussed in the literature. We consider the three possibilities in this chapter. When we discuss bounds, it is useful to consider the kinetic and interaction energies separately. This naturally brings in Lieb–Thirring [125, 126] and Lieb–Oxford [113, 120] inequalities, which provide lower bounds on these two functionals.

In Sect. 3.7 we recall the Hohenberg–Kohn theorem, which is another important abstract result in DFT. It turns out that its proof relies on some unique continuation problems for N-particle systems, which are not yet completely understood.

Notation Everywhere in this chapter $\mathbf{x} = (\mathbf{r}, \sigma)$ denotes both the space variable $\mathbf{r} \in \mathbb{R}^d$ and the spin variable $\sigma \in \mathbb{Z}_q$. Although the physical case corresponds to $d = 3$ and $q = 2$, it is sometimes useful to keep d and q general to better emphasize the role of the dimension and spin. To simplify our notation we use the convention that $\int_{\mathbb{R}^d \times \mathbb{Z}_q} f(\mathbf{x})\, d\mathbf{x} = \sum_{\sigma \in \mathbb{Z}_q} \int_{\mathbb{R}^d} f(\mathbf{r}, \sigma)\, d\mathbf{r}$. For N particles we use the notation $\mathbf{X} = (\mathbf{x}_1, \dots, \mathbf{x}_N) \in (\mathbb{R}^d \times \mathbb{Z}_q)^N$ with a similar convention for the integral $\int_{(\mathbb{R}^d \times \mathbb{Z}_q)^N}$.

We recall that the one-body density ρ_Ψ of an N-particle fermionic (normalized) wavefunction

$$\Psi \in \bigwedge_1^N L^2(\mathbb{R}^d \times \mathbb{Z}_q, \mathbb{C})$$

is defined by

$$\rho_\Psi(\mathbf{r}) := N \sum_{\sigma_1, \dots, \sigma_N \in \mathbb{Z}_q} \int_{\mathbb{R}^d} \cdots \int_{\mathbb{R}^d} |\Psi(\mathbf{r}, \sigma_1, \mathbf{r}_2, \sigma_2, \dots, \mathbf{r}_N, \sigma_N)|^2\, d\mathbf{r}_2 \cdots d\mathbf{r}_N.$$

The interpretation of ρ_Ψ is that it provides the average number of particles in space, without taking their spin into account.

3.2 Universal Functionals in Density Functional Theory

Following [99, 116] we express the ground state energy as a variational problem involving only the density, and we discuss some (known and unknown) mathematical properties of the associated functionals.

3.2.1 *The Levy–Lieb Universal Functional*

For completeness we consider general interaction potentials w in any space dimension d. This is useful to understand the particularities of the physical case at work in DFT, namely the Coulomb interaction $w(\mathbf{r}) = |\mathbf{r}|^{-1}$ in dimension $d = 3$.

Let

$$v_+ \in L^1_{\rm loc}(\mathbb{R}^d, \mathbb{R}), \qquad v_-, w \in L^p(\mathbb{R}^d, \mathbb{R}) + L^\infty(\mathbb{R}^d),$$

with w even, with $v_\pm \geqslant 0$, with $v := v_+ - v_-$ and with

$$p \begin{cases} = 1 & \text{when } d = 1, \\ > 1 & \text{when } d = 2, \\ = \frac{d}{2} & \text{when } d \geqslant 3. \end{cases} \tag{3.1}$$

Under the assumption (3.1), the N-body potential

$$W_N^{v,w}(\mathbf{r}_1, \dots, \mathbf{r}_N) := \sum_{j=1}^N v(\mathbf{r}_j) + \sum_{1 \leqslant j < k \leqslant N} w(\mathbf{r}_j - \mathbf{r}_k)$$

is infinitesimally $(-\Delta)$–form bounded from below, which means that

$$\int_{(\mathbb{R}^d \times \mathbb{Z}_q)^N} W_N^{v,w} |\Psi|^2 \geqslant -\varepsilon \int_{(\mathbb{R}^d \times \mathbb{Z}_q)^N} |\nabla \Psi|^2 - C_{N,\varepsilon} \int_{(\mathbb{R}^d \times \mathbb{Z}_q)^N} |\Psi|^2$$

for all Ψ and all $\varepsilon > 0$, with $C_{N,\varepsilon}$ a constant depending only on ε and N. From this we can deduce that the quadratic form associated with the symmetric N-body operator

$$H_N^{v,w} := \sum_{j=1}^N \left(-\frac{\Delta_{\mathbf{r}_j}}{2} + v(\mathbf{r}_j) \right) + \sum_{1 \leqslant j < k \leqslant N} w(\mathbf{r}_j - \mathbf{r}_k) \tag{3.2}$$

is closed on the energy space

$$Q(H_N^{v,w}) := \left\{ \Psi \in \bigwedge_1^N L^2(\mathbb{R}^d \times \mathbb{Z}_q) \ : \right.$$
$$\left. \int_{(\mathbb{R}^d\times\mathbb{Z}_q)^N} |\nabla\Psi(\mathbf{X})|^2 \, \mathrm{d}\mathbf{X} + \int_{\mathbb{R}^d} v_+(\mathbf{r})\, \rho_\Psi(\mathbf{r})\, \mathrm{d}\mathbf{r} < \infty \right\}. \tag{3.3}$$

This allows us to work with the associated *Friedrichs self-adjoint realization* of $H_N^{v,w}$, see [149, Sec. VIII.6] and [150, Sec. X.3]. Everywhere we work with fermions, that is, the wavefunction Ψ is assumed to be anti-symmetric with respect to exchanges of its variables. Note that the bosonic case is obtained when the number of spin states q is equal to N, see [121, Sec. 3.1.2–3.1.3].

The Hamiltonian (3.2) describes N fermionic particles evolving in $\mathbb{R}^d$, with q spin states, submitted to an external potential v and interacting via the pair potential w. An important physical quantity is the *ground state energy* of the system, which is obtained by minimizing the corresponding energy

$$\boxed{E_N[v] := \inf_{\substack{\Psi\in Q(H_N^{v,w}) \\ \int_{(\mathbb{R}^d\times\mathbb{Z}_q)^N} |\Psi|^2=1}} \left\langle \Psi, H_N^{v,w}\Psi \right\rangle}$$

where $\left\langle \Psi, H_N^{v,w}\Psi \right\rangle$ is understood in the sense of quadratic forms. By the variational principle, this is just the bottom of the spectrum of the operator $H_N^{v,w}$:

$$E_N[v] := \min \sigma\left(H_N^{v,w}\right)$$

in the fermionic subspace. We do not emphasize the interaction w in our notation since it will usually be fixed. When needed we will instead use the notation $E_N^w[v]$.

The main idea of [99, 116] is to replace the infimum over Ψ by a two-step minimization

$$\inf_\Psi (\cdots) = \inf_\rho \inf_{\substack{\Psi \\ \rho_\Psi=\rho}} (\cdots)$$

where we first minimize over the density ρ and then over all the wavefunctions having this prescribed density. This procedure requires us to identify the set of *N-representable densities*, that is, those arising from a Ψ in the form domain of $H_N^{v,w}$, a question that we now address.

The Hoffmann-Ostenhof inequality [81] states that

$$\sum_{j=1}^N \int_{(\mathbb{R}^d\times\mathbb{Z}_q)^N} |\nabla_j\Psi(\mathbf{X})|^2 \, \mathrm{d}\mathbf{X} \geqslant \int_{\mathbb{R}^d} \left|\nabla\sqrt{\rho_\Psi}(\mathbf{r})\right|^2 \mathrm{d}\mathbf{r} \tag{3.4}$$

for all (bosonic or fermionic) wavefunctions (see also Theorem 3.11 below). This inequality implies that we should restrict ourselves to densities such that

$$\sqrt{\rho} \in H^1(\mathbb{R}^d) = \left\{ f \in L^2(\mathbb{R}^d) \; : \; \nabla f \in L^2(\mathbb{R}^d) \right\}.$$

This turns out to be the optimal condition.

Theorem 3.1 (Representability of the One-Particle Density [116]) *Let $\rho \in L^1(\mathbb{R}^d, \mathbb{R}_+)$ be such that $\sqrt{\rho} \in H^1(\mathbb{R}^d)$ and $\int_{\mathbb{R}^d} \rho(\mathbf{r})\, \mathrm{d}\mathbf{r} = N \in \mathbb{N}$. Then there exists one normalized antisymmetric wavefunction $\Psi \in \bigwedge_1^N L^2(\mathbb{R}^d \times \mathbb{Z}_q, \mathbb{C})$ of finite kinetic energy, $\int_{(\mathbb{R}^d \times \mathbb{Z}_q)^N} |\nabla \Psi|^2 < \infty$, such that $\rho = \rho_\Psi$.*

The proof of the theorem is much easier for $q \geqslant N$, where the antisymmetry can be put entirely in the spin variables. One can just take

$$\Psi(\mathbf{X}) = \prod_{j=1}^N \sqrt{\frac{\rho(\mathbf{r}_j)}{N}} \, \frac{\det(\delta_j(\sigma_k))_{1 \leqslant j,k \leqslant N}}{\sqrt{N!}}.$$

When $q < N$ the proof in [116], inspired by March and Young [129] and Harriman [79], consists of taking a Slater determinant

$$\Psi(\mathbf{X}) = \frac{1}{\sqrt{N!}} \det\{\varphi_j(\mathbf{x}_k)\}$$

with the orbitals

$$\varphi_j(\mathbf{x}) = \sqrt{\frac{\rho(\mathbf{r})}{N}} \mathrm{e}^{i\theta_j(\mathbf{r})} \delta_0(\sigma) \tag{3.5}$$

where the phases θ_j are chosen to make the φ_j orthonormal. Although we know that there exist such phases (an explicit example will be given later in (3.44)), we have very bad control over their behavior in N. This will be discussed later in Sect. 3.4 when we consider the kinetic energy cost of introducing such phases.

At this point we have found the set of N-representable densities

$$\left\{ \rho \in L^1(\mathbb{R}^d, \mathbb{R}_+) \; : \; \int_{\mathbb{R}^d} \rho = N, \; \int_{\mathbb{R}^d} |\nabla \sqrt{\rho}|^2 < \infty \right\}. \tag{3.6}$$

Note that this is a convex set since $\rho \mapsto \int_{\mathbb{R}^d} |\nabla\sqrt{\rho}|^2$ is convex [118, Thm. 7.8]. With Theorem 3.1 at hand, we can rewrite the ground state energy as a minimization principle over ρ in this convex set:

$$\boxed{E_N[v] := \inf_{\substack{\sqrt{\rho}\in H^1(\mathbb{R}^d) \\ \int_{\mathbb{R}^d}\rho=N \\ \int_{\mathbb{R}^d} v_+\rho<\infty}} \left\{ F_{\rm LL}[\rho] + \int_{\mathbb{R}^d} v(\mathbf{r})\,\rho(\mathbf{r})\,\mathrm{d}\mathbf{r} \right\}} \tag{3.7}$$

where

$$F_{\rm LL}[\rho] := \inf_{\substack{\Psi\in\bigwedge_1^N L^2(\mathbb{R}^d\times\mathbb{Z}_q,\mathbb{C}) \\ \|\Psi\|_{L^2}=1 \\ \rho_\Psi=\rho}} \left\{ \frac{1}{2}\sum_{j=1}^N \int_{(\mathbb{R}^d\times\mathbb{Z}_q)^N} |\nabla_j\Psi(\mathbf{X})|^2\,\mathrm{d}\mathbf{X} \right.$$
$$\left. + \sum_{1\leqslant j<k\leqslant N} \int_{(\mathbb{R}^d\times\mathbb{Z}_q)^N} |\Psi(\mathbf{X})|^2 w(\mathbf{r}_j-\mathbf{r}_k)\,\mathrm{d}\mathbf{X} \right\} \tag{3.8}$$

is called the *Levy–Lieb functional*. It is the lowest possible (kinetic plus interaction) energy of a quantum system having the prescribed density ρ. This universal functional is the central object of DFT, since knowing it would allow us to compute the ground state energy of a system with any external potential v, by (3.7). In this chapter we will review what is known about $F_{\rm LL}$. But first we need to introduce two other universal functionals, which are obtained after convexifying $F_{\rm LL}$ in two different ways.

3.2.2 *Lieb's Universal Functional*

Note that $v \mapsto E_N[v]$ is concave (as seen from the variational principle, it is a minimization over Ψ of affine functions in v). More precisely, from (3.7) we see that E_N is the Legendre transform of $F_{\rm LL}$ on the convex set of N-representable densities. This naturally raises the question of whether $F_{\rm LL}$ is, conversely, the Legendre transform of E_N. This turns out to be *wrong* since $F_{\rm LL}$ is *not convex* [116]. It is therefore convenient to look at the convex hull (also called lower convex envelope)

$$F_{\rm L} := \mathrm{Conv}(F_{\rm LL}),$$

which is the Legendre transform of E_N. Here the convex hull means that it is the largest convex function below $F_{\rm LL}$. We always assume that $\int_{\mathbb{R}^d}\rho = N$. Another kind of convex hull with respect to N will be considered later. As proved in [116], the function $F_{\rm L}$ is explicit and given by a similar definition as in (3.8) but with mixed states instead of pure states:

$$\boxed{F_{\rm L}[\rho] := \inf_{\substack{\Gamma=\Gamma^*\geqslant 0\\ {\rm Tr}\,\Gamma=1\\ {\rm Tr}\,(-\Delta)\Gamma<\infty\\ \rho_\Gamma=\rho}} {\rm Tr}\left(H_N^{0,w}\Gamma\right).} \tag{3.9}$$

We recall that the density ρ_Γ of a mixed state Γ (a non-negative self-adjoint operator satisfying ${\rm Tr}\,(\Gamma) = 1$), diagonalized in the form

$$\Gamma = \sum_j \alpha_j |\Psi_j\rangle\langle\Psi_j|$$

with $\alpha_j \geqslant 0$ and $\sum_j \alpha_j = 1$, is defined by

$$\rho_\Gamma := \sum_j \alpha_j\, \rho_{\Psi_j}.$$

It is useful to know that the infimum is attained in (3.9), as well as for the Levy–Lieb functional in (3.8).

Theorem 3.2 (Existence of Optimal (Pure and Mixed) States [116]) *Let* $\rho \in L^1(\mathbb{R}^d, \mathbb{R}_+)$ *be such that* $\int_{\mathbb{R}^d} \rho = N \in \mathbb{N}$ *and* $\sqrt{\rho} \in H^1(\mathbb{R}^d)$. *Then the infima in* (3.8) *and* (3.9) *are attained.*

The proof uses the fact that a minimizing sequence Ψ_j (resp. Γ_j) is necessarily compact in $L^2(\mathbb{R}^d \times \mathbb{Z}_q, \mathbb{C})^N$ (resp. in the trace class), since the density is fixed, hence the sequence is tight.

Let us consider a density ρ and a corresponding minimizing N-particle mixed state Γ. Diagonalizing Γ in the form $\Gamma = \sum_j \alpha_j |\Psi_j\rangle\langle\Psi_j|$, we see that

$$F_{\rm L}[\rho] = \sum_j \alpha_j \left\langle \Psi_j, H_N^{0,w}\Psi_j\right\rangle \geqslant \sum_j \alpha_j\, F_{\rm LL}[\rho_{\Psi_j}]$$

and since the upper bound is obvious, we conclude that (3.9) can also be written in the form

$$F_{\rm L}[\rho] = \min_{\substack{\rho=\sum_j \alpha_j \rho_j\\ \sum_j \alpha_j=1\\ \sqrt{\rho_j}\in H^1(\mathbb{R}^d)\\ \int_{\mathbb{R}^d}\rho_j=N}} \sum_j \alpha_j\, F_{\rm LL}[\rho_j]. \tag{3.10}$$

This is the claimed convex hull of $F_{\rm LL}$.

Since an affine function always attains its minimum at an extreme point of a convex set, the ground state energy $E_N[v]$ is given by the same formula

$$E_N[v] = \inf_{\substack{\Gamma=\Gamma^*\geqslant 0\\ \operatorname{Tr}\Gamma=1\\ \operatorname{Tr}(-\Delta)\Gamma<\infty\\ \rho_\Gamma=\rho}} \operatorname{Tr}\left(H_N^{v,w}\Gamma\right) = \inf_{\substack{\sqrt{\rho}\in H^1(\mathbb{R}^d)\\ \int_{\mathbb{R}^d}\rho=N\\ \int_{\mathbb{R}^d} v_+\rho<\infty}} \left\{F_{\rm L}[\rho] + \int_{\mathbb{R}^d} v(\mathbf{r})\,\rho(\mathbf{r})\,d\mathbf{r}\right\}$$

as we had in (3.7) for pure states. From this discussion it seems more natural to work with the convex Lieb functional $F_{\rm L}$, instead of $F_{\rm LL}$. The following duality principle holds.

Theorem 3.3 (Duality [116]) *We have, with p as in* (3.1),

$$\begin{aligned} F_{\rm L}[\rho] &= \sup_{v\in L^p(\mathbb{R}^d)+L^\infty(\mathbb{R}^d)} \left\{E_N[v] - \int_{\mathbb{R}^d} v(\mathbf{r})\rho(\mathbf{r})\,d\mathbf{r}\right\} \\ &= \sup_{\substack{v\in L^p(\mathbb{R}^d)+L^\infty(\mathbb{R}^d)\\ H_N^{v,w}\geqslant 0}} \left\{-\int_{\mathbb{R}^d} v(\mathbf{r})\rho(\mathbf{r})\,d\mathbf{r}\right\}. \end{aligned} \tag{3.11}$$

In the second line the constant $E_N[v]$ has been included in v, hence the constraint that $H_N^{v,w} \geqslant 0$ in the operator sense.

We have seen in Theorem 3.2 that (3.8) and (3.9) are attained. On the other hand, the supremum in (3.11) will *not* be attained for most densities. Indeed, ρ would then be the density of a mixed ground state for the corresponding $E_N[v]$ but this set is believed to be very small. For instance, if ρ vanishes on a set of positive measure, the supremum cannot be attained with a v for which unique continuation holds on the whole space. This is discussed in Sect. 3.7.

The importance in applications of the convex formulation of DFT based on the functionals $F_{\rm LL}$ and $F_{\rm L}$ is reviewed in [80].

3.2.3 *The Grand Canonical Universal Functional*

At this step we have defined two universal functionals $F_{\rm LL}$ and $F_{\rm L}$ of the density ρ, which provide the same ground state energy $E_N[v]$ in the presence of an external potential v. Since Lieb's functional $F_{\rm L}$ is convex, it is to be preferred over $F_{\rm LL}$. The convexity implies the dual formula stated in Theorem 3.3.

In spite of its convexity, the functional $F_{\rm L}$ does not behave well with respect to the weak topology of $H^1(\mathbb{R}^d)$. This might lead to some difficulties in processes where some electrons are lost, e.g. for scattering. Indeed, if we have a sequence ρ_n such that $\sqrt{\rho_n} \rightharpoonup \sqrt{\rho}$ weakly but not strongly, then we may have $\int_{\mathbb{R}^d}\rho(\mathbf{r})\,d\mathbf{r} < N$. The integral does not even need to be an integer, in which case $F_{\rm L}[\rho]$ would not make any sense. For this reason it is natural to introduce a functional allowing for non-integer values of the particle number. The following grand-canonical version did not appear in [116], but was mentioned in [141] and was recently studied in [106, 108].

For us, a grand-canonical state (commuting with the particle number) will be a collection $\Gamma = (\Gamma_n)_{n\geqslant 0}$ of non-negative self-adjoint operators, each of them acting on the n-particle space $\bigwedge_1^n L^2(\mathbb{R}^d \times \mathbb{Z}_q)$, and such that

$$\Gamma_0 + \sum_{n\geqslant 1} \mathrm{Tr}\,(\Gamma_n) = 1.$$

Here Γ_0 is a number in $[0, 1]$ which gives the probability that there is no particle at all. The corresponding density is the sum

$$\rho_\Gamma = \sum_{n\geqslant 1} \rho_{\Gamma_n}$$

so that the average number of particles in the system is given by $\int_{\mathbb{R}^d} \rho_\Gamma = \sum_{n\geqslant 1} n\mathrm{Tr}\,(\Gamma_n)$. We define the grand-canonical universal functional as

$$\boxed{F_{\rm GC}[\rho] := \inf_{\substack{\sum_{n\geqslant 1}\mathrm{Tr}\,\Gamma_n\leqslant 1\\ \sum_{n\geqslant 1}\mathrm{Tr}\,(-\Delta)\Gamma_n<\infty\\ \sum_{n\geqslant 1}\rho_{\Gamma_n}=\rho}} \sum_{n\geqslant 1} \mathrm{Tr}\,\left(H_n^{0,w}\Gamma_n\right).} \tag{3.12}$$

In order to guarantee that the problem is well posed, we need some more assumptions on the interaction potential w. We assume that the system is *stable of the second kind* [152], that is, there exists a constant C such that

$$\forall n \geqslant 1, \qquad H_n^{0,w} \geqslant -Cn. \tag{3.13}$$

Inserting this in (3.12) implies

$$F_{\rm GC}[\rho] \geqslant -C \int_{\mathbb{R}^d} \rho(\mathbf{r})\,\mathrm{d}\mathbf{r}$$

hence the infimum in (3.12) is finite. A typical example is that of a non-negative interaction potential $w \geqslant 0$ such as Coulomb, or more generally a potential w which is classically stable of the second-kind [152], that is, which satisfies the same assumption as (3.13) with the kinetic energy removed:

$$\forall n \geqslant 2, \qquad \sum_{1\leqslant j<k\leqslant n} w(\mathbf{r}_j - \mathbf{r}_k) \geqslant -Cn \qquad \text{a.e. on } (\mathbb{R}^d)^n. \tag{3.14}$$

For a density with integer particle number $\int_{\mathbb{R}^d} \rho(\mathbf{r})\, d\mathbf{r} = N \in \mathbb{N}$ the grand canonical functional is the lowest of the three universal functionals:

$$F_{\rm GC}[\rho] \leqslant F_{\rm L}[\rho] \leqslant F_{\rm LL}[\rho].$$

The following can be shown similarly as for Theorem 3.2, using the stability assumption (3.13).

Theorem 3.4 (Existence of Optimal Grand-Canonical States) *Assume that the system is stable of the second kind as in* (3.13). *Let* $\rho \in L^1(\mathbb{R}^d, \mathbb{R}_+)$ *be such that* $\sqrt{\rho} \in H^1(\mathbb{R}^d)$. *Then the infimum in* (3.12) *is attained.*

From the existence of a minimizer we deduce as before that

$$F_{\rm GC}[\rho] = \min_{\substack{\rho=\sum_n \alpha_n \rho_n \\ \sum_n \alpha_n = 1 \\ \sqrt{\rho_n} \in H^1(\mathbb{R}^d) \\ \int_{\mathbb{R}^d} \rho_n = n}} \sum_n \alpha_n F_{\rm L}[\rho_n] = \min_{\substack{\rho=\sum_j \beta_j \rho_j \\ \sum_j \beta_j = 1 \\ \sqrt{\rho_j} \in H^1(\mathbb{R}^d) \\ \int_{\mathbb{R}^d} \rho_j \in \mathbb{N}}} \sum_j \beta_j F_{\rm LL}[\rho_j].$$

In other words, the grand canonical functional is also a convex hull of the original Levy–Lieb functional $F_{\rm LL}$, but convex combinations $\rho = \sum_j \beta_j \rho_j$ are considered with the ρ_j having an arbitrary number of particles. It is not required that all the ρ_j have the fixed number N of particles like for $F_{\rm L}$.

An interesting question is to determine whether an optimal grand canonical state $(\Gamma_n)_{n\geqslant 0}$ corresponding to a given ρ always satisfies $\Gamma_n \equiv 0$ for $n \geqslant n_{\max}$. In this case we say that $\Gamma = (\Gamma_n)_{n\geqslant 0}$ has a compact support in n. No result of this sort seems to have appeared in the literature up to now.

The following theorem asserts that $F_{\rm GC}$ is the weak-$*$ lower semi-continuous envelope of $F_{\rm L}$, in an appropriate sense.

Theorem 3.5 (Weak Lower Semi-Continuity) *We assume that*

$$0 \leqslant w \in L^p(\mathbb{R}^d) + L^\infty(\mathbb{R}^d)$$

with p as in (3.1) *and that w tends to 0 at infinity. The functional $F_{\rm GC}$ is the weak-$*$ lower semi-continuous closure of $F_{\rm L}$, in the following sense:*

(i) For any sequence $(\sqrt{\rho_j})_{j\geqslant 1} \subset H^1(\mathbb{R}^d)$ converging weakly in $\dot{H}^1(\mathbb{R}^d)$ to $\sqrt{\rho} \in H^1(\mathbb{R}^d)$, we have

$$F_{\rm GC}[\rho] \leqslant \liminf_{j\to\infty} F_{\rm GC}[\rho_j]. \tag{3.15}$$

(*ii*) *For any* $\sqrt{\rho} \in H^1(\mathbb{R}^d)$, *there exists a sequence* $(\sqrt{\rho_j})_{j\geqslant 1} \subset H^1(\mathbb{R}^d)$ *converging strongly to* $\sqrt{\rho}$ *in* $\dot{H}^1(\mathbb{R}^d) \cap L^p(\mathbb{R}^d)$ *for all* $2 < p < p^*$, *such that*

$$F_{\rm GC}[\rho] = \lim_{j\to\infty} F_{\rm L}[\rho_j]. \tag{3.16}$$

Here

$$p^* = \begin{cases} \infty & \text{in dimensions } d = 1, 2, \\ \frac{2d}{d-2} & \text{in dimensions } d \geqslant 3, \end{cases}$$

is the critical Sobolev exponent. Since we are not aware that a proof of Theorem 3.5 has been explicitly written anywhere, we provide the full argument later in Sect. 3.8. It is inspired by [102].

When the liminf on the right of (3.15) is finite (which we can always assume, otherwise the statement is void), then the Hoffmann-Ostenhof inequality (3.4) implies that $\sqrt{\rho_j}$ is bounded in the homogeneous Sobolev space $\dot{H}^1(\mathbb{R}^d)$. However $\int_{\mathbb{R}^d} \rho_j$ need not be bounded in general and this is why only the weak convergence in $\dot{H}^1(\mathbb{R}^d)$ was assumed. This plays an important role in (*ii*). Consider a ρ and an associated optimal grand-canonical state $\Gamma = (\Gamma_n)_{n\geqslant 0}$ for $F_{\rm GC}[\rho]$. If we have $\Gamma_{n_k} \neq 0$ for a sequence $n_k \to \infty$, then this means that infinitely many particles are needed to properly represent ρ grand-canonically. Although this is very unlikely to happen in practical situations, this can probably not be avoided for a general interaction w and a general ρ. Then we need a diverging number of particles

$$\int_{\mathbb{R}^d} \rho_j = N_j \to +\infty$$

in our canonical state associated with ρ_j, even if it has a bounded energy. On the other hand, if there exists one minimizer $(\Gamma_n)_{n\geqslant 0}$ for $F_{\rm GC}[\rho]$ which has a compact support in n, then (*ii*) holds with a sequence converging weakly in $H^1(\mathbb{R}^d)$, as will be clear from our proof. This is one reason why it is important to understand whether optimal states always have a compact support in n, as we have mentioned previously.

Next we discuss the dual formulation of $F_{\rm GC}$. For $v = v_+ - v_-$ with $v_- \in L^p(\mathbb{R}^d) + L^\infty(\mathbb{R}^d)$ and $v_+ \in L^1_{\rm loc}(\mathbb{R}^d)$, we find that the Legendre transform of $F_{\rm GC}$ is given by

$$E_\lambda^{\rm GC}[v] := \inf_{\substack{\sqrt{\rho}\in H^1(\mathbb{R}^d) \\ \int_{\mathbb{R}^d} \rho=\lambda \\ \int_{\mathbb{R}^d} v_+\rho<\infty}} \left\{ F_{\rm GC}[\rho] + \int_{\mathbb{R}^d} v\rho \right\} = \inf_{\substack{\sum_n \alpha_n=1 \\ \sum_n n\alpha_n=\lambda}} \sum_n \alpha_n E_n[v]. \tag{3.17}$$

For $\lambda = N \in \mathbb{N}$, we have $E_N^{\rm GC}[v] \leqslant E_N[v]$ but equality will in general not hold. If the function $n \mapsto E_n[v]$ is convex in the discrete sense, that is,

$$E_n[v] - E_{n-1}[v] \leqslant E_{n+1}[v] - E_n[v], \qquad \forall n \geqslant 1, \tag{3.18}$$

then it follows that

$$E_{N+\theta}^{\rm GC}[v] = (1-\theta)E_N[v] + \theta E_{N+1}[v]$$

for all $N \in \mathbb{N}$ and $\theta \in [0,1)$ [141]. In particular $E_N^{\rm GC}[v] = E_N[v]$. Therefore, if (3.18) holds, then the grand canonical functional $F_{\rm GC}[\rho]$ provides the same ground state energy in an external potential v as the canonical ones $F_{\rm LL}[\rho]$ and $F_{\rm L}[\rho]$. In physical terms the condition (3.18) means that the electron ionization energy is greater than or equal to the electron affinity. It is a famous conjecture that (3.18) holds for the Coulomb potential in dimension $d = 3$, for atomic or molecular external potentials v [5, 141]. A counterexample is provided in [116] for a different w. Note that (3.18) always holds for $w \equiv 0$.

If (3.18) does *not* hold, then $E_\lambda^{\rm GC}[v]$ is equal to the convex hull of $n \mapsto E_n[v]$. This amounts to considering the set $\mathcal{K} \subset \mathbb{N}$ of the points k such that

$$E_k[v] - E_{k-1}[v] = \max_{n=2,\dots,k} \left(E_n[v] - E_{n-1}[v]\right).$$

One obtains

$$E_N^{\rm GC}[v] = E_{n_1}[v] + \frac{E_{n_1}[v] - E_{n_2}[v]}{n_1 - n_2}(N - n_1),$$

where $n_1 < N < n_2$ are the two closest points in $\mathcal{K}$ on the left and right of N.

3.2.4 *Kohn–Sham Exchange Correlation*

In the previous sections we have explained the Levy–Lieb variational formulation of the ground state energy of the N-particle problem in terms of the density only, which is really in the spirit of DFT. Practitioners prefer to use an auxiliary set of N orthonormal functions $\Phi = (\varphi_1, \dots, \varphi_N)$, which describe N fictitious uncorrelated electrons, to build the desired density through the formula

$$\rho_\Phi(\mathbf{r}) = \sum_{n=1}^{N} \sum_{\sigma \in \mathbb{Z}_q} |\varphi_n(\mathbf{r},\sigma)|^2 = \rho_\Psi(\mathbf{r})$$

with the *Slater determinant*

$$\Psi(\mathbf{x}_1,\ldots,\mathbf{x}_N)=\frac{\det(\varphi_j(\mathbf{x}_k))}{\sqrt{N!}}. \tag{3.19}$$

This method provides a better representation of the kinetic energy, but it is much more costly from a computational point of view. We quickly explain this approach due to Kohn–Sham [93] here.

For a density ρ with $\int_{\mathbb{R}^d}\rho(\mathbf{r})\mathrm{d}\mathbf{r}=N\in\mathbb{N}$, we introduce the lowest kinetic energy of Slater determinants

$$T_{\mathrm{S}}[\rho]:=\min_{\substack{\varphi_1,\ldots,\varphi_N\in H^1(\mathbb{R}^d\times\mathbb{Z}_q,\mathbb{C})\\ \langle\varphi_i,\varphi_j\rangle=\delta_{ij}\\ \rho_\Phi=\rho}}\frac{1}{2}\sum_{j=1}^N\int_{\mathbb{R}^d\times\mathbb{Z}_q}|\nabla\varphi_j(\mathbf{x})|^2\,\mathrm{d}\mathbf{x} \tag{3.20}$$

(the min is attained for the same reason as in Theorem 3.2). We then add and subtract T_{S} from F_{LL}, which allows us to rewrite the N-particle ground state using N orbitals as

$$\begin{aligned}E_N[v]:=\inf_{\substack{\varphi_1,\ldots,\varphi_N\in H^1(\mathbb{R}^d\times\mathbb{Z}_q,\mathbb{C})\\ \langle\varphi_i,\varphi_j\rangle=\delta_{ij}\\ \int_{\mathbb{R}^d}\rho_\Phi v_+<\infty}}&\left\{\frac{1}{2}\sum_{j=1}^N\int_{\mathbb{R}^d\times\mathbb{Z}_q}|\nabla\varphi_j(\mathbf{x})|^2\,\mathrm{d}\mathbf{x}+\int_{\mathbb{R}^d}v(\mathbf{r})\rho_\Phi(\mathbf{r})\,\mathrm{d}\mathbf{r}\right.\\ &\left.+\frac{1}{2}\int_{\mathbb{R}^d}\int_{\mathbb{R}^d}w(\mathbf{r}-\mathbf{r}')\rho_\Phi(\mathbf{r})\rho_\Phi(\mathbf{r}')\,\mathrm{d}\mathbf{r}\,\mathrm{d}\mathbf{r}'+E_{\mathrm{xc}}[\rho_\Phi]\right\},\end{aligned} \tag{3.21}$$

where

$$E_{\mathrm{xc}}[\rho]:=F_{\mathrm{LL}}[\rho]-T_{\mathrm{S}}[\rho]-\frac{1}{2}\int_{\mathbb{R}^d}\int_{\mathbb{R}^d}w(\mathbf{r}-\mathbf{r}')\rho(\mathbf{r})\rho(\mathbf{r}')\,\mathrm{d}\mathbf{r}\,\mathrm{d}\mathbf{r}' \tag{3.22}$$

is called the *exchange-correlation energy*. From a mathematical point of view, the Kohn–Sham approach a priori requires the study of both $F_{\mathrm{LL}}[\rho]$ and $T_{\mathrm{S}}[\rho]$ as separated functionals. It is an interesting question to find a way to study $E_{\mathrm{xc}}[\rho]$ directly, without interpreting it as a difference. In chemistry one often relies on the *adiabatic connection formula* (see Remark 3.1 below), which however involves another kinetic energy functional $T[\rho]$ discussed later in Sect. 3.4.

Instead of using N uncorrelated electrons as the main variable, one may also use a one-particle density matrix γ, which is often called the *Kohn–Sham method with fractional occupations*. The method is similar but one has to subtract the lowest kinetic energy $T_{\mathrm{GC}}[\rho]$ of all possible one-particle density matrices, which is defined later in Sect. 3.4.

3.3 The Uniform Electron Gas and the Local Density Approximation

The universal functionals $F_{\rm LL}$, $F_{\rm L}$ and $F_{\rm GC}$ defined in the previous section allow us in principle to describe any fermionic system interacting via the potential w. But these functionals are of course not known exactly and finding them is essentially the same as solving the N-particle problem. One of the main purposes of DFT is to find reliable and efficient approximations. Here we discuss the most widely used of these approximations, called the *Local Density Approximation (LDA)* [46, 82, 93, 128, 136, 140], where they are replaced by purely local ones. The LDA is often considered as *"the mother of all approximations"* [142] and it yields surprisingly good results, even in cases where the density is not at all slowly varying [128, 136]. Its successors involving gradient corrections are even better and have become the standard in DFT calculations. In this section we only consider the Coulomb case in dimension $d = 3$ but we expect similar results for other potentials in all dimensions.

Of course, the functionals $F_{\rm LL}$, $F_{\rm L}$ and $F_{\rm GC}$ are not local at all. Two electrons at different places are always entangled and, furthermore, the Coulomb potential has a very long range so that electrons interact even when they are far apart. In the LDA one makes the assumption that the only non-local part is the Hartree term (the classical Coulomb energy of the density ρ) and one approximates the rest by a local function of ρ, that is, the integral of a function f depending only on the value $\rho(\mathbf{r})$ at $\mathbf{r}$:

$$F_{\rm LL,L,GC}[\rho] \approx \underbrace{\frac{1}{2}\iint_{\mathbb{R}^3\times\mathbb{R}^3} \frac{\rho(\mathbf{r})\,\rho(\mathbf{r}')}{|\mathbf{r}-\mathbf{r}'|}\,\mathbf{dr}\,\mathbf{dr}'}_{\substack{\textbf{non-local}\\ \text{classical Coulomb energy of } \rho}} + \underbrace{\int_{\mathbb{R}^3} f\big(\rho(\mathbf{r})\big)\,\mathbf{dr}.}_{\substack{\textbf{local}\\ f=\text{energy per unit vol.}\\ \text{of uniform electron gas}}} \tag{3.23}$$

The function f is chosen to be the energy per unit volume of an infinite gas of constant density ρ, called the *Uniform Electron Gas* (UEG), so that the approximation becomes exact when ρ is constant over a very large domain. Because the UEG is an infinite system it should not depend on whether it is defined canonically or grand-canonically. Hence the function f must be the same for the three functionals $F_{\rm LL}$, $F_{\rm L}$ and $F_{\rm GC}$.

The idea behind the LDA is as depicted in Fig. 3.1. After subtraction of the Hartree term, one splits the space into small boxes (of volume $\mathbf{dr}$) and assumes that the remaining energy is the sum of the local energies. In each little box, one replaces the density by a constant. One does not use the energy of the constant function in the small box, but rather the energy per unit volume of an infinite system having the corresponding uniform density, multiplied by the volume $\mathbf{dr}$ of the small box.

In the LDA the complicated Levy–Lieb functionals therefore get replaced by a new universal function $f : \mathbb{R}_+ \to \mathbb{R}$, which is much simpler since it only depends on one real parameter. But the function f is also not known exactly, and we will see that it displays a very rich structure.

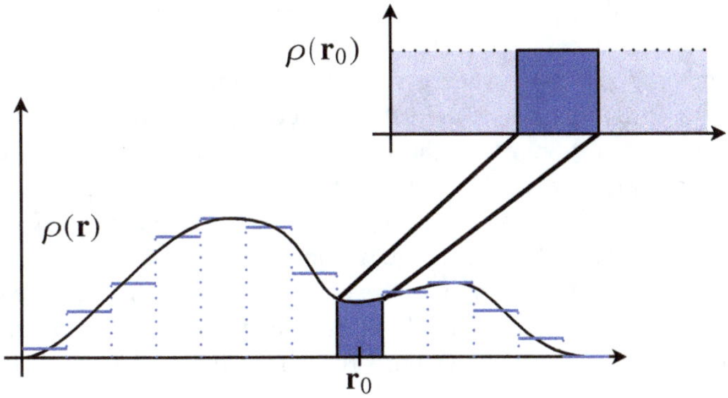

Fig. 3.1 Main idea of the Local Density Approximation in DFT. The Levy–Lieb energy (with the Hartree term subtracted) is replaced by the sum of the energies per unit volume of an infinite uniform gas with the local density $\rho(\mathbf{r}_0)$, times the volume $d\mathbf{r}$

In this section we report on the results in [108] where the LDA was rigorously justified for the first time. The proper regime is that of slowly varying densities, that is, densities ρ which are very flat on sufficiently large domains such that $f(\rho)$ becomes a good local approximation. To this end we start by defining the function f.

3.3.1 The Uniform Electron Gas

The uniform electron gas was rigorously defined in [106, 108] and it is obtained by assuming that the density is exactly constant over a large domain which grows so as to cover the whole space. The result is the following.

Theorem 3.6 (Uniform Electron Gas [106, 108]) *Let $w(\mathbf{r}) = |\mathbf{r}|^{-1}$ in dimension $d = 3$ and $q \geqslant 1$ be the number of spin states. Let $\rho_0 > 0$. Let Ω be a fixed open convex set of unit volume $|\Omega| = 1$. Let $\chi \in L^1(\mathbb{R}^3)$ be a radial non-negative function of compact support such that $\int_{\mathbb{R}^3} \chi = 1$ and $\int_{\mathbb{R}^3} |\nabla\sqrt{\chi}|^2 < \infty$. Then the following thermodynamic limit exists*

$$f(\rho_0) := \lim_{L\to\infty} L^{-3}\bigg(F_{\rm GC}\big[\rho_0 \mathbb{1}_{L\Omega} * \chi\big] - \frac{\rho_0^2}{2}\iint_{\mathbb{R}^3\times\mathbb{R}^3} \frac{(\mathbb{1}_{L\Omega} * \chi)(\mathbf{r})(\mathbb{1}_{L\Omega} * \chi)(\mathbf{r}')}{|\mathbf{r}-\mathbf{r}'|}\, d\mathbf{r}\, d\mathbf{r}'\bigg) \tag{3.24}$$

and does not depend on Ω and χ.

Electrons have $q = 2$ spin states but we have written the result with a general q for convenience. It is expected that the exact same result holds for the other two functionals F_{LL} and F_{L}, with the same limit $f(\rho_0)$. This has not yet been proved, except in the classical case where the kinetic energy is dropped (see Sect. 3.5).

The number $f(\rho_0)$ is the energy per unit volume (with the Hartree energy subtracted) of an infinite gas submitted to the constraint that its density is exactly constant over $\mathbb{R}^3$, $\rho(\mathbf{r}) \equiv \rho_0$. In the literature, $f(\rho_0)$ is often confused with the corresponding *Jellium energy*. In Jellium there is no constraint on ρ but one adds instead a uniform background of density ρ_0 which compensates the long range of the Coulomb potential [104, 119]. That the two models coincide has only been shown in the classical case [36, 107] so far, as we will mention in Sect. 3.5 below. In the quantum case the same result should hold, but the proof has not been written yet.

Here are some rigorously known properties of the function f.

Theorem 3.7 (Properties of f [72, 106, 108, 119]) *The function f is locally Lipschitz: There exists a constant C such that*

$$|f(\rho_1) - f(\rho_2)| \leqslant C\left(\max(\rho_1,\rho_2)^{\frac{1}{3}} + \max(\rho_1,\rho_2)^{\frac{2}{3}}\right)|\rho_1 - \rho_2|, \tag{3.25}$$

for all $\rho_1, \rho_2 \geqslant 0$. The function f satisfies the uniform bound

$$c_{\mathrm{TF}}(3)q^{-\frac{2}{3}}\rho^{\frac{5}{3}} - \frac{3}{5}\left(\frac{9\pi}{2}\right)^{\frac{1}{3}}\rho^{\frac{4}{3}} \leqslant f(\rho) \leqslant c_{\mathrm{TF}}(3)q^{-\frac{2}{3}}\rho^{\frac{5}{3}} - c_{\mathrm{D}}(1,3)q^{-\frac{1}{3}}\rho^{\frac{4}{3}} \tag{3.26}$$

for all $\rho \geqslant 0$, where $c_{\mathrm{TF}}(3) = (3/10)(6\pi^2)^{2/3}$ and $c_{\mathrm{D}}(1,3) = (3/4)\,(6/\pi)^{1/3}$ are respectively the Thomas–Fermi and Dirac constants, discussed later in Sects. 3.4 and 3.5. It behaves at small densities like

$$f(\rho) = c_{\mathrm{UEG}}(1,3)\rho^{\frac{4}{3}} + o\left(\rho^{\frac{4}{3}}\right)_{\rho\to 0^+} \tag{3.27}$$

where

$$-1.4508 \simeq -\frac{3}{5}\left(\frac{9\pi}{2}\right)^{\frac{1}{3}} \leqslant c_{\mathrm{UEG}}(1,3) \leqslant -1.4442$$

is the classical UEG energy discussed later in Sect. 3.5, and at large densities like

$$f(\rho) = c_{\mathrm{TF}}(3)q^{-\frac{2}{3}}\rho^{\frac{5}{3}} - c_{\mathrm{D}}(1,3)q^{-\frac{1}{3}}\rho^{\frac{4}{3}} + o\left(\rho^{\frac{4}{3}}\right)_{\rho\to\infty}. \tag{3.28}$$

The statement involves several constants that will be introduced in the next sections. It is believed that f is smooth except at finitely many points corresponding to phase transitions. In the case of spin-1/2 particles like electrons ($q = 2$), numerical simulations in [3, 20, 24, 47, 84, 88, 146, 175] indicate that there might be one or two such points. For a long time it was believed that the system can be a ferromagnetic Wigner crystal, a ferromagnetic fluid and a paramagnetic fluid. Recent results indicate that the ferromagnetic fluid phase might not exist [3, 84], however. More transitions could occur in the solid phase (for instance an anti-ferromagnetic crystal). In spite of the clear numerical evidence that there are phase transitions, proving it remains a very challenging open problem [10]. Several approximate formulas for the function f are used in DFT, including for instance the celebrated Perdew–Wang (PW92) functional [144].

3.3.2 The Local Density Approximation of $F_{\rm GC}$

We now state a result from [108], where the LDA was justified for the first time in the quantum case.

Theorem 3.8 (LDA for $F_{\rm GC}$ [108]) *Let $w(\mathbf{r}) = |\mathbf{r}|^{-1}$ in dimension $d = 3$ and $q \geqslant 1$ be the number of spin states. Then there exists a constant $C = C(q)$ such that*

$$\left|F_{\rm GC}[\rho] - \frac{1}{2}\iint_{\mathbb{R}^3\times\mathbb{R}^3} \frac{\rho(\mathbf{r})\rho(\mathbf{r}')}{|\mathbf{r}-\mathbf{r}'|}\,\mathrm{d}\mathbf{r}\,\mathrm{d}\mathbf{r}' - \int_{\mathbb{R}^3} f\left(\rho(\mathbf{r})\right)\mathrm{d}\mathbf{r}\right|$$

$$\leqslant \varepsilon\int_{\mathbb{R}^3}\left(\rho(\mathbf{r})+\rho(\mathbf{r})^2\right)\mathrm{d}\mathbf{r}+\frac{C(1+\varepsilon)}{\varepsilon}\int_{\mathbb{R}^3}|\nabla\sqrt{\rho}(\mathbf{r})|^2\,\mathrm{d}\mathbf{r}+\frac{C}{\varepsilon^{15}}\int_{\mathbb{R}^3}|\nabla\sqrt{\rho}(\mathbf{r})|^4\,\mathrm{d}\mathbf{r} \tag{3.29}$$

for every $\varepsilon > 0$ and every non-negative density $\rho \in L^1(\mathbb{R}^3)\cap L^2(\mathbb{R}^3)$ such that $\nabla\sqrt{\rho} \in L^2 \cap L^4(\mathbb{R}^3)$. Here f is the function defined in Theorem 3.6.

It is expected that the exact same result holds for the canonical functionals $F_{\rm LL}$ and $F_{\rm L}$, with of course the additional constraint that $\int_{\mathbb{R}^3}\rho \in \mathbb{N}$.

The last gradient term $\varepsilon^{-15}|\nabla\sqrt{\rho}|^4$ was chosen for simplicity but the same result actually holds with $\varepsilon^{1-4p}|\nabla\rho^\theta|^p$ instead, under the conditions that $p > 3$, $0 < \theta < 1$ and $2 \leqslant p\theta \leqslant 1 + p/2$. The constant C then depends on the chosen p and θ.

In addition to the large power of ε, which is an artifact of the proof in [108], the form of the error term is probably not optimal. It is reasonable to expect that the right side of (3.29) should only involve quantities like $\rho^{5/3}$, $\rho^{4/3}$, $|\nabla\sqrt{\rho}|^2$ and $|\nabla\rho^{1/3}|^2$ or perhaps $|\nabla\rho|$, which have the same scaling as the kinetic and Coulomb terms.

The inequality (3.29) holds for every density but it is useful only when the two gradient terms are much smaller than the first term,

$$\int_{\mathbb{R}^3}\left(|\nabla\sqrt{\rho}(\mathbf{r})|^2+|\nabla\sqrt{\rho}(\mathbf{r})|^4\right)\mathrm{d}\mathbf{r} \ll \int_{\mathbb{R}^3}\left(\rho(\mathbf{r})+\rho(\mathbf{r})^2\right)\mathrm{d}\mathbf{r},$$

so that after optimizing over ε one gets a small term on the right side of (3.29). One interesting case is when the density is given in terms of a fixed function ρ with $\int_{\mathbb{R}^3} \rho = 1$, which is rescaled in the manner

$$\rho_N(\mathbf{r}) = \rho(N^{-1/3}\mathbf{r}).$$

After optimizing over ε we obtain the following expansion of the grand-canonical Levy–Lieb energy:

$$F_{\mathrm{GC}}[\rho_N] = \frac{N^{\frac{5}{3}}}{2} \iint_{\mathbb{R}^3\times\mathbb{R}^3} \frac{\rho(\mathbf{r})\rho(\mathbf{r}')}{|\mathbf{r}-\mathbf{r}'|}\,\mathrm{d}\mathbf{r}\,\mathrm{d}\mathbf{r}' + N\int_{\mathbb{R}^3} f\big(\rho(\mathbf{r})\big)\,\mathrm{d}\mathbf{r} + O\left(N^{\frac{11}{12}}\right). \tag{3.30}$$

The first term is the trivial non-local Coulomb term, whereas the next term in the expansion is the LDA. It is an interesting open question to determine the next order correction, which is believed to be also local, of order $N^{1/3}$, and to involve gradients. The exact same result as (3.30) is expected for F_{LL} and F_{L}.

Remark 3.9 (LDA for the Exchange-Correlation Energy) Theorem 3.15 below is a result similar to Theorem 3.8 for the grand-canonical kinetic energy T_{GC} alone, and implies a corresponding bound for the difference of the two functionals. These two bounds justify the LDA for the (grand-canonical) exchange-correlation energy, as was defined in Sect. 3.2.4.

In the next two sections we study separately the kinetic energy functional and the classical interaction functional. We discuss known upper and lower bounds and derive the LDA for these functionals in a similar (but simpler) manner as for the full Levy–Lieb functional F_{GC}. Although the minimum of a sum is in general not the sum of the two minima, understanding the kinetic and interaction energies separately will give us useful information on the full functional, as explained in Sect. 3.6 below.

3.4 Kinetic Energy and Lieb–Thirring Inequalities

3.4.1 Three Kinetic Energy Functionals

We have introduced in (3.20) the lowest kinetic energy $T_{\mathrm{S}}[\rho]$ that can be reached with Slater determinants, for a given density ρ with $\int_{\mathbb{R}^3} \rho(\mathbf{r})\,\mathrm{d}\mathbf{r} = N$. We can define in a similar manner the lowest kinetic energy that can be reached with all possible wave functions

$$\boxed{T[\rho] := \min_{\substack{\Psi \in \bigwedge_1^N H^1(\mathbb{R}^d \times \mathbb{Z}_q, \mathbb{C}) \\ \|\Psi\|_{L^2}=1 \\ \rho_\Psi = \rho}} \frac{1}{2} \sum_{j=1}^N \int_{(\mathbb{R}^d \times \mathbb{Z}_q)^N} |\nabla_j \Psi(\mathbf{X})|^2 \, \mathrm{d}\mathbf{X}.} \tag{3.31}$$

This is nothing but $F_{\mathrm{LL}}^0[\rho]$, the Levy–Lieb functional with interaction $w \equiv 0$. Recall that T and T_{S} depend on the number of spin states q.

Since this is a non-interacting problem, one may think at first sight that minimizers will always be Slater determinants, that is, $T[\rho]$ and $T_{\mathrm{S}}[\rho]$ should coincide. But this is not true in general [116] and the best one can say for a general ρ is that $T_{\mathrm{S}}[\rho] \geqslant T[\rho]$.

There are two other natural kinetic functionals corresponding to $F_{\mathrm{L}}^0[\rho]$ and $F_{\mathrm{GC}}^0[\rho]$, respectively. For the first one the minimization is extended to mixed canonical states and for the second one to grand-canonical states. It turns out that these two are equal:

$$F_{\mathrm{L}}^0[\rho] = F_{\mathrm{GC}}^0[\rho].$$

The reason is that the kinetic energy can be expressed in terms of the one-particle density matrix γ and that the set of such matrices which are N-representable by a mixed state coincides with those which are representable by a grand-canonical state [27, 28]. By duality, this also follows from the fact that the inequality (3.18) always holds in the non-interacting case.

In order to explain all this in detail, we first recall that the one-particle density matrix γ_Ψ of a wavefunction Ψ is the self-adjoint operator acting on the one-particle space $L^2(\mathbb{R}^d \times \mathbb{Z}_q)$ with integral kernel

$$\gamma_\Psi(\mathbf{x}, \mathbf{y}) = N \int_{(\mathbb{R}^d \times \mathbb{Z}_q)^{N-1}} \Psi(\mathbf{x}, \mathbf{X}) \overline{\Psi(\mathbf{y}, \mathbf{X})} \, \mathrm{d}\mathbf{X}.$$

This gives

$$\frac{1}{2} \sum_{j=1}^N \int_{(\mathbb{R}^d \times \mathbb{Z}_q)^N} |\nabla_j \Psi(\mathbf{X})|^2 \, \mathrm{d}\mathbf{X} = \operatorname{Tr} \left(\frac{-\Delta}{2} \right) \gamma_\Psi$$

with the trace interpreted in the quadratic form sense. Every density matrix of an antisymmetric Ψ satisfies $0 \leqslant \gamma_\Psi = (\gamma_\Psi)^* \leqslant 1$ and $\operatorname{Tr}(\gamma_\Psi) = N$. When we consider mixed N-particle states we obtain the convex hull of the set of N-representable density matrices. This convex hull is definitely contained in the convex set

$$\left\{ \gamma = \gamma^* \ : \ 0 \leqslant \gamma \leqslant 1, \ \operatorname{Tr}(-\Delta)\gamma < \infty, \ \operatorname{Tr}(\gamma) = N \right\}.$$

But the extreme points of this set are the rank-N orthogonal projections with finite kinetic energy. Those are exactly the one-particle density matrices of the Slater determinants. Hence we must have equality of the two convex sets. By considering grand-canonical states the set will not increase further. See Theorem 3.26 for a related result.

This discussion leads us towards introducing the following kinetic energy functional

$$\boxed{T_{\rm GC}[\rho] := \min_{\substack{0\leqslant\gamma=\gamma^*\leqslant 1\\ {\rm Tr}\,(-\Delta)\gamma<\infty\\ \rho_\gamma=\rho}} {\rm Tr}\left(\frac{-\Delta}{2}\right)\gamma.} \tag{3.32}$$

We call it "grand-canonical" since $\int_{\mathbb{R}^d}\rho$ can now take any positive value. But for $\int_{\mathbb{R}^d}\rho\in\mathbb{N}$, this is just Lieb's canonical energy $T_{\rm GC}[\rho]=F_{\rm L}^0[\rho]$. In this case we also have

$$T_{\rm GC}[\rho]\leqslant T[\rho]\leqslant T_{\rm S}[\rho],\qquad \text{when}\quad \int_{\mathbb{R}^d}\rho\in\mathbb{N}.$$

The functional $T_{\rm GC}[\rho]$ is convex and it is the convex hull and the weak-$*$ semi-continuous closure of both $T[\rho]$ and $T_{\rm S}[\rho]$, similarly as in Theorem 3.5. Minimizers exist for these three functionals, as in Theorem 3.2. It suffices to take $w\equiv 0$ in all those theorems.

Remark 3.10 (N-Representability of the One-Particle Density Matrix) There are complicated constraints on a one-particle density matrix γ to ensure that it arises from an N-particle wavefunction Ψ. For instance, when $N=2$ then all the eigenvalues of γ must be of even multiplicity. Another example is that no γ of rank $N+1$ is N-representable. See [28, 62, 101] for these two examples and [2, 13, 91, 154] for more advanced results when $N\geqslant 3$.

Now we mention some known upper and lower bounds on the kinetic energy functionals. Lower bounds naturally involve the lowest functional $T_{\rm GC}[\rho]$. Upper bound should ideally involve $T_{\rm S}[\rho]$ but we will see that much more is known on $T_{\rm GC}[\rho]$.

3.4.2 *Lower Bounds: Hoffmann-Ostenhof and Lieb–Thirring Inequalities*

The first lower bound is the Hoffmann-Ostenhof inequality mentioned previously in (3.4) and which holds for the grand-canonical kinetic energy as well.

Theorem 3.11 (Hoffmann-Ostenhof Inequality [81]) *For every $\rho \geqslant 0$ such that $\sqrt{\rho} \in H^1(\mathbb{R}^d)$, we have*

$$T_{\rm GC}[\rho] \geqslant \frac{1}{2}\int_{\mathbb{R}^d} \left|\nabla\sqrt{\rho}(\mathbf{r})\right|^2 \,\mathrm{d}\mathbf{r}. \tag{3.33}$$

Using the Gagliardo–Nirenberg inequality [118]

$$\|u\|_{L^2(\mathbb{R}^d)}^{\frac{4}{d}} \int_{\mathbb{R}^d} |\nabla u(\mathbf{r})|^2 \,\mathrm{d}\mathbf{r} \geqslant c_{\rm GN}(d) \int_{\mathbb{R}^d} |u(\mathbf{r})|^{2+\frac{4}{d}} \,\mathrm{d}\mathbf{r}, \tag{3.34}$$

for $u = \sqrt{\rho}$, we obtain

$$T_{\rm GC}[\rho] \geqslant \frac{c_{\rm GN}(d)}{2N^{\frac{2}{d}}} \int_{\mathbb{R}^d} \rho(\mathbf{r})^{1+\frac{2}{d}} \,\mathrm{d}\mathbf{r}$$

and this is optimal for bosons. But for fermions this is not optimal at all. The Lieb–Thirring inequality states that one can replace the N-dependent prefactor by an N-independent one (or, rather, by a q-dependent constant where q is the number of spin states).

Theorem 3.12 (Lieb–Thirring [121, 125, 126]) *There exists a positive constant $c_{\rm LT}(d) > 0$ such that*

$$T_{\rm GC}[\rho] \geqslant q^{-\frac{2}{d}} c_{\rm LT}(d) \int_{\mathbb{R}^d} \rho(\mathbf{r})^{1+\frac{2}{d}} \,\mathrm{d}\mathbf{r} \tag{3.35}$$

for all $\rho \geqslant 0$ such that $\sqrt{\rho} \in H^1(\mathbb{R}^d)$.

Note the spin dependence in $q^{-2/d}$, which is compatible with the fact that the bosonic case is recovered when $q = N$. For particles like electrons we have $q = 2$ and the constant is N-independent.

For large fermionic systems the Lieb–Thirring inequality is an advantageous replacement for the Gagliardo–Nirenberg inequality, to which it reduces in the case $N = 1$ (in particular we always have $c_{\rm LT}(d) \leqslant c_{\rm GN}(d)/2$). Since its invention, the Lieb–Thirring inequality (3.35) has played a central role in the mathematical understanding of large fermionic systems. It was originally used to give a proof of stability of matter [112, 117, 121, 125] that is much shorter than the original proof of Dyson and Lenard [48]. Later the Lieb–Thirring inequality was generalized to systems at positive density [57, 58] where ρ is a local perturbation of a constant, and to the dynamic case where it extends Strichartz's inequality [59, 60].

The right side of (3.35) is related to the kinetic energy of the free Fermi gas. Indeed, we recall that the translation-invariant orthogonal projector

$$P_{\rho_0} = \mathbb{1}\left(-\Delta \leqslant 2\frac{d+2}{d} c_{\rm TF}(d) q^{-\frac{2}{d}} \rho_0^{\frac{2}{d}}\right) \tag{3.36}$$

has the constant density $\rho_{P_{\rho_0}} = \rho_0$ and the constant kinetic energy density $c_{\rm TF}(d)\, q^{-2/d} \rho_0^{1+2/d}$, where

$$c_{\rm TF}(d) = \frac{2\pi^2 d}{(d+2)} \left(\frac{d}{|\mathbb{S}^{d-1}|}\right)^{\frac{2}{d}} \tag{3.37}$$

is called the *Thomas–Fermi* constant.

The best constant $c_{\rm LT}(d)$ in (3.35) is unknown but it is definitely less than or equal to $c_{\rm TF}(d)$. This is seen by using the trial state $\gamma_R = \chi(\cdot/R) P_{\rho_0} \chi(\cdot/R)$ and taking the limit $R \to \infty$. The famous *Lieb–Thirring conjecture* [159] states that

$$c_{\rm LT}(d) = \min\left\{\frac{c_{\rm GN}(d)}{2}, c_{\rm TF}(d)\right\} = \begin{cases} \dfrac{c_{\rm GN}(d)}{2} & \text{for } d = 1, 2, \\ c_{\rm TF}(d) & \text{for } d \geqslant 3. \end{cases} \tag{3.38}$$

In other words, the conjecture states that the best constant is obtained either for the infinite non-interacting uniform electron gas, or for one isolated electron. This conjecture was investigated numerically in [98]. The proof of the conjecture (3.38) in dimension $d = 3$ would have a great impact since it would mean that the Thomas–Fermi–Dirac (TFD) energy is an exact lower bound to the many-particle problem [114], as we will mention later in Sect. 3.6. The Thomas–Fermi energy is the simplest functional in Density Functional Theory and knowing that it is an exact lower bound would simplify drastically many mathematical results, in addition to increasing its physical significance.

The best known estimate on $c_{\rm LT}(d)$ was recently proved in [56] and reads

$$\forall d \geqslant 1, \qquad \frac{c_{\rm LT}(d)}{c_{\rm TF}(d)} \geqslant (1.456)^{-\frac{2}{d}}.$$

It improves upon the previously best known result where 1.456 was replaced by 1.814 and which was proved in $d = 1$ in 1991 by Eden and Foias [49] and in $d \geqslant 2$ by Dolbeault et al. [45] in 2008. We refer to [54–56, 159] for a recent overview of other important results on the Lieb–Thirring inequality.

By duality, the Lieb–Thirring inequality implies a bound on the sum of the negative eigenvalues of a one-particle Schrödinger operator in an external potential v, denoted by functional calculus as $-\operatorname{Tr}(-\Delta/2 + v)_-$. Namely, we have by (3.7)

$$\begin{aligned} -\operatorname{Tr}(-\Delta/2 + v)_- &= \inf_{N \geqslant 0} E_N^0[v] \\ &= \inf_{\substack{\sqrt{\rho} \in H^1(\mathbb{R}^d) \\ \int_{\mathbb{R}^d} v_+ \rho < \infty}} \left\{ T_{\rm GC}[\rho] + \int_{\mathbb{R}^d} v(\mathbf{r})\, \rho(\mathbf{r})\, \mathrm{d}\mathbf{r} \right\} \end{aligned}$$

$$
\begin{aligned}
&\geqslant \inf_{\substack{\sqrt{\rho}\in H^1(\mathbb{R}^d)\\ \int_{\mathbb{R}^d} v_+\rho<\infty}} \left\{q^{-\frac{2}{d}}c_{\rm LT}(d)\int_{\mathbb{R}^d}\rho(\mathbf{r})^{1+\frac{2}{d}}\,{\rm d}\mathbf{r}+\int_{\mathbb{R}^d}v(\mathbf{r})\,\rho(\mathbf{r})\,{\rm d}\mathbf{r}\right\}\\
&= -\frac{2d^{\frac{d}{2}}q}{(d+2)^{1+\frac{d}{2}}c_{\rm LT}(d)^{\frac{d}{2}}}\int_{\mathbb{R}^d}v_-(\mathbf{r})^{1+\frac{d}{2}}\,{\rm d}\mathbf{r}, \qquad (3.39)
\end{aligned}
$$

where E_N^0 indicates that we take $w\equiv 0$. Since $T_{\rm GC}[\rho]=F_{\rm GC}^0[\rho]$ is the Legendre transform of $E_N^0[v]$, the inequality (3.39) is actually equivalent to the Lieb–Thirring inequality (3.35). The original proof of Lieb and Thirring [125, 126] actually showed (3.39) and it was only much later that Rumin [153] found a direct proof of (3.35).

The semi-classical constant (3.37) naturally occurs for slowly varying densities in the LDA regime, as we will see. Nam proved in [133] that one can replace the Lieb–Thirring (unknown) constant $c_{\rm LT}(d)$ by $c_{\rm TF}(d)$ at the expense of a gradient correction.

Theorem 3.13 (Nam's Lieb–Thirring Inequality with Gradient Correction [133]) *Let $q,d\geqslant 1$. There exists a universal constant $\kappa(d)$ (independent of the number of spin states q) such that*

$$
T_{\rm GC}[\rho]\geqslant q^{-\frac{2}{d}}c_{\rm TF}(d)(1-\varepsilon)\int_{\mathbb{R}^d}\rho(\mathbf{r})^{1+\frac{2}{d}}\,{\rm d}\mathbf{r}-\frac{\kappa(d)}{\varepsilon^{3+\frac{4}{d}}}\int_{\mathbb{R}^d}|\nabla\sqrt{\rho}(\mathbf{r})|^2\,{\rm d}\mathbf{r} \qquad (3.40)
$$

for all $0<\varepsilon<1$ and all $d\geqslant 1$.

This was the first step towards a proof of the validity of the LDA for the kinetic energy, to which we will come back soon.

Li and Yau proved in [111] a lower bound involving the optimal Thomas–Fermi constant in a bounded domain $\Omega\subset\mathbb{R}^d$:

$$
T_{\rm GC}[\rho]\geqslant c_{\rm TF}(d)|\Omega|^{-\frac{2}{3}}\left(\int_\Omega\rho(\mathbf{r})\,{\rm d}\mathbf{r}\right)^{\frac{5}{3}}\qquad\text{for all }\sqrt{\rho}\in H_0^1(\Omega) \qquad (3.41)
$$

(see also [118, Thm. 12.3] and [75, Lem. 9]). The bound is particularly useful for densities ρ which are (almost) constant over a domain Ω. For instance for $\rho(\mathbf{r})=\rho_0\mathbb{1}_\Omega*\chi(\mathbf{r})$ with ${\rm supp}(\chi)\subset B_1$ we find the exact lower bound

$$
T_{\rm GC}[\rho_0\mathbb{1}_\Omega*\chi(\mathbf{r})]\geqslant c_{\rm TF}(d)(\rho_0)^{\frac{5}{3}}\frac{|\Omega|^{\frac{5}{3}}}{|\Omega+B_1|^{\frac{2}{3}}}=c_{\rm TF}(d)(\rho_0)^{\frac{5}{3}}\left(|\Omega|-C|\Omega|^{\frac{2}{3}}\right)_+
$$

whereas (3.40) yields a worse error term.

3.4.3 Upper Bounds

For upper bounds one should ideally consider the larger functional $T_S[\rho]$. In dimension $d = 1$, March and Young [129, Eq. (9)] gave the proof of an estimate similar to (3.40) without the parameter ε in front of the gradient correction

$$T_S[\rho] \leqslant q^{-2}\frac{\pi^2}{6}\int_{\mathbb{R}} \rho(x)^3\,\mathrm{d}x + \frac{1}{2}\int_{\mathbb{R}} \left|(\sqrt{\rho})'(x)\right|^2\,\mathrm{d}x \tag{3.42}$$

where $\pi^2/6 = c_{\mathrm{TF}}(1)$. In the same paper they also state a result in 3D (for a constant $c > c_{\mathrm{TF}}(3)$) but the proof has a mistake [116, Sec. 5.B]. The bound (3.42) is proved by using as trial state the orbitals

$$\varphi_n(x) = \sqrt{\frac{\rho(x)}{N}}\exp\left(\frac{2in\pi}{N}\int_{-\infty}^{x}\rho(t)\,\mathrm{d}t\right) \tag{3.43}$$

(we take $q = 1$ for simplicity), where $n \in \mathbb{Z}$ and the phases are seen to make the φ_n orthonormal. Computing the kinetic energy of this trial state, one obtains

$$T_S[\rho] \leqslant 2\pi^2\frac{\sum n^2}{N^3}\int_{\mathbb{R}} \rho(x)^3\,\mathrm{d}x + \frac{1}{2}\int_{\mathbb{R}} \left|(\sqrt{\rho})'(x)\right|^2\,\mathrm{d}x.$$

Taking all the integers n less than or equal to $N/2$ in absolute value and using the precise behavior of the series gives the result for $q = 1$.

The method can be generalized to higher dimensions using a similar method, but the estimate has a bad behavior in N. The orbitals

$$\varphi_n(\mathbf{x}) = \sqrt{\frac{\rho(\mathbf{r})}{N}}\mathrm{e}^{i\theta_n(\mathbf{r})}\delta_0(\sigma), \qquad \theta_n(\mathbf{r}) = \frac{2n\pi}{N}\int_{-\infty}^{r_1}\int_{\mathbb{R}^{d-1}}\rho(t,\mathbf{r}')\,\mathrm{d}t\,\mathrm{d}\mathbf{r}' \tag{3.44}$$

were considered in [79, 116] and these are the phases which we already mentioned in (3.5). Using this trial state one obtains [116] for $q = 1$

$$T_S[\rho] \leqslant \left(\frac{2\pi^2}{3}N^2 + CN\right)\int_{\mathbb{R}^d} |\nabla\sqrt{\rho}(\mathbf{r})|^2\,\mathrm{d}\mathbf{r}. \tag{3.45}$$

An upper bound on $T_S[\rho]$ involving only $\int_{\mathbb{R}^d}|\nabla\sqrt{\rho}|^2$ has to have a constant diverging at least as fast as $N^{2/d}$, due to the Lieb–Thirring inequality. In [11, 176, 177] the optimal upper bound of this form was shown:

$$T_S[\rho] \leqslant CN^{\frac{2}{d}}\int_{\mathbb{R}^d} |\nabla\sqrt{\rho}(\mathbf{r})|^2\,\mathrm{d}\mathbf{r}. \tag{3.46}$$

The idea of the proof is to apply a deformation of the space in order to map ρ onto the constant density in a box, which is then represented by a usual Slater determinant made of plane waves. One would expect an upper bound on $T_S[\rho]$ involving both $\int_{\mathbb{R}^d}\rho^{1+2/d}$ and $\int_{\mathbb{R}^d}|\nabla\sqrt{\rho}|^2$, with coefficients independent of N as in (3.42), but this seems unknown at present. The periodic case was studied in [12].

Recently, an upper bound similar to (3.40) was proved in [108] for the grand-canonical functional $T_{GC}[\rho]$.

Theorem 3.14 (Upper Bound on $T_{GC}[\rho]$ [108]) *Let $d, q \geqslant 1$. There exists a constant $\kappa'(d)$ such that*

$$T_{GC}[\rho] \leqslant q^{-\frac{2}{d}} c_{TF}(d)\,(1+\varepsilon)\int_{\mathbb{R}^d}\rho(\mathbf{r})^{1+\frac{2}{d}}\mathrm{d}\mathbf{r} + \kappa'(d)\frac{1+\varepsilon}{\varepsilon}\int_{\mathbb{R}^d}|\nabla\sqrt{\rho}(\mathbf{r})|^2\,\mathrm{d}\mathbf{r} \tag{3.47}$$

for all $\varepsilon > 0$ and all $\rho \geqslant 0$ with $\sqrt{\rho}\in H^1(\mathbb{R}^d)$.

The main difficulty in the proof of (3.47) is the constraint that the one-particle density matrix must have the exact density ρ. One can provide rather good upper bounds if we allow the density to vary a bit. For instance, by using coherent states [114] the density ρ is replaced by $\rho * |f|^2$ where f is the profile used to build the coherent states (typically a Gaussian).

The proof of (3.47) instead relies on the following trial one-particle density matrix

$$\gamma = \int_0^\infty \sqrt{\eta\left(\frac{t}{\rho(\mathbf{r})}\right)}\;\mathbb{1}\left(-\Delta \leqslant 2\frac{d+2}{d}c_{TF}(d)\,q^{-\frac{2}{d}}t^{\frac{2}{d}}\right)\sqrt{\eta\left(\frac{t}{\rho(\mathbf{r})}\right)}\,\frac{\mathrm{d}t}{t}. \tag{3.48}$$

Here the two functions $\sqrt{\eta(t/\rho(\mathbf{r}))}$ are interpreted as multiplication operators, whereas the operator in the middle is the Fourier multiplier P_t introduced before in (3.36). The non-negative function η is chosen such that

$$\int_0^\infty \eta(t)\,\mathrm{d}t = 1, \qquad \int_0^\infty \eta(t)\,\frac{\mathrm{d}t}{t} \leqslant 1. \tag{3.49}$$

The main idea is to represent the density ρ by using the smooth "layer cake principle" [118, Thm. 1.13]

$$\rho(\mathbf{r}) = \int_0^\infty \eta\left(\frac{t}{\rho(\mathbf{r})}\right)\mathrm{d}t,$$

where we think of η as very concentrated around 1, and to then take the free Fermi gas P_t as in (3.36) on the support of $\eta(t/\rho)$, where ρ is very close to t. The measure $\mathrm{d}t/t$ in (3.48) ensures that $\rho_\gamma = \rho$ exactly. On the other hand the condition

$\int_0^\infty \eta(t)\,\mathrm{d}t/t \leqslant 1$ ensures that $0 \leqslant \gamma \leqslant 1$ and means that η must put slightly more weight on the right of 1 than on the left. Computing the kinetic energy of the trial state (3.48) and optimizing over η, one obtains (3.47).

We have explained the construction of the trial state (3.48) to emphasize how much easier it is to work in the grand-canonical setting. It is an important open problem to obtain a bound similar to (3.47) on $T_{\rm S}[\rho]$ or $T[\rho]$. For $T_{\rm S}[\rho]$ this amounts to understanding how to build N orthogonal orbitals with the prescribed density, and to obtain the lowest possible energy. This problem is somewhat related to the smooth Hobby–Rice problem. There one considers N arbitrary L^2-normalized functions $\varphi_1, \dots, \varphi_N \in H^1(\mathbb{R}^d, \mathbb{C})$ and looks for the minimal kinetic energy cost to orthonormalize them using only phases: $\varphi'_j = \varphi_j e^{i\theta_j}$. It was proved in [61, 97, 155] that such phases θ_j always exist, but known bounds involve $\|\nabla\theta_j\|_{L^1}$ which are not enough to deduce anything on the H^1 norm of the orbitals φ'_j. In view of (3.45), one would suspect that

$$\min_{\substack{|\varphi'_j|=|\varphi_j| \\ \langle\varphi'_j,\varphi'_k\rangle=\delta_{jk}}} \sum_{j=1}^N \int_{\mathbb{R}^d\times\mathbb{Z}_q} |\nabla\varphi'_j(\mathbf{x})|^2\,\mathrm{d}\mathbf{x} \leqslant C(N,d)\sum_{j=1}^N \int_{\mathbb{R}^d\times\mathbb{Z}_q} |\nabla|\varphi_j|(\mathbf{x})|^2\,\mathrm{d}\mathbf{x}$$

but this seems unknown at present. In (3.43) the reference orbitals are all equal to $\sqrt{\rho/N}$ but this is probably not the optimal choice for $T_{\rm S}[\rho]$ in dimension $d \geqslant 2$.

3.4.4 *Local Density Approximation for the Kinetic Energy*

From the lower bound (3.40) and the upper bound (3.47) we obtain the following result, which is similar to Theorem 3.8 but involves only quantities that all scale the same, namely like inverse-length squared.

Theorem 3.15 (Local Density Approximation of the Kinetic Energy [108, 133]) *Let $d, q \geqslant 1$. There exists a universal constant $C(d)$ such that*

$$\begin{aligned}&\left|T_{\rm GC}[\rho] - q^{-\frac{2}{d}} c_{\rm TF}(d)\int_{\mathbb{R}^d}\rho(\mathbf{r})^{1+\frac{2}{d}}\,\mathrm{d}\mathbf{r}\right| \\ &\quad\leqslant \varepsilon q^{-\frac{2}{d}}\int_{\mathbb{R}^d}\rho(\mathbf{r})^{1+\frac{2}{d}}\,\mathrm{d}\mathbf{r} + C(d)\left(1+\varepsilon^{-3-\frac{4}{d}}\right)\int_{\mathbb{R}^d}|\nabla\sqrt{\rho}(\mathbf{r})|^2\,\mathrm{d}\mathbf{r} \qquad (3.50)\end{aligned}$$

for all $\rho \geqslant 0$ with $\sqrt{\rho} \in H^1(\mathbb{R}^d)$ and all $\varepsilon > 0$.

In the regime where

$$\int_{\mathbb{R}^d}|\nabla\sqrt{\rho}(\mathbf{r})|^2\,\mathrm{d}\mathbf{r} \ll \int_{\mathbb{R}^d}\rho(\mathbf{r})^{1+\frac{2}{d}}\,\mathrm{d}\mathbf{r}$$

the optimization over ε gives a right side which is negligible compared to the left side. In this regime we can approximate the kinetic energy functional in the manner

$$T_{\rm GC}[\rho] \approx q^{-\frac{2}{d}} c_{\rm TF}(d) \int_{\mathbb{R}^d} \rho(\mathbf{r})^{1+\frac{2}{d}} \, d\mathbf{r}. \tag{3.51}$$

The right side is called the Thomas–Fermi kinetic energy and it is the simplest approximation to $T_{\rm GC}[\rho]$. If we fix a density ρ with $\int_{\mathbb{R}^d} \rho = 1$ and take $\rho_N(\mathbf{r}) = \rho(\mathbf{r}N^{-1/d})$, then we find from (3.50) that

$$T_{\rm GC}[\rho(\cdot N^{-1/d})] = Nq^{-\frac{2}{d}} c_{\rm TF}(d) \int_{\mathbb{R}^d} \rho(\mathbf{r})^{1+\frac{2}{d}} \, d\mathbf{r} + O\left(N^{\frac{2d+1}{2d+2}}\right).$$

From semi-classical analysis it is expected that for a sufficiently regular ρ the next term should be equal to

$$\frac{d-2}{6d} N^{1-\frac{2}{d}} \int_{\mathbb{R}^d} |\nabla\sqrt{\rho}(\mathbf{r})|^2 \, d\mathbf{r}, \tag{3.52}$$

which is called the *von Weizsäcker correction*, see [136, Sec. 6.7] and [128, pp. 89–90]. This is in reference to the historical work [171] for atoms in dimension $d = 3$ where however von Weizsäcker chose the coefficient $1/2$ instead of $1/18$.[1] The value of the prefactor in (3.52) was predicted in [83, 90, 94, 156, 170]. That the coefficient is negative in dimension $d = 1$ is related to the non-optimality of the Thomas-Fermi constant in the Lieb–Thirring inequality (3.35) and is well known in one-dimensional semi-classical analysis [21].

Even without having a clean upper bound like (3.47), it is reasonable to believe that

$$\lim_{N\to\infty} \frac{T[\rho(\cdot/N^{1/d})]}{N} = \lim_{N\to\infty} \frac{T_{\rm S}[\rho(\cdot/N^{1/d})]}{N} = q^{-\frac{2}{d}} c_{\rm TF}(d) \int_{\mathbb{R}^d} \rho(\mathbf{r})^{1+\frac{2}{d}} \, d\mathbf{r}$$

but this does not seem to be known at present. If the fixed density ρ is replaced by a well chosen locally constant density ρ_N converging to ρ, then this was proved in [71, Thm. 4].

3.4.5 Derivation from Levy–Lieb at Large Densities

In this section we show that our kinetic energy functionals can be obtained from the corresponding Levy–Lieb functionals in a proper limit of large densities.

[1] In order to recover Scott's correction in atoms, the coefficient must actually be taken equal to 0.083 [114].

For completeness, we consider a rather arbitrary interaction potential w in any dimension.

Theorem 3.16 (Convergence at High Density) *Let $w \in L^p(\mathbb{R}^d) + L^\infty(\mathbb{R}^d)$ with p as in* (3.1) *and $\rho \geqslant 0$ such that $\sqrt{\rho} \in H^1(\mathbb{R}^d)$. If $\int_{\mathbb{R}^d} \rho \in \mathbb{N}$ we have*

$$\lim_{\lambda\to\infty} \frac{F_{\rm LL}\big[\lambda^d \rho(\lambda\,\cdot)\big]}{\lambda^2} = T[\rho], \qquad \lim_{\lambda\to\infty} \frac{F_{\rm L}\big[\lambda^d \rho(\lambda\,\cdot)\big]}{\lambda^2} = T_{\rm GC}[\rho]. \tag{3.53}$$

If $\int_{\mathbb{R}^d} \rho \in \mathbb{R}_+$ and the additional classical stability assumption (3.14) *holds, we have*

$$\lim_{\lambda\to\infty} \frac{F_{\rm GC}\big[\lambda^d \rho(\lambda\,\cdot)\big]}{\lambda^2} = T_{\rm GC}[\rho]. \tag{3.54}$$

One can also prove the convergence of optimal states or even write the theorem in the form of Gamma convergence. In a similar manner, $T_{\rm S}[\rho]$ arises from the Hartree–Fock-type Levy–Lieb functional where one only minimizes over Slater determinants.

Since we have not found the proof in the literature, we provide it here for completeness.

Proof We start with $F_{\rm LL}$. By scaling we see that

$$\frac{F_{\rm LL}^{w}\big[\lambda^d \rho(\lambda\cdot)\big]}{\lambda^2} = F_{\rm LL}^{w_\lambda}\big[\rho\big]$$

with the new interaction potential $w_\lambda(\mathbf{r}) = \lambda^{-2} w(\mathbf{r}/\lambda)$. Our assumptions on w imply that w is infinitesimally $(-\Delta)$-form bounded, that is, $|w| \leqslant \varepsilon(-\Delta) + C_\varepsilon$ for all $\varepsilon > 0$. After scaling this implies

$$|w_\lambda| \leqslant \varepsilon(-\Delta) + \frac{C_\varepsilon}{\lambda^2}.$$

For the two-particle operator this gives

$$\sum_{1\leqslant j<k\leqslant N} |w_\lambda(\mathbf{r}_j - \mathbf{r}_k)| = \frac{1}{2}\sum_{j=1}^N \sum_{k\neq j} |w_\lambda(\mathbf{r}_j - \mathbf{r}_k)| \leqslant \frac{N-1}{2} \sum_{j=1}^N \left(-\varepsilon\Delta_j + \frac{C_\varepsilon}{\lambda^2}\right)$$

and we thus obtain

$$\frac{1-\varepsilon(N-1)}{2}\sum_{j=1}^{N}(-\Delta)_j - \frac{N(N-1)C_\varepsilon}{2\lambda^2} \leqslant H_N^{0,w_\lambda} \leqslant \frac{1+\varepsilon(N-1)}{2}\sum_{j=1}^{N}(-\Delta)_j + \frac{N(N-1)C_\varepsilon}{2\lambda^2}. \tag{3.55}$$

This yields the bound

$$(1-\varepsilon(N-1))T[\rho] - \frac{N(N-1)C_\varepsilon}{2\lambda^2} \leqslant \frac{F_{\rm LL}^w\big[\lambda^d\rho(\lambda\cdot)\big]}{\lambda^2} \leqslant (1+\varepsilon(N-1))T[\rho] + \frac{N(N-1)C_\varepsilon}{2\lambda^2}.$$

The limit (3.53) follows after taking first $\lambda \to \infty$ and then $\varepsilon \to 0$. For an explicit potential such as Coulomb we know how C_ε depends on ε and one can then give a quantitative bound.

For $F_{\rm L}$ the argument is exactly the same, with the same bound and $T[\rho]$ replaced by $T_{\rm GC}[\rho]$. For $F_{\rm GC}$ the above argument does not work due to the bad behavior in N. Instead, we rescale the stability assumption (3.14) on w and obtain

$$H_n^{0,w_\lambda} \geqslant \sum_{j=1}^{n}(-\Delta)_j - \frac{C}{\lambda^2}n,$$

which provides the lower bound

$$\frac{F_{\rm GC}^w\big[\lambda^d\rho(\lambda\cdot)\big]}{\lambda^2} \geqslant T_{\rm GC}[\rho] - \frac{C}{\lambda^2}\int_{\mathbb{R}^d}\rho(\mathbf{r})\,d\mathbf{r}.$$

For the upper bound we consider a fixed grand-canonical state $\Gamma = (\Gamma_n)_{n\geqslant 0}$ such that

$$\sum_{n\geqslant 1}{\rm Tr}\,(H_n^{0,0}\Gamma_n) \leqslant T_{\rm GC}[\rho] + \eta$$

for some small $\eta > 0$. From the proof in Sect. 3.8, we can assume that Γ has compact support: $\Gamma_n \equiv 0$ for $n \geqslant K$. Using the previous bound (3.55) in the canonical case, we obtain the bound

$$\frac{F_{\rm LL}^w\big[\lambda^d\rho(\lambda\cdot)\big]}{\lambda^2} \leqslant (1+\varepsilon(K-1))\big(T_{\rm GC}[\rho]+\eta\big) + \frac{K(K-1)C_\varepsilon}{2\lambda^2}.$$

The limit now follows after taking $\lambda \to \infty$, $\varepsilon \to 0$ and finally $\eta \to 0$. □

Remark 3.1 (Adiabatic Connection) For homogeneous potentials such as Coulomb, scaling ρ is the same as changing the strength of the interaction. This is the spirit of the *adiabatic connection formula*, which is often used in quantum chemistry to interpolate between the non-interacting and interacting problems [80]. Let us for instance discuss $F_{\rm L}^{w}[\rho]$ and the corresponding kinetic energy $T_{\rm GC}[\rho] = F_{\rm L}^{0}[\rho]$. We introduce a coupling constant t in front of w and look at the function $t \mapsto F_{\rm L}^{tw}[\rho]$. It is concave on $[0, 1]$ (and increasing if $w \geqslant 0$). It has left and right derivatives everywhere, which are given by the minimal and maximal values of the interaction energy among all the possible minimizers Γ_t of $F_{\rm L}^{tw}[\rho]$, by the Feynman-Hellmann theorem. These two derivatives are equal, except possibly on a countable subset of $[0, 1]$. We can express

$$F_{\rm L}^{w}[\rho] - T_{\rm GC}[\rho] = \int_0^1 \frac{\partial}{\partial t} F_{\rm L}^{tw}[\rho]\, {\rm d}t - \int_0^1 {\rm Tr}\left(\sum_{1\leqslant j<k\leqslant N} w(\mathbf{r}_j - \mathbf{r}_k)\right)\Gamma_t\, {\rm d}t$$

where Γ_t is any minimizer for $F_{\rm L}^{tw}[\rho]$. This is a formula for the direct plus exchange-correlation energy in Kohn-Sham theory with fractional occupations. It is sometimes useful to consider a general path $t \in [0, 1] \mapsto w_t$ in place of the simple linear switching, see [174] and [80, Sec. 2.4].

3.5 The Classical Interaction Energy and Lieb–Oxford Inequalities

In this section we study the Levy–Lieb functional with the kinetic energy dropped, which then becomes a purely classical problem.

3.5.1 A Multi-Marginal Optimal Transport Problem

In the classical problem there is no difference between fermions and bosons. In the canonical setting, the main variable is a symmetric probability density $\mathbb{P}(\mathbf{r}_1, \dots, \mathbf{r}_N)$ over $(\mathbb{R}^d)^N$, which in the quantum case corresponds to

$$\mathbb{P}(\mathbf{r}_1, \dots, \mathbf{r}_N) = \sum_{\sigma_1,\dots,\sigma_N \in \mathbb{Z}_q} |\Psi(\mathbf{r}_1, \sigma_1, \dots, \mathbf{r}_N, \sigma_N)|^2$$

for pure states and to an average of such quantities for mixed states. The sum over the spin variables occurs since the interaction potential has been assumed to be spin-independent. In general, $\mathbb{P}$ will not be absolutely continuous with respect to the Lebesgue measure, however. The problem is therefore better stated in the form

$$\boxed{F_{\rm SCE}[\rho] = \inf_{\substack{\mathbb{P}:\\ \rho_{\mathbb{P}}=\rho}} \int_{(\mathbb{R}^d)^N} \sum_{1\leqslant j<k\leqslant N} w(\mathbf{r}_j - \mathbf{r}_k)\, \mathrm{d}\mathbb{P}(\mathbf{r}_1, \ldots, \mathbf{r}_N)} \tag{3.56}$$

with the density

$$\rho_{\mathbb{P}}(\mathbf{r}) = N \int_{(\mathbb{R}^d)^{N-1}} \mathrm{d}\mathbb{P}(\mathbf{r}, \mathbf{r}_2, \ldots, \mathbf{r}_N).$$

The acronym SCE means *Strictly Correlated Electrons* [69, 160–164] since, as we will explain, the minimizing solution $\mathbb{P}$ is typically supported on a set of small dimension where the positions of the particles are highly dependent on each other. In general ρ can be a singular measure. In the worst case ρ is the sum of N Dirac deltas, in which case $\mathbb{P}$ has to be the symmetrized tensor product of these N deltas so that the locations of the particles are then completely fixed. For simplicity we will always assume that $\rho \in L^1(\mathbb{R}^d)$. Nevertheless, the minimizing $\mathbb{P}$ need not be a function.

There is a grand-canonical version of $F_{\rm SCE}$ which is stated in the form

$$\boxed{F_{\rm GSCE}[\rho] = \inf_{\substack{\mathbb{P}=(\mathbb{P}_n)_{n\geqslant 0}\\ \sum_{n\geqslant 0}\mathbb{P}_n((\mathbb{R}^d)^n)=1\\ \sum_{n\geqslant 1}\rho_{\mathbb{P}_n}=\rho}} \sum_{n\geqslant 2} \int_{(\mathbb{R}^d)^n} \sum_{1\leqslant j<k\leqslant n} w(\mathbf{r}_j - \mathbf{r}_k)\, \mathrm{d}\mathbb{P}_n(\mathbf{r}_1, \ldots, \mathbf{r}_n).} \tag{3.57}$$

This was introduced in [106] and further studied in [42, 107, 108]. For the problem to be well posed for all densities, w needs to satisfy the stability condition (3.14).

The two classical problems (3.56) and (3.57) belong to the class of *multi-marginal optimal transport* problems [33, 35, 41, 137, 162]. We only mention here a few striking results. The existence of a minimizing $\mathbb{P}$ for (3.56) follows by compactness arguments similar to Theorem 3.2, for a large class of interaction potentials including the Coulomb potential. The argument is the same in the grand-canonical case (3.57). It was proved in [30] that the infimum can be restricted to *Monge states*, which are the most correlated N-particle probability densities with one-particle density ρ and take the form

$$\mathbb{P}(\mathbf{r}_1, \ldots, \mathbf{r}_N) = \mathrm{Sym} \int_{\mathbb{R}^d} \delta_{\mathbf{y}}(\mathbf{r}_1)\delta_{T\mathbf{y}}(\mathbf{r}_2) \cdots \delta_{T^{N-1}\mathbf{y}}(\mathbf{r}_N)\, \frac{\rho(\mathbf{y})}{N}\, \mathrm{d}\mathbf{y}, \tag{3.58}$$

where $T : \mathbb{R}^d \to \mathbb{R}^d$ is a transport map such that $T\#\rho = \rho$ and $T^N = \mathrm{Id}$ and Sym denotes symmetrization. The formula means that the position $\mathbf{y} = \mathbf{r}_1$ of the first particle completely determines the positions $\mathbf{r}_2 = T\mathbf{r}_1, \ldots, \mathbf{r}_N = T^{N-1}\mathbf{r}_1$ of the other $N-1$ particles through the transport map T (and the picture is symmetrized with respect to the indices of the particles at the end). When moving the first particle (at the appropriate speed so as to build the desired density ρ) the other particles follow in a 'strictly correlated' way.

Even if the infimum in (3.56) is the same when restricted to Monge states, there might exist no Monge minimizer [32, 162]. Only when $N=2$, or in one dimension for all $N \geqslant 2$ [29] one can be sure that Monge minimizers exist. In fact, in dimension $d=1$ and for a positive interaction $w \geqslant 0$, the problem admits a minimizer $\mathbb{P}$ which does not depend on w at all! It is the Monge state with increasing transport map $T\#(\rho \mathbb{1}_{(r_{k-1},r_k)}) = \rho \mathbb{1}_{(r_k,r_{k+1})}$ where $r_0 = -\infty < r_1 < r_2 < \cdots < r_{N-1} < r_N = +\infty$ are chosen such that $\int_{r_i}^{r_{i+1}} \rho(r)\,\mathrm{d}r = 1$ [29]. The corresponding N-particle probability can also be expressed in the manner

$$\mathbb{P}(r_1,\ldots,r_N) = \operatorname{Sym} \int_0^1 \delta_{r_1(s)} \otimes \cdots \otimes \delta_{r_N(s)}\,\mathrm{d}s, \tag{3.59}$$

where $r_k(s) : [0,1] \to [r_{k-1}, r_k]$ is the inverse of the increasing function $r \mapsto s_k(r) = \int_{r_{k-1}}^{r} \rho(t)\,\mathrm{d}t$. This is displayed in Fig. 3.2. For instance, for the uniform density $\rho(r) = \mathbb{1}(0 \leqslant r \leqslant N)$ we have simply $T(y) = y$. The N points are placed on the lattice $(y+\mathbb{Z}) \cap [0, N)$ and their position is averaged over $s \in [0,1]$:

$$\mathbb{P}(r_1,\ldots,r_N) = \operatorname{Sym} \int_0^1 \delta_s(r_1)\delta_{1+s}(r_2)\cdots\delta_{N-1+s}(r_N)\,\mathrm{d}s. \tag{3.60}$$

This is called a *floating Wigner crystal* in Physics and Chemistry [9, 47, 107, 130], since the particles are exactly located on a lattice, whose position is varied. We will come back to this special state later in Sect. 3.5.4.

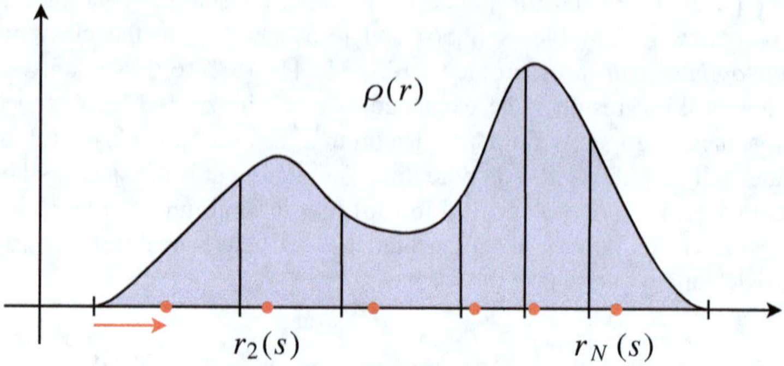

Fig. 3.2 Form (3.59) of the optimal Monge-type probability $\mathbb{P}$ in one dimension. The positions of all the particles are fixed by the position of the first particle and they are moved to the right at a proper speed so as to reproduce the desired density ρ

In [22] some interesting properties of the exact (Monge or not Monge) minimizer $\mathbb{P}$ of (3.56) were established. This includes the fact that the N particles have a positive distance to each other on the support of $\mathbb{P}$, for a repulsive interaction such as Coulomb. The dual formulation is similar to (3.11) and takes the form

$$F_{\rm SCE}[\rho] = \sup_{\substack{v\in C^0(\mathbb{R}^d)\\ \sum_{1\leqslant j<k\leqslant N} w(\mathbf{r}_j-\mathbf{r}_k)+\sum_{j=1}^N v(\mathbf{r}_j)\geqslant 0}} \left\{-\int_{\mathbb{R}^d} v(\mathbf{r})\rho(\mathbf{r})\,\mathrm{d}\mathbf{r}\right\}. \tag{3.61}$$

One important feature of the SCE problem is that there exists an optimal potential $v_{\rm SCE}$ solving the supremum in (3.61), under rather weak assumptions on the interaction potential w. The optimal $v_{\rm SCE}$ is called a *Kantorovich* potential. This is in stark contrast to the quantum case (3.11), where unique continuation drastically reduces the set of densities ρ for which the supremum is attained (see Sect. 3.7 below). Under the sole assumption that w is radial decreasing, diverges at the origin and is C^1 outside of the origin (like for the Coulomb potential $w(\mathbf{r}) = |\mathbf{r}|^{-1}$ in dimension $d = 3$), it was proved in [22] that there exists an optimal Kantorovich potential $v_{\rm SCE}$ which is bounded and Lipschitz. An optimal N-particle probability $\mathbb{P}$ must then be supported on the set

$$\operatorname{argmin}\left\{\sum_{1\leqslant j<k\leqslant N} w(\mathbf{r}_j-\mathbf{r}_k)+\sum_{j=1}^N v_{\rm SCE}(\mathbf{r}_j)\right\}.$$

In other words, the N particles should minimize the associated N-particle classical problem with the external potential $v_{\rm SCE}$. All densities $\rho \in L^1(\mathbb{R}^3)$ are v-representable in the classical case.

3.5.2 *Convergence of the Levy–Lieb Functional at Low Density*

We have seen above in Theorem 3.16 that the kinetic energy functional becomes dominant at large densities. Similarly, the interaction becomes dominant at low densities, provided that w has the right scaling at large distances. To simplify our exposition, from now on we restrict our discussion to power-law (Riesz) potentials

$$w(\mathbf{r}) = \frac{1}{|\mathbf{r}|^s}, \qquad 0 < s < \min(2, d).$$

The main result is the following.

Theorem 3.17 (Convergence at Low Density) *Let $w(\mathbf{r}) = |\mathbf{r}|^{-s}$ with $0 < s < \min(2, d)$. Let $\rho \geqslant 0$ such that $\sqrt{\rho} \in H^1(\mathbb{R}^d)$. If $\int_{\mathbb{R}^d} \rho \in \mathbb{N}$ we have*

$$\lim_{\lambda \to 0} \frac{F_{\rm LL}\big[\lambda^d \rho(\lambda \cdot)\big]}{\lambda^{1+\frac{s}{d}}} = \lim_{\lambda \to 0} \frac{F_{\rm L}\big[\lambda^d \rho(\lambda \cdot)\big]}{\lambda^{1+\frac{s}{d}}} = F_{\rm SCE}[\rho]. \tag{3.62}$$

If $\int_{\mathbb{R}^d} \rho \in \mathbb{R}_+$ we have

$$\lim_{\lambda \to 0} \frac{F_{\rm GC}\big[\lambda^d \rho(\lambda \cdot)\big]}{\lambda^{1+\frac{s}{d}}} = F_{\rm GSCE}[\rho]. \tag{3.63}$$

Note that the stability condition (3.13) is always satisfied for the positive potential $w(\mathbf{r}) = |\mathbf{r}|^{-s}$, hence the grand-canonical energies are well defined.

The proof is much more complicated than Theorem 3.16, since an optimizer $\mathbb{P}$ for $F_{\rm SCE}[\rho]$ or $F_{\rm GSCE}[\rho]$ will never have a finite kinetic energy, even under the assumption that $\sqrt{\rho} \in H^1(\mathbb{R}^d)$. The limit for $F_{\rm LL}$ was shown for $\int_{\mathbb{R}^d} \rho = 2$ with spin first by Cotar et al. in [33] and later extended to $\int_{\mathbb{R}^d} \rho = 3$ by Bindini and de Pascale in [8]. The limit for $F_{\rm L}[\rho]$ and all $\int_{\mathbb{R}^d} \rho \in \mathbb{N}$ was solved in [103] whereas the case of $F_{\rm LL}$ was finally treated in [34]. The proof for $F_{\rm GC}$ follows along the lines of [103]. The next order in λ was predicted in [70] and proved in some cases in [31].

To summarize, at large densities the Levy–Lieb functional behaves like the kinetic energy of non-interacting quantum particles, whereas at low density the particles tend to be very correlated and solve the corresponding classical problem.

3.5.3 Lieb–Oxford Inequality

We discuss here upper and lower bounds on the interaction energy, with an emphasis on lower bounds (Lieb–Oxford inequality).

The easiest upper bound is obtained by taking the decorrelated trial state $\mathbb{P} = (\rho/N)^{\otimes N}$, that is, independent particles distributed according to the density ρ. This gives the bound

$$F_{\rm SCE}[\rho] \leqslant \frac{1 - \frac{1}{N}}{2} \iint_{\mathbb{R}^{2d}} w(\mathbf{r} - \mathbf{r}')\rho(\mathbf{r})\rho(\mathbf{r}')\, d\mathbf{r}\, d\mathbf{r}'. \tag{3.64}$$

The right side is, up to the constant $1 - 1/N$, the classical energy of the density distribution ρ and it is a non-local term. The factor $1/N$ can be dropped for repulsive potentials.

It is relatively easy to prove a similar lower bound, under the additional assumption that w is continuous and has a non-negative Fourier transform, $\widehat{w} \geqslant 0$. In this case we have

$$\iint_{\mathbb{R}^{2d}} w(\mathbf{r}-\mathbf{r}')\,\mathrm{d}\eta(\mathbf{r})\,\mathrm{d}\eta(\mathbf{r}') = (2\pi)^{d/2}\int_{\mathbb{R}^d} \widehat{w}(k)|\widehat{\eta}(k)|^2\,\mathrm{d}k \geqslant 0$$

for every signed measure η. Taking $\eta = \sum_{j=1}^N \delta_{\mathbf{r}_j} - f$ and expanding we find the pointwise inequality on $(\mathbb{R}^d)^N$

$$\sum_{1\leqslant j<k\leqslant N} w(\mathbf{r}_j-\mathbf{r}_k) \geqslant \sum_{j=1}^N w*f(\mathbf{r}_j) - \frac{1}{2}\iint_{\mathbb{R}^{2d}} w(\mathbf{r}-\mathbf{r}')f(\mathbf{r})f(\mathbf{r}')\,\mathrm{d}\mathbf{r}\,\mathrm{d}\mathbf{r}' - \frac{w(0)N}{2}.$$

This is valid for all f and the last error term comes from the case $j = k$. Integrating against any state $\mathbb{P}$ with density ρ and taking $f = \rho$, we obtain the following lower bound:

$$F_{\mathrm{SCE}}[\rho] \geqslant \frac{1}{2}\iint_{\mathbb{R}^{2d}} w(\mathbf{r}-\mathbf{r}')\rho(\mathbf{r})\rho(\mathbf{r}')\,\mathrm{d}\mathbf{r}\,\mathrm{d}\mathbf{r}' - \frac{w(0)}{2}\int_{\mathbb{R}^d}\rho(\mathbf{r})\,\mathrm{d}\mathbf{r}, \quad \text{when } \widehat{w} \geqslant 0. \tag{3.65}$$

For a long-range potential the first term grows faster than N for most densities, hence the last error term is often much lower than the classical energy.

For Coulomb or other power-law potentials, the previous argument does not work since $w(0) = +\infty$. One solution is to regularize the potential at the origin but this also modifies the classical interaction energy. One can estimate the error under appropriate regularity assumptions on ρ. But Lieb [113] and then Lieb–Oxford [120] have proved a universal bound which has the right scaling behavior and does not require the potential to be smeared out. We state it for power-law potentials but the inequality is slightly more general.

Theorem 3.18 (Lieb–Oxford Inequality [4, 72, 113, 120, 121, 124, 127]) *Assume that $w(\mathbf{r}) = |\mathbf{r}|^{-s}$ with $0 < s < d$ in dimension $d \geqslant 1$. Then there exists a universal constant $c_{\mathrm{LO}}(s,d) > 0$ such that*

$$F_{\mathrm{GSCE}}[\rho] \geqslant \frac{1}{2}\iint_{\mathbb{R}^{2d}} \frac{\rho(\mathbf{r})\rho(\mathbf{r}')}{|\mathbf{r}-\mathbf{r}'|^s}\,\mathrm{d}\mathbf{r}\,\mathrm{d}\mathbf{r}' - c_{\mathrm{LO}}(s,d)\int_{\mathbb{R}^d}\rho(\mathbf{r})^{1+\frac{s}{d}}\,\mathrm{d}\mathbf{r}, \tag{3.66}$$

for every $\rho \in (L^1 \cap L^{1+\frac{s}{d}})(\mathbb{R}^d, \mathbb{R}_+)$.

From now on we always call $c_{\mathrm{LO}}(s,d)$ the *smallest* constant for which the inequality (3.66) is valid for all ρ. Note that $c_{\mathrm{LO}}(s,d)$ works for every particle number $\int_{\mathbb{R}^d}\rho$. If one adds the constraint that $\int_{\mathbb{R}^d}\rho = \lambda$ then the optimal constant depends on λ but it is non-decreasing and has the limit $c_{\mathrm{LO}}(s,d)$ when $\lambda \to \infty$.

Although only the case $s = 1$ and $d = 3$ was considered in the original papers [113, 120], the proof for $s = 1$ and $d = 2$ given in [4, 72, 124] extends to any $0 < s < d$ in any dimension, see [127, Lemma 16]. This proof involves the Hardy–Littlewood estimate for the maximal function M_ρ [73],

$$\|M_\rho\|_{L^{1+s/d}(\mathbb{R}^d)} \leqslant c_{\rm HL}(s,d)\,\|\rho\|_{L^{1+s/d}(\mathbb{R}^d)}$$

and, consequently, the best known estimate on $c_{\rm LO}(s,d)$ involves the unknown constant $c_{\rm HL}(s,d)$. A Lieb-Oxford bound was shown for $w(\mathbf{r}) = -\log|\mathbf{r}|$ in two dimensions in [110, Prop. 3.8]. In dimension $d = 1$, optimal Lieb-Oxford bounds are studied in [40].

In the 3D Coulomb case, $d = 3$ and $s = 1$, the best estimate known so far on the optimal Lieb-Oxford constant is

$$\boxed{1.4442 \leqslant c_{\rm LO}(1,3) \leqslant 1.5765.} \tag{3.67}$$

The upper constant was equal to 8.52 in [113], to 1.68 in [120] and later improved to 1.64 in [89]. The better value 1.58 was obtained very recently in [109]. The lower bound has been claimed in [100, 138] and only shown recently in [36, 108]. It will be discussed in the next section. It was conjectured in [100, 135, 148] that the best Lieb-Oxford constant is indeed about 1.44. It remains an important challenge to find the optimal constant in (3.66). Several of the most prominent functionals used in Density Functional Theory make use of the value of the Lieb-Oxford constant for calibration [100, 138, 139, 143, 166–168].

A different Lieb-Oxford inequality was recently proved in the 3D Coulomb case in [109]. It reads

$$\iint_{\mathbb{R}^{3N}} \left(\sum_{1\leqslant j<k\leqslant N} \frac{1}{|\mathbf{r}_j - \mathbf{r}_k|} \right) \mathrm{d}\mathbb{P}(\mathbf{r}_1,\dots,\mathbf{r}_N) - \frac{1}{2}\iint_{\mathbb{R}^6} \frac{\rho_{\mathbb{P}}(\mathbf{r})\rho_{\mathbb{P}}(\mathbf{r}')}{|\mathbf{r}-\mathbf{r}'|}\,\mathrm{d}\mathbf{r}\,\mathrm{d}\mathbf{r}' \geqslant -1.2490 \int_{\mathbb{R}^3} \rho_{\mathbb{P}}(\mathbf{r})^{\frac{4}{3}}\,\mathrm{d}\mathbf{r}, \tag{3.68}$$

under the additional assumption that $\mathbb{P}$ has *negative correlations*, which means

$$N(N-1)\iint_{\mathbb{R}^{3N-6}} \mathrm{d}\mathbb{P}(\mathbf{r},\mathbf{r}',\mathbf{r}_3,\dots,\mathbf{r}_N) \leqslant \rho_{\mathbb{P}}(\mathbf{r})\,\rho_{\mathbb{P}}(\mathbf{r}'), \qquad \text{for a.e. } \mathbf{r},\mathbf{r}' \in \mathbb{R}^3. \tag{3.69}$$

This condition is satisfied when $\mathbb{P}$ is the square of a Slater determinant (3.19), in which case the left side of (3.68) is called the *exchange energy* and the best constant is believed to be 1.09 [143]. But many other states satisfy the condition (3.69). In statistical mechanics, this is typical of gas phases [152] at high temperature. Since $1.25 < 1.44 < c_{\rm LO}(1,3)$, this means that such states cannot provide the optimal

Lieb-Oxford constant. In fact, we explain below how to obtain the lower bound 1.44 from a solid (periodic) phase.

The *indirect energy* is the equivalent of the exchange-correlation energy defined in the quantum case in Sect. 3.2.4:

$$E_{\mathrm{Ind}}[\rho] = F_{\mathrm{SCE}}[\rho] - \frac{1}{2}\iint_{\mathbb{R}^{2d}} w(\mathbf{r}-\mathbf{r}')\rho(\mathbf{r})\rho(\mathbf{r}')\,\mathrm{d}\mathbf{r}\,\mathrm{d}\mathbf{r}'.$$

For power-law interactions it is always negative and bounded from below by a constant times $\int_{\mathbb{R}^d} \rho^{1+s/d}$.

3.5.4 *Constant Densities and the Classical Uniform Electron Gas*

We discuss here the special case of densities which are constant over a finite set and the limit when this set fills the whole space. This is the classical equivalent of the *Uniform Electron Gas* discussed in the quantum case in Sect. 3.3.1 above. This special case will give us some lower bounds on the Lieb–Oxford constant $c_{\mathrm{LO}}(s,d)$, including the bound 1.44 in dimension $d = 3$ stated in (3.67).

The classical equivalent of Theorem 3.6 was proved in [106].

Theorem 3.19 (The Classical Uniform Electron Gas Energy [106]) *Assume that $w(\mathbf{r}) = |\mathbf{r}|^{-s}$ with $0 < s < d$ in dimension $d \geqslant 1$. Let $\rho_0 > 0$. Let Ω be a fixed open convex set of unit volume $|\Omega| = 1$. Then there exists a universal constant $c_{\mathrm{UEG}}(s,d) > 0$ such that*

$$\begin{aligned}
&\lim_{L\to\infty} L^{-d}\left(F_{\mathrm{GSCE}}[\rho_0 \mathbb{1}_{L\Omega}] - \frac{\rho_0^2}{2}\iint_{(L\Omega)^2}\frac{\mathrm{d}\mathbf{r}\,\mathrm{d}\mathbf{r}'}{|\mathbf{r}-\mathbf{r}'|^s}\right)\\
&\qquad = \lim_{\substack{L\to\infty\\ L^d\in\mathbb{N}/\rho_0}} L^{-d}\left(F_{\mathrm{SCE}}[\rho_0 \mathbb{1}_{L\Omega}] - \frac{\rho_0^2}{2}\iint_{(L\Omega)^2}\frac{\mathrm{d}\mathbf{r}\,\mathrm{d}\mathbf{r}'}{|\mathbf{r}-\mathbf{r}'|^s}\right)\\
&\qquad = c_{\mathrm{UEG}}(s,d)\rho_0^{1+\frac{s}{d}}.
\end{aligned} \tag{3.70}$$

In particular we obtain $c_{\mathrm{LO}}(s,d) \geqslant -c_{\mathrm{UEG}}(s,d)$.

The constant $c_{\mathrm{UEG}}(1,3)$ is the one which has appeared before in Theorem 3.7. At low density, the quantum UEG behaves like a classical gas by an equivalent of Theorem 3.17 for infinite systems [106].

Note that the classical canonical and grand-canonical functionals are known to give the same thermodynamic limit. In the quantum case this is not yet known. We have stated the theorem for a fixed domain Ω which is scaled but the same result holds for a general sequence Ω_L that has a regular boundary in the sense of Fisher [106].

Except in dimensions $d \in \{1, 8, 24\}$ and $\max(0, d-2) \leqslant s < d$, some special cases to which we will come back, the constant $c_{\rm UEG}(s, d)$ is unknown. In order to get upper bounds on $c_{\rm UEG}(s, d)$, we need to construct trial states. The idea is to use a *floating crystal* similar to (3.60), that is, to place the particles on a lattice and then average over translations to obtain a constant density.

Let $\mathscr{L} \subset \mathbb{R}^d$ be a lattice of normalized unit cell Q. We then only retain the points of the lattice intersecting the large cube $C_L = (-L/2, L/2)^d$ and average over the translations of this finite lattice over Q. This way we obtain a trial state which is constant over the union of the corresponding translates of Q. In general this is only an approximation of C_L but since the limit (3.70) is insensitive to the type of domains, this will not create any difficulty. The trial state is, therefore, given by

$$\mathbb{P}_{\mathscr{L},L} := \mathrm{Sym} \int_Q \bigotimes_{\ell \in \mathscr{L} \cap C_L} \delta_{\ell + \mathbf{y}} \, \mathrm{d}\mathbf{y}. \tag{3.71}$$

Then $\mathbb{P}_{\mathscr{L},L}$ has the constant density

$$\rho_{\mathbb{P}_{\mathscr{L},L}} = \mathbb{1}_{\Omega_L} \qquad \text{over the set} \qquad \Omega_L = \bigcup_{\ell \in \mathscr{L} \cap C_L} Q + \ell. \tag{3.72}$$

The state is as displayed in Fig. 3.3. Note that the energy of the probability measure $\mathbb{P}_{\mathscr{L},L}$ is simply the interaction of the lattice points, since the interaction potential is translation-invariant:

$$\int_{(\mathbb{R}^d)^N} \sum_{1 \leqslant j < k \leqslant N} \frac{1}{|\mathbf{r}_j - \mathbf{r}_k|^s} \, \mathrm{d}\mathbb{P}_{\mathscr{L},L} = \frac{1}{2} \sum_{\substack{\ell \neq \ell' \\ \in \mathscr{L} \cap C_L}} \frac{1}{|\ell - \ell'|^s}.$$

It is instructive to see first what happens in the short range case $s > d$. Then the energy per unit volume converges to

$$\lim_{L \to \infty} \frac{1}{2|\Omega_L|} \sum_{\substack{\ell \neq \ell' \\ \in \mathscr{L} \cap C_L}} \frac{1}{|\ell - \ell'|^s} = \frac{1}{2} \sum_{\ell \in \mathscr{L} \setminus \{0\}} \frac{1}{|\ell|^s} =: \zeta_{\mathscr{L}}(s), \qquad s > d. \tag{3.73}$$

The function on the right side is called the *Epstein Zeta function* [10, 17, 51] and it is the natural generalization to $\mathbb{R}^d$ of the usual Riemann Zeta function, which it coincides with when $d = 1$ (hence $\mathscr{L} = \mathbb{Z}$). It turns out that the limit in the long range case can be expressed with the (analytic extension) of $\zeta_{\mathscr{L}}$, for potentials decaying to zero at infinity faster than Coulomb. Something special is happening at $s = d - 2$.

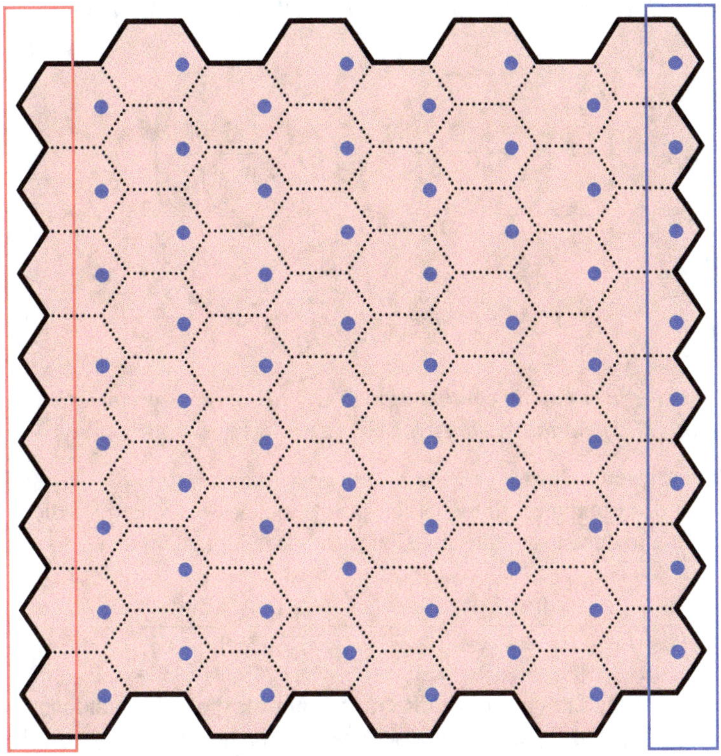

Fig. 3.3 A two-dimensional picture of the Jellium model. The dots represent the point particles, which are placed on a finite subset of a lattice $\mathscr{L}$. The colored set is the union of the corresponding unit cells and it represents a uniform background charge distribution of opposite charge. The indirect energy of the floating crystal is obtained after integrating the position of the lattice over the unit cell Q as in (3.79). When the lattice is not centered, this results in an excess of point charges on one side and an excess of background charge on the other side, indicated by the two rectangles. These large boundary charge fluctuations are responsible for the shift in Theorem 3.20 at $s = d-2$

Theorem 3.20 (Indirect Energy of the Floating Wigner Crystal [14–16, 96, 104, 105]) *Let $d-2 \leqslant s < d$ in dimension $d \geqslant 3$ and $0 < s < d$ in dimensions $d = 1, 2$. Let $\mathscr{L} \subset \mathbb{R}^d$ be a lattice with a normalized unit cell Q having no dipole and no quadrupole moment:*

$$\int_Q \mathbf{r}\, d\mathbf{r} = 0, \qquad \int_Q r_i r_j\, d\mathbf{r} = \frac{\delta_{ij}}{d} \int_Q |\mathbf{r}|^2\, d\mathbf{r}.$$

Then the indirect energy per unit volume of the floating Wigner crystal (3.71) *converges to*

$$\lim_{L\to\infty} |\Omega_L|^{-1}\left(\frac{1}{2}\sum_{\substack{\ell\neq\ell'\\ \in\mathscr{L}\cap C_L}}\frac{1}{|\ell-\ell'|^s}-\frac{1}{2}\iint_{(\Omega_L)^2}\frac{\mathrm{d}\mathbf{r}\,\mathrm{d}\mathbf{r}'}{|\mathbf{r}-\mathbf{r}'|^s}\right)$$

$$=\begin{cases}\zeta_{\mathscr{L}}(s) & \text{for } s>d-2,\\ \zeta_{\mathscr{L}}(d-2)+\dfrac{|\mathbb{S}^{d-1}|}{2d}\displaystyle\int_Q |\mathbf{r}|^2\,\mathrm{d}\mathbf{r} & \text{for } s=d-2,\end{cases} \tag{3.74}$$

where $\zeta_{\mathscr{L}}(s)$ *is the analytic continuation to* $\mathbb{C}\setminus\{d\}$ *of the Epstein Zeta function on the right of* (3.73)*, initially defined for* $\Re(s)>d$.

A similar result holds for $-1\leqslant s\leqslant 0$ in $d=1$ and $s=0$ in $d=2$ [96, 104].

The first divergent term in the lattice sum is the classical energy, which behaves like $N^{2-s/d}$ and depends on the shape of the chosen large domain C_L:

$$\frac{1}{2}\iint_{(\Omega_L)^2}\frac{\mathrm{d}\mathbf{r}\,\mathrm{d}\mathbf{r}'}{|\mathbf{r}-\mathbf{r}'|^s}\sim_{L\to\infty}\frac{L^{2d-s}}{2}\iint_{(C_1)^2}\frac{\mathrm{d}\mathbf{r}\,\mathrm{d}\mathbf{r}'}{|\mathbf{r}-\mathbf{r}'|^s}. \tag{3.75}$$

This is because the lattice sum is a Riemann sum for the corresponding integral at that scale. Replacing C_L by another set changes this macroscopic term. Note that the analytic extension of (3.75) is a $o(L^d)$ for $s>d$. This term probably exists in the short range case too, but it is lower order and it was not seen in the limit (3.73).

The theorem provides the next order term in the long range case $d-2<s<d$. This is an extensive quantity (of the order of the volume) which has a limit independent of the shape C_L. This limit is simply the analytic extension of the short range energy. This is compatible with our picture that the classical energy (3.75) is the leading term for $s<d$ but once it is removed, we are essentially back to (3.73).

At $s=d-2$ the picture changes. Another term of the order $L^{2(d-1)-s}$, which was lower order for all $s>d-2$, becomes relevant for the energy per unit volume at $s=d-2$ and dominates for $s<d-2$. As we will explain later, this is a kind of surface term.

In dimension $d=1$ we know from [29] that the floating crystal is optimal and provides the minimal classical energy at constant density. Therefore we deduce from Theorem 3.20 that

$$c_{\rm UEG}(s,1)=\zeta(s),\qquad \text{for all } 0<s<1. \tag{3.76}$$

In particular, $-c_{\rm LO}(s,1)\leqslant\zeta(s)$.

In higher dimensions, the floating crystal is not known to be an exact minimum, and furthermore there are several possible crystals. Therefore we only obtain the upper bound

$$-c_{\rm LO}(s,d)\leqslant c_{\rm UEG}(s,d)\leqslant \min_{\mathscr{L}}\zeta_{\mathscr{L}}(s),\qquad \text{for } \max(0,d-2)<s<d. \tag{3.77}$$

The minimum is over all lattices of normalized unit cell. It is expected that the last inequality should be an equality for some values of the dimension d including $d = 1, 2, 3$. So far this is only known in dimensions $d = 8$ and $d = 24$ [25, 145]. In dimension $d = 2$ the minimum on the right of (3.77) is known to be achieved by the triangular lattice [23, 43, 50, 131, 147] whereas in dimension $d = 3$, numerics indicates that it is achieved by the Body-Centered Cubic lattice (BCC) for $0 < s \leqslant 3/2$ and the Face-Centered Cubic lattice (FCC) for $3/2 \leqslant s < 3$ [10, 17, 68, 157].

The surprising jump of the energy per unit volume (3.74) in the Coulomb case $s = d-2$ was first discovered in 1979 by Hall [76] based on an unpublished remark by Plaskett in 1959. The conendrum raised by Hall was discussed in several papers in the 80s, see for instance [1, 39, 77, 78, 86, 134]. It was rediscovered in 1988 by Borwein et al. [15] and was recently revived and reformulated in [105, App. B]. It has indeed always been assumed in the Physics and Chemistry literature that the floating crystal is a good trial state for the UEG, and that $c_{\rm UEG}(1, 3)$ should even be equal to the BCC lattice energy, whose value is $\zeta_{\rm BCC}(1) \simeq -1.4442$ (see [26] and [68, p. 43]). This value is used in most DFT functionals based on the Uniform Electron Gas. But the Coulomb potential is exactly the one for which the floating crystal behaves badly, by Theorem 3.20.

Note that the jump exists and is unavoidable in 1D, where the floating crystal is known to be optimal, but it happens at the negative value $s = -1$. Indeed, for $w(r) = -|r|$ we have the expansion similar to Theorem 3.20

$$F_{\rm SCE}\big[\rho_0\mathbb{1}_{[0,L]}\big] = -\frac{L^3(\rho_0)^2}{2}\int_0^1\int_0^1 |x-y|\,{\rm d}x\,{\rm d}y + \frac{L(\rho_0)^2}{6} + O(1) \tag{3.78}$$

with $1/6 > -\zeta(-1) = 1/12$.

In order to better understand what is going on, it is useful to reinterpret the result in terms of the *Jellium model* [104, 119]. In this model there is no constraint on the electronic density but the particles interact with a compensating uniform background of opposite charge. At density one, the corresponding energy is defined by

$$\mathcal{E}_{\rm Jel}(\Omega,\mathbf{r}_1,\ldots,\mathbf{r}_N) = \sum_{1\leqslant j<k\leqslant N}\frac{1}{|\mathbf{r}_j-\mathbf{r}_k|^s} - \sum_{j=1}^N\int_\Omega \frac{{\rm d}\mathbf{r}'}{|\mathbf{r}_j-\mathbf{r}'|^s} + \frac12\iint_{\Omega\times\Omega}\frac{{\rm d}\mathbf{r}\,{\rm d}\mathbf{r}'}{|\mathbf{r}-\mathbf{r}'|^s}$$

where Ω is any measurable set of volume $|\Omega| = N$, representing the uniform background. A short calculation shows that the indirect energy of the floating crystal can be written in the form

$$\frac{1}{2}\sum_{\substack{\ell\neq\ell'\\ \in\mathscr{L}\cap C_L}}\frac{1}{|\ell-\ell'|^s}-\frac{1}{2}\iint_{(\Omega_L)^2}\frac{\mathrm{d}\mathbf{r}\,\mathrm{d}\mathbf{r}'}{|\mathbf{r}-\mathbf{r}'|^s}=\int_Q \mathcal{E}_{\rm Jel}\big(\Omega_L,(\mathscr{L}\cap C_L)+\mathbf{y}\big)\,\mathrm{d}\mathbf{y}. \tag{3.79}$$

In other words, it is the average of Jellium energies where the lattice points are moved over the fixed background Ω_L. In this interpretation it becomes clear why the averaging over $\mathbf{y}$ induces the shift in (3.74): moving the particles away from the center of the unit cells is not at all energetically favorable. When the particles are moved in one direction this creates a large excess of negative charges on one side and a corresponding excess of background charge on the opposite side (Fig. 3.3). These two opposite boundary charges have an interaction energy proportional to $(L^{d-1})^2/L^s = L^{2(d-1)-s}$ which is exactly of the order of the volume in the Coulomb case $s = d-2$ and grows faster for $s < d-2$. On the other hand, when the particles are placed exactly at the center of the unit cells, one recovers the analytic extension of the short range energy for all $s > d-4$.

Theorem 3.21 (Jellium Energy of the Clamped Wigner Crystal [10, 14–16, 96, 104, 105, 119]) *Assume that $d \geqslant 1$. Let $\mathscr{L}\subset\mathbb{R}^d$ be a lattice satisfying the same assumptions as in Theorem 3.20. Then the Jellium energy per unit volume of the Wigner crystal clamped at the center of the unit cells converges to*

$$\lim_{L\to\infty}\frac{\mathcal{E}_{\rm Jel}\big(\Omega_L,\mathscr{L}\cap C_L\big)}{|\Omega_L|}=\zeta_{\mathscr{L}}(s) \tag{3.80}$$

for all $\max(0,d-4) < s < d$.

The floating crystal is really not a good trial state for the UEG. The conundrum raised in Theorem 3.20 was recently resolved in [36, 107]. Cotar and Petrache managed to prove in [36] that the (unknown) UEG energy $c_{\rm UEG}(s,d)$ is always continuous for $0 < s < d$ and that it is equal to the (also unknown) Jellium energy for $d-2 \leqslant s < d$. The proof of continuity in s is very delicate and requires the use of advanced analytical techniques due to Fefferman and collaborators [52, 74, 85]. A short time later, the same result was obtained in [107] with a different and much simpler argument. Here we only explain this argument for the special case of the floating crystal, that is, we show how to modify the trial state (3.71) in order to cancel the shift appearing in (3.74).

The main idea of [107] is to immerse the crystal in a thin layer of fluid. In other words, the floating crystal is melted close to the boundary in order to reduce the large charge fluctuations. The fluid gets displaced with the crystal when the latter is averaged over translations. Think of a block of ice completely filling a container. In order to move the ice it is necessary to melt it close to the container walls.

To describe this procedure, let us denote by C_L' a slightly larger cubic container such that $\Omega_L + Q \subset C_L'$, where we recall that Ω_L is the union of the unit cells $Q+\ell$

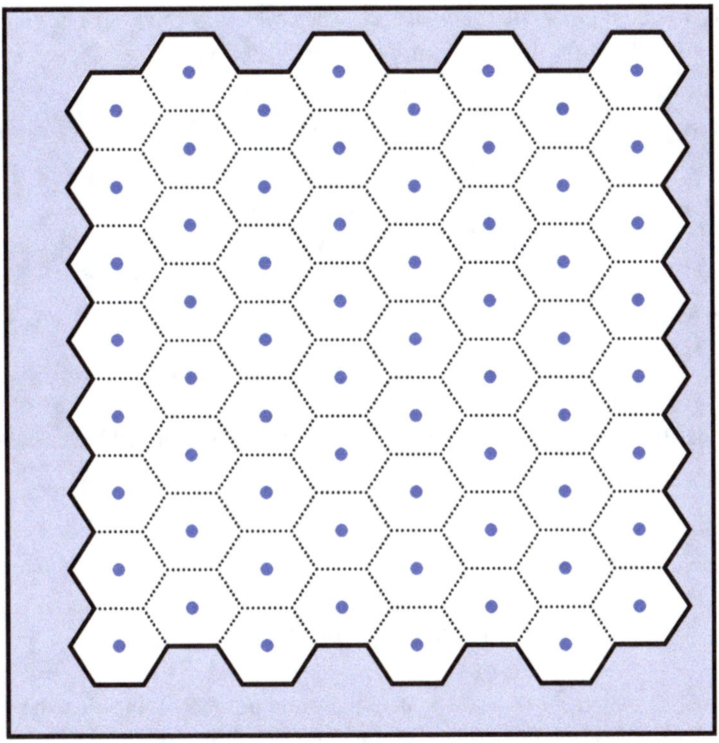

Fig. 3.4 A two-dimensional picture of the modified floating crystal (3.81) from [107]. The dots represent the point particles which are at the centers of hexagons of volume one. As the whole crystal block is translated by $\mathbf{y}$, the incompressible fluid gets displaced to fill the remaining space $C'_L \setminus (\Omega_L + \mathbf{y})$. The resulting density is only constant well inside the container

with $\ell \in \mathscr{L} \cap C_L$. We can take $C'_L = C_{L+\lambda}$, where λ is any fixed distance larger than the diameter of Q. We assume that the volume of the fluid $|C'_L \setminus \Omega_L| = M$ is an integer. It satisfies $M \leqslant CL^{d-1} \ll L^d$. The new trial state has the $N = |\Omega_L|$ particles on the floating crystal, translated by $\mathbf{y} \in Q$ as before in (3.71), together with M other particles forming an uncorrelated fluid in $C'_L \setminus (\Omega_L + \mathbf{y})$, the set remaining after we have subtracted the union of all the cells centered at the particle positions (Fig. 3.4):

$$\widetilde{\mathbb{P}}_{\mathscr{L},L} = \mathrm{Sym} \int_Q \bigotimes_{\ell \in \mathscr{L} \cap C_L} \delta_{\ell+\mathbf{y}} \otimes \left(\frac{\mathbb{1}_{C'_L \setminus (\Omega_L + \mathbf{y})}}{M} \right)^{\otimes M} \mathrm{d}\mathbf{y}. \tag{3.81}$$

Note that the state of the fluid is correlated with the position $\mathbf{y}$ of the crystal.

Theorem 3.22 (Indirect Energy of the Modified Floating Crystal [107]) *Let* $\max(0, d-4) < s < d$ *in dimension* $d \geqslant 1$ *and* $\mathscr{L} \subset \mathbb{R}^d$ *a lattice satisfying* *the same assumptions as in Theorem 3.20. Then the indirect energy per unit volume of the modified floating Wigner crystal* (3.81) *converges to*

$$\lim_{L\to\infty}|C'_L|^{-1}\Bigg(\int_{(\mathbb{R}^d)^N}\sum_{1\leqslant j<k\leqslant N}\frac{1}{|\mathbf{r}_j-\mathbf{r}_k|^s}d\widetilde{\mathbb{P}}_{\mathscr{L},L}$$
$$-\frac{1}{2}\iint_{\mathbb{R}^{2d}}\frac{\rho_{\widetilde{\mathbb{P}}_{\mathscr{L},L}}(\mathbf{r})\rho_{\widetilde{\mathbb{P}}_{\mathscr{L},L}}(\mathbf{r}')}{|\mathbf{r}-\mathbf{r}'|^s}\,d\mathbf{r}\,d\mathbf{r}'\Bigg)=\zeta_{\mathscr{L}}(s). \tag{3.82}$$

In particular, we obtain

$$-c_{\rm LO}(s,d)\leqslant c_{\rm UEG}(s,d)\leqslant\zeta_{\mathscr{L}}(s),\qquad \textit{for all}\ \max(0,d-4)<s<d.$$

For the BCC lattice in dimension $d=3$, one finds the claimed lower bound

$$c_{\rm LO}(1,3)\geqslant-\zeta_{\rm BCC}(1)\simeq 1.4442.$$

It is reasonable to conjecture that

$$e_{\rm UEG}(s,d)=\min_{\mathscr{L}}\zeta_{\mathscr{L}}(s)$$

for $d=1,2,3$ and all $0<s<d$, which amounts to saying that the uniform electron gas is always crystallized at zero temperature [10, 104]. The modified trial state (3.81) suggests that the system can only be a solid in the bulk. In a neighborhood of the boundary, it is probably a fluid because the particles have to be able to move sufficiently far away around the crystal to compensate the large charge fluctuations.

In this section we have explained some major difficulties encountered when trying to construct good trial states for the Uniform Electron Gas. Those are entirely due to the boundary, namely to the fact that we work with a finite piece of material in the physical space $\mathbb{R}^d$. If we set up the model on the torus, as is often done in practical calculations, these difficulties disappear [104, Sec. IV.C].

It is worth mentioning an estimate due to Lieb and Narnhofer [119] in the Coulomb case $s=1$ in dimension $d=3$ which states that

$$F_{\rm SCE}[\mathbb{1}_\Omega]\geqslant\frac{1}{2}\iint_{\Omega\times\Omega}\frac{d\mathbf{r}\,d\mathbf{r}'}{|\mathbf{r}-\mathbf{r}'|}-\frac{3}{5}\left(\frac{9\pi}{2}\right)^{\frac{1}{3}}|\Omega| \tag{3.83}$$

for *any* open set Ω of integer volume. The constant $(3/5)(9\pi/2)^{1/3}\simeq 1.4508$ is surprisingly close to the expected optimal value $-\zeta_{\rm BCC}(1)$ and it implies in any case that

$$\boxed{-1.4508\leqslant c_{\rm UEG}(1,3)\leqslant-1.4442.}$$

This is the bound that appeared in Theorem 3.7. For negatively-correlated states as in (3.69), the constant can be replaced by $(3/2)(\pi/6)^{1/3} \simeq 1.2090$ [109].

In dimension $d = 3$ for $s = 1$ it was conjectured in [135, 148] that the classical Uniform Electron Gas gives the optimal Lieb–Oxford constant, i.e., $c_{\rm LO}(1,3) = -c_{\rm UEG}(1,3)$.

Remark 3.23 (Determinantal Processes and the Dirac Constant) Let $\rho = \rho_0 \mathbb{1}_{C_L}$ with $N = \rho_0 L^d \in \mathbb{N}$. Instead of the floating crystal (3.71) and its modified version (3.81), one can consider the square of a Slater determinant as a trial state:

$$\mathbb{P}(\mathbf{r}_1, \ldots, \mathbf{r}_N) = \frac{1}{\sqrt{N!}} L^{-Nd} |\det \left(\mathrm{e}^{i\frac{2\pi}{L}\mathbf{k}_i \cdot \mathbf{r}_j} \mathbb{1}_{C_L}(\mathbf{r}_j) \right) |^2,$$

where $\mathbf{k}_1, \ldots, \mathbf{k}_N$ are N distinct points in $\mathbb{Z}^d$ (our trial state contains no spin). We find

$$\begin{aligned} F_{\rm SCE}[\rho_0 \mathbb{1}_{C_L}] \leqslant \frac{\rho_0^2}{2} \iint_{(C_L)^2} w(\mathbf{r} - \mathbf{r}')\, \mathrm{d}\mathbf{r}\, \mathrm{d}\mathbf{r}' \\ - \frac{1}{2} \iint_{(C_L)^2} w(\mathbf{r} - \mathbf{r}') \left| L^{-d} \sum_{j=1}^N \mathrm{e}^{i\frac{2\pi}{L}\mathbf{k}_j \cdot (\mathbf{r} - \mathbf{r}')} \right|^2 \mathrm{d}\mathbf{r}\, \mathrm{d}\mathbf{r}'. \end{aligned}$$

As for the free Fermi gas we choose all the points $\mathbf{k}_i$ in a given ball centered at the origin. For $w(\mathbf{r}) = |\mathbf{r}|^{-s}$ the second term behaves like $L^d \rho_0^{1+s/d} c_{\rm D}(s,d)$ where the *Dirac constant* [44]

$$c_{\rm D}(s,d) = \frac{1}{2(2\pi)^d} \int_{\mathbb{R}^d} \frac{|\widehat{\mathbb{1}_{B_{k_{\rm F}}}}(\mathbf{r})|^2}{|\mathbf{r}|^s}\, \mathrm{d}\mathbf{r}$$

is the exchange energy per unit volume of a free Fermi gas. Here the Fermi radius is $k_{\rm F} = \sqrt{2(d+2) c_{\rm TF}(d)/d}$. This proves that

$$-c_{\rm LO}(s,d) \leqslant c_{\rm UEG}(s,d) \leqslant -c_{\rm D}(s,d)$$

but this bound is worse than the floating crystal. The particles are not correlated enough. In dimension $d = 3$ with $s = 1$ one finds $c_{\rm D}(1,3) = (3/4)\,(6/\pi)^{1/3} \simeq 0.9305$. Recall that for an arbitrary determinantal point process, we have the better Lieb-Oxford inequality (3.68) from [109].

3.5.5 Local Density Approximation for the Classical Interaction Energy

We have discussed in the previous section the case of exactly constant densities and their limit of infinite volume. We now consider the case of slowly varying densities, which are assumed to be essentially constant over large sets.

First, we mention that the Lieb–Narnhofer bound (3.83) from [119] was generalized to arbitrary densities, in the form of lower bounds involving gradient-type corrections [7, 105]. For instance, the bound

$$F_{\mathrm{GSCE}}[\rho] \geqslant \frac{1}{2}\iint_{\mathbb{R}^6} \frac{\rho(\mathbf{r})\rho(\mathbf{r}')}{|\mathbf{r}-\mathbf{r}'|}\,\mathrm{d}\mathbf{r}\,\mathrm{d}\mathbf{r}' - \left(\frac{3}{5}\left(\frac{9\pi}{2}\right)^{\frac{1}{3}} + \varepsilon\right)\int_{\mathbb{R}^3} \rho(\mathbf{r})^{\frac{4}{3}}\,\mathrm{d}\mathbf{r} - \frac{0.001206}{\varepsilon^3}\int_{\mathbb{R}^3} |\nabla\rho(\mathbf{r})|\,\mathrm{d}\mathbf{r} \tag{3.84}$$

was shown to hold in [105] for any $\varepsilon > 0$.

The following result gives a quantitative estimate on the grand-canonical classical energy for slowly varying densities.

Theorem 3.24 (Local Density Approximation of the Classical Coulomb Energy [108]) *Consider the case $w(\mathbf{r}) = |\mathbf{r}|^{-1}$ in dimension $d = 3$. There exists a constant C such that*

$$\left|F_{\mathrm{GSCE}}(\rho) - \frac{1}{2}\iint_{\mathbb{R}^3\times\mathbb{R}^3} \frac{\rho(\mathbf{r})\rho(\mathbf{r}')}{|\mathbf{r}-\mathbf{r}'|}\,\mathrm{d}\mathbf{r}\,\mathrm{d}\mathbf{r}' - c_{\mathrm{UEG}}(1,3)\int_{\mathbb{R}^3} \rho(\mathbf{r})^{\frac{4}{3}}\,\mathrm{d}\mathbf{r}\right| \leqslant \varepsilon\int_{\mathbb{R}^3}\left(\rho(\mathbf{r}) + \rho(\mathbf{r})^{\frac{4}{3}}\right)\mathrm{d}\mathbf{r} + \frac{C}{\varepsilon^7}\int_{\mathbb{R}^3} |\nabla\rho^{\frac{1}{3}}(\mathbf{r})|^4\,\mathrm{d}\mathbf{r} \tag{3.85}$$

for every $\varepsilon > 0$ and every non-negative density $\rho \in L^1(\mathbb{R}^3) \cap L^{4/3}(\mathbb{R}^3)$ such that $\nabla\rho^{1/3} \in L^4(\mathbb{R}^3)$.

The gradient term can be replaced by $\varepsilon^{-b}\int_{\mathbb{R}^3} |\nabla\rho^\theta(\mathbf{r})|^p\,\mathrm{d}\mathbf{r}$ for any $p > 3$ and $0 < \theta < 1$ such that $\theta p \geqslant 4/3$, with $b = \max\{2p-1, (1+3\theta)p-4\}$.

If we fix a density ρ with $\int_{\mathbb{R}^3}\rho = 1$ and take $\rho_N(\mathbf{r}) = \rho(\mathbf{r}N^{-1/3})$, then we find that

$$F_{\mathrm{GSCE}}\big[\rho(\cdot N^{-1/3})\big] = \frac{N^{\frac{5}{3}}}{2}\iint_{\mathbb{R}^3\times\mathbb{R}^3} \frac{\rho(\mathbf{r})\rho(\mathbf{r}')}{|\mathbf{r}-\mathbf{r}'|}\,\mathrm{d}\mathbf{r}\,\mathrm{d}\mathbf{r}' + c_{\mathrm{UEG}}(1,3)N\int_{\mathbb{R}^3} \rho(\mathbf{r})^{\frac{4}{3}}\,\mathrm{d}\mathbf{r} + O\left(N^{\frac{5}{6}}\right). \tag{3.86}$$

Compare this expansion with the quantum case (3.30).

It is an open problem to prove an estimate similar to (3.85) for the canonical SCE functional $F_{\rm SCE}$. However, a non-quantitative convergence similar to (3.86) is known even for $w(\mathbf{r}) = |\mathbf{r}|^{-s}$ in all dimensions $d > s > 0$:

$$F_{\rm SCE}\big[\rho(\cdot N^{-1/d})\big] = \frac{N^{2-\frac{s}{d}}}{2}\iint_{\mathbb{R}^d\times\mathbb{R}^d}\frac{\rho(\mathbf{r})\rho(\mathbf{r}')}{|\mathbf{r}-\mathbf{r}'|^s}\,\mathrm{d}\mathbf{r}\,\mathrm{d}\mathbf{r}' \\ + c_{\rm UEG}(s,d)N\int_{\mathbb{R}^3}\rho(\mathbf{r})^{1+\frac{s}{d}}\,\mathrm{d}\mathbf{r} + o(N). \tag{3.87}$$

In the Coulomb case for $d = 3$, (3.87) was proved in [106] whereas general power-law potentials were covered in [37].

3.6 Upper and Lower Bounds on the Levy–Lieb Functionals

In the previous sections we have studied the kinetic and interaction energies separately and reviewed several known upper and lower bounds. Since the minimum of a sum is always greater than or equal to the sum of the minima, we easily obtain lower bounds on the full Levy–Lieb functional. For instance, putting together the Lieb–Thirring and Lieb–Oxford inequalities, we find

$$F_{\rm GC}[\rho] \geqslant c_{\rm LT}(d)q^{-\frac{2}{d}}\int_{\mathbb{R}^d}\rho(\mathbf{r})^{1+\frac{2}{d}}\,\mathrm{d}\mathbf{r} + \frac{1}{2}\iint_{\mathbb{R}^d\times\mathbb{R}^d}\frac{\rho(\mathbf{r})\rho(\mathbf{r}')}{|\mathbf{r}-\mathbf{r}'|^s}\,\mathrm{d}\mathbf{r}\,\mathrm{d}\mathbf{r}' \\ - c_{\rm LO}(s,d)\int_{\mathbb{R}^3}\rho(\mathbf{r})^{1+\frac{s}{d}}\,\mathrm{d}\mathbf{r},$$

for the interaction $w(\mathbf{r}) = |\mathbf{r}|^{-s}$ in dimension $d > s > 0$. The right side takes the same form as the *Thomas–Fermi–Dirac functional*, except for the values of the two constants in front of the terms $\rho^{1+2/d}$ and $\rho^{1+s/d}$. One would obtain the right constant in front of the term $\rho^{1+2/d}$ in dimension $d \geqslant 3$ if the Lieb–Thirring conjecture mentioned in Sect. 3.4.2 had been proved. We refer to [114] for a review of results on Thomas–Fermi-type functionals. From Nam's bound (3.40) one can replace the Lieb–Thirring constant by $(1-\varepsilon)c_{\rm TF}(d)$ at the expense of a (negative) gradient correction.

Upper bounds are more complicated because a trial state that works for $T[\rho]$ could be very bad for the classical energy and conversely. We have seen in Sect. 3.4 that the set of one-particle density matrices that are N-representable by a mixed states is exactly given by the operators $\gamma = \gamma^*$ such that $0 \leqslant \gamma \leqslant 1$ and $\mathrm{Tr}\,(\gamma) = N$. This is because any such operator is the convex combination of rank-N projections which correspond to Slater determinants. But for estimating the interaction energy we need some more information on the two-particle density.

An explicit convex combination which provided a bound on the two-particle density matrix was derived in [115]. The idea is the following. Assume for simplicity that $\gamma = \sum_{i=1}^{K} n_i |u_i\rangle\langle u_i|$ has finite rank K. By Horn's lemma [115] there exists a set of N orthonormal vectors $V^1, \ldots, V^N$ in $\mathbb{C}^K$ such that $\sum_{k=1}^{N} |V_i^k|^2 = n_i$ for all i. Define then the new orbitals $f_\theta^k = \sum_{j=1}^{K} e^{i\theta_j} V_j^k u_j$ where $\theta = (\theta_1, \ldots, \theta_K) \in (0, 2\pi)^K$ and the associated trial mixed state

$$\Gamma = \frac{1}{(2\pi)^K} \int_0^{2\pi} d\theta_1 \cdots \int_0^{2\pi} d\theta_K \, |f_\theta^1 \wedge \cdots \wedge f_\theta^N\rangle\langle f_\theta^1 \wedge \cdots \wedge f_\theta^N|.$$

A computation shows that its one-particle density matrix is exactly γ, whereas its two-particle density matrix is

$$\Gamma^{(2)} = \mathscr{A}(\gamma \otimes \gamma)\mathscr{A} - \sum_{1 \leqslant j < k \leqslant K} \left| \sum_{i=1}^{N} V_j^i \overline{V_k^i} \right|^2 |u_j \wedge u_k\rangle\langle u_j \wedge u_k|, \tag{3.88}$$

where $\mathscr{A}$ is the orthogonal projection onto the anti-symmetric two-particle subspace $L^2(\mathbb{R}^d \times \mathbb{Z}_q, \mathbb{C}) \wedge L^2(\mathbb{R}^d \times \mathbb{Z}_q, \mathbb{C})$. The first operator has the integral kernel

$$\mathscr{A}(\gamma \otimes \gamma)\mathscr{A}(\mathbf{x}_1, \mathbf{x}_2; \mathbf{y}_1, \mathbf{y}_2) = \gamma(\mathbf{x}_1, \mathbf{y}_1)\gamma(\mathbf{x}_2, \mathbf{y}_2) - \gamma(\mathbf{x}_1, \mathbf{x}_2)\gamma(\mathbf{y}_1, \mathbf{y}_2). \tag{3.89}$$

The operator $\mathscr{A}(\gamma \otimes \gamma)\mathscr{A}$ is exactly the two-particle density matrix of the unique quasi-free state over the Fock space that has the one-particle density γ [6]. Integrating against the potential w, the first term in (3.89) gives the classical energy (Hartree term) whereas the second gives the exchange term, which is non-positive for $w \geqslant 0$.

Since the last term in (3.88) is a non-positive operator, the following was obtained in [115] after using the density of finite rank operators.

Lemma 3.25 (Mixed Canonical States and Quasi-Free States [115]) *Let* $0 \leqslant \gamma = \gamma^* \leqslant 1$ *be a one-particle density matrix such that* $\mathrm{Tr}\,(\gamma) = N \in \mathbb{N}$. *Then there exists a mixed state* Γ *over the fermionic* N*-particle space* $\bigwedge_1^N L^2(\mathbb{R}^d \times \mathbb{Z}_q, \mathbb{C})$ *such that its one-particle density matrix is* γ *and its two-particle density matrix* $\Gamma^{(2)}$ *satisfies*

$$\Gamma^{(2)} \leqslant \mathscr{A}(\gamma \otimes \gamma)\mathscr{A} \tag{3.90}$$

in the operator sense.

Using the lemma for the trial state (3.48) employed in the proof of Theorem 3.14 and neglecting the exchange term, the following was derived in [108].

Theorem 3.26 (Upper Bound on $F_{\rm L}[\rho]$ [108]) *For $w(\mathbf{r}) = |\mathbf{r}|^{-s}$ in dimension $d > s > 0$, we have*

$$F_{\rm L}[\rho] \leqslant c_{\rm TF}(d)(1+\varepsilon)q^{-\frac{2}{d}} \int_{\mathbb{R}^d} \rho(\mathbf{r})^{1+\frac{2}{d}}\, d\mathbf{r} + \kappa'(d)\frac{1+\varepsilon}{\varepsilon}\int_{\mathbb{R}^d} |\nabla\sqrt{\rho}(\mathbf{r})|^2\, d\mathbf{r} \\ + \frac{1}{2}\iint_{\mathbb{R}^d\times\mathbb{R}^d} \frac{\rho(\mathbf{r})\rho(\mathbf{r}')}{|\mathbf{r}-\mathbf{r}'|^s}\, d\mathbf{r}\, d\mathbf{r}' \qquad (3.91)$$

for any $\varepsilon > 0$.

This time, the right side of (3.91) involves an energy functional that looks like the Thomas–Fermi–von Weizäcker energy [114]. It is an open problem to derive a similar upper bound for $F_{\rm LL}$.

3.7 The Hohenberg–Kohn Theorem

In this chapter we have mainly discussed the convex formulation of Density Functional Theory [80, 116] based on the universal functionals of Levy and Lieb. Another important result is the *Hohenberg-Kohn theorem* [82] which, in spite of its rather abstract character, is often cited as the main justification for the use of the density to replace the N-particle wavefunction. As we will explain, the necessary assumptions for the validity of this theorem are not yet fully understood mathematically. In fact, this result relies on the *unique continuation principle* which is not completely settled for N-particle Hamiltonians. Before stating the Hohenberg-Kohn theorem, we therefore start by discussing unique continuation in detail.

For simplicity we assume throughout the whole section that there is no spin:

$$q = 1.$$

Adding q does not change anything in the following results but it makes the notation a bit heavier.

3.7.1 Many-Body Unique Continuation

We refer for instance to [95] for a discussion on the importance of the unique continuation for the Hohenberg–Kohn theorem. Because unique continuation is a purely local property we allow here external potentials v whose positive part $v_+ = \max(v, 0)$ is only locally integrable. For simplicity, we assume that its negative part $v_- = \max(-v, 0)$ and the interaction potential w are infinitesimally $(-\Delta)$-form-bounded, as was done in the body of the chapter.

Definition 3.27 (Many-Body Unique Continuation) Let

$$v_+ \in L^1_{\rm loc}(\mathbb{R}^d, \mathbb{R}), \qquad v_-, w \in L^p(\mathbb{R}^d, \mathbb{R}) + L^\infty(\mathbb{R}^d)$$

with $v_\pm \geqslant 0$ and p satisfying (3.1). We say that the potentials $v = v_+ - v_-$ and w satisfy the *many-body unique continuation principle* if, for every integer $N \geqslant 1$, (the Friedrichs realization of) $H_N^{v,w}$ satisfies the unique continuation principle in its form domain: if we have $H_N^{v,w}\Psi = \lambda\Psi$ for some $\lambda \in \mathbb{R}$ and $\Psi \in Q(H_N^{v,w})$ with $|\{\Psi = 0\}| > 0$, then $\Psi \equiv 0$.

The equation $H_N^{v,w}\Psi = 0$ is understood in $Q(H_N^{v,w})'$, that is,

$$\frac{1}{2}\int_{\mathbb{R}^{dN}} \nabla\Phi(\mathbf{X})^* \cdot \nabla\Psi(\mathbf{X})\, {\rm d}\mathbf{X} + \int_{\mathbb{R}^{dN}} W_N^{v,w}(\mathbf{X})\Phi(\mathbf{X})^*\Psi(\mathbf{X})\, {\rm d}\mathbf{X} = 0$$

for every $\Phi \in Q(H_N^{v,w})$ or, equivalently, in the sense of distributions. Recall that the full N-body potential is defined by

$$W_N^{v,w}(\mathbf{r}_1, \dots, \mathbf{r}_N) := \sum_{j=1}^N v(\mathbf{r}_j) + \sum_{1\leqslant j<k\leqslant N} w(\mathbf{r}_j - \mathbf{r}_k).$$

Our formulation of unique continuation is one of the strongest possible, in that it only requires Ψ to vanish on a set of positive measure in order to deduce that $\Psi \equiv 0$. This is the property which is needed in the proof of the Hohenberg–Kohn theorem, as we will see. This is sometimes called "unique continuation on sets of positive measures". For $v_+ \in L^p_{\rm loc}(\mathbb{R}^d, \mathbb{R}^+)$ with p as in (3.1) it is shown in [38] that any $\Psi \in H^1_{\rm loc}(\mathbb{R}^{dN})$ vanishing on a set of positive measure and solving $H_N^{v,w}\Psi = \lambda\Psi$ must have a point $\mathbf{X}_0 \in \mathbb{R}^{dN}$ where it vanishes to infinite order, that is, such that

$$\forall\alpha > 0, \qquad \int_{|\mathbf{X}-\mathbf{X}_0|\leqslant\alpha} |\Psi(\mathbf{X})|^2\, {\rm d}\mathbf{X} = O(r^\alpha).$$

Unique continuation for functions vanishing to infinite order at one point is usually called "strong unique continuation". Many authors consider instead the "weak unique continuation" problem where Ψ is instead assumed to vanish on an open set, but this is not sufficient for the Hohenberg–Kohn theorem.

Unique continuation is a very well studied question. Note first that if we restrict our attention to the potentials for N electrons in a molecule, where

$$v(\mathbf{r}) = -\sum_{m=1}^M \frac{z_m}{|\mathbf{r} - \mathbf{R}_m|}, \qquad w(\mathbf{r} - \mathbf{r}') = \frac{1}{|\mathbf{r} - \mathbf{r}'|},$$

then any eigenfunction of $H_N^{v,w}$ is analytic outside of the singularities of the potential [132], which form a set of zero measure. Therefore it satisfies the unique continuation principle. However, restricting the theory to this very special, though physically relevant, class of potentials is not appropriate in density functional theory. In order to fully understand the density, it is necessary to allow the largest possible class of potentials.

In a famous work [87], Jerison and Kenig have proved that the (strong) unique continuation principle holds for $-\Delta+W$ in $\mathbb{R}^D$ under the sole assumption that $W \in L^p_{\rm loc}(\mathbb{R}^D)$ with p satisfying (3.1) and d replaced by D. This was then generalized by Koch and Tataru in [92] and many other authors. These results apply to the N-particle setting under the condition that

$$W_N^{v,w} \in L^{\frac{dN}{2}}_{\rm loc}(\mathbb{R}^{dN})$$

and this is valid for all $N \geqslant 2$ when

$$v, w \in L^p_{\rm loc}(\mathbb{R}^d) \text{ for all } 1 \leqslant p < \infty.$$

This is not far from asking that the potentials are locally bounded (in which case the result would follow for instance from the singular Carleman-type estimate proved in [151]). We see that L^p conditions are not well adapted to the N-particle problem, since they yield N-dependent constraints on v and w. More natural assumptions on v and w involve relative bounds with respect to the Laplacian, since such properties are easily propagated to all N. For instance if v and w are infinitesimally $(-\Delta)$-form bounded in $\mathbb{R}^d$,

$$\forall \varepsilon > 0, \qquad |v| + |w| \leqslant \varepsilon(-\Delta) + C_\varepsilon, \tag{3.92}$$

then so is the N-particle potential $W_N^{v,w}$ in $\mathbb{R}^{dN}$ for every N. It seems reasonable to conjecture that the many-body unique continuation principle holds under the sole assumption (3.92), but this is not known, even for $N = 1$. See [165] for a similar conjecture in the Kato class, which involves L^1 norms instead of L^2 norms.

Georgescu [67] and Schechter–Simon [158] have provided one of the first results for N-body systems with N-independent assumptions on v and w, but they required the wavefunction to vanish on an open set (weak unique continuation). Recently [63, 64], Garrigue has extended their result to cover the case of functions vanishing on a set of positive measure. His main assumption is that

$$|v_+|^2 \mathbb{1}_{B_R} + |v_-|^2 + |w|^2 \leqslant \varepsilon(-\Delta)^{\frac{3}{2}-\delta} + C_{\delta,\varepsilon,R}$$

for some $\delta > 0$ and all $\varepsilon, R > 0$. This condition in the one-body space $\mathbb{R}^d$ is inherited by $W_N^{v,w}$ in $\mathbb{R}^{dN}$ for every $N \geqslant 1$. After using the Sobolev inequality, the following result was shown in [64].

Theorem 3.28 (Unique Continuation for L^p Potentials [64]) *Any potentials v, w with $v_+ \in L^p_{\rm loc}(\mathbb{R}^d)$ and $v_-, w \in L^p(\mathbb{R}^d)+L^\infty(\mathbb{R}^d)$ with $p > \max(2, 2d/3)$ satisfy the many-body unique continuation property of Definition 3.27.*

Theorem 3.28 now covers Coulomb-type potentials. This is the best result known at the moment for many-body unique continuation. It is an important problem to generalize it to more singular potentials v.

3.7.2 Main Theorem and Some Open Problems

Let us now state the Hohenberg–Kohn theorem, which says that the density uniquely determines the potential, under the condition that unique continuation holds.

Theorem 3.29 (Hohenberg–Kohn) *Let $(v_1)_-, (v_2)_-, w \in L^p(\mathbb{R}^d)+L^\infty(\mathbb{R}^d)$ and $(v_1)_+, (v_2)_+ \in L^1_{\rm loc}(\mathbb{R}^d)$ with p as in (3.1) and assume that (v_1, w) or (v_2, w) satisfies the many-body unique continuation property of Definition 3.27. If there are two ground states Ψ_1 and Ψ_2 of, respectively, $H_N^{v_1,w}$ and $H_N^{v_2,w}$ so that $\rho_{\Psi_1} = \rho_{\Psi_2}$, then we have $v_1 = v_2 + C$ for some constant C.*

The following proof is essentially the one given in [63, 116].

Proof Changing v_2 into $v_2 - (E_N[v_2] - E_N[v_1])/N$ we can assume that the two ground state energies are equal. Note that the assumption $\rho_{\Psi_1} = \rho_{\Psi_2}$ implies that $\Psi_1 \in \mathcal{Q}(H_N^{v_2,w})$ and $\Psi_2 \in \mathcal{Q}(H_N^{v_1,w})$, the form domains of the two operators defined in (3.3). We can write

$$\begin{aligned}\left\langle \Psi_1, H_N^{v_1,w}\Psi_1\right\rangle &= \left\langle \Psi_1, H_N^{v_2,w}\Psi_1\right\rangle + \int_{\mathbb{R}^d} \rho_{\Psi_1}(\mathbf{r})\Big(v_1(\mathbf{r}) - v_2(\mathbf{r})\Big)\,\mathbf{dr}\\ &\geqslant \left\langle \Psi_2, H_N^{v_2,w}\Psi_2\right\rangle + \int_{\mathbb{R}^d} \rho_{\Psi_1}(\mathbf{r})\Big(v_1(\mathbf{r}) - v_2(\mathbf{r})\Big)\,\mathbf{dr}.\end{aligned}$$

Exchanging the two indices and using that the two densities are equal, we obtain that there is equality everywhere. In particular, Ψ_1 is a ground state for $H_N^{v_2,w}$, hence belongs to its operator domain and solves the equation

$$\left(H_N^{v_1,w} - H_N^{v_2,w}\right)\Psi_1 = \sum_{j=1}^N \big(v_1(\mathbf{r}_j) - v_2(\mathbf{r}_j)\big)\Psi_1 = 0$$

in the sense of distributions hence also almost everywhere. Due to the unique continuation principle (for either v_1 or v_2), we know that $|\{\Psi_1 = 0\}| = 0$ hence this implies that

$$\sum_{j=1}^N \big(v_1(\mathbf{r}_j) - v_2(\mathbf{r}_j)\big) = 0$$

for almost every $\mathbf{r}_1, \ldots, \mathbf{r}_N \in \mathbb{R}^d$. Integrating against $f^{\otimes N}$ with $f \in C_c^\infty(\mathbb{R}^d)$, we deduce that

$$N \int_{\mathbb{R}^d} \big(v_1(\mathbf{r}) - v_2(\mathbf{r})\big) f(\mathbf{r})\,\mathrm{d}\mathbf{r} = 0.$$

We obtain $v_1 = v_2$ a.e., as we wanted. □

Remark 3.30 (Mixed States) There is a similar Hohenberg–Kohn theorem for mixed states. That is, if we have two N-particle mixed states Γ_1 and Γ_2 supported on the ground state eigenspaces of $H_N^{v_1,w}$ and $H_N^{v_2,w}$ respectively, such that $\rho_{\Gamma_1} = \rho_{\Gamma_2}$, then $v_1 = v_2 + C$. The proof is similar.

From Theorem 3.28, we know that the Hohenberg–Kohn theorem holds when $p > \max(2, 2d/3)$. We now discuss some consequences of this result.

Let us consider a fixed interaction $w \in L^p(\mathbb{R}^d) + L^\infty(\mathbb{R}^d)$ (in DFT w is usually the Coulomb potential in dimension $d = 3$). We introduce the set of *v-representable densities*

$$\mathcal{R}_w := \Big\{\rho_\Psi \ :\ \Psi \text{ ground state of } H_N^{v,w} \text{ for some } (v, w) \\ \text{satisfying the many-body unique continuation in Definition 3.27}\Big\}.$$

The Hohenberg–Kohn theorem states that any $\rho \in \mathcal{R}_w$ arises from a *unique* potential v, up to a constant. We remark that the set $\mathcal{R}_w$ might be quite small. In fact, all the densities $\rho \in \mathcal{R}_w$ are positive in the following sense. If we had $|\{\rho = 0\}| > 0$ then the associated Ψ would vanish on a set of infinite measure, which contradicts the unique continuation principle.[2] It is an interesting question [65, 66] to determine how small $\mathcal{R}_w$ is.

That the set $\mathcal{R}_w$ might be quite small is not really a problem in density functional theory, since the sought-after ground state density of course always belongs to $\mathcal{R}_w$. However, the smallness of $\mathcal{R}_w$ creates some technical difficulties. For instance if we have $\rho \in \mathcal{R}_w$ with associated potential v then one cannot use the implicit function theorem to determine a potential for other densities in the neighborhood of ρ.

In DFT, one important question is to understand the dependence of $\mathcal{R}_w$ on the interaction potential w. In fact, in Kohn–Sham theory the goal is to replace the many-body problem with interaction w by a non-interacting eigenvalue problem. If we have $\rho \in \mathcal{R}_w \cap \mathcal{R}_0$ with $\rho = \rho_\Psi$ and $H_N^{v,w}\Psi = 0$, then we conclude that there exists a unique potential $v_{\rm KS}$, called the *Kohn–Sham potential*, and a normalized

[2] To include densities vanishing on a set of positive measure we have to allow v_+ to be infinite on such sets and rephrase the unique continuation principle accordingly.

ground state Ψ' such that $\rho_{\Psi'} = \rho$ and

$$H^{v+v_{\rm KS},0}\Psi' = \sum_{j=1}^{N}\left(-\frac{\Delta_{\mathbf{r}_j}}{2} + v(\mathbf{r}_j) + v_{\rm KS}(\mathbf{r}_j)\right)\Psi' = 0.$$

The spectrum of $H^{v+v_{\rm KS},0}$ is determined in terms of the one-particle Kohn–Sham operator $h_{\rm KS} = -\Delta/2 + v + v_{\rm KS}$. When its eigenvalues satisfy

$$\lambda_N(h_{\rm KS}) < \lambda_{N+1}(h_{\rm KS})$$

then Ψ' is unique up to a phase and equal to the Slater determinant

$$\Psi' = (N!)^{-1/2}\det(\varphi_i(\mathbf{x}_j)),$$

where the orbitals $\varphi_1, \ldots, \varphi_N$ are the N first eigenfunctions of $h_{\rm KS}$. Hence the interacting problem has been mapped onto a non-interacting problem. For this reason, it is desirable that $\mathcal{R}_w \cap \mathcal{R}_0$ is as large as possible, perhaps even equal to the whole set $\mathcal{R}_w$. This question does not seem to have been studied in detail. Understanding the Coulomb case is the main goal of the Kohn–Sham formulation of DFT. See [66] for numerical results in this direction.

In the original approach of [82, 93], the Hohenberg–Kohn theorem is used to define universal functionals. Namely, for every $\rho \in \mathcal{R}_w$, we know that there exists a unique potential v (up to constants) and an N-particle wavefunction Ψ such that $\rho = \rho_\Psi$ and Ψ is a ground state for $E_N[v]$. Should Ψ be non-degenerate, this defines a map $\rho \mapsto \Psi[\rho]$ and therefore one can define the universal energy functional by $\rho \mapsto \langle \Psi[\rho], H_N^{0,w}\Psi[\rho]\rangle$, and similarly for the kinetic and interaction energies. This approach is not satisfactory from a mathematical point of view, however, since the set $\mathcal{R}_w$ is essentially unknown.

3.8 Proof of Theorem 3.5

Note that since $w \geqslant 0$, we always have $F_{\rm GC}[\rho] \geqslant 0$. In particular, there is nothing to prove when $\rho = 0$. In addition, the Lieb–Thirring inequality (3.35) or the Hoffmann-Ostenhof inequality (3.4) imply that $F_{\rm GC}[\rho] > 0$ for $\rho \neq 0$.

For the proof we need to introduce the k-particle density matrices [102]

$$\Gamma^{(k)} = \sum_{n \geqslant k} \frac{n!}{(n-k)!}{\rm Tr}_{k+1\to n}[\Gamma_n]$$

of a grand-canonical state $\Gamma = (\Gamma_n)_{n\geqslant 0}$, where $\mathrm{Tr}_{k+1\to n}$ means the partial trace in the $n-k+1$ last variables. The energy can be expressed in terms of $\Gamma^{(1)}$ and $\Gamma^{(2)}$ only, as follows

$$\sum_{n\geqslant 1} \mathrm{Tr}\,(H_n^{0,w}\Gamma_n) = \frac{1}{2}\mathrm{Tr}\left((-\Delta)\Gamma^{(1)}\right) + \frac{1}{2}\mathrm{Tr}\left(w_{12}\Gamma^{(2)}\right).$$

Here $w_{12} \geqslant 0$ denotes the multiplication operator by $w(\mathbf{r}_1 - \mathbf{r}_2)$ on the two-particle space $L^2(\mathbb{R}^d \times \mathbb{Z}_q, \mathbb{C}) \wedge L^2(\mathbb{R}^d \times \mathbb{Z}_q, \mathbb{C})$.

We will also use the concept of localized states [102]. For a function $0 \leqslant \chi \leqslant 1$ on $\mathbb{R}^d$, the localized state $\Gamma_{|\chi}$ of a grand-canonical state Γ is characterized by the property that its density matrices are equal to $\Gamma^{(k)}_{|\chi} = \chi^{\otimes k}\Gamma^{(k)}\chi^{\otimes k}$ for all k.

Proof of (i) Let $\Gamma_j = (\Gamma_{j,n})_{n\geqslant 0}$ be a grand-canonical minimizer for $F_{\mathrm{GC}}[\rho_j]$, whose existence is guaranteed by Theorem 3.4. After extraction of a subsequence we may assume that $F_{\mathrm{GC}}[\rho_j]$ converges to a finite limit (if the limit is $+\infty$ there is nothing to show).

Since the second term is non-negative, the kinetic energy $\mathrm{Tr}\,(-\Delta)\Gamma_j^{(1)}$ must be uniformly bounded, hence $\sqrt{-\Delta}\Gamma_j^{(1)}\sqrt{-\Delta}$ is bounded in the trace-class. For fermions, $\Gamma_j^{(1)}$ is in addition bounded in operator norm by 1. Note however that we have no a priori bound on the number of particles $\mathrm{Tr}\,(\Gamma_j^{(1)}) = \int_{\mathbb{R}^d}\rho_j$, which could diverge. The idea is to use the *local* trace class topology instead.

By the Hoffmann-Ostenhof and Lieb–Thirring inequalities (3.4) and (3.35), $\sqrt{\rho_j}$ is bounded in $\dot{H}^1(\mathbb{R}^d) \cap L^{2+4/d}(\mathbb{R}^d)$. In particular ρ_j is bounded in L^1 on any finite ball, uniformly with respect to the center of the ball. This means that $\Gamma_j^{(1)}$ is locally uniformly bounded in the trace-class. Due to the kinetic energy bound, we can therefore assume, after extraction of a subsequence, that $\Gamma_j^{(1)}$ converges strongly locally in the trace class to some operator $\Gamma^{(1)}$ which is such that $\rho_{\Gamma^{(1)}} = \rho$, the weak limit of the sequence ρ_j. Since ρ is integrable by assumption, $\Gamma^{(1)}$ is indeed trace-class.

The argument is somewhat more complicated for $\Gamma_j^{(2)}$. Let $0 \leqslant \chi \leqslant 1$ be a function of compact support. Denoting the localized state by $\Gamma_{j|\chi}$, we know that $\Gamma^{(1)}_{j|\chi} = \chi\Gamma_j^{(1)}\chi$ is bounded in the trace class. By Yang's inequality [172, 173], we have $\|\Upsilon_n^{(2)}\| \leqslant Cn\mathrm{Tr}\,(\Upsilon_n)$ for any n-particle operator $\Upsilon_n = \Upsilon_n^* \geqslant 0$ and a universal constant C. This implies that

$$\left\|\chi^{\otimes 2}\Gamma_j^{(2)}\chi^{\otimes 2}\right\| = \left\|\Gamma^{(2)}_{j|\chi}\right\| \leqslant \sum_{n\geqslant 2}\left\|\Gamma^{(2)}_{j|\chi,n}\right\| \leqslant C\sum_{n\geqslant 2} n\,\mathrm{Tr}\,(\Gamma_{j|\chi,n}) \leqslant C\int_{\mathbb{R}^d}\chi^2\rho_j.$$

Hence $\Gamma_j^{(2)}$ is a locally bounded sequence of operators. No local bound on the trace is known, but this does not create any difficulty. Up to extraction of a subsequence, we can therefore assume that $\Gamma_j^{(2)} \rightharpoonup \Gamma^{(2)}$ weakly locally as operators. Since $w_{12} \geqslant 0$ by assumption, Fatou's lemma for operators now implies that

$$\liminf_{j\to\infty} F_{\rm GC}[\rho_j] \geqslant \frac{1}{2}\mathrm{Tr}\left((-\Delta)\Gamma^{(1)}\right) + \frac{1}{2}\mathrm{Tr}\left(w_{12}\Gamma^{(2)}\right). \tag{3.93}$$

We have thus shown that the energy is weakly lower semi-continuous when expressed in terms of the one and two-particle density matrices. Our next task is to go back to states in Fock space.

Following [102, Lemma 3], we know that there exists a state $\Gamma = (\Gamma_n)_{n\geqslant 0}$ on the Fock space which has the density matrices $\Gamma^{(1)}$ and $\Gamma^{(2)}$. The argument uses a different notion of weak convergence and goes as follows. First we extract a subsequence in the sense of weak-$*$ convergence over the local algebra of anti-commutation relations. The weak limit is in principle an abstract state over the local CAR but since its density is ρ, which is integrable over $\mathbb{R}^d$, the state is actually normal and arises from a grand-canonical state $\Gamma = (\Gamma_n)_{n\geqslant 0}$ [18, 19], which has the above density matrices, as we claimed. In particular, the right side of (3.93) is by definition $\geqslant F_{\rm GC}[\rho]$, which concludes the proof of (3.15). □

Proof of (ii) Let $\rho \geqslant 0$ be such that $\sqrt{\rho} \in H^1(\mathbb{R}^d)$ and let $\Gamma = (\Gamma_n)_{n\geqslant 0}$ be a grand-canonical state such that $\rho_\Gamma = \rho$ and

$$F_{\rm GC}[\rho] = \frac{1}{2}\mathrm{Tr}\left(-\Delta\Gamma^{(1)}\right) + \frac{1}{2}\mathrm{Tr}\left(w_{12}\Gamma^{(2)}\right).$$

By optimality, we must have

$$\mathrm{Tr}\left(H_n^{0,w}\Gamma_n\right) = \mathrm{Tr}\left(\Gamma_n\right) F_{\rm L}\left[\frac{\rho_{\Gamma_n}}{\mathrm{Tr}\left(\Gamma_n\right)}\right] \tag{3.94}$$

for all $n \geqslant 1$ such that $\Gamma_n \neq 0$ (otherwise we could decrease the energy by choosing another state, without changing the density). We split the rest of the argument into several steps. □

Step 1. Approximation by a State with $\rho_{\Gamma_{n_0}} \geqslant \alpha\rho$ In this first step we slightly modify Γ in order to guarantee that $\rho_{\Gamma_{n_0}} \geqslant \alpha\rho$ for some $\alpha > 0$ and some n_0, a property which will play a role in the next step. To this end we remark that there exists a state $\widetilde{\Gamma}$ with density $\rho_{\widetilde{\Gamma}} = \rho$ and which only lives over the N and $(N+1)$–particle subspaces, where N is the integer part of $\int_{\mathbb{R}^d}\rho$. If $\int_{\mathbb{R}^d}\rho \in \mathbb{N}$ we just take $\widetilde{\Gamma}$ to minimize $F_{\rm LL}[\rho]$. Otherwise, we can write $\int_{\mathbb{R}^d}\rho = N+\kappa$ with $\kappa \in (0,1)$ and we consider the state

$$\widetilde{\Gamma} = (1-\kappa)\Gamma_N + \kappa\Gamma_{N+1},$$

where Γ_N optimizes $F_{\rm LL}[N\rho/(N+\kappa)]$ and Γ_{N+1} optimizes $F_{\rm LL}[(N+1)\rho/(N+\kappa)]$. The total density is then

$$\rho_{\widetilde{\Gamma}} = \left((1-\kappa)\frac{N}{N+\kappa} + \kappa\frac{N+1}{N+\kappa}\right)\rho = \rho,$$

as desired. With the trial state $\widetilde{\Gamma}$ at hand, we consider the new state $\widetilde{\Gamma}_\varepsilon = (1-\varepsilon)\Gamma + \varepsilon\widetilde{\Gamma}$. This state has the exact density ρ and its energy converges to that of Γ when $\varepsilon \to 0$. This new state has the desired property that

$$\rho_{(\widetilde{\Gamma}_\varepsilon)_N} \geqslant \varepsilon\rho_{\widetilde{\Gamma}_N} = \frac{\varepsilon N(1-\kappa)}{N+\kappa}\rho$$

and an energy very close to that of Γ.

Without loss of generality, we can thus assume for the rest of the proof that we have a state Γ with the exact density $\rho_\Gamma = \rho$ and such that $\rho_{\Gamma_{n_0}} \geqslant \alpha\rho$ for some n_0 and $\alpha > 0$.

Step 2. Approximation by a State with Compact Support in n In case that Γ_n does not vanish for large n, we replace the state $\Gamma = (\Gamma_n)_{n\geqslant 0}$ by a new state $\Gamma' = (\Gamma'_n)$ so that $\Gamma'_n \equiv 0$ for n large enough, at the expense of a small error in the energy. Although it is possible to keep the exact density, we will here allow the density to vary a little.

First note that since $\rho \neq 0$, we have $F_{\rm GC}[\rho] > 0$ and therefore $\Gamma_0 < 1$ (otherwise the energy would vanish). This allows us to define Γ' by

$$\begin{cases} \Gamma'_0 = \Gamma_0 + \sum_{n\geqslant K+1} \operatorname{Tr}(\Gamma_n) & \text{for } n = 0, \\ \Gamma'_n = \Gamma_n & \text{for } 1 \leqslant n \leqslant K, \\ \Gamma'_n = 0 & \text{for } n \geqslant K+1. \end{cases}$$

In other words, we truncate the state and compensate the missing mass in the vacuum. The energy of Γ' is equal to $\sum_{n=1}^K \operatorname{Tr}(H_n^{0,w}\Gamma_n)$ and it converges to $F_{\rm GC}[\rho]$ as $K \to \infty$. Similarly, its density is equal to $\sum_{n=1}^K \rho_{\Gamma_n}$ and it converges to ρ in $L^1(\mathbb{R}^d)$. In addition, we have for the one-particle density matrix $\operatorname{Tr}(-\Delta)\gamma' \leqslant \operatorname{Tr}(-\Delta)\gamma$ which proves that

$$\int_{\mathbb{R}^d} |\nabla\sqrt{\rho_{\Gamma'}}|^2 \leqslant \sum_{n\geqslant 1}\int_{\mathbb{R}^d} |\nabla\sqrt{\rho_{\Gamma_n}}|^2 \leqslant \operatorname{Tr}(-\Delta)\gamma \tag{3.95}$$

by the Hoffmann-Ostenhof inequality (3.33). This gives the strong convergence of $\rho_{\Gamma'}$ to ρ in $L^p(\mathbb{R}^d)$ when $K \to \infty$ for $1 \leqslant p < p^*/2$ where p^* is the critical Sobolev exponent. Finally, we would like to prove the convergence

$$\nabla \sqrt{\sum_{n=1}^{K} \rho_{\Gamma_n}} \to \nabla \sqrt{\sum_{n=1}^{\infty} \rho_{\Gamma_n}} = \nabla \sqrt{\rho} \tag{3.96}$$

strongly in $L^2(\mathbb{R}^d)$, and this is where the first step helps. Indeed, we have the following lemma.

Lemma 3.31 *Let (ρ_j) be a sequence such that $\alpha\rho \leqslant \rho_j \leqslant \rho$ for some $\alpha > 0$ and $\rho_j(x) \to \rho(x)$ a.e., where $\sqrt{\rho} \in H^1(\mathbb{R}^d)$. If $\nabla\sqrt{\rho - \rho_j} \to 0$ strongly in $L^2(\mathbb{R}^d)$, then $\nabla\sqrt{\rho_j} \to \nabla\sqrt{\rho}$ strongly in $L^2(\mathbb{R}^d)$.*

Proof of of Lemma 3.31 We write $u_j = \sqrt{\rho_j}/\sqrt{\rho}$, which satisfies $\sqrt{\alpha} \leqslant u_j \leqslant 1$ by assumption and which converges almost everywhere to 1. Then the assumption means that

$$\nabla\sqrt{\rho - \rho_j} = \sqrt{1 - u_j^2}\,\nabla\sqrt{\rho} - \sqrt{\rho}\frac{u_j \nabla u_j}{\sqrt{1 - u_j^2}} \to 0$$

strongly in $L^2(\mathbb{R}^d)$. The first term goes to 0 by dominated convergence, hence we conclude that the second term tends to 0 in $L^2(\mathbb{R}^d)$. Then we have

$$\begin{aligned}\int_{\mathbb{R}^d} |\nabla\sqrt{\rho} - \nabla\sqrt{\rho_j}|^2 &= \int_{\mathbb{R}^d} \left|(1 - u_j)\nabla\sqrt{\rho} - \sqrt{\rho}\nabla u_j\right|^2 \\ &\leqslant 2\int_{\mathbb{R}^d} (1 - u_j)^2 |\nabla\sqrt{\rho}|^2 + 2\int_{\mathbb{R}^d} \rho |\nabla u_j|^2.\end{aligned}$$

The first term converges again to 0, whereas the second can be estimated by

$$\int_{\mathbb{R}^d} \rho|\nabla u_j|^2 \leqslant \frac{1-\alpha}{\alpha}\int_{\mathbb{R}^d} \frac{\rho u_j^2 |\nabla u_j|^2}{1 - u_j^2} \to 0,$$

and the lemma follows. □

In our case, the inequality (3.95) implies that $\sum_{n\geqslant 1}\int_{\mathbb{R}^d} |\nabla\sqrt{\rho_{\Gamma_n}}|^2$ is a convergent series. In addition, by the Cauchy–Schwarz inequality for series,

$$\int_{\mathbb{R}^d} \left|\nabla\sqrt{\sum_{n\geqslant K+1} \rho_{\Gamma_n}}\right|^2 \leqslant \sum_{n\geqslant K+1} \int_{\mathbb{R}^d} |\nabla\sqrt{\rho_{\Gamma_n}}|^2 \underset{K\to\infty}{\longrightarrow} 0.$$

Since we have for $K \geqslant n_0$

$$\sum_{n=1}^{K} \rho_{\Gamma_n} \geqslant \alpha\rho$$

for some $\alpha > 0$, the lemma implies the convergence (3.96).

At this step we have replaced $\Gamma = (\Gamma_n)_{n\geqslant 0}$ by a new state $\Gamma' = (\Gamma'_n)$ with compact support in n, and a close energy. In addition, $\|\rho_{\Gamma'} - \rho_\Gamma\|_{L^1(\mathbb{R}^d)}$ and $\left\|\nabla\sqrt{\rho_{\Gamma'}} - \nabla\sqrt{\rho_\Gamma}\right\|_{L^2(\mathbb{R}^d)}$ are small. To simplify our exposition we assume henceforth that Γ itself satisfies $\Gamma_n \equiv 0$ for $n \geqslant K+1$.

Step 3. Approximation by a State with Compact Support in Space Next we localize the state Γ in order to make it have a compact support in space. Let $\chi_R := \chi(\cdot/R)$ for some $\chi \in C_c^\infty(\mathbb{R}^d)$ satisfying $\chi(0) = 1$ and $0 \leqslant \chi \leqslant 1$. We consider the localized state $\Gamma_{|\chi_R}$ which has the density $\chi_R^2\rho$ and the energy

$$\frac{1}{2}\mathrm{Tr}\,((-\Delta)\chi_R\Gamma^{(1)}\chi_R) + \frac{1}{2}\mathrm{Tr}\,(w_{12}(\chi_R^2)^{\otimes 2}\Gamma^{(2)}).$$

Note that the space localization does not modify the support in n [102]. That is, the state $\Gamma_{|\chi_R}$ satisfies $(\Gamma_{|\chi_R})_n \equiv 0$ for $n \geqslant K+1$ (the same value as for Γ). The energy and the density converge strongly as $R \to \infty$ to that of Γ. Hence we can assume in the following that $\Gamma = (\Gamma_n)_{n\geqslant 0}$ has both a compact support in n and in space.

Step 4. Construction of the Canonical Sequence In the previous approximations we have replaced the initial grand-canonical state by a new state Γ which has an energy close to the minimal energy $F_{\mathrm{GC}}[\rho]$ and a density close to the initial density, in all the appropriate function spaces. In this step we finally construct the sequence ρ_j but it will only be close to ρ_Γ in the spaces mentioned in the statement of the theorem. We recall that $\Gamma_n \equiv 0$ for $n \geqslant K+1$.

Let $\varphi_\ell \in C_c^\infty(B_1)$ be K orthonormal functions in the unit ball $B_1 \subset \mathbb{R}^d$ and define $\varphi_{j,\ell}(\mathbf{r}) = j^{-d/2}\varphi_\ell(\mathbf{r}/j - \mathbf{v})$ where $\mathbf{v} \neq 0$ is any fixed vector in $\mathbb{R}^d$. These are K orthonormal functions in the translated and dilated ball $j\mathbf{v} + jB_1$. We then introduce the following K-particle mixed state

$$\Upsilon_j := \Gamma_0|S_{j,0}\rangle\langle S_{j,0}| + \sum_{n=1}^{K-1} \Gamma_n \wedge |S_{j,n}\rangle\langle S_{j,n}| + \Gamma_K$$

where

$$S_{j,n} := \varphi_{j,n+1} \wedge \cdots \wedge \varphi_{j,K}$$

is an $(K-n)$-particle Slater determinant. For j large enough the Γ_n's and $\varphi_{j,\ell}$ have disjoint support, hence

$$\mathrm{Tr}\,(\Upsilon_j)=\Gamma_0+\sum_{n=1}^{N}\mathrm{Tr}\,(\Gamma_n)=1,$$

as required. After a lengthy but straightforward calculation, one finds that Υ_j has the energy

$$\mathrm{Tr}\,(H_K^{0,w}\Upsilon_j)=\sum_{n=1}^{K}\mathrm{Tr}\,(H_n^{0,w}\Gamma_n)+\sum_{n=0}^{K-1}\left\langle S_{j,n}H_{K-n}^{0,w}S_{j,n}\right\rangle\mathrm{Tr}\,(\Gamma_n)$$
$$+\sum_{n=0}^{K-1}\int_{\mathbb{R}^d}\int_{\mathbb{R}^d}w(\mathbf{r}-\mathbf{r}')\rho_{\Gamma_n}(\mathbf{r})\,\rho_{S_{j,n}}(\mathbf{r}')\,\mathrm{d}\mathbf{r}\,\mathrm{d}\mathbf{r}'.$$

Under our assumptions on w the last two terms converge to 0 in the limit $j\to\infty$, hence the energy of Υ_j converges to that of Γ. In particular,

$$\limsup_{j\to\infty}F_{\mathrm{L}}[\rho_{\Upsilon_j}]\leqslant\lim_{j\to\infty}\mathrm{Tr}\,(H_K^{0,w}\Upsilon_j)=\sum_{n=1}^{K}\mathrm{Tr}\,(H_n^{0,w}\Gamma_n).$$

In addition, the density is

$$\rho_{\Upsilon_j}=\rho_\Gamma+\sum_{n=0}^{K-1}\mathrm{Tr}\,(\Gamma_n)\sum_{\ell=n+1}^{K}|\varphi_{j,\ell}|^2$$

and its square root converges strongly to $\sqrt{\rho_\Gamma}$ in $\dot{H}^1(\mathbb{R}^d)\cap L^p(\mathbb{R}^d)$ for all $2<p<p^*$, but not for $p=2$.

Step 5. Conclusion Using an $\varepsilon/2$ argument to justify the approximations made in Steps 1–3, we have managed to construct the sequence ρ_j mentioned in the statement and proved that it satisfies

$$\limsup_{j\to\infty}F_{\mathrm{L}}[\rho_j]\leqslant F_{\mathrm{GC}}[\rho].\tag{3.97}$$

Next we notice that the lower bound

$$\liminf_{j\to\infty}F_{\mathrm{L}}[\rho_j]\geqslant\liminf_{j\to\infty}F_{\mathrm{GC}}[\rho_j]\geqslant F_{\mathrm{GC}}[\rho]$$

follows from (i). Therefore we obtain the stated limit

$$\lim_{j\to\infty} F_{\rm L}[\rho_j] = F_{\rm GC}[\rho]$$

and this completes the proof of the theorem. □

Acknowledgments This project has received funding from the European Research Council (ERC) under the European Union's Horizon 2020 research and innovation programme (grant agreements MDFT No 725528 of M.L. and AQUAMS No 694227 of R.S.).

References

1. A. Alastuey and B. Jancovici. On the classical two-dimensional one-component Coulomb plasma. *J. Phys. France* **42**, 1–12 (1981).
2. M. Altunbulak and A. Klyachko. The Pauli principle revisited. *Comm. Math. Phys.* **282**, 287–322 (2008).
3. S. Azadi and N. D. Drummond, Low-density phase diagram of the three-dimensional electron gas. *Phys. Rev. B*, **105**, 245135 (2022).
4. V. Bach. Error bound for the Hartree-Fock energy of atoms and molecules. *Commun. Math. Phys.* **147**, 527–548 (1992).
5. V. Bach and L. Delle Site. On some open problems in many-electron theory. In: *Many-Electron Approaches in Physics, Chemistry and Mathematics*, edited by V. Bach and L. Delle Site, pp. 413–417, Mathematical Physics Studies, Springer International Publishing (2014).
6. V. Bach, E.H. Lieb, and J.P. Solovej. Generalized Hartree–Fock theory and the Hubbard model. *J. Statist. Phys.* **76**, 3–89 (1994).
7. R.D. Benguria, G.A. Bley and M. Loss. A new estimate on the indirect Coulomb energy. *Int. J. Quantum Chem.* **112**, 1579–1584 (2012).
8. U. Bindini and L. De Pascale. Optimal transport with Coulomb cost and the semiclassical limit of density functional theory. *J. Éc. polytech. Math.* **4**, 909–934 (2017).
9. R.F. Bishop and K.H. Lührmann. Electron correlations. II. Ground-state results at low and metallic densities. *Phys. Rev. B* **26**, 5523–5557 (1982).
10. X. Blanc and M. Lewin. The crystallization conjecture: A review. *EMS Surv. Math. Sci.* **2**, 255–306 (2015).
11. O. Bokanowski and B. Grébert. A decomposition theorem for wave functions in molecular quantum chemistry. *Math. Models Methods Appl. Sci.* **6**, 437–466 (1996).
12. O. Bokanowski, B. Grébert and N.J. Mauser. Local density approximations for the energy of a periodic Coulomb model. *Math. Models Methods Appl. Sci.* **13**, 1185–1217 (2003).
13. R. Borland and K. Dennis. The conditions on the one-matrix for three-body fermion wavefunctions with one-rank equal to six. *J. Phys. B* **5**, 7–15 (1972).
14. D. Borwein, J.M. Borwein, and R. Shail. Analysis of certain lattice sums. *J. Math. Anal. Appl.* **143**, 126–137 (1989).
15. D. Borwein, J.M. Borwein, R. Shail, and I.J. Zucker. Energy of static electron lattices. *J. Phys. A* **21**, 1519–1531 (1988).
16. D. Borwein, J.M. Borwein, and A. Straub. On lattice sums and Wigner limits. *J. Math. Anal. Appl.* **414**, 489–513 (2014).
17. J.M. Borwein, M.L. Glasser, R.C. McPhedran, J.G. Wan, and I.J. Zucker. *Lattice sums then and now.* Vol. 150 of Encyclopedia of Mathematics and its Applications, With a foreword by Helaman Ferguson and Claire Ferguson, Cambridge University Press, Cambridge (2013).

18. O. Bratelli and D.W. Robinson. *Operator Algebras and Quantum Statistical Mechanics. 1: C*- and W*-Algebras. Symmetry Groups. Decomposition of States.* 2nd ed. Texts and Monographs in Physics, Springer (2002).
19. O. Bratelli and D.W. Robinson. *Operator Algebras and Quantum Statistical Mechanics 2: Equilibrium States. Models in Quantum Statistical Mechanics.* 2nd ed. Texts and Monographs in Physics, Springer (2002).
20. S.G. Brush, H.L. Sahlin, and E. Teller. Monte carlo study of a one-component plasma. I. *J. Chem. Phys.* **45**, 2102–2118 (1966).
21. K. Burke. Lieb's most useful contribution to density functional theory? In: *The Physics and Mathematics of Elliott Lieb. The 90th Anniversary Volume I*, edited by R.L. Frank, A. Laptev, M. Lewin and R. Seiringer, ch. 7, pp. 131–142, EMS Press (2022).
22. G. Buttazzo, T. Champion and L. De Pascale. Continuity and estimates for multimarginal optimal transportation problems with singular costs. *Appl. Math. Optim.* **78**, 185–200 (2018).
23. J.W.S. Cassels. On a problem of Rankin about the Epstein zeta-function. *Proc. Glasgow Math. Assoc.* **4**, 73–80 (1959).
24. D.M. Ceperley and B.J. Alder. Ground State of the electron gas by a stochastic method. *Phys. Rev. Lett.* **45**, 566–569 (1980).
25. H. Cohn, A. Kumar, S.D. Miller, D. Radchenko and M. Viazovska. Universal optimality of the E_8 and Leech lattices and interpolation formulas. *Ann. of Math.* (2), **196**(3), 983–1082 (2022).
26. R.A. Coldwell-Horsfall and A.A. Maradudin. Zero-point energy of an electron lattice. *J. Math. Phys.* **1**, 395–404 (1960).
27. A. Coleman. Structure of fermion density matrices. *Rev. Modern Phys.* **35**, 668–689 (1963).
28. A. Coleman and V. Yukalov. *Reduced Density Matrices: Coulson's Challenge.* Springer Verlag (2000).
29. M. Colombo, L. De Pascale and S. Di Marino. Multimarginal optimal transport maps for one-dimensional repulsive costs. *Canad. J. Math.* **67**, 350–368 (2015).
30. M. Colombo and S. Di Marino. Equality between Monge and Kantorovich multimarginal problems with Coulomb cost. *Ann. Mat. Pura Appl.* **194**, 307–320 (2015).
31. M. Colombo, S. Di Marino and F. Stra. First order expansion in the semiclassical limit of the Levy–Lieb functional. *ArXiv:2106.06282* (2021).
32. M. Colombo and F. Stra. Counterexamples in multimarginal optimal transport with Coulomb cost and spherically symmetric data. *Math. Models Methods Appl. Sci.* **26**, 1025–1049 (2016).
33. C. Cotar, G. Friesecke, and C. Klüppelberg. Density functional theory and optimal transportation with Coulomb cost. *Comm. Pure Appl. Math.* **66**, 548–599 (2013).
34. C. Cotar, G. Friesecke and C. Klüppelberg. Smoothing of transport plans with fixed marginals and rigorous semiclassical limit of the Hohenberg-Kohn functional. *Arch. Ration. Mech. Anal.* **228**, 891–922 (2018).
35. C. Cotar, G. Friesecke and B. Pass. Infinite-body optimal transport with Coulomb cost. *Calc. Var. Partial Differ. Equ.* **54**, 717–742 (2015).
36. C. Cotar and M. Petrache. Equality of the jellium and uniform electron gas next-order asymptotic terms for Coulomb and Riesz potentials. *ArXiv:1707.07664v5 (version 5)* (2019).
37. C. Cotar and M. Petrache. Next-order asymptotic expansion for N-marginal optimal transport with Coulomb and Riesz costs. *Adv. Math.* **344**, 137–233 (2019).
38. D.G. de Figueiredo and J.-P. Gossez. Strict monotonicity of eigenvalues and unique continuation. *Comm. Partial Differential Equations* **17**, 339–346 (1992).
39. F.W. de Wette. Comments on the electrostatic energy of a Wigner solid. *Phys. Rev. B* **21**, 3751–3753 (1980).
40. S. Di Marino. In preparation (2019).
41. S. Di Marino, A. Gerolin and L. Nenna. Optimal Transportation Theory with Repulsive Costs. In: *Topological Optimization and Optimal Transport in the Applied Sciences*, ch. 9, 204–256, Radon Series on Computational and Applied Mathematics, De Gruyter (2017).
42. S. Di Marino, M. Lewin and L. Nenna. Grand-canonical optimal transport. *ArXiv:2201.06859* (2022).

43. P.H. Diananda. Notes on two lemmas concerning the Epstein zeta-function. *Proc. Glasgow Math. Assoc.* **6**, 202–204 (1964).
44. P.A.M. Dirac. Note on exchange phenomena in the Thomas atom. *Proc. Camb. Philos. Soc.* **26**, 376–385 (1930).
45. J. Dolbeault, A. Laptev, and M. Loss. Lieb–Thirring inequalities with improved constants. *J. Eur. Math. Soc. (JEMS)* **10**, 1121–1126 (2008).
46. R. Dreizler and E. Gross. *Density functional theory.* Springer, Berlin (1990).
47. N.D. Drummond, Z. Radnai, J.R. Trail, M.D. Towler and R.J. Needs. Diffusion quantum Monte Carlo study of three-dimensional Wigner crystals. *Phys. Rev. B* **69**, 085116 (2004).
48. F.J. Dyson and A. Lenard. Stability of matter. I. *J. Math. Phys.* **8**, 423–434 (1967).
49. A. Eden and C. Foias. A simple proof of the generalized Lieb–Thirring inequalities in one-space dimension. *J. Math. Anal. Appl.* **162**, 250–254 (1991).
50. V. Ennola. A lemma about the Epstein zeta-function. *Proc. Glasgow Math. Assoc.* **6**, 198–201 (1964).
51. P. Epstein. Zur Theorie allgemeiner Zetafunctionen. *Math. Ann.* **56**, 615–644 (1903).
52. C. Fefferman. The thermodynamic limit for a crystal. *Commun. Math. Phys.* **98**, 289–311 (1985).
53. E. Fermi. Un metodo statistico per la determinazione di alcune priorieta dell'atome. R*end. Accad. Naz. Lincei* **6**, 602–607 (1927).
54. R. Frank, A. Laptev, and T. Weidl. *Schrödinger operators: Eigenvalues and Lieb-Thirring inequalities*, Cambridge Studies in Advanced Mathematics, Cambridge University Press (2022).
55. R.L. Frank. The Lieb–Thirring inequality: Recent results and open problems. In: *Nine Mathematical Challenges: An Elucidation*, edited by A. Kechris, N. Makarov, D. Ramakrishnan and X. Zhu, Proc. Symp. Pure Math., Amer. Math. Soc., vol. 104, pp. 45–86 (2021).
56. R.L. Frank, D. Hundertmark, M. Jex and P.T. Nam. The Lieb–Thirring inequality revisited. *J. Eur. Math. Soc. (JEMS)* **23**, 2583–2600 (2021).
57. R.L. Frank, M. Lewin, E.H. Lieb and R. Seiringer. Energy Cost to Make a Hole in the Fermi Sea. *Phys. Rev. Lett.* **106**, 150402 (2011).
58. R.L. Frank, M. Lewin, E.H. Lieb and R. Seiringer. A positive density analogue of the Lieb–Thirring inequality. *Duke Math. J.* **162**, 435–495 (2012).
59. R.L. Frank, M. Lewin, E.H. Lieb and R. Seiringer. Strichartz inequality for orthonormal functions. *J. Eur. Math. Soc. (JEMS)* **16**, 1507–1526 (2014).
60. R.L. Frank and J. Sabin. Restriction theorems for orthonormal functions, Strichartz inequalities, and uniform Sobolev estimates. *Amer. Math. J.* **139**, 1649–1691 (2017).
61. F. Frick and M. Superdock. A nonlinear Lazarev–Lieb theorem: L^2-orthogonality via motion planning. *J. Topol. Anal.*, **14**(3), 569–585 (2022).
62. G. Friesecke. The multiconfiguration equations for atoms and molecules: charge quantization and existence of solutions. *Arch. Ration. Mech. Anal.* **169**, 35–71 (2003).
63. L. Garrigue. Unique continuation for many body Schrödinger operators and the Hohenberg–Kohn theorem. *Math. Phys. Anal. Geom.* **21**, Art. 27, 11 (2018).
64. L. Garrigue. Unique continuation for many-body Schrödinger operators and the Hohenberg–Kohn theorem. II. The Pauli Hamiltonian. *Doc. Math.* **25**, 869–898 (2020).
65. L. Garrigue. Some properties of the potential-to-ground state map in quantum mechanics. *Comm. Math. Phys.* **386**, 1803–1844 (2021).
66. L. Garrigue. Building Kohn-Sham potentials for ground and excited states. *Arch. Ration. Mech. Anal.* **245**, 949–1003 (2022).
67. V. Georgescu. On the unique continuation property for Schrödinger Hamiltonians. *Helv. Phys. Acta* **52**, 655–670 (1979).
68. G. Giuliani and G. Vignale. *Quantum Theory of the Electron Liquid.* Cambridge University Press (2005).
69. P. Gori-Giorgi and M. Seidl. Density functional theory for strongly-interacting electrons: perspectives for physics and chemistry. *Phys. Chem. Chem. Phys.* **12**, 14405–14419 (2010).

70. P. Gori-Giorgi, G. Vignale and M. Seidl. Electronic zero-point oscillations in the strong-interaction limit of density functional theory. *J. Chem. Theory Comput.* **5**, 743–753 (2009).
71. N. Gottschling and P.T. Nam. Convergence of Levy–Lieb to Thomas–Fermi density functional. *Calc. Var. Partial Differential Equations* **57**, Art. 146, 13 (2018).
72. G.M. Graf and J.P. Solovej. A correlation estimate with applications to quantum systems with Coulomb interactions. *Rev. Math. Phys.* **06**, 977–997 (1994).
73. L. Grafakos. *Classical Fourier analysis.* Springer-Verlag, New York (2008).
74. J.N. Gregg. The existence of the thermodynamic limit in Coulomb-like systems. *Comm. Math. Phys.* **123**, 255–276 (1989).
75. C. Hainzl, M. Lewin and J.P. Solovej. The thermodynamic limit of quantum Coulomb systems. Part II. Applications. *Advances in Math.* **221**, 488–546 (2009).
76. G.L. Hall. Correction to Fuchs' calculation of the electrostatic energy of a Wigner solid. *Phys. Rev. B* **19**, 3921–3932 (1979).
77. G.L. Hall. Response to "Comment on the average potential of a Wigner solid". *Phys. Rev. B* **24**, 7415–7418 (1981).
78. G.L. Hall and T.R. Rice. Wigner solids, classical Coulomb lattices, and invariant average potential. *Phys. Rev. B* **21**, 3757–3759 (1980).
79. J.E. Harriman. Orthonormal orbitals for the representation of an arbitrary density. *Phys. Rev. A* **24**, 680–682 (1981).
80. T. Helgaker and A.M. Teale. Lieb variation principle in density-functional theory. In: *The Physics and Mathematics of Elliott Lieb. The 90th Anniversary Volume I*, edited by R.L. Frank, A. Laptev, M. Lewin and R. Seiringer, ch. 22, pp. 527–559, EMS Press (2022).
81. M. Hoffmann-Ostenhof and T. Hoffmann-Ostenhof. Schrödinger inequalities and asymptotic behavior of the electron density of atoms and molecules. *Phys. Rev. A* **16**, 1782–1785 (1977).
82. P. Hohenberg and W. Kohn. Inhomogeneous electron gas. *Phys. Rev.* **136**, B864–B871 (1964).
83. A. Holas, P.M. Kozlowski, and N.H. March. Kinetic energy density and Pauli potential: dimensionality dependence, gradient expansions and non-locality. *J. Phys. A Math. Gen.* **24**, 4249–4260 (1991).
84. M. Holzmann and S. Moroni, Itinerant-electron magnetism: The importance of many-body correlations, *Phys. Rev. Lett.* **124**, 206404 (2020).
85. W. Hughes. Thermodynamics for Coulomb systems: a problem at vanishing particle densities. *J. Statist. Phys.* **41** (1985), 975–1013.
86. J. Ihm and M.L. Cohen. Comment on "Correction to Fuchs' calculation of the electrostatic energy of a Wigner solid". *Phys. Rev. B* **21**, 3754–3756 (1980).
87. D. Jerison and C.E. Kenig. Unique continuation and absence of positive eigenvalues for Schrödinger operators. With an appendix by E.M. Stein. *Ann. of Math. (2)*, **121**, 463–494 (1985).
88. M.D. Jones and D.M. Ceperley. Crystallization of the one-component plasma at finite temperature. *Phys. Rev. Lett.* **76**, 4572–4575 (1996).
89. G. Kin-Lic Chan and N.C. Handy. Optimized Lieb–Oxford bound for the exchange-correlation energy. *Phys. Rev. A* **59**, 3075–3077 (1999).
90. D. Kirzhnits. Quantum corrections to the Thomas–Fermi equation. *Soviet Phys. JETP* **5**, 64–71 (1957).
91. A.A. Klyachko. Quantum marginal problem and N-representability. *J. Phys. Conf. Series* **36**, 72–86 (2006).
92. H. Koch and D. Tataru. Carleman estimates and absence of embedded eigenvalues. *Commun. Math. Phys.* **267**, 419–449 (2006).
93. W. Kohn and L.J. Sham. Self-consistent equations including exchange and correlation effects. *Phys. Rev. (2)*, **140**, A1133–A1138 (1965).
94. M. Koivisto and M.J. Stott. Kinetic energy functional for a two-dimensional electron system. *Phys. Rev. B* **76**, 195103 (2007).
95. P.E. Lammert. In search of the Hohenberg–Kohn theorem. *J. Math. Phys.* **59**, 042110 (2018).
96. A.B. Lauritsen. Floating Wigner crystal and periodic jellium configurations. *J. Math. Phys.* **62**, no. 083305 (2021).

97. O. Lazarev and E.H. Lieb. A smooth, complex generalization of the Hobby–Rice theorem. *Indiana Univ. Math. J.* **62**, 1133–1141 (2013).
98. A. Levitt. Best constants in Lieb–Thirring inequalities: a numerical investigation. *J. Spectr. Theory* **4**, 153–175 (2014).
99. M. Levy. Universal variational functionals of electron densities, first-order density matrices, and natural spin-orbitals and solution of the v-representability problem. *Proc. Natl. Acad. Sci. U.S.A.* **76**, 6062–6065 (1979).
100. M. Levy and J.P. Perdew. Tight bound and convexity constraint on the exchange-correlation-energy functional in the low-density limit, and other formal tests of generalized-gradient approximations. *Phys. Rev. B* **48**, 11638–11645 (1993).
101. M. Lewin. Solutions of the multiconfiguration equations in quantum chemistry. *Arch. Ration. Mech. Anal.* **171**, 83–114 (2004).
102. M. Lewin. Geometric methods for nonlinear many-body quantum systems. *J. Funct. Anal.* **260**, 3535–3595 (2011).
103. M. Lewin. Semi-classical limit of the Levy–Lieb functional in Density Functional Theory. *C. R. Math. Acad. Sci. Paris* **356**, 449–455 (2018).
104. M. Lewin. Coulomb and Riesz gases: The known and the unknown. *J. Math. Phys.* **63**, 061101 (2022). Special collection in honor of Freeman Dyson.
105. M. Lewin and E.H. Lieb. Improved Lieb–Oxford exchange-correlation inequality with gradient correction. *Phys. Rev. A* **91**, 022507 (2015).
106. M. Lewin, E.H. Lieb and R. Seiringer. Statistical mechanics of the Uniform Electron Gas. *J. Éc. polytech. Math.* **5**, 79–116 (2018).
107. M. Lewin, E.H. Lieb and R. Seiringer. Floating Wigner crystal with no boundary charge fluctuations. *Phys. Rev. B* **100**, 035127 (2019).
108. M. Lewin, E.H. Lieb and R. Seiringer. The Local Density Approximation in Density Functional Theory. *Pure Appl. Anal.* **2**, 35–73 (2019).
109. M. Lewin, E.H. Lieb and R. Seiringer. Improved Lieb-Oxford bound on the indirect and exchange energies. *Lett. Math. Phys.*, **112**, Art. 92 (2022). Themed collection "Mathematical Physics and Numerical Simulation of Many-Particle Systems"; V. Bach and L. Delle Site (eds.).
110. M. Lewin, P.T. Nam, S. Serfaty and J.P. Solovej. Bogoliubov spectrum of interacting Bose gases. Comm. *Pure Appl. Math.* **68**, 413–471 (2015).
111. P. Li and S.T. Yau. On the Schrödinger equation and the eigenvalue problem. *Commun. Math. Phys.* **88**, 309–318 (1983).
112. E.H. Lieb. The stability of matter. *Rev. Mod. Phys.* **48**, 553–569 (1976).
113. E.H. Lieb. A lower bound for Coulomb energies. *Phys. Lett. A* **70**, 444–446 (1979).
114. E.H. Lieb. Thomas-Fermi and related theories of atoms and molecules. *Rev. Mod. Phys.* **53**, 603–641 (1981).
115. E.H. Lieb. Variational principle for many-fermion systems. *Phys. Rev. Lett.* **46**, 457–459 (1981).
116. E.H. Lieb. Density functionals for Coulomb systems. *Int. J. Quantum Chem.* **24**, 243–277 (1983).
117. E.H. Lieb. The stability of matter: from atoms to stars. *Bull. Amer. Math. Soc. (N.S.)* **22**, 1–49 (1990).
118. E.H. Lieb and M. Loss. *Analysis.* 2nd ed. Vol. 14 of Graduate Studies in Mathematics, American Mathematical Society, Providence, RI (2001).
119. E.H. Lieb and H. Narnhofer. The thermodynamic limit for jellium. *J. Stat. Phys.* **12**, 291–310 (1975).
120. E.H. Lieb and S. Oxford. Improved lower bound on the indirect Coulomb energy. *Int. J. Quantum Chem.* **19**, 427–439 (1980).
121. E.H. Lieb and R. Seiringer. *The Stability of Matter in Quantum Mechanics.* Cambridge Univ. Press (2010).
122. E.H. Lieb and B. Simon. Thomas–Fermi theory revisited. *Phys. Rev. Lett.* **31**, 681–683 (1973).

123. E.H. Lieb and B. Simon. The Thomas–Fermi theory of atoms, molecules and solids. *Adv. Math.* **23**, 22–116 (1977).
124. E.H. Lieb, J.P. Solovej and J. Yngvason. Ground states of large quantum dots in magnetic fields. *Phys. Rev. B* **51**, 10646–10665 (1995).
125. E.H. Lieb and W.E. Thirring. Bound on kinetic energy of fermions which proves stability of matter. *Phys. Rev. Lett.* **35**, 687–689 (1975).
126. E.H. Lieb and W.E. Thirring. Inequalities for the moments of the eigenvalues of the Schrödinger hamiltonian and their relation to Sobolev inequalities. In: *Studies in Mathematical Physics*, edited by E.H. Lieb, pp. 269–303, Princeton University Press (1976).
127. D. Lundholm, P.T. Nam, and F. Portmann. Fractional Hardy–Lieb–Thirring and related inequalities for interacting systems. *Arch. Ration. Mech. Anal.* **219**, 1343–1382 (2016).
128. S. Lundqvist and N. March (eds.) *Theory of the Inhomogeneous Electron Gas*. Physics of Solids and Liquids, Springer US (1983).
129. N.H. March and W.H. Young. Variational methods based on the density matrix. *Proc. Phys. Soc.* **72**, 182 (1958).
130. S. Mikhailov and K. Ziegler. Floating Wigner molecules and possible phase transitions in quantum dots. *Eur. Phys. J. B* **28**, 117–120 (2002).
131. H.L. Montgomery. Minimal theta functions. *Glasgow Math. J.* **30**, 75–85(1988).
132. J.C.B. Morrey. On the analyticity of the solutions of analytic non-linear elliptic systems of partial differential equations. I. Analyticity in the interior. *Amer. J. Math.* **80**, 198–218 (1958).
133. P.T. Nam. Lieb–Thirring inequality with semiclassical constant and gradient error term. *J. Funct. Anal.* **274**, 1739–1746 (2018).
134. B.R.A. Nijboer and T.W. Ruijgrok. On the energy per particle in three- and two-dimensional Wigner lattices. *J. Statist. Phys.* **53**, 361–382 (1988).
135. M.M. Odashima and K. Capelle. How tight is the Lieb–Oxford bound? *J. Chem. Phys.* **127**, 054106 (2007).
136. R. Parr and W. Yang. *Density-Functional Theory of Atoms and Molecules*. International Series of Monographs on Chemistry, Oxford University Press, USA (1994).
137. B. Pass. Multi-marginal optimal transport: theory and applications. *ESAIM Math. Model. Numer. Anal.* **49**, 1771–1790 (2015).
138. J.P. Perdew. Unified Theory of Exchange and Correlation Beyond the Local Density Approximation. In: *Electronic Structure of Solids '91*, edited by P. Ziesche and H. Eschrig, pp. 11–20, Akademie Verlag, Berlin (1991).
139. J.P. Perdew, K. Burke, and M. Ernzerhof. Generalized gradient approximation made simple. *Phys. Rev. Lett.* **77**, 3865–3868 (1996).
140. J.P. Perdew and S. Kurth. Density Functionals for Non-relativistic Coulomb Systems in the New Century. In: *A Primer in Density Functional Theory*, edited by C. Fiolhais, F. Nogueira and M.A.L. Marques, pp. 1–55, Springer Berlin-Heidelberg (2003).
141. J.P. Perdew, R.G. Parr, M. Levy and J.L. Balduz. Density-functional theory for fractional particle number: Derivative discontinuities of the energy. *Phys. Rev. Lett.* **49**, 1691–1694 (1982).
142. J.P. Perdew and K. Schmidt. Jacob's ladder of density functional approximations for the exchange-correlation energy. *AIP Conference Proceedings* **577**, 1–20 (2001).
143. J.P. Perdew and J. Sun.The Lieb-Oxford Lower Bounds on the Coulomb Energy, Their Importance to Electron Density Functional Theory, and a Conjectured Tight Bound on Exchange. In: *The Physics and Mathematics of Elliott Lieb. The 90th Anniversary Volume II*, edited by R.L. Frank, A. Laptev, M. Lewin and R. Seiringer, ch. 36, pp. 165–178, EMS Press (2022).
144. J.P. Perdew and Y. Wang. Accurate and simple analytic representation of the electron-gas correlation energy. *Phys. Rev. B* **45**, 13244–13249 (1992).
145. M. Petrache and S. Serfaty. Crystallization for Coulomb and Riesz interactions as a Consequence of the Cohn–Kumar Conjecture. *Proc. Am. Math. Soc.* **148**, 3047–3057 (2020).

146. E. Pollock and J. Hansen. Statistical mechanics of dense ionized matter. ii. equilibrium properties and melting transition of the crystallized one-component plasma. *Phys. Rev. A* **8**, 3110–3122 (1973).
147. R.A. Rankin. A minimum problem for the Epstein zeta-function. *Proc. Glasgow Math. Assoc.* **1**, 149–158 (1953).
148. E. Räsänen, S. Pittalis, K. Capelle, and C.R. Proetto. Lower bounds on the exchange-correlation energy in reduced dimensions. *Phys. Rev. Lett.* **102**, 206406 (2009).
149. M. Reed and B. Simon, *Methods of Modern Mathematical Physics. I. Functional analysis*, Academic Press (1972).
150. M. Reed and B. Simon, *Methods of Modern Mathematical Physics. II. Fourier analysis, self-adjointness*, Academic Press (1975).
151. R. Regbaoui. Strong uniqueness for second order differential operators. *J. Differential Equations* **141**, 201–217 (1997).
152. D. Ruelle. *Statistical mechanics. Rigorous results.* Singapore: World Scientific. London: Imperial College Press (1999).
153. M. Rumin. Balanced distribution-energy inequalities and related entropy bounds. *Duke Math. J.* **160**, 567–597 (2011).
154. M.B. Ruskai. N-representability problem: Particle-hole equivalence. *J. Mathematical Phys.* **11**, 3218–3224 (1970).
155. V. Rutherfoord. On the Lazarev–Lieb extension of the Hobby–Rice theorem. *Adv. Math.* **244**, 16–22 (2013).
156. L. Salasnich. Kirzhnits gradient expansion for a D-dimensional Fermi gas. *J. Phys. A* **40**, 9987–9992 (2007).
157. P. Sarnak and A. Strömbergsson. Minima of Epstein's zeta function and heights of flat tori. *Invent. Math.* **165**, 115–151 (2006).
158. M. Schechter and B. Simon. Unique continuation for Schrödinger operators with unbounded potentials. *J. Math. Anal. Appl.* **77**, 482–492 (1980).
159. L. Schimmer. The state of the Lieb–Thirring conjecture. In *The Physics and Mathematics of Elliott Lieb. The 90th Anniversary Volume II*, edited by R.L. Frank, A. Laptev, M. Lewin and R. Seiringer, ch. 39, pp. 253–275, EMS Press (2022).
160. M. Seidl. Strong-interaction limit of density-functional theory. *Phys. Rev. A* **60**, 4387–4395 (1999).
161. M. Seidl, T. Benyahia, D.P. Kooi, and P. Gori-Giorgi. The Lieb-Oxford bound and the optimal transport limit of DFT. In: *The Physics and Mathematics of Elliott Lieb. The 90th Anniversary Volume II*, edited by R.L. Frank, A. Laptev, M. Lewin and R. Seiringer, ch. 43, pp. 345–360, EMS Press (2022).
162. M. Seidl, S. Di Marino, A. Gerolin, L. Nenna, K.J.H. Giesbertz and P. Gori-Giorgi. The strictly-correlated electron functional for spherically symmetric systems revisited. *ArXiv:1702.05022* (2017).
163. M. Seidl, P. Gori-Giorgi and A. Savin. Strictly correlated electrons in density-functional theory: A general formulation with applications to spherical densities. *Phys. Rev. A* **75**, 042511 (2007).
164. M. Seidl, J.P. Perdew, and M. Levy. Strictly correlated electrons in density-functional theory. *Phys. Rev. A* **59**, 51–54 (1999).
165. B. Simon. Schrödinger semigroups. *Bull. Amer. Math. Soc. (N.S.)* **7**, 447–526 (1982).
166. J. Sun, J.P. Perdew and A. Ruzsinszky. Semilocal density functional obeying a strongly tightened bound for exchange. *Proceedings of the National Academy of Science* **112**, 685–689 (2015).
167. J. Sun, R.C. Remsing, Y. Zhang, Z. Sun, A. Ruzsinszky, H. Peng, Z. Yang, A. Paul, U. Waghmare, X. Wu, M.L. Klein and J.P. Perdew. Accurate first-principles structures and energies of diversely bonded systems from an efficient density functional. *Nature Chemistry* **8**, 831–836 (2016).
168. J. Sun, A. Ruzsinszky and J.P. Perdew. Strongly constrained and appropriately normed semilocal density functional. *Phys. Rev. Lett.* **115**, 036402 (2015).

169. L.H. Thomas. The calculation of atomic fields. *Proc. Camb. Philos. Soc.* **23**, 542–548 (1927).
170. M.-I. Trappe, Y. Len, H. Ng and B.-G. Englert. Airy-averaged gradient corrections for two-dimensional fermion gases. *Annals of Physics* **385**, 136–161 (2017).
171. C.F.v. Weizsäcker. Zur theorie der Kernmassen. *Z. Phys.* **96**, 431–458 (1935).
172. C.N. Yang. Concept of off-diagonal long-range order and the quantum phases of liquid He and of superconductors. *Rev. Modern Phys.* **34**, 694–704 (1962).
173. C.N. Yang. Some properties of the reduced density matrix. *J. Mathematical Phys.* **4**, 418–419 (1963).
174. W. Yang. Generalized adiabatic connection in density functional theory. *J. Chem. Phys.* **109**, 10107–10110 (1998).
175. F.H. Zong, C. Lin, and D.M. Ceperley. Spin polarization of the low-density three-dimensional electron gas. *Phys. Rev. E* **66**, 036703 (2002).
176. G. Zumbach and K. Maschke. New approach to the calculation of density functionals. *Phys. Rev. A* (3), **28**, 544–554 (1983).
177. G. Zumbach and K. Maschke. Erratum: "New approach to the calculation of density functionals". *Phys. Rev. A* (3), **29**, 1585–1587 (1984).

Chapter 4
The Strong-Interaction Limit of Density Functional Theory

Gero Friesecke, Augusto Gerolin, and Paola Gori-Giorgi

Abstract This is a comprehensive review of the strong-interaction limit of density functional theory. It covers the derivation of the limiting strictly correlated electrons (SCE) functional from exact Hohenberg–Kohn DFT, basic aspects of SCE physics such as the nonlocal dependence of the SCE potential on the density, equivalent formulations and the mathematical interpretation as optimal transport with Coulomb cost, rigorous results (including exactly soluble cases), approximations, numerical methods, integration into Kohn–Sham DFT (KS SCE), and applications to molecular systems, an example being that KS SCE, unlike the local density approximation or generalized gradient approximations, dissociates H_2 correctly. We have made an effort to make this review accessible to a broad audience of physicists, chemists, and mathematicians.

4.1 Introduction

The strong-interaction limit of DFT is the inhomogeneous low-density limit associated with the uniform coordinate scaling

$$\rho_\gamma(\mathbf{r}) = \gamma^3 \rho(\gamma \mathbf{r})$$

G. Friesecke (✉)
Department of Mathematics, Technische Universität München, Garching bei München, Germany
e-mail: gf@ma.tum.de

A. Gerolin (✉)
Department of Chemistry and Biomolecular Sciences, University of Ottawa, Ottawa, Canada

Department of Mathematics and Statistics, University of Ottawa, Ottawa, Canada

P. Gori-Giorgi (✉)
Department of Theoretical Chemistry, Vrije Universiteit Amsterdam, Amsterdam, The Netherlands

AI4Science Microsoft Research, Amsterdam, The Netherlands

E. Cancès, G. Friesecke (eds.), *Density Functional Theory*, Mathematics and Molecular Modeling, https://doi.org/10.1007/978-3-031-22340-2_4

of the single-particle density at fixed particle number, with $\gamma \to 0$. In this limit, the Levy–Lieb functional, which gives the minimum kinetic and interaction energy subject to the given density, has the leading order asymptotics

$$F_{\mathrm{LL}}[\rho_\gamma] \sim \gamma \, \mathrm{V}_{\mathrm{ee}}^{\mathrm{SCE}}[\rho],$$

and the corresponding optimal wavefunction Ψ_γ has the asymptotics

$$\sum_{s_1,\ldots,s_N \in \mathbb{Z}_2} |\Psi_\gamma(\mathbf{r}_1, s_1, \ldots, \mathbf{r}_N, s_N)|^2 \sim \gamma^{3N} \rho_N(\gamma \mathbf{r}_1, \ldots, \gamma \mathbf{r}_N),$$

where ρ_N solves the variational principle of having minimal Coulomb energy subject to the given density ρ and $\mathrm{V}_{\mathrm{ee}}^{\mathrm{SCE}}[\rho]$ denotes the resulting minimal energy.

This appears to be the only case in which one can obtain insight into how to extract information about the interaction energy directly from the density. As it turns out, in this limit none of the ingredients from the traditional "Jacob's ladder" of DFT approximations (local density, local density gradients, Kohn–Sham kinetic energy density, Hartree–Fock exchange, virtual orbitals) play any role. Instead, maps based on integrals not derivatives of the density appear. These maps are mathematically related to the field of optimal transport, and physically describe strictly correlated electrons (SCE). The SCE functional $\mathrm{V}_{\mathrm{ee}}^{\mathrm{SCE}}$ appearing above is the limiting Hartree-exchange-correlation functional.

While the strong-interaction limit is, of course, not reached in nature, it points the way towards the real physics happening in molecular systems containing strong correlations, without having to leave the realm of Kohn–Sham DFT. Two important examples whose physics is missed by Kohn–Sham DFT with semilocal or hybrid exchange-correlation functionals but captured correctly by integrating the SCE functional into Kohn–Sham DFT (KS SCE) are weakly charged nanosystems, see Fig. 4.11, and H_2 near the dissociation limit, see Fig. 4.14.

This chapter provides a self-contained introduction to this limit and its fascinating physics and mathematics which has been unearthed in the past two decades, and reviews the current state of the art.

4.2 The Many-Electron Schrödinger Equation and Universal Density Functional

In this section we quickly introduce the time-independent electronic Schrödinger equation and the exact reformulation of the ground state problem via a universal density functional.

4.2.1 The Many-Electron Schrödinger Equation

We consider a quantum mechanical system of N non-relativistic electrons (of mass m_e and charge $-e$), moving around classical nuclei with positions $\mathbf{R}_1, \ldots, \mathbf{R}_M \in \mathbb{R}^d$ and charges $Z_1 e, \ldots, Z_M e$ (Born–Oppenheimer approximation). Our main interest is in the physical space $\mathbb{R}^3$, but we consider the general space dimension $d \geq 1$ since it will be instructive to illustrate key properties of the strong interaction limit with lower dimensional examples. The electrons are described by a wave function $\Psi : (\mathbb{R}^d \times \mathbb{Z}_2)^N \to \mathbb{C}$ of N positions $\mathbf{r}_i \in \mathbb{R}^d$ and spin coordinates $s_i \in \{\uparrow, \downarrow\} = \mathbb{Z}_2$.

The Pauli exclusion principle states that the electronic wave function must be antisymmetric with respect to permutations of the electron coordinates,

$$\Psi(\mathbf{r}_{\sigma(1)}, s_{\sigma(1)}, \ldots, \mathbf{r}_{\sigma(N)}, s_{\sigma(N)}) = \operatorname{sign}(\sigma)\Psi(\mathbf{r}_1, s_1, \ldots, \mathbf{r}_N, s_N), \quad \sigma \in \mathfrak{S}_N, \tag{4.1}$$

where $\mathfrak{S}_N$ denotes the group of permutations of the indices $1, \ldots, N$. The set of square-integrable N-electron wave functions, $\{\Psi \in L^2((\mathbb{R}^d \times \mathbb{Z}_2)^N; \mathbb{C}) \;:\; (4.1)\}$, will be denoted $\bigwedge_{i=1}^N L^2(\mathbb{R}^d \times \mathbb{Z}_2; \mathbb{C})$. The one-body density of an electronic wave function $\Psi \in \bigwedge_{i=1}^N L^2(\mathbb{R}^d \times \mathbb{Z}_2; \mathbb{C})$ is defined by

$$\rho_\Psi(\mathbf{r}_j) = N \sum_{s_1,\ldots,s_N \in \mathfrak{S}_N} \int_{\mathbb{R}^{d(N-1)}} |\Psi(\mathbf{r}_1, s_1, \ldots, \mathbf{r}_N, \mathbb{Z}_2)|^2 \prod_{i \neq j} \mathrm{d}\mathbf{r}_i, \quad \forall j \in \{1, \ldots, N\}.$$

The energy $E[\Psi, v]$ of a fermionic state Ψ with external potential $v : \mathbb{R}^d \to \mathbb{R}$ is given, in atomic units $\hbar = m_e = e = 1$, by

$$E[\Psi, v] = T[\Psi] + V_{\mathrm{ee}}[\Psi] + V_{\mathrm{ne}}[\Psi, v], \tag{4.2}$$

where $T[\Psi]$ is the *kinetic energy*,

$$T[\Psi] = \frac{1}{2} \sum_{s_1 \in \mathbb{Z}_2} \int_{\mathbb{R}^d} \cdots \sum_{s_N \in \mathbb{Z}_2} \int_{\mathbb{R}^d} \sum_{i=1}^N |\nabla_{\mathbf{r}_i} \Psi(\mathbf{r}_1, s_1 \ldots, \mathbf{r}_N, s_N)|^2 \mathrm{d}\mathbf{r}_1 \ldots \mathrm{d}\mathbf{r}_N;$$

$V_{\mathrm{ee}}[\Psi]$ is the electron-electron interaction energy

$$V_{\mathrm{ee}}[\Psi] = \sum_{s_1 \in \mathbb{Z}_2} \int_{\mathbb{R}^d} \cdots \sum_{s_N \in \mathbb{Z}_2} \int_{\mathbb{R}^d} \sum_{1 \leqslant i < j < N}^{N} w(\mathbf{r}_i - \mathbf{r}_j) \, |\Psi(\mathbf{r}_1, s_1 \ldots, \mathbf{r}_N, s_N)|^2 \mathrm{d}\mathbf{r}_1 \ldots \mathrm{d}\mathbf{r}_N,$$

and $V_{\mathrm{ne}}[\Psi, v]$ is the electron-nuclei interaction energy,

$$V_{\mathrm{ne}}[\Psi, v] = \sum_{s_1 \in \mathbb{Z}_2} \int_{\mathbb{R}^d} \cdots \sum_{s_N \in \mathbb{Z}_2} \int_{\mathbb{R}^d} \sum_{i=1}^{N} v(\mathbf{r}_i) |\Psi(\mathbf{r}_1, s_1 \ldots, \mathbf{r}_N, s_N)|^2 d\mathbf{r}_1 \ldots d\mathbf{r}_N,$$

where $w : \mathbb{R}^d \to \mathbb{R}$ is an interaction potential satisfying $w(\mathbf{r}) = w(-\mathbf{r})$, so that the total interaction potential

$$V_{\mathrm{ee}}(\mathbf{r}_1, \ldots, \mathbf{r}_N) = \sum_{1 \le i < j \le N} w(\mathbf{r}_i - \mathbf{r}_j) \tag{4.3}$$

is symmetric.[1] Typically,

$$w(\mathbf{r}) = |\mathbf{r}|^{-1} \tag{4.4}$$

is the Coulomb electron repulsion and v is the Coulomb potential generated by M nuclei which are at positions $\mathbf{R}_\nu$ with charges Z_ν,

$$v(\mathbf{r}) = -\sum_{\nu=1}^{M} \frac{Z_\nu}{|\mathbf{r} - \mathbf{R}_\nu|}. \tag{4.5}$$

If additional fields are present, the external potential v contains extra terms.

The central quantity of interest is the *ground state energy* of the system. By the *Rayleigh–Ritz variational principle*, it is given by

$$E_0[v] = \inf\{E[\Psi, v] : \Psi \in \mathcal{W}^N\}, \tag{4.6}$$

where the infimum is taken over the class $\mathcal{W}^N$ of wavefunctions which are antisymmetric and have finite kinetic energy,

$$\mathcal{W}^N = \tag{4.7}$$

$$\left\{ \Psi \in \bigwedge_{i=1}^{N} H^1(\mathbb{R}^d \times \mathbb{Z}_2; \mathbb{C}) : \sum_{s_1,\ldots,s_N \in \mathbb{Z}_2} \int_{\mathbb{R}^{dN}} |\nabla \Psi|^2 d\mathbf{r}_1 \ldots d\mathbf{r}_N < +\infty, ||\Psi|| = 1 \right\}.$$

Here H^1 is the usual Sobolev space of square-integrable functions with square-integrable gradient, and $||\Psi||$ denotes the L^2 norm of Ψ. The ground state energy (4.6) is well defined whenever the potentials v and w are sufficiently

[1] We follow the usual convention to use the same letter V_{ee} both for the total interaction potential, a function on $\mathbb{R}^{dN}$, and the associated quadratic form, a functional on the wavefunction space $\mathcal{W}^N$.

regular so that the functional E is well defined on $H^1((\mathbb{R}^d \times \mathbb{Z}_2)^N)$. A simple sufficient condition in dimension $d = 3$ which encompasses (4.5), (4.4) is v, $w \in L^{3/2}(\mathbb{R}^3) + L^\infty(\mathbb{R}^3)$.

Whether or not the infimum in (4.6) is actually a minimum, that is, a minimizing Ψ exists, is much more subtle. For neutral or positively charged molecules in dimension $d = 3$ ((4.4), (4.5) with $Z = \sum_{i=1}^M Z_i > N - 1$) the answer is yes, as was proved by Zhislin [145] via a careful spectral analysis of the underlying Hamiltonian operator. For an alternative proof based on variational methods see Friesecke [48].

4.2.2 Universal Density Functional

In quantum mechanics, the absolute value squared $|\Psi(\mathbf{r}_1, s_1, \ldots, \mathbf{r}_N, s_N)|^2$ of a wave function $\Psi \in \mathcal{W}^N$ corresponds to an N-point probability distribution: it gives the probability density of finding the electrons at positions $\mathbf{r}_i \in \mathbb{R}^d$ with spins $s_i \in \mathbb{Z}_2$, $i \in \{1, \ldots, N\}$.

By integrating the N-point probability distribution over the spins, we obtain the N-point position density,

$$\pi_N^\Psi(\mathbf{r}_1, \ldots, \mathbf{r}_N) := \sum_{s_1,\ldots,s_N \in \mathbb{Z}_2} |\Psi(\mathbf{r}_1, s_1, \ldots, \mathbf{r}_N, s_N)|^2, \quad \Psi \in \mathcal{W}^N. \tag{4.8}$$

The single particle density $\rho_\Psi(\mathbf{r}_j)$ is then obtained by integrating out all but one electron position $\mathbf{r}_j \in \mathbb{R}^d$,

$$\rho_\Psi(\mathbf{r}_j) := N \int_{\mathbb{R}^{d(N-1)}} \pi_N^\Psi(\mathbf{r}_1, \mathbf{r}_2, \ldots, \mathbf{r}_j, \ldots, \mathbf{r}_N) \prod_{i \neq j} \mathrm{d}\mathbf{r}_i, \quad \forall\, j \in \{1, \ldots, N\}. \tag{4.9}$$

We denote by $\Psi \mapsto \rho$ the relation between Ψ and ρ given by Eqs. (4.8) and (4.9). This means that the wave function Ψ has single-electron density ρ.

Following the work of Hohenberg and Kohn [78], Levy [91] and Lieb [97] showed that the electronic ground state problem (4.6) can be recast as a minimization over single-electron densities ρ instead of many-electron wavefunctions Ψ:

$$E_0[v_{\mathrm{ne}}] = \inf_{\rho \in \mathcal{D}^N} \left\{ F_{\mathrm{LL}}[\rho] + N \int_{\mathbb{R}^d} v_{\mathrm{ne}}(\mathbf{r}) \rho(\mathbf{r}) \mathrm{d}\mathbf{r} \right\}, \tag{4.10}$$

with

$$F_{\mathrm{LL}}[\rho] = \min \left\{ T[\Psi] + V_{\mathrm{ee}}[\Psi] : \Psi \in \mathcal{W}^N, \Psi \mapsto \rho \right\}, \tag{4.11}$$

where F_{LL} is the Levy–Lieb functional. The above direct definition of F_{LL} by a constrained search replaced an earlier, indirect existence proof of a universal functional satisfying (4.10) [78]. The space $\mathcal{D}^N$ is defined as the set of densities ρ coming from a wave function $\Psi \in \mathcal{W}^N$ (i.e, $\Psi \mapsto \rho$), i.e., the N-representable one-particle densities. It can be fully characterized [97] and is given by

$$\mathcal{D}^N = \left\{\rho \in L^1(\mathbb{R}^d) : \rho \geqslant 0, \sqrt{\rho} \in H^1(\mathbb{R}^d), \int_{\mathbb{R}^d} \rho = N\right\}. \tag{4.12}$$

Also, it is known that the minimum in (4.11) is attained. For more details about these matters see Chap. 3 by Lewin et al.

4.3 The Strictly Correlated Electrons (SCE) Functional

4.3.1 Constrained-Search Definition

From the early days of DFT it has been clear that a useful approximation to the kinetic energy contribution in (4.11) is given by the functional

$$T_{\text{S,LL}}[\rho] = \min_{\Psi \in \mathcal{W}^N,\, \Psi \mapsto \rho} \langle \Psi | T | \Psi \rangle \tag{4.13}$$

and by its further approximation $T_{\text{S}}[\rho]$ obtained by Kohn and Sham [85] via restricting the above search to Slater determinants built from orthonormal spin orbitals,

$$T_{\text{S}}[\rho] = \min \left\{ \sum_{i=1}^{N} \sum_{s \in \mathbb{Z}_2} \int_{\mathbb{R}^d} \tfrac{1}{2} |\nabla \varphi_i(\mathbf{r}, s)|^2 \ : \ \varphi_i \in H^1(\mathbb{R}^d \times \mathbb{Z}_2; \mathbb{C})\ \forall i, \right.$$
$$\left. \langle \varphi_i | \varphi_j \rangle = \delta_{ij}\ \forall i, j,\ \sum_{i=1}^{N} \sum_{s \in \mathbb{Z}_2} |\varphi_i(\mathbf{r}, s)|^2 = \rho(\mathbf{r})\ \forall \mathbf{r} \right\}. \tag{4.14}$$

The natural analogue of $T_{s,LL}$ for the interaction energy contribution in (4.11) is the SCE functional

$$V_{\text{ee}}^{\text{SCE}}[\rho] = \inf_{\Psi \in \mathcal{W}^N,\, \Psi \mapsto \rho} \langle \Psi | V_{\text{ee}} | \Psi \rangle, \tag{4.15}$$

which was introduced by Seidl [122]. The acronym SCE stands for *strictly correlated electrons*, and will be explained shortly. As detailed in the next section, the functional (4.15) is a rigorous leading-order asymptotic limit of $F_{\text{LL}}[\rho]$ in the low-density regime, where interaction dominates, just as the kinetic functional (4.13) is a leading-order asymptotic limit at high density, where the kinetic energy dominates.

What is more, there also exists a natural analogue to T_S for interaction, which approximates, in the case of N electrons in the physical space $\mathbb{R}^3$, the high-dimensional minimization over wavefunctions on $3N$ dimensional space in (4.15) by a minimization over just N maps on $\mathbb{R}^3$; see Sect. 4.3.4.

4.3.2 Derivation as a Low-Density or Strong-Interaction Limit of the Levy–Lieb Functional

For any given N-particle density ρ on $\mathbb{R}^d$, consider its dilation obtained by uniform coordinate scaling

$$\rho^\gamma(\mathbf{r}) = \gamma^d \rho(\gamma \mathbf{r}),$$

where $\gamma > 0$ is a scaling factor. Note that this scaling preserves the total particle number,

$$\int_{\mathbb{R}^d} \rho^\gamma(\mathbf{r}) = \int_{\mathbb{R}^d} \rho(\mathbf{r}) = N.$$

We are interested in the small-γ regime, which corresponds to a low-density limit.

If Ψ is a wavefunction with density ρ, then the scaled wavefunction

$$\Psi^\gamma(\mathbf{r}_1, s_1, \ldots, \mathbf{r}_N, s_N) = \gamma^{\frac{dN}{2}} \Psi(\gamma \mathbf{r}_1, s_1, \ldots, \gamma \mathbf{r}_N, s_N)$$

has density ρ^γ. But as first noticed by Levy and Perdew [92], scaling does not commute with constrained search. Instead, by an elementary change of variables,

$$T[\Psi^\gamma] = \gamma^2 T[\Psi], \quad V_{\text{ee}}[\Psi^\gamma] = \gamma V_{\text{ee}}[\Psi],$$

and therefore

$$\begin{aligned} F_{\text{LL}}[\rho^\gamma] &= \min_{\Psi^\gamma \in \mathcal{W}_N,\, \Psi^\gamma \mapsto \rho^\gamma} \left\langle \Psi^\gamma | T + V_{\text{ee}} | \Psi^\gamma \right\rangle \\ &= \gamma \min_{\Psi \in \mathcal{W}_N,\, \Psi \mapsto \rho} \left\langle \Psi | \gamma T + V_{\text{ee}} | \Psi \right\rangle \\ &= \gamma^2 F^{1/\gamma}[\rho], \end{aligned} \tag{4.16}$$

where

$$F^{\lambda}[\rho] = \min_{\Psi \in \mathcal{W}_N,\, \Psi \mapsto \rho} \Big(T[\Psi] + \lambda V_{\mathrm{ee}}[\Psi] \Big) \tag{4.17}$$

is a Levy–Lieb functional with coupling constant λ. This suggests, assuming that the minimization in the second line of (4.16) commutes with taking the limit $\gamma \to 0$,

$$F_{\mathrm{LL}}[\rho^{\gamma}] \underset{\gamma \to 0}{\sim} \gamma\, V_{\mathrm{ee}}^{\mathrm{SCE}}[\rho] \tag{4.18}$$

or equivalently, by starting from the Levy–Lieb functional with coupling constant, Eq. (4.17), as done in [122, 127]

$$\lim_{\lambda \to \infty} \tfrac{1}{\lambda} F^{\lambda}[\rho] = V_{\mathrm{ee}}^{\mathrm{SCE}}[\rho]. \tag{4.19}$$

Mathematically, as pointed out in [127] it is not obvious whether the minimization in the second line of (4.16) commutes with passing to the limit $\gamma \to 0$ since the optimal wavefunction depends on γ. Nevertheless the above leading-order asymptotics can be rigorously justified; see Theorem 4.3 in the next section.

Repeating the calculation in (4.16) without the kinetic energy and replacing "min" by "inf" shows that

$$V_{\mathrm{ee}}^{\mathrm{SCE}}[\rho^{\gamma}] = \gamma V_{\mathrm{ee}}^{\mathrm{SCE}}[\rho], \tag{4.20}$$

whence the asymptotic result (4.18) can also be re-written as

$$F_{\mathrm{LL}}[\rho^{\gamma}] \underset{\gamma \to 0}{\sim} V_{\mathrm{ee}}^{\mathrm{SCE}}[\rho^{\gamma}]. \tag{4.21}$$

Off the low-density limit, we remark that $V_{\mathrm{ee}}^{\mathrm{SCE}}$ still provides a rigorous lower bound for the Levy–Lieb functional,

$$F_{\mathrm{LL}}[\rho] \geq V_{\mathrm{ee}}^{\mathrm{SCE}}[\rho] \quad \forall \rho \in \mathcal{D}^N. \tag{4.22}$$

This is a trivial consequence of the constrained-search Definitions (4.11) and (4.15) and the nonnegativity of the kinetic energy functional T. For typical atomic densities on $\mathbb{R}^3$, this lower bound is a significant improvement over the Lieb–Oxford bound with best known constant.

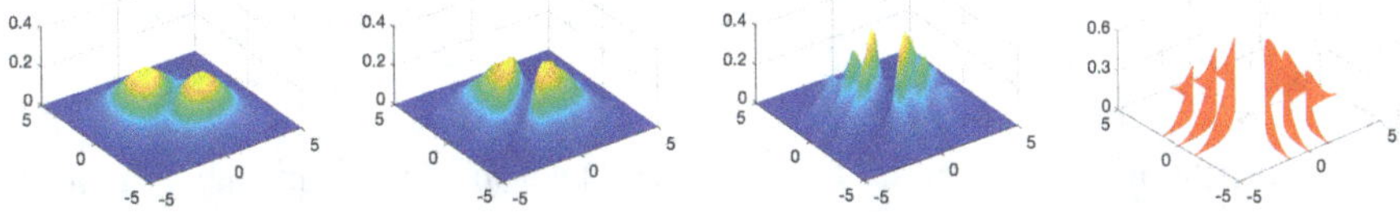

Fig. 4.1 Numerically computed ground state wave functions for $F^{\lambda}[\rho]$ in (4.17) for $N = 4$ and one-body density $\rho(r) = \frac{1}{2L}(1+\cos(\frac{\pi}{L}r))$, $r \in [-L, L]$ $(L = 5)$ for different values of λ: $\lambda = 0.1$, 1, 10, ∞. Shown: pair density $\sum_{s_1,s_2,s_3,s_4} \int dr_3\, dr_4 |\Psi^{\lambda}(r_1, s_1, r_2, s_2, r_3, s_3, r_4, s_4)|^2$. Picture from [21], see also [109] for a numerical approximation of (4.17) with $N = 2$. The pair density on the left is governed by exchange effects, whereas the one on the right is governed purely by Coulombic correlations

4.3.3 *Enlarging the Constrained Search to Probability Measures*

The variational principle underlying the definition of $V_{ee}^{SCE}[\rho]$ in (4.15),

$$\text{Minimize } \langle\Psi|V_{ee}|\Psi\rangle = \int_{\mathbb{R}^{dN}} V_{ee}(\mathbf{r}_1, \ldots, \mathbf{r}_N)\, \pi_N^{\Psi}(\mathbf{r}_1, \ldots, \mathbf{r}_N)\, d\mathbf{r}_1 \ldots d\mathbf{r}_N$$
$$\text{over } \{\Psi \in \mathcal{W}^N \ : \ \Psi \mapsto \rho\}, \tag{4.23}$$

with N-point density π_N^{Ψ} as in (4.8), typically has no minimizer. That is, no minimizing wavefunction $\Psi \in \mathcal{W}^N$ exists and the infimum in (4.15) is not attained.[2] Physically, this reflects the phenomenon that if $\Psi_\lambda[\rho]$ is a sequence of square-integrable functions depending on a parameter $\lambda > 0$ such that $\langle\Psi_\lambda|V_{ee}|\Psi_\lambda\rangle$ approaches the infimum in (4.15) as λ tends to infinity—prototypical is the Ψ_λ that minimizes $\langle\Psi|T + \lambda V_{ee}|\Psi\rangle$ subject to $\Psi \mapsto \rho$—then $|\Psi_\lambda|^2$ integrates to 1 but is typically concentrating on a lower dimensional subset, as depicted in Fig. 4.1.

This basic shortcoming of (4.23)—that wavefunctions which are closer and closer to being optimal in the constrained search (4.15) do not converge to any proper wavefunction—can be overcome as follows [14, 32]. First, interpret the variational principle (4.23) as a variational principle for the N-point density as suggested by the second expression in (4.23); second, enlarge the space of admissible N-point densities to the space $\mathcal{P}(\mathbb{R}^{dN})$ of *probability measures* on $\mathbb{R}^{dN}$ with density ρ. Then the constrained search becomes well-posed, that is, optimizers exist. See Theorem 4.1 below. This enlargement allows N-point densities to concentrate on lower dimensional subsets as in Fig. 4.1. The condition that a probability measure $\Pi \in \mathcal{P}(\mathbb{R}^{dN})$ has density ρ now means that Π has marginals equal to the density divided by the particle number, $\frac{\rho}{N}$:

[2] This is not cured by dropping the requirement in (4.7) that Ψ must have square-integrable gradient and requiring mere square-integrability, i.e. replacing $\bigwedge_{i=1}^{N} H^1(\mathbb{R}^d \times \mathbb{Z}_2; \mathbb{C})$ by $\bigwedge_{i=1}^{N} L^2(\mathbb{R}^d \times \mathbb{Z}_2; \mathbb{C})$.

$$\int_{(\mathbb{R}^d)^{j-1}\times A_j\times(\mathbb{R}^d)^{N-j}} \mathrm{d}\Pi = \int_{A_j} \frac{\rho}{N} \text{ for all } j \in \{1,\ldots,N\} \text{ and all open sets } A_j \text{ in } \mathbb{R}^d. \tag{4.24}$$

We denote the relation given by Eq. (4.24) by $\Pi \mapsto \rho$. This yields the variational principle

$$\text{Minimize } \int_{\mathbb{R}^{dN}} V_{\mathrm{ee}}(\mathbf{r}_1,\ldots,\mathbf{r}_N)\,\mathrm{d}\Pi(\mathbf{r}_1,\ldots,\mathbf{r}_N) \text{ over } \{\Pi \in \mathcal{P}(\mathbb{R}^{dN}) \,:\, \Pi \mapsto \rho\} \tag{4.25}$$

and the following enlarged-constrained-search definition of the SCE functional

$$V_{\mathrm{ee}}^{\mathrm{SCE}}[\rho] = \min_{\Pi\in\mathcal{P}(\mathbb{R}^{dN}),\,\Pi\mapsto\rho} \int_{\mathbb{R}^{dN}} V_{\mathrm{ee}}(\mathbf{r}_1,\ldots,\mathbf{r}_N)\,\mathrm{d}\Pi(\mathbf{r}_1,\ldots,\mathbf{r}_N). \tag{4.26}$$

This alternative definition of $V_{\mathrm{ee}}^{\mathrm{SCE}}[\rho]$ and the underlying enlarged variational problem (4.25) were introduced by Buttazzo et al. [14] and Cotar et al. [32], along with the insight that minimizers now exist (see Theorem 4.1 (1) below) and (4.26) is mathematically an *optimal transport problem* and can be usefully analyzed with methods from optimal transport theory (see Sect. 4.3.6). We call (4.25) the SIL variational principle, the acronym SIL standing for strong-interaction limit.

The notation in (4.26) ("min" instead of "inf"; using the same notation for the ensuing density functional even though a priori the right-hand side of (4.26) could be lower than that in (4.15) since the minimization is over a larger set) is justified because of:

Theorem 4.1 *Let ρ be any N-particle density in the class $\mathcal{D}^N$ (see* (4.12)*), and let $w(\mathbf{r}) = |\mathbf{r}|^{-1}$ be the Coulomb interaction.*

(1) *The minimum in* (4.26) *is attained; that is, there exists a minimizing probability measure Π.*
(2) [10, 32, 33] *The minimum value in* (4.26) *is equal to the infimum in* (4.15).

Statement (1) is a special case of general existence theorems in optimal transport theory. For a textbook account see [50]. Proofs of such results rely on Prokhorov's theorem from probability theory as well as on approximation and lower semi-continuity results for functionals of the form $\Pi \mapsto \int V_{\mathrm{ee}}\mathrm{d}\Pi$.

Statement (2), although plausible, is mathematically much more subtle. It rests on the nontrivial result that arbitrary symmetric probability measures $\Pi \in \mathcal{P}(\mathbb{R}^{dN})$ with marginal ρ can be approximated by N-point densities of quantum wavefunctions $\Psi \in \mathcal{W}^N$ with the *same* marginal. Note that such wavefunctions must be antisymmetric and must have a square-integrable gradient; but applying standard smoothing techniques from mathematics—such as mollification—to a given probability measure with marginal ρ does not preserve the marginal, nor does it yield the N-point density of an antisymmetric function. This result, and the ensuing statement (2), was first proved for $N = 2$ [32], and later extended to $N = 3$ [10] and general N [33] (see also [95] for a similar extension to general N allowing mixed states).

Remark 4.2 (Symmetrization) The minimum value in (4.26) is unchanged, and still attained, when the minimization over arbitrary probability measures with marginal ρ/N, $\{\Pi \in \mathcal{P}(\mathbb{R}^{dN}) : \Pi \mapsto \rho\}$, is restricted to *symmetric* probability measures with marginal ρ/N, where a probability measure $\Pi \in \mathcal{P}(\mathbb{R}^{dN})$ is said to be symmetric if

$$\int_{A_1\times\ldots\times A_N} d\Pi = \int_{A_{\sigma(1)}\times\ldots\times A_{\sigma(N)}} d\Pi \text{ for all open sets } A_1, \ldots, A_N \text{ in } \mathbb{R}^d \text{ and all permutations } \sigma.$$

This is because whenever Π is a probability measure in $\mathcal{P}(\mathbb{R}^{dN})$ with marginals ρ, Eq. (4.24), then so is its symmetrization $S_N\Pi$ defined by

$$(S_N\Pi)(A_1 \times \ldots \times A_N) = \frac{1}{N!}\sum_\sigma \Pi(A_{\sigma(1)} \times \ldots \times A_{\sigma(N)}), \tag{4.27}$$

the sum being over all permutations of $\{1, \ldots, N\}$; and the integral on the r.h.s. of (4.26) for Π agrees with that for $S_N\Pi$, thanks to the permutation symmetry of V_{ee}.

Next we rigorously justify the asymptotic relations (4.18), (4.19), (4.21) and complement them with an asymptotic result on the associated constrained-search wavefunctions.

Theorem 4.3 ([33])) *For any N-electron density ρ in the class $\mathcal{D}^N$ (see* (4.12)*), and with $w(\mathbf{r}) = |\mathbf{r}|^{-1}$ being the Coulomb interaction, the asymptotic results* (4.18), (4.19), (4.21) *hold. Moreover if $\Psi_\lambda[\rho]$ is any minimizer in the constrained-search definition of $F^\lambda[\rho]$ (see* (4.17)*), then every limit point*[3] *Π of the sequence of N-point densities $\pi^{\Psi_\lambda[\rho]}$ is a minimizer in the enlarged-search Definition* (4.26) *of $V_{ee}^{SCE}[\rho]$.*

[3] By a limit point Π of a sequence Π_λ of probability measures we mean a limit point in the sense of narrow convergence, that is, convergence of the integrals $\int f\, d\Pi_\lambda$ to $\int f\, d\Pi$ for any bounded continuous function f.

Proof of (4.18), (4.19), ***and*** (4.21) The proof, taken from [33], is easy, so we include it. We show (4.19), the other statements being equivalent. Fix ρ. First, pick any minimizer $\Psi^{\lambda}[\rho]$ in the constrained-search definition of $F^{\lambda}[\rho]$, then

$$\frac{1}{\lambda}F^{\lambda}[\rho] = \frac{1}{\lambda}\Big(T[\Psi^{\lambda}[\rho]] + \lambda V_{\text{ee}}[\Psi^{\lambda}[\rho]]\Big) \geq V_{\text{ee}}[\Psi^{\lambda}[\rho]] \geq V_{\text{ee}}^{\text{SCE}}[\rho], \tag{4.28}$$

that is, the SCE functional is a lower bound of the left-hand side. To show that it is also an asymptotic upper bound for large λ, we fix any positive number ϵ and pick a wavefunction $\widetilde{\Psi}[\rho]$ in $\mathcal{W}^N$ such that $V_{\text{ee}}[\widetilde{\Psi}[\rho]] \leq V_{\text{ee}}^{SCE}[\rho] + \epsilon$. It follows that

$$\frac{1}{\lambda}F^{\lambda}[\rho] \leq \frac{1}{\lambda}\Big(T[\widetilde{\Psi}[\rho]] + \lambda V_{\text{ee}}[\widetilde{\Psi}[\rho]]\Big).$$

Since $\widetilde{\Psi}$ belongs to $\mathcal{W}^N$, its kinetic energy $T[\widetilde{\Psi}]$ is finite, and so

$$\limsup_{\lambda\to\infty} \frac{1}{\lambda}F^{\lambda}[\rho] \leq V_{\text{ee}}[\widetilde{\Psi}[\rho]] \leq V_{\text{ee}}^{\text{SCE}}[\rho] + \epsilon.$$

Since $\epsilon > 0$ was arbitrary,

$$\limsup_{\lambda\to\infty} \frac{1}{\lambda}F^{\lambda}[\rho] \leq V_{\text{ee}}^{\text{SCE}}[\rho]. \tag{4.29}$$

Combining (4.28) and (4.29) yields (4.19). □

The above simple argument only shows that the asymptotic error in (4.19) is $o(1/\lambda)$, but does not give its order, which turns out to be $O(1/\lambda^{1/2})$, see Sect. 4.3.5.

4.3.4 The SCE Ansatz

The SIL variational principle (4.25) still requires minimization over a high-dimensional space of N-point probability measures.

Seidl [122] (see also [127]) proposed the following low-dimensional ansatz: we restrict minimization over N-point probability measures to minimization over singular probability measures of the special form

$$d\Pi(\mathbf{r}_1, \ldots, \mathbf{r}_N) = \frac{\rho(\mathbf{r}_1)}{N} \prod_{n=2}^{N} \delta\big(\mathbf{r}_n - f_{n-1}(\mathbf{r}_1)\big)d\mathbf{r}_1 \ldots d\mathbf{r}_N \tag{4.30}$$

where, for any $\mathbf{r}_1 \in \mathbb{R}^d$, $\delta\big(\mathbf{r}_n - f_{n-1}(\mathbf{r}_1)\big)$ denotes the delta function of $\mathbf{r}_n$ (alias Dirac measure) centered at $f_{n-1}(\mathbf{r}_1)$, and $f_1, \ldots, f_{N-1}$ are maps from $\mathbb{R}^d$ to $\mathbb{R}^d$. The singular densities (4.30) are concentrated on the d-dimensional set

$$\Omega_0 = \{(\mathbf{r}_1, \ldots, \mathbf{r}_N) \in \mathbb{R}^{dN} \ : \mathbf{r}_2 = f_1(\mathbf{r}_1), \ldots, \mathbf{r}_N = f_N(\mathbf{r}_1)\}. \tag{4.31}$$

From a physical point of view, such a density describes a state in which the position of one of the electrons, say $\mathbf{r}_1$, can be freely chosen according to the density ρ, but this then uniquely fixes the position of all the other electrons through the functions $f_2, \ldots, f_N$, that is, $\mathbf{r}_2 = f_1(\mathbf{r}_1)$ etc. Thus states of form (4.30) are called *strictly correlated states*, or SCE states for short. The f_i are called *co-motion functions* or *transport maps*.

The marginal constraint that Π must have marginals ρ, Eq. (4.24), turns into the following constraint on the maps f_n: the f_n must transport the density ρ to itself,

$$f_{i\sharp}\rho = \rho \quad \forall i \in \{2, \ldots, N\} \tag{4.32}$$

where, for any measurable map $f \ : \mathbb{R}^p \to \mathbb{R}^q$ and any measure μ on $\mathbb{R}^p$, the push-forward $f_\sharp\mu$ is the measure on $\mathbb{R}^q$ defined by

$$(f_\sharp\mu)(B) = \mu(f^{-1}(B)) \text{ for all open sets } B \text{ in } \mathbb{R}^q. \tag{4.33}$$

More explicitly, if $p = q$, μ is absolutely continuous with density ρ, f is a diffeomorphism, and the density of the push-forward $f_\sharp\mu$ is denoted by $f_\sharp\rho$, we have

$$(f_\sharp\rho)(\mathbf{r}') = |\text{det}Df^{-1}(\mathbf{r}')|\, \rho\big(f^{-1}(\mathbf{r}')\big).$$

By substituting this formula for the push-forward into (4.32) and changing variables $f^{-1}(\mathbf{r}') = \mathbf{r}$, the constraint (4.32) turns—provided the f_n are diffeomorphisms—into the following nonlinear first-order partial differential equation:

$$\rho(f_i(\mathbf{r})) = \frac{\rho(\mathbf{r})}{|\text{det}Df_i(\mathbf{r})|} \quad \forall i \in \{2, \ldots, N\}.$$

Plugging the ansatz (4.30) into the SIL variational principle (4.25) and integrating out the variables $\mathbf{r}_2, \ldots, \mathbf{r}_N$ yields the SCE variational principle

$$\text{Minimize} \int_{\mathbb{R}^d} V_{\text{ee}}\big(\mathbf{r}_1, f_1(\mathbf{r}_1), \ldots, f_{N-1}(\mathbf{r}_1)\big)\, \frac{\rho(\mathbf{r}_1)}{N}\, d\mathbf{r}_1 \tag{4.34}$$

$$\text{over maps } f_1, \ldots, f_{N-1} \in \mathcal{T}_\rho,$$

with the minimization being over maps in the admissible class

$$\mathcal{T}_\rho = \{f : \mathbb{R}^d \to \mathbb{R}^d : f \text{ measurable}, \ f_\sharp\rho = \rho\}. \tag{4.35}$$

Thanks to Theorem 4.4 (1) below, this yields a third construction of the SCE functional,

$$V_{\rm ee}^{\rm SCE}[\rho] = \inf_{f_1,\dots,f_{N-1}\in\mathcal{T}_\rho} \int_{\mathbb{R}^d} V_{\rm ee}\big(\mathbf{r}, f_1(\mathbf{r}),\dots,f_{N-1}(\mathbf{r})\big)\,\frac{\rho(\mathbf{r})}{N}\,d\mathbf{r}. \tag{4.36}$$

In the Coulomb case, (4.4), and denoting $f_0(\mathbf{r}) = \mathbf{r}$, we thus have

$$V_{\rm ee}^{\rm SCE}[\rho] = \inf_{f_1,\dots,f_{N-1}\in\mathcal{T}_\rho} \sum_{0\le i<j\le N-1} \int_{\mathbb{R}^d} \frac{1}{|f_i(\mathbf{r}) - f_j(\mathbf{r})|}\,\frac{\rho(\mathbf{r})}{N}\,d\mathbf{r}. \tag{4.37}$$

Physically, this means that one needs to minimize the mutual Coulomb repulsion of the co-motion functions. This construction of the SCE functional was introduced by Seidl [122]. A priori it is not clear, but was conjectured by Seidl, that it is equivalent to the original construction (4.15). This is now rigorously known (see Corollary 4.6 below).

The construction (4.36) should be considered the analogue for interaction of the classical Kohn–Sham kinetic energy functional T_S. Just as T_S is determined by N low-dimensional functions (the Kohn–Sham spin orbitals $\varphi_1,\dots\varphi_N : \mathbb{R}^3\times\mathbb{Z}_2 \to \mathbb{C}$), $V_{\rm ee}^{\rm SCE}$ is determined by $N-1$ low-dimensional maps (the co-motion functions or transport maps $f_1,\dots,f_{N-1} : \mathbb{R}^3 \to \mathbb{R}^3$), which can be easily stored on a computer. Moreover—like the Kohn–Sham orbitals—the co-motion functions are obtained by just minimizing a 3-dimensional integral.

The reader is warned, however, that the behavior of the SCE variational principle and its relationship to the SIL variational principle is subtle, and open questions remain. In particular, it is not known—except in special cases—whether minimizers in (4.37) exist. The following results have been rigorously proved.

Theorem 4.4 *Let $\rho : \mathbb{R}^d \to \mathbb{R}$ be any N-particle density in the class $\mathcal{D}^N$ (see* (4.12)*), and let $w(\mathbf{r}) = |\mathbf{r}|^{-1}$ be the Coulomb interaction.*

(1) *The infimum in* (4.36) *is equal to the minimum in* (4.26).
(2) *For two electrons ($N = 2$), and in arbitrary space dimension d, the infimum in* (4.36) *is attained; that is, there exists a minimizing map f_1. Moreover f_1 is unique, and the induced probability measure* (4.30) *is the unique minimizer of the SIL variational principle* (4.25).
(3) *In one space dimension ($d = 1$), and for arbitrary N, the infimum in* (4.36) *is attained; that is, there exist minimizing maps $f_1,\dots,f_{N-1}$. Moreover the symmetrization (see Remark 4.2) of the associated probability measure* (4.30) *is the unique symmetric minimizer of the SIL variational principle* (4.25).

Statement (1) is a consequence, pointed out in [28], of a general theorem of Ambrosio [5] and Pratelli [116] in optimal transport theory. For $N = 2$ or $d = 1$, use of the Ambrosio–Pratelli theorem can be avoided since the assertion follows from (2) respectively (3).

The existence of optimal maps in (2) and (3) is subtle and depends on special Coulombic features. For non-Coulombic counterexamples see Remark 4.7 below. In the Coulomb case, it is an open question whether the infimum in (4.36) is attained for general (physically reasonable) densities ρ when $d > 1$ and $N \geq 3$.

Statement (2) completely justifies Seidl's SCE ansatz for $N = 2$: the SCE problem

$$\text{Minimize} \int_{\mathbb{R}^d} \frac{1}{|\mathbf{r} - f_1(\mathbf{r})|} \text{ over maps } f_1 \in \mathcal{T}_\rho$$

has a unique minimizer and the associated SCE state

$$\mathrm{d}\Pi(\mathbf{r}_1, \mathbf{r}_2) = \frac{\rho(\mathbf{r}_1)}{2} \delta\big(\mathbf{r}_2 - f_1(\mathbf{r}_1)\big)\, \mathrm{d}\mathbf{r}_1 \mathrm{d}\mathbf{r}_2 \tag{4.38}$$

is the unique minimizer of the SIL problem

$$\text{Minimize} \int_{\mathbb{R}^d \times \mathbb{R}^d} \frac{1}{|\mathbf{r}_1 - \mathbf{r}_2|} \mathrm{d}\Pi(\mathbf{r}_1, \mathbf{r}_2) \text{ over } \Pi \mapsto \rho.$$

This was proved in [32], by modifying the analysis by Gangbo and McCann [55] of optimal transport with costs $w(\mathbf{r}, \mathbf{r}')$ which are convex or concave in the displacement $\mathbf{z} = \mathbf{r} - \mathbf{r}'$. Note that the Coulomb cost is neither: near any displacement $\mathbf{z}_0 \neq 0$, it is convex in radial direction and concave in all perpendicular directions. A simpler proof using Kantorovich duality (see Sect. 4.3.7) was suggested in [14], and made rigorous in [42]. The SCE map is given by

$$f_1(\mathbf{r}) = \mathbf{r} + \frac{\nabla u(\mathbf{r})}{|\nabla u(\mathbf{r})|^{3/2}}, \tag{4.39}$$

for some function $u : \mathbb{R}^d \to \mathbb{R}$ (Kantorovich potential). The notion of Kantorovich potential will be explained in Sect. 4.3.7. Equation (4.39) follows by solving Eq. (4.59) for f_1.

Statement (3), together with an explicit construction of the optimal maps given in Sect. 4.3.11, was suggested in the original paper by Seidl [122] on grounds of physical arguments, and was rigorously proved in [27] with the help of cyclical monotonicity methods from optimal transport theory. See Sect. 4.3.11 for more information.

The uniqueness statements in (2) and (3) are somewhat surprising: the optimal N-point densities arising from Levy–Lieb constrained search in the strongly interacting

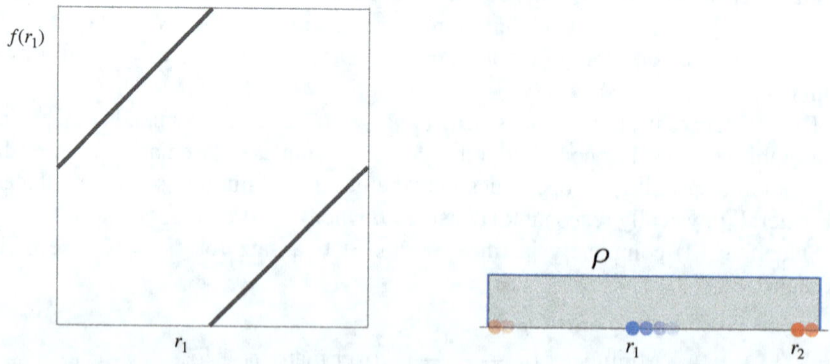

Fig. 4.2 SCE state of a two-electron system with homogeneous density in a one-dimensional domain. By Theorems 4.3 and 4.4, this state is an asymptotically exact approximation to the true quantum ground state at low density. Left: optimal co-motion function or transport map f. Right: position of the two electrons in the one-dimensional domain. The position of the second electron, r_2, is determined by that of the first electron, r_1, through the equation $r_2 = f(r_1)$, with the optimal f keeping the electrons at a constant distance, of half the domain size. The position of the first electron varies over the whole domain according to the density ρ (see Eq. (4.38)). As the first electron (depicted in blue) passes through the mid-point, the position of the second electron (depicted in red) jumps from the right end to the left end, causing a discontinuity of f

limit are always *unique* when either $N = 2$ or $d = 1$! No analogue holds off the strongly interacting limit.

Example 4.5 Consider a two-electron system with uniform density in a one-dimensional interval $[0, L]$. The unique minimizer $f_1 = f$ of the SCE variational principle (4.34) can be shown (see Sect. 4.3.11) to be

$$f(r_1) = \begin{cases} r_1 + \frac{L}{2} & \text{if } r_1 \le \frac{L}{2} \\ r_1 - \frac{L}{2} & \text{if } r_1 > \frac{L}{2}. \end{cases} \tag{4.40}$$

See Fig. 4.2.

By combining Theorems 4.4 and 4.1 we obtain:

Corollary 4.6 *Let* $\rho : \mathbb{R}^d \to \mathbb{R}$ *be any* N*-particle density in the class* $\mathcal{D}^N$ *(see* (4.12)*), and let* $w(\mathbf{r}) = |\mathbf{r}|^{-1}$ *be the Coulomb interaction. Then the infimum in* (4.36) *is equal to that in* (4.15)*.*

Proof This follows from the fact that both quantities are equal to the minimum value of the SIL variational principle (4.25), by Theorem 4.1 (2) respectively Theorem 4.4 (1). □

We remark that no proof is known which bypasses the SIL variational principle, even though the corollary was conjectured before the latter was introduced.

We close this introductory section on the SCE ansatz with some remarks.

Remark 4.7 (Nonattainment) For simple non-Coulombic counterexamples to attainment of the infimum in (4.36) for $N = 3$ even in one space dimension, see [49, 59]. For instance, one can take the uniform density in the interval $[0, 3]$ and the interaction potential $w(r) = r^4/4 - r^3/3$ [49]. Earlier more intricate counterexamples can be found in [107]. Such a nonattainment has the undesirable consequence that numerically computed optimal maps will necessarily exhibit wilder and wilder oscillations as the mesh is refined or the basis set approaches completeness, and fail to converge in any pointwise sense to actual optimal maps.

Remark 4.8 (Existence of Non-SCE Minimizers) For a Coulombic example for $N = 3$ in three space dimensions showing that the SIL variational principle can possess minimizers which are not of SCE form, see [113]. This example exhibits nonuniqueness and it is not known whether it also admits minimizers which are of SCE form.

Remark 4.9 (Alternative Formulations of the SCE Ansatz) Denoting $f_0(\mathbf{r}) = \mathbf{r}$, one can write the SCE ansatz (4.30) in the following form in which all coordinates $\mathbf{r}_1, \ldots, \mathbf{r}_N$ appear on an equal footing:

$$d\Pi(\mathbf{r}_1, \ldots, \mathbf{r}_N) = \int \frac{\rho(\mathbf{r})}{N} \prod_{n=1}^{N} \delta\big(\mathbf{r}_n - f_{n-1}(\mathbf{r})\big)\, d\mathbf{r}. \tag{4.41}$$

Also, one can work with the symmetrized form of this ansatz,

$$d\Pi(\mathbf{r}_1, \ldots, \mathbf{r}_N) = \frac{1}{N!} \sum_{\sigma} \int \frac{\rho(\mathbf{r})}{N} \prod_{n=1}^{N} \delta\big(\mathbf{r}_n - f_{\sigma(n-1)}(\mathbf{r})\big)\, d\mathbf{r} \tag{4.42}$$

(where σ runs over the permutations of the SCE map indices $0, \ldots, N - 1$); the symmetrization doesn't change the energy $\int V_{\text{ee}}\, d\Pi$, and the symmetrized form (4.42) minimizes the SIL problem (4.26) if and only if the unsymmetrized form (4.41) does, as was explained in Remark 4.2.

Remark 4.10 (Nonsmoothness of Optimal Maps) The reader might wonder why, in SCE theory, no differentiability and not even continuity is imposed on the competing maps (the maps in the admissible class (4.35) are merely required to be measurable). This is because optimal maps, when they exist, are typically discontinuous. This important effect can be understood intuitively from simple examples as in Fig. 4.2. As the first electron passes through the midpoint of the domain, the position of the second electron jumps from the right end of the domain to the left end, yielding the discontinuous map depicted in the Figure. For a radial density in three dimensions ($d = 3$), an analogous discontinuity occurs in that spheres near zero are mapped to spheres near infinity [32]. For general densities and general N, the presence of discontinuities across unknown surfaces makes Eq. (4.36) very challenging for numerical computations.

4.3.5 The Next Leading Term

So far we have treated the limit of the Levy–Lieb functional at infinite coupling strength λ (or, equivalently, at extreme low density). One could ask how this limit is approached, or, in other words, what is the next leading term in Eqs. (4.18)–(4.19).

The strategy employed in [68] to compute this next leading term relies on the assumption that the minimizer in (4.25) is of the SCE or Monge type, see the detailed discussion in the previous Sect. 4.3.4. Under this assumption, as shown in Sect. 4.3.8, the classical potential energy

$$E_{\text{pot}}(\mathbf{r}_1,\ldots,\mathbf{r}_N) = V_{\text{ee}}(\mathbf{r}_1,\ldots,\mathbf{r}_N) - \sum_{i=1}^{N} v_{\text{SCE}}(\mathbf{r}_i), \tag{4.43}$$

with $v_{\text{SCE}}(\mathbf{r})$ defined by Eqs. (4.59) and (4.64), attains its minimum on the manifold Ω_0 parametrized by the co-motion functions,

$$\Omega_0 = \{(\mathbf{r}_1,\ldots,\mathbf{r}_N) \in \mathbb{R}^{dN} : \mathbf{r}_1 = \mathbf{r}, \mathbf{r}_2 = \mathbf{f}_2(\mathbf{r}),\ldots,\mathbf{r}_N = \mathbf{f}_N(\mathbf{r})\}. \tag{4.44}$$

When λ in Eq. (4.17) is very large but finite, we can expect that the support of the minimizer in Eq. (4.17) be strongly localized around Ω_0, as illustrated by Fig. 4.1 in Sect. 4.3.3. We can then expand E_{pot} around its minimum through second order. The corresponding Hessian matrix $\mathbb{H}(\mathbf{r})$ evaluated on Ω_0, for any fixed $\mathbf{r}$, will have d zero eigenvalues (along the manifold Ω_0) and $dN - d$ positive eigenvalues. By using curvilinear coordinates along the manifold Ω_0 and orthogonal to it, the sought next leading term is determined by adding the kinetic energy to the second-order expansion of E_{pot}, which corresponds to the Hamiltonian of zero-point oscillations in the space orthogonal to Ω_0 [68]. The final result is that Eqs. (4.18)–(4.19) are extended to [66, 68]

$$F_{\text{LL}}[\rho^\gamma] \underset{\gamma\to 0}{\sim} \gamma\, V_{\text{ee}}^{\text{SCE}}[\rho] + \gamma^{3/2} F^{\text{ZPE}}[\rho] \tag{4.45}$$

$$F_\lambda \underset{\lambda\to\infty}{\sim} \lambda V_{\text{ee}}^{\text{SCE}}[\rho] + \sqrt{\lambda}\, F^{\text{ZPE}}[\rho], \tag{4.46}$$

where

$$F^{\text{ZPE}}[\rho] = \frac{1}{2}\int_{\mathbb{R}^d} \frac{\rho(\mathbf{r})}{N} \text{Tr}\left(\sqrt{\mathbb{H}(\mathbf{r})}\right). \tag{4.47}$$

In [71] this term has been computed explicitly for $N = 2$ electrons in 1d and it has been compared with accurate numerical calculations for the Levy-Lieb functional at very large λ, finding excellent agreement.

The intuition that the next term of the Levy–Lieb functional at infinite coupling strength λ should be given by zero-point oscillations around the manifold parametrized by the co-motion functions appeared for the first time in Seidl's seminal work [122]. He also carried out explicit calculations in 3D for the spherically-symmetric case with $N = 2$ electrons, using the co-motion function introduced in Sect. 4.3.12. This idea was extended to the general many-electron case in [68], where it was also found that Seidl's original calculation had a wrong factor 2. Very recently, a rigorous proof of Eqs. (4.45)–(4.47) for the many-electron $1d$ case has been provided by Colombo et al. [29].

4.3.5.1 The Fermionic Statistics

Equations (4.45)–(4.47) are the first-order correction due to kinetic energy in the large-λ (or $\hbar \to 0$) limit of the Levy–Lieb functional. This correction is still independent of the particle statistics. A natural question to ask is then at which order will the fermionic antisymmetry enter.

In Refs. [67, 68] it has been conjectured that the particle statistics enters in the $\lambda \to \infty$ limit at orders $\sim e^{-\sqrt{\lambda}}$. The physical intuition behind this idea is simply that the effect on the energy of antisymmetrization vanishes as the overlap between gaussians centered at each set of strictly-correlated positions (each $\mathbf{r}$ value in Ω_0). The scaling $\sqrt{\lambda}$ of such gaussians comes from the zero-point hamiltonian. This conjecture has been confirmed numerically [71] for the case of $N = 2$ electrons in 1D, again by comparison with accurate numerical calculations of the exact Levy-Lieb functional at large λ.

4.3.6 The Strongly Interacting Limit of DFT from the Point of View of Optimal Transport

We now introduce a fruitful interpretation of the strongly interacting limit of DFT as "optimal transport with Coulomb cost".

Optimal transport theory (see [50, 117, 119, 136] for textbook accounts) is concerned with the following two problems, introduced in special cases in fundamental work by Kantorovich [80] respectively Monge [108]:

(a) *Kantorovich optimal transport problem:* For given probability measures $\mu_1, \ldots, \mu_N$ defined on closed subsets $X_1, \ldots, X_N$ of $\mathbb{R}^d$, find a joint probability measure Π on the product space $X = X_1 \times \ldots \times X_N \subseteq \mathbb{R}^{Nd}$ which minimizes a cost functional

$$C[\Pi] = \int_X c(\mathbf{r}_1, \ldots, \mathbf{r}_N)\, d\Pi(\mathbf{r}_1, \ldots, \mathbf{r}_N)$$

subject to the marginal constraints

$$\int_{X_1 \times \ldots \times X_{i-1} \times A_i \times X_{i+1} \times \ldots \times X_N} \mathrm{d}\Pi = \int_{A_i} \mathrm{d}\mu_i \text{ for all measurable sets } A_i \subseteq X_i$$

$$\text{and all } i \in \{1, \ldots, N\}.$$

Here $c : X_1 \times \ldots \times X_N \to \mathbb{R} \cup \{+\infty\}$ is some given cost function, and the validity of the above constraint is denoted $\Pi \mapsto \mu_1, \ldots, \mu_N$.

(b) *Monge optimal transport problem:* For given probability measures $\mu_1, \ldots, \mu_N$ defined on measurable subsets $X_1, \ldots, X_N$ of $\mathbb{R}^d$ of positive volume which possess integrable densities $p_1, \ldots, p_N$ (i.e. $p_i \in L^1(X_i)$), and a cost function c as above, find measurable maps $f_1, \ldots, f_{N-1}$ with $f_i : X_1 \to X_{i+1}$ which minimize

$$I[f_1, \ldots, f_{N-1}] = \int_{X_1} c\big(\mathbf{r}_1, f_1(\mathbf{r}_1), \ldots, f_{N-1}(\mathbf{r}_N)\big)\, \mathrm{d}\mu_1$$

subject to the marginal constraints

$$f_{i\sharp}p_1 = p_{i+1} \quad \text{for } i \in \{1, \ldots, N-1\}.$$

This corresponds to making the ansatz

$$\mathrm{d}\Pi(\mathbf{r}_1, \ldots, \mathbf{r}_N) = \mathrm{d}\mu_1(\mathbf{r}_1)\delta\big(\mathbf{r}_2 - f_1(\mathbf{r}_1)\big) \cdots \delta\big(\mathbf{r}_N - f_{N-1}(\mathbf{r}_1)\big)\mathrm{d}\mathbf{r}_2 \ldots \mathrm{d}\mathbf{r}_N \tag{4.48}$$

or equivalently—using the notion of push-forward introduced in (4.33)—

$$\Pi = (\mathrm{id}, f_1, \ldots, f_N)_\sharp \mu_1 \tag{4.49}$$

in the Kantorovich problem, where id denotes the identity map $\mathrm{id}(\mathbf{r}_1) = \mathbf{r}_1$.

Example 4.11 (N Equal Marginals, Coulomb Cost) If we take

$$X_1 = \cdots = X_N = \mathbb{R}^d, \mu_1 = \cdots = \mu_N = \frac{\rho}{N}, c(\mathbf{r}_1, \ldots, \mathbf{r}_N) = \sum_{1 \le i < j \le N} \frac{1}{|\mathbf{r}_i - \mathbf{r}_j|}$$

the Kantorovich optimal transport problem is precisely the SIL variational problem, (4.25), and the Monge optimal transport problem is precisely the SCE variational problem, (4.34).

Thus the strongly interacting limit of DFT can be viewed as *optimal transport with Coulomb cost*. This viewpoint, introduced by Buttazzo et al. [14] and Cotar et al. [32], opened the door to much of the current understanding of the strong-interaction limit of DFT.

Example 4.12 (Two Unequal Marginals, Positive Power Cost) The prototype problem of classical optimal transport theory going back to [80, 108] is to instead take

$$N = 2, \quad X_1 = X_2 = \mathbb{R}^d, \quad c(\mathbf{r}_1, \mathbf{r}_2) = |\mathbf{r}_1 - \mathbf{r}_2|^p, \quad p \geq 1.$$

That is, one considers:

- only two marginals;
- unequal instead of equal marginals;
- a positive instead of a negative power of the Euclidean distance as cost.

Denoting $\mu_1 = \mu$, $\mu_2 = \nu$, $f_1 = T$, $\mathbf{r}_1 = x$, $\mathbf{r}_2 = y$, the Kantorovich problem then becomes

$$\text{Minimize } C[\Pi] = \int_{\mathbb{R}^d \times \mathbb{R}^d} |x - y|^p \mathrm{d}\Pi$$
$$(x, y) \text{ over } \Pi \in \mathcal{P}(\mathbb{R}^{2d}) \text{ subject to } \Pi \mapsto \mu,\ \nu \tag{4.50}$$

and the Monge problem becomes

$$\text{Minimize } I[T] = \int_{\mathbb{R}^d} |x - T(x)|^p \mathrm{d}\mu(x) \text{ over measurable maps } T\ :\ \mathbb{R}^d \to \mathbb{R}^d$$
$$\text{subject to } T_\sharp \mu = \nu. \tag{4.51}$$

The analogue of the SCE functional is the optimal cost as a functional of the two prescribed marginals,

$$C_{\text{opt}}[\mu, \nu] = \min\{C[\Pi]\ :\ \Pi \mapsto \mu,\ \nu\} = \inf\{I[T]\ :\ T_\sharp \mu = \nu\}.$$

Its p-th root, $W_p(\mu, \nu) = (C_{\text{opt}}[\mu, \nu])^{1/p}$, is the celebrated p-Wasserstein distance, which is a metric on the space of probability measures.

Thus the SCE functional can be thought of as the *Coulomb analogue of the Wasserstein distance*.

We remark that the motivation of Monge and Kantorovich for considering Example 4.12 came from civil engineering, respectively economics, and explains the name *optimal transport*: Monge thought of moving a given pile of sand on a construction site into a given hole in a way that minimizes the overall distance of transport, with $T(x)$ describing the target position of sand originally located at x and with pile and hole modelled, respectively, by μ and ν. Kantorovich thought of transporting some economic good, say steel, from producers (steel mines) to consumers (factories), at minimal transportation cost; $\Pi(x, y)$ then describes the density of goods transported from location x to location y, and is called a *transport plan*. In the latter context it is natural *not* to make the Monge ansatz

$$d\Pi(x, y) = d\mu(x)\delta\big(y - T(x)\big)dy$$

but instead allow one producer located at x to supply several consumers located at different positions y, i.e. consider the general problem (4.50).

The general question for which costs and marginals the Monge and Kantorovich problems are equivalent, i.e. the Kantorovich problem admits minimizers of Monge form, is not well understood. A sufficient condition [119] for $N = 2$ (and, say, compact convex sets X_1 and X_2 and continuously differentiable costs c) is that the marginal measure μ_1 is absolutely continuous and c satisfies the so-called twist condition that the map $\mathbf{r}_2 \mapsto \nabla_{\mathbf{r}_1} c(\mathbf{r}_1, \mathbf{r}_2)$ be injective for every r_1. For $N > 2$, generalized twist conditions have been studied by Pass [111, 114, 115]; unfortunately these are not satisfied for the Coulomb cost.

4.3.7 Dual Construction of the SCE Functional

We now introduce a fourth—dual—construction of the SCE functional.

A cornerstone principle of optimal transport theory, *Kantorovich duality*, says that the minimum of a given Kantorovich optimal transport problem (see Sect. 4.3.6) equals the supremum of an associated explicit dual problem. The general form of the dual is recalled in Appendix 4.6. For the SIL problem (4.25), the dual problem is the following (see Appendix 4.6 for a quick derivation from general OT theory): maximize the functional

$$J[u] = \sum_{i=1}^{N} \int_{\mathbb{R}^d} u(\mathbf{r})\,\rho(\mathbf{r})\,d\mathbf{r} \tag{4.52}$$

over potentials $u : \mathbb{R}^d \to \mathbb{R}$ which must satisfy the pointwise constraint

$$\sum_{i=1}^{N} u(\mathbf{r}_i) \le V_{\text{ee}}(\mathbf{r}_1, \ldots, \mathbf{r}_N) \;\; \forall \;\; (\mathbf{r}_1, \ldots, \mathbf{r}_N) \in \mathbb{R}^{dN}. \tag{4.53}$$

Maximization is over the admissible class

$$\mathcal{A} = \{u : \mathbb{R}^d \to \mathbb{R} \mid u \text{ bounded and measurable, } u \text{ satisfies (4.53)}\}. \tag{4.54}$$

This yields the following alternative definition of the SCE functional:

$$V_{ee}^{SCE}[\rho] = \sup_{u \in \mathcal{A}} \int_{\mathbb{R}^d} u(\mathbf{r})\, \rho(\mathbf{r})\, d\mathbf{r}. \tag{4.55}$$

This construction is due to Buttazzo, DePascale, and Gori-Giorgi [14]. Note that the optimization here is not over N-point densities, but over (suitable) external potentials u. Optimizers are called *Kantorovich potentials*. Heuristically, they can be thought of as Lagrange multipliers associated with the marginal constraints in the original problem (4.25). This is explained in our discussion of optimality conditions in Sect. 4.3.8.

It can be rigorously shown that the new construction yields, again, the SCE functional, and that optimal potentials exist:

Theorem 4.13 *Let* $\rho : \mathbb{R}^d \to \mathbb{R}$ *be any N-particle density in the class* $\mathcal{D}^N$ *(see* (4.12)*), and let* $w(\mathbf{r}) = |\mathbf{r}|^{-1}$ *be the Coulomb interaction. Then:*

(1) *[14] The supremum in* (4.55) *is equal to the minimum in* (4.26)*.*
(2) *[14, 38] The supremum in* (4.55) *is attained; that is, there exists a maximizing potential u in the class* (4.54)*.*
(3) *[13, 42] If, in addition,* $\rho > 0$ *everywhere, there exists a maximizing potential which is in addition Lipschitz continuous.*

Statement (1) follows directly from the general Kantorovich duality theorem of OT theory; see Appendix 4.6. The question of existence and regularity of optimal potentials is more delicate. Note that the Coulomb potential V_{ee} which upper-bounds $u(\mathbf{r}_1) + \cdots + u(\mathbf{r}_N)$ tends to plus infinity as the distance $\mathbf{r}_i - \mathbf{r}_j$ between any two position coordinates goes to zero; so one might a priori think that u's are favourable which also tend to plus infinity at certain places. But statement (2) in the above theorem says that this does not happen; the existence proof of bounded optimal potentials is due to [14] for $N = 2$ and to [38] for general N.

Quantum Analogue We remark that the dual construction of the SCE functional in Eq. (4.55) admits a quantum analogue. In [97], Lieb proposed an extension of the Levy–Lieb functional (4.11) to mixed states, i.e. $\mathrm{F_L} : \mathcal{D}_N \to \mathbb{R}$,

$$\mathrm{F_L}[\rho] = \tag{4.56}$$

$$\min \left\{ \mathrm{Tr}\left(-\frac{1}{2}\sum_{\mathrm{j}=1}^{\mathrm{N}} \Delta_{\mathbf{r}_\mathrm{j}} + \mathrm{V_{ee}}(\mathbf{r}_1, \ldots, \mathbf{r}_\mathrm{N}) \right) \Gamma \,:\, \Gamma = \Gamma^* \geqslant 0, \mathrm{Tr}(\Gamma) = 1, \Gamma \mapsto \rho \right\},$$

where Γ is an operator acting on the fermionic Hilbert space and, similarly to (4.9), $\Gamma \mapsto \rho$ denotes the relation $\rho(\mathbf{r}) = N \int_{\mathbb{R}^{dN}} \Gamma(\mathbf{r}, \mathbf{r}_2, \ldots, \mathbf{r}_N; \mathbf{r}, \mathbf{r}_2, \ldots, \mathbf{r}_N)\, d\mathbf{r}_2 \ldots d\mathbf{r}_N$. In [91], M. Levy introduced a similar functional requiring in addition that $\Gamma = |\psi\rangle\langle\psi|$ be a rank-one operator. An advantage of the Lieb functional F_L is that it is convex. Ignoring issues of rigor, (4.56) admits a dual formulation

$$F_{\mathrm{L}}[\rho] = \sup \left\{ \int_{\mathbb{R}^3} u(\mathbf{r})\rho(\mathbf{r})\,\mathrm{d}\mathbf{r} \,:\, \sum_{i=1}^{N} u(\mathbf{r}_i) \le -\frac{1}{2}\sum_{j=1}^{N} \Delta_{\mathbf{r}_j} + V_{\mathrm{ee}} \right\}, \tag{4.57}$$

with the above inequality understood in the sense of self-adjoint operators. For a rigorous discussion of Eq. (4.57) see Chap. 3 by Lewin et al. This equation is the quantum analogue (for mixed states) of the dual construction of the SCE functional. Note that because the right-hand side of the constraint on u now contains an additional positive term, the value of the supremum will be higher than in (4.55), as it should be.

4.3.8 Optimality Conditions

With the help of Kantorovich duality one obtains very interesting necessary conditions for solutions to the SIL variational principle (4.25). In particular, for optimizers of SCE (alias Monge) form one can express the gradient of the Kantorovich potential u in terms of the co-motion functions (alias transport maps).

We follow the rigorous presentation for general OT problems in [50], but specialize throughout to the SIL problem. For the benefit of less mathematically minded readers, we also include a heuristic derivation at the end of this section.

Theorem 4.14 (Optimality Conditions [50]) *Let $\rho : \mathbb{R}^d \to \mathbb{R}$ be any N-particle density in the class $\mathcal{D}^N$ (see (4.12)). Let $V_{\mathrm{ee}} : \mathbb{R}^{dN} \to \mathbb{R}\cup\{+\infty\}$ be any interaction potential which is symmetric, bounded from below, lower semi-continuous, and has the property that the minimum in (4.26) is finite. Suppose Π is a solution to the SIL problem (4.26), and u is a solution to the dual problem, i.e. a maximizer of J in the class $\mathcal{A}$.*

(1) *Π is zero outside the set*

$$\mathcal{M} = \{(\mathbf{r}_1, \ldots, \mathbf{r}_N) \in \mathbb{R}^{dN} \,:\, V_{\mathrm{ee}}(\mathbf{r}_1, \ldots, \mathbf{r}_N) - \sum_{i=1}^{N} u(\mathbf{r}_i) = min\}.$$

(2) *Π is an unconstrained minimizer (i.e., a minimizer on $\mathcal{P}(\mathbb{R}^{dN})$) of the modified functional*

$$\mathcal{L}[\Pi] = \int_{\mathbb{R}^{dN}} \Big(V_{\mathrm{ee}}(\mathbf{r}_1, \ldots, \mathbf{r}_N) - \sum_{i=1}^{N} u(\mathbf{r}_i) \Big)\, \mathrm{d}\Pi(\mathbf{r}_1, \ldots, \mathbf{r}_N).$$

(3) *At any point* $(\mathbf{r}_1, \ldots, \mathbf{r}_N)$ *in* $\mathcal{M}$ *where the function in* (1) *is differentiable with respect to* $\mathbf{r}_1$,

$$\nabla u(\mathbf{r}_1) = \nabla_{\mathbf{r}_1} V_{\text{ee}}(\mathbf{r}_1, \ldots, \mathbf{r}_N). \tag{4.58}$$

In particular, if Π *is of SCE form,* (4.30)*, and* $V_{\text{ee}}(\mathbf{r}_1, \ldots, \mathbf{r}_N)$ *is the Coulomb interaction* $\sum_{1 \leq i < j \leq N} \frac{1}{|\mathbf{r}_i - \mathbf{r}_j|}$,

$$\nabla u(\mathbf{r}) = -\sum_{i=1}^{N-1} \frac{\mathbf{r} - f_i(\mathbf{r})}{|\mathbf{r} - f_i(\mathbf{r})|^3}$$

$$\text{at any point } \mathbf{r} \text{ where } u \text{ is differentiable and } \rho(\mathbf{r}) > 0. \tag{4.59}$$

The physical and mathematical meaning of these results is as follows.

(1) says that the classical potential energy

$$E_{\text{pot}}(\mathbf{r}_1, \ldots, \mathbf{r}_N) = V_{\text{ee}}(\mathbf{r}_1, \ldots, \mathbf{r}_N) - \sum_i u(\mathbf{r}_i)$$

is minimal on the manifold of configurations which occur with nonzero probability under the optimal plan Π. In particular, when Π is of the SCE or Monge type, the classical potential energy is minimal on the manifold (4.31) parametrized by the co-motion functions. Besides being interesting in its own right, this underlies the derivation of the next leading term of the Levy–Lieb functional outlined in Sect. 4.3.5.

(3) says that the Kantorovich potential u is an *effective one-body potential emulating the many-body system*, in the following sense: its gradient at the point r is precisely the classical repulsive force exerted on an electron at $\mathbf{r}$ by the other electrons at positions $f_i(\mathbf{r})$. Equation (4.59) is called the *force equation*.

(2) can be viewed as an infinite-dimensional Lagrange multiplier rule, with any Kantorovich potential (i.e. any optimizer of the dual variational principle (4.55)) playing the role of a Lagrange multiplier associated with the constraint $\Pi \mapsto \rho$.

We remark that results of the above form have a long history in OT theory; for the two-marginal problem with interaction potential $|\mathbf{r}_1 - \mathbf{r}_2|$ respectively $|\mathbf{r}_1 - \mathbf{r}_2|^2$, (1) goes back to Kantorovich himself [80], while the differential version (3) and its usefulness were first realized by Knott and Smith [83].

Proof The following proof, taken from [50], is simple and illuminating, so we include it. By Kantorovich duality (in the form of Theorem 4.13 (1)),

$$0 = \int_{\mathbb{R}^{dN}} V_{\text{ee}} \, d\Pi - \sum_{i=1}^{N} \int_{\mathbb{R}^d} u(\mathbf{r}_i) \frac{\rho(\mathbf{r}_i)}{N} \, d\mathbf{r}_i.$$

Since Π has equal marginals $\frac{\rho}{N}$, $\int_{\mathbb{R}^d} u(\mathbf{r}_i)\,\frac{\rho(\mathbf{r}_i)}{N}\,d\mathbf{r}_i = \int_{\mathbb{R}^{dN}} u(\mathbf{r}_i)\,d\Pi(\mathbf{r}_1,\ldots,\mathbf{r}_N)$, and so

$$0 = \int_{\mathbb{R}^{dN}} \Big(V_{\mathrm{ee}}(\mathbf{r}_1,\ldots,\mathbf{r}_N) - \sum_i u(\mathbf{r}_i)\Big) d\Pi(\mathbf{r}_1,\ldots,\mathbf{r}_N).$$

But since u satisfies the constraint (4.149) at every point in $\mathbb{R}^{dN}$, the integrand is nonnegative. So the minimum value of the integrand must be zero and attained, and Π must vanish wherever the integrand is positive. This establishes (1) and (2). The elementary calculus fact that the gradient of a differentiable function vanishes at minimum points now yields (4.58). Finally, (4.59) follows since the point $\big(\mathbf{r}_1, f_1(\mathbf{r}_1),\ldots, f_{N-1}(\mathbf{r}_1)\big)$ belongs to $\mathcal{M}$ whenever the density ρ is positive at $\mathbf{r}_1$. □

We complete this section with a more heuristic derivation of the optimality conditions.

Heuristic derivation of Theorem 4.14 Let us re-write the SIL variational principle (4.25) in the form

$$\text{Minimize } C[\Pi] = \int_{\mathbb{R}^{dN}} V_{\mathrm{ee}}\,d\Pi \text{ subject to the constraints} \tag{4.60}$$
$$G^{(\mathbf{r}_1)}[\Pi] = \rho(\mathbf{r}_1)\;\forall \mathbf{r}_1 \in \mathbb{R}^d,$$

where $G^{(\mathbf{r}_1)}[\Pi]$ is the functional which assigns to an N-point probability measure Π the value of its single-particle density at the point $\mathbf{r}_1$, and where the minimization is over symmetric probability measures (see Remark 4.2). Let us now postulate the existence of a family of Lagrange multipliers $(\lambda(\mathbf{r}_1))_{\mathbf{r}_1\in\mathbb{R}^d}$, one for each $G^{(\mathbf{r}_1)}$, such that minimizers of C subject to the constraints $G^{(\mathbf{r}_1)}[\Pi] = \rho(\mathbf{r}_1)$ are unconstrained minimizers of the Lagrangian

$$\mathcal{L}[\Pi] = C[\Pi] - \int_{\mathbb{R}^d} \lambda(\mathbf{r}_1)\, G^{(\mathbf{r}_1)}[\Pi] d\mathbf{r}_1.$$

But since $G^{(\mathbf{r}_1)}[\Pi]$ is the one-body density of Π, and Π is symmetric,

$$\begin{aligned}\mathcal{L}[\Pi] &= \int_{\mathbb{R}^{dN}} \Big[V_{\mathrm{ee}}[\mathbf{r}_1,\ldots,\mathbf{r}_N] - \lambda(\mathbf{r}_1)\,N\Big] d\Pi(\mathbf{r}_1,\ldots.,\mathbf{r}_N)\\ &= \int_{R^{dN}} \Big[V_{\mathrm{ee}}(\mathbf{r}_1,\ldots,\mathbf{r}_N) - \sum_{i=1}^N \lambda(\mathbf{r}_i)\Big] d\Pi(\mathbf{r}_1,\ldots,\mathbf{r}_N),\end{aligned} \tag{4.61}$$

so the Lagrangian coincides with the functional in (2) with $\lambda = u$. It is clear that minimizers of the Lagrangian must be concentrated on the set of pointwise minimizers of the integrand, yielding (1). Statement (3) now follows as in the rigorous proof.

The above argumentation obtains the Kantorovich potential u quickly but non-rigorously as a Lagrange multiplier. In fact, with such a heuristic construction of u, statements (1) and (3) were already derived in [127] before the discovery of the *SCE theory/optimal transport* connection.

But readers are put on notice that there is no such thing as a general and rigorous Lagrange multiplier rule which would guarantee the existence of Lagrange multipliers for infinite-dimensional non-smooth problems like Levy–Lieb constrained search or its strongly interacting limit (4.25). In DFT (in its original form with both kinetic energy and electron repulsion present), the existence problem for Lagrange multipliers—i.e., the existence of one-body potentials which, when added to the Hamiltonian $T + V_{\rm ee}$, reduce a constrained search to an unconstrained search—is known as the *v-representability problem*. This is a longstanding open problem, see e.g. [76, 84, 90, 97, 133]. For variants of the problem at positive temperature respectively quantum lattices see [19, 20]; v-representability for a regularization of the exact Levy–Lieb functional is discussed in Chap. 5 by Kvaal.

4.3.9 *Solution of the Purely-Interacting v-Representability Problem*

We now show that in the strongly interacting limit the v-representability problem, alias the problem of existence of Lagrange multipliers for density functionals defined by constrained search, can be completely solved. As we will see, this fact follows by combining known results. We assume in this section that $w(\mathbf{r}) = |\mathbf{r}|^{-1}$ is the Coulomb interaction.

Recall that a density $\rho : \mathbb{R}^d \to \mathbb{R}$ is called

- *N-representable* if it comes from a wave function $\Psi \in \mathcal{W}^N$ (i.e. $\Psi \mapsto \rho$)
- *v-representable* if it comes from a minimizer of $\langle \Psi | T + V_{\rm ee} + \sum_i v(\mathbf{r}_i) | \Psi \rangle$ on $\mathcal{W}^N$ for some potential $v : \mathbb{R}^d \to \mathbb{R}$
- *non-interacting v-representable* if it comes from a minimizer of $\langle \Psi | T + \sum_i v(\mathbf{r}_i) | \Psi \rangle$ on $\mathcal{W}^N$ for some potential $v : \mathbb{R}^d \to \mathbb{R}$
- *purely-interacting v-representable* if it comes from a minimizer of $\int_{\mathbb{R}^{dN}} \big(V_{\rm ee} + \sum_i v(\mathbf{r}_i) \big) d\Pi$ on $\mathcal{P}(\mathbb{R}^{dN})$ for some potential $v : \mathbb{R}^d \to \mathbb{R}$.

Theorem 4.15 (N-Representability Implies Purely-Interacting v-Representability) *Any N-representable ρ, i.e. any ρ belonging to the class $\mathcal{D}^N$ (see (4.12)), is purely-interacting v-representable by some bounded measurable potential $v : \mathbb{R}^d \to \mathbb{R}$. Explicitly, the following choice will do:*

$$v = -u \tag{4.62}$$

where u is any bounded Kantorovich potential for ρ (see Theorem 4.13 for existence of the latter).

This result is quite remarkable, given that—to our knowledge—not much is known on the rigorous level off the strongly interacting limit.

Proof of Theorem 4.15 By Theorem 4.13 (2), there exists a bounded maximizer u of the dual functional, i.e. an associated Kantorovich potential. Let $v = -u$. By Theorem 4.1 (1), there exists a minimizer $\Pi[\rho]$ of the SIL variational problem (4.25). By Theorem 4.14 (2), this $\Pi[\rho]$ is a minimizer of $\int_{R^{dN}} \big(V_{\text{ee}} + \sum_i v(\mathbf{r}_i)\big)\mathrm{d}\Pi$ on $\mathcal{P}(\mathbb{R}^{dN})$. Since by construction Π has density ρ, it follows that v represents ρ. □

If in addition $\rho > 0$ everywhere, the above proof together with Theorem 4.13 (3) shows that ρ is even purely-interacting v-representable by some Lipschitz continuous potential.

4.3.10 Functional Derivative and SCE Potential

It is not difficult to deduce from Theorem 4.15 that when the density ρ is sufficiently nice (say, continuous and everywhere positive) and the Kantorovich potential $u[\rho]$ (i.e. the maximizer of the dual problem (4.55)) is unique, the SCE functional is functionally differentiable at ρ with functional derivative

$$\frac{\delta V_{\text{ee}}^{\text{SCE}}[\rho]}{\delta\rho}[\rho] = u[\rho] + \text{const} \tag{4.63}$$

where const is an arbitrary additive constant. Here for any functional F on $\mathcal{D}^N$ the functional derivative $\frac{\delta F}{\delta\rho}[\rho]$ at some density ρ (if it exists) is defined by the requirement

$$\frac{\mathrm{d}}{\mathrm{d}t} F[\rho + t\,\eta]\Big|_{t=0} = \int_{\mathbb{R}^d} \frac{\delta F}{\delta\rho}[\rho](\mathbf{r})\,\eta(\mathbf{r})\,\mathrm{d}\mathbf{r}$$

for all smooth mass-preserving localized perturbations $\eta : \mathbb{R}^d \to \mathbb{R}$ (mathematically: $\eta \in C_0^\infty(\mathbb{R}^d)$, $\int \eta = 0$), and is unique up to an additive constant. For an informal derivation of Eq. (4.63) see e.g. [22], and for a rigorous proof under suitable assumptions see [41].

As for any Hartree-exchange-correlation functional, the Hartree-exchange-correlation *potential* associated to the SCE functional is the functional derivative with additive constant chosen so that the potential vanishes at infinity, in our case

$$v_{\rm SCE}[\rho](\mathbf{r}) = u[\rho](\mathbf{r}) + C[\rho], \quad C[\rho] \text{ a constant that ensures} \\ \lim_{|\mathbf{r}|\to\infty} v_{\rm SCE}[\rho](\mathbf{r}) = 0. \tag{4.64}$$

This functional derivative is called the SCE potential.

To summarize: *the SCE potential for the strongly correlated limit of DFT agrees up to a shift with the Kantorovich potential from optimal transport theory.*

Assume now that the density is everywhere positive, that the ground state of (4.25) is an SCE state, and that

$$\text{the values } f_1(\mathbf{r}), \ldots, f_{N-1}(\mathbf{r}) \text{ stay in a bounded region as } |\mathbf{r}| \to \infty. \tag{4.65}$$

It then follows from (4.59) that the SCE potential has the correct asymptotic behavior

$$v_{\rm SCE}[\rho](\mathbf{r}) \sim_{|\mathbf{r}|\to\infty} \frac{N-1}{|\mathbf{r}|}. \tag{4.66}$$

By contrast, Hartree-exchange-correlation potentials for all semilocal functionals (LDA, GGAs) are well known to have the wrong asymptotics on physical (i.e. exponentially decaying) densities,

$$v_{\rm Hxc}^{\rm semiloc}[\rho](\mathbf{r}) \sim_{|\mathbf{r}|\to\infty} \frac{N}{|\mathbf{r}|}. \tag{4.67}$$

Open Problem Rigorously justify (4.65), and hence (4.66), for general densities ρ. Note that for $N = 2$ and radial densities, or any N and arbitrary densities in one dimension, assumption (4.65) follows from the explicit formulae for the f_i in [32] respectively [27] (Fig. 4.3).[4]

Example 4.16 Let $N = 2$, and let $\rho(r) = 2/\pi(1 + r^2)$ be the one-dimensional Lorenzian density, normalized so that $\int \rho = 2$. The co-motion function $f_1 = f$ can be computed explicitly and is given by $f(r) = -1/r$, see Example 4.18 in Sect. 4.3.11. The SCE potential must satisfy the differential equation (4.59) which in our case reads

$$v'_{\rm SCE}(r) = \frac{\operatorname{sgn}(r)}{\left[r - f(r)\right]^2} = \operatorname{sgn}(r)\,\frac{r^2}{(r^2+1)^2}.$$

The boundary condition $v_{\rm SCE}(r) \to 0$ for $r \to \infty$ (Eq. (4.64)) yields the solution

$$v_{\rm SCE}(r) = \frac{\operatorname{sgn}(r)}{2}\left[\arctan r - \frac{r}{r^2+1}\right] - \frac{\pi}{4}.$$

[4] Note added in proof: this problem has recently been solved in https://arxiv.org/abs/2210.07830.

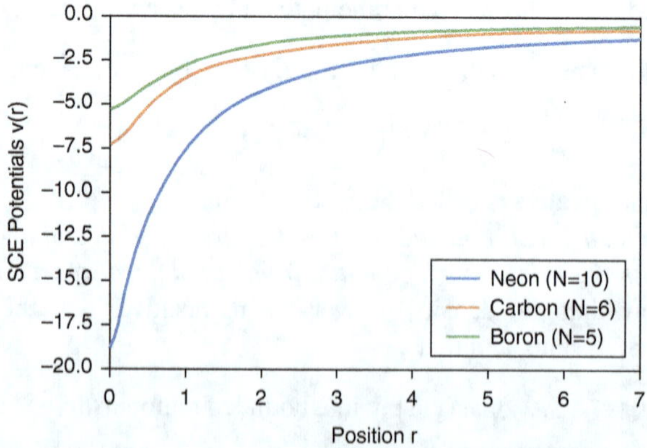

Fig. 4.3 SCE potentials $v_{\mathrm{SCE}}(r)$ corresponding to the (radially symmetric) densities of Neon ($N = 10$), Carbon ($N = 6$) and Boron ($N = 5$), Fig. 9 in [127]. Data obtained by the following steps: (i) compute the density $\rho(r)$ by an accurate full CI or quantum Monte Carlo computation; (ii) compute the SGS maps corresponding to ρ as described in Section 4.3.12 below; (iii) obtain the corresponding SCE potentials v_{SCE} via Eqs. (4.59) and (4.64)

4.3.11 Strictly Correlated Electrons in One Dimension

In one dimension the strong interaction limit (Eq. (4.25)) can be solved exactly. The minimizing probability measure is given by an SCE state (4.30) with explicit co-motion functions alias transport maps. The minimizer was found by Seidl himself in the original paper [122], on grounds of physical intuition. A proof of its optimality was found much later by Colombo, De Pascale and Di Marino [27].

Seidl's Construction For a given integrable density $\rho : \mathbb{R} \to \mathbb{R}$ with $\rho \geq 0$ and $\int \rho = N$, begin by choosing $f_1 : \mathbb{R} \to \mathbb{R}$ so that the amount of density between r and $f_1(r)$ is 1. Now choose f_2 so that the amount of density between $f_1(r)$ and $f_2(r)$ is again 1, and so on, i.e., denoting $f_0(r) = r$,

$$\int_{f_i(r)}^{f_{i+1}(r)} \rho(r')\,\mathrm{d}r' = 1 \tag{4.68}$$

for all $i = 0, \ldots, N-1$. For Eq. (4.68) to possess a solution $f_{i+1}(r)$ in $\mathbb{R} \cup \{+\infty\}$ we must have $\int_{f_i(r)}^{\infty} \rho \geq 1$; otherwise one needs to integrate first up to $+\infty$ and then onwards from $-\infty$ so as to obtain a total value of 1,

$$\int_{f_i(r)}^{\infty} \rho(r')\,\mathrm{d}r' + \int_{-\infty}^{f_{i+1}(r)} \rho(r')\,\mathrm{d}r' = 1. \tag{4.69}$$

Physically this means that, given that the first electron is at some position $x_1 = r$, all the other electrons at $x_2 = f_1(r), \ldots, x_N = f_{N-1}(r)$ are separated by an equal

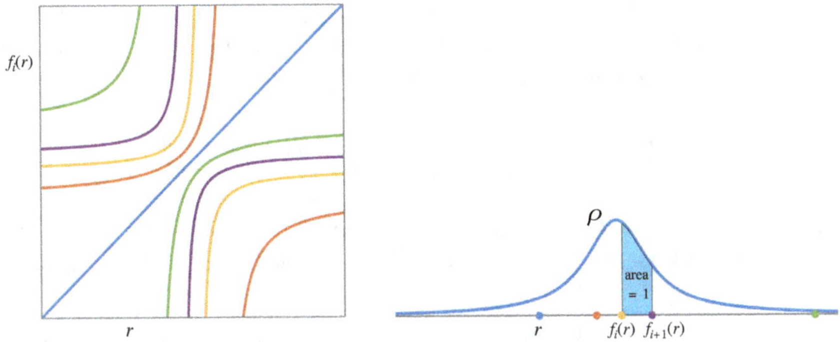

Fig. 4.4 SCE state of N electrons in one dimension. *Right:* Electron positions. Given that the first electron (depicted in blue) is at r, the positions of the other electrons are completely determined by the requirement that neighboring electrons are separated by an amount of density of 1 (blue area). The co-motion functions or transport maps f_i ($i = 1, \ldots, N-1$) of SCE theory are defined as the positions of the other electrons as a function of the first position r. The latter is distributed according to the given single-particle density ρ. *Left:* Graphs of the maps f_i, with $f_0(r) = r$ also shown. The figure corresponds to the Lorenzian density (Example 4.18) and $N = 5$

amount of density between nearest neighbors. See Fig. 4.4, right panel. As always for SCE states, the first electron position is distributed according to the given density ρ.

The above construction can be expressed concisely in terms of the cumulative distribution function

$$G_\rho(r) = \int_{-\infty}^{r} \frac{\rho(r')}{N}\,\mathrm{d}r' \tag{4.70}$$

and its generalized inverse

$$G_\rho^{-1}(y) = \inf\{r \in \mathbb{R} : G_\rho(r) > y\}. \tag{4.71}$$

(When ρ is continuous and everywhere positive, G_ρ^{-1} is just the usual inverse function; the above definition has the virtue that it works for any nonnegative integrable ρ with integral N.) Equations (4.68) and (4.69) now take the form $G_\rho(f_{i+1}(r)) - G_\rho(f_i(r)) = 1/N$ respectively $1 - G_\rho(f_i(r)) + G_\rho(f_{i+1}(r)) = 1/N$, so by solving for f_{i+1} in terms of f_i and using $f_0(r) = r$

$$f_i(r) = \begin{cases} G_\rho^{-1}\big(G_\rho(r) + \frac{i}{N}\big) & \text{if } G_\rho(r) \leqslant \frac{N-i}{N} \\ G_\rho^{-1}\big(G_\rho(r) + \frac{i}{N} - 1\big) & \text{otherwise,} \end{cases} \tag{4.72}$$

for $i \in \{1, \ldots, N-1\}$.

Optimality This construction is indeed optimal:

Theorem 4.17 *Let $w(r) = |r|^{-1}$. For any nonnegative integrable density $\rho : \mathbb{R} \to \mathbb{R}$ with $\int \rho = N$, the SCE state* (4.30) *with $f_1, \ldots, f_{N-1}$ given by the Seidl construction* (4.72) *is a minimizer of the SIL problem*

$$\text{Minimize} \int_{\mathbb{R}^N} V_{\text{ee}}(r_1, \ldots, r_N)\, \mathrm{d}\Pi(r_1, \ldots, r_N) \text{ over } \{\Pi \in \mathcal{P}(\mathbb{R}^N) : \Pi \mapsto \rho\}.$$

Moreover when ρ is everywhere positive, this minimizer is unique for $N = 2$, and its symmetrization (see Remark 4.2) is the unique symmetric minimizer for arbitrary N.

This theorem is due to [32] for $N = 2$ and to [27] for arbitrary N. Despite the intuitive nature of the optimizer, the proof is not elementary. It is based on a careful analysis of the structure of V_{ee}-cyclically monotone sets in $\mathbb{R}^N$, and strongly relies on both optimal transport theory and the ordering properties of the real line. Note that uniqueness cannot hold for $N \geq 3$ unless symmetry is required, as re-labelling the f_i then yields another solution. This is purely a mathematical, not a physical effect since solutions to the SIL problem arising as low-density limits of N-point densities of quantum wavefunctions (as described by Theorem 4.3) are always symmetric, corresponding to the symmetrization of the state (4.30) and (4.72).

Group Law Formula (4.72) implies an interesting group law for the co-motion functions, already noticed in [122]: the ith function is the i-fold composition of the first function with itself,

$$f_i = \underbrace{f_1 \circ \cdots \circ f_1}_{i \text{ times}},$$

and the N-fold composition of the first function gives the identity $f_0(r) = r$.

Explicit Examples The following examples further illustrate the nonlinear governing Eqs. (4.68)–(4.69), and may serve as useful benchmarks for numerical simulations in the strongly interacting limit (or close to it).

Example 4.5, ctd Consider a two-electron system with ρ being the uniform density in a one-dimensional interval $[0, L]$. In this case we have $G_\rho(r) = r/L$, and formula (4.72) readily yields the co-motion function (4.40). For its graph, see Fig. 4.2. Mathematically this map switches the right and left half of the interval; note that its composition with itself indeed gives the identity, as it must by the group law.

Example 4.18 ([68, 71]) Let $\rho(r) = N/\pi(1 + r^2)$ be the Lorenzian density, normalized so that $\int \rho = N$. In this case $G_\rho(r) = \frac{1}{\pi} \arctan r + \frac{1}{2}$ and so Eq. (4.68) for f_1 in the region $G_\rho(r) \leq \frac{N-1}{N}$ is, recalling the notation $f_0(r) = r$,

$$\arctan f_1(r) = \arctan r + \tfrac{\pi}{N}. \tag{4.73}$$

When $N = 2$ it follows that

$$f_1(r) = -\frac{1}{r} \tag{4.74}$$

(note that then the derivatives of both sides of (4.73) agree, as do their values at $r = 0$). From now on let us assume $N \geq 3$. In this case we can use the addition formula for the tangent, $\tan(x+y) = (\tan x + \tan y)/(1 - \tan x \tan y)$ for $x,\ y,\ x+y \notin \pi/2 + \mathbb{Z}$, and obtain

$$f_1(r) = \frac{r + t_1}{1 - t_1 r}, \qquad t_1 = \tan \tfrac{\pi}{N}. \tag{4.75}$$

In the region $G_\rho(r) > \frac{N-1}{N}$, or equivalently $\arctan r > \frac{\pi}{2} - \frac{\pi}{N}$, or equivalently (because $\tan(\frac{\pi}{2} - x) = 1/\tan x$) $r > 1/t_1$, Eq. (4.69) for f_1 is

$$\arctan f_1(r) - \left(-\tfrac{\pi}{2}\right) = \arctan r - \frac{\pi}{2} + \frac{\pi}{N},$$

that is to say $\arctan f_1(r) = \arctan r + \frac{\pi}{N} - \pi$. Using the addition formula for the tangent and $\tan x = \tan(x - \pi)$ we again find that f_1 is given by (4.75), so this formula describes f_1 on the whole real line. It remains to compute its i-fold composition f_i. Here we give a different derivation as compared to [68, 71]. Note that mathematically f_1 is a Moebius map, i.e. a map of the form $M_a(r) = (r + a)/(1 - ar)$. Using the (elementary to check) composition formula $M_a \circ M_b = M_{\frac{a+b}{1-ab}}$ and the addition formula for the tangent we find

$$f_i(r) = \frac{r + t_i}{1 - t_i r}, \qquad t_i = \tan \tfrac{i\pi}{N} \quad (i \in \{1, \ldots, N-1\}). \tag{4.76}$$

Moreover, setting $i = N$ in the above formula we recover the abstract fact that the N-fold composition of f_1 must be the identity. Hence the co-motion functions for the Lorenzian density form a discrete subgroup of the Moebius group. For the graph of these functions when $N = 5$ see Fig. 4.4.

4.3.12 *Radially Symmetric Densities*

When the one-body density ρ is radially symmetric, Seidl, Gori-Giorgi and Savin [127] conjectured an explicit minimizing probability measure in (4.25) of a radial-symmetry-preserving SCE form which is related to the explicit SCE state of one-dimensional systems.[5] Let us describe their conjecture in detail.

The starting point is the following reduction to a 1d problem with effective interaction.

Lemma 4.19 (Reduction to a 1d Problem, [9, 113]) *Let $\rho : \mathbb{R}^d \to \mathbb{R}$ be an integrable density with $\rho \geq 0$ and $\int \rho = N$ which is radially symmetric, that is, $\rho(\mathbf{r}) = \rho_0(|\mathbf{r}|)$ for some function ρ_0, and let*

$$\rho_{\rm rad}(r) = \omega_d r^{d-1} \rho_0(r),$$

where ω_d is the area of the unit sphere in $\mathbb{R}^d$ (for $d = 3$, $\omega_d = 4\pi$). Then the SCE functional defined by (4.26) *reduces to*

$$V_{\rm ee}^{\rm SCE}[\rho] = \min_{\eta \in \mathcal{P}([0,\infty)^N),\ \eta \mapsto \rho_{\rm rad}} \int_{[0,\infty)^N} V_{\rm ee}^{\rm rad}(r_1, \ldots, r_N)\, d\eta(r_1, \ldots, r_N), \tag{4.77}$$

where $V_{\rm ee}^{\rm rad}$ is the reduced Coulomb cost

$$V_{\rm ee}^{\rm rad}(r_1, \ldots, r_N) = \min \left\{ \sum_{1 \leqslant i < j \leqslant N} \frac{1}{|\mathbf{r}_j - \mathbf{r}_i|} \ :\ |\mathbf{r}_i| = r_i \ \forall i = 1, \ldots, N \right\}. \tag{4.78}$$

Moreover, $\Pi \in \mathcal{P}(\mathbb{R}^{dN})$ is a minimizer for the full SIL variational principle (4.25) *in d dimensions if and only if its radial projection $\Pi_{\rm rad}$, defined by*

$$\int_{A_1 \times \ldots \times A_N} d\Pi^{\rm rad}(r_1, \ldots, r_N) = \int_{\{|\mathbf{r}_1| \in A_1\} \times \ldots \times \{|\mathbf{r}_N| \in A_N\}} d\Pi(\mathbf{r}_1, \ldots, \mathbf{r}_N)$$

for all intervals $A_1, \ldots, A_N$, is a minimizer for the right-hand side of (4.77) *and $V_{\rm ee}(\mathbf{r}_1, \ldots, \mathbf{r}_N) = V_{\rm ee}^{\rm rad}(|\mathbf{r}_1|, \ldots, |\mathbf{r}_N|)$ Π-a.e.*

In [122, 127], the following interesting explicit state was conjectured to be optimal for the reduced problem in (4.77):

$$d\eta(r_1, \ldots, r_N) = \rho_{\rm rad}(r_1) \prod_{n=2}^{N} \delta(r_n - S^{(n)}(r_1)), \tag{4.79}$$

[5] The original conjecture concerned the physical case $d = 3$. Subsequently, two-dimensional models have also been considered in the literature [125, 127].

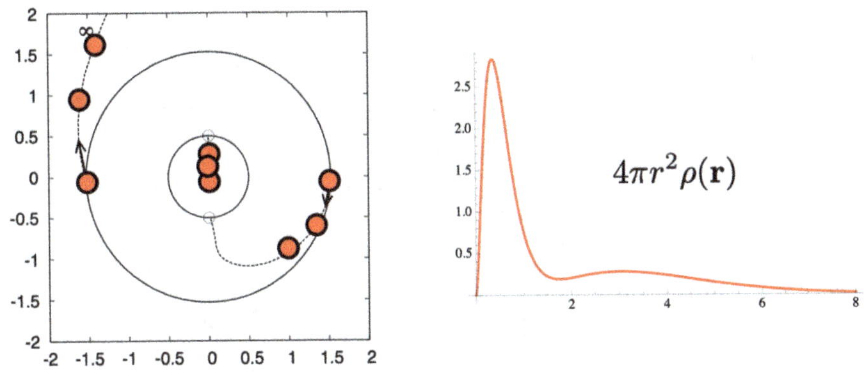

Fig. 4.5 SGS state (left panel) for the Lithium atom density (right panel). The extreme angular and radial correlation exhibited by this state is illustrated here while sending one of the electrons (the leftmost) to infinity

where $S^{(n)}$ denotes the n-fold composition $S \circ \cdots \circ S$ and $S : [0, \infty) \to [0, \infty)$ is defined as follows. Let $0 = a_0 < a_1 < \cdots < a_{N-1} < a_N = \infty$ be such that the intervals $A_n = [a_{n-1}, a_n)$ between successive a_n's carry equal mass, that is, $\int_{A_n} \rho^{\text{rad}} = 1$ for all n, and let $S|_{A_n}$ be the unique function such that

$$S|_{A_n} \text{ decreasing}, \quad S \text{ transports } \rho^{\text{rad}}|_{A_n} \text{ to } \rho^{\text{rad}}|_{A_{n+1}} \tag{4.80}$$

(with the convention $A_{N+1} = A_1$). In terms of the original SIL problem (4.25), this ansatz corresponds to the SCE ansatz (4.30) with maps satisfying the additional property

$$|f_n(\mathbf{r})| = S^{(n)}(|\mathbf{r}|)$$

for the above explicit S and with suitably chosen angles so that $\Pi \mapsto \rho$ and $V_{\text{ee}}(\mathbf{r}_1, \ldots, \mathbf{r}_N) = V_{\text{ee}}^{\text{rad}}(|\mathbf{r}_1|, \ldots, |\mathbf{r}_N|)$ Π-a.e. We call S the *SGS map*, and the probability measure η given by (4.79), (4.80) the *SGS state*. See Figs. 4.5 and 4.6.

The SGS state has been rigorously proved to be optimal in some specific cases.

Example 4.20 (SCE for Radially Symmetric Densities, [32, Theorem 4.10]) Let $N = 2$, and let ρ be a radially symmetric density on $\mathbb{R}^d$ such that $\rho(\mathbf{r}) > 0$ for all $\mathbf{r}$. Then the optimal co-motion function f is given by

$$f(\mathbf{r}) = s(|\mathbf{r}|)\frac{\mathbf{z}}{|\mathbf{z}|}$$

for some function $s : [0, +\infty) \to \mathbb{R}$ such that $s \leqslant 0$, s is increasing, $\lim_{r\to+\infty} s(z) = 0$, and $\lim_{r\to 0+} s(r) = -\infty$.

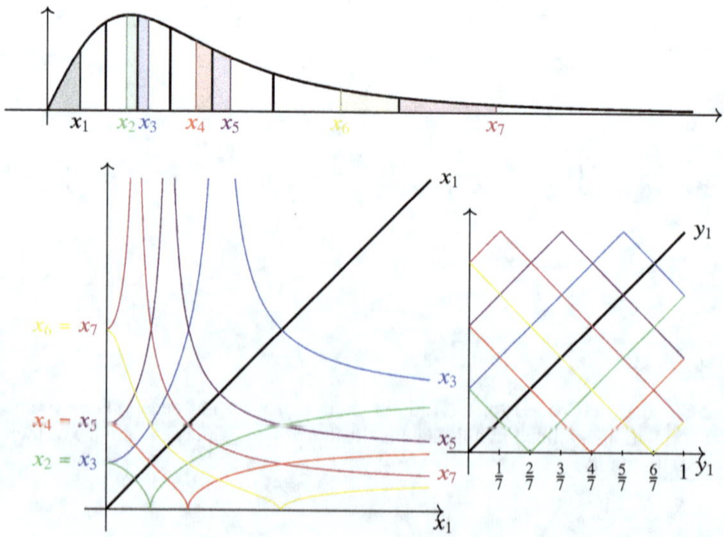

Fig. 4.6 SGS state when $N = 7$. Top: a radial measure $\rho_{\mathrm{rad}}(r)$. Bottom left: the maps $S, S^{(2)}, \ldots, S^{(6)} : [0, \infty) \to [0, \infty)$, plotted with colors green, blue, red, violet, yellow, and brown. Bottom right: the same graphs under a change of variables $y_1 = \int_0^{x_1} \frac{\rho_{\mathrm{rad}}}{N}$ which transports ρ_{rad} to the uniform density on the interval [0, 1]. Picture from [124]

The function s in the above example corresponds to minus the SGS map, i.e. $s = -S$ in the $N = 2$ case.

Optimality of the SGS state has also been proved for some special class of densities ρ when $N = d = 3$ [30, 57, 125] and $N = 3$ and $d = 2$ [11].

Recently, counterexamples of radially symmetric probability densities were found for which the SGS state is not optimal. The simplest one is a uniform density on a thin annulus:

Example 4.21 ([30], see also [11, 57, 124, 125] for related examples) Let $N = 3$. For sufficiently small $\varepsilon > 0$, and the density

$$\rho_\varepsilon^{\mathrm{rad}} = c_\varepsilon 1_{[1,1+\varepsilon]} \tag{4.81}$$

(with the constant c_ε chosen such that $\int \rho^{\mathrm{rad}} = 3$), the SGS state η_ε defined by (4.79)–(4.80) is not optimal for the variational problem (4.77).

This example illustrates that guessing the optimal SCE states can be a tricky business even for 1d problems, and makes it all the more remarkable that optimality of Seidl's guess for the 1d Coulomb problem is a rigorous theorem (Theorem 4.17). The proof of nonoptimality relies on a Taylor expansion of the reduced interaction $\mathrm{V}_{\mathrm{ee}}^{\mathrm{rad}}$ (defined in Eq. (4.78)) at the point (1, 1, 1) and on cyclical monotonicity methods from optimal transport theory.

While this counterexample disproves optimality of the SGS state in general, the density (4.81) is quite different from typical atomic densities and the following remains an interesting mathematical problem.

Open Problem Find sufficient conditions on radial densities ρ such that the SGS state is optimal for (4.77).

4.3.13 An Example with Irregular Co-motion Functions for Repulsive Harmonic Interactions

One of the more challenging properties of co-motion functions is that they are typically discontinuous. Here we give an extreme example with modified electron-electron interaction which is discontinuous *everywhere*, due to Di Marino et al. [42].

Example 4.22 Let d be arbitrary, $V_{ee}(\mathbf{r}_1, \ldots, \mathbf{r}_N) = -\sum_{1\leq i<j\leq N} |\mathbf{r}_i - \mathbf{r}_j|^2$ (repulsive harmonic interaction), and $N = 3$. Let $\rho = 3 \cdot 1_{[0,1]^d}$ (uniform density on a cube in $\mathbb{R}^d$). Then there exists a nowhere continuous map $T : [0, 1]^d \to [0, 1]^d$ which transports ρ to itself such that

$$d\Pi(\mathbf{r}_1, \mathbf{r}_2, \mathbf{r}_3) = \frac{\rho(\mathbf{r}_1)}{N}\delta(\mathbf{r}_2 - T(\mathbf{r}_1))\delta(\mathbf{r}_3 - T(T(\mathbf{r}_1)) \tag{4.82}$$

is an optimal probability measure for the SIL problem (4.25).

The map T is an explicit fractal map. For $d = 1$ it is depicted in Fig. 4.7 and constructed as the unique fixed point of the iteration

$$f \mapsto g(x) = \begin{cases} \frac{1}{3}f(3x) + \frac{1}{3} & \text{for } 0 \leqslant x < \frac{1}{3} \\ \frac{1}{3}f(3x-1) + \frac{2}{3} & \text{for } \frac{1}{3} \leq x < \frac{2}{3} \\ \frac{1}{3}f(3x-2) & \text{for } \frac{2}{3} \leqslant x < 1, \end{cases} \tag{4.83}$$

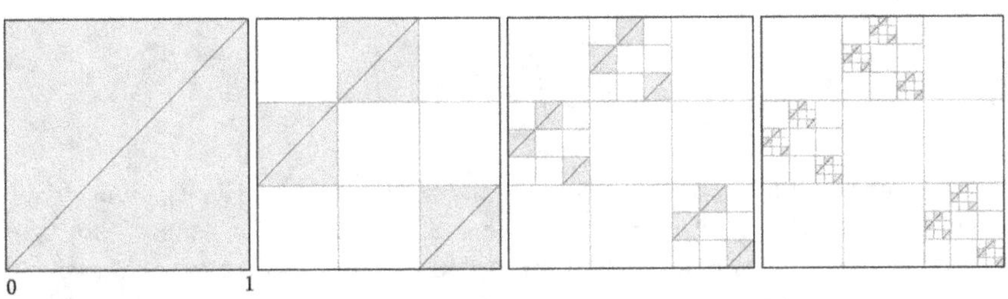

Fig. 4.7 Construction of the optimal map T in Example 4.22. The picture shows the graph (in blue) of the first few iterations of Eq. (4.83); each graph consists of three scaled copies of the previous one. The exact map is reached in the limit of infinitely many iterations

starting with $f(x) = x$. To see what the iteration is doing, divide $[0, 1]^2$ into a 3×3 grid of squares and put scaled copies of the graph of the original function into the two squares directly above the diagonal and the bottom right square. Optimality of the resulting fractal SCE state (4.82) is easy to see from the following special property of the repulsive harmonic cost which was first observed by Pass [112]: thanks to the identity $V_{ee} = |\mathbf{r}_1 + \ldots + \mathbf{r}_N|^2 - N\sum_{i=1}^N |\mathbf{r}_i|^2$ and the fact that the integral of the second term against a probability measure Π only depends on its marginal, the minimizers of the SIL problem are precisely the probability measures supported on the surface $\mathbf{r}_1 + \cdots + \mathbf{r}_N = 0$.

The above example and construction works for arbitrary N, see [42].

Open Problem Do such extreme examples also occur for the Coulomb interaction? Note that the repulsive harmonic interaction arises by locally Taylor-expanding the Coulomb interaction in angular direction.

4.3.14 Minimizers of the Discretized SIL Variational Principle Are Quasi-Monge States

We now come back to the important issue that the SIL variational principle (4.25) still requires minimization over a high-dimensional space of N-point probability measures, whereas the low-dimensional SCE ansatz (4.30) can fail to yield an optimizer of (4.25). One can ask whether some modified low-dimensional ansatz is enough to solve (4.25) exactly. In other words, can one achieve Seidl's original goal of solving the strongly interacting limit of DFT with a low-dimensional ansatz that can be easily stored on a computer?

For the discretization of (4.25) on a grid, Friesecke and Vögler [53] found a modified ansatz which achieves this, for arbitrary space dimensions, densities and interaction potentials:

$$\begin{aligned} d\Pi(\mathbf{r}_1, \ldots, \mathbf{r}_N) &= S_N \int_{\mathbb{R}^d} \alpha(\mathbf{r}) \prod_{n=1}^{N} \delta\big(\mathbf{r}_n - f_{n-1}(\mathbf{r})\big)\, d\mathbf{r} \qquad (4.84) \\ &= \frac{1}{N!}\sum_{\sigma} \int_{\mathbb{R}^d} \alpha(\mathbf{r}) \prod_{n=1}^{N} \delta\big(\mathbf{r}_n - f_{\sigma(n-1)}(\mathbf{r})\big)\, d\mathbf{r}, \end{aligned}$$

where σ runs over all permutations of the indices $0, \ldots, n-1$, α is some (free to choose) probability density on the single-particle space $\mathbb{R}^d$, and the f_n are maps from $\mathbb{R}^d$ to $\mathbb{R}^d$. States of this form are called *quasi-Monge states* or *quasi-SCE states*. With the specific choice $\alpha = \frac{\rho}{N}$, the quasi-Monge ansatz (4.84) reduces precisely to the SCE (alias Monge) ansatz in its symmetric form (4.42). The novelty is the additional freedom of choosing the auxiliary density α. For the quasi-Monge ansatz, the marginal constraint $\Pi \mapsto \rho$ takes, instead of the conditions $f_{n\sharp}\rho = \rho$

$(n = 0, \dots, N-1)$ (Eq. (4.32)), the form of a single condition,

$$\frac{1}{N}\sum_{n=0}^{N-1} f_{n\sharp}\alpha = \frac{\rho}{N}. \tag{4.85}$$

That is, the *average* push-forward of the auxiliary density α under the quasi-SCE maps must be the (suitably normalized) physical density.

Plugging the ansatz (4.84) into the SIL variational principle (4.25) and integrating out the variables $\mathbf{r}_2, \dots, \mathbf{r}_N$ yields the *quasi-Monge* or *quasi-SCE* variational principle

$$\text{Minimize } \int_{\mathbb{R}^d} V_{\text{ee}}(f_0(\mathbf{r}), \dots, f_{N-1}(\mathbf{r}))\,\alpha(\mathbf{r})\,d\mathbf{r} \text{ over probability densities } \alpha \text{ and maps } f_0, \dots, f_{N-1}, \tag{4.86}$$

with the minimization being subject to the constraint (4.85).

Theorem 4.23 (Justification of the Quasi-Monge Ansatz, [53]) *Let ρ be any discrete N-particle density on $\mathbb{R}^d$, that is to say $\rho(\mathbf{r}) = \sum_{i=1}^{\ell} \rho_i \delta(\mathbf{r} - \mathbf{a}_i)$ for some distinct discretization points $\mathbf{a}_i \in \mathbb{R}^d$ and some $\rho_i \geq 0$ with $\sum_i \rho_i = N$, and let $V_{\text{ee}} : \mathbb{R}^{dN} \to \mathbb{R} \cup \{+\infty\}$ be any interaction potential which is symmetric in the electron coordinates (e.g., the Coulomb interaction $V_{\text{ee}}(\mathbf{r}_1, \dots, \mathbf{r}_N) = \sum_{i<j} 1/|\mathbf{r}_i - \mathbf{r}_j|$). Then the SIL problem* (4.25) *possesses a minimizer which is a quasi-Monge state* (4.84). *Equivalently, it possesses a minimizer of the form* (4.88), *i.e., a superposition of at most ℓ symmetrized Dirac measures.*

This result rigorously reduces the number of unknowns from exponential to linear with respect to the number of electrons; more precisely, from ℓ^N (the dimension of the space of N-point probability measures supported on $\{\mathbf{a}_1, \dots, \mathbf{a}_\ell\}^N$) to $\ell \cdot (N+1)$ (ℓ unknowns for each of the N quasi-Monge maps, and another ℓ unknowns for the auxiliary density α).

The above result fails if the class of quasi-Monge states is narrowed to Monge (alias SCE) states, see [49]. For continuous ρ's, it is an open question whether the SIL problem always (or at least in the Coulomb case) admits minimizers of quasi-Monge form.

Proof of Theorem 4.23 (Following [53].) Let us explain the intuition and reasoning behind the quasi-Monge ansatz and Theorem 4.23, which comes from convex geometry. Before passing to a geometric viewpoint, we note that by the symmetry of V_{ee} the minimization in (4.25) can be restricted to symmetric probability measures (see Remark 4.2); moreover any symmetric probability measure $\Pi \in \mathcal{P}(\mathbb{R}^{dN})$ with $\Pi \mapsto \rho$ must be of the form

$$d\Pi(\mathbf{r}_1, \ldots, \mathbf{r}_N) = \sum_{i_1,\ldots,i_N=1}^{\ell} \gamma_{i_1 \ldots i_N} \delta(\mathbf{r}_1 - \mathbf{a}_{i_1}) \cdots \delta(\mathbf{r}_N - \mathbf{a}_{i_N}) d\mathbf{r}_1 \ldots d\mathbf{r}_N$$

for some symmetric tensor $(\gamma_{i_1 \ldots i_N}) \in \mathbb{R}^{\ell \times \ldots \times \ell}$ with nonnegative entries which sum to 1. Now geometrically, for fixed discretization points $a_1, \ldots, a_\ell$ the set of these probability measures is a finite-dimensional convex polytope; let us denote it by $\mathcal{P}_{\text{sym}}(\{\mathbf{a}_1, \ldots \mathbf{a}_\ell\}^N)$. The subset satisfying the marginal constraint $\Pi \mapsto \rho$, i.e.

$$\sum_{i_2,\ldots,i_N=1}^{\ell} \gamma_{i_1 i_2 \ldots i_N} = \rho_{i_1} \;\; \forall \; i_1 \in \{1, \ldots, \ell\}, \tag{4.87}$$

is also a convex polytope called the *Kantorovich polytope*; let us denote it by $\mathcal{P}_\rho(\{\mathbf{a}_1, \ldots \mathbf{a}_\ell\}^N)$. While general probability measures in these sets possess a huge number of coefficients which increases combinatorially with the number N of particles, the key point is that the *extreme points*[6] of these sets are very sparse, with only a small number of nonzero coefficients. The extreme points of $\mathcal{P}_\rho(\{\mathbf{a}_1, \ldots, \mathbf{a}_\ell\})$ are easily seen to be symmetrized products of delta functions, $S_N \delta(\mathbf{r}_1 - \mathbf{a}_{i_1}) \cdots \delta(\mathbf{r}_N - \mathbf{a}_{i_N})$, where S_N is the symmetrization operator. Now consider a subset of a convex polytope satisfying *one* linear constraint, geometrically: the intersection of the polytope with a hyperplane. It is geometrically expected (and not difficult to prove) that all extreme points of this new set are convex combinations of just two extreme points of the original polytope. Analogously, by a well-known result in convex geometry the intersection of a convex polytope with k hyperplanes has extreme points given by convex combinations of just $k+1$ of the original extreme points. Since the marginal condition (4.87) imposes $\ell - 1$ constraints (note that one of the ℓ constraints is redundant due to the sum of the $\gamma_{i_1 \ldots i_N}$ being 1), the extreme points of the Kantorovich polytope are convex combinations of just ℓ symmetrized delta functions, i.e., probability measures of the form

$$\sum_{\nu=1}^{\ell} \alpha_\nu S_N \delta(\mathbf{r}_1 - \mathbf{a}_{i_1^{(\nu)}}) \ldots \delta(\mathbf{r}_N - \mathbf{a}_{i_N^{(\nu)}}) \tag{4.88}$$

for some nonnegative coefficients α_ν. Defining the maps f_n by $f_{n-1}(\mathbf{a}_\nu) = \mathbf{a}_{i_n^{(\nu)}}$ yields that all extreme points are quasi-Monge states (4.84). Theorem 4.23 now follows from the general principle that the minimum of a linear functional (such as $\int V_{ee} \, d\Pi$) over a convex polytope is always attained at some extreme point. □

[6] These are the points that cannot be written as convex combinations of any other points in the set.

The quasi-Monge (or quasi-SCE) ansatz and Theorem 4.23 underlie the numerical method described in Sect. 4.4.5.3.

4.3.15 Entropic Regularization of the SCE Functional

We have seen in Fig. 4.1 and Sect. 4.3.8 that in the strongly interacting limit, the N-body density concentrates on the lower-dimensional manifold on which the classical effective potential energy $V_{ee}(\mathbf{r}_1, \ldots, \mathbf{r}_N) - \sum_i v_{SCE}(\mathbf{r}_i)$ is minimal. A regularization of the SCE functional which has nice mathematical properties and smears out the N-body density is the following:

$$\mathrm{V}_{ee}^{\tau}[\rho] = \inf_{\pi \in \mathcal{P}(\mathbb{R}^{dN}) \cap L^1(\mathbb{R}^{dN}),\, \pi \mapsto \rho} \mathcal{V}_{ee}[\pi] + \tau S[\pi]. \tag{4.89}$$

Here $\tau > 0$ is a small parameter, $\mathcal{V}_{ee}$ is the usual electron interaction energy, and S is (minus) the Shannon–Von Neumann entropy,

$$S[\pi] = \int_{\mathbb{R}^{dN}} \pi(\mathbf{r}_1, \ldots, \mathbf{r}_N)\Big(\log \pi(\mathbf{r}_1, \ldots, \mathbf{r}_N) - 1\Big)\mathrm{d}\mathbf{r}_1 \ldots \mathrm{d}\mathbf{r}_N. \tag{4.90}$$

As shown in Lemma 4.27 below, the negative part of the entropy density has finite integral under very mild conditions on ρ (e.g., finite first moment suffices), and so definitions (4.89)–(4.90) make rigorous sense. The existence of a minimizer in (4.89) can be obtained assuming that $\rho \log \rho \in L^1(\mathbb{R}^d)$ [60]. Physically, the right-hand side in (4.89) can be viewed as the free energy of N classical particles with interaction potential V_{ee} and density ρ at inverse temperature τ. But our goal here is not to model a physical system at finite temperature, but instead to approximate the SCE functional.

Figure 4.8 illustrates the effect of the entropy term in a two-electron example: the larger the regularization parameter τ, the more the minimizers π are spread out around the support of the SCE state. (Recall that by Theorem 4.4, when $N = 2$ the SIL variational principle is uniquely minimized by an SCE state.)

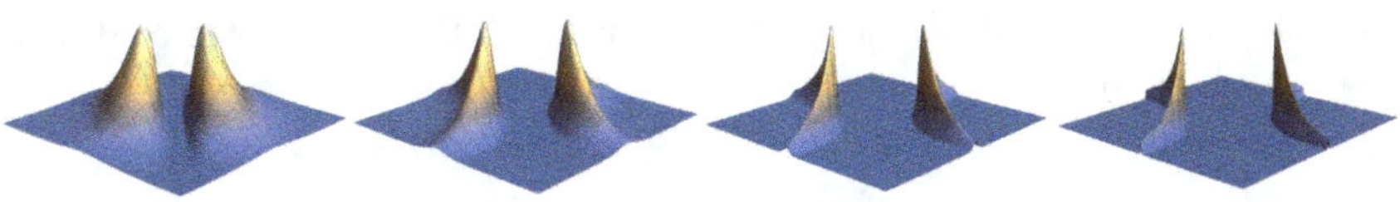

Fig. 4.8 Numerically computed optimal density $\pi^{\tau}(r_1, r_2)$ in (4.89) for two electrons in one dimension and the density $\rho(r) = c_0/\cosh r$, $r \in [-10, 10]$, for different values of the regularization parameter τ. Here c_0 is a normalization constant, and we have used the effective Coulomb interaction [7] $V_{ee}(r_1, r_2) = 1.07\, \mathrm{e}^{-|r_1 - r_2|/2.39}$

The corresponding regularization for the Wasserstein distance squared instead of the SCE functional (Example 4.12 instead of Example 4.11) in fact goes back to Erwin Schrödinger in 1931 [121], and had a completely different motivation: Schrödinger was looking for models for the "most likely" evolution law between two probability distributions of particle positions which have been empirically observed at different times, perhaps hoping to re-discover his—then still controversial—quantum mechanics in a novel way.

An equivalent entropic problem has been considered by Chayes et al. in [19, 20]. In their setting, the integration in the entropy functional S is not against the Lebesgue measure but against the product measure $\prod_{i=1}^{N} \rho(\mathbf{r}_i)/N$, which constitutes a natural model in classical statistical mechanics. The role of the reference measure will be explained below and can be understood via Eq. (4.93).

In optimal transport, entropic regularization became a popular basis for computational methods following an influential paper by Cuturi [34] in machine learning and Galichon and Salanié in economics [54]; see Sect. 4.4 for a computational algorithm. Regularization by entropies other than the Shannon–Von Neumann one is considered in [41, 99].

4.3.15.1 Basic Properties

Let us now informally discuss the basic properties of (4.89).

Unique Minimizer Assuming that $\rho \log \rho \in L^1(\mathbb{R}^d)$, a minimizer π^τ in (4.89) exists [60]. This is expected from the convexity of the functional $\mathcal{V}_{ee} + \tau S$. Since the functional is strictly convex on the domain where it is finite, minimizers must be unique.

Euler–Lagrange Equation; Form of Minimizer Assume a Lagrange multiplier rule as in (4.61). That is, assume the existence of Lagrange multipliers $(\lambda(\mathbf{r}_1))_{\mathbf{r}_1 \in \mathbb{R}^d}$ such that π^τ is the unconstrained minimizer of the Lagrangian

$$\mathcal{L}[\pi] = \int_{\mathbb{R}^{dN}} \Big(V_{ee}(\mathbf{r}_1, \ldots, \mathbf{r}_N) - \sum_{i=1}^{N} \lambda(\mathbf{r}_i) \Big) \pi(\mathbf{r}_1, \ldots, \mathbf{r}_N) \mathrm{d}\mathbf{r}_1 \ldots \mathrm{d}\mathbf{r}_N + \tau S[\pi].$$

Thus the function $\lambda(\mathbf{r})$ has the usual physical interpretation of DFT as minus the potential that enforces the density constraint, and will in the following be denoted $u^\tau(\mathbf{r})$. It follows that

$$0 = \frac{\mathrm{d}}{\mathrm{d}t}\Big|_{t=0} \mathcal{L}[\pi + t\eta]$$

for all variations η with $\int \eta = 0$ and $\pi \pm \eta \geq 0$. That is to say, $0 = \int [\big(V_{ee} - \sum_i u^\tau(\mathbf{r}_i)\big) + \tau \log \pi]\eta$ and therefore

$$V_{ee} - \sum_i u^\tau(\mathbf{r}_i) + \tau \log \pi = \text{const.}$$

By solving for π and adjusting u^τ by an additive constant, it follows that

$$\pi^\tau(\mathbf{r}_1, \ldots, \mathbf{r}_N) = \prod_{i=1}^{N} a^\tau(\mathbf{r}_i) \mathrm{e}^{-\frac{V_{ee}(\mathbf{r}_1,\ldots,\mathbf{r}_N)}{\tau}} \quad \text{with } a^\tau(\mathbf{r}_i) = \mathrm{e}^{\frac{1}{\tau} u^\tau(\mathbf{r}_i)}. \tag{4.91}$$

The function a^τ can—independently of its construction above with the help of Lagrange multipliers—be interpreted as an entropic weight function which makes the probability density π^τ satisfy the constraint $\pi^\tau \mapsto \rho$. Note that by this constraint and Eq. (4.91), a^τ must satisfy the following governing equation in which Lagrange multipliers no longer appear:

$$a^\tau(\mathbf{r}_j) \int_{\mathbb{R}^{d(N-1)}} \prod_{i \neq j} a^\tau(\mathbf{r}_i) \mathrm{e}^{-\frac{V_{ee}(\mathbf{r}_1,\cdots,\mathbf{r}_N)}{\tau}} \prod_{i \neq j} \mathrm{d}\mathbf{r}_i = \frac{\rho(\mathbf{r}_j)}{N} \quad \forall\, j \in \{1, \ldots, N\}. \tag{4.92}$$

The above equations constitute the so-called (multi-marginal) Schrödinger system. When the density ρ is Gaussian and the interaction potential w is taken to be the repulsive or attractive harmonic interaction, the entropically regularized problem can be solved exactly, see [58], for the one-dimensional case, and [39, 79, 103] for the general case. In [17], Carlier and Laborde showed the existence of a solution of the system (4.92) via an inverse function theorem argument by assuming that the one-body density ρ belongs to $L^\infty(\mathbb{R}^d)$.

Relative Entropy Formulation The functional $\mathcal{V}_{ee} + \tau S$ agrees up to an additive constant with the Kullback–Leibler divergence (or minus the relative entropy)[7] between π and a kernel function $\mathcal{K}$ of the electronic interaction V_{ee} [89]:

$$\mathcal{V}_{ee}[\pi] + \tau S[\pi] = \tau \mathrm{KL}(\pi | \mathcal{K}) - \tau \text{ with } \mathcal{K} = \mathrm{e}^{-V_{ee}/\tau}.$$

Thus the optimizer π^τ is the density with marginal ρ which has minimal relative entropy with respect to the kernel $\mathcal{K}$.

[7] The KL divergence between two nonnegative densities with possibly unequal mass is formally defined as $\mathrm{KL}(f|g) = \int f \log \frac{f}{g}$.

The Role of the Reference Measure In the literature, the entropy functionals which are typically studied replace integration against the Lebesgue measure in (4.90) by integration against the product of the marginals $\mu^{\otimes N} = \otimes_{i=1}^N \mu$ (or any other finite reference measure), where $\rho/N = \mu$. As shown in Lemma 1.5 in [40] (see also [60] for the Coulomb case), both problems are equivalent since the following identity holds

$$\mathrm{V}_{\mathrm{ee}}^{\tau}[\rho] = \inf_{\pi \mapsto \rho} \left\{ \mathcal{V}_{\mathrm{ee}}[\pi] + \tau \int_{\mathbb{R}^{dN}} \frac{\mathrm{d}\pi}{\mathrm{d}\mu} \left(\log \frac{\mathrm{d}\pi}{\mathrm{d}\mu} - 1 \right) \mathrm{d}\mu \right\} + \tau \int_{\mathbb{R}^d} \rho \log \frac{\rho}{N} \mathbf{dr}\,. \tag{4.93}$$

Therefore, whenever at least one side of the equality above is finite, the original variational problem from the definition of $\mathrm{V}_{\mathrm{ee}}^{\tau}[\rho]$ (Eq. (4.89)) and the variational problem defined on the right-hand side of (4.93) have the same minimizers.

Dual Formulation As for the exact (unregularized) strong-interaction limit of DFT, there is a dual variational principle for the Lagrange multiplier and an associated dual construction of $V_{\mathrm{ee}}^{\tau}[\rho]$. We have

$$\mathrm{V}_{\mathrm{ee}}^{\tau}[\rho] = \sup_u J[u], \tag{4.94}$$

where

$$J[u] = \int_{\mathbb{R}^d} u(\mathbf{r})\rho(\mathbf{r})\mathbf{dr} - \tau \int_{\mathbb{R}^{dN}} \mathrm{e}^{-\frac{1}{\tau}[V_{\mathrm{ee}}(\mathbf{r}_1,\ldots,\mathbf{r}_N) - \sum_i u(\mathbf{r}_i)]} \mathbf{dr}_1 \ldots \mathbf{dr}_N \tag{4.95}$$

and the supremum in (4.94) is over a suitable class of potentials. The second term in (4.95) can be viewed as a soft version of the inequality constraint $E_{\mathrm{pot}}(\mathbf{r}_1, \ldots, \mathbf{r}_N) = V_{\mathrm{ee}}(\mathbf{r}_1, \ldots, \mathbf{r}_N) - u(\mathbf{r}_1) + \ldots + u(\mathbf{r}_N) \geq 0$ in the unregularized theory (see (4.149)), as it penalizes deviations from this inequality. Indeed, via the Laplace principle we have that, whenever the second term in (4.94) is finite,

$$\begin{aligned} \lim_{\tau \to 0^+} -\tau \log \left(\int_{\mathbb{R}^{dN}} \mathrm{e}^{-\frac{1}{\tau}[V_{\mathrm{ee}}(\mathbf{r}_1,\ldots,\mathbf{r}_N) - \sum_i u(\mathbf{r}_i)]} \mathbf{dr}_1 \ldots \mathbf{dr}_N \right) \\ = \inf_{\mathbf{r}_1,\ldots,\mathbf{r}_N \in \mathbb{R}^d} \{E_{\mathrm{pot}}(\mathbf{r}_1, \ldots, \mathbf{r}_N)\}. \end{aligned}$$

In the discrete setting, this is precisely the LogSumExp formula. The existence of an optimizer u^{τ} for the dual problem and the representation formulae (4.91), (4.96) with this u^{τ} were proved in [40, 41] under the assumption that $\rho \log \rho \in L^1(\mathbb{R}^d)$ and V_{ee} is measurable and bounded.

Functional Derivative As in exact SCE theory, the functional derivative of the energy functional is formally given by the optimal potential in the dual problem, that is to say

$$\frac{\delta V^{\tau}_{\mathrm{ee}}[\rho]}{\delta\rho} = u^{\tau} + \text{const}, \tag{4.96}$$

where u^{τ} is the maximizer of (4.95) (assuming such a maximizer exists and is unique). As in SCE theory, a natural choice of the additive constant is to require $\lim_{|\mathbf{r}|\to\infty}\big(u^{\tau}(\mathbf{r}) + \text{const}\big) = 0$. The ensuing potential $v^{\tau} = u^{\tau} + \text{const}$ is then an approximation to the SCE potential.

4.3.15.2 Relation with the Levy–Lieb Functional

Just like the SCE functional itself, its entropic regularization is a rigorous lower bound of the exact functional, provided the regularization parameter τ is chosen suitably. More precisely:

Theorem 4.24 ([125]) *Let Ψ be any N-electron wavefunction in the space $\mathcal{W}^N$ (see (4.7)), or alternatively any bosonic wavefunction in $H^1(\mathbb{R}^{dN})$, and suppose $\Psi \mapsto \rho$. Let V_{ee} be the Coulomb interaction. Then the scaled Levy–Lieb functional defined in Eq.* (4.17) *satisfies*

$$\frac{F^{\lambda}[\rho]}{\lambda} \geq V^{\tau}_{\mathrm{ee}}[\rho] \quad \textit{with } \tau = \frac{\pi}{2\lambda}. \tag{4.97}$$

In particular, the original Levy–Lieb functional (4.11) *satisfies*

$$F_{\mathrm{LL}}[\rho] \geq V^{\pi/2}_{\mathrm{ee}}[\rho]. \tag{4.98}$$

This result is a consequence of the logarithmic Sobolev inequality (LSI). We include a proof, following Seidl et al. [125]. We begin by recalling a standard version of the LSI.

Theorem 4.25 (LSI, Corollary 7.3 in [70]) *Let $\nu \in \mathcal{P}(\mathbb{R}^n)$ such that $\nu(\mathbf{r}) = \mathrm{e}^{-V(\mathbf{r})}$ with $\mathrm{D}^2 V \geqslant \kappa \mathrm{Id}$. Then, for every locally integrable function $f \geqslant 0$ on $\mathbb{R}^n$ such that $f\nu \in \mathcal{P}(\mathbb{R}^n)$ we have that $\int_{\mathbb{R}^n} f \log f \mathrm{d}\nu \leqslant \frac{2}{\kappa}\int |\nabla\sqrt{f}|^2\, d\nu$.*

This implies the following LSI for the Lebesgue measure:

Corollary 4.26 (LSI for the Lebesgue Measure, [125]) *Let $f \geqslant 0$ be a function such that $\sqrt{f} \in H^1(\mathbb{R}^n)$ and $f \in \mathcal{P}(\mathbb{R}^n)$. Then $\int_{\mathbb{R}^n} f \log f \mathrm{d}\mathbf{r} \leqslant \frac{1}{\pi}\int_{\mathbb{R}^n} |\nabla\sqrt{f}|^2 d\mathbf{r}$.*

Proof of Corollary 4.26

1. In the LSI in Theorem 4.25, the requirement on f that $\int f\,\mathrm{d}\nu = 1$ can be relaxed to $0 < \int f\,\mathrm{d}\nu \le 1$. This follows by applying the LSI to f/α, $\alpha = \int f\,\mathrm{d}\nu$, and noting that the extra term $-(1/\alpha)\int f \log\alpha\,\mathrm{d}\nu$ on the left-hand side is ≥ 0.
2. Take $\nu_{\mathbf{r}_2}(\mathbf{r}_1) = \mathrm{e}^{-\pi|\mathbf{r}_1-\mathbf{r}_2|^2}$, then ν satisfies the assumption of the LSI with $\kappa = 2\pi$, and moreover $\int f\mathrm{d}\nu_{\mathbf{r}_2} \le 1$. Hence by the LSI,

$$\int f \log f\,\mathrm{d}\nu_{\mathbf{r}_2} \le \frac{1}{\pi}\int |\nabla\sqrt{f}|^2 \mathrm{d}\nu_{\mathbf{r}_2}.$$

3. Integrate over $\mathbf{r}_2$ and use that $\int \mathrm{e}^{-\pi|\mathbf{r}_1-\mathbf{r}_2|^2}\mathbf{dr}_2 = 1$. This yields the assertion.

□

Proof of Theorem 4.24 Let $\Psi \in \mathcal{W}^N$, $\Psi \mapsto \rho$, and let Π be its N-point position density (4.8). By a version of the Hoffmann-Ostenhof inequality [77],[8] $\sqrt{\Pi} \in H^1(\mathbb{R}^{dN})$ and

$$T[\Psi] \ge \frac{1}{2}\int_{\mathbb{R}^{dN}} |\nabla\sqrt{\Pi}|^2 \mathbf{dr}_1 \dots \mathbf{dr}_N.$$

This together with the LSI for the Lebesgue measure (Corollary 4.26) applied to π gives

$$\begin{aligned}\frac{1}{\lambda}T[\Psi] + V_{\mathrm{ee}}[\Psi] &\geqslant \frac{\pi}{2\lambda}\int_{\mathbb{R}^{dN}} \Pi \log \Pi\,\mathbf{dr}_1 \dots \mathbf{dr}_N + \int_{\mathbb{R}^{dN}} V_{\mathrm{ee}}\,\Pi\,\mathbf{dr}_1 \dots \mathbf{dr}_N \\ &= \frac{\pi}{2\lambda}\Big(S(\Pi) + 1\Big) + V_{\mathrm{ee}}[\Pi].\end{aligned}$$

Taking the infimum over $\Psi \in \mathcal{W}^N$ yields $F^\lambda[\rho]/\lambda \ge V_{\mathrm{ee}}^\tau[\rho] + \tau$, with τ as in the theorem.

□

Although Theorem 4.24 provides a lower bound for the Levy–Lieb functional (4.11), in practice this bound can be rather loose [58].

[8] Strictly speaking, this inequality and related ones are proved in [77] under the tacit assumption that $\sqrt{\Pi}$ (or related reduced quantities) belong to H^1 and can be differentiated by the chain rule. For further discussion of this point see Chap. 5 by Kvaal in this volume.

4.3.15.3 Well Definedness of Entropy and Convergence to the SCE Functional

We now show that the entropy is well defined under very mild conditions on ρ (e.g., finite first moment suffices), and that the entropically regularized functional $\mathrm{V}_{\mathrm{ee}}^{\tau}$ converges to the SCE functional when the regularization parameter tends to zero.

Note that a priori both the positive and the negative part of the integral (4.90) could be divergent; Lemma 4.27 excludes this for the negative part, and so the integral always has a well defined value in $\mathbb{R} \cup \{+\infty\}$.

Lemma 4.27 (Well-Definedness of Entropy and of the Regularized SCE Functional) *Let $\rho \in L^1(\mathbb{R}^d)$, $\rho \geq 0$, $\int \rho = N$, and assume ρ has finite first moment, that is to say $\int |\mathbf{r}|\rho(\mathbf{r})\mathrm{d}\mathbf{r} < \infty$. Let $\pi \in \mathcal{P}(\mathbb{R}^{dN}) \cap L^1(\mathbb{R}^{dN})$ with $\pi \mapsto \rho$. Then the negative part $(\pi \log \pi)_-$ has finite integral; more precisely, for some constant $A_\rho > 0$ which depends only on ρ but not on π*

$$\int (\pi \log \pi)_- \geq -A_\rho > -\infty,$$

where $f_-(\mathbf{r}) = \min\{f(\mathbf{r}), 0\}$ denotes the negative part of a function f. Hence S as defined by (4.90) *is well defined as a functional*

$$S : \{\pi \in \mathcal{P}(\mathbb{R}^{dN}) \cap L^1(\mathbb{R}^{dN}) : \pi \mapsto \rho\} \to \mathbb{R} \cup \{+\infty\},$$

and $V_{\mathrm{ee}}^{\tau}[\rho]$ as defined by (4.89) *is well defined as an element of $\mathbb{R} \cup \{+\infty\}$.*

The assumption that ρ has finite first moment cannot be omitted. For instance, for $N = 2$ and $d = 1$ the N-body density

$$\pi(r_1, r_2) = c_0 \prod_{i=1}^{2} \frac{1}{r_i(\log r_i)^2} \quad \text{on } [2, \infty)^2,$$

continued by zero to $\mathbb{R}^2$ and with c_0 chosen such that $\int \pi = 1$, belongs to $L^1(\mathbb{R}^2)$ but satisfies $\int (\pi \log \pi)_- = -\infty$, as the interested reader can check using that $\int_2^\infty \frac{1}{z|\log z|^\alpha} \mathrm{d}z = \infty$ for $\alpha = 1$ but $< \infty$ for $\alpha > 1$. In particular, in such a case the equivalence described in (4.93) does not necessarily hold.

The lemma implies that for any interaction potential V_{ee} on $\mathbb{R}^{dN}$ which is symmetric and bounded from below, such as the Coulomb interaction, $V_{\mathrm{ee}}^{\tau}[\rho]$ is well defined as an element of $\mathbb{R} \cup \{+\infty\}$.

Proof of Lemma 4.27 $\pi \log \pi$ is ≤ 0 precisely in the region $\Omega = \{\mathbf{r} \in \mathbb{R}^{dN} : \pi(\mathbf{r}) \leq 1\}$. Split Ω into $\Omega_< = \{\mathbf{r} \in \Omega : 0 \leq \pi(\mathbf{r}) < \mathrm{e}^{-(|\mathbf{r}_1|+\ldots+|\mathbf{r}_N|)}\}$ and $\Omega_> = \{\mathbf{r} \in \Omega : \pi(\mathbf{r}) \geq \mathrm{e}^{-(|\mathbf{r}_1|+\ldots+|\mathbf{r}_N|)}\}$. Since $g(z) = z \log z$ satisfies $|g(z)| \leq C\sqrt{z}$ in

[0, 1] for some constant C,

$$\begin{aligned}\int_{\mathbb{R}^{dN}} |(\pi \log \pi)_-| &= \int_{\Omega_<} |(\pi \log \pi)_-| + \int_{\Omega_>} |(\pi \log \pi)_-| \\ &\le C \int_{\Omega_<} \mathrm{e}^{-(|\mathbf{r}_1|/2+\ldots+|\mathbf{r}_N|/2)} \mathrm{d}\mathbf{r}_1 \ldots \mathrm{d}\mathbf{r}_N \\ &+ \int_{\Omega_>} \pi(\mathbf{r}_1, \ldots, \mathbf{r}_N)\Big(|\mathbf{r}_1| + \cdots + |\mathbf{r}_N|\Big) \mathrm{d}\mathbf{r}_1 \ldots \mathrm{d}\mathbf{r}_N \\ &\le C\Big(\int_{\mathbb{R}^d} \mathrm{e}^{-|\mathbf{r}_1|/2} \mathrm{d}\mathbf{r}_1\Big)^N + \int_{\mathbb{R}^d} \rho(\mathbf{r}_1)|\mathbf{r}_1| \mathrm{d}\mathbf{r}_1 =: A_\rho. \end{aligned} \tag{4.99}$$

By the assumption that ρ has finite first moment, the right-hand side is finite, completing the proof of the lemma. □

Finally, we prove that—as intuitively expected—the entropically regularized functional V_{ee}^{τ} converges to the exact SCE functional when the regularization parameter tends to zero. The corresponding Γ-convergence result was obtained in [60].

Theorem 4.28 *Let ρ be any N-electron density which belongs to the class $\mathcal{D}^N$ (see* (4.12)*) and has finite first moment, and let V_{ee} be the Coulomb interaction. Then*

$$\lim_{\tau \to 0} V_{\mathrm{ee}}^{\tau}[\rho] = V_{\mathrm{ee}}^{\mathrm{SCE}}[\rho]. \tag{4.100}$$

Proof We combine the upper bound on $V_{\mathrm{ee}}^{\tau}[\rho]$ from Theorem 4.24, the asymptotic result on $F^{\lambda}[\rho]/\lambda$ in Eq. (4.19) (see Theorem 4.3), and the lower bound from Lemma 4.27. By inequality (4.99) we have for any $\pi \mapsto \rho$

$$\tau S[\pi] \ge \tau \int_{\mathbb{R}^{dN}} (\pi \log \pi)_- \; - \; \tau \ge -\tau\Big(A_\rho + 1\Big)$$

and hence, by adding $\mathcal{V}_{\mathrm{ee}}[\pi]$ to both sides and taking the infimum over π

$$V_{\mathrm{ee}}^{\tau}[\rho] \ge V_{\mathrm{ee}}^{\mathrm{SCE}}[\rho] - \tau\Big(A_\rho + 1\Big).$$

Obviously this lower bound converges to $V_{\mathrm{ee}}^{\mathrm{SCE}}[\rho]$ as $\tau \to 0$. On the other hand, by Theorem 4.24 we have $V_{\mathrm{ee}}^{\tau}[\rho] \le F^{\lambda}[\rho]/\lambda$ and by Theorem 4.3 this upper bound also converges to $V_{\mathrm{ee}}^{\mathrm{SCE}}[\rho]$; hence so must $V_{\mathrm{ee}}^{\tau}[\rho]$. □

4.4 Numerical Methods and Approximations

The SCE functional cannot at the moment be accurately and efficiently computed for general three-dimensional densities and large N. But accurate numerical methods are available for small N or special situations, novel methods aimed at large N are under development, and less accurate approximations can already be computed for large N. We review these methods and approximations in this section, and their use within Kohn–Sham DFT in Sect. 4.5.

4.4.1 Numerical Methods Based on Co-motion Functions

Numerical implementations using co-motion functions were confined to the following cases:

- the exact maps are known: general N in one dimension (see Sect. 4.3.11);
- an explicit ansatz, able to get very close to the true minimum, exists: spherically symmetric (radial) case (see Sect. 4.3.12).

In addition, co-motion functions can be extracted from optimal plans in the case

- $N = 2$, for which the existence of the map is proven and there are 1-1 correspondences between map, optimal plan, and Kantorovich potential (see Eqs. (4.38) and (4.39)).

We review here and in the following section the implementation for these three classes of problems. Their use in combination with Kohn–Sham DFT is then discussed in Sect. 4.5.

4.4.1.1 One-Dimensional N-Electron Systems

The SCE functional has been implemented for one-dimensional (1D) many-electron systems using the exact co-motion functions (maps) of Seidl [122], which we reported and illustrated in Sect. 4.3.11. These applications typically aim at modeling physical systems in which electrons are confined in elongated traps (quantum wires): the interaction used is thus 3D Coulomb renormalized for small interparticle distances. The idea is that at long range the electrons feel the $1/|x|$ interaction, but at short range they can avoid each other due to the finite thickness of the wire, which is mimicked by removing the divergence at $x = 0$. For example, a widely used effective quasi-1D interaction is obtained by integrating the 3D Coulomb interaction over normalized gaussians in two of the three spatial directions [64], modeling harmonic confinement within a wire of thickness b,

$$v_{\text{ee}}^{\text{wire}}(x) = \frac{1}{4\pi\, b^2}\int_{-\infty}^{\infty} \mathrm{d}y \int_{-\infty}^{\infty} \mathrm{d}z \frac{\mathrm{e}^{-\frac{1}{4b^2}(x^2+b^2)}}{\sqrt{x^2+y^2+z^2}} = \frac{\sqrt{\pi}}{2\,b}\exp\left(\frac{x^2}{4\,b^2}\right)\operatorname{erfc}\left(\frac{|x|}{2\,b}\right). \tag{4.101}$$

This interaction is finite at $x = 0$, where it has a cusp, behaves as $1/|x|$ for large x and it is convex for $x \geq 0$. Other popular quasi-1D interactions are the soft Coulomb and the regularized Coulomb,

$$v_{\text{ee}}^{\text{soft}}(x) = \frac{1}{\sqrt{x^2+a^2}}, \tag{4.102}$$

$$v_{\text{ee}}^{\text{reg}}(x) = \frac{1}{|x|+a}. \tag{4.103}$$

Notice, however, that the 1D maps of Seidl [123] are exact only for interactions (costs) that are convex for $x \geq 0$ [27]. This means that when using $v_{\text{ee}}^{\text{soft}}(x)$, which is concave for $x \in [0, a/\sqrt{2}]$, the Seidl maps are not guaranteed to yield the true minimizer, as illustrated, for example, in Fig. 2 of Ref. [71].

Numerical realizations of the 1D Seidl maps are reported in Refs. [71, 72, 100–102, 104]. The implementation of the maps directly follows from Sect. 4.3.11: given a density $\rho(x)$ on a grid, the cumulant function $F_\rho(x)$ is evaluated on the same grid, and its inverse $F_\rho^{-1}(x)$ is simply obtained by swapping the columns. The grid can be restored by using a spline interpolation for $F_\rho^{-1}(x)$, and the maps are readily obtained. Numerical issues can appear in regions where the density is close to zero, with $F_\rho^{-1}(x)$ raising extremely steeply. An alternative method to obtain the 1D maps without the need to construct $F_\rho^{-1}(x)$ is discussed in Ref. [72].

4.4.1.2 Spherically Symmetric Densities

For spherically symmetric densities the radial SGS maps (4.79)–(4.80) conjectured in [122, 127] have been implemented in Refs. [63, 127] for the 3D case using numerical densities for atoms from He to Ne, and in Ref. [105] for the 2D case, where the SCE functional has been combined self-consistently with Kohn–Sham DFT to describe electrons confined in a parabolic potential at low density.

The construction of the radial maps is implemented as in the 1D case. However, the computational complexity is now higher due to the evaluation of the reduced radial cost of Eq. (4.77), which requires an angular minimization for given radial distances. For the two-dimensional case treated in Ref. [105], where the number of relative angles to minimize was equal to $N-1$, the procedure has been the following. For an initial non-degenerate radial configuration and given initial starting angles, the quasi-Newton Broyden–Fletcher–Goldfarb–Shanno (BFGS) algorithm was used to find the closest local minimum. Then the radial position of the "first" electron was changed in small discrete steps, the radial positions of the remaining electrons were computed using the SGS maps, and the angles were optimized using the BFGS algorithm, with starting angles taken from the previous step. This procedure rests on the assumption that the optimal angles change continuously with the radial configuration. The starting angles for the initial radial configuration can be chosen by using simulated annealing as a global optimization strategy. It should be stressed that the angular minimization does not need to be performed for the whole set N_{grid} of radial grid points. In fact, the N radial distances are periodic, as each circular shell $r \in [a_i, a_{i+1}]$ (with $a_i = F_\rho^{-1}(i)$, $i \in \mathbb{N}$), corresponds to the same physical situation,[127] simply describing a permutation of the set of distances occurring in the first shell $r \in [0, a_1]$. Thus, by keeping track of the minimizing angles, and by readapting the grid in every circular shell, it is possible to do the angular minimization only N_{grid}/N times rather than N_{grid} times.

4.4.2 Methods Based on Linear Programming

Direct discretization of the SIL variational principle (4.25) yields a linear program, which is numerically tractable when $N = 2$.

4.4.2.1 The $N = 2$ Case

For two-electron systems in 3D with general density, Chen et al. [22] have implemented a method to directly solve the SIL variational principle via linear programming and extract the co-motion function and the SCE potential from the SIL solution. They used this approach to compute the co-motion function and the KS-SCE binding curve of the H_2 molecule (see Figs. 4.9 and 4.14).

One truncates $\mathbb{R}^3$ to a bounded domain, discretizes it into ℓ finite regions $e_1, \dots, e_\ell$, and represents each element by a point $\mathbf{a}_\ell$ located at its barycenter. The single-particle density becomes a vector in $\mathbb{R}^\ell$ with components $\rho_\ell = \int_{e_\ell} \rho(\mathbf{r})\, d\mathbf{r}$. The two-particle density π is represented by a matrix $\gamma = (\gamma_{ij}) \in \mathbb{R}^{\ell\times\ell}$ with $\gamma_{ij} = \int_{e_i}\int_{e_j} d\pi(\mathbf{r}_1, \mathbf{r}_2)$, and the interaction $V_{\text{ee}}(\mathbf{r}_1, \mathbf{r}_2)$ becomes a matrix $(c_{ij}) \in \mathbb{R}^{\ell\times\ell}$ with $c_{ij} = \frac{1}{|\mathbf{a}_i - \mathbf{a}_j|}$. The SIL problem (4.25) then becomes

$$\min_{\gamma\in\mathbb{R}^{\ell\times\ell}} \sum_{1\le i,j\le\ell} c_{ij}\gamma_{ij} \tag{4.104}$$

$$\text{s/to} \quad \sum_{j=1}^{\ell} \gamma_{ij} = \frac{\rho_i}{2}, \; i = 1, \dots, \ell, \quad \sum_{i=1}^{\ell} \gamma_{ij} = \frac{\rho_j}{2}, \; j \in \{1, \dots, \ell\},$$

$$\gamma_{ij} \ge 0.$$

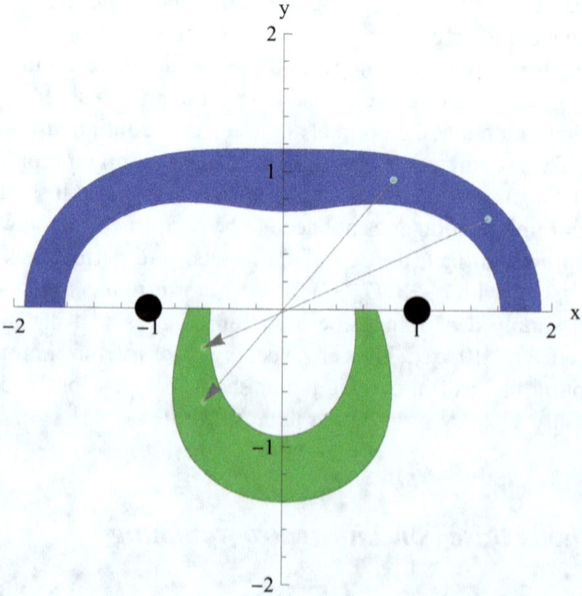

Fig. 4.9 Co-motion function for the H_2 molecule [22]. The blue region—corresponding to the points in a half plane adjacent to the molecular axis with density between 0.04 and 0.08—is mapped to the green region in the opposite half plane. The black dots indicate the positions of the nuclei

This is a standard linear programming problem of the form $\min_x f^T x$ subject to $Ax = b$, $x_k \geq 0$, where x is the vector containing the entries of γ. The solution can be obtained with a standard linear programming software (in [22], the authors used MOSEK). For a uniform discretization of the density, the number of degrees of freedom in the linear program would still be huge; instead an adaptive mesh was used in which all elements contain roughly the same amount of density, that is to say the mesh is much finer in the high-density region near the nuclei. (For automated generation of such a mesh, the finite element package PHK was used. See Chap. 10 by Dai and Zhou for more information about this package.) The solution to (4.104) entails an approximation to the co-motion function f at the barycenters $\{\mathbf{a}_i\}_{i=1}^{\ell}$, namely the barycenter of the image of $\mathbf{a}_i$ under the transport plan γ:

$$f^{(\ell)}(\mathbf{a}_i) = \sum_{j=1}^{\ell} \frac{\gamma_{ij}}{\rho_i/2} \mathbf{a}_j \; i \in \{1, \ldots, \ell\}, \tag{4.105}$$

where γ_{ij} can be regarded as the mass transported from $\mathbf{a}_i$ to $\mathbf{a}_j$ and the normalization factor $\rho_i/2$ guarantees that the barycentric weights sum to 1. Since, for $N = 2$, the optimal N-point density π for the continuum problem is unique and of SCE form (see (4.41) and Theorem 4.4), it follows that if the discretization is sufficiently fine, i.e. ℓ is large enough, $f^{(\ell)}$ is a good approximation to f. The resulting co-motion function for the H_2 molecule is depicted in Fig. 4.9.

4.4.2.2 The $N > 2$ Case and the Curse of Dimension

Since the above method uses a real-space discretization of the SIL variational principle whose unknown is the N-particle density on $\mathbb{R}^{3N}$, it is limited in practice to $N = 2$, to keep the number of computational degrees of freedom manageable. Indeed, for general N the N-particle density π must be represented by an order-N tensor $(\gamma_{i_1 \dots i_N}) \in \mathbb{R}^{\ell \times \dots \times \ell}$ with entries $\gamma_{i_1 \dots i_N} = \int_{e_{i_1} \times \dots \times e_{i_N}} \mathrm{d}\pi(\mathbf{r}_1, \dots, \mathbf{r}_N)$. Since π can be assumed to be symmetric (see Remark 4.2), γ can be assumed to be symmetric under permutation of indices and Eq. (4.104) becomes

$$\begin{aligned} &\min_{(\gamma_{i_1 \dots i_N}) \in \mathbb{R}^{\ell \times \dots \times \ell} \text{ symmetric}} \sum_{1 \le i_1, \dots, i_N \le \ell} V_{\mathrm{ee}}(a_{i_1}, \dots, a_{i_N}) \, \gamma_{i_1 \dots i_N} \qquad (4.106) \\ &\text{s/to} \quad \sum_{i_2, \dots, i_N = 1}^{\ell} \gamma_{i_1 i_2 \dots i_\ell} = \frac{\rho_{i_1}}{N}, \quad i_1 \in \{1, \dots, \ell\}, \\ &\gamma_{i_1 \dots i_N} \ge 0. \end{aligned}$$

This is still a linear program, but in ℓ^N (or, using symmetry, $\binom{N+\ell-1}{\ell+1}$) variables. For rigorous hardness results on problems of the form (4.106) with pairwise interaction see [4, 52].

4.4.3 *Methods Based on the Dual Formulation*

Mendl and Lin [104] have implemented a method for solving the dual formulation of the SCE functional, Eqs. (4.55) and (4.53), and applied it to the Beryllium atom, a four-electron quantum wire in 1D, and a model trimer in 3D. In the 3D case, they parametrized the (unknown) Kantorovich potential by a pseudocharge,

$$v(\mathbf{r}) = \int \frac{m(\mathbf{r}')}{|\mathbf{r} - \mathbf{r}'|} \mathrm{d}\mathbf{r}',$$

with m given by a small number of Gaussians and satisfying $\int m = N - 1$ to account for the asymptotic behavior $v(\mathbf{r}) \sim (N-1)/|\mathbf{r}|$ for large $|\mathbf{r}|$ (see (4.66)). They showed that the constrained maximization in (4.55), (4.53) is equivalent to a nested pair of unconstrained optimizations,

$$V_{\rm ee}^{\rm SCE}[\rho] = \sup_{v}\Big(\int v(\mathbf{r})\rho(\mathbf{r})\mathrm{d}\mathbf{r} + g[v]\Big) \text{ with} \tag{4.107}$$

$$g[v] = \min_{(\mathbf{r}_1,\dots,\mathbf{r}_N)\in\mathbb{R}^{dN}}\Big(V_{\rm ee}(\mathbf{r}_1,..,\mathbf{r}_N) - \sum_{i=1}^{N} v(\mathbf{r}_i)\Big). \tag{4.108}$$

The inner optimization for given v was implemented by a quasi-Newton method and the outer optimization via a gradient-free simplex algorithm. For the Beryllium atom, using just two Gaussians for m resulted in a relative error of the SCE energy of only 1.6% compared to the SCE energy obtained via the SGS co-motion functions for radially symmetric densities [127] as described in Sect. 4.4.1.2. Also, the obtained SCE potential was in good agreement with the one based on the radial co-motion functions.

As the authors point out, this approach is in practice limited to small systems, because the inner optimization is high-dimensional, nonlinear, and highly degenerate for the optimal v (recall that the set of minimizers is typically d-dimensional), and the outer optimization (4.107) is nonlinear and nonsmooth, and hence unsuitable for numerical optimization over a large number of degrees of freedom.

4.4.4 Multi-Marginal Sinkhorn Algorithm

In optimal transport, a standard computational method [35] is to pass to the entropic regularization (in our case, problem (4.89)) and solve the latter via the Sinkhorn algorithm. This is a simple and robust algorithm which goes back to Sinkhorn in the context of estimating Markov transition matrices [131]; it was introduced into two-marginal optimal transport in [34] and generalized to several marginals in [8]. The multi-marginal Sinkhorn algorithm with Coulomb cost was implemented by Benamou et al. [9] (see also [110]) to compute the SCE energy and potential for the He and Li atoms.

The multi-marginal Sinkhorn algorithm goes as follows; we state it here in the continuous setting. One starts from the exact form (4.91) of the optimizer. One now allows the N entropic weight functions $a_j(\mathbf{r}_j) = a^\tau(\mathbf{r}_j)$ in this form to be different (so as to be able to update them one by one). One updates them iteratively so as to enforce the j-th marginal constraint, (4.92) for j:

$$a_j(\mathbf{r}_j)\int_{\mathbb{R}^{d(N-1)}}\prod_{i\neq j} a_i(\mathbf{r}_i)\mathrm{e}^{-V_{\rm ee}(\mathbf{r}_1,\dots,\mathbf{r}_N)/\tau}\prod_{i\neq j}\mathrm{d}\mathbf{r}_i \stackrel{!}{=} \rho(\mathbf{r}_j)/N. \tag{4.109}$$

Solving for a_j yields an explicit formula for a_j in terms of the other a_i. Thus a single updating cycle consists of the N steps

$$a_j^{\text{new}}(\mathbf{r}_j) = \frac{\rho(\mathbf{r}_j)/N}{\left[\begin{array}{c}\int_{\mathbb{R}^{d(N-1)}} \prod_{i<j} a_i^{\text{new}}(\mathbf{r}_i) \prod_{i>j} a_i^{\text{old}}(\mathbf{r}_i) \\ \times e^{-V_{\text{ee}}(\mathbf{r}_1,\ldots,\mathbf{r}_N)/\tau} \prod_{i\neq j} d\mathbf{r}_i\end{array}\right]}, \; j = 1, \ldots, N. \tag{4.110}$$

One then repeats the cycle until convergence.

Convergence of the Sinkhorn algorithm is rigorously guaranteed under mild conditions on the interaction potential and the density (e.g., bounded potentials and $\rho \log \rho \in L^1$ are sufficient); see [131] for the discretized $N = 2$ case, [118] for the general $N = 2$ case, and [40] for $N \geq 2$. The (linear) rate of convergence for the Sinkhorn algorithm was obtained in [23, 46] in the $N = 2$ case, and in [15] for the multi-marginal Sinkhorn algorithm. For a two-electron example in dimension one computed with the Sinkhorn algorithm see Fig. 4.8.

In [9], Benamou et al. demonstrated that for the He atom (and the choice $\tau = 0.02$) the algorithm yields an accurate approximation to the SCE energy and the SCE potential compared to the (in this case rigorously justified) SGS map-based solution; the relative error of the potential in the L^∞ norm was only 0.4%. Moreover, for the Li atom the numerical Sinkhorn solution exhibited very good qualitative agreement with the SGS solution.

Some regularization is essential for the Sinkhorn approach. As τ approaches zero—so that the entropic regularization $V_{\text{ee}}^{\tau}[\rho]$ from (4.89) approaches the exact SCE functional (4.26)—the convergence speed of the algorithm also goes to zero (see e.g. [35, 50]), and numerical instabilities can appear associated with the extremely small order $e^{-1/\tau}$ of the integrand (see e.g. [8]).

The idea of regularization underlying the algorithm fits well into our DFT context as the optimal N-point density is smeared out anyway off the strongly interacting limit. However, a significant limitation from the point of view of DFT is the high-dimensionality of the integral in (4.109), (4.110). For a discretization of the one-body density by ℓ gridpoint values, the cost of a single integral evaluation for fixed $\mathbf{r}_j$ is $O(\ell^{N-1})$, limiting the method to small N.

4.4.5 *Towards Large N*

Very recently, some promising methods have been proposed which should, at least in principle, be suitable for tackling the case of large N. These have been demonstrated to show good performance on one-dimensional test examples where the Seidl solution from Sect. 4.3.11 is available for comparison. At the time of writing, it has yet to be demonstrated that any of these methods is capable of accurately computing the SCE energy for large N in three dimensions.

4.4.5.1 Semidefinite Convex Relaxation

The starting point of this method, introduced by Khoo and Ying [82], is the fact that the SIL problem 4.25 can, due to the fact that $V_{\rm ee}$ is a two-body potential (4.3), be reformulated as a minimization over N-representable 2-point probability measures:

$$V_{\rm ee}^{SCE}[\rho] = \min_{\substack{\Gamma\in\mathcal{P}(\mathbb{R}^d\times\mathbb{R}^d)\\ \Gamma\ N\text{-representable},\ \Gamma\mapsto\rho}} \binom{N}{2} \int_{\mathbb{R}^d\times\mathbb{R}^d} w_{\rm ee}(\mathbf{r}-\mathbf{r}')\,\mathrm{d}\Gamma(\mathbf{r},\mathbf{r}'). \qquad (4.111)$$

Here a two-point probability measure on $\mathbb{R}^d \times \mathbb{R}^d$ is called *N-representable* if it is the 2-marginal of a symmetric N-point probability measure on $\mathbb{R}^{dN}$. This two-body formulation of the SCE functional was introduced in [51], and is a direct adaptation of the well-known two-body reduced density matrix formulation [26] of the Rayleigh–Ritz variational principle (4.6) to the strongly correlated limit of DFT.

After discretization as described in Sect. 4.4.2.1, the two-point marginal becomes a matrix $\Gamma = (\Gamma_{ij}) \in \mathbb{R}^{\ell\times\ell}$, and N-representability means that Γ is obtained from some symmetric tensor $(\gamma_{i_1\ldots i_\ell}) \in \mathbb{R}^{\ell\times\ldots\times\ell}$ with nonnegative entries which sum to one by $\Gamma_{i_1i_2} = \sum_{i_3,\ldots,i_N} \gamma_{i_1i_2i_3\ldots i_N}$.

The extreme points of the set of discrete N-representable 2-marginals have been determined explicitly [53] (see [16, 82] for generalizations to 3-marginals respectively general k-marginals).

Theorem 4.29 ([53]) *The set of extreme points of the set $\mathcal{R}_2$ of discrete N-representable 2-marginals is*

$$\mathcal{R}_2^{\rm ext} =$$

$$\left\{\frac{N}{N-1}\lambda\lambda^T - \frac{1}{N-1}\mathrm{diag}(\lambda) : \lambda\in\mathbb{R}^\ell,\ \lambda_i\geq 0\,\forall i,\ \mathbf{1}^T\lambda=1,\ \lambda_i\in\{0,\tfrac{1}{N},\tfrac{2}{N},\ldots\}\right\}.$$

In particular, $\mathcal{R}_2$ is the convex hull of $\mathcal{R}_2^{\rm ext}$.

Here $\mathbf{1}$ denotes the vector in $\mathbb{R}^\ell$ with all components equal to 1. The discretized problem is then

$$\min_{\Gamma} \sum_{1\leq i,j\leq\ell} c_{ij}\Gamma_{ij} \qquad (4.112)$$

$$\text{s/to } \Gamma\in\mathcal{R}_2,\ \Gamma\mathbf{1} = \frac{\rho}{N}.$$

Khoo and Ying [82] introduced the following convex relaxation of this problem in which $\mathcal{R}_2$ is replaced by a slightly larger but simpler set:

$$\min_{\Gamma} \sum_{1\le i,j\le \ell} c_{ij}\Gamma_{ij} \tag{4.113}$$

$$\text{s/to } \Gamma \in \widetilde{\mathcal{R}_2} = \left\{ \frac{N}{N-1}\Lambda - \frac{1}{N-1}\text{diag}(\Lambda\mathbf{1}) \ :\ \Lambda_{ij} \ge 0 \forall i,j,\ \Lambda \succeq 0,\ \mathbf{1}^T\Lambda\mathbf{1}=1 \right\},$$

$$\Gamma\mathbf{1} = \frac{\rho}{N}.$$

Here $\Lambda \succeq 0$ means matrix positivity of Λ.

It is clear that $\widetilde{\mathcal{R}_2} \supset \mathcal{R}_2$, since $\widetilde{\mathcal{R}_2}$ is convex and—by inspection—contains the set of extreme points of $\mathcal{R}_2$ given in Theorem 4.29. A theoretical argument in support of the approximation (4.113) is:

Theorem 4.30 ([82]) *The extreme points of the true set $\mathcal{R}_2$ of discrete N-representable 2-marginals are still extreme points of $\widetilde{\mathcal{R}_2}$.*

Intuitively this means that, at least near the extreme points of the exact set $\mathcal{R}_2$ of N-representable 2-marginals, the relaxation is very tight.

Viewed as a minimization over Λ, (4.112) is a semidefinite program (SDP), i.e. a problem of minimizing a linear cost subject to finitely many linear equalities or inequalities and a matrix positivity constraint. It has been implemented in [82] using a uniform grid and the large-scale SDP solver SDPNAL+. For 1D problems with $N = 8$, up to $\ell = 1600$ gridpoints, and different one-body densities, the solutions reported in [82] are in excellent qualitative agreement with the pair density of the exact Seidl solution. The relative energy error compared to the unapproximated discrete problem (4.111) is estimated to be of the order of 10^{-2} to 10^{-4}, depending on the choice of one-body density. Also, (4.113) is solved for 6 electrons in 2D with a Gaussian density on a 10×10 grid.

Khoo and Ying [82] also give a dual formulation of the SDP (4.113) which yields an approximation to the Kantorovich potential. For 1D test problems with 8 electrons and 200 gridpoints, a relative accuracy of 10^{-2} to 10^{-3} in the L^2 norm is reported compared to the exact potential obtained from the Seidl solution and Eq. (4.59).

4.4.5.2 Langevin Dynamics with Moment Constraints

This approach was proposed by Alfonsi et al. [1, 2]. The idea is to only discretize the density constraint, but not the N-point density, and then use a stochastic particle method to simulate the many-electron density. One performs a Galerkin (or “moment”) discretization of the marginal constraint (4.24) by requiring only a fixed number M of integral constraints, of the form

$$\int_{\mathbb{R}^{Nd}} \varphi_m(\mathbf{r}_i)\, \mathrm{d}\gamma(\mathbf{r}_1, \ldots, \mathbf{r}_N) = \int_{\mathbb{R}^d} \varphi_m \, \mathrm{d}\mu \;\; \forall i = 1, \ldots, N, \; \forall m = 1, \ldots, M, \tag{4.114}$$

where $\mu = \rho/N$ is the prescribed single-particle density and $\varphi_1, \ldots, \varphi_M$ are suitable single-particle basis functions on $\mathbb{R}^d$. Moreover since the marginal constraint has been relaxed, one introduces a mild additional constraint on the class of admissible N-electron densities γ to prevent mass from escaping to infinity,

$$\int_{\mathbb{R}^{dN}} \sum_{i=1}^{N} \theta(|\mathbf{r}_i|)\, \mathrm{d}\gamma(\mathbf{r}_1, \ldots, \mathbf{r}_N) \le A \tag{4.115}$$

for some nonnegative increasing function $\theta : [0, \infty) \to [0, \infty)$ with $\theta(r) \to \infty$ $(r \to \infty)$ and some constant $A > 0$. The SIL problem (4.25) is now approximated by:

$$\text{Minimize} \int_{\mathbb{R}^{Nd}} V_{\mathrm{ee}} \, \mathrm{d}\gamma \text{ over } \gamma \in \mathcal{P}(\mathbb{R}^{Nd}) \text{ subject to (4.114) and (4.115).} \tag{4.116}$$

Under suitable assumptions on the basis functions, and for A chosen sufficiently large, the minimum value of (4.116) can be shown to converge to the SCE energy $\mathrm{V}_{\mathrm{ee}}^{\mathrm{SCE}}[\rho]$ as the number M of basis functions tends to infinity [2]. The key property of (4.116) opening the door to numerical methods is the following.

Theorem 4.31 ([2]) *Assume $\mu \in \mathcal{P}(\mathbb{R}^d)$, and suppose that the basis functions $\varphi_1, .., \varphi_M : \mathbb{R}^d \to \mathbb{R}$ are continuous, belong to $L^1(\mathrm{d}\mu)$, and satisfy the growth bound $|\varphi_m(\mathbf{r})| \le const(1+\theta(|\mathbf{r}|))^s$ for some $s \in (0, 1)$. Assume that $V_{\mathrm{ee}} : \mathbb{R}^{dN} \to \mathbb{R} \cup \{+\infty\}$ is nonnegative and $\int V_{\mathrm{ee}} \mathrm{d}\gamma$ is finite for some γ satisfying (4.114), and that A is sufficiently large. Then there exists a minimizer of (4.116) of the form $\mathrm{d}\gamma(\mathbf{r}_1, \ldots, \mathbf{r}_N) = \sum_{\nu=1}^{K} \alpha_\nu S_N \delta(\mathbf{r}_1 - \mathbf{a}_1^{(\nu)}) \ldots \delta(\mathbf{r}_N - \mathbf{a}_N^{(\nu)})$ for some $K \le M+2$, some coefficients $\alpha_\nu \ge 0$, and some $\mathbf{a}_i^{(\nu)} \in \mathbb{R}^d$.*

Thus a sparse ansatz for the many-electron density consisting of $K \le M + 2$ symmetrized Dirac measures (where M is the number of constraints discretizing the marginal condition) is sufficient. This result generalizes Theorem 4.23 from discrete problems to semi-discrete problems with continuous state space and discretized marginal constraint.

In order to numerically solve (4.116), in [1] a stochastic particle method in continuous state space has been implemented. More precisely, the authors use constrained overdamped Langevin dynamics in the potential V_{ee}, which is a natural stochastic evolution equation for minimizing V_{ee}, applied to weighted sums of K symmetrized Dirac measures moving on the constraint manifold (4.114). For 5 electrons in a one-dimensional interval and the regularized Coulomb interaction (4.103) with $a = 0.1$, up to $M = 40$ basis functions taken to be Legendre polynomials, and superpositions of up to $K = 10{,}000$ symmetrized Dirac measures, the method achieves good agreement with the Seidl solution described in Sect. 4.3.11. The implementation uses an iterative method to maintain the constraints (which are nonlinear in the particle positions), as well as judicious choices of the time steps, temperature profile, and numbers of symmetrized Diracs to balance accuracy and computational efficiency.

An attractive feature of this method besides its feasibility for large numbers of electrons is the fact that space is not discretized. In [1] simulations are reported for 100 electrons in three dimensions subject to 52 marginal constraints, again using superpositions of 10,000 symmetrized Dirac measures. At the time of writing, it remains an interesting open question to assess, in such situations, the accuracy of the model (4.116) and its numerical solutions.

4.4.5.3 Genetic Column Generation

This method was proposed recently by Friesecke et al. [52]. It directly solves the discretized SIL problem (4.106), by combining the sparse but exact quasi-SCE or quasi-Monge ansatz (see Theorem 4.23), the method of column generation from discrete optimization, and basic ideas from machine learning.

The idea is to alternate between solving the SIL problem on a small but otherwise unconstrained subset of the many-electron configuration space, and updating the subset based on the (primal and dual) SIL solution. Recall that after discretization, the many-electron density becomes a density γ on X^N, where $X = \{\mathbf{a}_1, \ldots, \mathbf{a}_\ell\}$ is a set of discretization points (e.g., a grid) for the single-electron configuration space $\mathbb{R}^d$. One now starts from the quasi-SCE or quasi-Monge ansatz in the form (4.88), which suffices to solve the discrete SIL problem (4.106) exactly (see Theorem 4.23), but—for computational reasons—allows a slightly larger number of delta functions:

$$\gamma(\mathbf{r}_1, \ldots, \mathbf{r}_N) = \sum_{\nu=1}^{\ell'} \alpha_\nu S_N \delta(\mathbf{r}_1 - \mathbf{r}_1^{(\nu)}) \ldots \delta(\mathbf{r}_N - \mathbf{r}_N^{(\nu)}), \quad \ell \le \ell' \le \beta\ell. \tag{4.117}$$

Here the $\mathbf{r}^{(\nu)} = (\mathbf{r}_1^{(\nu)}, \ldots, \mathbf{r}_N^{(\nu)})$ are arbitrary N-point configurations in X^N and $\beta > 1$ is a hyperparameter (taken to be 5 in [52]) which limits the number of N-point configurations to $O(\ell)$ instead of the naively required $O(\ell^N)$. To achieve a unique correspondence between symmetrized Diracs and N-point configurations one restricts the $\mathbf{r}^{(\nu)}$ to the sector $X^N_{\text{sym}} = \{(\mathbf{a}_{i_1}, \ldots, \mathbf{a}_{i_N}) \in X^N \, : \, i_1 \le \ldots \le i_N\}$, making the expansion coefficients α_ν in (4.117) unique.

The ansatz (4.117) involves two sets of degrees of freedom, the subset $\Omega = \{\mathbf{r}^{(1)}, \ldots, \mathbf{r}^{(\ell')}\}$ of the many-electron configuration space and the coefficient vector $(\alpha_1, \ldots, \alpha_{\ell'})$, which are updated alternatingly. For fixed Ω, the coefficient vector is governed by the SIL problem (4.106) restricted to the ansatz (4.117), which reads, using that $S_N\delta(\mathbf{r}_1 - \mathbf{r}_1^{(\nu)})\ldots\delta(\mathbf{r}_N - \mathbf{r}_N^{(\nu)})$ has single-particle density $\rho^{(\nu)}(\mathbf{r}) = \sum_{i=1}^N \delta(\mathbf{r} - \mathbf{r}_i^{(\nu)})$,

$$\min_{\alpha\in\mathbb{R}^{\ell'}} \sum_{\nu=1}^{\ell'} \alpha_\nu V_{ee}(\mathbf{r}^{\nu)}) \text{ s/to } \sum_{\nu=1}^{\ell'} \alpha_\nu \rho^{(\nu)}(\mathbf{a}_i) = \frac{\rho_i}{N}, \quad i \in \{1, \ldots, \ell\}, \quad \alpha_\nu \geq 0. \tag{4.118}$$

This is just a small linear program with an $\ell \times O(\ell)$ constraint matrix. Updating the set Ω is done in a simple but subtle manner, as standard methods would incur the curse of dimension (see below). One also uses the dual problem

$$\max_{u:X\to\mathbb{R}} \sum_{i=1}^{\ell} u(\mathbf{a}_i)\rho_i \text{ s/to } u(\mathbf{r}_1^{(\nu)}) + \cdots + u(\mathbf{r}_N^{(\nu)}) \leq V_{ee}(\mathbf{r}^{(\nu)}) \; \forall \nu \in \{1, \ldots, \ell'\}, \tag{4.119}$$

whose solution u is an approximation to the Kantorovich potential.

An updating cycle in the genetic column generation (`GenCol`) method goes as follows:

1. Given a set $\Omega \subset X_{\text{sym}}^N$ of N-particle configurations, update the primal solution α and the dual solution u by solving (4.118) and (4.119).
2. Given the updates α^{new} and u^{new}, update Ω by the following genetic learning method:

– pick a random "parent" configuration $\mathbf{r}^{(\nu)} \in \Omega$ satisfying $\alpha_\nu^{\text{new}} > 0$;
– create a random "child" $\mathbf{r}^* \in X_{\text{sym}}^N$ by moving one electron position to a nearest neighbor;
– repeat these steps until $u^{\text{new}}(\mathbf{r}_1^*) + \cdots + u^{\text{new}}(\mathbf{r}_N^*) > V_{ee}(\mathbf{r}^*)$ and set $\Omega^{\text{new}} = \Omega \cup \{\mathbf{r}^*\}$. (4.120)

Steps 1. and 2. are iterated until convergence, with the oldest configurations which do not contribute to the current optimal plan (i.e. satisfy $\alpha_\nu^{\text{new}} = 0$) being deleted from Ω whenever its size ℓ' exceeds the maximum allowed size $\beta\ell$.

The simple but powerful genetic learning aspect of the search rule in (4.120) is that only "successful" N-electron configurations in Ω (i.e. ones that contribute to the current optimal plan (4.117) with a nonzero coefficient α_ν) are allowed to bear offspring. Numerical observations and theoretical considerations show that this is essential for overcoming the curse of dimension. An unbiased random search of new configurations, or the updating step in the classical column generation method of solving the so-called pricing problem,[9] would merely turn the curse of dimension with respect to the size of the state space into a curse of dimension with respect to the number of search steps.

The rationale behind the acceptance criterion in (4.120) is that any new configuration $\mathbf{r}^*$ satisfying it represents a constraint of the full dual problem (Eq. (4.119) with the $\mathbf{r}^{(\nu)}$ being replaced by all configurations in X^N_{sym}) which the current dual solution u^{new} violates. Adding this configuration to the set Ω "cuts off" u^{new} from the optimization domain of the dual problem, yielding a new dual solution and an energy decrease. For a rigorous justification see [52].

Figure 4.10, taken from [52], shows the solution of the SIL problem (4.25) computed by the `GenCol` algorithm for 10 electrons in a 1D interval discretized by 100 gridpoints. In this example, the grid spacing is normalized to 1, the density is taken to be $\rho(x) = \text{const}(0.2 + \sin^2(\frac{x}{\ell+1}))$, and the interaction is the soft Coulomb potential (4.102) with $a = 0.1$. With the initial set of many-electron configurations chosen randomly, the algorithm always found the exact Seidl solution (see Sect. 4.3.11) of the discretized problem to machine precision using less than 7000 iterations and less than 5 samples per iteration. This means that only a tiny fraction of the configuration space was accessed. The energy decreased steadily at an exponential rate.

Tests reported in [52] on larger 1D systems with up to $N = 30$ electrons on 120 grid points (corresponding to a space of N-point densities of dimension $\ell^N \approx 2.4 \times 10^{62}$) show only a slow polynomial growth in N of the number of iterations required to find the exact solution to machine precision, with the average number of samples needed per iteration to satisfy the acceptance criterion remaining approximately constant.

Apart from its simplicity and efficiency in high dimensions, attractive features of the genetic column generation method are that after discretization no further approximations are made (and the discrete SIL problem is solved accurately), and that the method also provides the Kantorovich potential for use within Kohn–Sham DFT.

Tests for accurately discretized three-dimensional densities are not yet available at the time of writing.

4.4.6 Approximations

As explained, there are presently no efficient algorithms to solve the SCE problem in an exact or very accurate way for the general three-dimensional case. In the usual spirit of DFT, several approximations for the functional $V_{\text{ee}}^{\text{SCE}}[\rho]$ have

[9] which consists in our case in finding a configuration $\mathbf{r}^*$ which maximizes the difference $u^{\text{new}}(\mathbf{r}_1^*) + \cdots + u^{\text{new}}(\mathbf{r}_N^*) - V_{\text{ee}}(\mathbf{r}^*)$.

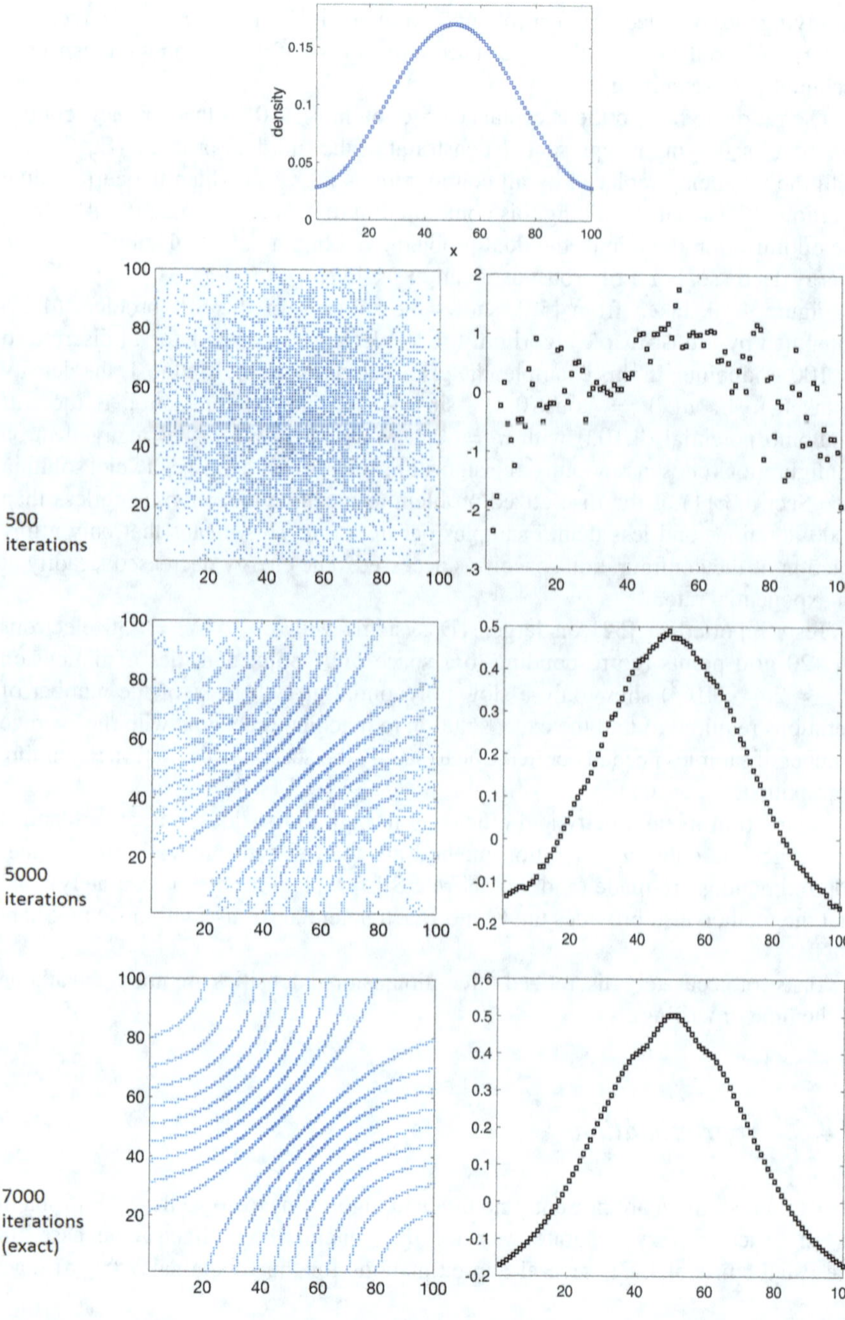

Fig. 4.10 Solution to the SIL problem (4.25) for 10 electrons in 1D with the `GenCol` algorithm [52]. *Top:* prescribed density. *Left:* Evolution under `GenCol` of the N-electron density from a random initial state, visualized via its two-point marginal (pair density). Gridpoints with nonzero values (i.e., "successful" configurations) are shown in blue, with larger markers indicating higher values. *Right:* Evolution of the Kantorovich potential. The final N-point density recovers Seidl's SCE state for the discretized problem with machine precision

been proposed and used in combination with Kohn–Sham DFT. We review the approximations in this section, and their use within Kohn–Sham DFT in Sect. 4.5.

4.4.6.1 Gradient Expansion: Point-Charge-Plus-Continuum Model (PC)

The first gradient expansion approximation (GEA) for the indirect energy functional $W_\infty[\rho] = V_{ee}^{SCE}[\rho] - U[\rho]$ has been proposed by Seidl et al. [128], and it is called the point-charge-plus continuum (PC) model,

$$W_\infty^{PC}[\rho] = \int d\mathbf{r} \left[A\, \rho(\mathbf{r})^{4/3} + B\, \frac{|\nabla\rho(\mathbf{r})|^2}{\rho(\mathbf{r})^{4/3}} \right], \tag{4.121}$$

where $A = -\frac{9}{10}(\frac{4\pi}{3})^{1/3}$ and $B = \frac{3}{350}(\frac{3}{4\pi})^{1/3}$. The model is built from the physical interpretation of $W_\infty[\rho]$ as the electrostatic energy of a system of perfectly correlated electrons with density ρ inside a classical background with the same charge density ρ of opposite sign [128]. Notice that the electrons are not allowed to relax in this fictitious external potential, as they are kept in the SCE state with the prescribed density. Only when the density is uniform is the energy of the SCE state the same as the one we would obtain by letting the electrons relax in the positive background external potential [96]. The idea of the PC model is that when the density is slowly varying the energy should be well approximated by surrounding each electron by a PC cell (given by the combined effect of the background and the remaining electrons) that neutralizes its charge and it is such that the electron plus its cell have zero dipole moment [128].

The PC approximation works rather well: for example, for the atomic densities from He to Ne, the values $W_\infty^{PC}[\rho]$ agree within 1% with the values obtained by using the radial co-motion functions (maps) described in Sect. 4.3.12, as shown in Table I of Ref. [127]. This is quite remarkable as, usually, gradient expansions for the exchange-correlation functionals fail in providing accurate quantitative results.

4.4.6.2 Generalized Gradient Approximations: The Modified PC Model

Although quantitatively accurate for the SIL energy, the main drawback of the PC model is that its functional derivative,

$$v_{xc}^{PC}(\mathbf{r}) = \frac{\delta W_\infty^{PC}[\rho]}{\delta\rho(\mathbf{r})}, \tag{4.122}$$

diverges to $-\infty$ in the tail of atomic and molecular densities [45], making a self-consistent Kohn–Sham calculation impossible. Moreover, the PC model fails for quasi-2D and quasi-1D systems [31] .

To overcome these problems, Constantin [31] has proposed a generalized gradient approximation (GGA) for $W_\infty[\rho]$, called the modified PC model (mPC), which reads

$$W_\infty^{\text{mPC}}[\rho] = A \int d\mathbf{r}\rho(\mathbf{r})^{4/3} \frac{1 + a\, s(\mathbf{r})^2}{1 + (a + 0.14)\, s(\mathbf{r})^2},$$
$$s(\mathbf{r}) = \frac{|\nabla\rho(\mathbf{r})|}{2(3\pi)^{1/3}\rho(\mathbf{r})^{4/3}}, \tag{4.123}$$

where A has the same value as in the original PC model, and $a = 2$. This approximation is less accurate for the SIL of atomic densities with respect to the original PC model (with errors around 9–10%), but has the advantage of a well-behaved functional derivative, and of achieving a physical description of the crossover from three to two dimensions.

4.4.6.3 Approximations with Some Non-Locality: The Non-local Radius (NLR) and the Shell Model

The PC and mPC are semilocal approximations, while, as we have seen, the exact SIL physics has an extreme non-local dependence on the density. Approximations that retain some (albeit limited) non-locality are the non-local radius (NLR) [140] and the shell models [6]. Both approximations use as key ingredient the spherically averaged density $\widetilde{\rho}(\mathbf{r}, u)$ around a given position $\mathbf{r}$, obtained by integrating out the angular dependence of $\mathbf{u}$,

$$\widetilde{\rho}(\mathbf{r}, u) = \int \rho(\mathbf{r} + \mathbf{u}) \frac{d\widehat{\mathbf{u}}}{4\pi}, \tag{4.124}$$

and, in analogy with the SCE structure for spherical densities conjectured in Ref. [127] and illustrated in Sect. 4.3.12, its cumulant

$$N_{\text{e}}(\mathbf{r}, u) = \int_0^u 4\pi\, x^2\, \widetilde{\rho}(\mathbf{r}, x)\, dx. \tag{4.125}$$

In the NLR model [140] the functional $W_\infty[\rho]$ is approximated as

$$W_\infty^{\text{NLR}}[\rho] = -\int d\mathbf{r}\, \rho(\mathbf{r}) \int_0^{R(\mathbf{r})} 2\pi\, \widetilde{\rho}(\mathbf{r}, u)\, u\, du, \tag{4.126}$$

where the radius $R(\mathbf{r})$ is defined by the condition that the underlying exchange-correlation hole be normalized:

$$N_{\text{e}}(\mathbf{r}, R(\mathbf{r})) = 1. \tag{4.127}$$

This simple approximation is less accurate than the PC and mPC models for the case of the uniform electron gas, giving a too high energy. For non-uniform densities, the NLR has the advantage, with respect to the PC and mPC models, of being exact for one-electron systems. For atomic densities, NLR makes errors, with respect to the SCE results of Ref. [127], of the order of 8–9% [140].

The shell model [6] substantially improves the NLR approximation, by making it exact for a uniform density, and reducing its error with respect to the SCE results for atomic densities by almost a factor of 10. While the NLR model approximates the exchange-correlation hole with a sphere depleting one electron from the spherically averaged density, the shell model adds a single positive oscillation, and reads

$$W_\infty^{\text{shell}}[\rho] = \int d\mathbf{r}\,\rho(\mathbf{r})\,2\pi\left(-\int_0^{u_s(\mathbf{r})}\widetilde{\rho}(\mathbf{r},u)\,u\,du + \int_{u_s(\mathbf{r})}^{u_c(\mathbf{r})}\widetilde{\rho}(\mathbf{r},u)\,u\,du\right), \tag{4.128}$$

where for all $\mathbf{r}$ we have $u_s = 0.849488\,u_c$, which is the condition needed to make the model exact for a uniform density. The value of $u_c(\mathbf{r})$ is then obtained again by the normalization condition,

$$2\,N_e(\mathbf{r}, 0.849488\,u_c(\mathbf{r})) - N_e(\mathbf{r}, u_c(\mathbf{r})) = 1. \tag{4.129}$$

4.5 Kohn–Sham Combined with the Strong-Interaction Limit

4.5.1 Kohn–Sham with the SCE Functional (KS SCE)

The Kohn–Sham scheme with the SCE functional (KS SCE) was first proposed and implemented in [100], and corresponds to a crude, but well-defined approximation for the HK functional,

$$F_{\text{KSSCE}}[\rho] = T_s[\rho] + \mathrm{V}_{\text{ee}}^{\text{SCE}}[\rho], \tag{4.130}$$

in which we replace the minimum of the sum of kinetic energy and electron-electron repulsion at fixed density, with the sum of the two minima. As such, the KS SCE will always provide a lower bound for the HK functional. When implemented self-consistently, the KS SCE scheme yields the usual KS equations with the Hartree-exchange-correlation potential given by the SCE or Kantorovich potential (written below for simplicity for a closed-shell system),

$$\begin{aligned} -\frac{1}{2}\nabla^2\varphi_i(\mathbf{r}) + (v_{\text{SCE}}(\mathbf{r},[\rho]) + v_{\text{ne}}(\mathbf{r}))\,\varphi_i(\mathbf{r}) &= \epsilon_i\,\varphi_i(\mathbf{r}), \\ \rho(\mathbf{r}) &= 2\sum_{i=1}^{N/2}|\varphi_i(\mathbf{r})|^2, \end{aligned} \tag{4.131}$$

where the SCE potential is equal to

$$v_{\rm SCE}(\mathbf{r},[\rho]) = u(\mathbf{r},[\rho]) + C[\rho], \tag{4.132}$$

with $u(\mathbf{r},[\rho])$ the maximizer in Eq. (4.55), and the constant $C[\rho]$ a shift that ensures $\lim_{\mathbf{r}\to\infty} v_{\rm SCE}(\mathbf{r}) = 0$, see Eq. (4.64). This shift is the same, in the $\lambda \to \infty$ limit of the density-fixed adiabatic connection, as the one introduced by Levy and Zahariev [94, 139]. If we want to compute the ground-state density and the ground state energy only, one could better work with u instead of $v_{\rm SCE}$, as with the former the energy becomes simply [22, 94, 139] the sum of the occupied orbital energies, $E_0 = 2\sum_{i=1}^{N/2} \epsilon_i$. The shift is needed if we want to estimate the ionisation potential $I = E_0^{N-1} - E_0^N$ from the highest occupied molecular orbital energy (HOMO), as $I = -\epsilon_{N/2}$ holds only when the exchange-correlation potential goes to zero far from the barycentre of nuclear charge [3, 93]. For further discussion of this point see Chap. 1 by Toulouse in this volume.

4.5.1.1 1D Case

The self-consistent KS SCE equations have been solved for 1D systems with the interaction $v_{\rm ee}^{\rm wire}(x)$ of Eq. (4.101) when the external potential is harmonic, $v_{\rm ne}(x) = \frac{1}{2}\omega^2 x^2$, [72, 100, 101], and with the soft Coulomb interaction $v_{\rm ee}^{\rm soft}(x)$ of Eq. (4.102) for model 1D atoms and molecules with 'nuclei' that attract the electrons with the same soft Coulomb potential [102]. At each KS iteration, the 1D co-motion functions [123] were computed numerically as explained in Sect. 4.4.1.1, and the potential $v_{\rm SCE}(x,[\rho])$ was obtained by simply integrating the force equation

$$v'_{\rm SCE}(x,[\rho]) = \sum_{i=2}^{N} w'(|x - f_i(x)|)\,\mathrm{sgn}(x - f_i(x)), \tag{4.133}$$

with boundary condition $v_{\rm SCE}(x \to \pm\infty, [\rho]) = 0$, and where $w(x)$ is the chosen 1D interaction (wire or soft Coulomb, see Sect. 4.4.1.1). In addition, at low density the highest occupied KS SCE eigenvalue gives a very accurate ionization energy of the system [101].

Harmonic External Potential

In Fig. 4.11 we show the self-consistent KS SCE densities for $N = 4$ electrons interacting with $v_{\rm ee}^{\rm wire}(x)$ of Eq. (4.101) when the external potential is harmonic, using scaled units in terms of $L = 2\,\omega^{-1/2}$, compared with accurate many-body results from configuration interaction (CI) and with KS within the local density approximation (LDA), provided for this interaction in Ref. [18]. We see that, as the system is driven to low density by reducing the strength of the harmonic

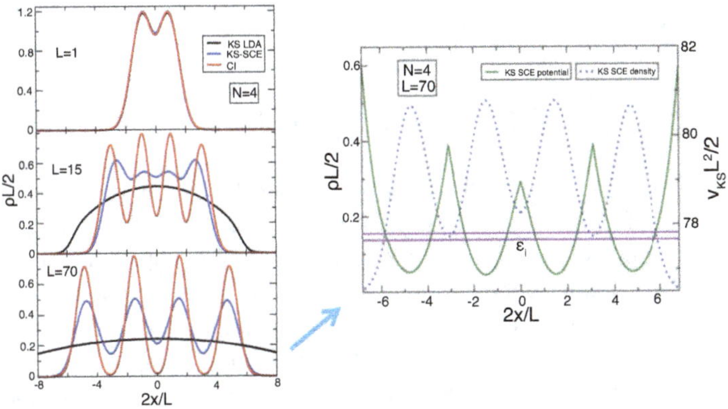

Fig. 4.11 Left: self-consistent KS SCE densities for $N = 4$ electrons interacting with $v_{\rm ee}^{\rm wire}(x)$ of Eq. (4.101) when the external potential is harmonic, $v_{\rm ne}(x) = \frac{1}{2}\omega^2 x^2$, compared with very accurate configuration interaction results (CI) and with KS LDA. Right: the total KS potential at self consistency, $v_{\rm KS} = v_{\rm ne} + v_{\rm SCE}$, for the most correlated case. The horizontal lines are the two eigenvalues of the occupied KS SCE orbitals. Results are in scaled units, where $L = 2\,\omega^{-1/2}$, and are taken from Ref. [101]

confinement (large L), the exact many-body solution undergoes a so-called "$2k_F \to 4k_F$" transition, in which the number of peaks in the density is doubled. At high density, in fact, the number of peaks is dictated by the number of occupied orbitals, $N/2$ for a closed shell system. At low density, we have an incipient Wigner molecular structure, in which the electrons are well separated. Notice that with the Coulomb interaction this Wigner molecular phase exhibits different properties than the simpler case of very short-range interactions, in which the physics can be captured by making the system spin-polarized (i.e. by occupying N orbitals instead of $N/2$). This is clearly illustrated in Ref. [142].

It is well-known that the local and semilocal approximations to the XC functional, as well as exact exchange, are not able to capture this "$2k_F \to 4k_F$" transition [134, 135] without introducing artificial symmetry breaking. This is also clearly shown by the KS LDA results of Fig. 4.11, which become very close in this limit to the Thomas–Fermi result (minus the external potential in the classically allowed region) predicting a too delocalized density. The KS SCE self-consistent density, although not quantitatively very accurate, has the correct qualitative behavior, with two peaks at high density and four at low density, and with the correct extension. The KS SCE HOMO energy is also very close to the exact many-body ionisation potential [100, 101]. In the right panel of Fig. 4.11 we show the total KS potential at self consistency, $v_{\rm KS} = v_{\rm ne} + v_{\rm SCE}$, for the most correlated case. The horizontal lines are the two occupied KS SCE eigenvalues. We see that the SCE functional is able to self-consistently build barriers that create classically forbidden regions inside the harmonic trap. Classically forbidden regions for the KS orbitals created by the Hartree-exchange-correlation potential seem to play a crucial role in describing strong correlation within KS DFT [12, 75, 144].

Model 1D Chemistry with Soft Coulomb Potential

In Ref. [102] the KS SCE method has been tested for model chemical systems in 1D, consisting of "nuclei" and electrons attracting each other with the soft-Coulomb potential (for the use of these 1D models to test DFT approximations, see also Refs. [74, 141]). While in the harmonic external potential we can drive the system to low density where the SCE becomes a very good approximation to the exact KS exchange-correlation functional, chemical systems (bound by the Coulomb external potential) are never in this regime. For this reason, KS SCE does not in general yield accurate results, with total energies that are way too low. An exception seems to be the good agreement between the eigenvalue of the highest occupied KS SCE orbital and the many-body chemical potential, as shown in Table 2 of Ref. [102].

4.5.1.2 2D Case

The circularly-symmetric 2D case of electrons interacting with the $1/r$ repulsion in the harmonic external potential has been studied with KS SCE in Ref. [105], using the SGS radial co-motion functions and the reduced radial cost of Eq. (4.78) implemented as described in Sect. 4.4.1.2. As in 1D, the aim is to model electrons strongly confined in one direction, found, for example, at the interface of semiconductor etherostructures. For this reason, the interaction remains the same as the 3D Coulomb one.

As discussed in Sect. 4.3.12, the SGS state defined by (4.79)–(4.80) is not guaranteed to yield the absolute minimum for the electron-electron interaction in a given radial density $\rho(r)$. Nonetheless, it can be proven [125] that, for a spherically-symmetric density, if we reduce the admissible class of maps $\mathcal{T}_\rho$ in the SCE functional (4.37) to a class $\mathcal{T}_\rho^{\mathrm{SGS}} \subset \mathcal{T}_\rho$ of maps given by the SGS ansatz defined in Eqs. (4.79)–(4.80) as an approximation for $\mathrm{V}_{\mathrm{ee}}^{\mathrm{SCE}}[\rho]$, even when the SGS maps are not optimal the functional derivative of this approximate $\mathrm{V}_{\mathrm{ee}}^{\mathrm{SCE}}[\rho]$ with respect to $\rho(r)$ still satisfies the force equation (written using the notation of Eq. (4.59)),

$$\nabla v_{\mathrm{SGS}}(\mathbf{r}) = -\sum_{i=2}^{N} \frac{\mathbf{r}-\mathbf{f}_i^{\mathrm{SGS}}(\mathbf{r})}{|\mathbf{r}-\mathbf{f}_i^{\mathrm{SGS}}(\mathbf{r})|^3}, \tag{4.134}$$

which we can integrate to obtain a potential $v_{\mathrm{SGS}}(r)$. In other words, the SGS maps provide a well-defined approximation to the exact SCE functional, with an easy to evaluate functional derivative, which, in turn, can be used in the KS equations.

In Fig. 4.12 we show the resulting KS SCE self-consistent radial density for $N = 3$ electrons for two low-density cases, compared with accurate Quantum

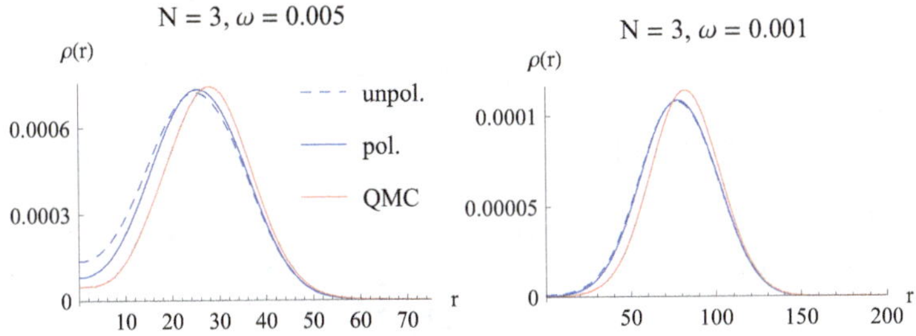

Fig. 4.12 Self-consistent radial KS SCE densities $\rho(r)$ for $N = 3$ electrons in 2D, with external potential $v_{\rm ne}(r) = \frac{1}{2}\omega^2 r^2$ [105], compared with accurate Quantum Monte Carlo (QMC) results [61, 73]. The KS SCE densities are shown for both the unpolarized (2 orbitals, of which only the lowest is doubly occupied) and spin polarized (3 different singly occupied KS orbitals) case

Monte Carlo (QMC) results from Refs. [61, 73]. The KS SCE calculations have been done for both the unpolarized case (2 KS orbitals, of which only the lowest is doubly occupied) and the spin-polarized case (3 different singly occupied KS orbitals). We see that the KS SCE densities are very close to the QMC ones, predicting the right shell structure with one peak. Total energies are in agreement with QMC within $\sim 4-6\%$ [105]. Notice that at such low densities it is very hard to even obtain converged results using KS with the local-spin density (LSD) approximation. We thus see that even if the SGS maps are not optimal for these densities (see [125]), they yield very good results when used in the self-consistent KS equations at low density. However, we have to mention that QMC predicts that at such small ω's the ground state is spin-polarized, while in KS SCE the unpolarized case always yields the lowest energy, due to the lack of any spin dependence in the SCE functional.

Figure 4.13 shows the self-consistent KS SCE total potential and density for $N = 10$ electrons (spin unpolarized) [105]. The green dashed curve is the energy of the highest occupied KS orbital. We clearly see, as in the 1D case of Fig. 4.11, that when the system is driven to low-density (small ω case), KS SCE is able to self-consistently create classically forbidden regions inside the trap.

4.5.1.3 3D Case

KS SCE has been tested on the anions of the He isoelectronic series [106] and on the dissociation curve of the H_2 molecule [22].

Anions of the He Isoelectronic Series

In this case, i.e., $N = 2$ electrons with $v_{\rm ne}(r) = -\frac{Z}{r}$, where Z is lowered until the system can no longer bind two particles, the co-motion function and the SCE potential are simply built following the original work of Seidl [123] (see

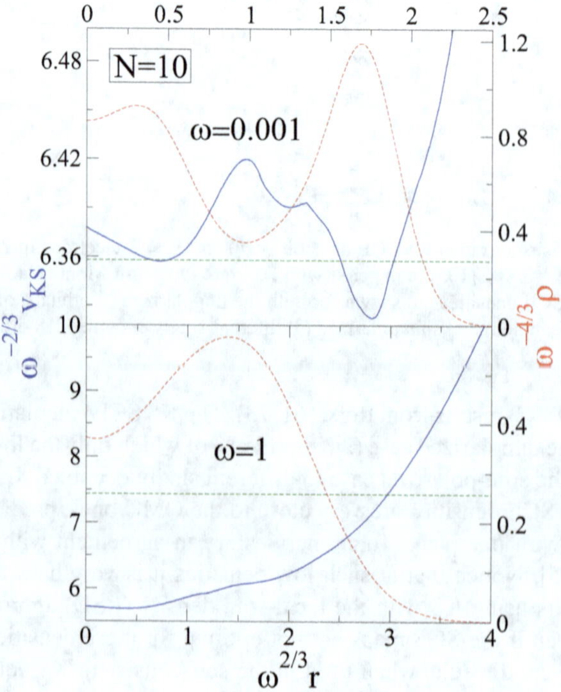

Fig. 4.13 Self-consistent KS SCE radial potential $v_{\mathrm{KS}}(r) = v_{\mathrm{SCE}}(r) + \frac{1}{2}\omega^2 r^2$ (blue solid line) and radial densities (red dashed line) for a strongly and weakly correlated case (top and bottom, respectively) of a 2D system composed of $N = 10$ electrons inside a circularly symmetric harmonic trap. The green dashed horizontal lines correspond to the energies of the highest occupied KS orbital. Notice the presence of classically forbidden regions inside the trap in the strongly correlated case ($\omega = 0.001$)

Example 4.20), which is a special case of the SGS maps. While very accurate wavefunction results predict [47] that one electron is lost by the system at a critical nuclear charge $Z_{\mathrm{crit}}^{\mathrm{exact}} \approx 0.911$, KS SCE binds two electrons down to $Z_{\mathrm{crit}}^{\mathrm{KSSCE}} \approx 0.7307$ [106]. This is because in the KS SCE case the two electrons can get much closer to the nucleus by perfectly avoiding each other, without raising the kinetic energy too much, which is only treated within KS.

The H_2 Molecule

The dissociation curve of the H_2 molecule has been computed within KS SCE in Ref. [22]. The result is shown in Fig. 4.14. To compute the self-consistent density and energy, an accurate adaptive three-dimensional finite element discretization was

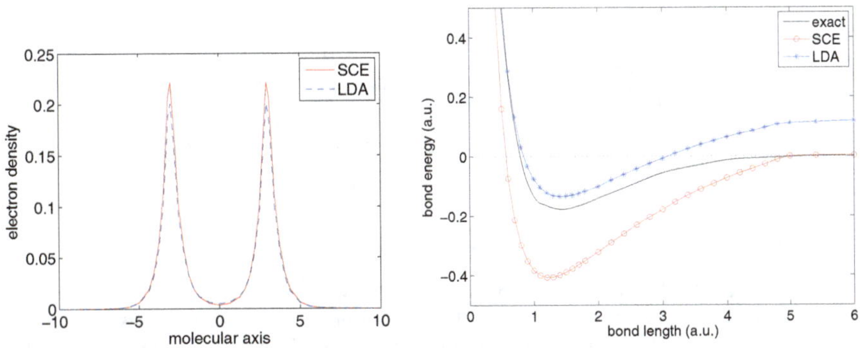

Fig. 4.14 *Right:* Dissociation curve of H_2 in KS SCE [22], that is, energy of H_2 minus twice the energy of the isolated H atom. For comparison, the KS LDA curve computed on the same mesh and the exact curve from Ref. [86] are also shown. Note that KS SCE, unlike the local density approximation, dissociates H_2 correctly. *Left:* Corresponding self-consistent KS SCE density and KS LDA density near dissociation

used and the SIL problem was solved using linear programming, as described in Sect. 4.4.2.1. The co-motion function for H_2 was then obtained from the SIL density via Eq. (4.105), and the SCE potential via the force Eqs. (4.59) and (4.64).

Not surprisingly, KS SCE predicts a binding energy that is way too low. A remarkable feature, though, is the ability of KS SCE to correctly dissociate the H_2 molecule, i.e., the molecular energy tends to twice the energy of the isolated H atom as the internuclear distance R becomes very large (see [22] for a rigorous proof). Local and semilocal approximations to the XC functionals are unable to do that, and exact exchange (or Hartree–Fock) perform even worse, unless we allow spin-symmetry breaking. Indeed, the extremely stretched H_2 molecule is often regarded as a severe test for XC functionals to check whether they are able to describe strong (or "static") correlation [24].

Although the SCE functional yields the exact energy when $R \to \infty$, we see that at large but finite R the KS SCE curve immediately start to deviate from the exact one. We can understand this error by making the following simple analysis. With the internuclear vector $\mathbf{R}$ directed along the x-axis, we can expand the electron-electron interaction at large R, which, without considering one-body terms and neglecting higher orders in R^{-1} yields

$$\frac{1}{\sqrt{(x_1 - x_2 - R)^2 + (y_1 - y_2)^2 + (z_1 - z_2)^2}} \sim \frac{2(x_1 - x_2)^2 - (y_1 - y_2)^2 - (z_1 - z_2)^2}{R^3}, \tag{4.135}$$

where the origins of $\mathbf{r}_1$ and $\mathbf{r}_2$ are placed on their respective nuclei. The SCE functional for large R then corresponds to the minimization of this interaction at

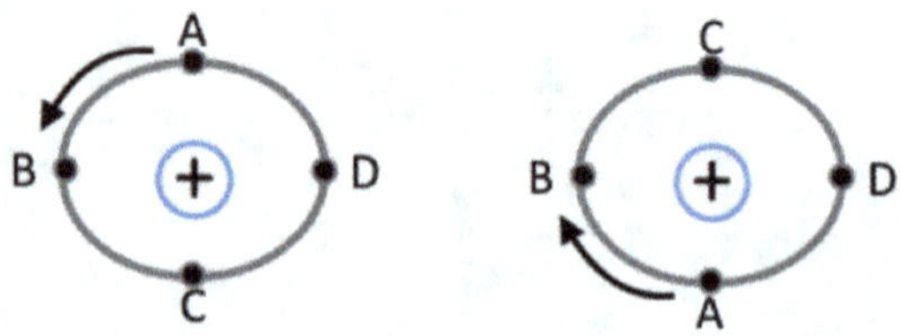

Fig. 4.15 When the distance R between the two atoms of the H_2 molecule gets very large, the optimal map describes the physics of perfectly coupled dipoles. The figure shows four pairs of electronic positions $\{\mathbf{r}, \mathbf{f}(\mathbf{r})\}$, labeled with the same letter A,B,C,D, with respect to the two positive nuclei

fixed one-body density (hence, the neglect of one-body terms that will not affect the minimizer). The SCE problem in this limit reduces then to the attractive harmonic cost [56] in the bond (x) direction and to the repulsive harmonic cost [42] in the two directions perpendicular to the bond axis. For large R, the optimal map will then approach the solution

$$f_x(\mathbf{r}) = x, \qquad f_y(\mathbf{r}) = -y, \qquad f_z(\mathbf{r}) = -z, \tag{4.136}$$

which corresponds to perfectly coupled dipoles (see Fig. 4.15). Such maps will give a finite (negative) expectation value for the r.h.s. of Eq. (4.135) even when the total density of the molecule is given by the sum of two spherical atomic densities, yielding an interaction energy that is too attractive, decaying as $\sim R^{-3}$ instead of the exact $\sim R^{-6}$. What is missing in the KS SCE approach is the raising in kinetic energy associated with the perfectly correlated dipoles of Fig. 4.15. A strategy to include the raise in kinetic energy in this asymptotic large-R regime is described in Ref. [88].

4.5.2 Interaction Strength Interpolation (ISI) Functionals

Another way to use the SIL in KS DFT is the interaction strength interpolation (ISI) construction, originally proposed by Seidl, Perdew and Levy (SPL) [130]. ISI is essentially the extension to non-uniform densities of Wigner's original idea [143] of approximating the energy of the uniform electron gas by interpolating between its high- and low-density asymptotics, which, by scaling, correspond to the weak- and strong-interaction limits, respectively.

The starting point is to use the Hellmann–Feynman theorem to write the exchange-correlation energy as an integral over the coupling-strength parameter λ of (4.17)[10]

$$E_{\rm xc}[\rho] = \int_0^1 W_\lambda[\rho], \tag{4.137}$$

where

$$W_\lambda[\rho] = V_{\rm ee}[\psi_\lambda[\rho]] - U[\rho], \tag{4.138}$$

with $\psi_\lambda[\rho]$ the minimizer in (4.17). The idea is then to construct approximations for the adiabatic connection integrand $W_\lambda[\rho]$ by interpolating between the $\lambda \to 0$ asymptotic expansion,

$$W_{\lambda\to 0} \sim E_{\rm x} + 2\,\lambda\, E_{\rm c}^{\rm GL2} + \cdots\,, \tag{4.139}$$

with $E_{\rm x}$ the exchange energy and $E_{\rm c}^{\rm GL2}$ the second-order Görling-Levy perturbation theory correlation energy [69], and the large-λ limit provided by the SIL, and possibly by the conjectured next leading term of Eq. (4.47),

$$W_{\lambda\to\infty} \sim W_\infty[\rho] + \frac{F^{\rm ZPE}[\rho]}{2\sqrt{\lambda}} + \cdots\,. \tag{4.140}$$

For example, SPL [130] proposed the following simple form to interpolate between the two limits, without using the term with $F^{\rm ZPE}[\rho]$:

$$W_\lambda^{\rm SPL} = W_\infty + \frac{E_{\rm x} - W_\infty}{\sqrt{1+2\lambda\chi}}\,, \tag{4.141}$$

with

$$\chi = \frac{2E_{\rm c}^{\rm GL2}}{W_\infty - E_{\rm x}}\,. \tag{4.142}$$

The SPL XC functional then reads

$$E_{\rm xc}^{\rm SPL} = (E_{\rm x} - W_\infty)\left[\frac{\sqrt{1+2\chi} - 1 - \chi}{\chi}\right] + E_{\rm x}. \tag{4.143}$$

[10] For further discussion see also Chap. 1 by Toulouse in this volume.

Several other interpolating functions that may or may not include $F^{\mathrm{ZPE}}[\rho]$ have been proposed in the literature, [43, 68, 98, 128, 129] and are reported, for example, in the appendix of Ref. [87].

4.5.2.1 Global Interpolations

Interpolations such as the one of Eq. (4.141) have been implemented and tested on several chemical systems by using for the $\lambda \to \infty$ limit the PC model of Sect. 4.4.6.1 (and its extension [128] to $F^{\mathrm{ZPE}}[\rho]$ when needed). In a practical calculation, KS orbitals with a given approximate semilocal or hybrid functional are used to compute the density ρ, the exchange energy E_{x}, and the second-order energy $E_{\mathrm{c}}^{\mathrm{GL2}}$, which are then fed into formulas such as (4.143) to obtain improved energies. The result is thus dependent on the chosen starting approximate functional used to generate the KS orbitals.

A basic problem of these *global* (in the sense that they are done on quantities that have been already integrated over all space) interpolations is the violation of size-consistency, i.e., if we take two different systems A and B that do not interact with each other, it is easy to verify from (4.143) that, in general,

$$E_{\mathrm{xc}}^{\mathrm{SPL}}(A+B) \neq E_{\mathrm{xc}}^{\mathrm{SPL}}(A) + E_{\mathrm{xc}}^{\mathrm{SPL}}(B), \tag{4.144}$$

an issue shared by all the other interpolation formulas proposed in the literature [25]. Notice that size-consistency of approximate electronic-structure methods is a very delicate issue when A and/or B have a degenerate ground state [65, 120]. Here we stress that even when degeneracy is not present, the fact that the input ingredients (E_{x}, $E_{\mathrm{c}}^{\mathrm{GL2}}$ and W_∞) enter in a non-linear way in the ISI formulas introduces anyway a size-consistency error. However, this error can be easily corrected [137]. In fact, the reason why size-consistency is crucial in chemistry is that we are interested in interaction energies rather than total energies. All we need to do is to set the limit of a molecular dissociation curve (when A and B are infinitely far apart) at the value given by the left-hand side of Eq. (4.144) rather than the one given by the right-hand side. Notice that both sides of this equation can be evaluated at exactly the same computational cost, as all that is needed is the input ingredients of the fragments A and B [137]. With this size-consistency correction it is possible to extract meaningful interaction energies from the ISI functionals [137].

The ISI functionals have been tested on several chemical data sets and systems [44, 62, 137]. They have been found to work reasonably well for interaction energies (especially of non-covalent systems) when Hartree–Fock orbitals (rather than KS ones) are used as input. This observation has triggered the study of the strong-interaction limit in Hartree–Fock theory [37, 126], which, in turn, has led to new interpolation schemes in this framework which are able to give very accurate results for a large variety of non-covalent interaction energies, ranging from small to medium-large systems [36].

If one wants to overcome the dependence on the input orbitals, one should evaluate the energy using the ISI functionals within a fully self-consistent KS scheme. For this, their functional derivative with respect to the density is needed, which is challenging due to the presence of second-order perturbation theory. Nonetheless, first attempts in the computation of the ISI functional derivatives have been reported in Refs. [45, 132], and self-consistent calculations are likely to appear soon.

4.5.2.2 Local Interpolations

Another possibility is to build the interpolations *locally*, in each point of space, by defining an energy density $w_\lambda(\mathbf{r}; [\rho])$ for the coupling-constant integrand $W_\lambda[\rho]$ of Eq. (4.138), writing $E_{xc}[\rho]$ as

$$E_{xc}[\rho] = \int d\mathbf{r}\, \rho(\mathbf{r}) \int_0^1 w_\lambda(\mathbf{r}; [\rho])\, d\lambda. \tag{4.145}$$

Energy densities are obviously not uniquely defined, and the only important requirement here is to use local quantities defined in the same way at weak and strong coupling. Some different choices for energy densities in the λ-interpolation context have been analyzed in [139], where it has been found that the electrostatic potential of the exchange-correlation hole[11] $h_{xc}^\lambda(\mathbf{r}_1, \mathbf{r}_2)$ seems to be the most suitable ,

$$w_\lambda(\mathbf{r}) = \frac{1}{2} \int \frac{h_{xc}^\lambda(\mathbf{r}, \mathbf{r}_2)}{|\mathbf{r} - \mathbf{r}_2|} d\mathbf{r}_2, \tag{4.146}$$

where $h_{xc}^\lambda(\mathbf{r}_1, \mathbf{r}_2)$ is defined in terms of the pair-density $P_2^\lambda(\mathbf{r}_1, \mathbf{r}_2)$ and the density ρ,

$$h_{xc}^\lambda(\mathbf{r}_1, \mathbf{r}_2) = \frac{P_2^\lambda(\mathbf{r}_1, \mathbf{r}_2)}{\rho(\mathbf{r}_1)} - \rho(\mathbf{r}_2), \tag{4.147}$$

with P_2^λ obtained from $\Psi_\lambda[\rho]$,

$$P_2^\lambda(\mathbf{r}, \mathbf{r}') = \tag{4.148}$$

$$N(N-1) \sum_{\sigma, \sigma', \sigma_3, \ldots, \sigma_N} \int |\Psi_\lambda(\mathbf{r}\sigma, \mathbf{r}', \sigma', \mathbf{r}_3, \sigma_3, \ldots, r_N, \sigma_N)|^2 d\mathbf{r}_3 \ldots d\mathbf{r}_N.$$

[11] For further discussion of this point, see Chap. 1 by Toulouse in this volume.

Local interpolations within this definition have been analysed and tested in Refs. [87, 138] on small systems, with mixed results.

4.6 Appendix: Kantorovich Duality

The dual construction of the SCE functional and potential (see Theorem 4.13) relies on Kantorovich duality. In this Appendix we give a precise mathematical statement of Kantorovich duality for multi-marginal optimal transport, and show how it implies Theorem 4.13 (1).

Recall the general Kantorovich optimal transport problem introduced in Sect. 4.3.6: for given marginal measures $\mu_1, \ldots, .\mu_N$ defined on closed subsets $X_1, \ldots, X_N$ of $\mathbb{R}^d$, minimize a cost functional

$$C[\gamma] = \int_{X_1 \times \ldots \times X_N} c(\mathbf{r}_1, \ldots, \mathbf{r}_N)\, \mathrm{d}\Pi(\mathbf{r}_1, \ldots, \mathbf{r}_N)$$

over probability measures Π on the product space $X_1 \times \ldots \times X_N$ subject to the marginal constraints

$$\int_{X_1 \times \ldots \times X_{i-1} \times A_i \times X_{i+1} \times \ldots \times X_N} \mathrm{d}\Pi = \int_{A_i} \mathrm{d}\mu_i \text{ for all measurable sets } A_i \subseteq X_i$$

$$\text{and all } i \in \{1, \ldots, N\}.$$

Here $c : X_1 \times \ldots \times X_N \to \mathbb{R} \cup \{+\infty\}$ is a given measurable cost function.

This problem is related to a certain dual variational problem: maximize the functional

$$J[u_1, \ldots, u_N] = \sum_{i=1}^{N} \int_{X_i} u_i \mathrm{d}\mu_i$$

over potentials $u_i : X_i \to \mathbb{R}$ $(i = 1, \ldots, N)$ which must satisfy the pointwise constraint

$$\sum_{i=1}^{N} u_i(\mathbf{r}_i) \leq c(\mathbf{r}_1, \ldots, \mathbf{r}_N) \;\; \forall \;\; (\mathbf{r}_1, \ldots, \mathbf{r}_N) \in X_1 \times \ldots \times X_N. \tag{4.149}$$

The following nontrivial statement, taken from the recent textbook [50], summarizes what is known in $\mathbb{R}^d$, and is general enough to cover the Coulomb cost.

Theorem 4.32 (Kantorovich Duality) *For given probability measures $\mu_1, \ldots, \mu_N$ defined on closed subsets $X_1, \ldots, X_N$ of $\mathbb{R}^d$, provided the cost function*

$c : X = X_1 \times \cdots \times X_N \to \mathbb{R} \cup \{+\infty\}$ is bounded from below and lower semi-continuous and the optimal cost is finite,

$$\inf_{\substack{\Pi \in \mathcal{P}(X) \\ \gamma \mapsto \mu_1, \dots, \mu_N}} \int_X c \, d\Pi = \sup_{(u_1, \dots, u_N) \in \mathcal{A}(c)} \sum_{i=1}^{N} \int_{X_i} u_i \, d\mu_i, \tag{4.150}$$

where $\mathcal{A}(c)$ is any of the following increasingly general sets of admissible potentials:

(1) $\mathcal{A}(c) = \{(u_1, \dots, u_N) : u_i \in C_0(X_i) \, \forall i,$ (4.149) *holds* $\forall \mathbf{r} \in X\}$
(2) $\mathcal{A}(c)$ *as in* (1), *with* $C_b(X_i)$ *in place of* $C_0(X_i)$
(3) $\mathcal{A}(c)$ *as in* (1), *with* $B(X_i) = \{u : X_i \to \mathbb{R} : u$ *bounded measurable*$\}$ *in place of* $C_0(X_i)$
(4) $\mathcal{A}(c) = \{(u_1, \dots, u_N) : u_i \in L^1(X_i; d\mu_i) \, \forall i,$ (4.149) *holds for* $\mu_1 \otimes \cdots \otimes \mu_N$*-a.e.* $\mathbf{r} \in X\}$.

Here we have used the standard notation $C_b(X_i)$ for the space of bounded continuous functions on X_i, and $C_0(X_i)$ for the space of decaying continuous functions on X_i (i.e. those u which in addition satisfy $u(\mathbf{r}_i) \to 0$ if $|\mathbf{r}_i| \to \infty$).

In the special case of two marginals defined on compact sets, cost functions c which are metrics (such as $|\mathbf{r}_1 - \mathbf{r}_2|$), and the choice (2) for the potentials, this fundamental result was discovered by Kantorovich [80]. A great many variants and modifications have subsequently appeared in the mathematics literature. Some of them replace the X_i by abstract spaces; many are worked out only for two marginals; almost all of them differ in the precise assumptions on the cost function and the class of admissible potentials. For instance, [117] (Theorems 2.1.4(b) and 2.1.1) and [81] cover bounded continuous cost functions and the class (3) for N marginals; [136] (Theorem 1.3) covers bounded-below lower semi-continuous cost functions and the class (4) for two marginals. Strictly speaking, none of the versions published prior to the discovery of the *optimal transport/SCE theory* connection applied directly to the multi-marginal Coulomb case, even though the underlying ideas essentially did. For the proof of Theorem 4.32 we refer the reader to [50].

Technical Remark From a functional analysis point of view, the natural class of admissible potentials in (4.150) is the smallest one, (1). This choice reflects the duality between potentials u_i and measures μ_i in the integral $\int u_i d\mu_i$; note that the linear hull of the space $\mathcal{P}(\mathbb{R}^d)$ of probability measures, that is, the space $\mathcal{M}(\mathbb{R}^d)$ of signed measures, is the dual of $C_0(\mathbb{R}^d)$. Enlarging this class from (1) to (2)–(4) has the virtue that the supremum of the dual problem is attained for increasingly general cost functions c.

Proof of Theorem 4.13 (1) using Theorem 4.32 Applying the Kantorovich duality theorem with X_i, μ_i, and c as in Example 4.11, and making the choice (3) for the class of admissible potentials, one obtains

$$\inf_{\substack{\Pi \in \mathcal{P}(\mathbb{R}^{Nd}) \\ \Pi \mapsto \rho}} \int_{\mathbb{R}^{Nd}} V_{ee} \, d\Pi =$$

$$\sup_{\substack{(u_1,\ldots,u_N):u_i \in B(\mathbb{R}^d)\,\forall i, \\ u_1(\mathbf{r}_1)+\ldots+u_N(\mathbf{r}_N) \leq V_{ee}(\mathbf{r}_1,\ldots,\mathbf{r}_N)\,\forall(\mathbf{r}_1,\ldots,\mathbf{r}_N)}} \sum_{i=1}^N \int_{\mathbb{R}^d} u_i(\mathbf{r}_i) \frac{\rho(\mathbf{r}_i)}{N} \, d\mathbf{r}_i.$$

The left-hand side is the enlarged-search definition (4.26) of the SCE functional $V_{ee}^{SCE}[\rho]$ (which, by Theorem 4.1, is equivalent to the original definition (4.15)). The right-hand side can be simplified. For any collection of potentials $(u_1, \ldots, u_N)$, the sum of the integrals on the right-hand side is preserved under the replacement $(u_1, \ldots, u_N) \mapsto (\bar{u}, \ldots, \bar{u})$, where $\bar{u}$ denotes the average $(u_1 + \cdots + u_N)/N$; moreover the constraint in (4.6) is also preserved, thanks to the symmetry of V_{ee}. Thus the right-hand side of (4.6) stays unaltered if the supremization is restricted to N equal potentials, $u_1 = \cdots = u_N = u$. But in this case the right-hand side reduces to that of (4.55), establishing Theorem 4.13 (1). □

Acknowledgments AG acknowledges partial support of his research by the Canada Research Chairs Program and Natural Sciences and Engineering Research Council of Canada. This work started when AG was at the Vrije Universiteit Amsterdam and has received funding from the European Union's Horizon 2020 research and innovation programme under the Marie Skłodowska-Curie grant agreement No. [795942].

References

1. A. Alfonsi, R. Coyaud, and V. Ehrlacher. Constrained overdamped langevin dynamics for symmetric multimarginal optimal transportation. *Mathematical Models and Methods in Applied Sciences* **32**(03), 403–455 (2022)
2. A. Alfonsi, R. Coyaud, V. Ehrlacher and D. Lombardi. Approximation of optimal transport problems with marginal moments constraints. *Math. Comp.* **90**, 689–737 (2021).
3. C.-O. Almbladh and U. von Barth. Exact results for the charge and spin densities, exchange-correlation and density-functional eigenvalues. *Phys. Rev. B* **31**, 3232–3244 (1985).
4. J. Altschuler and E. Boix-Adsera. Hardness results for Multimarginal Optimal Transport problems. *Discrete Optimization* **42**, 100669 (2021).
5. L. Ambrosio. Lecture notes on optimal transport problems. In: *Mathematical Aspects of Evolving Interfaces*, pp. 1–52. vol. 1812, Springer Lecture Notes in Mathematics (2003).
6. H. Bahmann, Y. Zhou and M. Ernzerhof. The shell model for the exchange-correlation hole in the strong-correlation limit. *J. Chem. Phys.* **145**, 124104 (2016).
7. T.E. Baker, E.M. Stoudenmire, L.O. Wagner, K. Burke and S.R. White. One-dimensional mimicking of electronic structure: The case for exponentials. *Phys. Rev. B* **91**, 235141 (2015). Err. **93** 119912 (2016).

8. J.-D. Benamou, G. Carlier, M. Cuturi, L. Nenna and G. Peyré. Iterative Bregman projections for regularized transportation problems. SIAM Journal on Scientific Computing, **37**(2), A1111–A1138 (2015).
9. J.-D. Benamou, G. Carlier and L. Nenna. A numerical method to solve multi-marginal optimal transport problems with coulomb cost. In: *Splitting Methods in Communication, Imaging, Science, and Engineering*, edited by R. Glowinski, S. Osher and W. Yin, pp. 577–601 Springer, Berlin-Heidelberg (2016).
10. U. Bindini and L. De Pascale. Optimal transport with Coulomb cost and the semiclassical limit of density functional theory. *J. Éc. polytech. Math.* **4**, 909–934 (2017).
11. U. Bindini, L. De Pascale and A. Kausamo. On Seidl-type maps for multi-marginal optimal transport with Coulomb cost. *arXiv preprint arXiv:2011.05063* (2020).
12. M.A. Buijse, E.J. Baerends and J. G. Snijders. Analysis of correlation in terms of exact local potentials: Applications to two-electron systems. *Phys. Rev. A* **40**, 4190–4202 (1989).
13. G. Buttazzo, T. Champion and L. De Pascale. Continuity and estimates for multimarginal optimal transportation problems with singular costs. *Applied Mathematics & Optimization* **78**, 185–200 (2018).
14. G. Buttazzo, L. De Pascale and P. Gori-Giorgi. Optimal-transport formulation of electronic density-functional theory. *Phys. Rev. A* **85**, 062502 (2012).
15. G. Carlier. On the linear convergence of the multi-marginal Sinkhorn algorithm. HAL Id: hal-03176512 (2021).
16. G. Carlier, G. Friesecke and D. Vögler. Convex geometry of finite exchangeable laws and de Finetti style representation with universal correlated corrections. *Probab. Theory Relat. Fields* (2022). https://doi.org/10.1007/s00440-022-01115-2
17. G. Carlier and M. Laborde. *SIAM Journal on Mathematical Analysis* **52**, 709–717 (2020).
18. M. Casula, S. Sorella and G. Senatore. Ground state properties of the one-dimensional coulomb gas using the lattice regularized diffusion monte carlo method. *Phys. Rev. B* 74, 245427 (2006).
19. J. Chayes and L. Chayes. On the validity of the inverse conjecture in classical density functional theory. *Journal of statistical physics* **36**, 471–488 (1984).
20. J. Chayes, L. Chayes and E.H. Lieb. The inverse problem in classical statistical mechanics. *Communications in Mathematical Physics* **93**, 57–121 (1984).
21. H. Chen and G. Friesecke. Pair densities in density functional theory. *Multiscale Modeling & Simulation* **13**, 1259–1289 (2015).
22. H. Chen, G. Friesecke and C.B. Mendl. Numerical methods for a Kohn–Sham density functional model based on optimal transport. *J. Chem. Theory Comput.* **10**, 4360–4368 (2014).
23. Y. Chen, T. Georgiou and M. Pavon. Entropic and displacement interpolation: a computational approach using the Hilbert metric. *SIAM Journal on Applied Mathematics* **76**, 2375–2396 (2016).
24. A. Cohen, P. Mori-Sánchez and W. Yang. Insights into current limitations of density functional theory. *Science* **321**, 792–794 (2008).
25. A.J. Cohen, P. Mori-Sánchez and W. Yang. *J. Chem. Phys.* **127**, 034101 (2007).
26. A.J. Coleman and V.I. Yukalov. *Reduced Density Matrices.* Lecture Notes in Chemistry Vol. 72, Springer (2000).
27. M. Colombo, L. De Pascale and S. Di Marino. Multimarginal optimal transport maps for one-dimensional repulsive costs. *Canad. J. Math.* **67**, 350–368 (2015).
28. M. Colombo and S. Di Marino. Equality between monge and kantorovich multimarginal problems with coulomb cost. *Annali di Matematica Pura ad Applicata*, 1–14 (2013).
29. M. Colombo, S. Di Marino and F. Stra. First order expansion in the semiclassical limit of the levy-lieb functional. *arXiv preprint arXiv:2106.06282* (2021).
30. M. Colombo and F. Stra. Counterexamples in multimarginal optimal transport with coulomb cost and spherically symmetric data. *Mathematical Models and Methods in Applied Sciences* **26**, 1025–1049 (2016).

31. L.A. Constantin. Correlation energy functionals from adiabatic connection formalism. *Phys. Rev. B* **99**, 085117 (2019).
32. C. Cotar, G. Friesecke, and C. Klüppelberg. Density functional theory and optimal transportation with coulomb cost. *Comm. Pure Appl. Math.* **66**, 548–99 (2013).
33. C. Cotar, G. Friesecke and C. Klüppelberg. Smoothing of transport plans with fixed marginals and rigorous semiclassical limit of the Hohenberg-Kohn functional. *Arch. Ration. Mech. An.* **228**, 891–922 (2018).
34. M. Cuturi. Sinkhorn distances: Lightspeed computation of optimal transport. In: *Advances in neural information processing systems 26 (NIPS 2013)*, edited by C.J. Burges, L. Bottou, M. Welling, Z. Ghahramani and K.Q. Weinberger, pp. 2292–2300, Curran Associates Inc., NY (2013).
35. M. Cuturi and G. Peyré. *Computational optimal transport.* Vol. 11, Now Publishers, Inc. (2019).
36. T.J. Daas, E. Fabiano, F. Della Sala, P. Gori-Giorgi and S. Vuckovic. Noncovalent interactions from models for the Møller–Plesset adiabatic connection. *The journal of physical chemistry letters* **12**, 4867–4875 (2021).
37. T.J. Daas, J. Grossi, S. Vuckovic, Z.H. Musslimani, D.P. Kooi, M. Seidl, K.J. Giesbertz and P. Gori-Giorgi. Large coupling-strength expansion of the Møller–Plesset adiabatic connection: From paradigmatic cases to variational expressions for the leading terms. *The Journal of chemical physics* **153**, 214112 (2020).
38. L. De Pascale. Optimal transport with coulomb cost. approximation and duality. *ESAIM: Math. Model. Numer. Anal.* **49**, 1643–1657 (2015).
39. E. del Barrio and J.-M. Loubes. The statistical effect of entropic regularization in optimal transportation. *arXiv:2006.05199* (2020).
40. S. Di Marino and A. Gerolin. An Optimal Transport approach for the Schrödinger bridge problem and convergence of Sinkhorn algorithm. *Journal of Scientific Computing* **85**(2), 1–28 (2020).
41. S. Di Marino and A. Gerolin. Optimal transport losses and Sinkhorn algorithm with general convex regularization. *arXiv preprint, arXiv:2007.00976* (2020).
42. S. Di Marino, A. Gerolin and L. Nenna. Optimal transport for repulsive costs. In: *Topological Optimization and Optimal Transport – In the Applied Sciences*, edited by M. Bergounioux, É. Oudet, M. Rumpf, G. Carlier, T. Champion and F. Santambrogio, chapter 9, De Gruyter (2017).
43. M. Ernzerhof. Construction of the adiabatic connection. *Chem. Phys. Lett.* **263**, 499 (1996).
44. E. Fabiano, P. Gori-Giorgi, M. Seidl, and F. Della Sala. Interaction-strength interpolation method for main-group chemistry: Benchmarking, limitations, and perspectives. *J. Chem. Theory. Comput.* **12**, 4885–4896 (2016).
45. E. Fabiano, S. Smiga, S. Giarrusso, T.J. Daas, F. Della Sala, I. Grabowski and P. Gori-Giorgi. Investigation of the exchange-correlation potentials of functionals based on the adiabatic connection interpolation. *J. Chem. Theory. Comput.* **15**, 1006–1015 (2019).
46. J. Franklin and J. Lorenz. On the scaling of multidimensional matrices. *Linear Algebra and its applications* **114**, 717–735 (1989).
47. D.E. Freund, B.D. Huxtable and J.D. Morgan. Variational calculations on the helium isoelectronic sequence. *Phys. Rev. A* **29**, 980–982 (1984).
48. G. Friesecke. The multiconfiguration equations for atoms and molecules: charge quantization and existence of solutions. *Archive for Rational Mechanics and Analysis*, **169**, 35–71 (2003).
49. A simple counterexample to the Monge ansatz in multi-marginal optimal transport, convex geometry of the set of Kantorovich plans, and the Frenkel–Kontorova model. *SIAM J. Math. Analysis* **51**, 4332–4355 (2019).
50. G. Friesecke. Lectures on optimal transport. *SIAM*, to appear (2022).
51. G. Friesecke, C.B. Mendl, B. Pass, C. Cotar and C. Klüppelberg. N-density representability and the optimal transport limit of the Hohenberg–Kohn functional. *J. Chem. Phys.* **139**, 164109 (2013).

52. G. Friesecke, A.S. Schulz and D. Vögler. Genetic column generation: Fast computation of high-dimensional multi-marginal optimal transport problems. *SIAM J. Sci. Comp.* **44**(3), A1632–A1654 (2022).
53. G. Friesecke and D. Vögler. Breaking the curse of dimension in multi-marginal kantorovich optimal transport on finite state spaces. *SIAM J. Math. Analysis* **50**, 3996–4019 (2018).
54. A. Galichon and B. Salanié. Matching with trade-offs: Revealed preferences over competing characteristics. CEPR Discussion Paper No. DP7858 (2010).
55. W. Gangbo and R.J. McCann. The geometry of optimal transportation. *Acta Math.* 177, 113–161 (1906).
56. W. Gangbo and A. Swiech. Optimal maps for the multidimensional monge-kantorovich problem. *Commun. Pure Appl. Math.* **51**, 23 (1998).
57. A. Gerolin. *Multi-marginal optimal transport and potential optimization problems for Schrödinger operators*. PhD thesis, Università degli studi di Pisa (2016).
58. A. Gerolin, J. Grossi, and P. Gori-Giorgi. Kinetic correlation functionals from the entropic regularisation of the strictly-correlated electrons problem. *Journal of Chemical Theory and Computation* **16**, 488–498 (2019).
59. A. Gerolin, A. Kausamo, and T. Rajala. Non-existence of optimal transport maps for the multi-marginal repulsive harmonic cost. *SIAM Journal on Mathematical Analysis* **51** (2019).
60. A. Gerolin, A. Kausamo, and T. Rajala. Multi-marginal Entropy-Transport with repulsive cost. *Calc. Var. PDEs* **59**, Art. 90 (2020).
61. A. Ghosal, A.D. Guclu, C.J. Umrigar, D. Ullmo and H.U. Baranger. N*ature Phys.* **2**, 336 (2006).
62. S. Giarrusso, P. Gori-Giorgi, F. Della Sala and E. Fabiano. Assessment of interaction-strength interpolation formulas for gold and silver clusters. *J. Chem. Phys.* **148**, 134106 (2018).
63. S. Giarrusso, S. Vuckovic, and P. Gori-Giorgi. Response potential in the strong-interaction limit of dft: Analysis and comparison with the coupling-constant average. *J. Chem. Theory Comput.* **14**, 4151–4167 (2018).
64. G.F. Giuliani and G. Vignale. *Quantum Theory of the Electron Liquid.* Cambridge University Press, New York (2005).
65. P. Gori-Giorgi and A. Savin. *J. Phys.: Conf. Ser.* **117**, 012017 (2008).
66. P. Gori-Giorgi and M. Seidl. Density functional theory for strongly-interacting electrons: perspectives for physics and chemistry. *Phys. Chem. Chem. Phys.* **12**, 14405–14419 (2010).
67. P. Gori-Giorgi, M. Seidl and G. Vignale. Density-functional theory for strongly interacting electrons. *Phys. Rev. Lett.* **103**, 166402 (2009).
68. P. Gori-Giorgi, G. Vignale and M. Seidl. Electronic zero-point oscillations in the strong-interaction limit of density functional theory. *J. Chem. Theory Comput.* **5**, 743–753 (2009).
69. A. Görling and M. Levy. Correlation-energy functional and its high-density limit obtained from a coupling-constant perturbation expansion. *Phys. Rev. B* **47**, 13105 (1993).
70. N. Gozlan and C. Léonard. Transport inequalities, a survey. *Markov Processes and Related Fields* **16**, 635–736 (2010).
71. J. Grossi, D.P. Kooi, K.J.H. Giesbertz, M. Seidl, A.J. Cohen, P. Mori-Sánchez and P. Gori-Giorgi. Fermionic statistics in the strongly correlated limit of density functional theory. *J. Chem. Theory Comput.* **13**, 6089–6100 (2017).
72. J. Grossi, Z. Musslimani, M. Seidl and P. Gori-Giorgi. Kohn–Sham equations with functionals from the strictly-correlated regime: Investigation with a spectral renormalization method. *Journal of Physics: Condensed Matter*. Epub ahead of print (2020).
73. A.D. Guclu, A. Ghosal, C.J. Umrigar and H.U. Baranger. *Phys. Rev. B* **77**, 041301 (2008).
74. N. Helbig, J.I. Fuks, M. Casula, M.J. Verstraete, M. Marques, I. Tokatly and A. Rubio. Density functional theory beyond the linear regime: Validating an adiabatic local density approximation. *Physical Review A* **83**, 032503 (2011).
75. N. Helbig, I.V. Tokatly, and A. Rubio. Exact Kohn–Sham potential of strongly correlated finite systems. *J. Chem. Phys.* **131**, 224105 (2009).
76. T. Helgaker, P. Jørgensen and J. Olsen. *Density-functional Theory: A Convex Treatment.* Wiley Blackwell (2016).

77. M. Hoffmann-Ostenhof and T. Hoffmann-Ostenhof. "Schrödinger inequalities" and asymptotic behavior of the electron density of atoms and molecules. *Phys. Rev. A* **16**, 1782–1785 (1977).
78. P. Hohenberg and W. Kohn. Inhomogeneous electron gas. *Phys. Rev. B* **136**, 864 (1964).
79. H. Janati, B. Muzellec, G. Peyré and M. Cuturi. Entropic optimal transport between unbalanced gaussian measures has a closed form. In *Advances in Neural Information Processing Systems*, vol. 33, edited by H. Larochelle, M. Ranzato, R. Hadsell, M.F. Balcan and H. Lin, pp.10468–10479, Curran Associates, Inc. (2020).
80. L.V. Kantorovich. On the transfer of masses. *Dokl. Akad. Nauk. SSSR.* **37**, 227 (1942).
81. H.G. Kellerer. Duality theorems for marginal problems. *Zeitschrift für Wahrscheinlichkeitstheorie und verwandte Gebiete* **67**, 399–432 (1984).
82. Y. Khoo and L. Ying. Convex relaxation approaches for strictly correlated density functional theory. *SIAM J. Sci. Comput.* **41**, B773–B795 (2019).
83. M. Knott and C.S. Smith. On the optimal mapping of distributions. *J. Optimization Theory and Appl.* **43**, 39–49 (1984).
84. W. Kohn. v-representability and density functional theory. *Physical review letters* **51**, 1596 (1983).
85. W. Kohn and L.J. Sham. Self-consistent equations including exchange and correlation effects. *Phys. Rev.* **140**, A1133–A1138 (1965).
86. W. Kolos and C.C.J. Roothaan. Accurate electronic wave functions for the h_2 molecule. *Rev. Mod. Phys.* **32**, 219–232 (1960).
87. D.P. Kooi and P. Gori-Giorgi. Local and global interpolations along the adiabatic connection of dft: a study at different correlation regimes. *Theoretical chemistry accounts* **137**, 1–12 (2018).
88. D.P. Kooi and P. Gori-Giorgi. A variational approach to London dispersion interactions without density distortion. *The journal of physical chemistry letters* **10**, 1537–1541 (2019).
89. C. Léonard. A survey of the Schrödinger problem and some of its connections with optimal transport. *Discrete Cont. Dyn.-A* **34**, 1533–1574 (2014).
90. M. Levy. Universal variational functionals of electron densities, first-order density matrices, and natural spin-orbitals and solution of the v-representability problem. *Proc. Natl. Acad. Sci.* **76**, 6062–6065 (1979).
91. M. Levy. Electron densities in search of hamiltonians. *Phys. Rev. A* **26**, 1200–1208 (1982).
92. M. Levy and J. Perdew. Hellman–Feynman, virial, and scaling requisites for the exact universal density functionals. Shape of the correlation potential and diamagnetic susceptibility for atoms. *Phys. Rev. A* **32**, 2010–2021 (1985).
93. M. Levy, J.P. Perdew and V. Sahni. Exact differential equation for the density and ionization energy of a many-particle system. *Phys. Rev. A* **30**, 2745–2748 (1984).
94. M. Levy and F. Zahariev. *Phys. Rev. Lett.* **113**, 113002 (2014).
95. M. Lewin. Semi-classical limit of the Levy–Lieb functional in Density Functional Theory. *C. R. Math.* **356**, 449–455 (2018).
96. M. Lewin, E.H. Lieb and R. Seiringer. Floating wigner crystal with no boundary charge fluctuations. *Physical Review B* **100**, 035127 (2019).
97. E.H. Lieb. Density functionals for Coulomb systems. *Int. J. Quantum. Chem.* **24**, 243–277 (1983).
98. Z.-F. Liu and K. Burke. Adiabatic connection in the low-density limit. *Phys. Rev. A* **79**, 064503 (2009).
99. D. Lorenz and H. Mahler. Orlicz space regularization of continuous optimal transport problems. *Appl. Math. Optim.* **85**, 14 (2022).
100. F. Malet and P. Gori-Giorgi. Strong correlation in Kohn–Sham density functional theory. *Phys. Rev. Lett.* **109**, 246402 (2012).
101. F. Malet, A. Mirtschink, J.C. Cremon, S.M. Reimann and P. Gori-Giorgi. Kohn–Sham density functional theory for quantum wires in arbitrary correlation regimes. *Phys. Rev. B* **87**, 115146 (2013).

102. F. Malet, A. Mirtschink, K.J.H. Giesbertz, L.O. Wagner and P. Gori-Giorgi. Exchange-correlation functionals from the strong interaction limit of dft: applications to model chemical systems. *Phys. Chem. Chem. Phys.* **16**, 14551–14558 (2014).
103. A. Mallasto, A. Gerolin, and H.Q. Minh. Entropy-regularized 2-Wasserstein distance between gaussian measures. *Information Geometry*, 1–35 (2021).
104. C.B. Mendl and L. Lin. Kantorovich dual solution for strictly correlated electrons in atoms and molecules. *Phys. Rev. B* **87**, 125106 (2013).
105. C.B. Mendl, F. Malet and P. Gori-Giorgi. Wigner localization in quantum dots from kohn-sham density functional theory without symmetry breaking. *Phys. Rev. B* **89**, 125106 (2014).
106. A. Mirtschink, C.J. Umrigar, J.D. Morgan III and P. Gori-Giorgi. Energy density functionals from the strong-coupling limit applied to the anions of the He isoelectronic series. *J. Chem. Phys.* **140**, 18A532 (2014).
107. A. Moameni and B. Pass. Solutions to multi-marginal optimal transport problems concentrated on several graphs. *ESAIM: Control Optim. Calc. Var.* **23**(2), 551–567 (2017).
108. G. Monge. *Mémoire sur la théorie des déblais et des remblais.* Histoire Acad. Sciences, Paris (1781).
109. P. Mori-Sánchez and A.J. Cohen. Exact density functional obtained via the Levy constrained search. *The journal of physical chemistry letters* **9**, 4910–4914 (2018).
110. L. Nenna. *Numerical methods for multi-marginal optimal transportation.* PhD thesis, Université Paris sciences et lettres (2016).
111. B. Pass. *Structural results on optimal transportation plans.* PhD thesis, University of Toronto (2011).
112. B. Pass. On the local structure of optimal measures in the multi-marginal optimal transportation problem. *Calculus of Variations and Partial Differential Equations* **43**, 529–536 (2012).
113. B. Pass. Remarks on the semi-classical Hohenberg–Kohn functional. *Nonlinearity*, **26**(9), 2731 (2013).
114. B. Pass. Multi-marginal optimal transport and multi-agent matching problems: uniqueness and structure of solutions. *Discrete Contin. Dyn. Syst.* **34**, 1623–1639 (2014).
115. B. Pass. Multi-marginal optimal transport: theory and applications. *ESAIM: Mathematical Modelling and Numerical Analysis* **49**(6), 1771–1790 (2015).
116. A. Pratelli. On the equality between Monge's infimum and Kantorovich's minimum in optimal mass transportation. *Annales de l'Institut Henri Poincare (B) Probability and Statistics* vol. **43**, 1–13 (2007).
117. S. Rachev and L. Rüschendorf. *Mass transportation problems.* Springer-Verlag, New York (1998).
118. L. Ruschendorf. Convergence of the iterative proportional fitting procedure. *The Annals of Statistics* **23**, 1160–1174 (1995).
119. F. Santambrogio. *Optimal Transport for Applied Mathematicians.* Progress in Nonlinear Differential Equations and Their Applications, Birkhäuser (2015).
120. A. Savin. Is size-consistency possible with density functional approximations? *Chem. Phys.* **356**, 91–97 (2009).
121. E. Schrödinger. *Über die umkehrung der naturgesetze.* Verlag Akademie der wissenschaften in kommission bei Walter de Gruyter u. Company (1931).
122. M. Seidl. Strong-interaction limit of density-functional theory. *Phys. Rev. A* **60**, 4387–4395 (1999).
123. M. Seidl. Adiabatic connection in density-function theory: Two electrons on the surface of a sphere. *Phys. Rev. A* **75**, 062506 (2007).
124. M. Seidl, S. Di Marino, A. Gerolin, L. Nenna, K.J. Giesbertz and P. Gori-Giorgi. The strictly-correlated electron functional for spherically symmetric systems revisited ii: Sgs conjecture. In preparation.
125. M. Seidl, S. Di Marino, A. Gerolin, L. Nenna, K.J. Giesbertz and P. Gori-Giorgi. The strictly-correlated electron functional for spherically symmetric systems revisited. *arXiv preprint, arXiv:1702.05022* (2017).

126. M. Seidl, S. Giarrusso, S. Vuckovic, E. Fabiano and P. Gori-Giorgi. Communication: Strong-interaction limit of an adiabatic connection in Hartree–Fock theory. *The Journal of Chemical Physics* **149**, 241101 (2018).
127. M. Seidl, P. Gori-Giorgi and A. Savin. Strictly correlated electrons in density-functional theory: A general formulation with applications to spherical densities. *Phys. Rev. A* **75**, 042511/12 (2007).
128. M. Seidl, J.P. Perdew and S. Kurth. Density functionals for the stron-interaction limit. *Phys. Rev. A* **62**, p. 012502 (2000).
129. M. Seidl, J.P. Perdew and S. Kurth. Simulation of all-order density-functional perturbation theory, using the second order and the strong-correlation limit. *Phys. Rev. Lett.* **84**, 5070–5073 (2000).
130. M. Seidl, J.P. Perdew and M. Levy. Strictly correlated electrons in density-functional theory. *Phys. Rev. A* **59**, 51–54 (1999).
131. R. Sinkhorn. A relationship between arbitrary positive matrices and doubly stochastic matrices. *The annals of mathematical statistics* **35**, 876–879 (1964).
132. S. Smiga and L.A. Constantin. Modified interaction-strength interpolation method as an important step toward self-consistent calculations. *Journal of chemical theory and computation* **16**, 4983–4992 (2020).
133. R. van Leeuwen. Density functional approach to the many-body problem: key concepts and exact functionals. *Adv. Quantum Chem.* **43**, 24–94 (2003).
134. D. Vieira. Spin-independent v-represetability of Wigner crystal oscillations in one-dimensional Hubbard chains: The role of spin-charge separation. *Phys. Rev. B* **86**, 075132 (2012).
135. D. Vieira and K. Capelle. Investigation of self-interaction corrections for an exactly solvable model system: ORbital dependence and electron localization. *J. Chem. Theory Comput.* **6**, 3319–3329 (2010).
136. C. Villani. *Topics in Optimal Transportation.* Grad. Stud. Math. **58**. Amer. Math. Soc., Providence (2003).
137. S. Vuckovic, P. Gori-Giorgi, F. Della Sala and E. Fabiano. Restoring size consistency of approximate functionals constructed from the adiabatic connection. *J. Phys. Chem. Lett.* **9**, 3137–3142 (2018).
138. S. Vuckovic, T.J.P. Irons, A. Savin, A.M. Teale and P. Gori-Giorgi. Exchange–correlation functionals via local interpolation along the adiabatic connection. *J. Chem. Theory Comput.* **12**, 2598–2610 (2016).
139. S. Vuckovic, M. Levy and P. Gori-Giorgi. Augmented potential, energy densities, and virial relations in the weak-and strong-interaction limits of DFT. *J. Chem. Phys.* **147**, 214107 (2017).
140. L.O. Wagner and P. Gori-Giorgi. Electron avoidance: A nonlocal radius for strong correlation. *Phys. Rev. A* **90**, 052512 (2014).
141. L.O. Wagner, E.M. Stoudenmire, K. Burke, and S.R. White. Reference electronic structure calculations in one dimension. *Phys. Chem. Chem. Phys.* **14**, 8581 (2012).
142. J.-J. Wang, W. Li, S. Chen, G. Xianlong, M. Rontani and M. Polini. Absence of Wigner molecules in one-dimensional few-fermion systems with short-range interactions. *Physical Review B* **86**, 075110 (2012).
143. E.P. Wigner. On the interaction of electrons in metals. *Phys. Rev.* **46**, 1002 (1934).
144. Z.-J. Ying, V. Brosco, G.M. Lopez, D. Varsano, P. Gori-Giorgi and J. Lorenzana. Anomalous scaling and breakdown of conventional density functional theory methods for the description of Mott phenomena and stretched bonds. *Phys. Rev. B* **94**, 075154 (2016).
145. G.M. Zhislin. Discussion of the spectrum of Schrödinger operators for systems of many particles. *Trudy Moskovskogo matematiceskogo obscestva* **9**, 81–120 (1960).

Chapter 5
Moreau–Yosida Regularization in DFT

Simen Kvaal

Abstract Moreau–Yosida regularization is introduced into the framework of exact DFT. Moreau–Yosida regularization is a lossless operation on lower semicontinuous proper convex functions over separable Hilbert spaces, and when applied to the universal functional of exact DFT (appropriately restricted to a bounded domain), gives a reformulation of the ubiquitous v-representability problem and a rigorous and illuminating derivation of Kohn–Sham theory.
The chapter comprises a self-contained introduction to exact DFT, basic tools from convex analysis such as sub- and superdifferentiability and convex conjugation, as well as basic results on the Moreau–Yosida regularization. The regularization is then applied to exact DFT and Kohn–Sham theory, and a basic iteration scheme based in the Optimal Damping Algorithm is analyzed. In particular, its global convergence established. Some perspectives are offered near the end of the chapter.

5.1 Introduction

In this chapter, we introduce Moreau–Yosida regularization into the exact DFT framework. Such regularization of convex optimization problems is lossless: Given a small $\epsilon > 0$, a non-smooth convex functional $F[\rho]$ is regularized into a differentiable convex functional ${}^{\epsilon}F[\rho]$. However, this smoothing is invertible, so that the information encoded is not lost. Thus, the regularization gives a *reformulation* of the convex optimization problem as a differentiable convex optimization problem. When applied to exact DFT, we gain insight into the ubiquitous v-representability problem and also a rigorous formulation of Kohn–Sham theory. This comes at a small price, as we need to describe densities and potentials using Banach spaces that are reflexive, which excludes, say, exact Coulomb potentials on infinite domains. This trade-off is acceptable, as no physical system is truly infinite in extent. Indeed,

S. Kvaal (✉)
Hylleraas Centre for Quantum Molecular Sciences, Department of Chemistry, University of Oslo, Oslo, Norway
e-mail: simen.kvaal@kjemi.uio.no

E. Cancès, G. Friesecke (eds.), *Density Functional Theory*, Mathematics and Molecular Modeling, https://doi.org/10.1007/978-3-031-22340-2_5

placing the system in a finite but large box is the only approximation being made in this chapter.

The motivation for introducing Moreau–Yosida regularization comes from two directions: First, DFT is committed to treating "all possible" potentials at once. For the exact interacting problem, this is surely overkill: After all we are mostly interested in the atomic Coulomb potentials, or at most well-behaved potentials modeling, say, harmonic traps. On the other hand, passage to the non-interacting Kohn–Sham system turns the problem upside down: The density is the known quantity, while the effective potential is the unknown function of the density. The standard approach is to differentiate the universal functional to obtain the exchange-correlation potential, but *this is not permitted in exact DFT*. Indeed, the large class of potentials implies that the universal functional will be highly non-smooth. We note in passing that in the strongly correlated limit of DFT, where one obtains the purely-interacting universal functional, the exchange-correlation potential can in fact be obtained rigorously by differentiation, as shown in Chap. 4 by Friesecke, Gerolin and Gori-Giorgi. The second motivation for Moreau–Yosida regularization is that formal Kohn–Sham theory relies on densities to be both interacting and non-interacting v-representable. This problem vanishes into thin air in the regularized Kohn–Sham formulation.

While Moreau–Yosida regularization resolves some problems, it still leaves open the question of how to approximate the regularized universal functional ${}^{\epsilon}F[\rho]$ in a systematic manner. On the other hand, the method of Moreau–Yosida regularization may also be used to model density functional approximations that are already convex.

This chapter has the following structure. In Sect. 5.2 we give a brief introduction to the convex formulation of exact DFT as given by Lieb [16]. Some results from the theory of quadratic forms over Hilbert spaces are needed, and we will refer the reader to the excellent literature that exists. The article [27] by Simon lays out the formulation of quantum mechanics using quadratic forms, and the monograph [25] by Schmüdgen gives further details, including the representation theorem of bounded-below closed quadratic forms. We include some definitions and theorems from the analysis of convex functions over Banach spaces, and also give proofs whenever deemed instructive. For an excellent and accessible introduction to convex analysis, see the short monograph [29] by van Tiel. After introducing exact DFT, we proceed to introduce the only approximation we need, that is, truncation of infinite space $\mathbb{R}^3$ to a bounded domain $\Omega \subset \mathbb{R}^3$. This allows the formulation of exact DFT in a Hilbert space setting, or more generally in a reflexive Banach space setting.

In Sect. 5.3 we introduce the Moreau–Yosida regularization of convex problems over Hilbert spaces. For a more detailed exposition in the Hilbert space setting, see the monograph [1] by Bauschke and Combettes. Most results can be wholly or partially generalized to convex functions over reflexive Banach spaces, see the monograph [2] by Barbu and Precupanu. In our exposition, we include some important definitions and results, with complete proofs in many cases due to the central nature of the material. In the following Sect. 5.4, we apply Moreau–Yosida regularization to box truncated exact DFT. In particular, we detail the rigorous derivation of Kohn–Sham theory in the regularized setting, including a basic analysis of the Moreau–Yosida Kohn–Sham optimal damping algorithm (MYKSODA). Our treatment is adapted from Refs. [9, 10, 17, 20, 21].

Finally, in Sect. 5.5 we provide a conclusion and discuss opportunities for future research.

The author gratefully acknowledges helpful feedback from Andre Laestadius and Markus Penz, and also Gero Friesecke who additionally provided the complete proof of the fact that finite-kinetic energy N-electron wavefunctions have densities with square roots in the Sobolev space $H^1(\mathbb{R}^3)$, see Theorem 5.2.

5.2 Exact DFT

In this section, we give a brief outline of elements of exact DFT, paying special attention to the regularity, or lack thereof, of the universal functional and ground-state energy. We then regularize exact DFT to a finite but large subdomain $\Omega \subset \mathbb{R}^3$. Our focus is on molecular electronic systems.

5.2.1 *The Variational Ground-State Problem*

In this section, we use the definitions

$$L^2_N := \bigwedge_{i=1}^{N} L^2(S), \quad S = \mathbb{R}^3 \times \mathbb{Z}_2\,,$$

$$H^k_N := H^k(S^N) \cap L^2_N\,.$$

Here, $H^k(S^N)$ denotes the set of elements of $L^2(S^N)$ whose 2^N spatial components are elements of the standard k'th order Sobolev space $H^k(\mathbb{R}^{3N})$, i.e., each spatial component has square-integrable weak partial derivatives up to order k [25]. The starting point for DFT is the ground-state problem of N interacting electrons in an external potential v on variational form, i.e.,

$$E[v] := \inf\{\mathcal{E}_0[\psi] + \mathcal{V}[\psi] \mid \psi \in \mathcal{W}_N\}\,, \tag{5.1}$$

with $\mathcal{W}_N = \{\psi \in H^1_N \mid \|\psi\|_2 = 1\}$ consisting of the normalized finite kinetic-energy wavefunctions, and with

$$\mathcal{E}_0[\psi] = \frac{1}{2}\langle \nabla\psi, \nabla\psi\rangle + \langle \psi, W\psi\rangle,$$

$$\mathcal{V}[\psi] = \langle \psi, V\psi\rangle = \int \rho_\psi(\mathbf{r})v(\mathbf{r})\,\mathrm{d}^3\mathbf{r}\,.$$

Here, V is the many-electron operator corresponding to the potential v, and ρ_ψ is the one-electron density of ψ. In order to make connection with more standard formulations of quantum mechanics, the variational minimization in terms of quadratic forms in Eq. (5.1) needs to be connected with the spectral theory of self-adjoint operators. Intuitively, the kinetic energy part of the variational formulation is obtained by integration by parts, enlarging the domain of the kinetic energy operator term $T = -\frac{1}{2}\nabla^2$ in the full Hamiltonian. However, the transition back and forth between a self-adjoint operator and a quadratic form is subtle, and for some potentials the domain of the Hamiltonian *operator* is not H^2_N due to singular behavior of v. For this exposition, we let the following be sufficient: Whenever $v \in L^{3/2}(\mathbb{R}^3) + L^\infty(\mathbb{R}^3)$, $\mathcal{V}$ will be relatively bounded with respect to $\mathcal{E}_0$, which means that as a perturbation it is sufficiently gentle to allow the sum of the forms to be well-defined with domain H^1_N by the KLMN theorem [25, 27]. The fundamental representation theorem of closed semibounded quadratic forms [25] guarantees the existence of a unique self-adjoint operator $\widehat{H}[v] : D[v] \to L^2$ such that for every ψ in the operator domain $D[v]$, we have $\mathcal{E}_0[\psi] + \mathcal{V}[\psi] = \langle \psi, \widehat{H}[v]\psi\rangle$. We point out that $D[v] \subset H^1_N$ is dense in L^2_N, and that $v \in L^2(\mathbb{R}^3) + L^\infty(\mathbb{R}^3)$ is sufficient to guarantee $D[v] = H^2_N$, a fundamental result of Kato [8]. However, for the stronger singularities present in $L^{3/2}(\mathbb{R}^3) + L^\infty(\mathbb{R}^3)$ it may happen that $D[v]$ becomes a proper subset of H^2_N. On the other hand, the *form domain* is always H^1_N for every $v \in L^{3/2}(\mathbb{R}^3) + L^\infty(\mathbb{R}^3)$. In any case, the infimum in Eq. (5.1) will be the bottom, i.e., infimum, of the spectrum of $\widehat{H}[v]$, and hence the connection is complete.

5.2.2 Densities

Central to DFT is the density $\rho_\psi \in L^1(\mathbb{R}^3)$ associated with a (normalized) $\psi \in L^2_N$.

Definition 5.1 For any $\psi \in L^2_N$, we define the density $\rho_\psi : \mathbb{R}^3 \to [0, +\infty)$ almost everywhere (a.e.) by the formula

$$\rho_\psi(\mathbf{r}) := N \sum_{s\in\mathbb{Z}_2} \int_{S^{N-1}} |\psi((\mathbf{r}, s), \mathbf{x}_2, \cdots, \mathbf{x}_N)|^2 \,\mathrm{d}\mathbf{x}_2 \cdots \mathrm{d}\mathbf{x}_N\,.$$

For a given $\rho \in L^1(\mathbb{R}^3)$, we write $\psi \mapsto \rho$ if $\rho_\psi = \rho$.

It is immediate that $\rho_\psi \geqslant 0$ a.e., and that $\|\rho_\psi\|_1 = N\|\psi\|^2$. In fact, we have a complete characterization of densities that come from elements $\psi \in \mathcal{W}_N$. In the proof of the following theorem, the proof that the density of $\psi \in \mathcal{W}_N$ satisfies $\rho_\psi^{1/2} \in H^1(\mathbb{R}^3)$ is due to G. Friesecke.

Theorem 5.2 **(N-Representable Densities)** *Let $\mathcal{D}^N$ be the set of measurable functions $\rho : \mathbb{R}^3 \to \mathbb{R}$ that satisfy $\rho \geqslant 0$ a.e., $\rho^{1/2} \in H^1(\mathbb{R}^3)$ with $\|\rho\|_1 = N$. Then for every $\rho \in \mathcal{D}^N$ there is a $\psi \in \mathcal{W}_N$ such that $\psi \mapsto \rho$. Conversely, for every $\psi \in \mathcal{W}_N$, $\rho_\psi \in \mathcal{D}^N$. Moreover, $\mathcal{D}^N$ is convex.*

Proof For the construction of a $\psi \in \mathcal{W}_N$ with $\rho_\psi = \rho$ for a given $\rho \in \mathcal{D}^N$, see [16, Theorem 1.2]. For the converse, we must prove that $\rho_\psi^{1/2} \in H^1(\mathbb{R}^3)$.

Let $\psi \in \mathcal{W}_N$, and let $\rho = \rho_\psi$. Differentiating, we obtain

$$\nabla\rho(\mathbf{r}) = N \sum_{s\in\mathbb{Z}_2} \int_{S^{N-1}} (\psi^*\nabla_1\psi + \psi\nabla_1\psi^*)\,\mathrm{d}\mathbf{x}_2\cdots\mathrm{d}\mathbf{x}_N\,.$$

Application of Cauchy–Schwarz gives

$$|\nabla\rho(\mathbf{r})|^2 \leqslant 8\rho(\mathbf{r})t(\mathbf{r}), \tag{5.2}$$

where

$$t(\mathbf{r}) = \frac{N}{2} \sum_{s\in\mathbb{Z}_2} \int_{S^{N-1}} |\nabla_1\psi|^2\,\mathrm{d}\mathbf{x}_2\cdots\mathrm{d}\mathbf{x}_N \in L^1(\mathbb{R}^3)$$

is the kinetic-energy density. Consequently, $\rho \in W^{1,1}(\mathbb{R}^3)$, the standard Sobolev space of $L^1(\mathbb{R}^3)$ functions with first-order weak derivatives in $L^1(\mathbb{R}^3)$.

Let $G(\mathbf{r}) = \exp(-|\mathbf{r}|^2)$, and consider $\rho_\epsilon(\mathbf{r}) = \rho(\mathbf{r}) + \epsilon G(\mathbf{r})$, with $\epsilon > 0$ a small parameter. We first prove that $\rho_\epsilon^{1/2} \to \rho^{1/2}$ in $L^2(\mathbb{R}^3)$ as $\epsilon \to 0$. It is clear that $|\rho_\epsilon^{1/2} - \rho^{1/2}| \to 0$ pointwise. By the elementary inequality $(a-b)^2 \leqslant 2(a^2+b^2)$ we have, for all $\epsilon \leqslant 1$,

$$|\rho_\epsilon^{1/2} - \rho^{1/2}|^2 \leqslant 2(\rho + \epsilon G + \rho) \leqslant 2G + 4\rho \in L^1(\mathbb{R}^3).$$

By the dominated convergence theorem, the integral of the left-hand side converges to zero, and hence $\rho_\epsilon^{1/2} \to \rho^{1/2}$ in $L^2(\mathbb{R}^3)$ as claimed.

We next prove that $\rho_\epsilon^{1/2}$ is bounded in $H^1(\mathbb{R}^3)$ for $\epsilon \leqslant 1$. Let $h(z) = (z + \epsilon G(\mathbf{r}))^{1/2}$, where $z \in [0, +\infty)$ and $\mathbf{r}$ is in some compact subset of $\mathbb{R}^3$. The function h is continuously differentiable with bounded derivative. By the chain rule for Sobolev functions, which says that (1) the composition of a C^1 function with bounded derivative and a function in $W^{1,1}_{\mathrm{loc}}(\mathbb{R}^3)$ is again in $W^{1,1}_{\mathrm{loc}}(\mathbb{R}^3)$, and (2) that its partial derivatives can be computed with the usual chain rule [6, Ch. 5, Exercise 17], we obtain $(\rho + \epsilon G)^{1/2} \in W^{1,1}(\mathbb{R}^3)_{\mathrm{loc}}$ with $\nabla(\rho + \epsilon G)^{1/2} =$

$\frac{1}{2}(\rho+\epsilon G)^{-1/2}(\nabla\rho+\epsilon\nabla G)$. Key now, is that this expression is uniformly bounded in $L^2(\mathbb{R}^3)$. Indeed,

$$|\nabla\rho_\epsilon^{1/2}|^2 \leq \frac{|\nabla\rho|^2+|\epsilon\nabla G|^2}{2(\rho+\epsilon G)} \leq \frac{4\rho t}{\rho+\epsilon G}+\epsilon\frac{|\nabla G|^2}{2G} \in L^1(\mathbb{R}^3). \tag{5.3}$$

Thus, $\rho_\epsilon^{1/2}$ is bounded in $H^1(\mathbb{R}^3)$, as claimed. By weak sequential compactness of bounded sets in the Hilbert space $H^1(\mathbb{R}^3)$ there exists a sequence $\{\epsilon_k\} \subset [0,1]$ such that $\rho_{\epsilon_k}^{1/2} \to u$ weakly in $H^1(\mathbb{R}^3)$. Since $\rho_{\epsilon_k}^{1/2} \to \rho^{1/2}$ strongly in $L^2(\mathbb{R}^3)$, $u=\rho^{1/2}$, and thus $\rho^{1/2} \in H^1(\mathbb{R}^3)$.

For convexity of $\mathcal{D}^N$, we first note that $\rho \in W^{1,1}(\mathbb{R}^3)$, since $\rho^{1/2} \in \mathcal{D}^N$ implies that $\rho=\varphi^2$ for a unique $\varphi \in H^1$, $\varphi \geqslant 0$ almost everywhere, and thus $\|\nabla\rho\|_1 = \|2\varphi\nabla\varphi\|_1 \leqslant 2\|\varphi\|_2\|\nabla\varphi\|_2 < +\infty$. Let $\lambda \in [0,1]$, $\rho_1,\rho_2 \in \mathcal{D}^N$, and set $\rho = \lambda\rho_1+(1-\lambda)\rho_2$. We need to check that $\rho^{1/2} \in H^1(\mathbb{R}^3)$, which is obtained by repeating the regularization argument carried out for $\rho=\rho_\psi \in W^{1,1}(\mathbb{R}^3)$. For that, we need a bound on $|\nabla\rho(\mathbf{r})|^2$ in lieu of Eq. (5.2). We begin with the inequality

$$|\nabla\rho|^2 \leqslant 4\lambda^2\varphi_1^2|\nabla\varphi_1|^2+4(1-\lambda)^2\varphi_2^2|\nabla\varphi_2|^2+8\lambda(1-\lambda)\varphi_1\varphi_2\nabla\varphi_1\cdot\nabla\varphi_2\,,$$

and apply Young's inequality, i.e., $2\varphi_1\varphi_2\nabla\varphi_2\cdot\nabla\varphi_2 \leqslant \varphi_1^2|\nabla\varphi_1|^2+\varphi_2^2|\nabla\varphi_2|^2$, and then rearrange to get

$$|\nabla\rho|^2 \leqslant 4\rho(\lambda|\nabla\varphi_1|^2+(1-\lambda)|\nabla\varphi_2|^2)\,,$$

which is precisely what is needed. □

Remark 5.3 The need for some regularization to an everywhere positive density comes from the fact that the commonly used formula $\nabla\rho^{1/2}=\frac{1}{2}\rho^{-1/2}\nabla\rho$ only makes sense if $\{\mathbf{r}\in\mathbb{R}^3 \mid \rho(\mathbf{r})=0\}$ has zero measure. When this can be verified, such as for $\rho=\rho_\psi$ everywhere strictly positive, the proof of the above result is somewhat simpler [16, Theorem 1.1].

Throughout this section, we let $X := L^1(\mathbb{R}^3)\cap L^3(\mathbb{R}^3)$, a Banach space with norm $\|\rho\|_X := \max(\|\rho\|_1,\|\rho\|_3)$. We note that the dual is $X^* = L^{3/2}(\mathbb{R}^3)+L^\infty(\mathbb{R}^3)$ with norm $\|v\|_{X^*}=\inf\{\|f\|_\infty+\|g\|_{3/2} \mid f+g=v\}$, also a Banach space [18].

Proposition 5.1 $\mathcal{D}^N \subset X$.

Proof We need only show that $\rho \in L^3(\mathbb{R}^3)$. But $H^1(\mathbb{R}^3)\subset L^6(\mathbb{R}^3)$ by the Sobolev embedding theorem, i.e. $\rho\in\mathcal{D}^N$ implies $\|\rho^{1/2}\|_6=\|\rho\|_3^{1/2}<+\infty$. □

5.2.3 Constrained Search and Skew Conjugate Pairs

Having a well-defined space of densities, we now introduce *constrained search* into Eq. (5.1), i.e.,

$$E[v] = \inf_{\rho \in X} \left\{ F_{\mathrm{LL}}[\rho] + \int v\rho \right\}, \tag{5.4}$$

where the Levy–Lieb functional is defined for any measurable ρ by the expression

$$\begin{aligned} F_{\mathrm{LL}}[\rho] &:= \inf_{\psi \mapsto \rho} \{\mathcal{E}_0[\psi]\} \\ &= \inf \left\{ \mathcal{E}_0[\psi] \;\middle|\; \psi \in \mathcal{W}_N,\ \rho_\psi = \rho \text{ a.e.} \right\}. \end{aligned}$$

We use the convention that $F_{\mathrm{LL}}[\rho] = +\infty$ whenever $\rho \notin \mathcal{D}^N$.

We now have that $E : X^* \to \mathbb{R}$ is given as a pointwise infimum over a nonempty family of continuous affine functions. Such functions are automatically concave and upper semicontinuous.

For any pair $(v, \rho) \in X^* \times X$ we have the inequality $E[v] \leqslant F_{\mathrm{LL}}[\rho] + \int v\rho$. Rearranging, we can define a new function $F : X \to \mathbb{R} \cup \{+\infty\}$ by the expression

$$F[\rho] := \sup_{v \in X^*} \left\{ E[v] - \int v\rho \right\}, \tag{5.5a}$$

which is then automatically *convex* and *lower* semicontinuous, satisfying $F[\rho] \leqslant F_{\mathrm{LL}}[\rho]$. As we will show, one also has

$$E[v] = \inf_{\rho \in X} \left\{ F[\rho] + \int v\rho \right\}. \tag{5.5b}$$

For an extended-valued map such as F, we define the *effective domain* $\mathrm{dom}(F)$ to be the points where F is real-valued.

Definition 5.4 (Skew Conjugation) Let B be a Banach space with dual B^*, consisting of the continuous linear functionals on B. Denote by $\langle \cdot, \cdot \rangle$ the dual pairing. For any $f : B \to \mathbb{R} \cup \{+\infty\}$ not identically $+\infty$ (f is then called proper) we define the (skew) concave conjugate functional $f^\wedge : B^* \to \mathbb{R} \cup \{-\infty\}$ by

$$f^\wedge[x^*] := \inf_{x \in B} \left\{ f[x] + \langle x^*, x \rangle \right\}.$$

For any $g : B^* \to \mathbb{R} \cup \{-\infty\}$, we define the (skew) convex conjugate functional $g^\vee : B \to \mathbb{R} \cup \{+\infty\}$ by

$$g^\vee[x] := \sup_{x^* \in B^*} \left\{ g[x^*] - \langle x^*, x \rangle \right\}.$$

A pair of functionals (f, g) satisfying $f = g^\vee$ and $g = f^\wedge$ are said to be a *skew-conjugate pair of functionals*.

The above introduced concept of (skew) convex/concave conjugate functionals and its notation is slightly unconventional [23], but useful in DFT. The *conventional* convex conjugate (Legendre–Fenchel transform) of $f : B \to \mathbb{R} \cup \{+\infty\}$ is the function $f^* : B^* \to \mathbb{R} \cup \{\pm\infty\}$ given by

$$f^*[x^*] := \sup_{x \in B} \left\{ \langle x^*, x\rangle - f[x] \right\}. \tag{5.6}$$

We introduce the class $\Gamma_0(B)$ of proper convex lower semicontinuous functions $f : B \to \mathbb{R} \cup \{+\infty\}$, meaningful for the Banach space B but also for general topological vector spaces such as B^*_w, the dual space B^* equipped with the weak-$*$ topology. A central fact is that the Legendre–Fenchel transformation is a bijection between these two classes of functions. This result gives insight into the role of convex conjugation in DFT.

Theorem 5.5 *The Legendre–Fenchel transform $f \mapsto f^*$ is a bijection between $\Gamma_0(B)$ and $\Gamma_0(B^*_w)$, and $(f^*)^* = f$, as well as $(g^*)^* = g$ for any $g \in \Gamma_0(B^*_w)$.*

Proof We mention two facts. First, $f \in \Gamma_0(B)$ if and only if $f \in \Gamma_0(B_w)$, where B_w is B equipped with the weak topology. This follows from f being lower semicontinuous if and only if f has closed convex sublevel sets. By Barbu and Precupanu [2, Proposition 1.73], norm-closed convex sets are weakly closed and vice versa. Second, a fact from functional analysis [24, Theorem IV.20] is that the dual of B_w is B^*_w, and that $(B^*_w)^* = B_w$.

From [2, Corrolary 2.21], if V is a locally convex topological vector space, $f : V \to \mathbb{R} \cup \{+\infty\}$ is proper if and only if $f^* : V^* \to \mathbb{R} \cup \{+\infty\}$ is proper. Since f^* is the supremum of a family of lower semicontinuous functions, $f \in \Gamma_0(B_w)$ implies $f^* \in \Gamma_0(B^*_w)$. By reflexivity the same argument gives that $g \in \Gamma_0(B^*_w)$ implies $g^* \in \Gamma_0(B_w)$. The biconjugation theorem (Fenchel–Moreau) for locally convex spaces [2, Theorem 2.22] can for reflexive locally convex topological spaces be phrased as $f \in \Gamma_0(V)$ implies $(f^*)^* = f$. Hence, the conjugation is a bijection between $\Gamma_0(B_w)$ and $\Gamma_0(B^*_w)$ and vice versa. Together with $\Gamma_0(B) = \Gamma_0(B_w)$, the bijection of $\Gamma_0(B)$ and $\Gamma_0(B^*_w)$ has been established. □

Theorem 5.5 can be reformulated in terms of skew conjugation. In the following, we write $f \in -\Gamma_0(B)$ if $-f \in \Gamma_0(B)$, and so on.

Proposition 5.2 *Let B be a Banach space.*

1. *Let $f \in \Gamma_0(B)$. Then $f^\wedge \in -\Gamma_0(B^*_w)$, and $f^\wedge[y] = -f^*[-y]$.*
2. *Let $g \in -\Gamma_0(B^*_w)$. Then $g^\vee \in \Gamma_0(B)$ and $g^\vee[x] = (-g)^*[-x]$.*
3. *The map $f \mapsto f^\wedge$ is a bijection between $\Gamma_0(B)$ and $-\Gamma_0(B^*_w)$, and the map $g \mapsto g^\vee$ is a bijection between $-\Gamma_0(B^*_w)$ and $\Gamma_0(B)$.*

Proof Proof of 1:

$$f^\wedge[y] = \inf_{x\in B}\{f[x] + \langle y, x\rangle\} = \inf_{x\in B}\{-(-f[x]) - \langle -y, x\rangle\}$$
$$= -\sup_{x\in B}\{\langle -y, x\rangle - f[x]\} = -f^*[-y] .$$

Proof of 2:

$$g^\vee[x] = \sup_{y\in B^*}\{g[y] - \langle y, x\rangle\} = \sup_{y\in B^*}\{\langle y, -x\rangle - (-g[y])\}$$
$$= (-g)^*[-x] .$$

Proof of 3: Follows from 1, 2, and Theorem 5.5. □

It is readily seen that $F = E^\vee = (F^\wedge)^\vee = (F_{\mathrm{LL}}^\wedge)^\vee$, and thus that $F = (F_{\mathrm{LL}}^\wedge)^\vee$. F is distinct from F_{LL}, as the latter function is not convex. We have the following characterization:

Theorem 5.6 *For any $f : B \to \mathbb{R}\cup\{+\infty\}$, $f^{**} = (f^\wedge)^\vee \in \Gamma_0(B)$ is the largest convex lower semicontinuous minorant of f, often called the convex envelope or closed convex hull of f.*

Proof See Theorem 6.15 in Ref. [29]. □

A central result in Lieb's analysis (with a proof attributed to Simon), is the following:

Theorem 5.7 *$F = F_{\mathrm{DM}}$, the density-matrix constrained-search functional defined by*

$$F_{\mathrm{DM}}[\rho] := \inf_{\gamma\mapsto\rho}\left\{\mathrm{Tr}(\widehat{H}[0]\gamma)\right\},$$

where the infimum extends over all trace-class operators $\mathcal{D}^N_{op}$ on the form $\gamma = \sum_k^\infty \lambda_k \left|\psi_k\right\rangle\left\langle\psi_k\right|$, with $0 \leqslant \lambda_k \leqslant 0$, $\sum_k \lambda_k = 1$, and $\{\psi_k\} \subset \mathcal{W}_N$ an L^2-orthonormal sequence. The notation $\gamma \mapsto \rho$ means that $\sum_k \lambda_k \rho_{\psi_k} = \rho$ almost everywhere. The effective domain of F_{DM} is $\mathcal{D}^N$.

Proof See Theorem 4.4 in Ref. [16], in which it is proven that F_{DM} is lower semicontinuous. Since it is also convex, we must have $F = F_{\mathrm{DM}}$ by Theorem 5.6. □

The ground-state problem can be written as a minimization over the set $\mathcal{D}^N_{\mathrm{op}}$ of density operators as

$$E_{\mathrm{DM}}[v] := \inf_{\gamma\in\mathcal{D}^N_{\mathrm{op}}} \mathrm{Tr}(\widehat{H}[v]\gamma) = \sum_k \lambda_k \left(\mathcal{E}_0[\psi_k] + \mathcal{V}[\psi_k]\right). \tag{5.7}$$

Proposition 5.3 *Equations* (5.7) *and* (5.1) *define the same function,* $E[v] = E_{\mathrm{DM}}[v]$ *for all* $v \in X^*$.

Proof Clearly, $E_{\mathrm{DM}}[v] \leqslant E[v]$, since $\mathcal{E}_0[\psi] + \mathcal{V}[\psi] = \mathrm{Tr}(\widehat{H}[v] \,|\psi\rangle\langle\psi|)$, i.e., the search domain is larger in Eq. (5.7). On the other hand, $E_{\mathrm{DM}}[v] \geqslant E[v]$, since for any $\gamma = \sum_k \lambda_k \,|\psi_k\rangle\langle\psi_k|$, $\mathrm{Tr}(\widehat{H}[v]\gamma) = \sum_k \lambda_k[\mathcal{E}_0[\psi_k] + \mathcal{V}[\psi_k]] \geqslant \sum_k \lambda_k E[v] = E[v]$. □

Both F_{LL} and F_{DM} have the important property that they are *expectation-valued*, i.e., that the infimums in their definition are attained [10] as expectation values of unique states:

Theorem 5.8 *For every* $\rho \in \mathcal{D}^N$, *there exists a unique* $\psi_\rho \in \mathcal{W}_N$ *such that* $F_{\mathrm{LL}}[\rho] = \langle\psi_0|\widehat{H}[0]|\psi_0\rangle$, *and a unique* $\gamma_\rho = \sum_k \lambda_k \,|\psi_k\rangle\langle\psi_k|$ *such that* $F_{\mathrm{DM}}[\rho] = \mathrm{Tr}(\gamma\widehat{H}[0]) = \sum_k \lambda_k\langle\psi_0, \widehat{H}[0]\psi_0\rangle$, *with* $\psi_k \in \mathcal{W}_N$.

Proof Theorem 3.3 and Corollary 4.5(ii) in Ref. [16]. □

5.2.4 Sub- and Superdifferentiability

Equations (5.5a) and (5.5b) cannot in general be differentiated to find a critical point condition, even if this is routinely done in the physics literature, see for example the classic monograph [22] by Parr and Yang. Indeed, neither F nor E are differentiable in general. However, for convex optimization problems, the weakest useful notion of differentiability is not the usual Gâteaux or Fréchet differentiability, but that of subdifferentiability.

Definition 5.9 Let B and C be topological vector spaces. Let $f : B \to \mathbb{R} \cup \{+\infty\}$, and let $x \in B$. The *subdifferential* of f at x is the set

$$\underline{\partial} f[x] = \left\{ y \in B^* \mid \forall x' \in B,\ f[x] + \langle y, x' - x\rangle \leqslant f[x'] \right\}.$$

The elements are called *subgradients*. Similarly, the *superdifferential* of $g : C \to \mathbb{R} \cup \{-\infty\}$ at x is the set

$$\overline{\partial} g[x] = \left\{ y \in C^* \mid \forall x' \in C,\ g[x] + \langle y, x' - x\rangle \geqslant g[x'] \right\}.$$

The elements are called *supergradients*.

In other words, the subdifferential is the set of slopes of tangent functionals of f at x that are nowhere above the graph of f, i.e., below-supporting tangent functionals. Similarly, the superdifferential consists of the set of slopes of above-supporting tangent functionals. Note, that for $f \in \Gamma_0(B^*_w)$, $C = B^*$ with the weak-$*$ topology, and thus $\overline{\partial} f[x] \subset B$.

Important properties of the subdifferential are summarized in the following proposition, whose proof is so easy we skip it:

Proposition 5.4 *Suppose* $f \in \Gamma_0(B)$, $g = f^\wedge \in -\Gamma_0(B^*_w)$ *(such that* $f = g^\vee$*) form a skew-conjugate pair of functionals. Then the following hold:*

1. *The subdifferential* $\underline{\partial} f : B \to 2^{B^*}$ *is a monotone operator, i.e., for all* $x_1, x_2 \in B$*, and for all* $y_i \in \underline{\partial} f[x_i] \subset B^*$,
$$\langle y_1 - y_2, x_1 - x_2 \rangle \geqslant 0 . \tag{5.8}$$
The superdifferential $\overline{\partial} g : C \to 2^{C^*}$ *is similarly monotone, i.e., for all* $x_1, x_2 \in C$*, and for all* $y_i \in \overline{\partial} g[x_i] \subset C^*$,
$$\langle y_1 - y_2, x_1 - x_2 \rangle \leqslant 0 . \tag{5.9}$$
If f (g) *is additionally strictly convex (concave), then Eq.* (5.8) *[*(5.9)*] holds strictly if* $x_1 \neq x_2$.
2. *Let* $x \in B$ *and* $y \in B^*$ *be given. Then Fenchel's inequality holds,*
$$g[x] - f[y] \leqslant \langle y, x \rangle .$$
3. *The following are equivalent:*
    a. $g[y] - f[x] = \langle y, x \rangle$.
    b. $-y \in \underline{\partial} f[x]$.
    c. $x \in \overline{\partial} g[y]$.
    d. $f[x] + \langle y, x \rangle \leqslant f[x'] + \langle y, x' \rangle$ *for all* $x' \in B$.
    e. $g[y] - \langle y, x \rangle \geqslant g[y'] - \langle y', x \rangle$ *for all* $y' \in B^*$.

Proof Easy exercise. □

The equivalence of 3b and 3c is particularly important, showing that two optimization problems are equivalent. Applied to exact DFT, we obtain the equivalence
$$-v \in \underline{\partial} F[\rho] \iff \rho \in \overline{\partial} E[v] \iff E[v] = F[\rho] + \int v\rho ,$$
giving the critical point conditions of Eqs. (5.5a) and (5.5b). We also have the following result, which relates the optimality condition to ground states of density operator form [10]:

Proposition 5.5 *Let* $\rho \in \mathcal{D}^N$ *and* $v \in X^*$ *be given. Define* $G[v] \subset \mathcal{D}^N_{op}$ *as the set of minimizers for Eq.* (5.7).

1. $G[v] = \mathrm{co}\{|\psi\rangle\langle\psi| \mid \psi \in \mathcal{W}_N,\ E[v] = \langle \psi, \widehat{H}[v]\psi \rangle\}$*, the convex hull of pure-state ground-state density operators.*
2. $E[v] = F[\rho] + \int v\rho$ *if and only if there is a* $\gamma \in G[v]$*, with* $\gamma \mapsto \rho$.

Proof Proof of 1: Clearly the stated convex hull is a subset of $G[v]$, since for $\gamma = \sum_k \lambda_k |\psi_k\rangle\langle\psi_k|$ with all the ψ_k ground-states, $\mathrm{Tr}(\widehat{H}[v]\gamma) = E[v]$. On the other hand, assume that γ is not in this convex hull. Then for at least one ψ_k in its decomposition, $\langle\psi_k, \widehat{H}[v]\psi_k\rangle > E[v]$, and $\mathrm{Tr}(\widehat{H}[v]\gamma) > E[v]$. Proof of 2: If: If $G[v]$ is empty, then $E[v] < F[\rho_\gamma] + \int \rho_\gamma v$ for any γ. Let $\gamma \in G[v]$, $\gamma \mapsto \rho$. Then $E[v] = \mathrm{Tr}(\widehat{H}[v]\gamma) = \mathrm{Tr}(\widehat{H}[0]\gamma) + \int \rho v$. Now, by definition $F[\rho] \leqslant \mathrm{Tr}(\widehat{H}[0]\gamma)$, but clearly equality must hold, otherwise we obtain a contradiction. Thus $E[v] = F[\rho] + \int \rho v$. Only if: Since F is expectation-valued (Theorem 5.8), there is a $\gamma \mapsto \rho$ such that $F[\rho] = \mathrm{Tr}(\widehat{H}[0]\gamma)$, and hence $E[v] = \mathrm{Tr}(\widehat{H}[0]\gamma) + \int \rho v = \mathrm{Tr}(\widehat{H}[v]\gamma)$. Thus, $\gamma \in G[v]$. □

5.2.5 Regularity of E and F

We next state some regularity results for convex functions in $\Gamma_0(B)$, with B being a Banach space throughout this section.

Definition 5.10 (Gâteaux and Fréchet Differentiability) A proper functional $f : B \to \mathbb{R} \cup \{+\infty\}$ is called *Gâteaux differentiable* at a point $x \in \mathrm{dom}(f)$ if, for all h in B, the directional derivative $f'(x; h) = \lim_{t\to 0^+} t^{-1}[f(x+th) - f(x)]$ exists, such that $f'(x; h) = \langle\nabla f(x), h\rangle$ for some $\nabla f(x) \in B^*$ called the *Gâteaux derivative*. If additionally

$$f(x+h) = f(x) + \langle\nabla f(x), h\rangle + o(\|h\|)$$

as $\|h\| \to 0$, then f is said to be *Fréchet differentiable* at x.

Note that we require the directional derivative to be a continuous linear functional. Some authors do not require the directional derivative to be continuous or even not linear in the definition of Gâteaux differentiability.

Theorem 5.11 *Let $f \in \Gamma_0(B)$, and $x \in B$. If f is Gâteaux differentiable at x then $\underline{\partial} f(x) = \{g\}$, a singleton, where $g = \nabla f(x_0)$, the Gâteaux derivative. Conversely, if f is continuous at x_0 and $\underline{\partial} f(x_0)$ is a singleton, then f is Gâteaux differentiable at x_0, and $\{\nabla f(x_0)\} = \underline{\partial} f(x_0)$.*

Proof See Proposition 2.40 in Ref. [2]. □

Remark 5.12 Note the continuity requirement in the converse statement. The condition that the subgradient must be a singleton is not sufficient to guarantee Gâteaux differentiability.

The following theorem demonstrates that convexity together with local boundedness above is quite a strong assumption on an $f \in \Gamma_0(B)$.

Theorem 5.13 *Suppose $f \in \Gamma_0(B)$, with B a Banach space.*

1. *If f is locally bounded above near $x \in B$, then f is locally Lipschitz continuous near x, and $\underline{\partial} f(x)$ is nonempty.*
2. *If f is defined on a convex open set $C \subset B$, and is locally bounded above near* some *$x \in C$, then f is locally Lipschitz near* all *of $x \in C$.*

Proof Proof of 1: Let $x \in B$ be given, and let B be the closed ball of radius $\delta > 0$ around x. Suppose $f(x) \leqslant M$ in B. Let $x + h \in B$ be arbitrary. Since $x = (x+h)/2 + (x-h)/2$, convexity of f gives $f(x) \leqslant f(x+h)/2 + f(x-h)/2$. Thus

$$f(x+h) \geqslant 2f(x) - f(x-h) \geqslant 2f(x) - M,$$

which implies

$$|f(x+h)| \leqslant \max\{M, M - 2f(x)\} \leqslant |M| + 2|f(x)| =: M'.$$

Let $B' \subset B$ be a slightly smaller concentric ball of radius $\delta - \epsilon$. Let $y_1, y_2 \in B'$. Consider the point

$$z = y_1 + \frac{\epsilon}{\|y_1 - y_2\|}(y_1 - y_2).$$

Now $\|z - x\| \leqslant \delta$ so that $x \in B$, and y_2 lies in the open interval (y_1, z). Explicitly, y_2 is given by the convex combination

$$y_2 = \frac{\epsilon}{\epsilon + \|y_2 - y_1\|} y_1 + \frac{\|y_2 - y_1\|}{\epsilon + \|y_2 - y_1\|} z.$$

Convexity of f now gives

$$f(y_2) \leqslant \frac{\epsilon}{\epsilon + \|y_2 - y_1\|} f(y_1) + \frac{\|y_2 - y_1\|}{\epsilon + \|y_2 - y_1\|} f(z),$$

which after rearrangement gives

$$f(y_2) - f(y_1) \leqslant (f(z) - f(y_2))\|y_2 - y_1\| \leqslant \frac{2M'}{\epsilon}\|y_2 - y_1\|.$$

Repeating the argument with y_1 and y_2 interchanged gives the desired Lipschitz continuity.

The existence of subgradients near x is a consequence of the geometric form of the Hahn–Banach theorem, see, e.g., Theorem 1.36 in Ref. [2]. For the complete proof, see Proposition 2.36 in Ref [2].

Proof of 2: Let B be a ball of radius δ around $x \in C$ on which f is locally bounded above by a constant M. Let $y \in C$. There exists a $z \in C$ such that $y =$

$\lambda z + (1-\lambda)x$ for some $\lambda \in [0,1]$. Now, it is straightforward to see, that for all $x' \in C$ such that $\|x' - y\| \leqslant (1-\lambda)\delta$, $f(x') \leqslant \max\{M, f(z)\}$. Thus, f is locally bounded above near y, and by (1), locally Lipschitz near y. Since y was arbitrary, we are done. □

We now discuss the behavior of the functionals $E \in -\Gamma_0(X^*_w)$ and $F \in \Gamma_0(X)$ in terms of classical differentiability and subdifferentials. Note that E is upper semicontinuous in the weak-∗ topology on X^*, and indeed by Barbu and Precupanu [2, Theorem 2.16] it is continuous in the stronger norm topology on X^* since E is everywhere defined and clearly upper semicontinuous as the infimum of a nonempty family of affine functions. However, we can say even more:

Theorem 5.14

1. *The map $E : X^* \to \mathbb{R}$ is locally Lipschitz continuous.*
2. *Suppose $v \in X^*$. Then E is Gâteaux differentiable at v if and only if $\hat{H}[v]$ has a smallest eigenvalue, and all normalized eigenvectors belonging to this eigenvalue share the same density.*

Proof Proof of 1: E is everywhere finite. In order to apply Theorem 5.13, we need to show that E is locally bounded above at *some* point in X^*, and we choose $v = 0$. The proof is adapted from Ref. [16].

Let $\|v\|_{X^*} < L/6$, where L is the constant in the Sobolev embedding of $H^1(\mathbb{R}^3)$ in $L^6(\mathbb{R}^3)$ ($\|\nabla f\|_2^2 \geqslant L\|f\|_6^2$). Write $v = u + w$ with $u \in L^{3/2}$ and $w \in L^\infty$, and note that $\|u\|_{3/2} + \|w\|_\infty < L/6$. Using the fact that the two-electron repulsion operator is positive, we get

$$
\begin{aligned}
E[v] &= \inf_{\psi \in \mathcal{W}_N} \frac{1}{2}\|\nabla\psi\|_2^2 + \langle\psi, W\psi\rangle + \int \rho_\psi v \\
&\geqslant \inf_{\psi \in \mathcal{W}_N} \frac{1}{2}\|\nabla\psi\|_2^2 + \int \rho_\psi v \\
&\geqslant \inf_{\psi \in \mathcal{W}_N} \frac{1}{2}\|\nabla\psi\|_2^2 - N\|w\|_\infty - \|\rho_\psi^{1/2}\|_6^2\|u\|_{3/2} \\
&\geqslant \inf_{\psi \in \mathcal{W}_N} \frac{1}{2}\|\nabla\psi\|_2^2 - N\|w\|_\infty - L^{-1}\|\nabla\rho^{1/2}\|_2^2\|u\|_{3/2}\,.
\end{aligned}
$$

Using Eq. (5.3), we get

$$
E[v] \geq \inf_{\psi \in \mathcal{W}_N} \frac{1}{2}\|\nabla\psi\|_2^2 - \frac{1}{6}NL - \frac{1}{6}(2\|\nabla\psi\|_2^2 + g) \geq -\frac{1}{6}NL - \frac{1}{6}g,
$$

where $g = \int (2G)^{-1}|\nabla G|^2 < +\infty$, $G(\mathbf{r}) = \exp(-|\mathbf{r}|^2)$. Hence, E is locally bounded above at $0 \in X^*$.

Proof of 2: Since E is continuous, Theorem 5.11 tells us that E is Gâteaux differentiable at $v \in X^*$ if and only if the superdifferential is a singleton. It follows from Proposition 5.5 that $\overline{\partial} E[v]$ is the convex hull of all densities ρ_ψ of all ground-state wavefunctions of $\widehat{H}[v]$. Thus, the superdifferential is a singleton if and only if all ground-state densities are the same. □

The universal functional F is quite badly behaved. Our discussion is mostly based on Ref. [13], in which many more details can be found. Since F is only defined on elements $\rho \in X$ for which $\int \rho = N$, it is clear that F cannot be Gâteaux differentiable, since any change in the particle number $\int \rho$ would give infinities. On the other hand, it could be differentiable in a more restricted sense, e.g., we could consider F as a function on the mentioned affine space and study directional derivatives, or even on the smaller space aff dom F (the affine hull of the effective domain). The following discussion shows that F is singular in these cases, too.

Theorem 5.15 *Let $X_N = \{\rho \in X \mid \int \rho = N\}$, an affine closed space of codimension 1. Let $X_N^+ \subset X_N$ be the subset of those elements that satisfy $\rho \geqslant 0$ a.e.*

1. *The map $F : X \to \mathbb{R} \cup \{+\infty\}$ has effective domain $\mathcal{D}^N$, i.e., $F[\rho] < +\infty$ if and only if $\rho \in \mathcal{D}^N$.*
2. *$\mathcal{D}^N$ is dense in X_N^+.*
3. *$\operatorname{aff} \mathcal{D}^N$ has empty algebraic interior in $\operatorname{aff} X_N^+ = X_N$. The algebraic interior of a subset $A \subset X_+$ consists of the points $\rho \in A$ such that, for every line $\ell \subset X_N$ through ρ, $\ell \cap A$ contains a line segment with ρ in its interior.*

Proof Proof of 1: The domain of F_{DM} is the convex hull of the domain of F_{LL}, which is $\mathcal{D}^N$, a convex set.

Proof of 2: For any $\sigma \in X_N^+$, we must find $\rho_\epsilon \in \mathcal{D}^N$ such that $\|\rho_\epsilon - \sigma\| \to 0$ as $\epsilon \to 0$. This can be done using standard mollification arguments: Let $g_\epsilon \in C_c^\infty(\mathbb{R}^3)$ be a mollifier, and define $\varphi = \rho^{1/2}/N^{1/2}$, $\psi_\epsilon = g_\epsilon * \varphi$, $\varphi_\epsilon = \psi_\epsilon/\|\psi_\epsilon\|$, $\rho_\epsilon = N\varphi_\epsilon^2$. Now, $\varphi_\epsilon \in H^1$, $\|\varphi_\epsilon\|_2^2 = 1$, and moreover $\varphi_\epsilon \geqslant 0$ a.e., so $\rho_\epsilon \geqslant 0$ a.e., and hence $\rho_\epsilon \in \mathcal{D}^N$. Furthermore, $\varphi_\epsilon \to \varphi$ in H^1, and $\|\rho_\epsilon - \rho\|_3 \leqslant C\|\varphi_\epsilon - \varphi\|_{H^1}^2$ by the same Sobolev embedding used in Theorem 5.2. Similarly, $\|\rho_\epsilon - \rho\|_1 \to 0$.

Proof of 3, adapted from Ref. [13]: Let $\rho_0 \in \mathcal{D}^N$ be given. We need to find a line segment $\{\rho_0 + s\delta\rho \mid s \in [0, 1]\} \in \operatorname{aff} \mathcal{D}^N$ such that $\rho_0 + s\delta\rho \notin \mathcal{D}^N$ for any $s > 0$.

Let $\sigma \in C^\infty(\mathbb{R}^3)$ be such that $\sigma(\mathbf{r}) \in [0, 1]$, $\sigma(\mathbf{r}) = 0$ outside the ball around the origin with unit volume, $\sigma(\mathbf{r}) \geqslant 1/2$ inside the ball around the origin with volume $1/2$ (implying that $\int \sigma \geqslant 1/4$). Furthermore, we require that $\sigma^{1/2} \in H^1$. Find a sequence B_n of balls of volume $\mu(B_n) = 2^{-n}$ with centers $\mathbf{r}_n$ satisfying $|\mathbf{r}_n - \mathbf{r}_m| \geqslant 1$ for all n, m. Additionally we require that the measure of $\{\rho_0 \geqslant 2^{-3n}\} \cap B_n \leqslant \mu(B_n)/4 = 2^{-n-2}$ (by taking a subsequence if necessary). Such a sequence exists, since otherwise ρ_0 cannot have a finite integral.

We now define

$$\delta\rho(\mathbf{r}) = \tau(\mathbf{r}) - \sum_n 2^{-2n}\sigma(2^n(\mathbf{r} - \mathbf{r}_n)),$$

where τ is smooth, nonnegative and with support that does not intersect any of the B_n, such that $\int \delta\rho = 0$. Moreover, we require that $\tau^{1/2} \in H^1$. Now, $\delta\rho \in \operatorname{aff} \mathcal{D}^N$.

On any given B_n, there is a region (in the inner part) of positive measure where

$$\rho(\mathbf{r}) \leqslant 2^{-3n} - s2^{-2n-1}.$$

Clearly, for any given $s > 0$ we can take n sufficiently large so that these values are negative. Thus, $\rho \notin \mathcal{D}^N$ for any $s > 0$. □

Remark 5.16 The construction of the direction $\delta\rho$ that immediately exits $\mathcal{D}^N$ exploits the requirement that $\rho \geqslant 0$ almost everywhere, and thus a pathology of the *domain* as opposed to the behavior of F *on the domain*. It is a fact that the restriction of F to $\mathcal{D}^N$ (with the X-topology) is everywhere discontinuous in the sense that we can construct a sequence ρ_n in $\mathcal{D}^N$ which is X-convergent to some $\rho \in X_N^+$, but for which $F[\rho_n] \to +\infty$. Such a construction is outlined in Ref. [13]. This also indicates that the topology of X is not really that well suited for DFT, since unlike F, the chosen topology on X is insensitive to density gradients.

5.2.6 *The Conjecture of Hohenberg and Kohn*

The last discussion in this section pertains to the structure of $\underline{\partial} F[\rho]$, and this is the domain of the Hohenberg–Kohn theorem [7, 16, Theorem 3.2]: That the external potential v is a unique function of the density ρ, up to a constant shift. This is a very appealing notion, because if ρ determines v, then it determines also $\widehat{H}[v]$ and hence the ground-state wavefunction ψ, and hence all physical observables as functions $\Omega[\rho]$. The system density is elevated to a basic variable, replacing the wavefunction as the state parameter of the quantum mechanics of N-electron systems. Unfortunately, the Hohenberg–Kohn theorem as originally stated is a *conjecture*, since its proof has one step which is not rigorous: One divides the Schrödinger equation by ψ pointwise, but this requires the unique continuation property, see [16, Theorem 3.2]. At the time of writing it is still a partially open question if the Hohenberg–Kohn conjecture is true for potentials $v \in X^*$, although Garrigue has established the unique continuation property for potentials $v \in L^p_{\mathrm{loc}}(\mathbb{R}^3)$ with $p > 2$, which include the Coulomb potentials. See also the discussion by Lammert [15].

Conjecture 5.17 (Hohenberg–Kohn) *The density $\rho \in \mathcal{D}^N$ determines the potential $v \in X^*$ up to an additive constant, i.e., we have either*

$$\underline{\partial} F[\rho] = \{-v + \mu \mid \mu \in \mathbb{R}\} \quad \textit{for some } v \in X^*,$$

or

$$\underline{\partial} F[\rho] = \emptyset ,$$

whenever there are no potentials for which $\rho \in \mathcal{D}^N$ is a ground-state density.

The concept of v-representability has received a lot of attention in the DFT literature.

Definition 5.18 (Ensemble v-Representability) We say that $\rho \in X$ is *(ensemble) v-representable* if there is a $v \in X^*$ such that $\rho = \rho_\gamma$ for some $\gamma \in G[v]$, i.e., that γ is a ground-state density operator of v. Equivalently,

$$E[v] = F[\rho] + \int v\rho ,$$

or, $-v \in \underline{\partial} F[\rho]$, or $\rho \in \overline{\partial} E[v]$. The set of (ensemble) v-representable densities is thus $\mathcal{B}_N := \text{dom}(\underline{\partial} F)$ (defined as those $\rho \in X$ for which $\underline{\partial} F[\rho] \neq \emptyset$).

Remark 5.19 The concept of (ensemble) v-representability introduced here is slightly less general than the one used in the classical DFT literature. Here, $\rho \in L^1$ is (ensemble) v-representable if it is the ground-state density of "some potential" $v : \mathbb{R}^3 \to \mathbb{R}$, with the function space otherwise unspecified. For example, a gaussian density is the ground-state density of a harmonic potential (for $N = 1$), but this potential is not in X^*. Lammert [14] distinguishes between (ensemble) v-representability and (ensemble) X^*-representability, which is identical to our concept.

One of the classical problems of DFT is to characterize the set $\mathcal{B}_N$, as this is the effective domain of the classical Hohenberg–Kohn functional F_{HK}, defined by restricting the effective domain of F to $\mathcal{B}_N$. The motivation for this study has been that since one wishes to differentiate F_{HK}, the set $\mathcal{B}_N$ needs to be sufficiently well-behaved. In particular, it needs to have an algebraic interior for directional derivatives to make sense. Since we have $F \leqslant F_{\text{HK}}$ and in particular $\text{dom}(F_{\text{HK}}) \subset \text{dom}(F)$, the algebraic interior of $\mathcal{B}_N$ must be empty. However, we can say the following based on the Brøndsted–Rockafellar theorem [1, Theorem 16.45], which implies that for $f \in \Gamma_0(B)$ (with B a Banach space), $\text{dom}(\underline{\partial} f)$ is dense in $\text{dom}(f)$.

Proposition 5.6 *$\mathcal{B}_N$ is dense in $\mathcal{D}^N$.*

Proof Simply note that $\mathcal{B}_N = \text{dom}(\underline{\partial} F)$ and $\mathcal{D}^N = \text{dom}(F)$. □

The Hohenberg–Kohn conjecture is often taken to be the foundation of DFT in the sense that it implies the existence of a universal Hohenberg–Kohn functional F_{HK}, but the present author considers this a red herring: Our developments so far have not relied on it, and there is actually not much extra to gain from elevating the conjecture to a theorem, since the mapping $\rho \mapsto v$ would be very ill-behaved, see the discussion by Lammert in Ref. [13]. First, not all $v \in X^*$ have ground states.

Second, not all $\rho \in \mathcal{D}^N$ can be ground-state densities ("v-representable"), and we have little knowledge of how the one-dimensional affine space $\underline{\partial} F[\rho]$ changes with ρ.

5.2.7 Some Remarks on Exact DFT

The previous section sets up exact DFT as a convex optimization problem, demonstrating a symmetry between the ground-state energy map $E[v]$ and the "universal" functional $F[\rho]$. The framework differs significantly from the "traditional" treatment that was initiated by the publications of Hohenberg, Kohn, and Sham [7, 11].

The framework has both strengths and weaknesses from the point of view of formulating the quantum physics of N-electron systems in external fields. First, it is a mathematically rigorous formulation of exact DFT that pinpoints some of the deficiencies of the "traditional" formulation of DFT, in terms of a well-established body of results.

Second, the potential space X^* feels both unnecessarily large and too small at the same time, containing not only the physical Coulomb potentials, but also a host of "wild" potentials that would never appear in actual physical problems, while lacking some obvious interesting potentials, like the harmonic oscillator. At the same time, the constrained-search formula (5.4) is valid for the latter, which pose no particular difficulties for quantum theory. Actually, Eq. (5.4) is valid for much more general potentials. For example, for $v = v_+ - v_- \in X^*$ with $v_\pm \in X^*$ almost everywhere positive, one may modify v_+ or v_0 (but not both) to arbitrary elements of $L^1_{\text{loc}}(\mathbb{R}^3)$ (even if we may not have a link to a self-adjoint $\widehat{H}[v]$ in such cases). It seems fortuitous that X^* happens to contain the most interesting potentials, namely the Coulomb potentials, as X is chosen for its capacity to hold $\mathcal{D}^N$.

Third, the usual norm topology on X is not so suitable for DFT. It seems difficult to introduce a "well-behaved" Banach space of densities with a dual that contains correspondingly "mostly" interesting potentials. Thus, the standard convex analytic setting could be refined to allow for a better description of exact DFT.

Finally, it seems hard to connect the above treatment of exact DFT with the development of density-functional approximations. Indeed, it is not clear that the existing functional approximations are approximations to F_{DM} (and one may suspect that they are not).

At this point, it is worthwhile to mention alternative formulations of DFT, in particular the coarse-grained formulation of Lammert [12, 14]. Coarse-graining can be motivated by the fact that (a) experimental resolution is not infinite, and (b) the nucleus is not a point particle, so that the singular Coulomb potential is not even exact. The space $\mathbb{R}^3$ is covered by a disjoint set of cells of uniformly bounded volume, such as a rectangular grid of uniform cells, and one considers equivalence classes $\boldsymbol{\rho}$ of those $\rho \in \mathcal{D}^N$ that have the same average in each cell. The potentials are taken to be constant over each cell. In this setting, every strictly positive $\boldsymbol{\rho}$ is shown to be ensemble v-representable. Moreover, the universal functional becomes much more well-behaved, in particular it is Gâteaux differentiable. Furthermore, Lammert demonstrates that certain limits conforming with Lieb's theory are obtained as the resolution is increased. In summary, Lammert's coarse-grained DFT represents a regularization alternative to Moreau–Yosida regularization which we study in Sect. 5.4

5.2.8 Box Truncated Exact DFT

One basic problem with the spaces X and X^* is that they are nonreflexive, and Moreau–Yosida regularization has the most powerful effect in a Hilbert space setting. We therefore consider the N-electron problem in a open, connected and bounded domain $\Omega \subset \mathbb{R}^3$, assuming Dirichlet boundary conditions on the Schrödinger equation. All the results from exact DFT carry over to this situation. On the other hand, the finite domain allows some simplifications and stronger statements. In particular, the long-range part of Coulomb potentials, being the source of the nonreflexiveness, disappear. In fact, we may now include harmonic potentials and other unbounded potentials in the class of external potentials. Moreover, every Hamiltonian $\widehat{H}[v]$ will have a ground state.

We begin with a simple lemma.

Lemma 5.20 *Let $X(\Omega) := L^1(\Omega) \cap L^3(\Omega)$. Then $X(\Omega) \subset L^2(\Omega)$, for Ω bounded or unbounded, and also $L^2(\Omega) \subset X(\Omega)'$. For Ω bounded, we have $L^2(\Omega) \subset L^1(\Omega)$, with continuous embedding. Furthermore, $L^2(\Omega) \subset X(\Omega)^*$ with continuous embedding.*

Proof Let $u \in X(\Omega)$, i.e., $\|u\|_1$ and $\|u\|_3$ are both finite.

$$\|u\|_2^2 = \int_{|u|\leqslant 1} |u|^2 + \int_{|u|>1} |u|^2 \leqslant \int_{|u|\leqslant 1} |u| + \int_{|u|>1} |u|^3 < +\infty .$$

Thus $u \in L^2(\Omega)$. It is a standard fact that $L^q(\Omega) \subset L^p(\Omega)$ for $1 \leqslant p < q < +\infty$ with continuous embedding for bounded Ω, proven by a simple application of Hölder's inequality. In particular, $L^2(\Omega) \subset L^1(\Omega)$. Moreover, $L^2(\Omega) \subset L^{3/2}(\Omega) \subset X(\Omega)^*$, and since $\|u\|_{3/2} \leqslant \|u\|_{X^*(\Omega)}$, the embedding is continuous. □

The significance of Lemma 5.20 is that it makes sense to consider F_{DM} as a function of $\rho \in L^2(\Omega)$, since $\mathcal{D}^N(\Omega) \subset L^2(\Omega)$ (with an obvious definition of $\mathcal{D}^N(\Omega)$):

Proposition 5.7 *The functional $F_{\mathrm{DM}} : L^2(\Omega) \to \mathbb{R} \cup \{+\infty\}$ is lower semicontinuous.*

Proof Suppose $\rho_n \to \rho$ in $L^2(\Omega)$. Then $\rho_n \to \rho$ in $L^1(\Omega)$ by Lemma 5.20, and $F[\rho] \leqslant \liminf_n F_{\mathrm{DM}}[\rho_n]$ by lower semicontinuity of $F_{\mathrm{DM}} : L^1(\Omega) \to \mathbb{R} \cup \{+\infty\}$. □

The second consequence of Lemma 5.20 is that it makes sense to consider the ground-state energy of the operator $\widehat{H}[v]$ for $v \in L^2(\Omega)$, i.e., the map $E : L^2(\Omega) \to \mathbb{R}$ is meaningful. It is also readily seen that Coulomb potentials are in $L^2(\Omega)$. We summarize this as a theorem:

Theorem 5.21 *For $\Omega \subset \mathbb{R}^3$ bounded, $E : L^2(\Omega) \to \mathbb{R}$ and $F_{\mathrm{DM}} : L^2(\Omega) \to \mathbb{R} \cup \{+\infty\}$ form a skew-conjugate pair of functionals. Moreover, for any $v \in L^2(\Omega)$ there exists a $\rho \in L^2(\Omega)$ such that $E[v] = F_{\mathrm{DM}}[\rho] + \langle v, \rho \rangle_2$ (i.e., $\overline{\partial} E[v] \neq \emptyset$).*

Proof We only need to prove the existence of a ground-state density for any $v \in L^2(\Omega)$. For $v \in L^2(\Omega)$, the self-adjoint N-electron Hamiltonian associated with the Hamiltonian quadratic form has domain $H^2_{N,0}(\Omega)$ [8, Theorem 1]. The Rellich–Kondrachov theorem implies that this Hamiltonian has a compact resolvent, and thus a purely discrete spectrum. In particular, the ground-state energy along with a ground-state eigenvector $\psi \in H^1_{N,0}(\Omega)$ exists for any external potential $v \in L^2(\Omega)$, and thus ρ_ψ is a minimizer for $E[v] = \inf_\rho F_{\mathrm{DM}}[\rho] + \langle v, \rho \rangle_2$. □

One may ask, why not use the unbounded domain $\Omega = \mathbb{R}^3$ when considering $L^2(\Omega)$ as the density and potential space? In this case convergence in L^2 does not imply L^1 convergence, and one opens up the possibility that F_{DM} is not lower semicontinuous, so that the Lieb functional F is strictly different from F_{DM}. This is, at the very least, a conceptual problem. Indeed, from Ref [10] we have the following:

Theorem 5.22 *Let $E : L^2(\mathbb{R}^3) \to \mathbb{R}$ be defined by*

$$E[v] = \inf_{\rho \in L^2(\mathbb{R}^3)} \{F_{\mathrm{DM}}[\rho] + \langle v, \rho \rangle\},$$

and let $F = E^\vee$.

1. *$E[v] \leqslant 0$. If $v \geqslant 0$ almost everywhere, then $E[v] = 0$.*
2. *$F[0] = 0$, and hence F_{DM} is not lower semicontinuous.*
3. *$0 \in \underline{\partial} F[0]$ and $0 \in \overline{\partial} E[0]$, but there are no ground-state wavefunctions for the Hamiltonian $\widehat{H}[0]$.*

Proof Proof of 1: Writing $v = v_+ - v_-$, with $v_\pm \in L^2(\mathbb{R}^3)$ almost everywhere positive, we obtain

$$E[v] = \inf_\rho \{F_{\mathrm{DM}}[\rho] + \langle v_+, \rho \rangle - \langle v_-, \rho \rangle\} \leqslant \inf_\rho \{F_{\mathrm{DM}}[\rho] + \langle v_+, \rho \rangle\} = E[v_+].$$

It is therefore sufficient to show that $E[v_+] \leqslant 0$.

Let $v \geqslant 0$ almost everywhere. Let $\lambda > 0$ be arbitrary and let $\Omega_{k,\lambda}$ with $k \in \mathbb{N}$ be disjoint cubes of side length λ such that $\cup_k \Omega_{k,\lambda} = \mathbb{R}^3$. Each $\Omega_{k,\lambda}$ can be obtained by translation of $\Omega_{1,\lambda}$. Since $v \in L^2(\mathbb{R}^3)$,

$$\int_{\mathbb{R}^3} v(\mathbf{r})^2 \, d\mathbf{r} = \sum_k \int_{\Omega_{k,\lambda}} v(\mathbf{r})^2 \, d\mathbf{r} < +\infty,$$

implying that $\int_{\Omega_{k,\lambda}} v(\mathbf{r})^2\,d\mathbf{r} \to 0$ as $k \to \infty$. Let $\psi \in \mathcal{W}_N$ have smooth components with support in $\Omega_{1,1}^N$ so that $\rho_\psi \in C_c^\infty(\mathbb{R}^3)$ has support contained in $\Omega_{1,1}$. By translating ψ (denoting the result by ψ_k), we can ensure that the support of ψ_k is inside $\Omega_{k,1}^N$ and

$$F_{\mathrm{DM}}(\rho_{\psi_k}) = F_{\mathrm{DM}}(\rho_\psi) \leqslant \langle\psi|T + W|\psi\rangle \equiv \langle T\rangle + \langle W\rangle,$$

independently of k. We obtain

$$E(v) \leqslant \inf_k \left(\langle T\rangle + \langle W\rangle + \int_{\Omega_{k,1}} v(\mathbf{r})\,\rho_k(\mathbf{r})\,d\mathbf{r} \right) = \langle T\rangle + \langle W\rangle,$$

where we have used the fact that

$$\int_{\Omega_{k,1}} v(\mathbf{r})\,\rho_k(\mathbf{r})\,d\mathbf{r} \leqslant \left(\int_{\Omega_{k,1}} v(\mathbf{r})^2\,d\mathbf{r} \right)^{1/2} \|\rho\|_2 \to 0$$

as $k \to \infty$. We now increase the size of the boxes $\Omega_{k,\lambda}$ by varying $\lambda > 1$. By dilating ψ in the manner

$$\psi(\mathbf{r}_1, \cdots) \to \lambda^{-3N/2}\psi(\lambda^{-1}\mathbf{r}_1, \cdots),$$

the support is still inside $\Omega_{1,\lambda}^N$ and the density is scaled as $\rho_\psi(\mathbf{r}) \to \lambda^{-3}\rho_\psi(\lambda^{-1}\mathbf{r})$. We obtain the scaling

$$\langle T\rangle + \langle W\rangle \to \lambda^{-2}\langle T\rangle + \lambda^{-1}\langle W\rangle.$$

By repeating the above argument for $\lambda = 1$ and letting $\lambda \to \infty$, we obtain $E[v] \leqslant 0$. On the other hand, $E[v] \geqslant 0$ since the Hamiltonian $\widehat{H}[v]$ is positive, yielding $E[v] = 0$. Proof of 2: $F_{\mathrm{DM}}[0] = +\infty$, but $F[0] = \sup_v E[v] = 0$. Thus $F[0] = (F_{\mathrm{DM}}^\wedge)^\vee[0] < F_{\mathrm{DM}}[0]$, and $F_{\mathrm{DM}} \notin \Gamma_0(L^2(\mathbb{R}^3))$. Proof of 3: Easy. □

5.3 Moreau–Yosida Regularization

In this section we introduce Moreau–Yosida regularization of convex lower-semicontinuous functions over separable Hilbert spaces. The Moreau–Yosida regularization of a convex optimization problem is *invertible*. Hence, we trade a possibly non-smooth optimization problem for a smooth and indeed quite well-behaved one. The invertibility implies that the solution of the exact problem is connected to the solution of the regularized problem. In this exposition, we present elementary proofs of most statements due to their central nature in this chapter. Most proofs are adapted from the excellent monograph [1] by Bauschke and Combettes, to which the reader is pointed for further details.

5.3.1 The Moreau Envelope

In this section, $(\mathcal{H}, \langle\cdot,\cdot\rangle)$ is a generic separable real Hilbert space. By the usual identification of $\mathcal{H}$ and $\mathcal{H}^*$ and the identification of weak and weak-$*$ topologies, we have $\Gamma_0(\mathcal{H}) = \Gamma_0(\mathcal{H}^*_w)$. Our central objects of study is the Moreau envelope ${}^\epsilon f \in \Gamma_0(\mathcal{H})$ of a convex lower semicontinuous function $f \in \Gamma_0(\mathcal{H})$ and the associated proximal mapping $\mathrm{Prox}_{\epsilon f} : \mathcal{H} \to \mathcal{H}$, introduced by Moreau in Ref [19]. The name Yosida is also attached to the formalism: In his proof of the Hille–Yosida theorem in Ref. [30] characterizing the generators of strongly continuous one-parameter semigroups, Yosida introduced a certain approximation of the resolvent of maximal monotone operators $A : \mathcal{H} \to 2^{\mathcal{H}}$. Moreau's Theorem [19, 26] connects the two concepts in the case where $A = \underline{\partial} f$, see Remark 5.26.

Definition 5.23 (Moreau Envelope) Let $\epsilon > 0$ be given, and let $f : \mathcal{H} \to \mathbb{R} \cup \{+\infty\}$ with nonempty $\mathrm{dom}(f)$. The *Moreau envelope* ${}^\epsilon f : \mathcal{H} \to \mathbb{R} \cup \{\pm\infty\}$ is defined by infimal convolution with the function $x \mapsto (1/2\epsilon)\|x\|^2$,

$$ {}^\epsilon f[x] = \min_{z\in\mathcal{H}} \left(f[z] + \frac{1}{2\epsilon}\|x - z\|^2 \right). \tag{5.10}$$

The unique minimizer Eq. (5.10) for $\epsilon = 1$ is defined as $\mathrm{Prox}_f\, x$, i.e.,

$$ {}^1 f[x] = f[\mathrm{Prox}_f\, x] + \frac{1}{2}\|x - \mathrm{Prox}_f\, x\|^2. \tag{5.11}$$

The map Prox_f is called *the proximal mapping*, and for general $\epsilon > 0$ we have

$$ {}^\epsilon f[x] = f[\mathrm{Prox}_{\epsilon f}\, x] + \frac{1}{2\epsilon}\|x - \mathrm{Prox}_{\epsilon f}\, x\|^2, \tag{5.12}$$

that is, the unique minimizer is in general $\mathrm{Prox}_{\epsilon f}\, x$.

Proposition 5.8 *Definition 5.23 makes sense, i.e., a minimizer of Eq.* (5.10) *exists and is unique.*

Proof The case $\epsilon = 1$ defines $\mathrm{Prox}_f\, x$, and we leave it to the reader to show that the minimizer is $\mathrm{Prox}_{\epsilon f}$ in general. Let $x \in \mathcal{H}$ be given, and let $h_x[z] := f[x] + (1/2\epsilon)\|z - x\|^2$. Let $\{z_n\} \subset \mathcal{H}$ be a minimizing sequence for the infimal convolution at x, i.e., $h_x[z_n] \to {}^\epsilon f[x]$. Such a sequence exists since $\mathrm{dom}(f) \neq \emptyset$. The sequence must be bounded, since $h_x[z_n] \to +\infty$ whenever $\|z_n\| \to +\infty$. By taking a subsequence if necessary, we can assume that z_n converges weakly to some $z_* \in \mathcal{H}$. We note that h_x is sequentially weakly lower semicontinuous (any lower semicontinuous convex function is sequentially weakly lower semicontinuous by [1, Theorem 9.1], and thus

$$h_x[z_*] \leqslant \liminf_n h_x[z_n] = {}^{\epsilon}f[x].$$

Thus, the infimum in the infimal convolution is attained at z_*. To show uniqueness, we use that strictly convex functions have unique minima: Suppose z'_* is a different minimizer. Since h_x is strictly convex, $h_x[\lambda z_* + (1-\lambda)z'_*] < \lambda h_x[z_*] + (1-\lambda)h_x[z'_*]$ for $\lambda \in (0, 1)$. But then $h_x[\lambda z_* + (1-\lambda)z'_*] < {}^{\epsilon}f[x]$, a contradiction. □

Remark 5.24 In the proof of Proposition 5.8, we used that $\mathcal{H}$ is a Hilbert space (more precisely, reflexivity is used) when picking a weakly convergent subsequence. Thus, in the general nonreflexive setting, the proximal mapping may not be well-defined.

5.3.2 *The Proximal Mapping*

Some important properties of the proximal mapping are the following:

Proposition 5.9 *Let $f \in \Gamma_0(\mathcal{H})$ and let $\epsilon > 0$ be given. The following holds:*

1. $p = \mathrm{Prox}_{\epsilon f}\, x \iff$ *for all $x' \in \mathcal{H}$ we have*

$$\epsilon^{-1}\langle x' - p, x - p\rangle + f[p] \leqslant f[x']. \tag{5.13}$$

2. $p = \mathrm{Prox}_{\epsilon f}\, x \iff \epsilon^{-1}(x - p) \in \underline{\partial} f[p]$.
3. $\mathrm{Prox}_{\epsilon f} : \mathcal{H} \to \mathcal{H}$ *is* firmly nonexpansive *[1, Section 4], i.e., for all $x, x' \in \mathcal{H}$,*

$$\|\mathrm{Prox}_{\epsilon f}\, x - \mathrm{Prox}_{\epsilon f}\, x'\|^2 + \|(\mathrm{Id} - \mathrm{Prox}_{\epsilon f})x - (\mathrm{Id} - \mathrm{Prox}_{\epsilon f})x'\|^2 \leqslant \|x - x'\|^2.$$

In particular $\mathrm{Prox}_{\epsilon f}$ *and* $\mathrm{Id} - \mathrm{Prox}_{\epsilon f}$ *are both Lipschitz continuous with constant* 1.

4. *If $x \in \mathrm{dom}(f)$ and $\epsilon \to 0^+$, then*

$$\|x - \mathrm{Prox}_{\epsilon f}\, x\|^2 = O(\epsilon).$$

Proof Proof of 1: Suppose $p = \mathrm{Prox}_f\, x$, and let x' be arbitrary. Set $z = \lambda x' + (1 - \lambda)p$. We have

$$f[p] = \left(\min_{x' \in \mathcal{H}} f[x'] + \frac{1}{2}\|x - x'\|^2\right) - \frac{1}{2}\|x - p\|^2,$$

so that

$$
\begin{aligned}
f[p] &\leqslant f[z] + \frac{1}{2}\|x - z\|^2 - \frac{1}{2}\|x - p\|^2 \\
&\leqslant \lambda f[x'] + (1-\lambda) f[p] + \frac{1}{2}\|x - p + \lambda(p - x')\|^2 - \frac{1}{2}\|x - p\|^2 \\
&= \lambda f[x'] + (1-\lambda) f[p] + \frac{1}{2}\lambda^2\|p - x'\|^2 - \lambda\langle p - x, p - x'\rangle .
\end{aligned}
$$

Hence, for $\lambda > 0$,

$$
f[p] \leqslant f[x'] + \frac{1}{2}\lambda\|p - x'\|^2 - \langle p - x, p - x'\rangle.
$$

Since this holds for all $\lambda \in (0, 1)$, we obtain Eq. (5.13) as $\lambda \to 0$, and substituting ϵf for f. The converse statement is easy: We rearrange Eq. (5.13), to get

$$
\begin{aligned}
f[p] + \frac{1}{2\epsilon}\|x - p\|^2 &\leqslant f[x'] + \frac{1}{2\epsilon}\|x - p\|^2 + \epsilon^{-1}\langle x - p, p - x'\rangle \\
&\leqslant f[x'] + \frac{1}{2\epsilon}\|x - p\|^2 + \epsilon^{-1}\langle p - x, p - x'\rangle \\
&\quad + \frac{1}{2\epsilon}\|p - x'\|^2 \\
&= f[x'] + \frac{1}{2\epsilon}\|x - x'\|^2.
\end{aligned}
$$

Since x' was arbitrary, we conclude that $p = \operatorname{Prox}_{\epsilon f} x$.

Proof of 2: Follows directly from 1 and the definition of the subgradient.

Proof of 3: Assume $p = \operatorname{Prox}_{\epsilon f} x$ and $p' = \operatorname{Prox}_{\epsilon f} x'$. From 1 we get

$$
\epsilon^{-1}\langle p' - p, x - p\rangle + f[p] \leqslant f[p']
$$

and

$$
\epsilon^{-1}\langle p - p', x' - p'\rangle + f[p'] \leqslant f[p] .
$$

Adding these inequalities and rearranging, we get

$$
\epsilon^{-1}\langle p - p', x - p - (x' - p')\rangle \geqslant 0 . \tag{5.14}
$$

Subtracting $\|x - x'\|^2/2\epsilon$ from each side and rearranging yields the equivalent condition of firm nonexpansiveness,

$$\|p - p'\|^2 + \|(x - p) - (x' - p')\|^2 \leqslant \|x - x'\|^2, \tag{5.15}$$

which also shows that $\mathrm{Prox}_{\epsilon f}$ and $\mathrm{Id} - \mathrm{Prox}_{\epsilon f}$ are both Lipschitz with constant 1.

Proof of 4: It is clear from the definition of the Moreau envelope that ${}^{\epsilon}f[x] \leqslant f[x]$. Let $\mu = \sup_{\epsilon>0} {}^{\epsilon}f[x] \leqslant f[x] < +\infty$, and thus,

$$\forall \epsilon > 0, \quad f[\mathrm{Prox}_{\epsilon f}\, x] + \frac{1}{2\epsilon}\|x - \mathrm{Prox}_{\epsilon f}\, x\|^2 \leqslant \mu\,.$$

Consider the map $h_x[z] := f[z] + \|x - z\|^2/2$. Then for all $\epsilon \in (0, 1)$, $\mathrm{Prox}_{\epsilon f}\, x$ is in the sublevel set $\{z \in \mathcal{H} \mid |h_x[z] < \mu\}$. This set is bounded. Hence $\nu := \sup_{\epsilon\in(0,1)} \|\mathrm{Prox}_{\epsilon f}\, x\| < +\infty$. We next use the fact that any $f \in \Gamma_0(\mathcal{H})$ possesses a continuous affine minorant [1, Theorem 9.19], i.e., there exist $\omega \in \mathcal{H}$ and $a \in \mathbb{R}$ such that $\langle \omega, \cdot\rangle + a \leqslant f$. Therefore,

$$\mu \geqslant \langle \omega, \mathrm{Prox}_{\epsilon f}\, x\rangle + a + \frac{1}{2\epsilon}\|x - \mathrm{Prox}_{\epsilon f}\, x\|^2 \geqslant -\nu\|\omega\| + a + \frac{1}{2\epsilon}\|x - \mathrm{Prox}_{\epsilon f}\, x\|^2.$$

Rearranging gives $\|x - \mathrm{Prox}_{\epsilon f}\, x\|^2 \leqslant 2\epsilon(\mu + \nu\|\omega\| - a) = O(\epsilon)$ as $\epsilon \to 0^+$. □

The Moreau envelope of a function in $\Gamma_0(\mathcal{H})$ has several nice properties, summarized in the following proposition.

Proposition 5.10 *Let $f \in \Gamma_0(\mathcal{H})$ and let $\epsilon > 0$ be given. The following holds:*

1. ${}^{\epsilon}f \in \Gamma_0(\mathcal{H})$, *and* $\mathrm{dom}({}^{\epsilon}f) = \mathcal{H}$.
2. *For every $\delta > \epsilon$, and for all $x \in \mathcal{H}$,*

$$\inf f \leqslant {}^{\delta}f[x] \leqslant {}^{\epsilon}f[x] \leqslant f[x]\,.$$

3. *For every $x \in \mathcal{H}$, ${}^{\epsilon}f[x] \to f[x]$ from below as $\epsilon \to 0^+$ (even if $x \neq \mathrm{dom}(f)$).*
4. *${}^{\epsilon}f$ is Fréchet differentiable, with derivative*

$$\nabla^{\epsilon}f[x] = \epsilon^{-1}(x - \mathrm{Prox}_{\epsilon f}\, x)\,.$$

The derivative is Lipschitz continuous with constant ϵ^{-1}.

Proof Proof of 1: The domain is already established. We show convexity. With differentiability shown in 3, it follows that ${}^{\epsilon}f \in \Gamma_0(\mathcal{H})$. Let $h_{x'}[x] = f[x'] + \|x - x'\|^2/2\epsilon$, which is convex. Thus, $h_{x'}[\lambda x_1 + (1 - \lambda)x_2] \leqslant \lambda h_{x'}[x_1] + (1 - \lambda)h_{x'}[x_2]$. Taking the infimum with respect to x' on both sides yields that ${}^{\epsilon}f$ is convex.

Proof of 2: Easy.

Proof of 3: As in the proof of Proposition 5.9(4), set $\mu = \sup_{\epsilon>0} {}^{\epsilon}f[x]$. From 2, ${}^{\epsilon}f[x] \to \mu \leqslant f[x]$ from below as $\epsilon \to 0^+$. Therefore, we can assume that $\mu < +\infty$ and demonstrate that $\lim_{\epsilon\to 0^+} {}^{\epsilon}f[x] \geqslant f[x]$. From Proposition 5.9(4) we get

$$\begin{aligned}\lim_{\epsilon\to 0^+} {}^{\epsilon}f[x] &= \lim_{\epsilon\to 0^+} f[\operatorname{Prox}_{\epsilon f} x] + \frac{1}{2\epsilon}\|x - \operatorname{Prox}_{\epsilon f} x\|^2 \\ &\geqslant \liminf_{\epsilon\to 0^+} f[\operatorname{Prox}_{\epsilon f} x] \geqslant f[x],\end{aligned}$$

where we used that f is lower semicontinuous and that $\operatorname{Prox}_{\epsilon f} x \to x$.

Proof of 4: Let $x, x' \in \mathcal{H}$, $p = \operatorname{Prox}_{\epsilon f} x$, $p' = \operatorname{Prox}_{\epsilon f} x'$. Using the definition of $\operatorname{Prox}_{\epsilon f}$ and Eq. (5.13), we derive two inequalities:

$$\begin{aligned}{}^{\epsilon}f[x'] - {}^{\epsilon}f[x] &= f[p'] - f[p] + \frac{1}{2\epsilon}(\|x' - p'\|^2 - \|x - p\|^2) \\ &\geqslant \epsilon^{-1}\langle p' - p, x - p\rangle + \frac{1}{2\epsilon}(\|x' - p'\|^2 - \|x - p\|^2) \\ &= \frac{1}{2\epsilon}\Big(\|x' - p' - x + p\|^2 + 2\langle x' - x, x - p\rangle\Big) \\ &\geqslant \epsilon^{-1}\langle x' - x, x - p\rangle.\end{aligned}$$

A similar calculation gives the second bound ${}^{\epsilon}f[x'] - {}^{\epsilon}f[x] \leqslant \epsilon^{-1}\langle x' - x, x' - p'\rangle$. Combining these two inequalities gives

$$\begin{aligned}0 &\leqslant {}^{\epsilon}f[x'] - {}^{\epsilon}f[x] - \epsilon^{-1}\langle x' - x, x - p\rangle \\ &\leqslant \epsilon^{-1}\langle x' - x, x' - p'\rangle - \epsilon^{-1}\langle x' - x, x - p\rangle \\ &= \epsilon^{-1}\langle x' - x, (x' - p') - (x - p)\rangle.\end{aligned}$$

Using Cauchy–Schwarz and the firm nonexpansiveness condition (Proposition 5.9(3)), we obtain

$$\begin{aligned}0 &\leqslant {}^{\epsilon}f[x'] - {}^{\epsilon}f[x] - \epsilon^{-1}\langle x' - x, x - p\rangle \\ &\leqslant \epsilon^{-1}\|x' - x\|\|(x' - p') - (x - p)\| \leqslant \epsilon^{-1}\|x' - x\| \\ &\quad \Big(\|x' - x\|^2 - \|p' - p\|^2\Big)^{1/2} \\ &\leqslant \epsilon^{-1}\|x' - x\|^2.\end{aligned}$$

It follows that

$$\lim_{x'\to x} \|x' - x\|^{-1}({}^{\epsilon}f[x'] - {}^{\epsilon}f[x] - \epsilon^{-1}\langle x' - x, x - p\rangle) = 0,$$

which proves that ${}^{\epsilon}f$ is Fréchet differentiable, with

$$\nabla^{\epsilon}f[x] = \epsilon^{-1}(x - \mathrm{Prox}_{\epsilon f}\, x)\,.$$

Since $\mathrm{Id} - \mathrm{Prox}_{\epsilon f}$ has Lipschitz constant 1, $\nabla^{\epsilon}f$ has Lipschitz constant ϵ^{-1}. □

We obtain a simple, but interesting fact from the Fréchet derivative from Proposition 5.10(4) and the variational characterization of the proximal mapping in Proposition 5.9(2): While $f \in \Gamma_0(\mathcal{H})$ may not be differentiable, the gradient of ${}^{\epsilon}f$ is always a subgradient at $\mathrm{Prox}_{\epsilon f}\, x$.

Corollary 5.25 *Let* $f \in \Gamma_0(\mathcal{H})$*, and* $\epsilon > 0$*. Then* $\nabla^{\epsilon}f[x] \in \underline{\partial} f[\mathrm{Prox}_{\epsilon f}\, x]$*.*

Remark 5.26 Moreau's Theorem [19, 26] states that, for $f \in \Gamma_0(\mathcal{H})$,

$${}^{\epsilon}f[x] = f[{}^{\epsilon}Jx] + \frac{\epsilon}{2}\|{}^{\epsilon}Ax\|^2$$

is convex and Fréchet differentiable with $\nabla^{\epsilon}f[x] = {}^{\epsilon}A$. Here, ${}^{\epsilon}A = \epsilon^{-1}(\mathrm{Id} - {}^{\epsilon}J)$ is the Yosida approximation to the resolvent ${}^{\epsilon}J = (\mathrm{Id} + \epsilon A)^{-1}$ of the maximal monotone operator $A = \underline{\partial} f$. Thus, ${}^{\epsilon}J = \mathrm{Prox}_{\epsilon f}$ (which is the only statement in Moreau's Theorem we have not proved), and we also have the compelling identity $\underline{\partial}^{\epsilon} f = {}^{\epsilon}\underline{\partial} f$.

5.3.3 Conjugate of the Moreau Envelope

We next discuss the skew concave conjugate of the Moreau envelope and its properties.

Proposition 5.11 *Let* $f \in \Gamma_0(\mathcal{H})$*, and let* $\epsilon > 0$*. Then*

$$({}^{\epsilon}f)^{\wedge}[x] = f^{\wedge}[x] - \frac{1}{2}\epsilon\|x\|^2,$$

which is strictly (indeed strongly) concave, and

$$\overline{\partial}({}^{\epsilon}f)^{\wedge}[x] = \overline{\partial} f^{\wedge}[x] - \epsilon x. \tag{5.16}$$

Proof Let $g = ({}^{\epsilon}f)^{\wedge}$, and let $y \in \mathcal{H}$. We first note that

$${}^{\epsilon}f[x] = \inf_{x_1 + x_2 = x}\left\{ f[x_1] + \frac{1}{2\epsilon}\|x_2\|^2 \right\}.$$

By the definition of the conjugate,

$$\begin{aligned} g[y] &= \inf_x \left\{ \langle y, x\rangle + \inf_{x_1+x_2} f[x_1] + \frac{1}{2\epsilon}\|x_2\|^2 \right\} \\ &= \inf_x \inf_{x_1+x_2=x} \left\{ \langle y, x\rangle + f[x_1] + \frac{1}{2\epsilon}\|x_2\|^2 \right\} \\ &= \inf_{x_1}\inf_{x_2} \left\{ \langle y, x_1\rangle + f[x_1] + \langle y, x_2\rangle + \frac{1}{2\epsilon}\|x_2\|^2 \right\} \\ &= f^\wedge[y] - \frac{\epsilon}{2}\|y\|^2, \end{aligned}$$

where we have used that if $\varphi[x] = \|x\|^2/2\epsilon$, then $\varphi^\wedge[y] = -\epsilon\|y\|^2/2$, which is left as an exercise.

To establish Eq. (5.16), we appeal to [29, Theorem 5.38], which, in the Hilbert space setting, states that if $f_1, f_2 \in \Gamma_0(\mathcal{H})$ and there is a point in $\mathrm{dom}(f_1)\cap\mathrm{dom}(f_2)$ where f_1 is continuous, then $\underline{\partial}(f_1 + f_2)[x] = \underline{\partial} f_1[x] + \underline{\partial} f_2[x]$ for every $x \in \mathcal{H}$. (The inclusion $\underline{\partial} f_1[x] + \underline{\partial} f_2[x] \subset \underline{\partial}(f_1 + f_2)[x]$ is easy to prove, but the converse inclusion does not hold in general.) Our result is established by noting that $\varphi, -f^\wedge \in -\Gamma_0(\mathcal{H})$, and that φ is everywhere continuous. □

Remark 5.27 Proposition 5.11 implies that the Moreau–Yosida regularization is lossless, i.e., for any $\epsilon > 0$

$$f = \left(({}^\epsilon f)^\wedge + \frac{\epsilon}{2}\|\cdot\|^2 \right)^\vee,$$

an explicit formula for the inverse of the regularization.

5.4 Moreau–Yosida Regularized Exact DFT

We apply Moreau–Yosida regularization to exact DFT in a box domain, as outlined in Sect. 5.2. The treatment in this section closely follows that of Ref. [9]. We describe Moreau–Yosida Kohn–Sham (MYKS) theory, and set up a basic self-consistent field (SCF) iteration, an abstract algorithm for the solution of the Kohn–Sham problem. We describe the Moreau–Yosida Kohn–Sham Optimal Damping Algorithm (MYKSODA), and prove a weak convergence result from Ref. [17].

In this section, we assume $\Omega = [-\ell, \ell]^3 \subset \mathbb{R}$ is a finite box, and we let $\mathcal{H} = L^2(\Omega)$ throughout. We omit all specifications of Ω in symbols like $\mathcal{D}^N(\Omega)$ for brevity. We let the Moreau–Yosida parameter $\epsilon > 0$ be fixed unless otherwise stated.

5.4.1 *Regularized Universal and Ground-State Energy Functionals*

We consider the Moreau envelope ${}^{\epsilon}F : \mathcal{H} \to \mathbb{R}$ and its concave skew conjugate. Thus, the central functionals of this section are

$$
\begin{aligned}
{}^{\epsilon}F[\rho] &= F[\mathrm{Prox}_{\epsilon F}\,\rho] + \frac{1}{2\epsilon}\|\rho - \mathrm{Prox}_{\epsilon F}\,\rho\|^2,\\
{}^{\epsilon}E[v] &:= E[v] - \frac{1}{2}\epsilon\|v\|^2.
\end{aligned}
$$

These are related via skew conjugation,

$$
\begin{aligned}
{}^{\epsilon}E[v] &= \min_{\rho\in\mathcal{H}} {}^{\epsilon}F[\rho] + \langle v, \rho\rangle,\\
{}^{\epsilon}F[\rho] &= \max_{v\in\mathcal{H}} {}^{\epsilon}E[v] - \langle v, \rho\rangle.
\end{aligned}
$$

The original infimum and supremum are a minimum and a maximum, since any $v \in \mathcal{H}$ has a ground state by Theorem 5.21, and since ${}^{\epsilon}E$ is strongly concave. The map ${}^{\epsilon}E$ is defined *not* as the Moreau envelope of E, but is instead related to the unregularized energy E by an explicit and easy-to-evaluate function of $v \in \mathcal{H}$. By Proposition 5.10, ${}^{\epsilon}F$ is everywhere Fréchet differentiable with $\nabla {}^{\epsilon}F[\rho] = \epsilon^{-1}(\rho - \mathrm{Prox}_{\epsilon F}\,\rho) \in \underline{\partial} F[\mathrm{Prox}_{\epsilon F}\,\rho]$. Thus $\mathrm{Prox}_{\epsilon F}\,\rho \in \mathcal{B}_N$. The converse is also true, which gives the remarkable fact that $\mathcal{B}_N$ is the range of the proximal mapping, so that the v-representability problem, i.e., the problem of characterizing those $\rho \in \mathcal{D}^N$ that are ground-state densities of a given $v \in \mathcal{H}$, is traded for the problem of evaluating ${}^{\epsilon}F$, since differentiating this function is not an issue.

Proposition 5.12 *The proximal mapping* $\mathrm{Prox}_{\epsilon F} : \mathcal{H} \to \mathcal{B}_N$ *is onto.*

Proof That $\mathrm{Prox}_{\epsilon F}^{-1} = \mathrm{Id} + \epsilon\underline{\partial} F$ comes directly from Proposition 5.9(2). By definition of range and domain, $\mathrm{ran}\, A = \mathrm{dom}\, A^{-1}$ for any set-valued operator $A : \mathcal{H} \to 2^{\mathcal{H}}$ [1, Definition 12.23], it follows immediately that $\mathrm{ran}(\mathrm{Prox}_{\epsilon F}) = \mathrm{dom}(\mathrm{Id} + \underline{\partial} F) = \mathcal{B}_N$. □

Having established that $\mathrm{ran}(\mathrm{Prox}_{\epsilon F}) = \mathcal{B}_N$, we ask what is the preimage of some $\rho_\epsilon \in \mathcal{B}_N$? The Hohenberg–Kohn Conjecture states that $\underline{\partial} F[\rho_0]$ is unique up to an additive constant. Assuming the conjecture to hold, we obtain an interesting property of the proximal mapping for a fixed $\epsilon > 0$.

Proposition 5.13 *Let* $\rho_0 \in \mathcal{B}_N$ *be the ground-state density of* $v \in \mathcal{H}$ *(which exists since any* v *has a ground-state in the box-truncated setting). Assume Conjecture 5.17, i.e.,*

$$
\underline{\partial} F[\rho_0] = \{-v + \mu \mid \mu \in \mathbb{R}\}\,.
$$

Then, $\rho \in \mathrm{Prox}_{\epsilon F}^{-1} \rho_0$ *is unique up to a constant shift, i.e.,*

$$\mathrm{Prox}_{\epsilon F}^{-1}[\rho_0] = \{\rho + \nu \mid \nu \in \mathbb{R}\}\,.$$

Proof Any potential for which $\mathrm{Prox}_{\epsilon F}\, \rho = \rho_0$ is a ground state is of the form $v = -\epsilon^{-1}(\rho - \rho_0) + \mu$, with $\mu \in \mathbb{R}$, and $\rho \in \mathrm{Prox}_{\epsilon F}^{-1}[\rho_0]$. Rearranging, $\rho = -\epsilon v + \rho_0 - \epsilon\mu$, from which the result follows. □

We can view $\mathrm{Prox}_{\epsilon F}$ as a kind of nonlinear "projection" that takes a $\rho \in \mathcal{H}$ and maps it to an (ensemble) v-representable density $\rho_\epsilon = \mathrm{Prox}_{\epsilon F}\, \rho$. Moreover, any (ensemble) v-representable density can be reached from some $\rho \in \mathcal{H}$. However, it is not true in general that $\mathrm{Prox}_{\epsilon F}^{\circ 2} = \mathrm{Prox}_{\epsilon F}$, so it is not a true projection.

The fact that $\mathrm{dom}({}^{\epsilon}F) = \mathcal{H}$ may seem like a severe shortcoming of the formalism, as $\mathcal{H}$ contains mostly "unphysical densities". On the other hand, for a given $\rho \in \mathcal{H}$, it is $\mathrm{Prox}_{\epsilon F}\, \rho$ which is the *physical* density. We therefore call general elements of $\mathcal{H}$ "quasidensities".

Definition 5.28 Let $\rho \in \mathcal{H}$ be given. We define the *proximal density* $\rho_\epsilon[\rho] := \mathrm{Prox}_{\epsilon F}\, \rho$, the resolvent of $\underline{\partial} F$, and the *proximal potential* $v_\epsilon[\rho] := -{}^{\epsilon}\underline{\partial} F[\rho] = -\epsilon^{-1}(\mathrm{Id} - \mathrm{Prox}_{\epsilon F})\rho$, with the Yosida approximation ${}^{\epsilon}\underline{\partial} F = \nabla^{\epsilon} F$ to the resolvent of $\underline{\partial} F$, cf. Remark 5.26.

Remark 5.29 The proximal density and potential satisfies

$$-\, v_\epsilon[\rho] \in \underline{\partial} F[\rho_\epsilon[\rho]] \iff \rho_\epsilon[\rho] \in \overline{\partial} E[v_\epsilon[\rho]]\,,$$

demonstrating how the proximal mapping generates a v-representable density *and* a corresponding potential via the Yosida approximation.

The regularized functionals are approximations to the exact functionals in the sense of Proposition 5.10. For example, for every quasidensity $\rho \in \mathcal{D}^N$, ${}^{\epsilon}F[\rho] \to F[\rho]$ from below as $\epsilon \to 0^+$. Moreover, from Proposition 5.9(4), we get $\rho_\epsilon \to \rho$. On the other hand, if $\rho \notin \mathcal{D}^N$, then ${}^{\epsilon}F[\rho] \to +\infty$.

5.4.2 Regularized Kohn–Sham Theory

Like in traditional Kohn–Sham theory, we set up a fictitious non-interacting system with an effective potential $v_{\mathrm{eff}} \in \mathcal{H}$ to be determined in a self-consistent manner. However, whereas in traditional Kohn–Sham the interacting and non-interacting systems are required to have the same *physical* densities, regularized Kohn–Sham theory requires the systems to have the same *quasidensities*. The Moreau–Yosida regularization resolves the two major issues with traditional Kohn–Sham theory: First, that ρ needs to be both interacting and non-interacting v-representable, resolved by $\rho \in \mathcal{H}$ being v-representable for any interaction strength in the Moreau–

Yosida regularized formulation. Second, that F is not at all differentiable, which makes the set-up of the self-consistent field equations non-rigorous, resolved by ${}^{\epsilon}F$ being continuously Fréchet differentiable by Proposition 5.10(4).

5.4.2.1 Adiabatic Connection

To prepare for Kohn–Sham theory, we briefly discuss the *adiabatic connection*, which relates the interacting N-electron problem to a non-interacting one. We introduce a connection parameter $\lambda \in [0, 1]$, and consider the modified energy functional on $\mathcal{W}_N$ given by

$$\mathcal{E}_0^{\lambda}[\psi] = \frac{1}{2}\langle \nabla\psi, \nabla\psi\rangle + \lambda\langle\psi, W\psi\rangle,$$

implying that the ground-state energy is a function of λ as well as the potential,

$$E^{\lambda}[v] := \inf\{\mathcal{E}_0^{\lambda}[\psi] + \mathcal{V}[\psi] \mid \psi \in \mathcal{W}_N\}.$$

At $\lambda = 0$, the problem is "solvable" in the sense that the N-electron Schdrödinger equation becomes separable, a key motivation behind Kohn–Sham theory. Similarly, we obtain a λ-dependent universal functional,

$$F^{\lambda}[\rho] = \inf_{\gamma\mapsto\rho} \mathrm{Tr}(\widehat{H}[0]^{\lambda}\gamma)\,.$$

The connection to spectral theory of self-adjoint operators in Sect. 5.2.1 does not depend on choosing $\lambda = 1$. Indeed, any $\lambda \in \mathbb{R}$ is acceptable. Moreover, the present discussion is valid for both the case $\Omega = \mathbb{R}^3$ and Ω a finite box.

Proposition 5.14 *Fix $v \in X^*$ and $\rho \in \mathcal{D}^N$. The maps from $\mathbb{R}$ to $\mathbb{R}$ defined by $\lambda \mapsto E^{\lambda}[v]$ and $\lambda \mapsto F^{\lambda}[\rho]$ are everywhere finite, concave, left and right differentiable, and almost everywhere differentiable.*

Proof We need to show concavity of $\lambda \mapsto E^{\lambda}[v]$. This follows from $E^{\lambda}[v] = \inf_{\gamma\in\mathcal{D}^N_{\mathrm{op}}}\{A_\gamma + \lambda B_\gamma\}$ with $A_\gamma = \mathrm{Tr}((T+V)\gamma)$ and $B_\gamma = \mathrm{Tr}(W\gamma)$, i.e., a pointwise infimum of a family of affine functions $\lambda \mapsto \mathrm{Tr}(\widehat{H}_0[v] + \lambda\mathrm{Tr}(\widehat{W}\gamma)$. Concavity of $\lambda \mapsto F^{\lambda}[\rho]$ follows immediately. It is clear that the both the maps are finite for all $\lambda \in \mathbb{R}$. By [29, Theorem 1.6], concave functions are everywhere left and right differentiable on the interior of their domains, and by Theorem 1.8 in the same reference, almost everywhere differentiable. □

One immediate complication that arises in the adiabatic connection is that the set of ensemble v-representable densities may change with λ, i.e., the set-valued map $\lambda \mapsto \mathcal{B}_N^{\lambda} \subset \mathcal{D}^N$ is non-trivial. There are no known results that relate the different $\mathcal{B}_N^{\lambda}$.

Turning to the Moreau–Yosida regularization in the domain $\Omega = [-\ell, \ell]^3$, we obtain λ-dependent functionals ${}^\epsilon E^\lambda : \mathcal{H} \to \mathbb{R}$ and ${}^\epsilon F^\lambda : \mathcal{H} \to \mathbb{R}$. Fix $\widehat{\rho} \in \mathcal{H}$, a target density. For any $\lambda \in \mathbb{R}$, there exists a unique proximal potential,

$$v_\epsilon^\lambda[\widehat{\rho}] = -\nabla^\epsilon F^\lambda[\widehat{\rho}] .$$

The regularized ground-state energy of that potential is

$${}^\epsilon E^\lambda[v_\epsilon^\lambda] = {}^\epsilon F^\lambda[\widehat{\rho}] + \langle v_\epsilon^\lambda, \widehat{\rho} \rangle,$$

related to the *unregularized* energy as

$$E^\lambda[v_\epsilon^\lambda] = {}^\epsilon E^\lambda[v_\epsilon^\lambda] + \frac{\epsilon}{2} \| v_\epsilon^\lambda \|^2 .$$

Associated with $\widehat{\rho} \in \mathcal{H}$ is also the proximal density $\rho_\epsilon^\lambda[\widehat{\rho}] = \mathrm{Prox}_{\epsilon F^\lambda}[\widehat{\rho}]$, being the true ground-state density of the proximal potential,

$$E^\lambda[v_\epsilon^\lambda] = F^\lambda[\rho_\epsilon^\lambda] + \langle v_\epsilon^\lambda, \rho_\epsilon^\lambda \rangle .$$

This is the adiabatic connection in the Moreau–Yosida formulation of DFT. However, we leave open the perhaps important question of the regularity properties of the maps $(\widehat{\rho}, \lambda) \mapsto v_\epsilon^\lambda[\widehat{\rho}]$ and $(\widehat{\rho}, \lambda) \mapsto \rho_\epsilon^\lambda[\widehat{\rho}]$.

5.4.2.2 Kohn–Sham Decomposition

Let $v_{\text{ext}} \in \mathcal{H}$ be the external potential for which we wish to determine $E^1[v_{\text{ext}}]$ and its corresponding ground-state density, known to exist by Theorem 5.21. (If the ground state of $\hat{H}[v_{\text{ext}}]$ is degenerate, there will be a convex set of solutions.) By Proposition 5.11, this problem is in a simple manner connected to the Moreau–Yosida regularized problem by Proposition 5.11,

$$\overline{\partial} E^1[v_{\text{ext}}] = \overline{\partial}^\epsilon E^1[v_{\text{ext}}] + \epsilon v_{\text{ext}} ,$$

and hence,

$$\rho \in \overline{\partial}^\epsilon E^1[v_{\text{ext}}] \iff \rho_\epsilon^1 = \rho - \epsilon v_{\text{ext}} \in \overline{\partial} E^1[v_{\text{ext}}].$$

We connect this problem at $\lambda = 1$ to the non-interacting problem at $\lambda = 0$, where now the quasidensity is known, whereas the corresponding potential $v_{\text{eff}} \in \mathcal{H}$ is unknown,

$$\rho \in \overline{\partial}^\epsilon E^0[v_{\text{eff}}] \iff \rho_\epsilon^0 = \rho - \epsilon v_{\text{eff}} \in \overline{\partial} E^0[v_{\text{eff}}].$$

We know that the potential $v_{\text{eff}} \in \mathcal{H}$ exists, since ${}^{\epsilon}F^0$ is differentiable at ρ. By Proposition 5.10, we have the stationary conditions

$$-v_{\text{ext}} = \nabla^{\epsilon} F^1[\rho]\,, \tag{5.17}$$

$$-v_{\text{eff}} = \nabla^{\epsilon} F^0[\rho]\,. \tag{5.18}$$

We introduce the *Hartree-exchange correlation energy* ${}^{\epsilon}E_{\text{Hxc}}[\rho] := {}^{\epsilon}F^1[\rho] - {}^{\epsilon}F^0[\rho]$, which is continuously differentiable with Lipschitz continuous gradient ${}^{\epsilon}v_{\text{Hxc}}[\rho] := \nabla^{\epsilon} E_{Hxc}[\rho]$, called the Hartree-exchange correlation potential. The condition (5.17) becomes

$$v_{\text{eff}} = v_{\text{ext}} + {}^{\epsilon}v_{\text{Hxc}}[\rho]\,. \tag{5.17'}$$

The idea now is that it is E_{Hxc}, and not ${}^{\epsilon}F^1$, which is available, at least as some kind of approximation. Equations (5.18) and (5.17') form a self-consistent field problem where the density ρ and potential v_{eff} are the unknowns.

This motivates the following abstract algorithm:

Algorithm 1: Basic MYKS-SCF iteration scheme

Fix $\epsilon > 0$.

1: Choose an initial guess $\rho_0 \in \mathcal{H}$.

2: For $i = 0, 1, \ldots,$ iterate:

1. Set $v_{\text{eff},i+1} = v_{\text{ext}} + {}^{\epsilon}v_{\text{Hxc}}[\rho_i]$.
2. Solve the unregularized noninteracting problem $\widetilde{\rho}_{i+1} \in \overline{\partial} E^0[v_{\text{eff},i+1}]$, and set $\rho_{i+1} = \widetilde{\rho}_{i+1} - \epsilon v_{i+1}$.
3. If $v_{\text{eff},i+1} = v_{\text{ext}} + {}^{\epsilon}v_{\text{Hxc}}[\rho_{i+1}]$, terminate.

3: Return $\rho_{\epsilon}^1 = \rho_{i+1} + \epsilon v_{\text{ext}}$ as the (physical) ground-state density, and $E^1[v_{\text{ext}}] = F^0[\rho_{i+1}] + {}^{\epsilon}E_{\text{Hxc}}[\rho_{i+1}] + \epsilon \|v_{\text{ext}}\|^2/2$ as the ground-state energy.

Remark 5.30 In Step 2.1 of Algorithm 1, the problem $\widetilde{\rho}_{i+1} \in \overline{\partial} E^0[v_{\text{eff},i+1}]$ is equivalent to solving the eigenvalue equation of a one-electron operator $\widehat{h} = -\nabla^2/2 + v_{\text{eff},i+1}$. By the Rellich–Kondrachov theorem, this operator has domain $H_0^2(\Omega)$, a compact resolvent and hence a purely discrete spectrum. The first N eigenfunctions $\varphi_k \in H_0^2(\Omega)$ define a solution $\widetilde{\rho}_{i+1} = \sum_{k=1}^N |\varphi_k|^2$. Any degeneracies in the spectrum lead to a finite number of solutions, the convex hull of which is $\overline{\partial} E^0[v_{\text{eff},i+1}]$. In particular, the superdifferential is always nonempty, so that the noninteracting problem can always be solved.

Remark 5.31 The formal limit $\epsilon \to 0^+$ in Algorithm 1 gives the traditional Kohn–Sham self-consistent field iterations for exact DFT. Of course, the Hartree-exchange correlation potential is not rigorously defined in this case.

5.4.3 Optimal-Damping Algorithm

Simple SCF iterations often suffer from bad convergence properties, with phenomena like asymptotic oscillations. This particular behavior can be traced to the fact that each SCF iteration takes too large a step, so that one "jumps over" a minimum. The optimal-damping algorithm (ODA) for the Hartree–Fock SCF iterations were introduced by Cancès and Le Bris in Refs. [3, 5], in which the "bare" SCF density update is damped by a factor in (0, 1]. In this section, we apply the idea of optimal damping for extended Kohn–Sham iterations from Ref. [4] to modify Algorithm 1, leading to the Moreau–Yosida regularized Kohn–Sham optimal-damping algorithm (MYKSODA). This algorithm was introduced in Ref. [17], where the convergence of the energy to an upper bound, Theorem 5.32, was proven. A stronger convergence proof for a finite-dimensional setting, Theorem 5.33, was introduced in Refs. [20, 21]. We here present the full proof here with some additional details and corrections.

The ODA modification of Algorithm 1 consists of a relaxation of the density step, i.e., not setting $\rho_{i+1} = \rho'_{i+1}$, but instead $\rho_{i+1} = \rho_i + t_i \rho'_{i+1}$ for some $t_i \in (0, 1]$. In Theorem 5.33, the step lengths t_i are actually not constrained, but convergence is still guaranteed.

Algorithm 2: MYKSODA

Fix $\epsilon > 0$.

1: Set $v_{\text{eff},0} = v_{\text{ext}}$, and $\rho_0 \in \overline{\partial}^\epsilon E^0[v_{\text{ext}}]$.

2: For $i = 0, 1, \ldots,$ iterate:

1. Set $v_{\text{eff},i+1} = v_{\text{ext}} + {}^\epsilon v_{\text{Hxc}}[\rho_i]$.
2. Solve for $\widetilde{\rho}_{i+1} \in \underline{\partial} E^0[v_{\text{eff},i+1}]$, and set $\rho'_{i+1} = \widetilde{\rho}_{i+1} - \epsilon v_{\text{eff},i+1}$.
3. Choose $t_i \in (0, 1]$ such that $\rho_{i+1} = \rho_i + t_i(\rho'_{i+1} - \rho_i)$ satisfies

$$\langle \nabla^\epsilon F[\rho_{i+1}] + v_{\text{ext}}, \rho'_{i+1} - \rho_i \rangle \leqslant 0\,. \tag{5.19}$$

4. If $v_{\text{eff},i+1} = v_{\text{ext}} + {}^\epsilon v_{\text{Hxc}}[\rho_{i+1}]$, terminate.

3: Return $\rho^1_\epsilon = \rho_{i+1} + \epsilon v_{\text{ext}}$ as the (physical) ground-state density, and $E^1[v_{\text{ext}}] = F^0[\rho_{i+1}] + {}^\epsilon E_{\text{Hxc}}[\rho_{i+1}] + \epsilon \|v_{\text{ext}}\|^2/2$ as the ground-state energy.

The MYKSODA algorithm is weakly convergent in the sense that the energy converges to an upper bound:

Theorem 5.32 *In Algorithm 2, the sequence of energy estimates*

$$e_i := {}^\epsilon F[\rho_i] + \langle v_{\text{ext}}, \rho_i \rangle$$

is strictly descending and hence convergent. Denoting the limit by $e[v_{\mathrm{ext}}]$, *we have*

$$e[v_{\mathrm{ext}}] + \epsilon \|v_{\mathrm{ext}}\|^2/2 \geqslant E^1[v_{\mathrm{ext}}]\,,$$

that is, an upper bound property of the computed energy limit.

Proof It is clear that step 2.2 produces $\rho'_{i+1} \in \underline{\partial}^{\epsilon} E^0[v_{\mathrm{eff},i+1}]$. We need to check that step 2.3 is well-defined. To that end, consider the function $G[\rho] := {}^{\epsilon}F^1[\rho] + \langle v_{\mathrm{ext}}, \rho\rangle$, which is to be minimized by the iterations. We note that step 2.1 can be rewritten as $v_{\mathrm{eff},i+1} + \nabla^{\epsilon} F^0[\rho_i] = v_{\mathrm{ext}} + \nabla^{\epsilon} F^1[\rho_i]$. The directional derivative of G at ρ in the direction $\Delta\rho_i := \rho'_{i+1} - \rho_i$ is $G'[\rho; \Delta\rho_i] = \langle \nabla^{\epsilon} F^1[\rho] + v_{\mathrm{ext}}, \rho'_{i+1} - \rho_i\rangle$, which gives

$$G'[\rho_i; \rho'_{i+1} - \rho_i] = \langle \nabla^{\epsilon} F^0[\rho_i] + v_{\mathrm{eff},i+1}, \rho'_{i+1} - \rho_i\rangle.$$

If the termination criterion was met at the previous iteration, $G'[\rho_i; \rho'_{i+1} - \rho_i] = 0$, and step 2.3 will not take place. Otherwise, since $\rho_i \in \bar{\partial}^{\epsilon} E^0[-\nabla^{\epsilon} F^0[\rho_i]]$ (see Remark 5.29), write

$$\sigma_{i+1} := \rho'_{i+1} + \epsilon v_{\mathrm{eff},i+1} \in \bar{\partial} E^0[v_{\mathrm{eff},i+1}],$$

$$\tau_i := \rho_i + \epsilon \nabla^{\epsilon} F^0[\rho_i] \in \bar{\partial} E^0[-\nabla^{\epsilon} F[\rho_i]].$$

This allows us to rewrite the directional derivative as

$$G'[\rho_i; \rho'_{i+1} - \rho_i] = \langle v_{\mathrm{eff},i+1} + \nabla^{\epsilon} F^0[\rho_i], \sigma_{i+1} - \tau_i\rangle - \epsilon \|v_{\mathrm{eff},i+1} + \nabla^{\epsilon} F^0[\rho_i]\|^2.$$

By monotonicity of the superdifferential (Propositions 5.4 and 1), the bracket is always nonpositive, which by step 2.1 in Algorithm 2 gives

$$G'[\rho_i; \Delta\rho_i] \leqslant -\epsilon^{-1} \|v_{\mathrm{eff},i+1} + \nabla^{\epsilon} F^0[\rho_i]\|^2 = -\epsilon \|v_{\mathrm{ext}} + \nabla^{\epsilon} F^1[\rho_i]\|^2. \qquad (5.20)$$

Since now $G[\rho_i]$ is strictly decreasing and convex in the direction $\Delta\rho_i$ ($t \mapsto G[\rho_i + t\Delta\rho_i]$ is *strongly* convex at $t = 0$), it follows that there is a maximum step length $t_i > 0$ such that $e_{i+1} = G[\rho_{i+1}] < G[\rho_i] = e_i$, and such that Eq. (5.19) holds, i.e., $G'[\rho_{i+1}, \rho'_{i+1} - \rho_i] \leqslant 0$. Since $e_i \geqslant \inf_{\rho} G[\rho] = {}^{\epsilon}E^1[v_{\mathrm{ext}}]$ for all i, it follows that the monotonically decreasing sequence e_i converges to an upper bound of ${}^{\epsilon}E^1[v_{\mathrm{ext}}] = E^1[v_{\mathrm{ext}}] - \epsilon \|v_{\mathrm{ext}}\|^2/2$. □

In the above, the step lengths $t_i \in (0, 1]$ were not specified. It is clear that any stronger convergence properties of the algorithm depend crucially on these. The final result of this section, paraphrased from Refs. [20, 21], exploits the definition of the Moreau–Yosida regularization at every step. The key idea is the following interpretation of Proposition 5.10(4): At every $x \in \mathcal{H}$, ${}^{\epsilon}f$ is tangent to the regularization parabola $h_x(z) := f(\mathrm{Prox}_{\epsilon f}\, x) + \|z - \mathrm{Prox}_{\epsilon f}\, x\|^2/(2\epsilon)$, i.e., evaluated at x, both have gradient $\epsilon^{-1}(x - \mathrm{Prox}_{\epsilon f}\, x)$ and their graphs are touching at x. Moreover, by definition ${}^{\epsilon}f \leqslant h_z$. The step sizes t_i are now chosen such that ρ_{i+1} minimizes the regularization parabola associated with $G[\rho_i]$ along the search direction $\Delta\rho_i$. However, one must assume that $\mathcal{H}$ is finite-dimensional for this to guarantee convergence. A finite-dimensional $\mathcal{H}$ arises, for example, when the N-electron Hilbert space is discretized by a finite-dimensional basis generated by a finite set of single-particle functions, and does not change the properties of E or F in any way relevant for the current discussion.

Theorem 5.33 *Assume that $\mathcal{H}$ is finite-dimensional. Then there exists a sequence of step lengths $\{t_i\}_i$ for Algorithm 2 such that $\{\rho_i\}$ has energies $e_i = {}^{\epsilon}F[\rho_i] + (v_{\text{ext}}, \rho_i)$ converging to the exact result ${}^{\epsilon}E[v_{\text{ext}}]$. Moreover, $\{\rho_i\}$ is the union of convergent subsequences $\rho_{i_n} \to \sigma \in \mathcal{H}$, such that $\sigma \in \bar{\partial}^{\epsilon}E[v_{\text{ext}}]$ is an exact ground-state quasidensity, and $v_{\text{eff},i} \to v_{\text{ext}} + {}^{\epsilon}v_{\text{Hxc}}[\sigma]$.*

Proof Let $p_i := \mathrm{Prox}_{\epsilon F}\, \rho_i$. Rewritten in terms of this proximal point, we have

$$\begin{aligned} G[\rho_i] &= F[p_i] + (v_{\text{ext}}, \rho_i) + \frac{1}{2\epsilon}\|\rho_i - p_i\|^2 \\ &= F[p_i] + (v_{\text{ext}}, p_i) - \frac{\epsilon}{2}\|v_{\text{ext}}\|^2 + \frac{1}{2\epsilon}\|\rho_i - p_i + \epsilon v_{\text{ext}}\|^2. \end{aligned} \tag{5.21}$$

The regularization parabola associated with $G[\rho]$ is similarly obtained, noting that the linear term (v_{ext}, ρ) of G is easily absorbed,

$$h_{\rho_i}[\sigma] := F[p_i] + (v_{\text{ext}}, p_i) - \frac{\epsilon}{2}\|v_{\text{ext}}\|^2 + \frac{1}{2\epsilon}\|\sigma - p_i + \epsilon v_{\text{ext}}\|^2. \tag{5.22}$$

We note that the global minimum of h_{ρ_i} is at $p_* := p_i - \epsilon v_{\text{ext}}$, and that $\nabla h_{\rho_i}[\rho_i] = \nabla G[\rho_i]$, by the remarks preceding the theorem. Let $f(t) = G[\rho_i + t\Delta\rho_i]$, which by Eq. (5.20) satisfies $f'(0) < 0$. At all t, $f(t) \leqslant g(t) := h_{\rho_i}[\rho_i + t\Delta\rho]$. Its derivative is

$$g'(t) = \epsilon^{-1}(\rho_i - p_* + t\Delta\rho, \Delta\rho), \quad g'(0) < 0.$$

Define $t_i > 0$ uniquely (and hence $\rho_{i+1} = \rho_i + t_i\Delta\rho_i$) by the condition $g'(t_0) = 0$, which gives

$$t_i = -\|\Delta\rho_i\|^{-2}(\rho_i - p_*, \Delta\rho),$$

and the equivalent orthogonality condition $\langle \rho_{i+1} - p_*, \rho_{i+1} - \rho_i \rangle = 0$. Geometrically, $\rho_{i+1} - \rho_i$ is the orthogonal projection of $\rho_* - \rho_i$ onto the search line. A brief calculation gives

$$\begin{aligned}\langle \nabla G[\rho_i], \Delta\rho_i \rangle &= g'(0) = t_i^{-1} \langle \nabla^\epsilon F[\rho_i] + v_{\text{ext}}, \rho_{i+1} - \rho_i \rangle \\ &= t_i^{-1}\epsilon^{-1} \langle \rho_i - \rho_{i+1} + \rho_{i+1} - p_*, \rho_{i+1} - \rho_i \rangle \\ &= t_i^{-1}\epsilon^{-1} \langle \rho_i - \rho_{i+1}, \rho_{i+1} - \rho_i \rangle \\ &= -t_i \epsilon^{-1} \|\Delta\rho_i\| \|\rho_{i+1} - \rho_i\| = -\frac{\|\Delta\rho_i\|^2 t_i^2}{\epsilon}.\end{aligned} \tag{5.23}$$

Let $m_i := h_{\rho_i}[\rho_{i+1}]$ be the minimum of the parabola section. We have $e_{i+1} \leqslant m_i$. Subtracting Eq. (5.22) with $\sigma = \rho_{i+1}$ from Eq. (5.21) and using orthogonality, we obtain

$$\frac{t_i^2 \|\Delta\rho_i\|^2}{2\epsilon} = \frac{1}{2\epsilon} \|\rho_{i+1} - \rho_i\|^2 = e_i - m_i \leqslant e_i - e_{i+1} \to 0.$$

We now show that ρ_i converges. By assumption, $\text{dom}(F)$ is bounded in $\mathcal{H}$. It is easy to see that ${}^\epsilon F[\rho] = O(\|\rho\|^2)$ as $\|\rho\| \to +\infty$. Since $e_i = {}^\epsilon F[\rho_i] + \langle v_{\text{ext}}, \rho_i \rangle \leqslant e_1$ for all i, $\{\rho_i\}$ (which is not a subset of $\text{dom}(F)$!) is bounded in $\mathcal{H}$. This again implies, by Proposition 5.104, that $\|\nabla^\epsilon F^\lambda(\rho_i)\|$ is bounded, hence $\{v_{\text{eff},i}\}$ is bounded. Equation (5.20) combined with Eq. (5.23) gives

$$t_i^{-1} \|\rho_{i+1} - \rho_i\| \|\Delta\rho_i\| = \|\Delta\rho_i\|^2 t_i^2 \geqslant \epsilon^2 \|\nabla^\epsilon F[\rho_i] + v_{\text{ext}}\|^2 = \epsilon^2 \|\nabla^\epsilon F^0[\rho_i] + v_{\text{eff},i}\|^2.$$

Since $\|\Delta\rho_i\| t_i \to 0$, we now have $\|\nabla^\epsilon F^0[\rho_i] + v_{\text{eff},i}\| \to 0$ and $\|\nabla^\epsilon F[\rho_i] + v_{\text{ext}}\| \to 0$. The latter implies that $v_{\text{ext}} = -\lim_{i\to+\infty} \nabla^\epsilon F[\rho_i]$. Let $\sigma \in \mathcal{H}$ be an accumulation point of $\{\rho_i\}$, which is guaranteed to exist by the Bolzano–Weierstrass theorem. There is a convergent subsequence $\rho_{i_n} \to \sigma$ for which we have $v_{\text{ext}} = -\lim_n \nabla^\epsilon F[\rho_{i_n}] = -\nabla^\epsilon F[\sigma]$. But then $\sigma \in \bar{\partial}^\epsilon E[v_{\text{ext}}]$ is a ground-state quasidensity of the exact regularized problem. We next obtain $\lim_{n\to+\infty} v_{\text{eff},i_n+1} = -\lim_{n\to+\infty} \nabla^\epsilon F^0[\rho_{i_n}] = -\nabla^\epsilon F^0[\sigma] = v_{\text{ext}} + {}^\epsilon v_{\text{Hxc}}[\sigma]$. Since the accumulation point was arbitrary, the sequence $\{\rho_i\}$ splits into subsequences converging to different ground-state densities σ, as claimed. □

Remark 5.34 The step length choices in the proof of Theorem 5.33 are not practical (if one can consider the MYKS scheme practical at all), since they require the whereabouts of the proximal point $\text{Prox}_{\epsilon F}\, \rho_i$. But knowledge of that would make the Moreau–Yosida regularization approach redundant.

5.4.4 Density-Functional Approximations

So far our discussion has pertained to the *exact* ground-state energy functional $E^\lambda \in \Gamma_0(\mathcal{H})$, which is an abstract setting far from actual numerical calculations. We now make some simple observations that open up the possibility of a rigorous treatment of Kohn–Sham theory for model density functionals [22], including the application and further analysis of the MYKSODA iteration scheme.

The Kohn–Sham decomposition relied on a family $\mathcal{U} = \{E^\lambda \in -\Gamma_0(\mathcal{H}) \mid \lambda \in [0, 1]\}$. This family was such that $\overline{\partial} E^\lambda[v] \neq \emptyset$ for any $v \in \mathcal{H}$, and generates the family $\{{}^\epsilon E^\lambda : \mathcal{H} \to \mathbb{R} \mid \lambda \in [0, 1]\}$ and the corresponding Moreau envelopes $\{{}^\epsilon F^\lambda : \mathcal{H} \to \mathbb{R} \mid \lambda \in [0, 1]\}$. Moreover, the idea was present that solving for $\rho \in \overline{\partial}^\epsilon E^0[v]$ for arbitrary $v \in \mathcal{H}$, i.e., the noninteracting problem, was in some sense easy compared to solving for $\rho^1_\epsilon \in \overline{\partial}^\epsilon E^1[v_{\text{ext}}]$. A property of the family $\mathcal{U}$ that was *not* used in the Kohn–Sham decomposition was the concavity in the parameter $\lambda \in [0, 1]$.

In Algorithm 1, the basic MYKS-SCF iteration scheme, the existence of solutions $\rho \in \overline{\partial}^\epsilon E^0[v]$ for arbitrary $v \in \mathcal{H}$ was essential, but otherwise no additional properties of $\mathcal{U}$ were used. In Algorithm 2, MYKSODA, and Theorem 5.32, the same is true.

This motivates the introduction of an approximation, or model, ${}^\epsilon\widetilde{E}_{\text{Hxc}}$ such that ${}^\epsilon\widetilde{F}^1 = {}^\epsilon F^0 + {}^\epsilon\widetilde{E}_{\text{Hxc}} \in \Gamma_0(\mathcal{H})$ is continuously differentiable, replacing the exact Moreau–Yosida regularization in Algorithms 1 and 2. Of course, *how* to obtain such models is a different matter.

A second approach is to consider the *exact* λ-dependent Hartree-exchange correlation functional $E^\lambda_{\text{Hxc}} = F^\lambda - F^0$, a functional with domain $\mathcal{D}^N$ but of otherwise unknown composition, and introduce a model functional $\widetilde{E}^\lambda_{\text{Hxc}} : \mathcal{D}^N \to \mathbb{R}$ and corresponding model universal functionals, viz.,

$$\widetilde{F}^\lambda[\rho] := F^0[\rho] + \widetilde{E}^\lambda_{\text{Hxc}}[\rho].$$

This family may or may not be convex, but suppose for simplicity that it is, and that $\widetilde{E}^\lambda_{\text{Hxc}}$ is in some sense "cheap" but also sufficiently regular as to keep $\widetilde{F}^\lambda \in \Gamma_0(\mathcal{H})$. The family induces a model energy $\widetilde{E}^\lambda \in -\Gamma_0(\mathcal{H})$ such that $\widetilde{E}^0 = E^0$, retaining the ease of solution in the non-interacting limit.

The next step is to consider the Moreau–Yosida regularized model Hartree-exchange correlation energy,

$${}^\epsilon\widetilde{E}^\lambda_{\text{Hxc}}[\rho] := {}^\epsilon\widetilde{F}^\lambda[\rho] - {}^\epsilon F^0[\rho] = {}^\epsilon(F^0 + \widetilde{E}^\lambda_{\text{Hxc}})[\rho] - {}^\epsilon F^0[\rho].$$

The main problem is now to identify how and when ${}^\epsilon\widetilde{E}^\lambda_{\text{Hxc}}$ can be easily calculated given that it involves a non-trivial convolution, which may negate the ease of evaluation of $\widetilde{E}^\lambda_{\text{Hxc}}[\rho]$.

5.5 Conclusion

In this chapter, we have reformulated exact DFT as introduced by Lieb in terms of the Moreau–Yosida regularization. Our starting point was the N-electron problem. Not considered were grand-canonical ensembles at finite or zero temperature, which adds some interesting aspects to the formalism [22]. Resulting in a compelling reformulation of DFT, including a rigorous formulation of Kohn–Sham theory, Moreau–Yosida regularization still leaves open the question of how to find functional approximations to the Hartree-exchange correlation energy.

The mathematical understanding of Moreau–Yosida regularized Kohn–Sham theory is in its early stages. For example, Theorem 5.33 requires a finite-dimensional density space. Stronger and more general results are of course highly desired, and could turn Moreau–Yosida Kohn–Sham theory into a practical tool when combined with approximate Hartree-exchange correlation functionals.

In current-density functional theory (CDFT), the ground-state energy $E[v, \mathbf{A}]$ of an N-electron system in the external potential v and magnetic potential $\mathbf{A}$ is considered, leading to both the density ρ and the current density $\mathbf{j} : \mathbb{R}^3 \to \mathbb{R}^3$ being variables in the constrained-search functional [28]. This requires a more general setting than that of a Hilbert space of densities. In Ref. [17], Moreau–Yosida regularized DFT was introduced in the abstract setting of a reflexive Banach space of densities, thereby accommodating CDFT and demonstrating that the regularization approach may also be important in other density-functional settings than the standard one considered in this chapter.

References

1. H.H. Bauschke and P.L. Combettes. *Convex Analysis and Monotone Operator Theory in Hilbert Spaces*. Springer Verlag, New York, Heidelberg, Berlin (2011).
2. V. Barbu and T. Precupanu. *Convexity and Optimization in Banach Spaces*. Springer Monographs in Mathematics. Springer, Dortrecht, Heidelberg, London, New York, 4th edition (2012).
3. E. Cancès. SCF algorithms for HF electronic calculations. In: *Lecture Notes in Chemistry*, volume 74, chapter 2. Springer, Heldelberg (2000).
4. E. Cancès. Self-consistent field algorithms for Kohn–Sham models with fractional occupation numbers. *The Journal of Chemical Physics* **114**, 10616–10622 (2001).
5. E. Cancès and C. Le Bris. Can we outperform the DIIS approach for electronic structure calculations? *International Journal of Quantum Chemistry* **79**, 82–90 (2000).
6. L.C. Evans. *Partial Differential Equations*, volume 19 of *Graduate Studies in Mathematics*. AMS (1998).
7. P. Hohenberg and W. Kohn. Inhomogeneous electron gas. *Physical Review* **136**, B864–B871 (1964).
8. T. Kato. Fundamental properties of Hamiltonian operators of Schrödinger type. *Transactions of the American Mathematical Society* **70**, 195–211 (1951).
9. S. Kvaal, U. Ekström, A.M. Teale, and T. Helgaker. Differentiable but exact formulation of density-functional theory. *The Journal of Chemical Physics* **140**, 18A518 (2014).

10. S. Kvaal and T. Helgaker. Ground-state densities from the Rayleigh–Ritz variation principle and from density-functional theory. *The Journal of Chemical Physics* **143**, 184106 (2015).
11. W. Kohn and L.J. Sham. Self-consistent equations including exchange and correlation effects. *Physical Review* **140**, A1133–A1138 (1965).
12. P.E. Lammert. Coarse-grained V representability. *The Journal of Chemical Physics* **125**, 074114 (2006).
13. P.E. Lammert. Differentiability of Lieb functional in electronic density functional theory. *International Journal of Quantum Chemistry* **107**, 1943–1953 (2007).
14. P.E. Lammert. Well-behaved coarse-grained model of density-functional theory. *Physical Review A* **82**, 012109 (2010).
15. P.E. Lammert. In search of the Hohenberg–Kohn theorem. *Journal of Mathematical Physics* **59**, 042110 (2018).
16. E.H. Lieb. Density Functionals for Coulomb Systems. *International Journal of Quantum Chemistry* **24**, 243–277 1983.
17. A. Laestadius, M. Penz, E.I. Tellgren, M. Ruggenthaler, S. Kvaal and T. Helgaker. Generalized Kohn–Sham iteration on Banach spaces. *The Journal of Chemical Physics* **149**, 164103 (2018).
18. T.-S. Liu and J.-K. Wang. Sums and intersections of Lebesgue spaces. *Mathematica Scandinavica* **23**, 241–251 (1968).
19. J.J. Moreau. Proximité et dualité dans un espace hilbertien. *Bulletin de la Société Mathématique de France* **93**, 273–299 (1965).
20. M. Penz, A. Laestadius, E.I. Tellgren, M. Ruggenthaler and P.E. Lammert. Erratum: Guaranteed Convergence of a Regularized Kohn-Sham Iteration in Finite Dimensions [Phys. Rev. Lett. **123**, 037401 (2019)]. *Physical Review Letters* **125**, 249902 (2020).
21. Markus Penz, Andre Laestadius, Erik I. Tellgren and Michael Ruggenthaler. Guaranteed Convergence of a Regularized Kohn–Sham Iteration in Finite Dimensions. *Physical Review Letters* **123**, 037401 (2019).
22. R.G. Parr and W. Yang. *Density-funtional Theory of Atoms and Molecules*. Oxford University Press, New York (1989).
23. R.T. Rockafellar. A general correspondence between dual minimax problems and convex programs. *Pacific Journal of Mathematics* **25**, 597–611 (1968).
24. M. Reed and B. Simon. *Methods of Modern Mathematical Physics I: Functional Analysis*. Academic Press (1980).
25. K. Schmüdgen. *Unbounded Self-adjoint Operators on Hilbert Space*. Graduate Texts in Mathematics. Springer (2012).
26. R.E. Showalter. *Monotone Operators in Banach Space and Nonlinear Partial Differential Operators*. American Mathematical Society, Providence, RI (1991).
27. B. Simon. Hamiltonians Defined as Quadratic Forms. *Communications in Mathematical Phyics* **21**, 192–210 (1971).
28. G. Vignale and M. Rasolt. Density-Functional Theory in Strong Magnetic Fields. *Physical Review Letters* **59**, 2360–2363 (1987).
29. J. van Tiel. *Convex Analysis: An Introductory Text*. John Wiley & Sons Ltd., Chichester (1984).
30. K. Yosida. On the differentiability and the representation of one-parameter semi-group of linear operators. *Journal of the Mathematical Society of Japan* **1**, 15–21 (1948).

Chapter 6
Thermodynamic Limits of Electronic Systems

David Gontier, Jianfeng Lu, and Christoph Ortner

Abstract We review thermodynamic limits and scaling limits of electronic structure models for condensed matter. We discuss several mathematical ways to implement these limits in three models of increasing chemical complexity and mathematical difficulty: (1) Thomas–Fermi-like models; (2) Hartree–Fock-like models; and (3) Kohn–Sham density functional theory models.

6.1 Introduction

The goal of thermodynamic limits, as introduced in the 1960's [23, 39], is to obtain mathematical models for infinite systems of particles. The overarching strategy is to study systems with a finite number of particles (which can be described efficiently by well-posed mathematical models), to let the number of particles go to infinity while filling the space, and to pass to the limit in the governing equations in order to obtain a limit model. The purpose of the present chapter is to review results of this kind in the context of electronic structure models in condensed matter.

JL was supported in part by National Science Foundation under grant NSF-DMS1454939 and NSF-CHE2037263

CO was supported by ERC Starting Grant 335120 and EPSRC Grant EP/R043612/1

D. Gontier (✉)
CEREMADE, Université Paris-Dauphine, PSL University, Paris, France
e-mail: gontier@ceremade.dauphine.fr

J. Lu
Departments of Mathematics, Physics, and Chemistry, Duke University, Durham, NC, USA
e-mail: jianfeng@math.duke.edu

C. Ortner
Department of Mathematics, University of British Columbia, Vancouver, Canada
e-mail: christoph.ortner@ubc.ca

E. Cancès, G. Friesecke (eds.), *Density Functional Theory*, Mathematics and Molecular Modeling, https://doi.org/10.1007/978-3-031-22340-2_6

Two prototypical applications of thermodynamic limits are (1) to justify models of the *energy per unit cell* of a homogeneous crystal (infinite periodic system); (2) to obtain models for the formation energy of a crystalline defects without artefacts due to the boundary conditions. In this chapter, we review different models and mathematical methods to treat both of these scenarios. Extensive references will be provided throughout the chapter.

In both cases, one would like to describe an infinite system of electrons in a potential generated by an infinite collection of nuclei at positions $\mathcal{R} \subset \mathbb{R}^3$. In most studies, $\mathcal{R}$ is a periodic lattice (we write $\mathcal{R} = \mathbb{L}$ in this case), or a perturbation of it, describing for instance a crystal with a defect, or a deformed crystal (there are some studies for amorphous solids, in which case $\mathcal{R}$ is a random set [12]). In order to highlight the main ideas of the thermodynamic limit, we restrict ourselves to the simple case of a periodic crystal with one nucleus of charge 1 per unit cell. We denote by m_{a} the charge density of a single nucleus, which we take to be smooth to avoid some technical details: $m_{\mathrm{a}} \in C^\infty(\mathbb{R}^3)$ with compact support and $\int m_{\mathrm{a}} = 1$. The total nuclear density of the crystal is then given by

$$m^{\mathrm{nuc}}(\mathbf{r}) := \sum_{\mathbf{R}\in\mathcal{R}} m_{\mathrm{a}}(\mathbf{r} - \mathbf{R}). \tag{6.1}$$

In order to approximate this infinite distribution of charges, we consider a sequence of finite systems that *converges* to the infinite one: we choose $\mathcal{R}_N \subset \mathcal{R}$ a finite subset of size $|\mathcal{R}_N| = N$, and study the finite electronic problem with N electrons, in the external potential generated by N nuclei arranged along $\mathcal{R}_N$. The total nuclear density is

$$m_N^{\mathrm{nuc}}(\mathbf{r}) := \sum_{\mathbf{R}\in\mathcal{R}_N} m_{\mathrm{a}}(\mathbf{r} - \mathbf{R}). \tag{6.2}$$

Given this finite distribution of charges, m_N^{nuc}, one formulates a variational problem to equilibrate the electrons,

$$I(\mathcal{R}_N) := \inf_{\gamma_N} E(m_N^{\mathrm{nuc}}; \gamma_N),$$

where the infimum is taken over all admissible states γ_N representing a system of N electrons, and E describes the energy of finite electronic systems. Usually, γ_N represents the electron density or density matrix. One then aims to make various statements about the limits of the energy, $I(\mathcal{R}_N)$ and the optimal electron variable γ_N^0. Examples of important questions in this context include:

- *Does the sequence $N^{-1} I(\mathcal{R}_N)$ converge to some limit $W(\mathcal{R})$ as $N \to \infty$ and $\mathcal{R}_N \to \mathcal{R}$?* In this case, $W(\mathcal{R})$ would correspond to an average energy per electron or energy per unit volume.

- *Does the sequence of minimisers γ_N^0 have a limit γ^0 as $N \to \infty$?* The limiting object would describe an infinite sea of electrons in a crystal. Which equations are satisfied by the limiting object γ^0?
- If $\mathscr{R} \approx \mathscr{R}'$, then *does the energy difference $\delta I_N := I(\mathscr{R}_N) - I(\mathscr{R}'_N)$ have a limit δI?* If $\mathscr{R}$ describes a crystal and $\mathscr{R}'$ the same crystal with a local defect, then δI is the defect formation energy.

In the following sections, we focus on the case where the energy E is given by one of the following three models: the Thomas–Fermi–von Weizsäcker model in Sect. 6.2, the (reduced) Hartree–Fock model in Sect. 6.3, as well as Kohn–Sham density functional theory models, in Sect. 6.4.

In what follows, the energy of an N-electron state γ_N is denoted by $E(\gamma_N)$. The infimum of this energy is $I_N = \inf_{\gamma_N} E(\gamma_N)$, and represents the ground state energy of an N-electron system. The energy per unit electron is $W_N = N^{-1} I_N$. For the orbital-free TFW model the N-electron state is given by the electron density ρ_N.

Remark 6.1 Throughout this review we are technically misapplying the term "thermodynamic limit", but we do so in a way that is consistent with its usage in the analysis literature. In a strict sense, the thermodynamic limit was introduced to describe many-particle systems in a limit where boundary effects can be neglected, and to employ the law of large numbers, large deviation theory and ergodic theory as a transition from microscopic states to macroscopic states (variational principles, PDEs etc). A key goal of this framework was to model the situation when the corresponding thermodynamic functions (pressure, free energy, susceptibility, magnetisation, etc) can have singularities which appear at the critical value of the intensive parameter (temperature, chemical potential, etc). We refer to [23, 39] for detailed treatments of the subject. The connection between the present review and the classical usage of the term "thermodynamic limit" is the study of the *many-particle limit* in which boundary and domain size effects can be ignored, however there is no (genuine) statistical mechanics aspect.

6.2 The Thomas–Fermi–von Weizsäcker Model

Thomas–Fermi models describe electronic structure purely in terms of the electron density and electrostatic potential, and can therefore be interpreted as a system of two nonlinear PDEs. In this setting there is a mature theory and general results on the structure of the model and in particular the thermodynamic limit. The original Thomas–Fermi model, while attractive due to its simplicity, does not allow for the existence of molecules [31]. We will therefore focus on the Thomas–Fermi–von Weizsäcker (TFW) model [42]. Our presentation is primarily based on the monograph [14], but also incorporates more recent results [3, 35].

6.2.1 TFW Model for a Cluster

We consider N nuclei at locations $\mathscr{R}_N$ and with total charge m_N^{nuc}, see Eq. (6.2). The non-dimensionalised TFW energy, parametrised by $\mathscr{R}_N$ as a functional of the electron density ρ, is given by

$$E^{\text{TFW}}(\mathscr{R}_N,\rho) := \underbrace{\int_{\mathbb{R}^3}\left(c_{\text{W}}|\nabla\sqrt{\rho}|^2 + c_{\text{TF}}\,\rho^{5/3}\right)}_{\text{kinetic energy}} \underbrace{+\tfrac{1}{2}D(\rho - m_N^{\text{nuc}}, \rho - m_N^{\text{nuc}})}_{\text{Coulomb interaction}}, \tag{6.3}$$

where the Coulomb quadratic form is defined by

$$D(f,g) := \iint_{(\mathbb{R}^3)^2} \frac{f(\mathbf{r})g(\mathbf{r}')}{|\mathbf{r}-\mathbf{r}'|}\,\mathrm{d}\mathbf{r}\,\mathrm{d}\mathbf{r}'. \tag{6.4}$$

The first two terms of (6.3) represent the kinetic energy, and the third term is the Coulomb energy. This term can further be split into

$$\tfrac{1}{2}D(\rho - m_N^{\text{nuc}}, \rho - m_N^{\text{nuc}}) = \tfrac{1}{2}D(\rho,\rho) - D(\rho, m_N^{\text{nuc}}) + \tfrac{1}{2}D(m_N^{\text{nuc}}, m_N^{\text{nuc}}).$$

The first term is the direct term, or Hartree term, and describes the mean-field self-interaction of the electrons. The second is the electron-nuclei Coulomb interaction and the last term is the nuclei-nuclei one. Since we fixed the lattice $\mathscr{R}$ beforehand, the last term is constant, and does not play role in the minimisation problem. In addition, c_{W} and c_{TF} are positive physical constants that are irrelevant from a mathematical perspective; hence, for the sake of notational convenience, we set them to $c_{\text{W}} = c_{\text{TF}} = 1$.

The charge-neutral electronic ground state is obtained by solving

$$I^{\text{TFW}}(\mathscr{R}_N) := \inf\left\{E^{\text{TFW}}(\mathscr{R}_N,\rho),\quad \rho \geqslant 0,\ \textstyle\int_{\mathbb{R}^3}\rho = N,\ \sqrt{\rho}\in H^1(\mathbb{R}^3)\right\}. \tag{6.5}$$

A direct computation shows that $\rho \mapsto \int_{\mathbb{R}^3}|\nabla\sqrt{\rho}|^2$ is convex, which is a key ingredient to obtain the following result (see [2] for the proof).

Proposition 6.1 *There exists a unique minimiser ρ_N of (6.5). In addition, $\rho_N > 0$.*

It can then be readily checked, at least formally, that the minimiser satisfies the Euler–Lagrange equation

$$-\frac{\Delta\sqrt{\rho_N}}{\sqrt{\rho_N}} + \tfrac{5}{3}\rho_N^{2/3} - (m_N^{\text{nuc}} - \rho_N) * \frac{1}{|\cdot|} = \theta_N,$$

for some Lagrange multiplier $\theta_N \in \mathbb{R}$ associated with the charge neutrality constraint $\int_{\mathbb{R}^3} \rho = N$. It now becomes convenient to make the transformation $\rho_N = u_N^2$, where we may again assume that $u_N > 0$, and to introduce the total electrostatic potential

$$V_N^{\text{tot}} := (m_N^{\text{nuc}} - \rho_N) * \frac{1}{|.|} - \theta_N$$

to obtain the Euler–Lagrange system

$$-\Delta u_N + \tfrac{5}{3} u_N^{7/3} - V_N^{\text{tot}} u_N = 0, \tag{6.6a}$$

$$-\Delta V_N^{\text{tot}} = 4\pi \left(m_N^{\text{nuc}} - u_N^2\right). \tag{6.6b}$$

We have absorbed the Lagrange multiplier θ_N into the electrostatic potential V_N^{tot}, shifting it by a constant, which in particular implies that we need not have $V_N^{\text{tot}}(\mathbf{r}) \to 0$ as $|\mathbf{r}| \to \infty$. See [2, 14, 30] for the details of this argument.

In the remainder of our treatment of the TFW model we review results establishing the convergence of the electron ground state (u_N, V_N^{tot}) as $N \to \infty$, as the nuclei configuration $\mathscr{R}_N$ grows. To establish this limit, a convenient function space setting is provided by the spaces (we denote by $B_R(\mathbf{r}) := \{\mathbf{r}' \in \mathbb{R}^3,\ |\mathbf{r}' - \mathbf{r}| < R\}$)

$$H_{\text{unif}}^k := \left\{v \in H_{\text{loc}}^k(\mathbb{R}^3) \;\middle|\; \sup_{\mathbf{r} \in \mathbb{R}^d} \|v\|_{H^k(B_1(\mathbf{r}))} < \infty\right\}.$$

6.2.2 *Thermodynamic Limit Model*

To pass to the thermodynamic limit $N \to \infty$ we begin with an *infinite* collection of (smeared) nuclei at positions $\mathscr{R} \subset \mathbb{R}^3$. Here, $\mathscr{R}$ need not be a periodic lattice. Since the energy of an infinite system is not well-defined, the associated electronic ground state cannot be immediately characterised by an analogue of the variational problem (6.5). However, the nonlinear PDE representation (6.6) has a straightforward generalisation. Indeed, let $m^{\text{nuc}}(\mathbf{r}) := \sum_{\mathbf{R} \in \mathscr{R}} m_{\text{a}}(\mathbf{r} - \mathbf{R})$, then it is natural to suppose that the electronic ground state for the nuclei arrangement $\mathscr{R}$ is given by $\rho = u^2$, where (u, V^{tot}) solves

$$-\Delta u + \tfrac{5}{3} u^{7/3} - V^{\text{tot}} u = 0, \tag{6.7a}$$

$$-\Delta V^{\text{tot}} = 4\pi \left(m^{\text{nuc}} - u^2\right). \tag{6.7b}$$

To justify this model we will establish its well-posedness and show that it indeed arises as the thermodynamic limit of (6.5) (or, equivalently, (6.6)).

To that end, we need to impose restrictions on the configuration $\mathscr{R}$. We assume that $\mathscr{R}$ describes roughly uniformly distributed matter, and in particular contains no clusters with arbitrary high densities, and no holes of arbitrary large volume. More precisely, we require that there exist constants $c_{1,2}, C_{1,2} > 0$ such that

$$\forall \mathbf{r} \in \mathbb{R}^3,\ R > 0, \qquad c_1 R^3 - c_2 \leqslant \#\big(\mathscr{R} \cap B_R(\mathbf{r})\big) \leqslant C_1 R^3 + C_2. \tag{6.8}$$

This condition is equivalent to (H1) and (H2) in [14]. One of the main results of [14] is the well-posedness of (6.7).

Theorem 6.2 (Well-Posedness [14, Thm 6.10]) *Under the condition* (6.8)*, there exists a unique pair* $(u, V^{\rm tot}) \in H^4_{\rm unif} \times H^2_{\rm unif}$, *with* $u \geqslant 0$, *solving* (6.7)*. Moreover,* $\inf u > 0$.

The majority of the monograph [14] is devoted to the proof of Theorem 6.2. Let us recall some key ideas: A crucial observation is that the linear operator

$$L_N \varphi := -\Delta\varphi + \big(\tfrac{5}{3} u_N^{4/3} - V_N^{\rm tot}\big)\varphi,$$

which is a kind of linearisation of (6.6a), is non-negative. This already hints at the existence of a strong stability property. Indeed, adapting this observation (see, for example, the proof of [14, Lemma 5.3]), the following result is shown in [35, Thm. 3.1], closely following variants of the same result in [14, Sec. 5.3] and [3].

Lemma 6.3 (Stability and Uniqueness [35, Thm. 3.1]) *Let* $\mathscr{R}, \mathscr{R}_* \subset \mathbb{R}^3$ *satisfy* (6.8)*, let* $m_1^{\rm nuc}, m_2^{\rm nuc}$ *be associated nuclear charge densities, and suppose that* $(u, V^{\rm tot}), (u_*, V_*^{\rm tot}) \in H^4_{\rm unif} \times H^2_{\rm unif}$ *are corresponding solutions to* (6.7) *with* $\inf u_i > 0$. *Then, there exist constants* $C \geq 0$ *and* $\alpha > 0$, *depending only on* $m_{\rm a}$ *and on the constants in* (6.8)*, such that*

$$|u(\mathbf{r}) - u_*(\mathbf{r})| + |V^{\rm tot}(\mathbf{r}) - V_*^{\rm tot}(\mathbf{r})| \leqslant C\left(\int_{\mathbb{R}^3} \mathrm{e}^{-\alpha|\mathbf{r}-\mathbf{z}|} \left|m^{\rm nuc} - m_*^{\rm nuc}\right|^2(\mathbf{z})\,\mathrm{d}\mathbf{z}\right)^{1/2}. \tag{6.9}$$

Lemma 6.3 immediately implies uniqueness of solutions to (6.7), but it is much stronger in that it also provides a pointwise stability that quantifies the dependence of the local electronic structure on the far-field. We will return to this result in Sect. 6.2.3.

To establish the existence of solutions, we use a thermodynamic limit argument. At the same time, this also justifies the model (6.7). To that end, we specify a sequence of clusters approximating $\mathscr{R}$: let $\mathscr{R}_N \subset \mathscr{R}$ and $r_N \uparrow \infty, c > 0$ such that

$$B_{r_N}(\mathbf{0}) \cap \mathscr{R} \subset \mathscr{R}_N \subset B_{r_N + c}(\mathbf{0}) \cap \mathscr{R}. \tag{6.10}$$

For each N, Proposition 6.1 yields the existence and uniqueness of an electronic ground state $(u_N, V_N^{\rm tot})$ solving (6.6).

The stability result stated in Lemma 6.3 already hints at the possibility of uniform a priori estimates on the solutions (u_N, φ_N), and indeed one can prove that

$$\|u_N\|_{H^4_{\mathrm{unif}}} + \|V_N^{\mathrm{tot}}\|_{H^2_{\mathrm{unif}}} \leqslant C, \tag{6.11}$$

where C depends only on m_{a} and on the constants in (6.8), see [14, Prop. 3.5] for the (involved and technical) details. A key technical step estimating the Lagrange multiplier, which we have hidden, is due to [40]. A summary of the proof, providing also quantitative estimates, can be found in [35, Prop. 6.1].

With the a priori estimate (6.11) in hand, we may extract a subsequence $(u_{N_j}, V_{N_j}^{\mathrm{tot}}) \rightharpoonup (u, V^{\mathrm{tot}})$ weakly in H^2_{loc} (we say $f_j \rightharpoonup f$ weakly in H^k_{loc} if $f_j|_D \rightharpoonup f|_D$ weakly in $H^k(D)$ for all bounded domains D) and it is straightforward to deduce that the limit satisfies the PDE (6.7). Since the limit is unique, it follows that the entire sequence converges. We obtain the following result.

Theorem 6.4 (Convergence) *Let $\mathscr{R}_N, \mathscr{R}$ satisfy* (6.10) *and* (6.8)*, and let $(u_N, V_N^{\mathrm{tot}}) \in H^4_{\mathrm{loc}} \times H^4_{\mathrm{loc}}$ and $(u, V^{\mathrm{tot}}) \in H^4_{\mathrm{loc}} \times H^4_{\mathrm{loc}}$ be the corresponding solutions of* (6.6) *and* (6.7) *respectively. Then*

$$(u_N, V_N^{\mathrm{tot}}) \rightharpoonup (u, V) \qquad \textit{weakly in } H^4_{\mathrm{loc}} \times H^4_{\mathrm{loc}}.$$

In particular, the convergence is locally uniform.

6.2.3 Discussion

We conclude this section with a series of further remarks about possible extensions and consequences of the results.

(1) Convergence rates: In Lemma 6.3 the conditions on the second solution u_* can be weakened. This allows us to prove that "well inside" the approximate domain $\mathscr{R}_N$, the solutions $(u_N, V_N^{\mathrm{tot}})$ and (u, V^{tot}) are exponentially close. Specifically, in [35, Proposition 4.1] it is shown that there are constants $C > 0$ and $\alpha > 0$ independent of $\mathbf{r} \in B_{r_N}$ and $N \in \mathbb{N}$ such that

$$\forall N \in \mathbb{N},\ \forall \mathbf{r} \in \mathbb{R}^3, \quad \left|u_N(\mathbf{r}) - u(\mathbf{r})\right| + \left|V_N^{\mathrm{tot}}(\mathbf{r}) - V^{\mathrm{tot}}(\mathbf{r})\right| \leqslant C\mathrm{e}^{-\alpha\, \mathrm{dist}(\mathbf{r}, \partial B_{r_N}(0))}.$$

Choosing $r_N' \ll r_N$ this readily translates into the convergence rate

$$\|u_N - u\|_{L^\infty(B_{r_N'})} + \|V_N^{\mathrm{tot}} - V^{\mathrm{tot}}\|_{L^\infty(B_{r_N'})} \leqslant C\mathrm{e}^{-\alpha(r_N - r_N')}.$$

The same argument also shows that boundary effects decay exponentially into the bulk of the cluster and justifies the common usage of buffer regions in electronic structure calculations.

(2) *Surfaces and sheets:* Our assumption (6.8) expressly disallows configurations with large sections of vacuum, in particular surfaces and 2D materials. Indeed, not only the mathematics but also the underlying physics changes in such situations. We refer to [3, 6, 32] for related results that go beyond this limitation.

(3) *The Dirac correction:* The Thomas–Fermi–Dirac–von Weizsäcker model adds an additional correction term to the energy functional,

$$E^{\mathrm{TFDW}}(\mathscr{R}_N,\rho) = \int_{\mathbb{R}^3}\left(|\nabla\sqrt{\rho}|^2+\rho^{5/3}\right)+\tfrac{1}{2}D(\rho-m_N^{\mathrm{nuc}},\rho-m_N^{\mathrm{nuc}})\underbrace{-c\int_{\mathbb{R}^3}\rho^{4/3}}_{\text{Dirac exchange}},$$

where the additional term can be interpreted as a model for the exchange of energy of the electrons. The additional challenge is that E^{TFDW} is no longer convex. We are unaware of an in-depth treatment of this model, but refer to [14, Sec 3.6.3] for a discussion of possible avenues and [20] for results on the related Cauchy–Born scaling limit for this model.

(5) *Further orbital-free DFT models:* Most orbital-free DFT models used in computations for practical materials have a more complicated functional form for kinetic energy and exchange-correlation energy, see e.g., [43, 44]. It is shown in [4] that the Wang–Teter kinetic energy [43] is not bounded from below, and thus the thermodynamic limit is ill-posed. The more complicated density-dependent orbital free kinetic-energy functionals, such as the Wang–Govind–Carter functional [44], are yet to be mathematically understood.

(6) *Charge screening:* The stability result Lemma 6.3 clearly shows that interaction in the TFW model is exponentially localised, *despite* the presence of the long-range Coulomb interaction. This can be interpreted as a very general screening result. Consider two configurations $\mathscr{R}, \mathscr{R}_*$ satisfying (6.8), and which coincide outside a large ball of radius $r > 0$: $\mathscr{R}\setminus B_r = \mathscr{R}_*\setminus B_r$. Let $(u, V^{\mathrm{tot}}), (u_*, V_*^{\mathrm{tot}})$ be the corresponding solutions. Then Lemma 6.3 implies

$$|u(\mathbf{r})-u_*(\mathbf{r})|+|V^{\mathrm{tot}}(\mathbf{r})-V_*^{\mathrm{tot}}(\mathbf{r})| \leqslant Ce^{-\alpha|\mathbf{r}|}. \tag{6.12}$$

For instance, if $\mathscr{R}_*$ contains more atoms than $\mathscr{R}$, one might expect the potentials to satisfy $V_*^{\mathrm{tot}}(\mathbf{r}) \approx V^{\mathrm{tot}}(\mathbf{r}) + Q/|\mathbf{r}|$ as $\mathbf{r}\to\infty$, where Q would be the extra effective charge. However, according to (6.12), this is not the case: the extra charge is completely screened. This is a very general fact in TFW theory, see [9, 35] for details.

One can also take into account the relaxation of the configuration $\mathscr{R}_*$ due to the presence of the defect. In this case, instead of an exponential decay, we obtain (see [17])

$$|u(\mathbf{r}) - u_*(\mathbf{r})| + |V^{\text{tot}}(\mathbf{r}) - V_*^{\text{tot}}(\mathbf{r})| \lesssim |\mathbf{r}|^{-2},$$

and, since $|\mathbf{r}|^{-2} = o(|\mathbf{r}|^{-1})$, we deduce that charges are screened, see [9] and [35, Thm. 4.1] for the details.

6.2.4 Scaling Limit

A question related to but distinct from the thermodynamic limit arises when considering the derivation of a continuum model for elastic material response from an underlying electronic structure model. Such scaling limits for the TFW model were first studied in [5], but the following discussion builds on the results of [35]. Specifically, the stability and locality estimate (6.3) yields stronger and quantitative results. In addition, for the sake of consistency with the KS-DFT case in Sect. 6.4, we consider a periodic instead of an infinite-domain setting.

6.2.4.1 Spatial Decomposition of Energy

In preparation we first mention another useful consequence of the stability and locality estimate found in Lemma 6.3, which is also of independent interest: Let $\mathscr{R} \subset \mathbb{R}^3$ be a finite configuration of nuclei or an infinite configuration satisfying (6.8) and let (u, V^{tot}) be the associated solutions to (6.6), then we define the energy density

$$\mathcal{E}(\mathscr{R}; \cdot) = |\nabla u|^2 + u^{10/3} + \frac{1}{8\pi}|\nabla V^{\text{tot}}|^2.$$

If $\mathscr{R}$ is finite then one may readily check [35, Eq. 4.18] that $I^{\text{TFW}}(\mathscr{R}) = \int_{\mathbb{R}^3} \mathcal{E}(\mathscr{R}; \mathbf{r})\, d\mathbf{r}$. We may therefore think of

$$I^{\text{TFW}}(\mathscr{R}, \Omega) := \int_\Omega \mathcal{E}(\mathscr{R}; \mathbf{r})\, d\mathbf{r}$$

as the energy stored in a compact sub-domain $\Omega \subset \mathbb{R}^3$. This intuition is further supported by the following result, proven in [35, Proof of Thm. 4.2], which is closely related to (6.3): there exist constants C, γ such that

$$\forall \mathbf{R} \in \mathscr{R},\ \forall \mathbf{r} \in \mathbb{R}^3, \qquad \left| \frac{\partial \mathcal{E}(\mathscr{R}; \mathbf{r})}{\partial \mathbf{R}} \right| \leqslant C e^{-\alpha|\mathbf{r}-\mathbf{R}|}. \tag{6.13}$$

In [35, Sec. 4.4] this observation is used to demonstrate exponential locality of interatomic forces in the TFW model. In the following section we use it for an easy derivation of the Cauchy–Born scaling limit.

6.2.4.2 The Cauchy–Born Scaling Limit

Consider a periodic arrangement $\mathscr{R} = B\mathbb{Z}^3$ of nuclei, where $B \in \mathbb{R}^{3\times 3}$ is a non-singular matrix, and let (u, V^{tot}) describe the corresponding TFW ground-state. Then, uniqueness of solutions to (6.6) implies that they must observe the same periodicity. In particular, we can define the Cauchy–Born energy function, which represents the energy stored in $\Omega_B := B[0, 1)^3$, that is

$$W^{\text{cb}}(B) := I^{\text{TFW}}(B\mathbb{Z}^3, \Omega_B) := \int_{\Omega_B} \mathcal{E}(\mathscr{R}; \mathbf{r})\, d\mathbf{r}.$$

A deformed configuration of the crystal is described by a continuum deformation field $Y(\mathbf{x}) = \mathbf{x} + U(\mathbf{x})$ where U is smooth and $\mathscr{R}$-periodic. We assume that U is chosen such that Y is bijective, i.e. a proper deformation.

Given a parameter $\epsilon > 0$ describing the inverse length-scale over which the deformation varies we define a deformed crystalline configuration by

$$\mathscr{R}^\epsilon := \left\{ Y_\epsilon(\mathbf{x}) := \epsilon^{-1} Y\big(\epsilon \mathbf{x}\big) \mid \mathbf{x} \in B\mathbb{Z} \right\}.$$

This definition encodes the *Cauchy–Born hypothesis* that nuclei in a crystal follow the continuum deformation field. An example of such an atomistic configuration shadowing a continuum field is given in Fig. 6.1(left). It must be emphasised that this is a simplifying assumption that is only approximately valid in specific deformation regimes and for simple crystals; see [22, 36] for in-depth discussions.

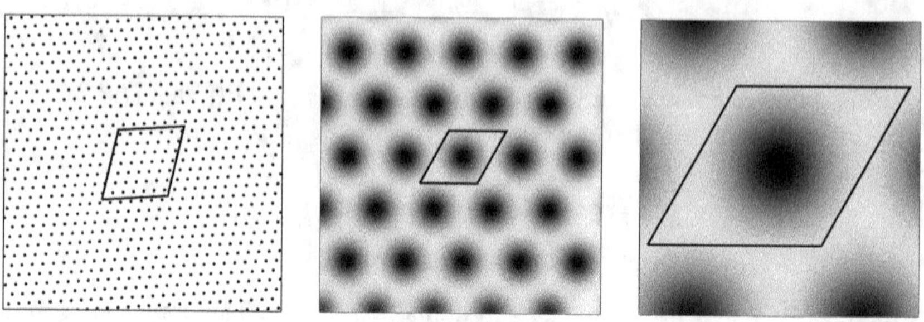

Fig. 6.1 Illustration of the scaling limit. Left: a deformation of a homogeneous crystal varying slowly relative to the scale of atoms; Center: the blow-up of a small section is nearly homogeneous; Right: The near-homogeneous section can be approximately represented by a single unit cell

In the following we concern ourselves with the *scaling limit* $\epsilon \to 0$ of the stored elastic energy per unit undeformed volume. To that end, let $\Omega_\epsilon := \epsilon^{-1}\Omega$ then $\mathscr{R}^\epsilon$ has periodic cells Ω_ϵ or alternatively also $Y_\epsilon(\Omega_\epsilon)$. That is, we may write the elastic energy per unit volume as

$$E^\epsilon := |\Omega^\epsilon|^{-1} I^{\mathrm{TFW}}(\mathscr{R}^\epsilon, \Omega_\epsilon) = \frac{1}{|\Omega^\epsilon|}\int_{\Omega^\epsilon} \mathcal{E}(\mathscr{R}^\epsilon, \mathbf{r})\,\mathrm{d}\mathbf{r} = \frac{\epsilon^3}{\det B}\int_{Y_\epsilon(\Omega_\epsilon)} \mathcal{E}(\mathscr{R}^\epsilon, \mathbf{r})\,\mathrm{d}\mathbf{r}.$$

We remark that the electronic coordinates $\mathbf{r}$ belong to the deformed space, i.e. it is natural to write $\mathbf{r} = Y_\epsilon(\mathbf{x})$.

The key observation now is that the locality estimate (6.9) on the electronic structure and its extension to the energy density (6.13) suggest that to predict the value of $\mathcal{E}(\mathscr{R}^\epsilon, \mathbf{r})$ for a non-uniform deformation varying at the macroscopic scale, it is not required to be aware of the global configuration $\mathscr{R}^\epsilon$ but it is sufficient to know the *local deformation* near $\mathbf{r}$. This is illustrated by Fig. 6.1.

To make this precise, let $\mathbf{n} \in \mathbb{L}$, $t_\mathbf{n} = Y_\epsilon(\mathbf{n})$ and $F_\mathbf{n} = \nabla Y_\epsilon(\mathbf{n})$, then

$$Y_\epsilon(\mathbf{x}) = t_\mathbf{n} + F_\mathbf{n}(\mathbf{x} - \mathbf{n}) + O(\epsilon),$$

for $\mathbf{x}$ in any bounded neighbourhood of $\mathbf{n}$. A Taylor expansion of $\mathcal{E}$ with respect to $\mathscr{R}^\epsilon$, employing (6.13), implies that for $\mathbf{r} \in Y_\epsilon(\mathbf{n} + \Omega)$ we have

$$\mathcal{E}(\mathscr{R}^\epsilon, \mathbf{r}) = \mathcal{E}\big(F\mathbb{Z}^d, \mathbf{r}\big) + O(\epsilon).$$

After integrating over one cell in the undeformed crystal, which in deformed coordinates becomes $Y_\epsilon(\mathbf{n} + \Omega)$, and making a further elementary approximation, we obtain

$$\int_{Y_\epsilon(\mathbf{n}+\Omega)} \mathcal{E}(\mathscr{R}^\epsilon, \mathbf{r})\,\mathrm{d}\mathbf{r} = W(F_\mathbf{n}) + O(\epsilon).$$

Finally, after summing over all such unit cells, using the scale-invariance of the deformation gradient one obtains the following convergence result. The second order error $O(\epsilon^2)$ is obtained by a more careful exploitation of the point symmetry in simple crystal lattices.

Theorem 6.5 *Let $Y \in C^4_{\mathrm{per}}(\Omega)$, then*

$$E^\epsilon(Y) = \int_\Omega W^{\mathrm{cb}}(\nabla_Y(\mathbf{x}))\,\mathrm{d}\mathbf{x} + O(\epsilon^2) \qquad \textit{as } \epsilon \to 0.$$

The presentation of this section follows unpublished notes. Related results using different techniques were first presented in [5, Theorem 5, case (i)]. The paper [5] also considers several generalisations, including domains with boundaries and alternative scaling regimes.

6.3 The Reduced Hartree–Fock Model

We now focus on the reduced Hartree–Fock (rHF) model. In this model, a fermionic system with N electrons is described by a one-body density matrix γ, which is a self-adjoint operator on $L^2(\mathbb{R}^3)$ satisfying the Pauli principle $0 \le \gamma \le 1$, and with trace N.

Together with the spectral theorem, this implies that γ is of the form

$$\gamma = \sum_{i=1}^{\infty} n_i |\varphi_i\rangle\langle\varphi_i|, \quad \text{with} \quad 1 \ge n_1 \ge n_2 \ge \cdots \ge 0, \quad \text{and} \quad \mathrm{Tr}(\gamma) = \sum_{i=1}^{\infty} n_i = N.$$

Here, the functions $(\varphi_i)_i$ form an orthonormal basis of eigenfunctions in $L^2(\mathbb{R}^3)$, and are called the *orbitals*, and the numbers $0 \le n_i \le 1$ are the *occupation numbers*. To such a one-body density matrix, we can associate its density $\rho_\gamma(\mathbf{r}) := \gamma(\mathbf{r}, \mathbf{r}) = \sum_i n_i |\varphi_i|^2(\mathbf{r})$.

In the potential generated by the nuclei at $\mathscr{R}_N \subset \mathbb{R}^3$, the rHF energy of a state γ is

$$E^{\mathrm{rHF}}(\mathscr{R}_N, \gamma) := \frac{1}{2}\mathrm{Tr}(-\Delta\gamma) + \frac{1}{2}D(\rho_\gamma - m_N^{\mathrm{nuc}}, \rho_\gamma - m_N^{\mathrm{nuc}}), \tag{6.14}$$

where m_N^{nuc} is the total nuclear density defined in (6.1). Compared with (6.3), we see that the kinetic energy part $\int |\nabla\sqrt{\rho}|^2 + \rho^{5/3}$ has been replaced by

$$\frac{1}{2}\mathrm{Tr}(-\Delta\gamma) := \frac{1}{2}\sum_{i=1}^{\infty} n_i \|\nabla\varphi_i\|^2_{L^2(\mathbb{R}^3)}.$$

In particular, this model is no longer a function of the density ρ, but of the one-body density matrix γ. The energy of the configuration $\mathscr{R}_N$ is given by the minimisation problem

$$I^{\mathrm{rHF}}(\mathscr{R}_N) := \inf\left\{E^{\mathrm{rHF}}(\mathscr{R}_N, \gamma),\ 0 \le \gamma = \gamma^* \le 1,\ \mathrm{Tr}(\gamma) = N\right\}. \tag{6.15}$$

Closely related to the rHF model is the Hartree–Fock (HF) model, where the exchange term is considered. This term is a correction to the direct Hartree energy, and is due to the fermionic nature of the particles. The HF model reads

$$E^{\mathrm{HF}}(\mathscr{R}_N, \gamma) := E^{\mathrm{rHF}}(\mathscr{R}_N, \gamma) - \frac{1}{2}\iint_{(\mathbb{R}^3)^2} \frac{|\gamma(\mathbf{r}, \mathbf{r}')|^2}{|\mathbf{r} - \mathbf{r}'|} \mathrm{d}\mathbf{r}\mathrm{d}\mathbf{r}'. \tag{6.16}$$

Since the HF model is not convex in γ, we only have partial results for it, and most of the following facts only hold for the rHF model.

In the thermodynamic limit, we consider a regular periodic lattice $\mathbb{L} \subset \mathbb{R}^3$, and a sequence of arrangements $\mathscr{R}_N \subset \mathbb{L}$ with $|\mathscr{R}_N| = N$, and satisfying (6.10). We want to study the energy per unit cell $N^{-1} I^{\mathrm{rHF}}(\mathscr{R}_N)$ as N goes to infinity.

The finite electron model (6.15) was introduced and studied by Solovej [41]. The existence of an optimiser γ_N^0 is provided here. The thermodynamic limit was latter studied by Catto, Le Bris and Lions in a series of papers. In [13], the authors announced their results, later proved in [15] (for the models presented here) and [16] (for the pure-state version of these problems, i.e. when γ is further constrained to be a rank-N projector). They prove the thermodynamic limit for the rHF model:

$$\lim_{N\to\infty} \frac{1}{N} I^{\mathrm{rHF}}(\mathscr{R}_N) = I_{\mathrm{per}}^{\mathrm{rHF}} + \frac{\mathfrak{m}}{2}, \tag{6.17}$$

where $I_{\mathrm{per}}^{\mathrm{rHF}}$ can be characterised by a minimisation periodic problem, that we describe in the next section, and $\mathfrak{m}$ is the Madelung constant, see (6.18) below. They conjectured that a similar result should hold for the HF model. Finally, they proved that the limiting problem $I_{\mathrm{per}}^{\mathrm{rHF}}$ (and its HF counterpart $I_{\mathrm{per}}^{\mathrm{HF}}$) is indeed well-posed. We discuss this point in the next section.

6.3.1 The Periodic Model

In order to write the limiting periodic model, as introduced by Catto, Le Bris and Lions, we define the set of periodic one-body density matrices (recall that in our simple setting, we expect one electron per unit cell)

$$\mathcal{P}_{\mathrm{per}} := \left\{ 0 \le \gamma = \gamma^* \le 1, \ \forall \ell \in \mathbb{L}, \ \tau_\ell \gamma = \gamma \tau_\ell, \ \underline{\mathrm{Tr}}\, \gamma = 1 \right\},$$

where τ_ℓ is the translation operator $\tau_\ell f(\mathbf{x}) := f(\mathbf{x} - \ell)$. Such a periodic density matrix has an $\mathbb{L}$-periodic density $\rho_\gamma(\mathbf{r}) := \gamma(\mathbf{r}, \mathbf{r})$. Its trace per unit cell $\underline{\mathrm{Tr}}\, \gamma$ is defined by

$$\underline{\mathrm{Tr}}\, \gamma := \mathrm{Tr}\, (\mathbb{1}_\Omega \gamma \mathbb{1}_\Omega) = \int_\Omega \rho_\gamma,$$

where Ω is a unit cell associated to the lattice $\mathscr{R} = \mathbb{L}$. A periodic density matrix $\gamma \in \mathcal{P}_{\mathrm{per}}$ has an $\mathbb{L}$-periodic density $\rho_\gamma(\mathbf{x}) := \gamma(\mathbf{x}, \mathbf{x}) \in L^1_{\mathrm{per}}(\Omega)$. We let G be the $\mathbb{L}$-periodic Coulomb kernel, solution to

$$-\Delta G := 4\pi \sum_{\mathbf{R} \in \mathbb{L}} \left(\delta_{\mathbf{R}} - |\Omega|^{-1} \right), \quad \text{and} \quad \int_\Omega G = 0,$$

and we introduce the periodic Coulomb quadratic form $D_{\rm per}(\cdot,\cdot)$ defined for periodic functions by (compare with (6.4))

$$D_{\rm per}(f,g) := \iint_{(\Omega)^2} f(\mathbf{r})g(\mathbf{r}')G(\mathbf{r}-\mathbf{r}')\mathrm{d}\mathbf{r}\mathrm{d}\mathbf{r}'.$$

The Madelung constant appearing in (6.17) is defined to be

$$\mathfrak{m} := \lim_{\mathbf{r}\to\mathbf{0}}\left(G(\mathbf{r}) - \frac{1}{|\mathbf{r}|}\right). \tag{6.18}$$

Note that since $F(\mathbf{r}) := G(\mathbf{r}) - |\mathbf{r}|^{-1}$ satisfies $\Delta F = 0$ on Ω, the function F is indeed smooth on Ω, hence has a well-defined value at $\mathbf{r} = \mathbf{0}$. The Madelung constant somehow describes the mismatch between the full space Coulomb kernel $|\mathbf{r}|^{-1}$ and the periodic kernel $G(\mathbf{r})$ (which can a priori be defined up to a constant).

With these notations, the limit $W_{\rm per}^{\rm rHF}$ for the perfect crystal is defined as the minimisation problem [15]

$$W_{\rm per}^{\rm rHF} := \inf\left\{E_{\rm per}^{\rm rHF}(\mathbb{L},\gamma),\ \gamma\in\mathcal{P}_{\rm per}\right\}, \tag{6.19}$$

where the energy per unit cell $E_{\rm per}^{\rm rHF}$ is

$$E_{\rm per}^{\rm rHF}(\mathbb{L},\gamma) := \frac{1}{2}\underline{\rm Tr}\,(-\Delta\gamma) + \frac{1}{2}D_{\rm per}(\rho_\gamma - m_{\rm per}^{\rm nuc}, \rho_\gamma - m_{\rm per}^{\rm nuc}), \tag{6.20}$$

and where $m_{\rm per}^{\rm nuc}(\mathbf{x}) := \sum_{\ell\in\mathbb{L}} m_{\rm a}(\mathbf{x}-\ell)$ is the periodic nuclear density. Comparing this expression with (6.14), we see that all terms have been "normalised" to take into account the periodicity of the infinite system.

The fact that $I_{\rm per}^{\rm rHF}$ is a well-posed problem was proved by Catto et al. in [15]. Later in [7], Cancès, Deleurence and Lewin proved that the minimiser γ satisfies the Euler–Lagrange equations

$$\gamma = \mathbb{1}\left(H_\gamma \le \varepsilon_F\right),\quad H_\gamma := -\frac{1}{2}\Delta + (\rho_\gamma - m_{\rm per}^{\rm nuc}) * G.$$

Here, $\varepsilon_F \in \mathbb{R}$ is the *Fermi level*. The operator H_γ is the mean-field one-body Hamiltonian of the crystal, which is a self-adjoint operator that commutes with $\mathbb{L}$-translations. Its spectral properties are well understood thanks to the Bloch transform [37, Chapter XIII], and its spectrum is composed of bands and gaps. When ε_F is in a gap, the crystal is an insulator, while when ε_F is in a band, it is a metal.

6.3.2 Supercell Methods, and Periodic Thermodynamic Limits

Once the periodic problem (6.19) has been written and justified, it is possible to understand its properties from other approaches. In [7] (see also [18]), Cancès, Deleurence and Lewin proved that this problem was also the limit of another thermodynamic limit. Their idea was to start directly with a periodic problem on the large supercell $\Omega_L := L\Omega$ with $N = L^3$ electrons, and take the limit $L \to \infty$. In other words, instead of working with one-body density matrices γ acting on the whole space $L^2(\mathbb{R}^3)$, they looked at one-body density matrices acting on the *supercell* $L^2_{\text{per}}(\Omega_L)$. We therefore define

$$\mathcal{P}^L_{\text{per}} := \left\{ \gamma \text{ acting on } L^2_{\text{per}}(\Omega_L),\ 0 \le \gamma = \gamma^* \le 1,\ \mathrm{Tr}_L\, \gamma = L^3 \right\},$$

where we set for simplicity $\mathrm{Tr}_L := \mathrm{Tr}_{\mathcal{S}(L^2_{\text{per}}(\Omega_L))}$. A one-body operator $\gamma \in \mathcal{P}^L_{\text{per}}$ has an $L\mathbb{L}$-periodic density $\rho_\gamma \in L^1_{\text{loc}}$. We also define the $L\mathbb{L}$-periodic Coulomb kernel as $G_L(\mathbf{x}) := L^{-1} G(L^{-1}\mathbf{x})$, and the L-periodic Coulomb quadratic form defined for $L\mathbb{L}$-periodic functions by

$$D_L(f, g) := \iint_{(\Omega_L)^2} f(\mathbf{r}) g(\mathbf{r}) G_L(\mathbf{r} - \mathbf{r}') \mathrm{d}\mathbf{r} \mathrm{d}\mathbf{r}'.$$

The supercell model is given by a periodic minimisation problem of the form

$$I^{\text{rHF}}_{\text{per},L}(\mathcal{R}^L) := \inf \left\{ E^{\text{rHF}}_{\text{per},L}(\mathcal{R}^L, \gamma),\ \gamma \in \mathcal{P}^L_{\text{per}} \right\}, \tag{6.21}$$

with the supercell energy

$$E^{\text{rHF}}_{\text{per},L}(\mathcal{R}^L, \gamma) := \frac{1}{2} \mathrm{Tr}_L\, (-\Delta_L \gamma) + \frac{1}{2} D_L(\rho_\gamma - m^{\text{nuc}}_L, \rho_\gamma - m^{\text{nuc}}_L). \tag{6.22}$$

Here, $\mathcal{R}^L$ is an $L\mathbb{L}$-periodic lattice (for instance $\mathcal{R}^L = \mathbb{L}$, or a deformation of it, see below), and m^{nuc}_L is the nuclear density $m^{\text{nuc}}_L := \sum_{\mathbf{R} \in \mathcal{R}_L} m_{\text{a}}(\mathbf{x} - \mathbf{R})$. In the case $\mathcal{R}^L = \mathbb{L}$, there are L^3 nuclei and electrons per supercell.

Even in the perfect crystal case, that is when $\mathcal{R}^L = \mathbb{L}$, the problems (6.19)–(6.20) and (6.21)–(6.22) differ. In (6.19), the minimisation is performed for γ acting on the whole space $L^2(\mathbb{R}^3)$, while in (6.21), it is performed for γ acting on the supercell $L^2(\Omega_L)$. These two types of operators cannot be compared, and it is not obvious a priori that there is a link between the two problems. Still, both operators give $\mathbb{L}$-periodic densities, which can be compared. This important fact allows us to prove the convergence [7]

$$\lim_{L \to \infty} \frac{1}{L^3} I^{\text{rHF}}_{\text{per},L}(\mathbb{L}) = W^{\text{rHF}}_{\text{per}}. \tag{6.23}$$

The result was later refined in [25], where the authors proved that, in the insulating case (see [10] for the metallic case), the convergence is exponential, in the sense that there exist constants $C \in \mathbb{R}$ and $\alpha > 0$ such that

$$\left| W_{\text{per}}^{\text{rHF}} - \frac{1}{L^3} I_{\text{per},L}^{\text{rHF}}(\mathbb{L}) \right| + \|\rho_{\text{per}}^0 - \rho_{\text{per, L}}^0\|_{L^\infty} \leq C e^{-\alpha L}, \tag{6.24}$$

where ρ_{per}^0 and $\rho_{\text{per},L}^0$ are the electronic densities of the periodic and supercell minimisers respectively, seen here as $\mathbb{L}$-periodic functions. This exponential convergence comes from the analyticity of the Bloch representation.

This means that the full space problem $W_{\text{per}}^{\text{rHF}}$ can be well-approximated by the supercell model $I_{\text{per},L}^{\text{rHF}}$. This latter problem can be studied efficiently from a numerical point of view, thanks to the Bloch transform. As noticed in [25], the supercell model corresponds exactly to a uniform discretisation of the Brillouin zone, as described in a famous paper by Monkhorst [34]. For metallic systems, the exact rate of convergence is unknown in the general case (see [10] for details).

Remark 6.6 When studying supercell methods for non-convex problems, symmetry breaking may happen (see e.g. [27, 38]). In this case, the density of the $L\mathbb{L}$-periodic problem may not be $\mathbb{L}$-periodic, and the periodic problem may not be the limit of supercell models.

To sum up, the energy per unit cell $W_{\text{per}}^{\text{rHF}}$ is the limit of two different sequences, namely

$$W_{\text{per}}^{\text{rHF}} = \lim_{N\to\infty} \left(\frac{1}{N} I^{\text{rHF}}(\mathscr{R}_N) \right) - \frac{\mathfrak{m}}{2}, \quad \text{and} \quad W_{\text{per}}^{\text{rHF}} = \lim_{L\to\infty} \left(\frac{1}{L^3} I_{\text{per},L}^{\text{rHF}}(\mathbb{L}) \right).$$

In the first limit, the crystal is seen as the limit of finite systems. This is the correct physical limit, as a real crystal is indeed always finite. However, we expect the convergence to be slow, due to boundary effects. On the other hand, the second limit has no real physical meaning, but gives an exponential rate of convergence in the insulating case.

6.3.3 *Local Defects in Crystals, in the Reduced Hartree–Fock Model*

We now discuss how to define the energy of a defect inside a crystal. We would like to define this energy as the difference between the energy of a crystal with a defect, and the energy of the crystal without the defect. Unfortunately, these two quantities are infinite. Also, the model with defect does not have an underlying periodicity, hence there is no notion of *energy per unit cell* in this case. One way to define the energy of a defect is through a thermodynamic limit procedure.

Let $\mathscr{R} := \mathbb{L}$ be the arrangement of nuclei for the perfect crystal, and let $\mathscr{R}_*$ be the one for the crystal with (local) defect, that is such that $\mathscr{R}$ and $\mathscr{R}_*$ coincide outside a ball of radius $r > 0$. The nuclear charge of the defect is therefore

$$\nu := m_*^{\mathrm{nuc}} - m^{\mathrm{nuc}} = \sum_{\mathbf{R} \in \mathscr{R}_* \setminus \mathscr{R}} m_{\mathrm{a}}(\cdot - \mathbf{R}) - \sum_{\mathbf{R} \in \mathscr{R} \setminus \mathscr{R}_*} m_{\mathrm{a}}(\cdot - \mathbf{R}).$$

For $L \in \mathbb{N}^*$, we can consider $\mathscr{R}_*^L$, the $L\mathbb{L}$-periodic arrangement which is equal to $\mathscr{R}_*$ on Ω_L (note that $\mathscr{R}^L = \mathscr{R} = \mathbb{L}$). In [7], the authors consider the supercell energy of the defect ν, defined by

$$J_L(\nu) := I_{\mathrm{per},L}^{\mathrm{rHF}}(\mathscr{R}_*^L) - I_{\mathrm{per},L}^{\mathrm{rHF}}(\mathbb{L}).$$

Here there is a slight complication: since we do not know a priori how many electrons should be in the system, we should not fix the number of electrons, but rather the Fermi level (grand canonical ensemble). To keep this presentation simple, we do not comment on this point, but refer to [7] for details.

Although the two quantities $I_{\mathrm{per},L}^{\mathrm{rHF}}(\mathscr{R}_*^L)$ and $I_{\mathrm{per},L}^{\mathrm{rHF}}(\mathbb{L})$ diverge to infinity with rate $O(L^3)$, the difference of the two quantities stays finite in the limit, and we can define the energy of the defect as

$$J_\infty(\nu) := \lim_{L \to \infty} J_L(\nu).$$

In [26], the authors prove that the corresponding rate of convergence is $O(L^{-1})$. This slow rate of convergence is due to the spurious interaction between the defect and its periodic images, a fact predicted in [29, 33]. This makes the supercell method quite a poor numerical method in this case.

It turns out that the limit $\mathcal{J}_\infty(\nu)$ can be characterised as a minimisation problem on a set of "defect" operators. It is unclear whether this last problem can be tackled directly with efficient numerical methods (see also [8]).

6.4 Scaling Limit for Kohn–Sham DFT

In this section, we discuss Kohn–Sham models. For a finite system with N electrons described by a one-body density matrix γ, the Kohn–Sham density functional takes the form

$$\mathcal{E}^{\mathrm{KS}}(\mathscr{R}_N, \gamma) := \frac{1}{2}\mathrm{Tr}(-\Delta\gamma) + \frac{1}{2}D(\rho_\gamma - m_N^{\mathrm{nuc}}, \rho_\gamma - m_N^{\mathrm{nuc}}) + E_{\mathrm{xc}}[\rho_\gamma], \qquad (6.25)$$

where m_N^{nuc} is the total nuclear density defined in (6.1). Compared with the reduced Hartree–Fock model (6.14), the Kohn–Sham model includes the exchange-correlation energy $E_{xc}[\rho_\gamma]$, where we have adopted the notation for an LDA or GGA type functional, and thus it can be explicitly written in terms of ρ_γ as

$$\text{(LDA)} \qquad E_{xc}[\rho_\gamma] = \int_{\mathbb{R}^3} \epsilon_{xc}(\rho_\gamma(\mathbf{r}))\mathrm{d}\mathbf{r}, \qquad \text{or}$$

$$\text{(GGA)} \qquad E_{xc}[\rho_\gamma] = \int_{\mathbb{R}^3} \epsilon_{xc}\big(\rho_\gamma(\mathbf{r}), |\nabla\sqrt{\rho_\gamma(\mathbf{r})}|^2\big)\mathrm{d}\mathbf{r}.$$

Even the simplest exchange-correlation functionals used in practice have complicated expressions, and hence will not be given explicitly here. Most of them are non-convex in ρ, such as, for instance the Dirac exchange term $E_{\mathrm{x}}^{\mathrm{Dirac}}[\rho] = -C_D \int \rho(\mathbf{x})^{4/3}\mathrm{d}\mathbf{x}$. Thus, even the existence of a minimiser to the Kohn–Sham DFT problem becomes a difficult question. The existence of minimisers for LDA functionals has been proved in [1, 24, 28], while for GGA-type functionals, it remains open with only preliminary results available (see the case of $N = 1$ in [1]).

We will henceforth assume LDA-type exchange-correlation functionals. The variation of the functional (6.25) gives rise to the Kohn–Sham equations for $\gamma = \sum_{i=1}^{N} |\psi_i\rangle\langle\psi_i|$, where ψ_i are the Kohn–Sham orbitals, solutions to

$$H^{\mathrm{KS}}[\rho_\gamma]\psi_i = E_i\psi_i \quad \text{where} \quad H^{\mathrm{KS}}[\rho] := -\frac{1}{2}\Delta + V_{\mathrm{H}}[\rho] + V_{\mathrm{xc}}[\rho] \tag{6.26}$$

with ρ_γ the density associated with γ and the Hartree and exchange-correlation potentials respectively given by

$$V_{\mathrm{H}}[\rho] = (\rho - m_N^{\mathrm{nuc}}) * \frac{1}{|\cdot|} \quad \text{and} \quad V_{\mathrm{xc}}[\rho] := \epsilon_{\mathrm{xc}}'(\rho(\cdot)).$$

The Kohn–Sham equations (6.26) form a set of nonlinear eigenvalue problems, as the effective Hamiltonian operator $H^{\mathrm{KS}}[\rho_\gamma]$ depends on the solution γ. We remark that in general there is no guarantee that the Kohn–Sham orbitals of the minimisers of (6.25) correspond to the lowest N eigenvalues of the self-consistent Hamiltonian, though in practice this is often assumed and known as the *Aufbau principle*.

Due to the non-convexity and hence possible symmetry breaking, see Remark 6.6, the thermodynamic limit of Kohn–Sham DFT with exchange-correlation functionals is very challenging and not much progress has been made.

To understand the behaviour of electronic structure in materials, we take a typical starting point of modelling in materials science—the periodic Kohn–Sham model with supercell Ω. This can be formulated using the density matrix similar to the periodic Hartree–Fock model discussed in Sect. 6.3.1. A periodic Kohn–Sham energy is of the form

$$E_{\mathrm{per}}^{\mathrm{KS}}(\mathbb{L}, \gamma) = E_{\mathrm{per}}^{\mathrm{rHF}}(\mathbb{L}, \gamma) + E_{\mathrm{per}}^{\mathrm{xc}}[\rho_\gamma],$$

where the rHF energy $E_{\text{per}}^{\text{rHF}}$ was defined in (6.20). This is the rHF model with the addition of a periodic exchange-correlation energy. One could follow the lines of Sect. 6.3.1 to study the thermodynamic limit. We can also consider the following alternative formulation, presented in [21], and that we present now.

The self-consistent Kohn–Sham eigenvalue problem (6.26) can be reformulated as a fixed point equation for the density

$$\rho(\mathbf{r}) = \mathcal{F}^{\text{KS}}[\rho](\mathbf{r}) := \left[\frac{1}{2\pi \mathrm{i}} \oint_{\mathscr{C}} \left(\lambda - H^{\text{KS}}[\rho]\right)^{-1} \mathrm{d}\lambda\right](\mathbf{r}, \mathbf{r}), \tag{6.27}$$

where $\mathscr{C}$ is a contour in the resolvent set separating the first N eigenvalues of H^{KS} from the rest of the spectrum (assuming a spectral gap). The right-hand side of (6.27) denotes the diagonal of the kernel of the density matrix viewed as an integral operator.

6.4.1 The Periodic Kohn–Sham DFT Model

For a periodic system with Bravais lattice $\mathbb{L}$, we can write a similar equation. We introduce the periodic Kohn–Sham Hamiltonian associated with some $\mathbb{L}$-periodic density ρ, given by

$$H_{\text{per}}^{\text{KS}}[\rho] = -\frac{1}{2}\Delta + V_{\text{H,per}}[\rho] + V_{\text{xc}}[\rho],$$

where the periodic Hartree potential solves

$$-\Delta V_{\text{H,per}}[\rho] = 4\pi(\rho - m^{\text{nuc}})$$

with periodic boundary condition, where m^{nuc} is understood as a background charge density given by the nuclei (to be specified below). As the potential is periodic, the Bloch–Floquet theory applies to the Hamiltonian. In particular, the spectrum of $H_{\text{per}}^{\text{KS}}[\rho]$ has a band structure. For each $\mathbf{k} \in \Omega^*$, the first Brillouin zone, the Bloch waves solve the eigenvalue problem

$$\left(\frac{1}{2}\left(-\mathrm{i}\nabla + \mathbf{k}\right)^2 + V_{\text{H,per}}[\rho] + V_{\text{xc}}[\rho]\right) u_{n,\mathbf{k}}(\mathbf{x}) = E_{n,\mathbf{k}} u_{n,\mathbf{k}}(\mathbf{x}), \quad \forall n = 1, 2, \ldots,$$

with periodic boundary condition on Ω. The spectrum is given by

$$\sigma(H_{\text{per}}^{\text{KS}}[\rho]) = \bigcup_n \bigcup_{\mathbf{k} \in \Omega^*} E_{n,\mathbf{k}}.$$

This is known as the band structure, see Fig. 6.2 for an illustration.

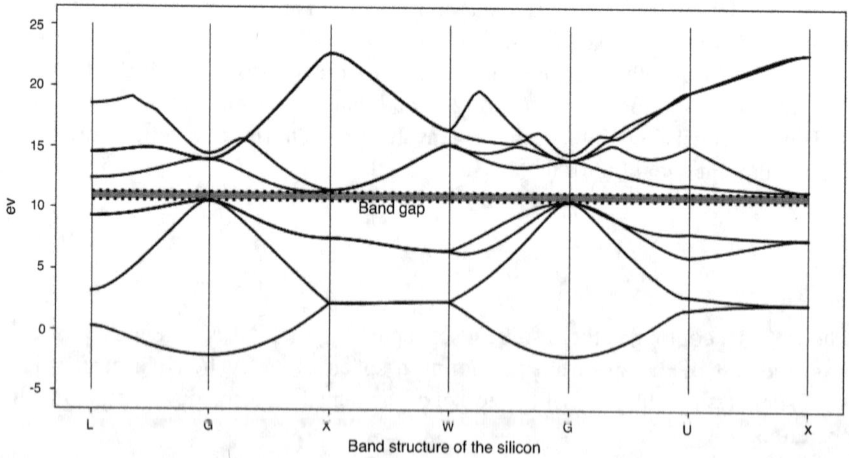

Fig. 6.2 Schematic band structure of crystalline silicon, along various lines connecting high-symmetry points in the first Brillouin zone Ω^*. The first 4 bands are occupied and separated by a band gap from the higher bands

If the first N bands are occupied and there exists a gap between the occupied and unoccupied spectrum (in physical terms, the system is an insulator), the Kohn–Sham map can be generalised to the periodic setting as

$$\mathcal{F}^{\mathrm{KS}}_{\mathrm{per}}[\rho](\mathbf{x}) := \left[\frac{1}{2\pi\mathrm{i}}\oint_{\mathscr{C}} \left(\lambda - H^{\mathrm{KS}}_{\mathrm{per}}[\rho]\right)^{-1} d\lambda\right](\mathbf{x},\mathbf{x}), \tag{6.28}$$

where the contour $\mathscr{C}$ lies in the resolvent set and separates the occupied and unoccupied spectra. For the periodic Kohn–Sham model, we thus recast the problem as a fixed point equation

$$\rho = \mathcal{F}^{\mathrm{KS}}_{\mathrm{per}}[\rho]. \tag{6.29}$$

Using the electron density ρ as the basic variable is more convenient in studying the scaling limit than the Kohn–Sham orbitals (Bloch waves), which will be discussed next.

6.4.2 *Scaling Limit for the Periodic Model*

Starting from the fixed point equation (6.29), valid for instance for a periodic configuration $\mathbb{L}$, we would like to find other solutions when the crystal is deformed.

We consider the electronic structure of an elastically deformed system in the scaling limit where the lattice parameter goes to 0. To set up the atomic configuration, we assume a lattice structure for the undeformed system. The atoms are located at $\varepsilon\mathbb{L}$, where $\mathbb{L}$ is a Bravais lattice, and the lattice parameter ε will serve as the scaling parameter in the limit, which can be understood as the ratio of the lattice parameter and the characteristic length scale of the system.

For simplicity, we assume that Ω coincides with the unit cell of $\mathbb{L}$. Thus, in Ω, atoms are located at $\Omega \cap \varepsilon\mathbb{L}$. The system consists of ε^{-3} atoms and correspondingly $N\varepsilon^{-3}$ electrons, where N is the number of valence electrons per atom.

Fix a smooth function $u : \mathbb{R}^3 \to \mathbb{R}^3$ of the form $u(\mathbf{x}) = B\mathbf{x} + u_{\mathrm{per}}(\mathbf{x})$, where B is a 3×3 matrix and u_{per} is periodic with respect to Ω. The deformed atom locations are

$$\mathbf{Y}_i^\varepsilon = \mathbf{X}_i^\varepsilon + u(\mathbf{X}_i^\varepsilon), \qquad \mathbf{X}_i^\varepsilon \in \varepsilon\mathbb{L}.$$

Correspondingly the background charge distribution is given by

$$m^{\mathrm{nuc},\varepsilon}(\mathbf{y}) = \sum_{\mathbf{X}_i \in \mathbb{L}} m_{\mathrm{a}}^\varepsilon(\mathbf{y} - \mathbf{Y}_i^\varepsilon),$$

where $m_{\mathrm{a}}^\varepsilon$ is the rescaled version of the charge contribution from each individual atom (recall that the lattice parameter is scaled to ε):

$$m_{\mathrm{a}}^\varepsilon = \varepsilon^{-3} m_{\mathrm{a}}(\cdot/\varepsilon).$$

As the lattice parameter is scaled to be ε, the Kohn–Sham Hamiltonian needs to be rescaled correspondingly as

$$H_{\varepsilon,u}^{\mathrm{KS}}[\rho] = -\frac{\varepsilon^2}{2}\Delta + V_{\mathrm{H}}^\varepsilon[\rho] + V_{\mathrm{xc}}[\rho],$$

where the Hartree potential $V_{\mathrm{H}}^\varepsilon$ solves

$$-\Delta V_{\mathrm{H}}^\varepsilon = 4\pi\varepsilon(\rho - m^{\mathrm{nuc},\varepsilon}).$$

Thus the electron density of the deformed system is determined by the fixed point of the Kohn–Sham map

$$\rho(\mathbf{r}) = \mathcal{F}_{\varepsilon,u}^{\mathrm{KS}}[\rho](\mathbf{x}) := \left[\frac{1}{2\pi \mathrm{i}} \oint_{\mathscr{C}} \left(\lambda - H_{\varepsilon,u}^{\mathrm{KS}}[\rho]\right)^{-1} \mathrm{d}\lambda\right](\mathbf{r}, \mathbf{r}). \tag{6.30}$$

In order to make sense of the Kohn–Sham map defined in (6.27) and (6.30), we require a gap between the occupied and unoccupied spectrum, and thus we make the following assumption for the undeformed system. The gap of the effective Hamiltonian of the perturbed system follows from a perturbation argument.

Assumption 6.7 (Insulating Undeformed System) There exists a Ω-periodic $\rho_0 \in C^\infty(\mathbb{R}^3)$, that is positive and uniformly bounded away from zero, such that

- The spectrum of the Hamiltonian $H^{\mathrm{KS}}_{\varepsilon=1,0}[\rho_0]$ has a positive gap between the occupied and unoccupied spectra.
- ρ_0 is a fixed point of the Kohn–Sham map:

$$\rho_0(\mathbf{r}) = \mathscr{F}^{\mathrm{KS}}_{\varepsilon=1,0}[\rho_0](\mathbf{r}) = \left[\frac{1}{2\pi \mathrm{i}} \oint_{\mathscr{C}} \left(\lambda - H^{\mathrm{KS}}_{\varepsilon=1,0}[\rho_0]\right)^{-1} \mathrm{d}\lambda\right](\mathbf{r}, \mathbf{r}),$$

where $\mathscr{C}$ is a contour in the resolvent set enclosing the occupied spectrum.

6.4.3 Cauchy–Born Rule for Electronic Structure

The question of the scaling limit is to characterise the electron density, as a solution to the Kohn–Sham equation (6.30), when the deformation is elastic in the sense that the deformation gradient is not too large. This is motivated by the Cauchy–Born rule for passing from atomistic models to elastic models, where the analogous question for DFT is to pass from electronic structure models to continuum elastic models. The scaling limit for Thomas–Fermi–von Weizsäcker model was studied by Blanc, Le Bris and Lions in [5], see Sect. 6.2.4.

For the Kohn–Sham type models, the scaling limit was proved by E and Lu in [19, 21] under the stability conditions on the level of linear response of the undeformed system. In order to state the stability assumptions, let us introduce the linearised Kohn–Sham map for the undeformed system

$$(\mathcal{L}_0 w) = \left[\frac{\delta \mathscr{F}^{\mathrm{KS}}_{\varepsilon=1,0}[\rho_0]}{\delta \rho}(w)\right].$$

It has been established [21] that $\mathcal{L}_0$ is a bounded linear operator on the space $\mathcal{X}_n := \dot{H}^{-1}_{\mathrm{per}}(n\Omega) \cap H^2_{\mathrm{per}}(n\Omega)$ for every $n \in \mathbb{N}$, where $H^2_{\mathrm{per}}(n\Omega)$ stands for the periodic Sobolev space with square integrable second derivatives and $\dot{H}^{-1}_{\mathrm{per}}(n\Omega)$ is the homogeneous Sobolev space with index -1 on the domain $n\Omega$.

Assumption 6.8 (Stability of Charge Density Wave Response) For every $n \in \mathbb{N}$, the operator $I - \mathcal{L}_0$ is uniformly invertible as an operator on $\mathcal{X}_n$.

Physically the stability assumption states that the undeformed crystal is stable with respect to spontaneous charge density wave perturbation at every scale. In particular, this prevents the possibility of symmetry breaking as $\varepsilon \to 0$.

We also define the macroscopic permittivity tensor for the undeformed crystal as

$$\mathsf{E}_0 = \frac{1}{2}(\mathsf{A}_0 + \mathsf{A}_0^*) + \frac{1}{4\pi}\mathsf{I},$$

where the 3×3 matrix A_0 is given by

$$\mathsf{A}_{0,\alpha\beta} := -2\Re \sum_{i}^{\text{occ}} \sum_{a}^{\text{unocc}} \int_{\Omega^*} \frac{\langle u_{a,\mathbf{k}}, i\partial_{\mathbf{k}_\alpha} u_{i,\mathbf{k}}\rangle \langle u_{a,\mathbf{k}}, i\partial_{\mathbf{k}_\beta} u_{i,\mathbf{k}}\rangle^*}{E_{i,\mathbf{k}} - E_{a,\mathbf{k}}} \frac{\mathrm{d}\mathbf{k}}{|\Omega^*|}$$
$$- \big\langle g_\alpha, \delta_\rho V_{\text{eff}} (I - \mathcal{L}_0)^{-1} g_\beta \big\rangle$$

where

$$g_\alpha(\mathbf{r}) := 2\Re \sum_{i}^{\text{occ}} \sum_{a}^{\text{unocc}} \int_{\Omega^*} \frac{\langle u_{a,\mathbf{k}}, i\partial_{\mathbf{k}_\alpha} u_{i,\mathbf{k}}\rangle}{E_{i,\mathbf{k}} - E_{a,\mathbf{k}}} u_{i,\mathbf{k}}^*(\mathbf{r}) u_{a,\mathbf{k}}(\mathbf{r}) \frac{\mathrm{d}\mathbf{k}}{|\Omega^*|},$$

and $\delta_\rho V_{\text{eff}}$ is the linearisation of the effective potential operator: $V_{\text{eff}}[\rho] = V_{\text{H}}[\rho] + V_{\text{xc}}[\rho]$ at ρ_0 for the undeformed crystal. The dielectric permittivity for the reduced Hartree–Fock theory has been studied in [11].

Assumption 6.9 (Stability of Dielectric Response) The macroscopic permittivity tensor for the undeformed crystal E_0 is positive definite.

The main result of [21] establishes the Cauchy–Born rule for the electronic structure.

Theorem 6.10 ([21, Thm. 5.1]) *Under Assumptions 6.7, 6.8 and 6.9, if the deformation gradient is sufficiently small, then there exists a ρ_u^ε satisfying the Kohn–Sham fixed point equation* (6.30)*, and furthermore, ρ_u^ε can be locally approximated by the Cauchy–Born rule:*

$$\|\rho_u^\varepsilon - \varepsilon^{-3}\rho_{CB}(\mathbf{r}/\varepsilon; \nabla u(\mathbf{r}))\|_{L^\infty} \lesssim \varepsilon^{1/2} \|\rho_u^\varepsilon\|_{L^\infty},$$

where $\rho_{CB}(\cdot; A)$ is the electron density of a homogeneous deformed system with $u(\mathbf{x}) = A\mathbf{x}$ (which is well-defined provided $|A|$ is not too large).

The main technical ingredients of the proof of the theorem is a two-scale analysis of the linearised Kohn–Sham map. As a part of the analysis, the effective potential and the macroscopic dielectric response of the deformed crystal can also be characterised, we refer the readers to [21] for details.

References

1. A. Anantharaman and E. Cancès. Existence of minimizers for Kohn–Sham models in quantum chemistry. *Ann. Inst. Henri Poincaré* **26**, 2425–2455 (2009).
2. R. Benguria, H. Brezis and E.H. Lieb. The Thomas–Fermi–von Weizsäcker theory of atoms and molecules. *Commun. Math. Phys.* **79**, 167–180 (1981).
3. X. Blanc. Unique solvability of a system of nonlinear elliptic PDEs arising in solid state physics. *SIAM J. Math. Anal.* **38**, 1235–1248 (2006).

4. X. Blanc and E. Cancès. Nonlinear instability of density-independent orbital-free kinetic-energy functionals. *J. Chem. Phys.* **122**, 214106 (2005).
5. X. Blanc, C. Le Bris and P.-L. Lions. From molecular models to continuum mechanics. *Arch. Ration. Mech. Anal.* **164**, 341–381 (2002).
6. E. Cancès, L. Cao and G. Stoltz. Removing a slab from the Fermi sea: the reduced Hartree-Fock model. *Nonlinearity* **33**, 156–195 (2020).
7. E. Cancès, A. Deleurence and M. Lewin. A new approach to the modeling of local defects in crystals: the reduced Hartree–Fock case. *Commun. Math. Phys.* **281**, 129–177 (2008).
8. E. Cancès, A. Deleurence and M. Lewin. Non-perturbative embedding of local defects in crystalline materials. *J. Phys. Condensed Matter* **20**, 294213 (2008).
9. E. Cancès and V. Ehrlacher. Local defects are always neutral in the Thomas–Fermi–von weiszäcker theory of crystals. *Arch. Ration. Mech. Anal.* **202**, 933–973 (2011).
10. E. Cancès, V. Ehrlacher, D. Gontier, A. Levitt and D. Lombardi. Numerical quadrature in the Brillouin zone for periodic Schrödinger operators. *Numer. Math.* **144**, 479–526 (2020).
11. E. Cancès and M. Lewin. The dielectric permittivity of crystals in the reduced Hartree–Fock approximation. *Arch. Ration. Mech. Anal.* **1**, 139–177 (2010).
12. E. Cancès, S. Lahbabi and M. Lewin. Mean-field models for disordered crystals. *J. Math. Pures Appl.* **100**, 241–274 (2013).
13. I. Catto, C. Le Bris and P.-L. Lions. Sur la limite thermodynamique pour des modèles de type Hartree et Hartree–Fock. *C. R. Acad. Sci. Paris* **327**, 259–266 (1998).
14. I. Catto, C. Le Bris and P.-L. Lions. *The mathematical theory of thermodynamic limits: Thomas–Fermi type models*. Oxford University Press (1998).
15. I. Catto, C. Le Bris and P.-L. Lions. On the thermodynamic limit for Hartree–Fock type models. *Ann. Inst. H. Poincaré (C)* **18**, 687–760 (2001).
16. I. Catto, C. Le Bris and P.-L. Lions. On some periodic Hartree-type models for crystals. *Ann. Inst. H. Poincaré (C)* **19**, 143–190 (2002).
17. H. Chen, F. Q. Nazar and C. Ortner. Geometry equilibration of crystalline defects in quantum and atomistic descriptions. *Math. Models Meth. Appl. Sc.* **29**, 419–492 (2019).
18. A. Deleurence. *Modélisation mathématique et simulation numérique de la structure électronique de cristaux en présence des défauts ponctuels*. PhD thesis, Paris Est (2008).
19. W.E and J. Lu. The electronic structure of smoothly deformed crystals: Wannier functions and the Cauchy–Born rule. *Arch. Ration. Mech. Anal.* **199**, 407–433 (2011).
20. W. E and J. Lu. Stability and the continuum limit of the spin-polarized Thomas–Fermi-Dirac–von Weizsäcker model. *Journal of Mathematical Physics* **53**, 115615 (2012).
21. W. E and J. Lu. *The Kohn–Sham equation for deformed crystals*. Memoir Amer. Math. Soc. 221 (2013).
22. W. E and P. Ming. Cauchy–Born rule and the stability of crystalline solids: Static problems. *Arch. Ration. Mech. Anal.* **183**, 241–297 (2007).
23. H.-O. Georgii. *Gibbs measures and phase transitions*. De Gruyter Studies in Mathematics Vol. 9, Walter de Gruyter (2011).
24. D. Gontier. Existence of minimizers for Kohn–Sham within the local spin density approximation. *Nonlinearity* **28**, 57–76 (2014).
25. D. Gontier and S. Lahbabi. Convergence rates of supercell calculations in the reduced Hartree–Fock model. *ESAIM: Math. Model. Num. Anal.* **50**, 1403–1424 (2016).
26. D. Gontier and S. Lahbabi. Supercell calculations in the reduced Hartree–Fock model for crystals with local defects. *Appl. Math. Res. Express*, 1–64 (2016).
27. D. Gontier, M. Lewin and F. Nazar. The nonlinear Schrödinger equation for orthonormal functions: Existence of ground states. *Arch. Ration. Mech. Anal.* **240**, 1203–1254 (2021).
28. C. Le Bris. *Quelques problèmes mathématiques en chimie quantique moléculaire*. PhD thesis, Ecole Polytechnique (1993).
29. M. Leslie and M. Gillan. The energy and elastic dipole tensor of defects in ionic crystals calculated by the supercell method. *J. Phys. C* **18**, 973–982 (1985).

30. E.H. Lieb. Thomas-Fermi and related theories of atoms and molecules. In: *The Stability of Matter: From Atoms to Stars: Selecta of Elliott H. Lieb*, edited by W. Thirring, pp. 259–297, Springer, Berlin-Heidelberg (1997).
31. E.H. Lieb and B. Simon. The Thomas–Fermi theory of atoms, molecules and solids. *Adv. Math.* **23**, 22–116 (1977).
32. J. Lu, V. Moroz and C.B. Muratov. Orbital-Free density functional theory of Out-of-Plane charge screening in graphene. *J. Nonlinear Sci.* **25**, 1391–1430 (2015).
33. G. Makov and M. Payne. Periodic boundary conditions in *ab initio* calculations. *Phys. Rev. B* **51**, 4014–4022 (1995).
34. H. Monkhorst and J. Pack. Special points for Brillouin-zone integrations. *Phys. Rev. B* **13**, 5188–5192 (1976).
35. F.Q. Nazar and C. Ortner. Locality of the Thomas-Fermi-von Weizsäcker equations. *Arch. Ration. Mech. Anal.* **224**, 817–70 (2017).
36. C. Ortner and F. Theil. Justification of the Cauchy–Born approximation of elastodynamics. *Arch. Ration. Mech. Anal.* **207**, 1025–1073 (2013).
37. M. Reed and B. Simon. *Methods of Modern Mathematical Physics. Analysis of Operators, Vol. IV*. Academic Press (1978).
38. J. Ricaud. Symmetry breaking in the periodic Thomas–Fermi–Dirac–von Weizsäcker model. Ann. H. Poincaré **19**, 3129–3177 (2018).
39. D. Ruelle. *Statistical mechanics: Rigorous results*. World Scientific (1999).
40. J.P. Solovej. Universality in the Thomas–Fermi–von Weizsäcker model of atoms and molecules. *Commun. Math. Phys.* **129**, 561–598 (1990).
41. J.P. Solovej. Proof of the ionization conjecture in a reduced Hartree–Fock model. *Invent. Math.* **104**, 291–311 (1991).
42. C.F. Von Weizsäcker. Zur theorie de kernmassen. *Z. Phys.* **96**, 431 (1935).
43. L.-W. Wang and M.P. Teter. Kinetic-energy functional of the electron density. *Phys. Rev. B* **45**, 13196 (1992).
44. Y.A. Wang, N. Govind and E.A. Carter. Orbital-free kinetic-energy density functionals with a density-dependent kernel. *Phys. Rev. B* **60**, 16350 (1999).

Chapter 7
Numerical Methods for Kohn–Sham Models: Discretization, Algorithms, and Error Analysis

Eric Cancès, Antoine Levitt, Yvon Maday, and Chao Yang

Abstract This chapter aims at presenting the main algorithms used to solve the Kohn–Sham models, as well as the current state-of-the-art of the numerical analysis of these models and algorithms.

7.1 Introduction

There are several avatars of the Kohn–Sham model, each characterized by

- the type of systems it describes: molecules or clusters, perfect crystals, disordered materials,
- the choice of the exchange-correlation functional: local, semilocal, nonlocal, orbital-dependent (see Chap. 1 by Toulouse in this volume),
- the electrons explicitly taken into account: all electrons vs pseudo-electron models (see Sect. 7.4),
- the underlying variational model: the standard Kohn–Sham models derived from the Levy-Lieb formulation of DFT, the extended Kohn–Sham models derived from the Valone–Lieb construction of DFT, the positive-temperature Kohn–Sham models [50].

E. Cancès (✉) · A. Levitt
CERMICS, Ecole des Ponts and Inria Paris, Marne-la-Vallée, France
e-mail: eric.cances@enpc.fr; antoine.levitt@inria.fr

Y. Maday
Sorbonne Université, CNRS, Université Paris Cité, Laboratoire Jacques-Louis Lions (LJLL), Paris, France

Institut Universitaire de France, Paris, France
e-mail: yvon.maday@sorbonne-universite.fr

C. Yang
Computational Research Division, Lawrence Berkeley National Laboratory, Berkeley, CA, USA
e-mail: cyang@lbl.gov

E. Cancès, G. Friesecke (eds.), *Density Functional Theory*, Mathematics and Molecular Modeling, https://doi.org/10.1007/978-3-031-22340-2_7

The most appropriate discretization method and the algorithm best suited to solve the resulting discretized problem depend on these four criteria.

The molecular orbital formulation of the standard Kohn–Sham model for finite systems, and the most common types of exchange-correlation functionals are presented in Sect. 7.2.1. The standard Kohn–Sham model can be reformulated in terms of the (one-body) density matrix. The density matrix formalism is also the right one to set the extended and finite-temperature Kohn–Sham models, introduced in Sect. 7.2.2. All Kohn–Sham type models are variational models: some energy (or free energy) functional is minimized over some "nice" set, which can be a manifold of N-tuples of orthonormal vectors (molecular orbital formulation of the standard Kohn–Sham model), a manifold of orthogonal projectors (density matrix formulation of the standard Kohn–Sham model), or the convex hull of a manifold of orthogonal projectors (extended and finite-temperature Kohn–Sham models). The geometries of these objects are discussed in Sect. 7.2.3.

Kohn–Sham models are used in both quantum chemistry and condensed matter physics (we will use this terminology here to refer to solid-state physics, the physics of liquids, and materials science all together). The models and numerical techniques used in these two fields are closely related, but however significantly different.

Quantum chemistry aims at modeling molecules, and understanding their properties and reactions. Most chemical reactions of practical interest take place in the liquid phase, and involve solvated molecules. For some systems, ignoring the solvent effects and considering the molecule of interest in vacuum is a sufficient approximation. For others, the solvent has to be taken into account. Explicit solvent models, in which each and every solvent molecule in a large simulation box are dealt with at the same level of theory as the solute molecule of interest, are commonly used in classical molecular dynamics, but are much too computationally expensive for DFT. A much cheaper alternative is to use implicit solvent models, in which the solute molecule is placed in a molecular-shape cavity and the solvent is modeled by a continuous dielectric medium of permittivity equal to the zero-frequency dielectric permittivity of the solvent. We refer to [138] for reviews of implicit solvent models. The rate-limiting computation is then the electronic state of a molecule in the environment of the solvent, which is mathematically and computationally similar to isolated molecules in vacuum. We therefore limit ourselves to this case in this chapter. The resulting equations are most often discretized in atomic basis sets; this approach is presented in Sect. 7.3.1.

In contrast, models used in condensed matter physics deal with extended systems containing virtually an infinite number of nuclei and electrons. The latter models can be seen as thermodynamic limits of the former models when the number of particles tends to infinity. Of course, it is not possible to simulate an infinite number of individual particles on a computer. There are two usual ways to circumvent this difficulty:

1. if the system is periodic (perfect crystals), a Bloch–Floquet transform can be used to reduce the problem with infinitely many particles set on the whole space

to a family of problems with finitely many particles set on the periodic cell with periodic boundary conditions (see e.g. [121]);
2. infinite aperiodic systems ("real" crystals, involving defects, amorphous solids, liquids) are usually dealt with using the supercell method: a chunk of the system is placed in a large rectangular box and appropriate boundary conditions are imposed on the Kohn–Sham orbitals to get rid of surface effects. This approach is also used in the field of homogenization, where the supercell is called a representative volume element (RVE).

It follows that in practice Kohn–Sham problems encountered in condensed matter physics are set on periodic cells, which either have a physical meaning (as for perfect crystals) or are chosen for numerical convenience (as in the supercell method). For this reason, it is natural to discretize them in plane-wave basis sets, in which Kohn–Sham orbitals and electronic densities are approximated by a finite linear combination of Fourier modes. This discretization method is presented in Sect. 7.3.2. Notably, plane-wave discretization methods perform extremely poorly on genuine all-electron Kohn–Sham models because of the singularities of the Coulomb potential generated by the nuclei. Most Kohn–Sham simulations in the condensed phase make use of pseudopotentials: only valence electrons are explicitly dealt with, valence Kohn–Sham orbitals (which have cusps at nuclear positions) are replaced by smoother valence pseudo-orbitals, and the local singular Kohn–Sham potential, by a smoother but nonlocal pseudopotential. This approximation is discussed in Sect. 7.4.

Whatever the system, the exchange-correlation functional and the discretization method that is used, the resulting discretized Kohn–Sham models can be formulated as constrained optimization problems. The generic methodology of constrained optimization proves too general, and the specific simple form of the constraints (which form Riemannian manifolds) allow for efficient custom algorithms, termed direct minimization algorithms. The Euler–Lagrange equations of the minimization problem can be recast as a nonlinear eigenvector problem, which has an appealing physical interpretation as a mean-field equation. Algorithms based on a fixed-point solution of these equations are known as self-consistent field (SCF) algorithms. In both cases, two challenges of different natures have to be faced: (i) ensure convergence toward a (hopefully) global minimum, and (ii) accelerate local convergence, which can be very slow due to ill-conditioning.

Algorithms to solve Kohn–Sham problems in atomic basis sets on the one hand, and in plane-wave basis sets on the other hand, have been developed almost independently by chemists and physicists respectively. This can be explained by the fact that the corresponding problems are quite different from a numerical point of view. Indeed, the performance of a given algorithm heavily depends on the physical properties of the system as well as on the discretization methods. Problems of global convergence are usually more severe in quantum chemistry, where bond breaking, open shells and limited basis sets contribute to the existence of local minima. These problems are relatively less common in the more rigid solid-state context. Similarly, ill-conditioning comes from different sources: overcompleteness of the basis set and degeneracies in quantum chemistry, low and high-frequency modes in condensed-matter. Finally, the problems caused by large systems are also different, and treated using specific methodologies. We will present the quantum chemistry perspective (atomic basis sets) in Sect. 7.5.3, and the materials science perspective (planewave basis sets) in Sect. 7.5.4.

Let us emphasize that very few algorithms for solving the Kohn–Sham model are well understood mathematically. Sometimes, convergence results have only been proved for the special cases of Hartree and Hartree–Fock models, for which the energy functional is quadratic in the density matrix. The mathematical understanding of most Kohn–Sham models and algorithms is only partial.

Likewise, error analysis for Kohn–Sham models still is in its infancy. The a priori error analysis of the Kohn–Sham LDA model was carried out in [24] for the plane-wave discretization of the supercell model. The proof was adapted in [41] to cover the cases of any variational discretization of the Kohn–Sham LDA model in a bounded domain with Dirichlet boundary conditions. The construction of a posteriori error estimators for Kohn–Sham, and more generally for electronic structure models, is a major challenge in molecular simulation. The a priori error analysis and the few results available to date on a posteriori error analysis are discussed in Sect. 7.6.

7.2 Mathematical Structures of Discretized Kohn–Sham Problems

For simplicity, we limit ourselves to non-magnetic spin-unpolarized systems, in which Kohn–Sham orbitals can be chosen real-valued and are occupied by two electrons (one with spin-up and one with spin-down).

7.2.1 Molecular Orbital Formulation of the Standard Kohn–Sham Model

The standard Kohn–Sham model for a molecule or a cluster (finite number of electrons and nuclei) is

$$E_0 = \inf\left\{ E^{\mathrm{KS}}(\Phi),\ \Phi = (\varphi_1, \ldots, \varphi_N) \in (H^1(\mathbb{R}^3))^N,\ \int_{\mathbb{R}^3} \varphi_i \varphi_j = \delta_{ij} \right\}, \tag{7.1}$$

where N is the total number of electron pairs, and E^{KS} the Kohn–Sham energy functional given by

$$E^{\mathrm{KS}}(\Phi) := \int_{\mathbb{R}^3} \sum_{i=1}^{N} |\nabla \varphi_i|^2 + \int_{\mathbb{R}^3} \rho_\Phi V_{\mathrm{ext}} + \frac{1}{2} \int_{\mathbb{R}^3} \int_{\mathbb{R}^3} \frac{\rho_\Phi(\mathbf{r})\, \rho_\Phi(\mathbf{r}')}{|\mathbf{r} - \mathbf{r}'|} \, d\mathbf{r} \, d\mathbf{r}' + E_{\mathrm{xc}}(\rho_\Phi). \tag{7.2}$$

In the above expression, V_{ext} is the external potential generated by the M nuclei of atomic charges $z_k \in \mathbb{N}^*$ and positions $\mathbf{R}_k \in \mathbb{R}^3$, ρ_Φ the electronic density associated with the orbitals Φ, and E_{xc} the exchange-correlation energy functional:

$$V_{\mathrm{ext}}(\mathbf{r}) := -\sum_{k=1}^{M} \frac{z_k}{|\mathbf{r} - \mathbf{R}_k|}, \qquad \rho_\Phi(\mathbf{r}) = 2 \sum_{i=1}^{N} |\varphi_i(\mathbf{r})|^2. \tag{7.3}$$

The factor 2 in (7.3) is due to the fact that each φ_i hosts two electrons. The expression of E_{xc} depends on the chosen approximate exchange-correlation functional. We will limit ourselves here to some simple examples: the Hartree (also called reduced Hartree–Fock, rHF) model for which $E_{\mathrm{xc}}^{\mathrm{rHF}} = 0$, the local density approximation (LDA), and the Xα model. We refer to the contributions of Lewin, Lieb, and Seiringer (Chap. 3) for a mathematical analysis of the LDA, and to the contribution by Toulouse (Chap. 1) for a detailed description of the many approximate exchange-correlation functionals that have been proposed by physicists and chemists. The LDA exchange-correlation functional is given by

$$E_{\mathrm{xc}}^{\mathrm{LDA}}(\rho) := \int_{\mathbb{R}^3} e_{\mathrm{xc}}^{\mathrm{HEG}}(\rho(\mathbf{r}))\, d\mathbf{r}, \tag{7.4}$$

where $e_{\mathrm{xc}}^{\mathrm{HEG}} : \mathbb{R}_+ \to \mathbb{R}$ is the exchange-correlation energy density of the homogeneous electron gas. The function $e_{\mathrm{xc}}^{\mathrm{HEG}}$ is not known explicitly. Several approximations of this function have been proposed, based on theoretically obtained asymptotic expansions in the high-density regime, and quantum Monte Carlo (QMC) simulations at lower densities [38]. The current reference approximation (PW92) has been proposed by Perdew and Wang [111]. In mathematical and numerical studies, it is convenient to replace the LDA functional by the Xα functional, which is explicit and shares the most important features of the LDA exchange-correlation functional:

$$E_{\mathrm{xc}}^{\mathrm{X}\alpha}(\rho) := -C_{\mathrm{D}} \int_{\mathbb{R}^3} \rho(\mathbf{r})^{4/3}\, d\mathbf{r},$$

where C_{D} is the Dirac constant.

In GGA (generalized gradient approximations), the exchange-correlation functional is of the form

$$E_{\mathrm{xc}}^{\mathrm{GGA}}(\rho) := \int_{\mathbb{R}^3} e_{\mathrm{xc}}^{\mathrm{GGA}}(\rho(\mathbf{r}), |\nabla\rho(\mathbf{r})|)\, \mathrm{d}\mathbf{r}, \tag{7.5}$$

where $e_{\mathrm{xc}}^{\mathrm{GGA}} : \mathbb{R}_+ \times \mathbb{R}_+ \to \mathbb{R}$ is a function such that $e_{\mathrm{xc}}^{\mathrm{GGA}}(\overline{\rho}, 0) = e_{\mathrm{xc}}^{\mathrm{HEG}}(\overline{\rho})$ and whose dependence in $|\nabla\rho(\mathbf{r})|$ is constructed by picking up a suitable functional form depending on a few parameters, and fitting these parameters using exact conditions and/or empirical data. The most popular GGA functionals are the PBE [110] and PBE0 [2] functionals.

Meta-GGA and hybrid functionals involve respectively the kinetic energy density

$$\tau_\Phi(\mathbf{r}) := 2\sum_{i=1}^{N} |\nabla\varphi_i(\mathbf{r})|^2$$

and the one-body density matrix

$$\gamma_\Phi(\mathbf{r}, \mathbf{r}') := \sum_{i=1}^{N} \varphi_i(\mathbf{r})\varphi_i(\mathbf{r}'),$$

and are therefore not explicit in the density ρ_Φ. In order to include all the above exchange-correlation functionals in a simple unified framework, we redefine the Kohn–Sham energy functional as

$$\begin{aligned} E^{\mathrm{KS}}(\Phi) : = &\int_{\mathbb{R}^3} \sum_{i=1}^{N} |\nabla\varphi_i|^2 + \int_{\mathbb{R}^3} \rho_\Phi V_{\mathrm{ext}} + \frac{1}{2}\int_{\mathbb{R}^3}\int_{\mathbb{R}^3} \frac{\rho_\Phi(\mathbf{r})\,\rho_\Phi(\mathbf{r}')}{|\mathbf{r}-\mathbf{r}'|}\, \mathrm{d}\mathbf{r}\, \mathrm{d}\mathbf{r}' \\ &+ \mathcal{E}_{\mathrm{xc}}(\gamma_\Phi), \end{aligned} \tag{7.6}$$

where we allow $\mathcal{E}_{\mathrm{xc}}$ to depend explicitly on the density matrix (note that $\tau_\Phi(\mathbf{r})$ can be easily obtained from γ_Φ). Another advantage of this framework is that it contains Hartree–Fock as a special case:

$$\begin{aligned} E^{\mathrm{HF}}(\Phi) : = &\int_{\mathbb{R}^3} \sum_{i=1}^{N} |\nabla\varphi_i|^2 + \int_{\mathbb{R}^3} \rho_\Phi V_{\mathrm{ext}} + \frac{1}{2}\int_{\mathbb{R}^3}\int_{\mathbb{R}^3} \frac{\rho_\Phi(\mathbf{r})\,\rho_\Phi(\mathbf{r}')}{|\mathbf{r}-\mathbf{r}'|}\, \mathrm{d}\mathbf{r}\, \mathrm{d}\mathbf{r}' \\ &+ \mathcal{E}_{\mathrm{xHF}}(\gamma_\Phi), \end{aligned} \tag{7.7}$$

where $\mathcal{E}_{\mathrm{xHF}}(\gamma)$ is the exact-exchange functional defined as

$$\mathcal{E}_{\mathrm{xHF}}(\gamma) := -\int_{\mathbb{R}^3}\int_{\mathbb{R}^3} \frac{|\gamma(\mathbf{r}, \mathbf{r}')|^2}{|\mathbf{r}-\mathbf{r}'|}\, \mathrm{d}\mathbf{r}\, \mathrm{d}\mathbf{r}'.$$

Once the model is set, variational discretization methods consist in choosing a set of N_b linearly-independent functions $\Xi := (\xi_1, \ldots, \xi_{N_b}) \in (H^1(\mathbb{R}^3))^{N_b}$ and solving the finite-dimensional minimization problem

$$E_0^{\Xi} = \inf \left\{ E^{\mathrm{KS}}(\Phi),\ \Phi = (\varphi_1, \ldots, \varphi_N) \in V_\Xi,\ \int_{\mathbb{R}^3} \varphi_i \varphi_j = \delta_{ij} \right\}, \tag{7.8}$$

where $V_\Xi = \mathrm{span}(\xi_1, \ldots, \xi_{N_b}) \subset H^1(\mathbb{R}^3)$. Denoting by $X \in \mathbb{R}^{N_b \times N}$ the matrix collecting the coefficients of the orbitals φ_i in the basis ξ_μ i.e.

$$\forall 1 \leq i \leq N, \quad \varphi_i(\mathbf{r}) = \sum_{\mu=1}^{N_b} X_{\mu i} \xi_\mu(\mathbf{r}),$$

problem (7.8) can be rewritten as

$$E_0^{\Xi} = \inf \left\{ E^{\Xi}(X),\ X \in \mathcal{M}_{\mathrm{MO}}^{\Xi} \right\}, \tag{7.9}$$

where

$$E^{\Xi}(X) = 2\mathrm{Tr}(h^{\Xi} X X^T) + 2\mathrm{Tr}\left(J^{\Xi}(XX^T) XX^T \right) + \mathcal{E}_{\mathrm{xc}}^{\Xi}(XX^T), \tag{7.10}$$

and

$$\mathcal{M}_{\mathrm{MO}}^{\Xi} := \left\{ X \in \mathbb{R}^{N_b \times N} \mid X^T S^{\Xi} X = I_N \right\}, \tag{7.11}$$

where I_N is the rank-N identity matrix. The matrices $S^{\Xi} \in \mathbb{R}_{\mathrm{sym}}^{N_b \times N_b}$ and $h^{\Xi} \in \mathbb{R}_{\mathrm{sym}}^{N_b \times N_b}$ (here and in the sequel, $\mathbb{R}_{\mathrm{sym}}^{n \times n}$ stands for the vector space of $n \times n$ real symmetric matrices) are the overlap and one-electron Hamiltonian matrix

$$\forall 1 \leq \mu, \nu \leq N_b, \quad S_{\mu,\nu}^{\Xi} = \int_{\mathbb{R}^3} \xi_\mu \xi_\nu, \quad h_{\mu,\nu}^{\Xi} = \frac{1}{2} \int_{\mathbb{R}^3} \nabla \xi_\mu \cdot \nabla \xi_\nu + \int_{\mathbb{R}^3} V_{\mathrm{ext}} \xi_\mu \xi_\nu. \tag{7.12}$$

The functions $J^{\Xi} : \mathbb{R}_{\mathrm{sym}}^{N_b \times N_b} \to \mathbb{R}_{\mathrm{sym}}^{N_b \times N_b}$ and $\mathcal{E}_{\mathrm{xc}}^{\Xi} : \mathbb{R}_{\mathrm{sym}}^{N_b \times N_b} \to \mathbb{R}$ are defined by

$$\forall P \in \mathbb{R}_{\mathrm{sym}}^{N_b \times N_b}, \quad J^{\Xi}(P) = A^{\Xi} : P, \quad \mathcal{E}_{\mathrm{xc}}^{\Xi}(P) = \mathcal{E}_{\mathrm{xc}} \left(\sum_{\mu,\nu=1}^{N_b} P_{\mu,\nu} \xi_\mu \otimes \xi_\nu \right),$$

where A^{Ξ} is the 4th-order tensor of two-electron integrals:

$$A_{\kappa\lambda\mu\nu}^{\Xi} := (\kappa\lambda|\mu\nu) := \int_{\mathbb{R}^3} \int_{\mathbb{R}^3} \frac{\xi_\kappa(\mathbf{r}) \xi_\lambda(\mathbf{r}) \xi_\mu(\mathbf{r}') \xi_\nu(\mathbf{r}')}{|\mathbf{r} - \mathbf{r}'|} \, d\mathbf{r} \, d\mathbf{r}'. \tag{7.13}$$

Recall that the notation $A^\Xi : P$ stands for the double contraction

$$[A^\Xi : P]_{\kappa\lambda} = \sum_{\mu,\nu=1}^{N_b} A^\Xi_{\kappa\lambda\mu\nu} P_{\mu\nu}.$$

Let us finally make a few remarks on the mathematical properties of the minimization problem (7.9). First, the function $E^\Xi : \mathbb{R}^{N_b\times N} \to \mathbb{R}$ is continuous and the minimization set $\mathcal{M}^\Xi_{\rm MO}$ is a non-empty compact set of $\mathbb{R}^{N_b\times N}$, which guarantees the existence of a minimizer to (7.9). Second, problem (7.9) is gauge invariant: denoting by $O(N)$ the group of $N \times N$ real orthogonal matrices, we have

$$\forall X \in \mathcal{M}^\Xi_{\rm MO}, \quad \forall U \in O(N), \quad XU \in \mathcal{M}^\Xi_{\rm MO} \quad \text{and} \quad E^\Xi(XU) = E^\Xi(X). \tag{7.14}$$

In particular, if X_0 is a minimizer of (7.10), then X_0U also is a minimizer of (7.10). Uniqueness of the minimizer up this gauge invariance is an open problem.

7.2.2 *Density Matrix Formulations of the Various Kohn–Sham Models*

We see from (7.10) that the discrete Kohn–Sham energy functional is in fact a function of the symmetric matrix $P = XX^T$:

$$E^\Xi(X) = \mathcal{E}^\Xi(XX^T),$$

where $\mathcal{E}^\Xi : \mathbb{R}^{N_b\times N_b}_{\rm sym} \to \mathbb{R}$ is defined by

$$\mathcal{E}^\Xi(P) = 2\mathrm{Tr}(h^\Xi P) + 2\mathrm{Tr}\left(J^\Xi(P)P\right) + \mathcal{E}^\Xi_{\rm xc}(P). \tag{7.15}$$

This is not surprising since the matrix $P = XX^T$ is in fact the matrix in the tensor basis $\xi_\mu \otimes \xi_\nu$ of the one-body reduced density matrix γ_Φ:

$$\begin{aligned}\gamma_\Phi(\mathbf{r},\mathbf{r}') &= \sum_{i=1}^N \varphi_i(\mathbf{r})\varphi_i(\mathbf{r}') \\ &= \sum_{i=1}^N \left(\sum_{\mu=1}^{N_b} X_{\mu i}\xi_\mu(\mathbf{r})\right)\left(\sum_{\nu=1}^{N_b} X_{\nu i}\xi_\nu(\mathbf{r}')\right) = \sum_{\mu,\nu=1}^{N_b} P_{\mu\nu}(\xi_\mu\otimes\xi_\nu)(\mathbf{r},\mathbf{r}').\end{aligned}$$

The gauge invariance (7.14) simply originates from the fact that X and XU for $U \in O(N)$ generate the same density matrix. Note that in the Hartree–Fock setting, the energy functional can be written as

$$\mathcal{E}^{\mathrm{HF},\Xi} = 2\mathrm{Tr}(h^{\Xi}P) + \mathrm{Tr}\left(G^{\Xi}(P)P\right), \tag{7.16}$$

with

$$[G^{\Xi}(P)]_{\mu\nu} = \sum_{\kappa,\lambda=1}^{N_b} \left(2A^{\Xi}_{\kappa\lambda\mu\nu} - A^{\Xi}_{\kappa\nu\mu\lambda}\right) P_{\kappa\lambda}. \tag{7.17}$$

It is also easy to check that when X spans the minimization set $\mathcal{M}^{\Xi}_{\mathrm{MO}}$ arising in (7.9) (and defined in (7.11)), the matrix $P = XX^T$ spans the set

$$\mathcal{M}^{\Xi}_{\mathrm{DM}} := \left\{P \in \mathbb{R}^{N_b\times N_b}_{\mathrm{sym}} \mid PS^{\Xi}P = P,\ \mathrm{Tr}(S^{\Xi}P) = N\right\}.$$

This set $\mathcal{M}_{\mathrm{DM}}$ is the set of rank-N S-orthogonal projectors in the N_b-dimensional real space.

We therefore have

$$E^{\Xi}_0 = \inf\left\{\mathcal{E}^{\Xi}(P),\ P \in \mathcal{M}^{\Xi}_{\mathrm{DM}}\right\}, \tag{7.18}$$

which corresponds to the variational approximation in the basis Ξ of the density matrix formulation of the continuous standard Kohn–Sham model:

$$E_0 = \left\{\mathcal{E}^{\mathrm{KS}}(\gamma),\ \gamma \in \mathcal{S}(L^2(\mathbb{R}^3)),\ \gamma^2 = \gamma,\ \mathrm{Tr}(\gamma) = N,\ \mathrm{Tr}(-\Delta\gamma) < \infty\right\}, \tag{7.19}$$

where $\mathcal{S}(L^2(\mathbb{R}^3))$ is the vector space of bounded self-adjoint operators on $L^2(\mathbb{R}^3)$ and

$$\mathcal{E}^{\mathrm{KS}}(\gamma) = \mathrm{Tr}(-\Delta\gamma) + \int_{\mathbb{R}^3}\rho_\gamma V_{\mathrm{ext}} + \frac{1}{2}\int_{\mathbb{R}^3}\int_{\mathbb{R}^3}\frac{\rho_\gamma(\mathbf{r})\,\rho_\gamma(\mathbf{r}')}{|\mathbf{r}-\mathbf{r}'|}\,d\mathbf{r}\,d\mathbf{r}' + E_{\mathrm{xc}}(\gamma),$$

ρ_γ denoting twice the density of the trace-class operator γ, i.e. $\rho_\gamma(\mathbf{r}) = 2\gamma(\mathbf{r},\mathbf{r})$, where $\gamma(\mathbf{r},\mathbf{r}')$ is the kernel of the operator γ. We refer to [122] for an introduction to the theory of trace-class operators.

Let us now turn to the discretized extended and finite-temperature Kohn–Sham models. These models are more conveniently formulated in the density matrix formalism. For both of them the minimization set is the convex hull of $\mathcal{M}^{\Xi}_{\mathrm{DM}}$, that is

$$\mathrm{Conv}(\mathcal{M}^{\Xi}_{\mathrm{DM}}) = \left\{P \in \mathbb{R}^{N_b\times N_b}_{\mathrm{sym}} \mid PS^{\Xi}P \le P,\ \mathrm{Tr}(S^{\Xi}P) = N\right\}.$$

The inequality $PS^{\Xi}P \leq P$ is to be understood in the sense of square symmetric matrices:

$$\text{for any } A, B \in \mathbb{R}^{n\times n}_{\text{sym}}, \quad A \leq B \quad \text{means} \quad \forall x \in \mathbb{R}^n,\ x^T A x \leq x^T B x.$$

Note that any matrix $P \in \mathrm{Conv}(\mathcal{M}^{\Xi}_{\mathrm{DM}})$ can be decomposed as

$$P = \sum_{i=1}^{N_b} n_i \Phi_i \Phi_i^T, \quad \text{where} \quad \Phi_i \in \mathbb{R}^{N_b}, \quad \Phi_i^T S^{\Xi} \Phi_j = \delta_{ij}, \quad 0 \leq n_i \leq 1,$$

where n_i is called the occupation number of the orbital Φ_i. The orbital Φ_i is called fully occupied if $n_i = 1$, partially occupied if $0 < n_i < 1$ and unoccupied, or virtual if $n_i = 0$. The discretized extended Kohn–Sham model is defined as

$$\inf\left\{\mathcal{E}^{\Xi}(P),\ P \in \mathrm{Conv}(\mathcal{M}^{\Xi}_{\mathrm{DM}})\right\}, \tag{7.20}$$

and corresponds to the discretization in the basis set Ξ of the continuous problem

$$E_0 = \left\{\mathcal{E}(\gamma),\ \gamma \in \mathcal{S}(L^2(\mathbb{R}^3)),\ 0 \leq \gamma \leq 1,\ \mathrm{Tr}(\gamma) = N,\ \mathrm{Tr}(-\Delta\gamma) < \infty\right\}, \tag{7.21}$$

obtained from Valone–Lieb formulation of DFT [94, 141]. The minima E_0 in (7.19) and (7.21) are equal for the exact Levy-Lieb and Valone–Lieb DFT functionals, but differ in general if approximate exchange-correlation functionals and finite basis sets are used.

A discretized finite-temperature Kohn–Sham model can be defined as

$$\inf\left\{\mathcal{E}^{\Xi}(P) + 2k_{\mathrm{B}}T\,\mathrm{Tr}(P\ln P + (1-P)\ln(1-P)),\ P \in \mathrm{Conv}(\mathcal{M}^{\Xi}_{\mathrm{DM}})\right\}, \tag{7.22}$$

where T is the temperature and k_{B} the Boltzmann constant. The functional

$$S(P) := -2k_{\mathrm{B}}\mathrm{Tr}(P\ln P + (1-P)\ln(1-P))$$

is the fermionic entropy for spin 1/2-unpolarized systems, favoring configurations with fractional occupation numbers. The functional $\mathcal{E}^{\Xi}(P) - TS(P)$ can therefore be interpreted as a free energy. It should be mentioned that while the above problem always has a minimizer, its continuous counterpart diverges: the infimum of the free energy $\mathcal{E}(\gamma) + 2k_{\mathrm{B}}T\ \mathrm{Tr}(\gamma\ln\gamma + (1-\gamma)\ln(1-\gamma))$ over the set of admissible density matrices is equal to $-\infty$, due to the presence of a continuous spectrum in the underlying Kohn–Sham Hamiltonian. On the other hand, the continuous finite-temperature Kohn–Sham model is well-defined for confined systems as well as, more interestingly, for periodic crystals, in the sense that the free-energy per unit volume has a minimizer. In both settings, the extended Kohn–Sham model is the limit of the finite-temperature Kohn–Sham model when the temperature goes to zero.

7.2.3 Geometrical Structure of the Kohn–Sham Problems

The minimization sets $\mathcal{M}_{\rm MO}^{\Xi}$ and $\mathcal{M}_{\rm DM}^{\Xi}$ in (7.9) and (7.18) have simple geometrical structures. They are respectively diffeomorphic to the smooth manifolds

$$\mathcal{M}_{\rm MO} := \left\{X \in \mathbb{R}^{N_b \times N} \mid X^T X = I_N\right\},$$

$$\mathcal{M}_{\rm DM} := \left\{P \in \mathbb{R}^{N_b \times N_b}_{\rm sym} \mid P^2 = P,\ {\rm Tr}(P) = N\right\},$$

through the linear transforms

$$X \mapsto (S^{\Xi})^{-1/2} X \quad \text{and} \quad P \mapsto (S^{\Xi})^{-1/2} P (S^{\Xi})^{-1/2}. \tag{7.23}$$

Note that $\mathcal{M}_{\rm MO}$ and $\mathcal{M}_{\rm DM}$ are themselves representations in $\mathbb{R}^{N_b \times N}$ and $\mathbb{R}^{N_b \times N_b}_{\rm sym}$ respectively of the Stiefel and Grassmann manifolds $V_N(\mathbb{R}^{N_b})$ and $G(N, N_b)$ of orthogonal bases and subspaces widely studied in differential geometry (see e.g. [54] and references therein).

In order to simplify the presentation, we will assume in the rest of the chapter that the basis set Ξ is orthonormal. The general case of a non-orthonormal basis can be recovered by using the change of variable (7.23) (we will however make some comments on the additional difficulties arising when the overlap matrix S^{Ξ} is ill-conditioned). The minimization sets then are the manifold $\mathcal{M}_{\rm MO}$ (orbital formulation) or $\mathcal{M}_{\rm DM}$ (density matrix formulation). Still in order to simplify the notation, we will set

$$E(X) := E^{\Xi}(X) \quad \text{and} \quad \mathcal{E}(P) = \mathcal{E}^{\Xi}(P),$$

the functions $E : \mathbb{R}^{N_b \times N} \to \mathbb{R}$ and $\mathcal{E} : \mathbb{R}^{N_b \times N_b}_{\rm sym} \to \mathbb{R}$ being related by the equality

$$\forall X \in \mathbb{R}^{N_b \times N}, \quad E(X) = \mathcal{E}(XX^T).$$

Let us recall some of the basic properties of the sets $\mathcal{M}_{\rm MO}$ and $\mathcal{M}_{\rm DM}$ that will be useful in our analysis. Let $g : \mathbb{R}^{N_b \times N} \to \mathbb{R}^{N \times N}_{\rm sym}$ be the polynomial map defined by $g(X) = X^T X - I_N$, so that $\mathcal{M}_{\rm MO} = g^{-1}(\{0\})$. It is easily checked that for each $X \in \mathcal{M}_{\rm MO}$,

$$g'(X) : \mathbb{R}^{N_b \times N} \ni A \mapsto X^T A + A^T X \in \mathbb{R}^{N \times N}_{\rm sym} \tag{7.24}$$

is surjective. This proves that g is a submersion [88], and therefore that $\mathcal{M}_{\rm MO}$ is a smooth submanifold of $\mathbb{R}^{N_b\times N}$ of dimension $N_bN - N(N+1)/2$. The tangent space to $\mathcal{M}_{\rm MO}$ at some $X \in \mathcal{M}_{\rm MO}$ is

$$T_X\mathcal{M}_{\rm MO} = \operatorname{Ker}\left(g'(X)\right) = \left\{A \in \mathbb{R}^{N_b\times N} \mid g'(X)(A) = X^TA + A^TX = 0\right\}. \tag{7.25}$$

Similar computations show that $\mathcal{M}_{\rm DM}$ is a smooth submanifold of $\mathbb{R}^{N_b\times N_b}_{\rm sym}$ of dimension $N(N_b - N)$, and that its tangent space at $P \in \mathcal{M}_{\rm DM}$ is the vector space of matrices

$$T_P\mathcal{M}_{\rm DM} = \left\{M \in \mathbb{R}^{N_b\times N_b}_{\rm sym} \mid PMP = (1-P)M(1-P) = 0\right\}.$$

If $P \in \mathcal{M}_{\rm DM}$, the range and kernel of P are respectively of dimension N and N_b-N, and are orthogonal. They are respectively called the occupied and virtual subspaces. Introducing an orthogonal matrix $U_0 \in \mathbb{R}^{N_b\times N_b}$ whose first N columns form an orthonormal basis of $\operatorname{Ran}(P)$ and the last $N_b - N$ an orthonormal basis of $\operatorname{Ker}(P)$, we have

$$P = U_0\begin{pmatrix} I_N & 0_{N\times(N_b-N)} \\ 0_{(N_b-N)\times N} & 0_{(N_b-N)\times(N_b-N)}\end{pmatrix}U_0^T, \tag{7.26}$$

and—dropping the subscripts indicating the sizes of the null matrices to simplify the notation—

$$T_P\mathcal{M}_{\rm DM} = \left\{U_0\begin{pmatrix} 0 & M_{\rm ov} \\ M_{\rm ov}^T & 0\end{pmatrix}U_0^T,\ M_{\rm ov} \in \mathbb{R}^{N\times(N_b-N)}\right\}, \tag{7.27}$$

where the subscript ov stands for occupied-virtual. The map $\Phi_P : \mathbb{R}^{N\times(N_b-N)} \to \mathcal{M}_{\rm DM}$ defined by

$$\Phi_P(A_{\rm ov}) := U_0\exp\left(-\mathcal{A}(A_{\rm ov})\right)\begin{pmatrix} I_N & 0 \\ 0 & 0\end{pmatrix}\exp\left(\mathcal{A}(A_{\rm ov})\right)U_0^T, \tag{7.28}$$

$$\mathcal{A}(A_{\rm ov}) := \begin{pmatrix} 0 & A_{\rm ov} \\ -A_{\rm ov}^T & 0\end{pmatrix}, \tag{7.29}$$

is surjective and its restriction to a neighborhood of 0 defines a local map of $\mathcal{M}_{\rm DM}$ in the neighborhood of P.

7.2.4 First-Order Optimality Conditions

As we have seen in the previous sections, the standard Kohn–Sham model boils down after discretization to an equality constrained optimization problem, which admits two equivalent natural formulations

- the orbital formulation (7.9) (i.e. an optimization problem on a Stiefel manifold);
- the density matrix formulation (7.18) (i.e. an optimization problem on a Grassmann manifold).

The orbital formulation (7.9) of the standard Kohn–Sham model is an equality constrained optimization problem of the form

$$\inf\left\{E(X),\ X \in \mathbb{R}^{N_b \times N},\ g(X) = 0\right\}, \tag{7.30}$$

where $g : \mathbb{R}^{N_b \times N} \to \mathbb{R}^{N \times N}_{\mathrm{sym}}$ is defined in (7.24) (recall that we assume an orthonormal basis and have set $E(X) := E^{\Xi}(X)$). As seen in the previous section, $\mathcal{M}_{\mathrm{MO}}$ is a smooth submanifold of $\mathbb{R}^{N_b \times N}$ and $T_X\mathcal{M}_{\mathrm{MO}}$ is defined in (7.25). Assuming that the function $P \mapsto \mathcal{E}^{\Xi}_{\mathrm{xc}}(P)$ is differentiable at each point of $\mathcal{M}_{\mathrm{DM}}$, the energy functional E is differentiable at each point of $\mathcal{M}_{\mathrm{MO}}$. Endowing $\mathbb{R}^{N_b \times N}$ with the Frobenius inner product defined by $(X_1, X_2)_{\mathrm{F}} := \mathrm{Tr}(X_1^T X_2)$, its gradient is given by

$$\nabla E(X) = 4H(XX^T)X, \tag{7.31}$$

where

$$H(P) = h^{\Xi} + 2J^{\Xi}(P) + \nabla\mathcal{E}^{\Xi}_{\mathrm{xc}}(P)$$

is the Kohn–Sham matrix (also called mean-field Hamiltonian) associated with the density matrix $P = XX^T$. The first-order optimality condition associated with (7.30) is then

$$\nabla E(X) \in T_X\mathcal{M}_{\mathrm{MO}}{}^{\perp} = \left(\mathrm{Ker}(g'(X))\right)^{\perp} = \mathrm{Ran}(g'(X)^*), \tag{7.32}$$

where $g'(X)^* : \mathbb{R}^{N \times N}_{\mathrm{sym}} \to \mathbb{R}^{N_b \times N}$ is the adjoint of $g'(X)$. A simple calculation shows that

$$\forall \Lambda \in \mathbb{R}^{N \times N}_{\mathrm{sym}}, \quad g'(X)^*(\Lambda) = 2X\Lambda. \tag{7.33}$$

Putting together (7.31)–(7.33) we finally obtain the Euler–Lagrange equations associated with (7.9):

$$\begin{cases} \text{seek } (X, \Lambda) \in \mathbb{R}^{N_b \times N} \times \mathbb{R}^{N\times N}_{\text{sym}} \text{ such that} \\ H(P)X = X\Lambda \\ X^T X = I_N \\ P = XX^T. \end{cases} \tag{7.34}$$

Interestingly, the gauge invariance (7.14) can be used to simplify the search for critical points of (7.9). Consider a solution (X, Λ) to the Euler–Lagrange equations (7.34). Since the Lagrange multiplier Λ of the constraint $g(X) = 0$ is a symmetric matrix, it can be diagonalized in an orthonormal basis:

$$\Lambda = U \operatorname{diag}(\varepsilon_1, \ldots, \varepsilon_N)\, U^T,$$

where $U \in O(N)$ is an orthogonal matrix and $\varepsilon_1 \leq \cdots \leq \varepsilon_N$ are the N eigenvalues of Λ, counted with their multiplicities. Observing that $(XU)^T(XU) = I_N$, $(XU)(XU)^T = XX^T = P$, and $H(P)(XU) = XU\operatorname{diag}(\varepsilon_1, \ldots, \varepsilon_N)$, we obtain that, up to replacing X by XU, all the solutions to the Euler–Lagrange equations (7.34) can be obtained by orthogonal transforms from the solutions to the so-called Kohn–Sham equations

$$\begin{cases} \text{seek } (\widetilde{X}_1, \ldots, \widetilde{X}_N) \in \mathbb{R}^{N_b \times N} \text{ and real numbers } \varepsilon_1 \leq \cdots \leq \varepsilon_N \text{ such that} \\ H(P)\widetilde{X}_i = \varepsilon_i \widetilde{X}_i \\ \widetilde{X}_i^T \widetilde{X}_j = \delta_{ij} \\ P = \sum_{i=1}^{N} \widetilde{X}_i \widetilde{X}_i^T, \end{cases} \tag{7.35}$$

where $\widetilde{X}_1, \ldots, \widetilde{X}_N$ are the N columns of the matrix XU. In the following, we drop the tildes for simplicity of notation.

Remark In the general case of a possibly non-orthonormal basis, the Kohn–Sham equations read

$$\begin{cases} \text{seek } (X_1, \ldots, X_N) \in \mathbb{R}^{N_b \times N} \text{ and real numbers } \varepsilon_1 \leq \cdots \leq \varepsilon_N \text{ such that} \\ H(P)X_i = \varepsilon_i S^{\Xi} X_i \\ X_i^T S^{\Xi} X_j = \delta_{ij} \\ P = \sum_{i=1}^{N} X_i X_i^T. \end{cases} \tag{7.36}$$

The Kohn–Sham equations therefore have the structure of a nonlinear generalized eigenvalue (or rather eigenvector) problem since the Kohn–Sham matrix $H(P)$ depends on the vector space spanned by its low-energy generalized eigenvectors X_i's, $1 \leq i \leq N$, through the density matrix P.

Remark The equations above are the discrete versions of the continuous Kohn–Sham equations, which are derived from the Euler–Lagrange equations of the variational problem (7.1) in the same way as above, and which read, in the LDA case, as

$$-\frac{1}{2}\Delta\varphi_i + V_{\mathrm{ext}}\varphi_i + \int_{\mathbb{R}^3} \frac{\rho_\Phi(\mathbf{r}')}{|\mathbf{r}-\mathbf{r}'|}\mathrm{d}\mathbf{r}' + V_{\mathrm{xc}}[\rho]\varphi_i = \varepsilon_i\varphi_i, \qquad \int_{\mathbb{R}^3} \varphi_i\varphi_j = \delta_{ij}, \tag{7.37}$$

where $V_{\mathrm{xc}}[\rho](\mathbf{r}) = \frac{\mathrm{d}e_{\mathrm{xc}}^{\mathrm{HEG}}}{\mathrm{d}\rho}(\rho(\mathbf{r}))$.

It can be checked that any critical point P of the density matrix formulation of the standard Kohn–Sham model (7.18) is such that

$$P = \sum_{i=1}^{N} X_i X_i^T \quad \text{with} \quad H(P)X_i = \varepsilon_i X_i, \quad X_i^T X_j = \delta_{ij}, \quad \varepsilon_1 \leq \cdots \leq \varepsilon_N.$$

In addition, the above conditions are equivalent to

$$P^2 = P, \quad \mathrm{Tr}(P) = N, \quad H(P)P - PH(P) = 0. \tag{7.38}$$

Conditions (7.38) also read

$$P \in \mathcal{M}_{\mathrm{DM}} \quad \text{and} \quad [H(P), P] = 0, \tag{7.39}$$

or equivalently

$$P \in \mathcal{M}_{\mathrm{DM}} \quad \text{and} \quad [[H(P), P], P] = 0. \tag{7.40}$$

Indeed, considering a generic element P in $\mathcal{M}_{\mathrm{DM}}$, and using the representation (7.26) of P, we have

$$H(P) = U_0 \begin{pmatrix} H_{\mathrm{oo}} & H_{\mathrm{ov}} \\ H_{\mathrm{ov}}^T & H_{\mathrm{vv}} \end{pmatrix} U_0^T,$$

$$[H(P), P] = U_0 \begin{pmatrix} 0 & -H_{\mathrm{ov}} \\ H_{\mathrm{ov}}^T & 0 \end{pmatrix} U_0^T,$$

$$[[H(P), P], P] = U_0 \begin{pmatrix} 0 & H_{\mathrm{ov}} \\ H_{\mathrm{ov}}^T & 0 \end{pmatrix} U_0^T.$$

The matrix $4H_{\rm ov} \in \mathbb{R}^{N\times(N_b-N)}$ can be identified with the gradient at $A_{\rm ov} = 0$ of the function $A_{\rm ov} \mapsto \mathcal{E}(\Phi_P(A_{\rm ov}))$, that is, of the discretized Kohn–Sham functional in the local map Φ_P defined by (7.28)–(7.29). Likewise, the matrix $2[[H(P), P], P]$ is the orthogonal projection of the gradient $\nabla\mathcal{E}(P) = 2H(P)$ on $T_P\mathcal{M}_{\rm DM}$, the tangent space to the manifold $\mathcal{M}_{\rm DM}$ at P, for the Frobenius inner product. The geometrical interpretation of the equivalent conditions (7.39) and (7.40) is that the gradient at $A_{\rm ov} = 0$ of the Kohn–Sham energy functional in the local map Φ_P vanishes, and that the gradient of $\mathcal{E}$ at P is orthogonal to the manifold $\mathcal{M}_{\rm DM}$ at this point.

The reader familiar with the Hartree–Fock model may have noticed that the discretized Kohn–Sham equations (7.35) have a similar structure as the discretized Hartree–Fock equations, which is not surprising since the minimizing sets and the gauge invariance properties are the same for the two models. There is however an important difference between the two models. For the (spin-unrestricted) Hartree–Fock model, it can be proved that if $\Phi = (\varphi_1, \ldots, \varphi_N)$ is a minimizer of the Hartree–Fock model, then $\varepsilon_1 \le \cdots \le \varepsilon_N$ are the lowest N eigenvalues of the Hartree–Fock Hamiltonian (*Aufbau* principle), and that there is a positive gap between ε_N and ε_{N+1} (no unfilled-shell property [9]). In addition, any minimizer of the Hartree–Fock energy functional on the convex hull $\mathrm{Conv}(\mathcal{M}_{\rm DM})$ is on $\mathcal{M}_{\rm DM}$. For Kohn–Sham models, these properties do not hold true for all molecular systems.

Let us now turn to the extended and finite-temperature Kohn–Sham models. We assume here that the functional $P \mapsto \mathcal{E}^{\Xi}(P)$ is differentiable at each point of $\mathrm{Conv}(\mathcal{M}_{\rm DM})$. Since $\mathrm{Conv}(\mathcal{M}_{\rm DM})$ is a non-empty closed convex subset of $\mathbb{R}^{N_b\times N_b}_{\rm sym}$, the first-order optimality conditions for (7.20) are given by the Euler inequality

$$\forall P' \in \mathrm{Conv}(\mathcal{M}_{\rm DM}), \quad \nabla\mathcal{E}(P)\cdot(P'-P) \ge 0,$$

which also reads

$$\forall P' \in \mathrm{Conv}(\mathcal{M}_{\rm DM}), \quad \mathrm{Tr}(H(P)(P'-P)) \ge 0. \tag{7.41}$$

It is well-known (see e.g. [25]) that any $P \in \mathrm{Conv}(\mathcal{M}_{\rm DM})$ for which (7.41) holds is of the form

$$P = \mathbb{1}_{(-\infty,\mu)}(H(P)) + \delta \quad \text{with} \quad \delta^T = \delta,\ 0 \le \delta \le 1_{\{\mu\}}(H(P)),\ \mathrm{Tr}(D) = N. \tag{7.42}$$

Otherwise stated, any critical point P of (7.20) is of the form

$$P = \sum_{i=1}^{N_b} n_i X_i X_i^T \quad \text{with} \quad H(P)X_i = \varepsilon_i X_i, \quad X_i^T X_j = \delta_{ij}, \quad \varepsilon_1 \le \cdots \le \varepsilon_{N_b} \tag{7.43}$$

where the occupation numbers of the Kohn–Sham orbitals X_i satisfy

$$\left| \begin{array}{ll} n_i = 1 & \text{if } \varepsilon_i < \mu \\ 0 \leq n_i \leq 1 & \text{if } \varepsilon_i = \mu \\ n_i = 0 & \text{if } \varepsilon_i > \mu \end{array} \right. \quad \text{and} \quad \sum_{i=1}^{N_b} n_i = N, \tag{7.44}$$

where μ is the Lagrange multiplier of the constraint $\mathrm{Tr}(P) = N$, called the Fermi level. In words, the Kohn–Sham energy levels below the Fermi level μ are fully occupied, the ones above the Fermi level are unoccupied, while the ones just at the Fermi level can be partially occupied. Two cases can be encountered:

1. either there is a gap between the Nth and $(N+1)$st energy levels (i.e. if $\varepsilon_N < \varepsilon_{N+1}$), in which case $P \in \mathcal{M}_{\mathrm{DM}}$ is a critical point of the standard Kohn–Sham problem (7.18) satisfying the *Aufbau* principle, and μ can be any number in the range $(\varepsilon_N, \varepsilon_{N+1})$;
2. or μ is a degenerate energy level of $H(P)$, in which case the occupation numbers of the Kohn–Sham orbitals at the Fermi level are not determined by the first-order optimality condition and are not known a priori; if P is a minimizer of (7.41), they can be obtained numerically by considering the second-order optimality condition (or higher-order conditions in case of degeneracies).

Lastly, the first-order optimality conditions associated with the finite-temperature Kohn–Sham model (7.22) inferred from Euler's inequality are

$$P = f_\beta(H(P) - \mu), \quad \mathrm{Tr}(P) = N, \tag{7.45}$$

where $\beta = \frac{1}{k_B T}$ is the inverse temperature, f_β the Fermi-Dirac function ($f_\beta(\varepsilon) = (1 + e^{\beta\varepsilon})^{-1}$) and μ the Lagrange multiplier of the constraint $\mathrm{Tr}(P) = N$. In terms of Kohn–Sham orbitals, (7.45) can be reformulated as (7.43), together with

$$n_i = f_\beta(\varepsilon_i - \mu), \quad \sum_{i=1}^{N_b} n_i = N. \tag{7.46}$$

Formally, the condition (7.44) on the occupation numbers for the extended Kohn–Sham model is the limit when T goes to zero of the condition (7.46) on the occupation numbers for the finite-temperature Kohn–Sham model.

7.2.5 *Quantities of Interest*

We have above explained how to define the ground-state energy of a molecular system in the standard Kohn–Sham approximation in a given atomic configuration

$\mathbf{R} = (\mathbf{R}_1, \ldots, \mathbf{R}_M) \in \mathbb{R}^{3M}$ and for a given basis set Ξ as a problem of the form

$$E_0(\mathbf{R}) = \inf \{\mathcal{E}(\mathbf{R}, P),\ \ P \in \mathcal{M}_{\mathrm{DM}}\}, \tag{7.47}$$

where $\mathbf{R} \in \mathbb{R}^{3M}$ are the atomic positions, and dropping the explicit dependence on Ξ for simplicity.

That computation is by itself of limited interest. It however opens the door to a variety of useful analyses, grouped under the umbrella term of "properties". For instance, the force exerted on nucleus k by the electrons is given by

$$F_k(\mathbf{R}) = -\frac{\partial E_0}{\partial \mathbf{R}_k}(\mathbf{R}).$$

Computing forces allows for the efficient optimization of geometries, and for molecular dynamics (integrating the classical equations of motion for the nuclei).

We now detail how to compute these forces. For simplicity, we will assume that the overlap matrix S^{Ξ} is set to the identity for all $\mathbf{R}$. When this is not the case (as for instance with atom-centered basis functions), then additional terms, called Pulay forces, appear, which can easily be computed through Lagrange multipliers [25].

Assume that the variational problem (7.47) has a unique global minimizer $P(\mathbf{R})$ that depends smoothly on $\mathbf{R}$. Then we can compute

$$\frac{\partial E_0}{\partial \mathbf{R}_k}(\mathbf{R}) = \frac{\partial \mathcal{E}}{\partial \mathbf{R}_k}(\mathbf{R}, P(\mathbf{R})) + \frac{\partial \mathcal{E}}{\partial P}(\mathbf{R}, P(\mathbf{R})) \cdot \frac{\partial P}{\partial \mathbf{R}_k}(\mathbf{R}).$$

This formula can be greatly simplified by noticing that, since $P(\mathbf{R})$ minimizes $\mathcal{E}$, by the first-order optimality conditions, the linear functional $\frac{\partial \mathcal{E}}{\partial P}(\mathbf{R}, P(\mathbf{R})) = H(\mathbf{R}, P(\mathbf{R}))$ vanishes on the tangent space to $\mathcal{M}_{\mathrm{DM}}$ at $P(\mathbf{R})$, to which $\frac{\partial P}{\partial \mathbf{R}_k}(\mathbf{R})$ belongs. It follows then that

$$\begin{aligned}\frac{\partial E_0}{\partial \mathbf{R}_k}(\mathbf{R}) &= \frac{\partial \mathcal{E}}{\partial \mathbf{R}_k}(\mathbf{R}, P(\mathbf{R})) \\ &= 2\mathrm{Tr}\left(\frac{\partial h}{\partial \mathbf{R}_k} P(\mathbf{R})\right),\end{aligned} \tag{7.48}$$

which can be computed easily (it is simply the discrete version of the classical electrostatic force exerted on nucleus k by the electron cloud). The above statement, that the derivatives of the energy with respect to external parameters do not involve the derivatives of variational quantities, is known as the (generalized) Hellmann–Feynman theorem.

In this fashion, one can compute first derivatives of the energy with respect to other external parameters, such as an external field, or cell parameters in the case of solids (giving access to stresses). Other properties of interest, such as force constants (the second derivatives of the energy with respect to atomic positions at equilibrium geometry), polarizabilities (the derivatives of the polarization with respect to an external electric field) and even more complex quantities (piezoelectric properties, anharmonic effects ...) can be accessed through perturbation theory, as we outline now. Consider the problem of computing $\delta P = \frac{\partial P}{\partial \mathbf{R}_k} \cdot \delta \mathbf{R}_k$, the variation δP in the density matrix caused by a displacement $\delta \mathbf{R}_k$ of nucleus k. Starting from the equation

$$[H(\mathbf{R}, P(\mathbf{R})), P(\mathbf{R})] = 0$$

we get

$$[H, \delta P] = \left[P, \left(\frac{\partial H}{\partial P} \cdot \delta P + \frac{\partial H}{\partial \mathbf{R}_k} \cdot \delta \mathbf{R}_k\right)\right]. \tag{7.49}$$

We now decompose this equation in an orthonormal basis $(X_n)_{n=1,\dots,N_b}$ of eigenvectors of H: $HX_n = \varepsilon_n X_n$ with $\varepsilon_1 \leq \cdots \leq \varepsilon_{N_b}$. We assume a gap $\varepsilon_N < \varepsilon_{N+1}$. In this basis,

$$\delta P = \sum_{n,m=1}^{N_b} \delta P_{mn} |X_m\rangle\langle X_n|.$$

Differentiating the equation $P^2 = P$, we obtain that $\delta P_{mn} = 0$ whenever $1 \leq m \leq N < n$ or $1 \leq n \leq N < m$, and using (7.49), we obtain

$$\delta P_{mn} = \frac{\left\langle X_m, \left[P, \left(\frac{\partial H}{\partial P} \cdot \delta P + \frac{\partial H}{\partial \mathbf{R}_k} \cdot \delta \mathbf{R}_k\right)\right] X_n\right\rangle}{\varepsilon_m - \varepsilon_n}. \tag{7.50}$$

Since the right-hand side depends on δP, this is a self-consistent linear equation, called the Dyson equation. It is the linearized version of the nonlinear self-consistent equation $P = \mathbb{1}_{(-\infty,\mu)}(H(P))$, and can be solved using the same tools.

This methodology can be extended in several directions:

- other types of perturbations (such as the derivative of the density with respect to an external field, or second derivatives of the energy) can be computed in the same way;
- equation (7.50) can be written in terms of orbitals (the resulting equation sometimes being called the Sternheimer equation [136]), avoiding a full diagonalization of the Hamiltonian;
- higher-order properties can be computed. A notable guiding principle is Wigner's $2n + 1$ rule, which generalizes the Hellmann–Feynman theorem: to compute the $(2n)$th and $(2n + 1)$st derivatives of the energy, one only needs to compute the first nth derivatives of the density matrix;
- the methodology can be applied to time-dependent models, resulting in frequency-dependent response properties.

We refer to [13] for the condensed-matter perspective, and to [109] for the quantum chemistry one.

7.3 Discretization Methods

The current most popular software is based on one of the following two discretization methods: Gaussian-type orbitals (GTOs), and planewaves (PW). The former is the method of choice in quantum chemistry for molecules and clusters. The latter is widely used in solid-state and condensed-matter physics.

GTOs were first introduced by Boys [19] in 1950 to discretize the Hartree–Fock model and have been at the origin of the rapid development of quantum chemistry in the second half of the 20th century. PW basis sets are natural discretization methods for solving Kohn–Sham-type equations on rectangular boxes with periodic boundary conditions [44, 56, 71, 86], as is the case for periodic solids (perfect crystals), and for supercell methods used to simulate disordered crystals, amorphous solids, and liquids.

Other approaches include general-purpose methods for discretizing partial differential equations such as (extended) finite element [80, 147], finite difference [40, 58, 64], spline [105], and wavelet methods [8, 63], as well as specific methods: other kinds of atomic orbitals (Slater-type orbitals [133], numerical atomic orbitals [134]), Augmented plane waves (APW)-type orbitals [4, 42, 129, 132], periodic sinc function basis [131], adaptive local-basis sets [95, 148]. For the sake of brevity, we focus in this chapter on GTO and PW methods. Other types of discretization methods are dealt with in subsequent chapters.

7.3.1 *Atomic Basis Sets for Molecules*

In atomic basis sets methods, each chemical element A is equipped with a finite number of rapidly decaying L^2-normalized basis functions $(\xi_\mu^{\mathrm{A}})_{1 \le \mu \le N_{\mathrm{A}}} \in H^1(\mathbb{R}^3)$ centered at the origin such that $\mathrm{Span}(\xi_\mu^{\mathrm{A}},\ 1 \le \mu \le N_{\mathrm{A}})$ is rotationally invariant. An atomic orbital basis set is a collection $((\xi_\mu^z)_{1 \le \mu \le N_z})_z$, where z spans the periodic table of the elements, or usually only a part of it (e.g. the first four rows, from hydrogen to xenon). Consider now a molecular system consisting of M nuclei of atomic numbers $z_1, \dots z_M$ in a configuration $(\mathbf{R}_1, \dots, \mathbf{R}_M) \in (\mathbb{R}^3)^M$. The basis set used to carry out a Kohn–Sham calculation in the atomic basis set $((\xi_\mu^z)_{1 \le \mu \le N_z})_z$ is then

$$\left(\xi_\mu^{z_k}(\cdot - \mathbf{R}_k),\ 1 \le k \le M,\ 1 \le \mu \le N_{z_k}\right).$$

Note that in atomic basis sets, the basis functions are attached to the nuclei and therefore move with them: for a given molecular system and a given atomic orbital basis set $((\xi_\mu^z)_{1\le\mu\le N_z})_z$, the discretization basis depends on the configuration (i.e. of the positions of the nuclei). As already mentioned in Sect. 7.2.5, this gives rise to the so-called Pulay forces [118] in molecular dynamics.

For consistency with atomic symmetries, the functions ξ_μ^z usually are of the form

$$\xi_\mu^z(\mathbf{r}) = Y_l^m\left(\frac{\mathbf{r}}{|\mathbf{r}|}\right)|\mathbf{r}|^l f(|\mathbf{r}|), \tag{7.51}$$

where $(Y_l^m)_{l\ge0,\,-l\le m\le l}$ are the real spherical harmonics (which forms an orthonormal basis of $L^2(\mathbb{S}^2)$), and $f\ :\ \mathbb{R}_+\ \to\ \mathbb{R}$ a smooth, fast decaying function. Note that if $(\xi_\mu^z)_{1\le\mu\le N_z}\ \in\ H^1(\mathbb{R}^3)$ contains a function of the form (7.51), then $Y_l^{m'}\left(\frac{\mathbf{r}}{|\mathbf{r}|}\right)|\mathbf{r}|^l f(|\mathbf{r}|)\ \in\ \mathrm{Span}((\xi_\mu^z)_{1\le\mu\le N_z})$ for all $-l\ \le\ m'\ \le\ l$ in order to fulfill rotation invariance. Recall that $|\mathbf{r}|^l Y_l^m\left(\frac{\mathbf{r}}{|\mathbf{r}|}\right)$ is a homogenous harmonic polynomial of total degree l.

In most quantum chemistry codes, the functions ξ_μ^z are Gaussian-type orbitals (GTOs), namely finite linear combinations of polynomials times Gaussian functions, that is

$$f(r) = \sum_{n=1}^{n_g} c_n \mathrm{e}^{-\alpha_n r^2},$$

with $\alpha_n > 0$ and $c_n \in \mathbb{R}$. A few quantum chemistry codes use Slater-types orbitals (STOs) for which

$$f(r) = \sum_{n=1}^{n_g} c_n \mathrm{e}^{-\alpha_n r}.$$

Compared to GTOs, STOs better account for the cusps of the Kohn–Sham orbitals at the nuclear positions as well as their decay rates at infinity. On the other hand, GTOs have a tremendous advantage over STOs: as pointed out by Boys [19], all the integrals in (7.12)–(7.13) can be computed explicitly for GTOs using three key properties of Gaussian functions:

1. the product of two Gaussian functions is a Gaussian function;
2. the Fourier transform of a Gaussian function is a Gaussian function;
3. the product of a Gaussian function by a homogeneous polynomial of order l is a linear combination of partial derivatives of order l of this Gaussian function, and vice-versa.

As a consequence, the kinetic, nuclei-electrons interaction, and Coulomb terms can be computed explicitly for GTOs. Only the exchange-correlation term must be computed on a grid. In most quantum chemistry packages, the quadrature points are obtained by unions of atomic quadrature grids. More precisely,

- each chemical element A is associated with a spherical quadrature grid $(\mathbf{r}_g^A)_{1\le g\le N_A}$ which is itself the tensor product of a radial grid by a set of Lebedev points on the unit sphere $\mathbb{S}^2$, and associated weights $(w_g^A)_{1\le g\le N_A}$;
- the quadrature grid for a molecular system with M nuclei of type A_k located at $\mathbf{R}_k$, $1\le k\le M$, is the union of all the points $\mathbf{r}_{k,g} := \mathbf{r}_g^{A_k}+\mathbf{R}_k$ for $1\le k\le M$ and $1\le g\le N_{A_k}$. The weight of the quadrature point $\mathbf{r}_{k,g}$ is given by the formula

$$w_{k,g} = w_g^{A_k} \frac{\prod_{l\neq k} f(\zeta_{k,g}^{kl})}{\sum_{m=1}^{M}\prod_{l\neq m} f(\zeta_{k,g}^{ml})}$$

where

$$\zeta_{k,g}^{ml} = \frac{|\mathbf{r}_{k,g}-\mathbf{R}_m| - |\mathbf{r}_{k,g}-\mathbf{R}_l|}{|\mathbf{R}_m-\mathbf{R}_l|},$$

and $f : [-1,1]\to\mathbb{R}_+$ is a smooth function satisfying $f(-1)=0$, $f(0)=1/2$ and $f(1)=0$. Usual choices include Becke and Stratmann–Scuseria–Frisch (SSF) partitioning functions. As a matter of example, the former is defined as

$$f(\zeta) = \frac{1}{2}\,(1-g(g(g(\zeta)))) \quad \text{where} \quad g(\zeta) = \frac{1}{2}\zeta(3-\zeta^2).$$

For very large molecular systems, Hamiltonian and overlap matrices are sparse due to the extremely fast decay of Gaussian atomic orbitals. In addition, so are density matrices, at least for insulators, and metals at finite temperature. This phenomenon, sometimes called the nearsightedness principle in the physics and chemistry literature [115], originates from the decay properties of the Green kernel of the resolvent of Schrödinger operators. This allows the use of sparse linear algebra methods (see [16] and references therein). The main bottleneck is then the building of the Kohn–Sham matrix

$$H(P) = h^{\Xi} + 2A^{\Xi} : P + \nabla E_{\text{xc}}^{\Xi}(P)$$

for the current density matrix P. The first term h^{Ξ} is simple to deal with since its $O(N_b)$ non-zero entries (at numerical precision) are explicit and easy to compute for Gaussian orbitals. The third term is local and can be computed in $O(N_b)$ operations also. The truly nonlocal second term involves the Coulomb integrals

$$A^{\Xi}_{\kappa\lambda\mu\nu} := (\kappa\lambda|\mu\nu) := \int_{\mathbb{R}^3}\int_{\mathbb{R}^3} \frac{\xi_\kappa(\mathbf{r})\xi_\lambda(\mathbf{r})\xi_\mu(\mathbf{r}')\xi_\nu(\mathbf{r}')}{|\mathbf{r}-\mathbf{r}'|}\,d\mathbf{r}\,d\mathbf{r}'.$$

The evaluation of the Coulomb term $A^{\Xi} : P$ can be accelerated using Fast Multipole Methods (FMMs). The original FMM for point charge distributions, due to Greengard and Rokhlin [72, 73], must be adapted to handle Gaussian-polynomial charge distributions. This has been done by several authors from the 90s on. We will not detail these technicalities here for the sake of brevity and refer the interested reader to [39] and references therein.

7.3.2 Plane-Wave Discretization

We examine now the plane-wave method. Although it was originally formulated for computing properties of crystals and is still mostly used for that purpose, for pedagogical purposes we introduce it here in the case of an isolated neutral molecule.

Note that the solutions φ_i of the self-consistent Eq. (7.37) for negative ε_i are exponentially localized (with characteristic length $1/\sqrt{-\varepsilon_i}$). It is therefore justified, with exponentially small error, to limit ourselves to a large box, which we will take for simplicity to be $\Omega = \left[-\frac{L}{2}, \frac{L}{2}\right]^3$. We will replace the space $\mathbb{R}^3$ by Ω equipped with the topology of a torus (therefore imposing periodic boundary conditions). We can then expand orbitals φ_i in the orthonormal Fourier basis $\{e_{\mathbf{K}}\}_{\mathbf{K}\in\mathcal{R}^*_L}$, with

$$e_{\mathbf{K}}(\mathbf{r}) = \frac{1}{\sqrt{|\Omega|}} e^{i\mathbf{K}\cdot\mathbf{r}}$$

and $\mathcal{R}^*_L = \frac{2\pi}{L}\mathbb{Z}^3$. We then have the expansion

$$\varphi_i(\mathbf{r}) = \sum_{\mathbf{K}\in\mathcal{R}^*_L} c_{i\mathbf{K}} e_{\mathbf{K}}(\mathbf{r}).$$

The kinetic, potential and exchange-correlation terms in (7.2) and (7.4)–(7.5) adapt easily by simply truncating the integrals to Γ. To approximate the Hartree term in a way that leads to simple computations, we replace the Coulomb kernel $\frac{1}{|\mathbf{r}|}$ by the periodic Coulomb kernel

$$G_L(\mathbf{r}) = \sum_{\mathbf{K}\in\mathcal{R}^*_L\setminus\{0\}} \frac{4\pi}{|\mathbf{K}|^2} e_{\mathbf{K}}(\mathbf{r}). \tag{7.52}$$

Then, we seek to minimize the total energy

$$E_L(\Phi) = \int_\Omega \left(\sum_{i=1}^N |\nabla \varphi_i|^2 + V\rho_\Phi \right) + \frac{1}{2} \int_{\Omega\times\Omega} \rho_\Phi(\mathbf{r})\rho_\varphi(\mathbf{r}')G_L(\mathbf{r}-\mathbf{r}')\,\mathbf{dr}\,\mathbf{dr}' + E_{\text{xc}}(\rho_\Phi)$$

variationally by limiting the discretization space $\{e_{\mathbf{K}}\}_{\mathbf{K}\in\mathcal{R}_L^*}$ to the subspace

$$\mathcal{X} = \text{Span}\left(e_{\mathbf{K}}, \mathbf{K} \in \mathcal{R}^*_{E_{\text{cut}}}\right) \quad \text{with} \quad \mathcal{R}^*_{E_{\text{cut}}} = \left\{ \mathbf{K} \in \mathcal{R}_L^*, \frac{1}{2}|\mathbf{K}|^2 \le E_{\text{cut}} \right\}$$

where $E_{\text{cut}} > 0$ is a truncation parameter that controls the maximum kinetic energy allowed in the system.

The resulting energy as a function of the coefficients $c_{i\mathbf{K}}$ is

$$\begin{aligned} E_{L,E_{cut}}(c) := &\sum_{\substack{i=1,\dots,N\\ \mathbf{K},\mathbf{K}'\in\mathcal{R}^*_{E_{\text{cut}}}}} \left(|\mathbf{K}|^2\delta_{KK'} + 2\widehat{V}_{\mathbf{K}-\mathbf{K}'} \right) \overline{c_{i\mathbf{K}}} c_{i\mathbf{K}'} \\ &+ 8\pi \sum_{\substack{i,j=1,\dots,N\\ \mathbf{K},\mathbf{K}'\in\mathcal{R}^*_{E_{\text{cut}}}\\ \mathbf{K}\neq\mathbf{K}'}} \frac{\overline{c_{i\mathbf{K}}} c_{i\mathbf{K}'} \overline{c_{jK}} c_{jK'}}{|\mathbf{K}'-\mathbf{K}|^2} \\ &+ E_{\text{xc}}\left(2 \sum_{\substack{i=1,\dots,N\\ \mathbf{K},\mathbf{K}\in\mathcal{R}^*_{E_{\text{cut}}}}} \overline{c_{i\mathbf{K}}} c_{i\mathbf{K}'} e_{\mathbf{K}'-\mathbf{K}}(\mathbf{r}) \right), \end{aligned}$$

where

$$\widehat{V}_{\mathbf{K}} = \frac{1}{\sqrt{|\Omega|}} \int_\Omega e^{-i\mathbf{K}\cdot\mathbf{r}} V(\mathbf{r})\mathbf{dr}$$

are the Fourier coefficients of the periodic extension of V.

Naively evaluated, these terms scale quadratically with both N and the number of plane waves $|\mathcal{R}^*_{E_{\text{cut}}}|$. However, many of these computations are convolutions, arising from pointwise multiplication in real space. These convolutions can be evaluated efficiently in real space, using the discrete convolution theorem: the cyclic convolution of two arrays can be computed by a discrete Fourier transform, evaluated efficiently using fast Fourier transforms (FFTs). To avoid aliasing effects arising from the cyclic convolution, zero-padding is used: the discrete Fourier transforms are performed on a Cartesian grid that contains all the $\mathbf{K} + \mathbf{K}'$, for $\mathbf{K}, \mathbf{K}' \in \mathcal{R}^*_{E_{\text{cut}}}$.

It is important to note that this procedure allows us to treat the kinetic, potential and Hartree terms *exactly*: for any given set of coefficients $c_{i\mathbf{K}}$, the energy terms computed in this way are the exact energy terms of the orbitals $\varphi_i(\mathbf{r}) = \sum_{\mathbf{K}\in\mathcal{R}^*_{E_{\text{cut}}}} c_{i\mathbf{K}} e_{\mathbf{K}}(\mathbf{r})$. Therefore, for the rHF model, the plane-wave method (for a given L) is *variational*: the ground state energy decreases with E_{cut}. However, the exchange-correlation term is a non-polynomial function of the density, and cannot be evaluated exactly in this fashion. In practice, this term is approximated by an integration on the same real-space grid as for the other terms.

The numerical analysis of this method and of the related Gross–Pitaevskii equation is discussed in Sect. 7.6. Note that this method is based on an expansion of the orbitals φ_i in a Fourier basis. In order for this to be effective, the orbitals need to be smooth, because of the equivalence between smoothness in real space and decay in reciprocal space. However, the singularity of the Coulomb potential imposes *cusps* on the φ_i: for instance, the first eigenfunction of the Hydrogen atom is proportional to $\mathrm{e}^{-|\mathbf{r}|}$. Further, even if we remove these cusps (for instance, by mollifying the Coulomb potential), the φ_i still need to oscillate to satisfy the orthogonality conditions $\varphi_i \perp \varphi_j$, and these oscillations need a large number of plane waves to represent properly. The plane-wave discretization is therefore not applicable directly to atomic systems; in practice, these problems are remedied through the use of the pseudopotential approximation.

7.4 Pseudopotentials

The pseudopotential method is a technique developed in solid-state physics to modify V_{ext} so that electrons close to a nucleus are treated as part of an ionic core. Only wavefunctions associated with valence electrons are to be computed. We refer to [65] and references therein for an overview of the pseudopotential method in practice. A very similar method is used in quantum chemistry, where it goes by the name of "effective core potential"; we refer to [49] for details. To simplify the discussion, we assume an isolated system of non-interacting electrons (we ignore the Hartree and exchange-correlation terms). With appropriate modifications, pseudopotentials can be extended to treat the Kohn–Sham equations.

Reintroducing spin temporarily, recall that the states of an atomic radial Hamiltonian $H_{\text{at}} = -\frac{1}{2}\Delta + V_{\text{at}}$ can be labelled as $\varphi_{n\ell m\sigma}$:

$$\begin{aligned} H_{\text{at}}\varphi_{n\ell m\sigma} &= \varepsilon_{n\ell}\varphi_{n\ell m\sigma} \\ \varphi_{n\ell m\sigma}(\mathbf{r}) &= \frac{R_{n\ell}(|\mathbf{r}|)}{|\mathbf{r}|} Y_{\ell m}\left(\frac{\mathbf{r}}{|\mathbf{r}|}\right), \end{aligned}$$

where $n \geq 1$, $\ell \geqslant 0$, $m = -\ell, \ldots, \ell$ and $\sigma \in \{\alpha, \beta\}$ are the principal, azimuthal, magnetic and spin quantum numbers. The functions $Y_{\ell m}$ are the real spherical harmonics. The functions $R_{n\ell}$ are solutions of the radial Schrödinger equation

$$-\frac{1}{2}R''_{n\ell}(r) + \left(\frac{\ell(\ell+1)}{2r^2} + V_{\mathrm{at}}(r)\right) R_{n\ell}(r) = \varepsilon_{n\ell} R_{n\ell}(r)$$

for $r > 0$, with $R_{n\ell}(0) = 0$ and $\int_0^\infty |R_{n\ell}|^2 = 1$.

There are $2(2\ell + 1)$ available states $(\varphi_{n\ell m\sigma})_{m=-\ell,\ldots,\ell,\sigma=\{\alpha,\beta\}}$ for a given energy level $\varepsilon_{n\ell}$. The electronic state of an atom is conventionally given as a sequence of terms of the form $(n + \ell)\ell^k$, meaning that k of the $4\ell + 2$ available states with a given (n, ℓ) are occupied. The angular momentum quantum number ℓ is labelled using letters: s, p, d, f, etc. For instance, silicon has 14 electrons, and its electronic structure is $1s^2\ 2s^2 2p^6\ 3s^2 3p^2$. This means: two electrons in the $\varphi_{n=1,\ell=0,m=0,\sigma=\{\alpha,\beta\}}$ orbitals, two electrons in the $\varphi_{n=2,\ell=0,m=0,\sigma=\{\alpha,\beta\}}$ orbitals, six electrons in the $\varphi_{n=1,\ell=1,m=\{-1,0,1\},\sigma=\{\alpha,\beta\}}$ orbitals, etc. The spatial extension $\langle r\rangle := \int_{\mathbb{R}^3} |\mathbf{r}||\varphi|^2(\mathbf{r})d\mathbf{r}$ and energies ε of these states, computed using the PBE density functional [110], are given Table 7.1, and the orbitals are plotted Fig. 7.1 (left panel).

Table 7.1 Energy ε and spatial extension $\langle r\rangle$ for the orbitals of the silicon atom at the PBE level. Data from the `atomic` code included in the Quantum Espresso distribution. For comparison, the energy of a typical covalent bond is about 0.1 Ha, and the interatomic distance in bulk silicon at ambient conditions is about 4.5 Bohr

State	(n, ℓ)	$\varepsilon_{n\ell}$ (Ha)	$\langle r\rangle$ (Bohr)
$1s^2$	$(1, 0)$	−64.4	0.11
$2s^2$	$(2, 0)$	−5.10	0.57
$2p^6$	$(1, 1)$	−3.51	0.54
$3s^2$	$(3, 0)$	−0.40	2.17
$3p^2$	$(2, 1)$	−0.15	2.79

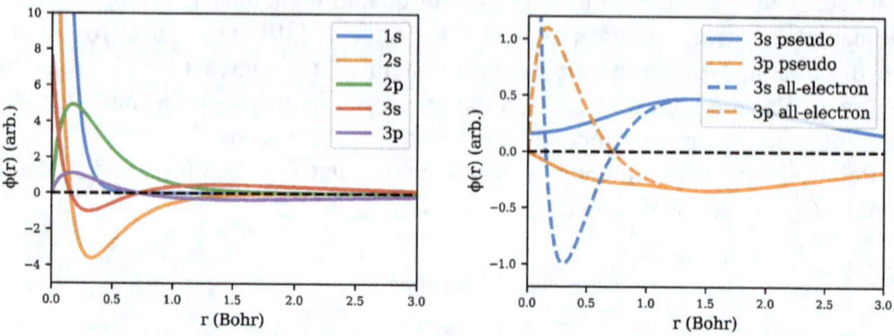

Fig. 7.1 Orbitals of the Si atom at the PBE level (left), and pseudized $3s$ and $3p$ orbitals (obtained with use of the pseudo-potential) (right). The pseudized orbitals are nodeless, and free of cusps

From the data in Table 7.1, it can be seen that electrons can be clearly separated into two classes. The 10 electrons in the $1s$, $2s$ and $2p$ configurations are *core* electrons. They are mainly localized close to the nucleus, with a low energy. This means that they react very little to their environment, as can be shown using the following heuristic argument. Assume that atom A has only one orbital φ_A, with energy E_A, and that atom B has only one orbital φ_B, with energy $E_B > E_A$. When in the diatomic configuration, the total Hamiltonian, expanded in the basis of the φ_A and φ_B, has the approximate form

$$\begin{pmatrix} E_A & \alpha \\ \alpha & E_B \end{pmatrix}$$

where the coupling term α is due to the overlap between φ_A and φ_B. The eigenvalues of this matrix are

$$E_{\pm} = \frac{E_A + E_B}{2} \pm \sqrt{\left(\frac{E_B - E_A}{2}\right)^2 + \alpha^2}.$$

In order to have $E_{\pm}$ significantly different from the isolated energies E_A and E_B (which leads to chemical bonding), we need α to be at least of comparable magnitude to $E_B - E_A$. In other words, to get a bond between two atoms, there needs to be orbitals with a significant overlap which are close in energy.

To a very good approximation, we can therefore consider the $1s$, $2s$ and $2p$ orbitals of silicon as *core* orbitals, that are frozen: they are the same in any chemical environment. By contrast, the $3s$ and $3p$ orbitals are *valence* orbitals, and bind to the orbitals of other atoms.

It follows from this splitting that representing all the φ_i is wasteful, since the core electrons do not contribute significantly to chemical bonding. On the other hand, the valence electrons still feel the influence of the core electrons, and have strong oscillations near the nuclei. Combined with the cusp at the nucleus, this makes them very hard to represent in a plane wave basis. A further complication is that the core electrons of heavy atoms, having a high velocity, are subject to non-negligible relativistic effects. These impact the valence orbitals indirectly through the orthogonalization requirements and the mean-field potential. All these difficulties are remedied through pseudopotentials, at the cost of an approximation.

To introduce pseudopotentials, we first split the energy states $\varepsilon_{n\ell}$ of the atom into core and valence orbitals. We ignore spin in the following discussion, and label the orbitals only by $\varphi_{n\ell m}$. We take $\ell = 0, \ldots, \ell_{\max}$ to be set the of angular momenta for which there is a valence orbital, $\mathcal{I}_\ell = \{n \in \mathbb{N}, \varepsilon_{n\ell} \text{ is a valence state}\}$ and $n_{\text{core},\ell} = \min \mathcal{I}_\ell - 1$ the number of core orbitals with angular momentum ℓ. For instance, for the Silicon atom a natural choice is to set the $1s$, $2s$ and $2p$ orbitals as core, and the $3s$ and $3p$ orbitals as valence: $\ell_{\max} = 1, \mathcal{I}_0 = \{3\}, \mathcal{I}_1 = \{2\}$, $n_{\text{core},0} = 2$, $n_{\text{core},1} = 1$.

The pseudopotential method then replaces the atomic potential V_{at} by a pseudopotential $\widetilde{V}_{\text{at}}$ (usually a nonlocal operator) such that the lowest energy eigenfunctions of the pseudo-Hamiltonian $\widetilde{H}_{\text{at}} = -\frac{1}{2}\Delta + \widetilde{V}_{\text{at}}$, the pseudo-orbitals $\widetilde{\varphi}_{n\ell m}$ for $\ell = 0, \dots, \ell_{\max}, n = 1, \dots, |\mathcal{I}_\ell|$, are smooth and match the valence orbitals $\varphi_{n+n_{\text{core},\ell},\ell m}$ of the real Hamiltonian $H_{\text{at}} = -\frac{1}{2}\Delta + V_{\text{at}}$ outside the core of the atom. In silicon, this means for instance that the new $1s$ state of $\widetilde{H}$ matches the $3s$ state of H outside a chosen cutoff radius r_c (see Fig. 7.1 (right panel)). Since the decay rate of an eigenstate with a negative eigenvalue ε is $\sqrt{-\varepsilon}$, it follows that the eigenvalues of $\widetilde{H}_{\text{at}}$ should match those of H_{at}:

$$\widetilde{H}_{\text{at}}\widetilde{\varphi}_{n\ell m} = \varepsilon_{n+n_{\text{core},\ell},\ell}\, \widetilde{\varphi}_{n\ell m}$$

for $\ell = 0, \dots, \ell_{\max}, n = 1, \dots, |\mathcal{I}_\ell|, m = -\ell, \dots, \ell$.

Once this procedure is established for an atom, all the individual potentials of the atoms in a molecule can be replaced ("pseudized") in this fashion, and the pseudo-Hamiltonian of the molecule is then solved.

Setting aside the construction of these pseudopotentials for a moment, two main issues have to be considered. First, *transferability*: to what extent do the results of a pseudopotential computation on a molecule match the results of the corresponding reference (all-electron) computation? Second, *softness*: how many Fourier modes are needed to represent the pseudo-orbitals accurately? Pseudopotentials seek an optimal compromise between these two competing objectives. Transferability is usually assessed empirically, by comparing the properties of a molecular system to those obtained using an all-electron computation. It is improved by choosing a small cutoff radius r_c and ensuring that all relevant electrons are included as valence. Softness is ensured by choosing a smooth pseudopotential $\widetilde{V}_{\text{at}}$ and by the fact that the lowest energy eigenfunction of a radial local Hamiltonian is nodeless, which reduces oscillations (see Fig. 7.1).

Building pseudopotentials is often done in two parts: given a local potential $V_{\text{at}}(\mathbf{r}) = -\frac{Z}{|\mathbf{r}|}$, first construct smooth pseudo-orbitals $\widetilde{\varphi}_{n\ell m}$, and second construct a pseudopotential $\widetilde{V}_{\text{at}}$ such that the eigenstates of $\widetilde{H}$ are the $\widetilde{\varphi}_{n\ell m}$.

The radial part $\widetilde{R}_{n\ell}$ of the pseudo-orbitals $\widetilde{\varphi}_{n\ell m}$ should match the original orbitals outside of a cutoff radius:

$$\widetilde{R}_{n\ell}(r) = R_{n+n_{\text{core},\ell},\ell} \qquad \text{for } |r| \geqslant r_c.$$

As eigenstates of $\widetilde{H}$, they should also be orthonormal. Orbitals with different angular momenta are automatically orthogonal; the orbitals with the same angular momenta must satisfy

$$\langle \widetilde{R}_{m\ell}, \widetilde{R}_{n\ell} \rangle_{L^2(\mathbb{R}^+)} = \delta_{mn},$$

which implies the "generalized norm-conservation condition"

$$\int_0^{r_c} \overline{\widetilde{R}_{n\ell}(r)}\widetilde{R}_{n'\ell}(r)\mathrm{d}r = \int_0^{r_c} \overline{R_{n+n_{\mathrm{core},\ell},\ell}(r)}R_{n'+n_{\mathrm{core},\ell},\ell}(r)\mathrm{d}r \tag{7.53}$$

for all $m, n \in \mathcal{I}_\ell$. Finally, they should be smooth, and, as low-energy eigenfunctions, have the correct amount of nodes.

Historically, most pseudopotentials have been constructed assuming the pseudization of only one orbital per angular momentum: $|\mathcal{I}_\ell| = 1$ for $\ell = 0, \dots, \ell_{\max}$. This is for instance reasonable in the silicon example above. For simplicity, we denote these orbitals and pseudo-orbitals by $\varphi_{\ell m}$ and $\varphi_{\ell m}$, with energies ε_ℓ. The Troullier–Martins procedure [139] then postulates a simple functional form for $\widetilde{R}_\ell$, and adjusts its parameters to match the values and the derivatives up to fourth order of $\widetilde{R}_\ell$ and R_ℓ at r_c. The RRKJ procedure expands $\widetilde{R}_\ell$ in a basis of Bessel functions, and minimizes their kinetic energy above a certain cutoff energy to ensure smoothness [120].

Once the $\widetilde{\varphi}_{\ell m}$ are constructed, a potential $\widetilde{V}_{\mathrm{at}}$ is constructed. A solution is to first choose the local radial potential $\widetilde{V}_\ell$ so that $\widetilde{\varphi}_{\ell m}$ is a solution of

$$\left(-\frac{1}{2}\Delta + \widetilde{V}_\ell\right)\widetilde{\varphi}_{\ell m} = \varepsilon_\ell \widetilde{\varphi}_{\ell m}.$$

This can be done by simply solving the equation above for $\widetilde{V}_\ell$. Then, we can choose the semilocal form

$$\widetilde{V}_{\mathrm{at}} = \sum_{\ell=0}^{\ell_{\max}} P_\ell \widetilde{V}_\ell P_\ell, \tag{7.54}$$

where $P_\ell = \sum_{m=-\ell,\dots,\ell} |Y_{\ell m}\rangle\langle Y_{\ell m}|$ is the projector on angular momentum ℓ. This operator automatically has the $\widetilde{\varphi}_{\ell m}$ as eigenstates with correct eigenvalues, completing the pseudo-potential.

This procedure however has a major defect: the operator $\widetilde{V}_{\mathrm{at}}$ is a nonlocal operator whose evaluation in plane-wave codes requires the computation of a dense matrix. This problem was solved by Kleinmann and Bylander [82], who realized that the above procedure was wasteful; the most economical operator that will produce the $\widetilde{\varphi}_{\ell m}$ as eigenstates is the low-rank form

$$\widetilde{V}_{\mathrm{at}}^{\mathrm{KB}} = \widetilde{V}_{\mathrm{loc}} + \sum_{\ell=0}^{\ell_{\max}} \sum_{m=-\ell}^{\ell} \frac{|\beta_{\ell m}\rangle\langle\beta_{\ell m}|}{\langle\beta_{\ell m}, \widetilde{\varphi}_{\ell m}\rangle} \tag{7.55}$$

where $\widetilde{V}_{\mathrm{loc}}$ is an arbitrary radial local potential and

$$\beta_{\ell m} = \left(\varepsilon_\ell - \left(-\frac{1}{2}\Delta + \widetilde{V}_{\mathrm{loc}}\right)\right)\widetilde{\varphi}_{\ell m}.$$

By construction, $(-\frac{1}{2}\Delta + \widetilde{V}_{\mathrm{at}}^{\mathrm{KB}})\widetilde{\varphi}_{\ell m} = \varepsilon_\ell \widetilde{\varphi}_{\ell m}$, as desired. The potential $\widetilde{V}_{\mathrm{loc}}$ is usually chosen to remove one of the ℓ components from the sum.

Note that we have only insisted on the fact that the $\widetilde{\varphi}_{\ell m}$ are eigenfunctions of $\widetilde{H}_{\mathrm{at}}$ at energies ε_ℓ, but have said nothing of the other states; it may very well happen that $\widetilde{H}_{\mathrm{at}}$ has states with lower energies ("ghosts"), invalidating the pseudopotential. This is usually treated heuristically, by varying parameters in the pseudopotential construction to avoid the appearance of ghosts.

The procedure outlined above is the basis for the construction of the historically successful Troullier–Martins and RRKJ pseudopotentials. More modern pseudopotentials try to reproduce more than one state per angular momentum. This is especially useful for highly-localized valence orbitals (like the $2p$ orbitals of oxygen), or semicore orbitals, which are not well isolated from the valence orbitals. In this case, one cannot use the semilocal form (7.54), because no single potential V_ℓ can reproduce all the pseudo-orbitals at the required energies. However, the form (7.55) naturally generalizes to this case: we seek a low-rank potential such that $\widetilde{H}_{\mathrm{at}}\widetilde{\varphi}_{n\ell m} = \varepsilon_{n\ell}\widetilde{\varphi}_{n\ell m}$ for $i = 1, \ldots, n_\ell$, and obtain

$$\widetilde{V}_{\mathrm{at}}^{\mathrm{KB}} = \widetilde{V}_{\mathrm{loc}} + \sum_{\ell=0}^{\ell_{\max}} \sum_{n=1}^{n_\ell} \sum_{m=-\ell}^{\ell} B_{nn'\ell m} |\beta_{n\ell m}\rangle\langle\beta_{n\ell m}|, \tag{7.56}$$

where the $\beta_{n\ell m}$ are defined as above, and

$$B_{nn'\ell m} = \langle\widetilde{\varphi}_{n\ell m}, \beta_{n'\ell m}\rangle.$$

From the generalized norm-conservation property (7.53), $\langle\widetilde{\varphi}_{n\ell m}, \widetilde{\varphi}_{n'\ell m}\rangle = \delta_{nn'}$, and it follows that B is Hermitian.

The scheme above was used by Hamann [74], who constructed pseudo-orbitals satisfying the generalized norm-conservation property (7.53). That work built on an earlier method of Vanderbilt, the "ultra-soft pseudopotentials", which relaxed the constraint (7.53) by introducing a modified inner product in which (7.53) holds, resulting in a generalized eigenvalue problem.

An alternative to norm-conserving or ultra-soft pseudopotentials is the projector-augmented wave (PAW) method of Blöchl [18], which uses a transformation that allows the recovery of the original all-electron orbitals. This is especially useful for probing properties that depend on fine details of the electrons close to the nuclei, such as nuclear magnetic resonance effects [112], but comes at the price of a more complex formalism, the response computations in particular being much more cumbersome.

Also notable is the "black-box" approach used in [70, 76], where a very simple functional form for the local and nonlocal part of the pseudopotential is postulated, and its parameters adjusted by least-squares fitting to a number of desirable properties, such as the eigenvalues ε_i and norm conservation.

Despite being extremely successful in practice, the various pseudopotential methods remain uncontrolled approximations which are not systematically improvable. For a given density functional, they are often the major source of variation between computational codes [89]. Their error analysis is still underdeveloped; see [30] for the existence of norm-conserving pseudopotentials which optimally reproduce the first-order Stark effect, and [51] for an analysis of the PAW method for a particular one-dimensional system.

7.5 Algorithms

The development of algorithms for solving Kohn–Sham problems in quantum chemistry, on the one hand, and condensed matter physics, on the other hand, has followed two different paths. Until the mid 90s and the introduction of hybrid functionals, DFT was barely used in quantum chemistry, while the Hartree–Fock model was very popular, on its own—in the early days—or as a first step toward correlated methods (Møller–Plesset perturbation theory, coupled cluster, quantum Monte Carlo…). From a mathematical viewpoint, the Hartree–Fock and zero-temperature standard Kohn–Sham model are very similar: they are minimization problems on the same sets (orbitals or density matrix formulations), and the energy functionals only differ by the exchange-correlation term, which is in some sense a "small" perturbation of the other three terms (kinetic energy, electrons-nuclei interaction, and Hartree term). It was therefore natural for quantum chemists to try and adapt to the Kohn–Sham setting the many algorithms for Hartree–Fock problems discretized in Gaussian basis sets developed since the 50s.

In parallel, in the mid 60s physicists started to develop algorithms for solving Kohn–Sham LDA and GGA models for solids discretized in plane-wave basis sets, and came up with specific techniques such as Anderson's density mixing [5], Thomas–Fermi charge mixing [119], Kerker's preconditioning in Fourier [81] or real [130] spaces, etc.

From a numerical point of view, solving the Kohn–Sham problem for molecules in a Gaussian basis set or for solids in a plane-wave basis set are quite different games. In particular,

- the self-consistent Hamiltonians in quantum chemistry tend to have a gap between occupied and virtual states, whereas that is not the case for metallic solids and numerical devices such as an artificial finite temperature are needed;
- the Kohn–Sham Hamiltonian of a large molecule is sparse and can be stored in memory for a Gaussian basis set, while it is dense and too large to be stored in memory for a plane-wave basis set (only matrix-vector products can be computed in practice);

- the number of discretization functions per electron pair is of the order of 2 to 10 for Gaussian basis sets, and typically 100 or more for plane-wave basis sets;
- Gaussian basis sets tend to become overcomplete when the size of the basis set increases (i.e. the overlap matrix becomes ill-conditioned), while plane-wave bases remain orthonormal when the energy cut-off increases.

This can explain why the two communities barely interacted until the beginning of the 21st century. The situation is rapidly changing since quantum chemists and materials scientists now share many common interests: hybrid functionals originally developed in quantum chemistry are now used in materials science, quantum chemists simulate extended systems to study, for example, heterogeneous catalysis...

In the following, we will consider the zero-temperature standard Kohn–Sham problem, which can be formulated either as

$$\min_{X \in \mathbb{R}^{N_b \times N},\ X^T X = I} E(X) \tag{7.57}$$

or

$$\min_{P \in \mathbb{R}^{N_b \times N_b}_{\rm sym},\ P^2 = P,\ {\rm Tr} P = N} \mathcal{E}(P), \tag{7.58}$$

where we recall that $E(X) = \mathcal{E}(XX^T)$. Both formulations are useful in different contexts. The first stores N orbitals of size $N_{\rm b}$, and is therefore more economical. However, it breaks a natural symmetry of the problem. Indeed, the energy does not depend on the orbitals themselves, but rather on the subspace they span:

$$\forall X \in \mathcal{M}_{\rm MO}, \quad \forall U \in O(N), \quad XU \in \mathcal{M}_{\rm MO} \quad \text{and} \quad E(XU) = E(X).$$

Density matrices quotient this gauge invariance automatically, and are therefore more convenient to work with theoretically. They are also useful in formulating linear scaling methods. However, they require storing and operating with an $N_{\rm b} \times N_{\rm b}$ matrix, which is prohibitively expensive when large basis sets are used. It is often most convenient to formulate and analyze the algorithms using density matrices, but implement them using orbitals and related quantities (such as the density ρ and self-consistent potential V). We follow this viewpoint in this presentation.

We will first describe in Sects. 7.5.1 and 7.5.2 the different algorithms used to solve the above problem in its density matrix formulation, without consideration of their community of origin, implementation or computational cost. Then, in Sects. 7.5.3 and 7.5.4, we will consider their implementation in the quantum chemistry and solid-state contexts respectively.

In order to lighten the notation, we assume throughout this section that the discretization basis Ξ is orthonormal, and we will omit the superscript Ξ in the expressions of the core Hamiltonian h^{Ξ}, the linear map G^{Ξ} defined in (7.17), the discretized Hartree–Fock energy functional $\mathcal{E}^{\rm HF,\Xi}$, etc.

7.5.1 *Self-Consistent Field Algorithms*

7.5.1.1 Simple SCF and Damping

As mentioned above, the main algorithms for solving the Kohn–Sham problem in quantum chemistry were originally developed for the Hartree–Fock (HF) model. In the HF setting, the energy functional $\mathcal{E}^{\rm HF}$ is given by (7.16). When the no-unfilled shell property holds true (which is always the case for spin-unrestricted models, see Sect. 7.2.4), the ground state density matrix satisfies

$$P^{\rm GS} = \mathcal{F}_{\rm HF-DM}(P^{\rm GS}), \tag{7.59}$$

where

$$\mathcal{F}_{\rm HF-DM}(P) = \operatorname{argmin}\left\{ \operatorname{Tr}(H^{\rm HF}(P)P'),\ P' \in \mathcal{M}_{\rm DM} \right\} = \mathbb{1}_{(-\infty,\mu)}(H^{\rm HF}(P)),$$

where $H^{\rm HF}(P) := \nabla\mathcal{E}^{\rm HF}(P) = h + G(P)$ is the Fock matrix (the HF Hamiltonian matrix) and μ a Fermi level, which depends on P and is chosen so that $\operatorname{Tr}(\mathcal{F}_{\rm HF-DM}(P)) = N$. One way to compute $\mathcal{F}_{\rm HF-DM}(P)$ for a given $P \in \mathcal{M}_{\rm DM}$, is to assemble the matrix $H^{\rm HF}(P)$, diagonalize it, and compute the orthogonal projector on the space spanned by N orthonormal eigenvectors associated with the lowest N eigenvalues (counting multiplicities). It is efficient for small or moderate size basis sets, but other methods must be used for large systems and basis sets. The function $\mathcal{F}_{\rm HF-DM}(P)$ is not uniquely defined if the Nth and $(N+1)$st eigenvalues of $H^{\rm HF}(P)$ are equal, but it is in the neighborhood of any ground state density matrix by virtue of the no-unfilled shell property.

Self-consistent field algorithms for HF are iterative algorithms aiming at solving (7.59). The simplest of them is the Roothaan algorithm [125] defined by

$$P_{k+1} = \mathcal{F}_{\rm HF-DM}(P_k). \tag{7.60}$$

This algorithm, as well as the level-shifting algorithm [127], has been analyzed in [28] from a mathematical point of view. Introducing the symmetric functional $E^{\rm HF} : \mathbb{R}^{N_b\times N_b}_{\rm sym} \times \mathbb{R}^{N_b\times N_b}_{\rm sym} \to \mathbb{R}$ defined by

$$E^{\rm HF}(P,P') = \operatorname{Tr}(hP) + \operatorname{Tr}(hP') + \operatorname{Tr}\left(G(P)P'\right),$$

it can be easily checked, using the symmetry property $\operatorname{Tr}\left(G(P)P'\right) = \operatorname{Tr}\left(G(P')P\right)$, that, in the case when the sequence of density matrices generated by the Roothaan algorithm is well defined, it satisfies for all $k \in \mathbb{N}$,

$$P_{2k+1} = \operatorname{argmin}\left\{ E^{\rm HF}(P,P_{2k}),\ P \in \mathcal{M}_{\rm DM} \right\},$$

$$P_{2k+2} = \operatorname{argmin}\left\{ E^{\rm HF}(P_{2k+1},P'),\ P \in \mathcal{M}_{\rm DM} \right\}.$$

In other words, the simple SCF scheme amounts to an alternate minimization of the functional E^{HF} on $\mathcal{M}_{\text{DM}} \times \mathcal{M}_{\text{DM}}$. It behaves as follows:

- either the whole sequence $(P_k)_{k\in\mathbb{N}}$ converges to the global minimizer of the Hartree–Fock model;
- or the sequences of even and odd iterates $(P_{2k})_{k\in\mathbb{N}}$ and $(P_{2k+1})_{k\in\mathbb{N}}$ both converge, but to different values P_{even} and P_{odd}, none of them being a minimizer of the Hartree–Fock energy functional.

The latter behavior is called *charge sloshing* in the physics and chemistry literature: when k is large (close to convergence), the electronic density ρ_{P_k} moves from one part of the molecule for k even to another one for k odd (e.g. from a functional group to another one) and back at the next iteration. This phenomenon is also encountered for isolated atoms where charge sloshing between e.g. $3d$ and $4s$ shells can be observed.

Let us now turn to the adaptation of this algorithm to the Kohn–Sham model. We assume that at each step of the algorithm under consideration there is a gap between the Nth and $(N+1)$st eigenvalues of the Kohn–Sham Hamiltonian $H(P_k) := \nabla\mathcal{E}(P_k)$, uniformly bounded from below by a positive constant independent of k (uniform well-posedness—UWP—assumption). The Roothaan algorithm for Kohn–Sham, also called the simple SCF algorithm in the physics literature, reads

$$P_{k+1} = \mathcal{F}_{\text{DM}}(P_k), \tag{7.61}$$

where, under the UWP assumption,

$$\mathcal{F}_{\text{DM}}(P) = \operatorname{argmin}\left\{\operatorname{Tr}(H(P)P'),\ P' \in \mathcal{M}_{\text{DM}}\right\} = \mathbb{1}_{(-\infty,\mu)}(H(P)),$$

where μ is chosen as above. In the special case of the Hartree model, for which the exchange-correlation energy functional is set to zero, the energy is quadratic in the density matrix and this algorithm behaves as in the HF setting (convergence or oscillation between two states which are not solutions to the problem). For a non-quadratic exchange-correlation functional, the iterates can oscillate between more than two states, or even display chaotic behavior for some systems [37].

A way to prevent charge sloshing is to use simple mixing, for which the iterates are defined as

$$P_{k+1} = \mathcal{F}_{\text{DM}}(\widetilde{P}_k), \quad \widetilde{P}_{k+1} = (1-\alpha)\widetilde{P}_k + \alpha P_{k+1}, \tag{7.62}$$

where $\widetilde{P}_0 = P_0$ and $0 < \alpha < 1$ is a fixed mixing parameter ($\alpha = 1$ corresponds to the simple SCF scheme, in which case $\widetilde{P}_k = P_k$ for each k). Note that unlike the P_k's, the iterates $\widetilde{P}_{k+1}$ in (7.62) do not live on $\mathcal{M}_{\text{DM}}$ but in its convex hull $\text{Conv}(\mathcal{M}_{\text{DM}})$. From a geometrical viewpoint, $\mathcal{F}_{\text{DM}}(\widetilde{P}_k)$ is the point P on $\mathcal{M}_{\text{DM}}$, and also actually in $\text{Conv}(\mathcal{M}_{\text{DM}})$, for which the slope at $t = 0$ of the function

$$t \mapsto \mathcal{E}\big((1-t)\widetilde{P}_k + tP\big)$$

is minimal, linking the mixing scheme with the Frank–Wolfe algorithm popular in convex optimization [108]. In particular, this implies that the energy $\mathcal{E}(\widetilde{P}_k)$ decreases along the iterations, for $0 < \alpha$ small enough. Under the UWP assumption that there is a uniform positive gap between the Nth and $(N+1)$st eigenvalues of the Kohn–Sham Hamiltonians $H(\widetilde{P}_k) := \nabla\mathcal{E}(\widetilde{P}_k)$ and regularity assumptions on $\mathcal{E}$, it can be shown that this algorithm "numerically converges" (in the sense that $P_{k+1} - P_k \to 0$ and $\widetilde{P}_k - P_k \to 0$) and that the limit (if the iteration actually converges) is a critical point—a local minimizer in practice—of $\mathcal{E}$ on $\mathcal{M}_{\rm DM}$ as well as a critical point—a local minimizer in practice—of $\mathcal{E}$ in $\mathrm{Conv}(\mathcal{M}_{\rm DM})$ [25].

Remark 7.1 For a majority of chemical systems, it is observed that the UWP assumption is satisfied for k large enough, and that the iterates $\widetilde{P}_k$ numerically converge toward a local, hopefully global, minimizer of $\mathcal{E}$ on $\mathcal{M}_{\rm DM}$. However, for some systems, the iterates $\widetilde{P}_k$ numerically converge toward a local minimizer of $\mathrm{Conv}(\mathcal{M}_{\rm DM})$ which does not belong to $\mathcal{M}_{\rm DM}$, in other words to a local minimizer of the extended Kohn–Sham model with fractional occupation numbers at the Fermi level, while the iterates P_k do not converge in general. It can be argued that for such systems, the standard Kohn–Sham problem is not appropriate since the Kohn–Sham energy can be lowered by allowing fractional occupancies.

The optimal damping algorithm [22] (ODA), also originally introduced in the HF setting [27] based on ideas in [93], is a variant of the simple mixing algorithm using an optimal step α_k:

$$P_{k+1} = \mathcal{F}_{\rm DM}(\widetilde{P}_k), \quad \widetilde{P}_{k+1} = (1-\alpha_k)\widetilde{P}_k + \alpha_k P_{k+1}, \tag{7.63}$$

with

$$\alpha_k \in \mathrm{argmin}\left\{\mathcal{E}\big((1-\alpha)\widetilde{P}_k + \alpha P_{k+1}x\big),\ \alpha \in [0,1]\right\}. \tag{7.64}$$

A proof of numerical convergence of the ODA—still under a UWP assumption—can be read in [25]. Stronger convergence results require more information, such as second-order conditions [37] or analyticity [90]. Note that Remark 7.1 is also valid for ODA.

Remark 7.2 Implementing the ODA algorithm for LDA or GGA functionals does not require density matrices to be stored: it suffices to store in memory electronic densities and Kohn–Sham orbitals. This is of course very important in practice since, for example, density matrices cannot be stored for plane-wave calculations on large systems.

Even when it converges, the solutions to the self-consistent Kohn–Sham equations obtained by the simple SCF scheme sometimes correspond to critical points which are local, non-global, minima, or to critical points which do not satisfy the *Aufbau* principle. This situation is sometimes encountered with GTO basis sets. Minimization based methods such as the ODA seem to be free from such problems.

The main limitation of the ODA is that it may converge very slowly in the vicinity of the solution. Acceleration methods are then needed.

7.5.1.2 Anderson–Pulay Acceleration Schemes

Consider a function $g : \mathcal{X} \to \mathcal{X}$ where $\mathcal{X}$ is a Euclidean (or Hermitian) space and the fixed-point problem: find x_* in $\mathcal{X}$ such that

$$g(x_*) = x_*. \tag{7.65}$$

Even when the convergence of the Picard iterations $x_{k+1} = g(x_k)$ locally converge, convergence may be very slow. The original formulation of the Anderson acceleration scheme [5], introduced in 1965, is:

$$x_{k+1} = \sum_{i=0}^{m_k} \theta_{k,i,*} g(x_{k-i}), \tag{7.66}$$

where the coefficients $\theta_{k,i,*}$ satisfy $\sum_{i=0}^{m_k} \theta_{k,i,*} = 1$ and are obtained by solving

$$(\theta_{k,i,*}) = \operatorname*{argmin}_{\theta \in \mathbb{R}^{m_k+1},\ \sum_{i=0}^{m_k} \theta_i = 1} \left\| \sum_{i=0}^{m_k} \theta_i (g(x_{k-i}) - x_{k-i}) \right\|_{\mathcal{X}}, \tag{7.67}$$

which has the structure of a linear least-square problem. When $m_k = k$ for all k, the scheme is called Anderson acceleration without truncation. An additional mixing step can be added, in which case the iterates are defined as

$$x_{k+1} = \beta_k \left(\sum_{i=0}^{m_k} \theta_{k,i,*} x_{k-i} \right) + (1 - \beta_k) \left(\sum_{i=0}^{m_k} \theta_{k,i,*} g(x_{k-i}) \right), \tag{7.68}$$

where $\beta_k > 0$ is allowed to vary with k (the original Anderson acceleration scheme corresponds to $\beta_k = 1$ for all k).

When applied to the affine function $g(x) : (A+I)x - b$ where $x, b \in \mathbb{R}^n$ and $A \in \mathbb{R}^{n\times n}$, the Anderson method is equivalent to the GMRES algorithm [126] for solving the linear system $Ax = b$, provided Anderson's iterates do not stagnate [124, 142]. It can thus be interpreted as a nonlinear version of GMRES (although less robust than GMRES in the linear case). It can also be interpreted as an extrapolation method [20].

Anderson acceleration has been rediscovered a number of times under various names, corresponding to different interpretations: "multisecant Broyden" and "inversion in the subspace". Indeed, it has recently been understood that the Anderson acceleration scheme can be seen as a special instance of a family of algorithms called Anderson–Pulay schemes [43]. This family also includes

Pulay's Direct Inversion in the Iterative Space (DIIS) schemes [116, 117] originally introduced to solve the Hartree–Fock equations (7.59), and widely used as well to solve the Kohn–Sham equations. This approach is often referred to as the *Pulay mixing* in the condensed matter physics literature. Note that Pulay's first DIIS scheme [116] is in fact identical to Anderson acceleration scheme, while Pulay's second DIIS [117] (also called C-DIIS, for commutator-DIIS) does not fit into Anderson's framework. In their most general formulation, Anderson–Pulay schemes are designed to accelerate local convergence toward a point x_* of a function g, which may, as for Hartree–Fock and Kohn–Sham models, take its values in a differentiable submanifold $\mathcal{M}$ of the Euclidean space $\mathcal{X}$. We therefore assume, following [43], that

A1. $\mathcal{X}$ is a Euclidean space, $\mathcal{M}$ is a differentiable submanifold of $\mathcal{X}$, $x_* \in \mathcal{M}$, U is an open neighborhood of x_* in $\mathcal{X}$, and $g : U \to \mathcal{M}$ is a function of class C^2 satisfying $g(x_*) = x_*$;
A2. There exist a Euclidean space $\mathcal{Y}$, and a C^2 function $f : U \to \mathcal{Y}$ satisfying the following properties: $f(x_*) = 0$ and there exists a constant $\sigma > 0$ such that

$$\forall x \in U \cap \mathcal{M}, \quad \sigma \|x - x_*\|_{\mathcal{X}} \leq \|f(x)\|_{\mathcal{Y}}. \tag{7.69}$$

From Assumption A2, we have that for all $x \in U \cap \mathcal{M}$, $f(x) = 0$ if and only if $x = x_*$. The function f can in fact be interpreted as a residual function: $\sigma^{-1}\|f(x)\|_{\mathcal{Y}}$ provides an upper bound of the distance from x to the solution x_*.
In Pulay's schemes, $\mathcal{X} = \mathbb{R}^{N_b \times N_b}_{\text{sym}}$, $\mathcal{M} = \mathcal{M}_{\text{DM}}$, g is the Hartree–Fock or Kohn–Sham map $g = \mathcal{F}_{\text{HF-DM}}$ or $g = \mathcal{F}_{\text{DM}}$, and the residual function f is defined as

- the natural residual map $f(P) = g(P) - P$ in the first scheme [116];
- the commutator $f(P) = [H(P), P]$ in the (more efficient) second scheme [117]. This is indeed a legitimate residual map in view of (7.39).

There exist several formulations of the Anderson–Pulay scheme differing by the way the previous iterates are taken into account. In the fixed-depth Anderson–Pulay scheme implemented in most electronic structure codes, the depth m_k is chosen equal to $m_k = \min(k, m_{\max})$, where $m_{\max}$ a fixed number, typically $m_{\max} = 20$, chosen empirically to optimize the performance of the algorithm in both computational time and memory requirement. Restarted versions [12, 114] have also been proposed. We detail here the recently introduced adaptive-depth formulation of the Anderson–Pulay scheme [43], which seems to outperform the previous versions:

1. **Initialization and first iteration:** x_0, $m_{\max} \in \mathbb{N}^*$, $\delta \geq 0$ and $\eta > 0$ being given, set

$$r_0 = f(x_0), \quad x_1 = g(x_0), \quad r_1 = f(x_1), \quad s_1 = r_1 - r_0, \quad k = 1, \quad m_k = 1.$$

2. **Subsequent iterations:** while $\|r_k\|_2 > \eta$ do

   a. solve

$$(\alpha_1^{(k)}, \ldots, \alpha_{m_k}^{(k)}) = \underset{(\alpha_1, \ldots, \alpha_{m_k}) \in \mathbb{R}^{m_k}}{\operatorname{argmin}} \left\| r_k - \sum_{i=1}^{m_k} \alpha_i s_{k-m_k+i} \right\|_2 \tag{7.70}$$

   and set

$$\widetilde{x}_{k+1} = x_k - \alpha_i^{(k)} \sum_{i=1}^{m_k} (x_{k-m_k+i} - x_{k-m_k+i-1})$$

   b. set

$$x_{k+1} = g(\widetilde{x}_{k+1}), \quad r_{k+1} = f(x_{k+1}), \quad s_{k+1} = r_{k+1} - r_k,$$

   c. set m_{k+1} the largest integer $m \leq \min(m_k + 1, m_{\max})$ such that for $k+1-m \leq i \leq k$, $\delta \|r_i\|_2 < \|r_{k+1}\|_2$
   d. set $k = k + 1$.

Taking $\delta = 0$ gives back the usual fixed-depth scheme. For $\delta > 0$, the depth m_{k+1} is adjusted so that the residuals at the previous m_{k+1} residuals are not too large compared to the current residual r_{k+1}. In view of (7.69), this ensures that the previous m_{k+1} iterates are not too far away from the current iterate x_{k+1}. This scheme can be proved to converge superlinearly under some assumptions on f and g [43].

Numerically, this procedure is often very efficient, and can converge quickly even if the underlying fixed-point iteration $P_{k+1} = \mathcal{F}_{\mathrm{DM}}(P_k)$ is divergent. It however lacks in robustness when started far away from a solution.

7.5.1.3 Newton and Quasi-Newton Methods for the SCF Equations

Problem (7.59) can also be seen as a nonlinear equation

$$R(P^{\mathrm{GS}}) := \mathcal{F}_{\mathrm{DM}}(P^{\mathrm{GS}}) - P^{\mathrm{GS}} = 0,$$

to which the standard Newton method can be applied:

$$P_{k+1} = P_k - D_k,$$

where D_k solves the correction equation

$$(1 - J_k) D_k = \mathcal{F}_{\mathrm{DM}}(P_k) - P_k \tag{7.71}$$

and J_k is the Jacobian of $\mathcal{F}_{\mathrm{DM}}$ at P_k.

However, a practical limitation of Newton's method is the high cost of evaluating the Jacobian and solving the Newton correction equation (7.71). One workaround is to use a Jacobian-free Krylov–Newton technique [83] and solve (7.71) iteratively using, for example, the GMRES algorithm [126]. The matrix vector multiplication of the form $J_k D$, which is required in each GMRES iteration, can be computed using methods similar to those used to compute response properties (see Sect. 7.2.5), or simply approximated by the finite difference formula

$$\frac{\mathcal{F}_{\mathrm{DM}}(P_k + tD) - \mathcal{F}_{\mathrm{DM}}(P_k)}{t},$$

for an appropriately chosen small scalar t. The finite difference calculation requires one additional function evaluation of $\mathcal{F}_{\mathrm{DM}}$ per GMRES step. Therefore, even though Newton's method may exhibit quadratic convergence, each Newton step may be expensive if the number of GMRES steps required to solve the correction equation is large. Numerical experiments presented in [62] show that without a good preconditioner, too many GMRES iterations are needed to ensure convergence.

Quasi-Newton methods can also be used, where the inverse Jacobian $(1 - J_k)^{-1}$ is approximated by a simpler matrix B_k. For instance, the simple mixing scheme above corresponds to $B_k = \alpha I$. More sophisticated schemes can be obtained by using variants of Broyden's method [21, 79, 135], which builds the B_k from previous iterates. In such an approach, B_k is obtained by performing a sequence of low-rank modifications to some initial approximation B_0 of the Jacobian inverse using a recursive formula. One particular choice among several is [57, 103]

$$\begin{aligned} &\min_{B} \frac{1}{2}||B - B_{k-1}||_F^2 \\ &\text{s.t. } \Upsilon_k = BY_k, \end{aligned} \tag{7.72}$$

where B_{k-1} is the approximation to the Jacobian inverse constructed in the $(k-1)$st Broyden step. The matrices Υ_k and Y_k above are defined as

$$\Upsilon_k = (s_k, s_{k-1}, \ldots, s_{k-\ell}), \quad Y_k = (y_k, y_{k-1}, \ldots, y_{k-\ell}), \tag{7.73}$$

where s_j and y_j are defined by

$$s_j = \mathrm{vec}\left[P_j - P_{j-1}\right] \text{ and } y_j = \mathrm{vec}\left[R(P_j) - R(P_{j-1})\right],$$

respectively. Here the residual matrices are given by $R(P) = \mathcal{F}_{\mathrm{DM}}(P) - P$, and the vec $[\cdot]$ notation is used to turn a matrix into a vector by stacking its columns on top of each other. The equation $\Upsilon_k = BY_k$ is the multisecant condition and imposes the consistency of the inverse Jacobian with the observed relationships between the P_j and $R(P_j)$.

It is easy to show that the solution to (7.72) is

$$B_k = B_{k-1} + (\Upsilon_k - B_{k-1}Y_k)Y_k^\dagger, \tag{7.74}$$

where $Y_k^\dagger$ denotes the pseudo-inverse of Y_k. If B_{k-1} is reset to αI, we obtain Anderson's method

$$\mathrm{vec}\,[P_{k+1}] = \mathrm{vec}\,[P_k] + \alpha(I - Y_k Y_k^\dagger)\mathrm{vec}\,[R(P_k)] - \Upsilon_k Y_k^\dagger\,\mathrm{vec}\,[R(P_k)]\,, \tag{7.75}$$

commonly used in KSDFT solvers [5].

7.5.2 Direct Minimization

Instead of solving the HF equation (7.59) by an SCF algorithm, the HF ground state can be obtained by direct minimization of the HF energy functional. A quadratically-convergent algorithm, consisting in obtaining P_{k+1} by performing one Newton step in the local map Φ_{P_k} defined by (7.28)–(7.29), was proposed by Bacskay in [10]. Trust region methods were introduced to the Hartree–Fock setting in [60, 137].

Because the Kohn–Sham problem is a constrained minimization problem as indicated in Sect. 7.2.3, we can try to solve the optimization problem directly. The simplest type of optimization procedure is the projected gradient algorithm

$$\widetilde{P}_{k+1} = P_k - \alpha[[H(P_k), P_k], P_k], \tag{7.76}$$

where $H(P_k) = \nabla\mathcal{E}(P_k)$ and where $[[H(P_k), P_k], P_k]$ is the projected gradient on the tangent space to $\mathcal{M}_{\mathrm{DM}}$ at P_k, and α is a line search parameter chosen to reduce $\mathcal{E}(P)$ along the projected gradient. A simple and often effective linear search strategy is the Barzilai–Borwein (BB) strategy [15]. As soon as P moves away from P_k along the projected gradient, it violates the idempotency constraint. To bring $\widetilde{P}_{k+1}$ back to the manifold $\mathcal{M}_{\mathrm{DM}}$ and obtain P_{k+1}, a retraction (also called purification) step is performed. The algorithms proposed in [31, 128, 149] are examples of this approach.

As usual in optimization methods, several strategies can be employed to enhance convergence. First, the quality of the search step can be improved using a preconditioner [3, 107], although the design of good preconditioners is less explored than for SCF algorithms [7]. Second, information from past iterations can be used to obtain curvature information, using for instance the conjugate gradient or the L-BFGS algorithm [1]. Finally, Newton's method can be used, yielding the QC-SCF algorithm [10].

Lastly, minimization algorithms can be combined with the SCF idea. The basic representative of this class is the ODA algorithm described above, where the SCF iteration is used to generate a search direction in which an energy minimization is performed. An extension of that idea is the EDIIS algorithm, in which the minimization is performed on a subspace using all past iterates rather than just a segment [87].

7.5.3 Implementation for Molecular Systems

The Gaussian basis sets used in most quantum chemistry packages are in some sense reduced bases: a very small number of basis functions per atom are optimized on a few reference simple systems (mostly isolated atoms) and are then used to simulate a whole variety of molecular systems. As a consequence, it is possible to simulate systems containing a few hundred electrons in basis sets with a few thousand elements. Density matrices can then be stored in memory and optimized dense linear algebra packages can be used to efficiently solve a generalized eigenvalue problem at each iteration of the SCF cycle.

For most molecular systems and small to medium size basis sets, an efficient iterative strategy to solve the Kohn–Sham equations is [87]:

- to make a few iterations of the EDIIS algorithm to ensure a decrease of the energy;
- to switch to DIIS to obtain fast convergence when some coarse convergence threshold is reached.

On the other hand, existing algorithms usually perform poorly for very large basis sets, giving rise to ill-conditioned overlap matrices ("overcomplete" Gaussian basis sets). This observation deserves a detailed mathematical analysis that has not yet been made to our knowledge.

When the Fermi level is a degenerate eigenvalue of the Kohn–Sham Hamiltonian, all the above described algorithms break down. An extension of Bacskay's quadratically convergent algorithm [10] was introduced in [26]. It works well for small and medium size-systems but, as for all Newton-type algorithms, cannot be applied to large systems.

7.5.4 Application to Condensed Matter

We first note that condensed-matter models are often formulated using a finite temperature $T = 1/\beta$ (setting to 1 the value of the Boltzmann constant k_{B}). This is usually not to capture physical effects due to the influence of the temperature on electronic properties (usually negligible at room temperature since the energy scale of the electrons is much larger than that of thermal fluctuations), but a numerical device to accelerate the convergence of properties of metallic infinite systems with respect to the supercell size. Indeed, metals present strong long-range correlations, and the convergence of properties with respect to the supercell size L is slow ($1/L$ for naive methods). Adding an artificial temperature introduces an exponential locality, and improves this to exponential convergence [36], at the price of slightly distorted properties.

As mentioned in Sects. 7.2.2 and 7.2.4, the fixed-point equation at finite temperature takes the form

$$P = \mathcal{F}_{\mathrm{DM}}(P) := \left[I + \mathrm{e}^{\beta(H(P)-\mu I)}\right]^{-1},$$

where the Fermi level μ is chosen such that $\mathrm{Tr}\mathcal{F}_{\mathrm{DM}}(P) = N$. The fixed-point mapping $\mathcal{F}_{\mathrm{DM}}$ is well-defined even in the absence of a gap, as there is no notion of "occupied" or "virtual" states (but only partially occupied states).

Planewave expansion, finite difference and finite element methods are often used to discretize the Kohn–Sham problem for condensed matter systems. In these approaches, the number of degrees of freedom per electron pair is often over 100. The dimension of the Hamiltonian N_b is therefore much larger than the number of occupied states N. For large systems that contain many atoms (in a unit cell), it is generally not practical to form and work with the density matrix P directly, and only orbitals, densities, and potentials are manipulated directly. This feature makes the implementation of SCF and minimization algorithms somewhat different from those discussed in the previous section. We will highlight some of these differences in this section.

7.5.4.1 Charge Density Mixing and Preconditioning

When the exchange-correlation functional is an explicit function of the charge density ρ (i.e. for LDA or GGA), the ground state density ρ^{GS} satisfies a Kohn–Sham equation of the form

$$\rho^{\mathrm{GS}} = \mathcal{F}_{\mathrm{KS}}(\rho^{\mathrm{GS}}). \tag{7.77}$$

Consequently, SCF iteration and acceleration techniques can be designed to find the fixed point of $\mathcal{F}_{\mathrm{KS}}$.

At a finite temperature $T = 1/\beta$, and in a real-space grid representation, $\mathcal{F}_{\mathrm{KS}}$ has the form

$$\mathcal{F}_{\mathrm{KS}}(\rho) = \mathrm{diag}\left(\left[I + \mathrm{e}^{\beta(H[\rho]-\mu I)}\right]^{-1}\right). \tag{7.78}$$

Here diag$(\cdot)$ denotes the diagonal elements of a matrix, and I is the identity matrix. The map $\mathcal{F}_{\mathrm{KS}}$ is well defined even if $H[\rho]$ has degenerate eigenvalues near the Fermi level μ, which is chosen to ensure that

$$\mathrm{Tr}\left(\left[I + \mathrm{e}^{\beta(H[\rho]-\mu I)}\right]^{-1}\right) = N. \tag{7.79}$$

The Hamiltonian $H[\rho]$ is the discrete counterpart of the continuous Kohn–Sham Hamiltonian

$$-\frac{1}{2}\Delta + v_{\mathrm{KS}}[\rho],$$

where the Kohn–Sham potential $v_{\mathrm{KS}}[\rho]$ is the sum of the fixed nuclear (pseudo-) potential and the density-dependent Hartree and exchange-correlation potentials:

$$v_{\mathrm{KS}}[\rho] = V_{\mathrm{nuc}} + v_{\mathrm{H}}[\rho] + v_{\mathrm{xc}}[\rho], \quad v_{\mathrm{H}}[\rho](\mathbf{r}) = \int_{\Omega} \rho(\mathbf{r}')G_L(\mathbf{r}-\mathbf{r}')\,\mathrm{d}\mathbf{r}',$$
$$v_{\mathrm{xc}}[\rho] = \frac{\partial E_{\mathrm{xc}}[\rho]}{\partial \rho},$$

G_L being the periodic Couloumb kernel defined in (7.52).

In practice, the solution (ρ, μ) to the coupled Eqs. (7.78) and (7.79) is obtained in an alternating fashion. For a fixed μ, we may update ρ by a simple SCF iteration

$$\rho_{k+1} = \mathcal{F}_{\mathrm{KS}}(\rho_k), \tag{7.80}$$

starting from an initial guess ρ_0. For each fixed ρ_k, a bisection method can be used to find the solution to (7.79) since the right-hand side of (7.79) is monotonic with respect to μ.

The SCF iteration can be viewed as a fixed point iteration applied to the Kohn–Sham map. To analyze the convergence of the SCF iteration and develop acceleration strategies, it is useful to examine the Jacobian associated with $\mathcal{F}_{\mathrm{KS}}$, which we denote by $\mathcal{J}_{\mathrm{KS}}[\rho] := \partial\mathcal{F}_{\mathrm{KS}}[\rho]/\partial\rho$.

Introducing the potential-to-density map

$$\mathcal{F}_{\mathrm{P}\to\mathrm{D}}(v) = \mathrm{diag}\left(\left[I + \mathrm{e}^{\beta(T+v-\mu I)}\right]^{-1}\right),$$

where T is the kinetic-energy matrix, we have $\mathcal{F}_{\mathrm{KS}}(\rho) = \mathcal{F}_{P\to D}(v_{\mathrm{KS}}[\rho])$ and it therefore follows from the chain rule that

$$\mathcal{J}_{\mathrm{KS}}[\rho] = \frac{\partial\mathcal{F}_{P\to D}}{\partial v}[v_{\mathrm{KS}}[\rho]] \cdot \frac{\partial v_{\mathrm{KS}}}{\partial \rho}[\rho], \tag{7.81}$$

where the first term on the right-hand side of (7.81) is the functional derivative of $\mathcal{F}_{P\to D}$. Its value at the ground-state Kohn–Sham potential is the irreducible (or independent-particle) polarizability

$$\chi_0 := \frac{\partial\mathcal{F}_{P\to D}}{\partial v}[v_{\mathrm{KS}}[\rho^{\mathrm{GS}}]]$$

of the system. At the continuous level, its kernel is given by

$$\chi_0(\mathbf{r}, \mathbf{r}') = 4 \sum_{n<m} \frac{f_m - f_n}{\varepsilon_m - \varepsilon_n} \varphi_m(\mathbf{r})\varphi_n(\mathbf{r})\varphi_m(\mathbf{r}')\varphi_n(\mathbf{r}'), \tag{7.82}$$

where the ε_m are the eigenvalues of the Kohn–Sham Hamiltonian $-\frac{1}{2}\Delta + v_{\mathrm{KS}}[\rho^{\mathrm{GS}}]$, φ_m an orthonormal basis of associated real-valued eigenfunctions, and

$$f_m = f(\varepsilon_m) \quad \text{with} \quad f(\varepsilon) = (1 + \mathrm{e}^{\beta(\varepsilon-\mu)})^{-1} \tag{7.83}$$

is the occupation number associated with ε_m (by convention, $(f_m - f_n)/(\varepsilon_m - \varepsilon_n) = f'(\varepsilon_m)$ if $\varepsilon_m = \varepsilon_n$).

The second term in (7.81) is the functional derivative of $v_{\mathrm{KS}}[\rho]$, and its kernel is given by

$$\frac{\partial v_{\mathrm{KS}}}{\partial \rho}[\rho](\mathbf{r}, \mathbf{r}') = G_L(\mathbf{r} - \mathbf{r}') + \frac{\partial v_{\mathrm{xc}}[\rho](\mathbf{r})}{\partial \rho(\mathbf{r}')}, \tag{7.84}$$

where G_L is again the periodic Coulomb kernel defined in (7.52). The second term on the right-hand side, which is called the exchange-correlation kernel, is the functional derivative of the exchange-correction potential with respect to ρ, which (at least formally) defines a self-adjoint operator. We denote it by $K_{\mathrm{xc}}(\mathbf{r}, \mathbf{r}')$.

Subtracting the exact solution ρ to (7.77) from both sides of (7.80), and performing a Taylor expansion of $\mathcal{F}_{\mathrm{KS}}$ at ρ_k up to the first order yields

$$\Delta\rho_{k+1} \approx \mathcal{J}_{\mathrm{KS}}(\rho_k)\Delta\rho_k, \tag{7.85}$$

where $\mathcal{J}_{\mathrm{KS}}(\rho_k)$ is the Jacobian of $\mathcal{F}_{\mathrm{KS}}$ evaluated at ρ_k. Therefore, when ρ_k is sufficiently close to the exact solution, the fixed point iteration (7.80) converges if the spectral radius of $\mathcal{J}_{\mathrm{KS}}(\rho_k)$, denoted by $\sigma(\mathcal{J}_{\mathrm{KS}}(\rho_k))$, is less than 1 for all k's. However, this condition does not usually hold. In this case, the simple mixing scheme introduced in (7.62) can be used to update ρ by

$$\rho_{k+1} = (1 - \alpha)\rho_k + \alpha\mathcal{F}_{\mathrm{KS}}(\rho_k), \tag{7.86}$$

where $0 < \alpha < 1$ is a mixing parameter. From linear stability analysis, when ρ_k is sufficiently close to the exact solution ρ^{GS}, simple mixing will lead to convergence if

$$\sigma(I - \alpha\epsilon) < 1, \tag{7.87}$$

where $\epsilon = I - \mathcal{J}_{\mathrm{KS}}(\rho^{\mathrm{GS}})$ is known as the dielectric operator (function) in the physics literature. It can be shown that, when ρ^{GS} is an energy minimum, the eigenvalues of $\mathcal{J}_{\mathrm{KS}}(\rho^{\mathrm{GS}})$ are positive, and therefore it is always possible to choose an α so that (7.87) holds [37]. However, such a choice may lead to slow convergence, depending on the conditioning of ϵ.

To analyze this it is convenient to neglect the K_{xc} term, an often reasonable assumption known as the *random phase approximation*, $\epsilon \simeq \epsilon_{RPA} := 1 - \chi_0 v_H$. The eigenvalues of this operator are easily seen to be greater than one, so that ill-conditioning can only appear through large eigenvalues (for more complicated situations where the random phase approximation is too crude, such as systems close to magnetic phase transitions, small eigenvalues of ϵ can also be problematic). These large eigenvalues can be caused either by large eigenvalues of χ_0, which are by (7.82) related to small gaps, or by large eigenvalues of v_H. From the expression (7.52) of G_L in Fourier space, it can easily be seen that the divergences of v_H are related to long-wavelength modes. Both these types of divergences can lead to the phenomenon of *charge sloshing*, wherein the density undergoes large oscillations during the iterations.

The behavior of ϵ for large systems in the random phase approximation is markedly different depending on its physical nature. In insulators, the long-wavelength divergence of the Coulomb potential is compensated by χ_0, so that ϵ_{RPA} is bounded independently of the system size [29]. It follows that charge sloshing is absent, and such systems are usually found to converge quickly. On the other hand, for metals, χ_0 does not compensate for the divergence of v_H, and charge sloshing hampers convergence. A simple model for this divergence is the case of the homogeneous electron gas, where v_{KS} is constant. In this case, χ_0 is a multiplication operator in Fourier space, characterized by its multiplier $\chi_0^{HEG}(q)$, which can be computed explicitly [66]. It can in particular be seen that $\chi_0^{HEG}(q)$ approaches a negative constant $-\zeta$ as $q \to 0$, where ζ is proportional to the inverse square of the Thomas–Fermi screening length. ϵ^{HEG} is therefore a multiplication operator in Fourier space, such that $\epsilon^{HEG}(q) \simeq 1 + \frac{4\pi\zeta}{|q|^2}$ for low q.

In the case of the homogeneous electron gas, the appropriate iteration scheme is then the preconditioned iteration

$$\rho_{k+1} = \rho_k + K(\mathcal{F}_{KS}(\rho_k) - \rho_k), \tag{7.88}$$

with K a fixed preconditioner, given as a multiplication in Fourier space by

$$K(q) = \frac{|q|^2}{|q|^2 + 4\pi\zeta},$$

a choice known as Kerker preconditioning [81]. When combined with Anderson–Pulay acceleration, this has been found to be very efficient in practice for a wide range of metals, and is the default iteration scheme in most plane-wave codes. A mathematical proof of the independence of the convergence rate of (7.88) on the system size in the reduced Hartree–Fock model can be found in [91]. For general systems, and in particular those containing hybrid insulating/conducting character, finding an efficient preconditioner is still a challenge; see [96, 151] and [144] for a review. Very recently, an inexpensive and parameter-free preconditioner based on the local density of states has been introduced [77]. It is based on a physically motivated approximation to the independent-particle susceptibility operator, and its efficiency has been demonstrated on several heterogeneous systems such as clusters and surfaces.

7.5.4.2 Iterative Diagonalization

An accelerated fixed point iteration for seeking the solution of (7.77) requires evaluating the Kohn–Sham map $\mathcal{F}_{\text{KS}}$ at each ρ_k. This evaluation is often performed by making use of spectral decomposition $H(\rho_k) = U\Lambda U^*$, where U contains the eigenvectors of $H(\rho_k)$ and $\Lambda = \text{diag}(\varepsilon_1, \varepsilon_2, \ldots, \varepsilon_{N_b})$ contains the corresponding eigenvalues. Using such a decomposition, we can write $\mathcal{F}_{\text{KS}}(\rho_k)$ as

$$\mathcal{F}_{\text{KS}}(\rho_k) = \sum_{m=1}^{N_b} f_m |\varphi_m|^2, \tag{7.89}$$

where f_m is the occupation number defined in (7.83), and φ_m is the mth column of U.

Because the dimension of H (N_b) can be extremely large, using full diagonalization algorithms for dense matrices implemented in LAPACK [6], ScaLAPACK [17] or ELPA [102] can be prohibitively expensive. For large basis sets such as planewaves and discretization methods based on finite elements and finite differences, N_b is typically much larger than the number of required eigenpairs, which is N (at zero temperature), or slightly larger (in the usual range of positive temperatures). Iterative eigensolvers such as the Davidson–Liu algorithm [46, 99] or the locally optimal block preconditioned conjugate gradient (LOBPCG) algorithm [84] are then preferred. These iterative eigensolvers can take advantage of an efficient representation of H that allows the multiplication of H by one or multiple vectors to be performed efficiently, through fast Fourier transforms (plane waves) or sparse matrix techniques (finite elements, finite differences).

One additional advantage of using iterative diagonalization methods is that approximate eigenvectors from the kth SCF iteration can be used as the starting guess of eigenvectors for the $(k+1)$st SCF iteration so that only a few iterative diagonalization steps may be needed per SCF iteration. This is certainly true when ρ_n has nearly converged. But even when ρ_n is far from the solution at the beginning of the SCF procedure, only a few iterative diagonalization steps are needed to yield approximate eigenpairs. This is due to the fact that in the first few SCF iterations, the error in nonlinear Eq. (7.77) is still relatively large. Hence, as long as the error in $\mathcal{F}_{\text{KS}}$ evaluation is less than the difference between ρ and $\mathcal{F}_{\text{KS}}(\rho)$ even when the latter is evaluated exactly, the convergence of the SCF procedure can still be achieved in subsequent iterations.

Another way to evaluate $\mathcal{F}_{\mathrm{KS}}(\rho)$ is to apply a subspace iteration to $T_d(H[\rho])$ where $T_d(t)$ is a d-th degree scaled and shifted Chebyshev polynomial of the second kind that satisfies the property

$$\begin{cases} |T_d(t)| \gg 1 \quad \text{for } \varepsilon_1 \leqslant t \leqslant \varepsilon_N, \\ |T_d(t)| \leqslant 1 \ \text{for } \varepsilon_{N+1} < t \leqslant \varepsilon_{N_b}. \end{cases}$$

This technique was pioneered in [152, 153], and is now widely used in a number of software tools. At zero temperature, the subspace iteration converges to an approximate invariant subspace spanned by eigenvectors associated with the lowest N eigenvalues. The vectors u_i, $i = 1, 2 \ldots, N$, used in (7.89) do not need to be eigenvectors. Any basis of the invariant subspace suffices. At finite temperature, only a small fraction of the eigenpairs of H corresponding to eigenvalues slightly below and above μ are computed using the method presented in [11, 106].

For large problems consisting of many atoms, the cost of an iterative diagonalization procedure is dominated by the computation required to obtain orthonormalized basis vectors of the desired subspace and extract eigenvalue and eigenvector approximations from such a subspace. The complexity of this computation is typically $O(N^3)$. Furthermore, some of these computations do not scale to a large number of processors on a massively parallel computer. The recent effort in developing spectrum slicing algorithms has the potential to address this issue, but requires solving difficult interior eigenvalue problems.

Methods for reducing the $O(N^3)$ have also developed since the 90s. These methods often try to approximate the matrix function $\left[1 + \mathrm{e}^{\beta(H[\rho]-\mu)}\right]^{-1}$ directly through either polynomial or rational expansions [14, 67–69, 85, 92, 98]. Some of these methods can scale linearly with respect to N, although with a large prefactor. Details of these methods can be found in the contribution by Lin et al. (Chap. 8).

7.5.4.3 Computing the Hartree and Fock (Exact) Exchange Potential

When the Kohn–Sham equations are discretized by GTOs for molecular systems, both the Hartree and the Fock (exact) exchange potentials are computed by using the two-electron integrals (7.13) evaluated in advance or on the fly.

When the problem is discretized by a large basis set such as planewaves, the Hartree potential $v_{\mathrm{H}}[\rho]$ is obtained by solving the periodic Poisson problem $-\Delta v_{\mathrm{H}}[\rho] = 4\pi\rho$ in a unit cell by using FFT. In a continuous formulation, the non-local Fock exchange potential is defined (at zero temperature for simplicity) as

$$v_{\mathrm{X}}(\mathbf{r}, \mathbf{r}') = -\sum_{i=1}^{N} \varphi_i(\mathbf{r})\varphi_i(\mathbf{r}')G_L(\mathbf{r} - \mathbf{r}'), \tag{7.90}$$

where $\varphi_i(\mathbf{r})$, $i = 1, 2, \ldots, N$ are the occupied Kohn–Sham orbitals, and $G_L(\mathbf{r}, \mathbf{r}')$ is the periodic Coulomb kernel defined in (7.52). When an iterative diagonalization method is used to compute the desired eigenpairs, we need to find a v_X which can be efficiently applied to a set of N wavefunctions, i.e.,

$$\int_\Omega v_X(\mathbf{r},\mathbf{r}')\psi_j(\mathbf{r}')\,\mathrm{d}\mathbf{r}' = -\sum_{i=1}^{N}\varphi_i(\mathbf{r})\int_\Omega \varphi_i(\mathbf{r}')\psi_j(\mathbf{r}')G_L(\mathbf{r},\mathbf{r}')\,\mathrm{d}\mathbf{r}'. \tag{7.91}$$

For each j, N integrals on the right-hand side of the above equation can be evaluated by performing N cyclic convolutions via FFTs. Since $j = 1, 2, \ldots, N$, the total number of FFTs that need to be performed is $O(N^2)$. An efficient low rank approximation technique called interpolative separable density fitting (ISDF) [78] has recently been developed to reduce the number of FFTs to a small multiple of N by constructing a set of numerical auxiliary basis functions that nearly span the same subspace as $\varphi_i(\mathbf{r})\varphi_j(\mathbf{r})$, for $i, j = 1, 2, \ldots, N$. Furthermore, by recognizing that the application of v_X to $\{\psi_j\}$, $j = 1, 2, \ldots, N$, only needs to be correct in the space spanned by $\{\psi_j\}$, further reduction in complexity can be achieved in the ISDF construction of the v_X operator through a technique called the adaptively compressed exchange operator [97]. The technique of using a small set of auxiliary basis functions to compactly represent pair product basis functions (to reduce the cost of the Fock exact exchange calculation) is also known as resolution of identity or density fitting in general. (See the review article [123] and the references therein.)

7.5.4.4 Direct Minimization

The algorithms in Sect. 7.5.2 can be adapted to the condensed matter setting by using orbitals. The basic gradient algorithm then takes the form

$$X^{(k+1)} = X^{(k)} - \alpha R(X^{(k)}),$$

where $R(X^k)$ is the residual that takes the form of

$$R(X) = H(XX^T)X - X(X^T H(XX^T)X), \tag{7.92}$$

followed by an orthogonalization step; choices include standard orthogonalization methods (based on the Gram–Schmidt, Cholesky or singular value decomposition algorithms), as well as the so-called WY update [143].

However, the separation of line search and the orthonormalization procedure used to keep X feasible is suboptimal. A better strategy is to carry out these two tasks simultaneously in a constrained optimization framework.

In order to do that, we express X^{k+1} as

$$X^{(k+1)} = X^{(k)}G_1 + K^{-1}R(X^{(k)})G_2 + (X^{(k)} - X^{(k-1)})G_3, \tag{7.93}$$

where G_1, G_2 and G_3 are $N \times N$ real matrices to be determined, and where $K \in \mathbb{R}_{\text{sym}}^{N_b \times N_b}$ is a preconditioner. We choose these matrices by solving the following constrained minimization problem

$$\begin{aligned} &\min_G E(YG) \\ \text{s.t. } &G^T Y^T Y G = I_N, \end{aligned} \tag{7.94}$$

where $Y = (X^{(k)}, K^{-1}R(X^{(k)}), X^{(k)} - X^{(k-1)})$ and $G = (G_1^T \; G_2^T \; G_3^T)^T$. This approach is referred to as a direct constrained minimization (DCM) method in [146]. It is a nonlinear extension of the locally optimal block preconditioned conjugate gradient (LOBPCG) method [84]. When P_{k-1} term is not included in (7.93), the DCM algorithm can be viewed as a nonlinear extension of the Davidson–Liu algorithm [99] for solving a linear eigenvalue problem. The problem (7.94) is of reduced size, that can be solved using stabilized SCF algorithms for instance; we refer to [146] for details.

Faster convergence can be achieved by a Riemannian Newton's method [1, 150] that preserves the orthonormality constraint while minimizing the total energy objective function. The key step in the Newton method is to compute the Newton search direction by solving the following equation

$$D^2 E(X^{(k)})[\Delta X^{(k)}] + \nabla E(X^{(k)}) = \mathbf{0}, \tag{7.95}$$

where $D^2 E(X^{(k)})$ is the Hessian of E at $X^{(k)}$ restricted to the manifold $\mathcal{M}_{\text{MO}}$ associated with the orthonormality constraint in (7.57), and $\nabla E(X^{(k)})$ is the projected gradient along the tangent of this constraint. Although it is not practical to write down the Hessian and solve the Newton equation to high accuracy, the Hessian does have a special structure that allows one to multiply it by a vector efficiently. This structure is described in [62, 100]. An iterative method can be used to solve the Newton equation to obtain a search direction $\Delta X^{(k)}$ that satisfies

$$\begin{aligned} \|D^2 E(X^{(k)})[\Delta X^{(k)}] + \nabla E(X^{(k)})\|_F &\leqslant \sigma \|\nabla E(X^{(k)})\|_F, \\ \text{Tr}(\nabla E(X^{(k)})^T (\Delta X^{(k)})) &\leqslant -\sigma \|\Delta X^{(k)}\|_F^2, \end{aligned} \tag{7.96}$$

where σ is a dynamically chosen parameter. Even though Newton's method can yield quadratic convergence, each step of the method requires a large linear system of equations to be solved. Without a good preconditioner, the cost for obtaining a good solution can be high. Constructing a good preconditioner is not trivial.

The approach discussed above is applicable to Kohn–Sham models at zero temperature. At finite temperature, we minimize an objective function similar to the one given in (7.22). In addition to optimizing with respect to the wavefunction X, we also need to optimize the occupation numbers. An efficient algorithm for solving this optimization problem is given in [140]. Similar algorithms which optimize (7.22) with respect X and the occupation numbers simultaneously can be found in [61, 104].

7.5.4.5 Stopping Criteria

When an iterative algorithm is designed to solve (7.77), a natural stopping criterion at step k is

$$\frac{\|\rho^{(k)} - \mathcal{F}_{\mathrm{KS}}(\rho^{(k)})\|_2}{\|\rho^{(k)}\|_2} < \tau, \tag{7.97}$$

where τ is some user specified error tolerance. A similar criterion can be used to check the convergence of the potential that depends on ρ. Although such a criterion does not guarantee the convergence of ρ to the ground state density, it does allow us to check the self-consistency of the approximate solution.

In a direct minimization approach, we can terminate an iterative procedure if the Frobenius norm of the residual $R(X)$ defined in (7.92) is sufficiently small. Again, such a stopping criterion does not guarantee X to be a global minimizer. But it ensures that X is a self-consistent approximation.

In the physics community, the difference between the Kohn–Sham total energy evaluated at an approximate X and an estimation known as the Harris–Foulke energy functional [59, 75] defined in the real space as

$$E_{\text{Harris–Foulke}}(\rho) = \mathrm{Tr}[H(\rho_{\mathrm{in}})P_{\mathrm{out}}] - \frac{1}{2}\int_\Omega v_{\mathrm{H}}[\rho_{\mathrm{in}}(\mathbf{r})]\rho_{\mathrm{in}}(\mathbf{r})d\mathbf{r}$$
$$- \int_\Omega v_{\mathrm{xc}}[\rho_{\mathrm{in}}(\mathbf{r})]\rho(\mathbf{r})d\mathbf{r} + E_{\mathrm{xc}}[\rho_{\mathrm{out}}],$$

where ρ_{in} is the input density, and ρ_{out} and P_{out} are the output density and density matrix, is often used in a termination criterion. When $\rho_{\mathrm{in}} = \rho_{\mathrm{out}}$ is the solution to (7.77), $E_{\text{Harris–Foulke}}(\rho)$ is identical to the Kohn–Sham total energy. However, in general, these two energy functions are different. When ρ_{in} converges to the ground state density, the difference between the two becomes smaller.

7.6 Error Analysis

The mathematical analysis of the error between the exact solution to the Kohn–Sham model under consideration and the numerical result obtained by discretization and an approximately converged iterative algorithm encompasses two frameworks: the a priori analysis, and the a posteriori analysis.

The a priori analysis aims at answering two questions:

1. **algorithmic error:** does the solution algorithm converge and how fast does it converge? The problems we are interested in this chapter are indeed highly non-linear and thus iterative methods are needed; the question is then to understand if the solution algorithm converges when the number of iterations tends to infinity

and—if so—to estimate the number of iterations necessary to reach a given accuracy. A related side question is to understand the fundamental reasons why different algorithms converge, and to identify key quantities such as the condition number of a specific matrix or the Lipschitz constant of some contraction map which control the convergence rates of these algorithms. Such an analysis then allows us to improve the existing algorithms by modifying them by acting on these key quantities, for example through the design of preconditioners or accelerated algorithms. The knowledge accumulated during the a priori analysis may also lead to reformulations of the problem which make it fall into a class of problems for which efficient and robust algorithms already exist (e.g. write it as a smooth convex minimization problem);

2. **discretization error:** the goal is to prove that, eventually, when the number of degrees of freedom uniformly[1] increases to infinity, then the discrete solution converges to the exact solution in some sense. In this case, the discretization method is called systematically improvable. Such an analysis generally provides a theoretical convergence rate, that is, a lower bound of the asymptotic speed of convergence of the error when the number of degrees of freedom uniformly increases to infinity. The a priori analysis helps us to choose the more appropriate numerical discretization schemes for a given problem by highlighting the link between the convergence rate of a given discretization scheme and properties of the exact solution such as its regularity. For instance, fixed (low) order discretization methods (e.g. finite elements) are generally indicated if the exact solution is globally not very regular, whereas high (or infinite) order methods (e.g. spectral or planewave methods) are generally preferred when the exact solution and its derivatives are uniformly bounded by well-controlled constants.

The *a priori analysis* allows us to build a first, mathematically consistent, numerical strategy (discretization and iterative algorithm) for solving the problem under consideration, based on some qualitative analysis: characterization of the class of problems this specific problem belongs to, regularity of the exact solution...

Let us mention the existence of other contributions (not analyzed here) to the error between the exact solution and the result of the numerical simulation, in particular (i) finite arithmetics errors, which can become significant if part of the calculation is done in single precision (which is of upmost importance for hybrid architectures involving GPUs), and (ii) implementation errors (bugs, which are almost inevitable in software consisting of hundreds of thousands lines of code). The model error, due to the fact that the exchange-correlation functional used in the calculation is an approximation of the true one, is of course important as well, but is currently out of reach of the mathematical and numerical analysis. The

[1] In this context, *uniformly* means that the discretization scheme is such that there exists a non-increasing function $f : \mathbb{N} \to \mathbb{R}_+$ going to zero at infinity such that for any function v in the natural function space $\mathcal{Y}$ on which the problem is set, for instance the space $H^1_{\rm per}(\Omega)$ for the simple model considered in Sect. 7.6.1, the distance from v to its best approximation in a discretization space with $N_{\rm dof}$ degrees of freedom is bounded by $f(N_{\rm dof})\|v\|_{\mathcal{Y}}$.

situation is different for wavefunction methods (e.g. configuration interaction or coupled cluster) for which estimating the model error should be possible since the approximate ground state wavefunction generated by these methods can be used to define a residual for the exact N-body Schrödinger equation. Some preliminary works in this direction are in progress.

In contrast with the a priori analysis, the a posteriori analysis aims at answering more practical questions: what is the size of the error between the exact (unknown) solution and the discrete solution that comes out of the numerical simulation and what can be done to improve the accuracy of the numerical result? By "size" we mean here an upper (ε^+) bound and a lower (ε^-) bound of some appropriate and relevant norm of the error such that $\varepsilon^+ - \varepsilon^-$ is much smaller than ε^+. Improved versions of this analysis further provide "indicators" that help us to decide what to do if these upper and lower bounds reveal that the accuracy is not sufficient to answer the question motivating the simulation. Examples are local a posteriori error indicators pointing out where to refine the mesh in finite element methods. This leads to adaptive schemes and to non-uniform solution-dependent quasi-optimal meshes. Other examples are iterative procedures in which the discretization and algorithmic components of the error are separately estimated at each iteration. This allows one to adaptively refine the discretization when the SCF or minimization algorithm gets closer to convergence, thus avoiding costly calculations in large discretization basis when the iterates are far from convergence. More generally, error balancing consists in designing numerical schemes which minimize the computational cost necessary to reach a given accuracy, or, conversely, minimize the error reachable at a given computational cost. A posteriori error estimators of the various components of the error are a key tool for the design of black-box error balancing strategies.

Note that most of the time, what we are looking for is not the knowledge of the global state (e.g. the Kohn–Sham ground state density matrix) but the computation of some properties or *quantities of interest* (e.g. the forces acting on the nuclei) that are evaluated from the numerical approximation of the global state. These quantities of interest are generally expressed as functionals that are continuous with respect to some norms. These are the norms for which the a posteriori error analysis has to be performed. Thus, estimates of the error on the global state for these norms automatically give rise to estimates of the error on the quantities of interest. Note that obtaining such bounds on these quantities of interest generally requires the introduction of auxiliary dual problems to obtain convergence rates consistent with the ones observed in numerical simulations, similar to the classical Aubin–Nitsche methodology in finite element methods [55].

7.6.1 A Priori Error Analysis of a Simple Model

For the sake of pedagogy, let us explain the strategy on the simple example of a one-dimensional periodic Gross–Pitaevskii model. The equations have the same form as those arising in Kohn–Sham DFT, with several important simplifications: the state of the system is described by a single orbital, the model is strictly convex in the density, and the *Aufbau* principle always holds.

Let $\Omega := (-\frac{1}{2}, \frac{1}{2})$ be the unit cell of the periodic lattice $\mathcal{R} = \mathbb{Z}$ of $\mathbb{R}$. The Sobolev spaces of 1-periodic functions (or distributions) are defined for all $s \in \mathbb{R}$ by

$$H^s_{\rm per}(\Omega) = \left\{ v \in H^s_{\rm loc}(\mathbb{R}) \mid v \text{ is 1-periodic} \right\}.$$

For each $s \in \mathbb{R}$, $H^s_{\rm per}(\Omega)$ is a Hilbert space, whose norm $\|\cdot\|_{H^s}$ can be easily expressed in terms of Fourier coefficients. Denoting by $e_K(x) = \mathrm{e}^{iKx}$, $K \in 2\pi\mathbb{Z}$, the 1-periodic Fourier modes, any $v \in H^s_{\rm per}(\Omega)$ can be expanded as

$$v = \sum_{K \in 2\pi\mathbb{Z}} \widehat{v}_K e_K,$$

and we can choose the following equivalent norm for $H^s_{\rm per}(\Omega)$:

$$\|v\|^2_{H^s} = \sum_{K \in 2\pi\mathbb{Z}} (1 + K^2)^s |\widehat{v}_K|^2. \tag{7.98}$$

We also have

$$L^2_{\rm per}(\Omega) = \left\{ v \in L^2_{\rm loc}(\mathbb{R}) \mid v \text{ is 1-periodic} \right\} = H^0_{\rm per}(\Omega), \quad \|v\|^2_{L^2_{\rm per}} = \int_\Omega v^2 = \|v\|^2_{H^0_{\rm per}}.$$

The one-dimensional periodic Gross–Pitaevskii problem reads

$$E_0 = \inf \left\{ E^{\rm GP}(v),\ v \in H^1_{\rm per}(\Omega),\ \int_\Omega v^2 = 1 \right\}, \tag{7.99}$$

where the energy functional $E^{\rm GP} : H^1_{\rm per}(\Omega) \to \mathbb{R}$ is given by

$$E^{\rm GP}(v) = \frac{1}{2} \int_\Omega |\nabla v|^2 + \frac{1}{2} \int_\Omega V v^2 + \frac{1}{4} \int_\Omega v^4, \tag{7.100}$$

where $V : \mathbb{R} \to \mathbb{R}$ is a 1-periodic bounded function. Problem (7.99) can be reformulated as

$$E_0 = \inf \left\{ \mathcal{E}^{\rm GP}(\rho),\ \rho \geq 0,\ \sqrt{\rho} \in H^1_{\rm per}(\Omega),\ \int_\Omega \rho = 1 \right\}, \tag{7.101}$$

where we have introduced the function $\mathcal{E}^{\rm GP}(\rho) := E^{\rm GP}(\sqrt{\rho})$.

The energy $\mathcal{E}^{\rm GP}$ is strictly convex and smooth, and it is easy to prove that problem (7.101) has a unique solution ρ_0, which is continuous and positive on $\mathbb{R}$—hence bounded away from zero since it is 1-periodic–, so that problem (7.99) has exactly two solutions: $u_0 = \sqrt{\rho_0}$ and $-u_0$; we refer e.g. to [23] for a detailed proof of these results. The Euler–Lagrange equation associated to this problem involves an eigenvalue λ_0 (the Lagrange multiplier of the normalization constraint) and reads:

$$-\Delta u_0 + V u_0 + u_0^3 = \lambda_0 u_0, \quad \|u_0\|_{L^2_{\rm per}} = 1. \tag{7.102}$$

The variational formulation of the above problem is

$$\forall v \in H^1_{\rm per}(\Omega), \quad \int_\Omega \nabla u_0 \nabla v + \int_\Omega V u_0 v + \int_\Omega u_0^3 v = \lambda_0 \int_\Omega u_0 v, \quad \|u_0\|_{L^2_{\rm per}} = 1. \tag{7.103}$$

Note that a solution u_0 to the nonlinear eigenvalue problem (7.102) may not be a minimizer of (7.99), but, as in the linear case, it will be if one of the following conditions is fulfilled:

- (u_0, λ_0) is a solution to (7.102) and λ_0 is the smallest eigenvalue of the nonlinear problem

$$-\Delta v + V v + v^3 = \mu v, \qquad \|v\|_{L^2_{\rm per}} = 1;$$

- (u_0, λ_0) is a solution to (7.102) and λ_0 is the smallest eigenvalue to the bounded-below self-adjoint operator with compact resolvent $-\Delta + V + u_0^2$.

The planewave approximation of this problem in the discretization space

$$\mathcal{X}_N := \mathrm{Span}\,(e_K,\ K \in 2\pi\mathbb{Z},\ |K| \le N)$$

(see Sect. 7.3.2 for the 3D setting) is given by

$$E_{0,N} = \inf\left\{E^{\rm GP}(v),\ v \in \mathcal{X}_N,\ \int_\Omega v^2 = 1\right\}, \tag{7.104}$$

which, obviously, has—at least—one solution. This states the existence of a solution $u_{0,N}$ satisfying the constraint $\|u_{0,N}\|_{L^2_{\rm per}} = 1$, $\int_\Omega u_0 u_{0,N} \ge 0$ (up to replacing $u_{0,N}$ by $-u_{0,N}$), and the variational equation

$$\forall v \in \mathcal{X}_N, \quad \int_\Omega \nabla u_{0,N} \nabla v + \int_\Omega V u_{0,N} v + \int_\Omega u_{0,N}^3 v = \lambda_{0,N} \int_\Omega u_{0,N} v. \tag{7.105}$$

From convexity arguments, there exists a positive constant α such that

$$E_{0,N} - E_0 \ge \alpha \|u_{0,N} - u_0\|^2_{H^1_{\rm per}}. \tag{7.106}$$

Then let $\Pi_{0,N}$ be the orthogonal projector from $H^s_{\text{per}}(\Omega)$ to its finite-dimensional subspace $\mathcal{X}_N$:

$$\forall v \in H^s_{\text{per}}(\Omega), \quad \Pi_{0,N} v = \sum_{K \in 2\pi\mathbb{Z},\ |K| \leq N} \widehat{v}_K e_K.$$

We deduce from the smoothness of the functional E^{GP} that

$$E_{0,N} - E_0 \leq E^{GP}\big(\Pi_{0,N} u_0\big) - E_0 \leq C \|u_0 - \Pi_{0,N} u_0\|_{H^1_{\text{per}}}$$

for some constant $C \in \mathbb{R}_+$ independent of N. Together with (7.106), this implies that $\|u_{0,N} - u_0\|_{H^1_{\text{per}}}$ converges to 0 as N tends to infinity. This also implies that, in any small enough neighborhood of u_0, there exists a unique solution $u_{0,N}$ to (7.104), provided N is large enough.

Our goal is now to compare the error between u_0 and $u_{0,N}$ (measured in the $H^1_{\text{per}}(\Omega)$-norm) on the one hand, and the error between λ_0 and $\lambda_{0,N}$ on the other hand, with the best fit error in $\mathcal{X}_N$ realized by $\Pi_{0,N} u_0$. To do this, we reformulate the problem (7.105) solved by $(u_{0,N}, \lambda_{0,N})$ as a nonlinear equation $F_N(u_{0,N}, \lambda_{0,N}) = 0$, with $F_N : H^1_{\text{per}}(\Omega) \times \mathbb{R} \to H^{-1}_{\text{per}}(\Omega) \times \mathbb{R}$ defined by

$$F_N(v, \mu) = \left(\mu v - \Pi_{0,N}\left[-\Delta v + V v + v^3\right], \int_\Omega v^2 - 1\right). \tag{7.107}$$

This has the property that $F_N(u_{0,N}, \lambda_{0,N}) = 0$, and furthermore

$$F_N(u_0, \lambda_0) = (\lambda_0(u_0 - \Pi_{0,N} u_0), 0),$$

which is small. Now we can link the residual $F_N(u_0, \lambda_0)$ to the error $(u_0, \lambda_0) - (u_{0,N}, \lambda_{0,N})$ by the expansion

$$F_N(u_0, \lambda_0) \simeq F_N(u_{0,N}, \lambda_{0,N}) + DF_N(u_{0,N}, \lambda_{0,N})[(u_0, \lambda_0) - (u_{0,N}, \lambda_{0,N})]. \tag{7.108}$$

By showing that $[DF_N(u_{0,N}, \lambda_{0,N})]^{-1}$ is uniformly bounded with respect to N (which follows from quite tedious arguments, see [52] for details), and controlling higher order terms, one obtains that there exists a constant $C \in \mathbb{R}_+$ such that for N large enough

$$\|u_0 - u_{0,N}\|_{H^1_{\text{per}}} + |\lambda_0 - \lambda_{0,N}| \leq C \|u_0 - \Pi_N u_0\|_{H^1_{\text{per}}}. \tag{7.109}$$

In the end, the convergence rate is thus only related to the Sobolev regularity of the eigenfunction u_0, which itself depends of the regularity of the external potential V: if $V \in H^\sigma_{\text{per}}(\Omega)$ for some $\sigma \geq 0$, then $u_0 \in H^{\sigma+2}_{\text{per}}(\Omega)$ by elliptic regularity arguments. Since for all $s > 1$, we have

$$\|v - \Pi_N v\|_{H^1_{\rm per}} = \left(\sum_{K \in 2\pi\mathbb{Z},\ \frac{1}{2}|K|^2 > N} (1+K^2)|\widehat{v}_K|^2 \right)^{1/2} \leq \frac{\|v\|_{H^s_{\rm per}}}{(1+2N)^{\frac{s-1}{2}}},$$

we obtain that if $V \in H^\sigma_{\rm per}(\Omega)$ for some $\sigma \geq 0$, then

$$\|u_0 - u_{0,N}\|_{H^1_{\rm per}} \leq CN^{-\frac{\sigma+1}{2}}, \tag{7.110}$$

for some constant $C \in \mathbb{R}_+$ independent of N. The above estimate derived from (7.109) is optimal.

In contrast, the estimate on the eigenvalue error obtained from (7.109) does not scale optimally. Indeed, in the linear case

$$-\Delta\widetilde{u}_0 + V\widetilde{u}_0 = \widetilde{\lambda}_0\widetilde{u}_0, \tag{7.111}$$

a quadratic convergence rate of the eigenvalue can be established following the Hellmann–Feynman type arguments in Sect. 7.2.5.

Using estimates *à la* Aubin–Nitsche in negative Sobolev norms, this approach can be extended to the nonlinear case of the Gross–Pitaevskii equation, yielding error estimates of the form

$$|\lambda_{0,N} - \lambda_0| \leq C'\, N^{-(\sigma+1)}. \tag{7.112}$$

It can also be established that

$$E_{0,N} - E_0 \leq C''\, N^{-(\sigma+1)}. \tag{7.113}$$

Similarly, other quantities of interest computed from $u_{0,N}$ and $\lambda_{0,N}$ are at hand and (7.110) and (7.112) allow us to derive a priori error estimates for these outputs.

Note that it is impossible to get accurate values of the errors from the above estimates since the constants C, C' and C'' in (7.110), (7.112) and (7.113) are not explicit, depending on quantities such as the gap that are not known a priori. This motivates the next section.

7.6.2 A Posteriori Error Analysis of a Simple Model

We follow up with the same problem (7.102), but now, as explained at the beginning of this section, we aim at an accurate numerical value for the size of the error between the exact solution (or quantity of interest) and the computed approximate solution (or computed QOI), and not a rate of convergence as the number of degrees of freedom of the discrete problem and the number of iterations of the SCF or minimization algorithm increase.

The general strategy to derive a posteriori error bounds is somewhat similar to the one for a priori error bounds described in the previous section. The difference is that it involves a nonlinear function F whose root is the exact solution: $F(u_0, \lambda_0) = 0$. Now the distance between some approximate solution (u, λ) and the exact solution (u_0, λ_0) is related to the size of $F(u, \lambda)$. Indeed, similarly as in (7.108), we have

$$F(u, \lambda) \simeq F(u_0, \lambda_0) + DF(u_0, \lambda_0)[(u, \lambda) - (u_0, \lambda_0)] \tag{7.114}$$

and $(u - u_0, \lambda - \lambda_0)$ behaves as $DF(u_0, \lambda_0)^{-1} F(u, \lambda)$, using this time the fact that $F(u_0, \lambda_0) = 0$.

The definition of $F : H^1_{\text{per}}(\Omega) \times \mathbb{R} \to H^{-1}_{\text{per}}(\Omega) \times \mathbb{R}$ is

$$F(v, \mu) = \left(\mu v - (-\Delta v + V v + v^3), \int_\Omega v^2 - 1 \right), \tag{7.115}$$

and we have, assuming that (u, λ) is close enough to (u_0, λ_0) and neglecting higher-order terms,

$$\begin{aligned}
\|u - u_0\|_{H^1_{\text{per}}} &\le \|(u, \lambda) - (u_0, \lambda_0)\|_{H^1_{\text{per}} \times \mathbb{R}} \\
&\le \left\| [DF(u_0, \lambda_0)]^{-1} F(u, \lambda) \right\|_{H^1_{\text{per}} \times \mathbb{R}} \\
&\le \left\| [DF(u_0, \lambda_0)]^{-1} \right\|_{\mathcal{L}(H^1_{\text{per}} \times \mathbb{R}, H^{-1}_{\text{per}} \times \mathbb{R})} \|F(u, \lambda)\|_{H^{-1}_{\text{per}} \times \mathbb{R}} \\
&= \left\| [DF(u_0, \lambda_0)]^{-1} \right\|_{\mathcal{L}(H^1_{\text{per}} \times \mathbb{R}, H^{-1}_{\text{per}} \times \mathbb{R})} \|\lambda u + \Delta u - V u - u^3\|_{H^{-1}_{\text{per}}}.
\end{aligned}$$

Here, the approximate solution (u, λ) can be $(u_{0,N}, \lambda_{0,N})$ (converged numerical solution in the discretization space $\mathcal{X}_N$) for N large enough, but the argument also holds for the kth iterate of a SCF or minimization algorithm in some (possibly adaptively refined) discretization space.

Note that $\|\lambda u + \Delta u - V u - u^3\|_{H^{-1}_{\text{per}}}$ can be easily evaluated in the Fourier representation using (7.98). To turn the above inequalities into a practical tool to evaluate $\|u - u_0\|_{H^1_{\text{per}}}$, we need an accurate bound for the operator norm of $DF(u_0, \lambda_0)^{-1}$ or an estimate for $\left\| [DF(u_0, \lambda_0)]^{-1} F(u, \lambda) \right\|_{H^1_{\text{per}} \times \mathbb{R}}$. For the sake of brevity, we will not detail this quite technical point here and refer the interested reader to [52]. The conclusion is that

$$\|u - u_0\|_{H^1_{\text{per}}} \le C \|\lambda u + \Delta u - V u - u^3\|_{H^{-1}_{\text{per}}},$$

with a constant $C \in \mathbb{R}_+$ which can be accurately estimated; this is a fundamental difference between a priori and a posteriori error bounds. As previously mentioned, the evaluation of this H^{-1}-norm of the residual $\lambda u + \Delta u - Vu - u^3$ is rather simple in the Fourier representation. Actually, the analysis can be improved to decompose the error into two terms: a discretization error due to the fact that u lives in a finite-dimensional discretization space $\mathcal{X}_N$; an algorithmic error corresponding to the fact that the SCF or minimization algorithm has not yet converged to $(u_{0,N}, \lambda_{0,N})$. The former component of the error vanishes in the limit $N \to \infty$, the latter when the iterative algorithm has converged in the current discretization space. This decomposition leads to adaptive algorithms in which the discretization space is increased as the SCF iterations proceed. We refer to [52] for more details.

7.6.3 Extensions to the Kohn–Sham Problem

Let us continue now to the a posteriori analysis of the Hartree–Fock and Kohn–Sham problems. The point is to write these problems again under the formal form $F = 0$, and an approximation like $F_{\text{cut}} = 0$ on the manifold

$$\mathcal{M} = \left\{ \Phi = (\varphi_1, \ldots, \varphi_N) \in (H^1(\mathbb{R}^3))^N, \int_{\mathbb{R}^3} \varphi_i \varphi_j = \delta_{ij} \right\}. \tag{7.116}$$

Due to the gauge invariance (7.14) of the Hartree–Fock or Kohn–Sham energy, the ground state problems do not have a unique solution and thus DF, at the ground state, is not invertible: it has a Kernel within the tangent space to $\mathcal{M}$. More precisely, for any $\Phi = (\varphi_1, \ldots, \varphi_N)^T \in \mathcal{M}$ we denote by $T_\Phi \mathcal{M}$ the tangent space at Φ to $\mathcal{M}$. It reads

$$T_\Phi \mathcal{M} = \left\{ (\psi_1, \ldots, \psi_N)^T \in (H^1_{\text{per}}(\Gamma))^N \mid \forall 1 \leq i, j \leq N, \int_\Gamma \varphi_i \psi_j + \psi_i \varphi_j = 0 \right\}.$$

Then, if we denote by

$$\Phi^{\perp\!\!\perp} = \left\{ \Psi = (\psi_1, \ldots, \psi_N)^T \in (H^1_{\text{per}}(\Gamma))^N \mid \forall 1 \leq i, j \leq N, \int_\Gamma \varphi_i \psi_j = 0 \right\}$$

the set of all orbitals that are strongly orthogonal to Φ, we have (see e.g. Lemma 4 in [101] and [53]) that

$$T_\Phi \mathcal{M} = \mathcal{A}\Phi \oplus \Phi^{\perp\!\!\perp},$$

where $\mathcal{A} = \left\{ A \in \mathbb{R}^{N \times N} \mid A^T = -A \right\}$ is the space of the $N \times N$ antisymmetric real matrices. The gauge invariance implies that $\mathcal{A}\Phi^0$ is a subset of the Kernel to DF. It is reasonable to assume that this Kernel is restricted to $\mathcal{A}\Phi^0$. Now, if this is indeed the case, due to the fact that the problem we are considering is a minimization problem, the second order condition further states

$$\forall W \in T_{\Phi^0}\mathcal{M}, \quad a_{\Phi^0}(W, W) \geq 0,$$

where for all $\Psi = (\psi_1, \ldots, \psi_N)^T$ and $\Upsilon = (\upsilon_1, \ldots, \upsilon_N)^T$ in $(H^1_{\text{per}}(\Gamma))^N$,

$$\begin{aligned} a_{\Phi^0}(\Psi, \Upsilon) &= \frac{1}{4} E^{\text{KS}''}(\Phi^0)(\Psi, \Upsilon) - \sum_{i=1}^N \epsilon_i^0 \int_\Gamma \psi_i \upsilon_i \\ &= \sum_{i=1}^N \langle (\mathcal{H}^{\text{KS}}_{\rho^0} - \epsilon_i^0)\psi_i, \upsilon_i \rangle_{H^{-1}_{\text{per}}, H^1_{\text{per}}} + 4 \sum_{i,j=1}^N D_\Gamma(\varphi_i^0 \psi_i, \varphi_j^0 \upsilon_j) \\ &\quad + 4 \sum_{i,j=1}^N \int_\Gamma \frac{d^2 e_{\text{xc}}^{\text{LDA}}}{d\rho^2}(\rho_{\text{c}} + \rho^0) \varphi_i^0 \psi_i \varphi_j^0 \upsilon_j. \end{aligned}$$

Under the previous assumption that the Kernel to DF is restricted to $\mathcal{A}\Phi^0$, this implies that a_{Φ^0} is positive definite on $\Phi^{0,\perp\!\!\!\perp}$, so that, using compactness arguments (see e.g. Proposition 1 in [101]), one can deduce a stronger statement: a_{Φ^0} is coercive on $\Phi^{0,\perp\!\!\!\perp}$ (for the H^1_{per} norm), i.e. the previous assumption is equivalent to assuming that there exists a positive constant c_{Φ^0} such that[2]

$$\forall \Psi \in \Phi^{0,\perp\!\!\!\perp}, \quad a_{\Phi^0}(\Psi, \Psi) \geq c_{\Phi^0} \|\Psi\|^2_{H^1_{\text{per}}}. \tag{7.117}$$

This coercivity constant appears in the upper bound of $[DF(\Phi^0, \epsilon^0)]^{-1}$ so that, similarly as above, there exists an orthogonal matrix U in $O(N)$ such that

$$\|U\Phi_N^0 - \Phi^0\|_{H^1} \leq \frac{C}{c_{\Phi^0}} \|F(\Phi_N^0, \epsilon_N^0)\|_{H^{-1}}.$$

This orthogonal matrix U can be identified: let us denote by $M_{\Phi_N^0, \Phi^0}$ the $N \times N$ matrix with entries

$$[M_{\Phi_N^0, \Phi^0}]_{ij} = \int \Phi_{Ni}^0 \Phi_j^0.$$

If $M_{\Phi_N^0, \Phi^0}$ is invertible, then $U = M^T_{\Phi_N^0, \Phi^0}(M_{\Phi_N^0, \Phi^0} M^T_{\Phi_N^0, \Phi^0})^{-1/2}$.

[2] Note that, as is proven in [CCM], in the simplified linear framework: $h = -\frac{1}{2}\Delta + V_{\text{local}} + V_{\text{nl}}$, the coercivity condition (7.117) is satisfied if and only if there is a gap $c_{\Phi^0} > 0$ between the lowest Nth and $(N+1)$st eigenvalues of h.

Pushing the analysis further, it is proven in [101] that, by using the coercivity property (7.117) we can actually solve a simple linear problem, the solution of which is named the reconstructed error, that can be used to improve the accuracy of the initial solution Φ_N^0. This approach is of a similar nature as the two grids method in [32] or [145] or the perturbative methodology in [33].

A last remark stands on the relative simplicity of the periodic framework since the use of a Fourier basis allows us to refer to the Fourier coefficients definition of the norms that makes the H^{-1}-norm above rather easily to evaluate. In the non-periodic case, where such an easy evaluation of the dual norm of the residual is not feasible, for instance with a local basis of finite element approximations, people classically refer to local reconstruction of the residual by solving local problems over patches of mesh elements that allows bounds to be provided on the dual norm of the residual in a computable way. We refer to [113] and [47, 48] for an original approach to this theory of equilibrated fluxes, and to [34, 35] and [45] for applications to eigenvalue problems.

After these elements on "*a posteriori analysis*", we can proceed to "*a priori analysis*".

We follow here the same lines as in the first subsection above; we start by showing the existence of a discrete solution in a neighborhood of a solution Φ^0.

The analysis needs to be slightly more subtle since Φ^0 is not unique due to the already mentioned gauge invariance. We have thus to work in the quotient space with respect to this gauge condition.

First, we remark that any element $\Psi_N in \mathcal{X}_N^N \cap \mathcal{M}^{\Phi_{N_c}}$ can be written as

$$\Psi_N = \Pi_N(\Phi^0) + \mathcal{S}(W_N)[\Pi_N(\Phi^0)] + W_N, \tag{7.118}$$

where $W_N \in \mathcal{X}_N^N \cap [\Pi_N(\Phi^0)]^{\perp\!\!\perp}$ satisfies $0 \leq M_{W_{N_c},W_{N_c}} \leq 1_N$ and where, for any W, $\mathcal{S}(W) = (1_N - M_{W,W})^{1/2} - 1_N$ is a $N \times N$ symmetric matrix. This provides an interesting parametrization $\mathcal{X}_N^N \cap \mathcal{M}^{\Phi_{N_c}}$ which allows us to translate the original discrete minimization Kohn–Sham problem into a minimization problem set over $\mathcal{X}_N^N \cap [\Pi_N(\Phi^0)]^{\perp\!\!\perp}$, which is easier to work with, and show that there exists a unique solution close to $\Pi_N(\Phi^0)$ thanks to the coercivity of a_{Φ^0}. We refer to [24] for the details on this property.

This allows even more because, following the same arguments involving the ground state energy as in the first subsection above, an estimate that states

$$\|\Phi_N^0 - \Phi^0\|_{H^1}^2 \leq \|[\Pi_N(\Phi^0)] - \Phi^0\|_{H^1} \tag{7.119}$$

already shows some convergence that can further be used in the next step.

Once a solution to the discrete problem is set, one can actually perform the a priori analysis, i.e. the analysis of the error between Φ^0 and Φ_N^0.

The basic tool is again related to the precise estimate of the norm of $[DF_N((\Phi_N^0, \epsilon_N^0)]^{-1}$ which is derived from the above estimate (7.117) on the norm of $[DF(\Phi^0, \epsilon^0)]^{-1}$ and subtle inequalities to evaluate the non linear terms thanks to (7.119).

We thus obtain

$$\|\Phi_N^0 - \Phi^0\|_{H^1} \leq CN^{-\sigma}, \tag{7.120}$$

for some constant $C \in \mathbb{R}_+$ independent of N and σ related to the regularity of Φ^0 (see [24] for more results in this situation).

Acknowledgments This project has received funding from the European Research Council (ERC) under the European Union's Horizon 2020 research and innovation programme (grant agreement EMC2 No 810367), and the Center for Applied Mathematics for Energy Research Applications (CAMERA) funded by U.S. Department of Energy, Office of Science, Advanced Scientific Computing Research under Contract No. DE-AC02-05CH11231.

References

1. P.-A. Absil, R. Mahony and R. Sepulchre. *Optimization Algorithms on Matrix Manifolds.* Princeton University Press, Princeton (2008).
2. C. Adamo and V. Barone. Toward reliable density functional methods without adjustable parameters: The PBE0 model. *The Journal of Chemical Physics* **110**(13), 6158–6170 (1999).
3. F. Alouges and C. Audouze. Preconditioned gradient flows for nonlinear eigenvalue problems and application to the Hartree–Fock functional. *Numerical Methods for Partial Differential Equations*, **25**(2), 380–400 (2009).
4. C. Ambrosch-Draxl. Augmented planewave methods. *Physica Scripta*, **T109**, 48 (2004).
5. D.G. Anderson. Iterative procedures for nonlinear integral equations. *J. ACM* **12**(4), 547–560 (1965).
6. E. Anderson, Z. Bai, C. Bischof, S. Blackford, J. Demmel, J. Dongarra, J. Du Croz, A. Greenbaum, S. Hammarling, A. McKenney and D. Sorensen. *LAPACK Users' Guide (third ed.).* SIAM, Philadelphia (1999).
7. P.-M. Anglade and X. Gonze. Preconditioning of self-consistent-field cycles in density-functional theory: The extrapolar method. *Phys. Rev. B* **78**, 045126 (2008).
8. T.A. Arias. Multiresolution analysis of electronic structure: semicardinal and wavelet bases. *Rev. Mod. Phys.* **71**, 267–311 (1999).
9. V. Bach, E.H. Lieb, M. Loss and J.P. Solovej. There are no unfilled shells in unrestricted Hartree–Fock theory. *Phys. Rev. Lett.* **72**(19), 2981–2983 (1994).
10. G.B. Bacskay. A quadratically convergent Hartree–Fock (qc-scf) method. application to closed shell systems. *Chemical Physics* **61**(3), 385 – 404 (1981).
11. A.S. Banerjee, L. Lin, P. Suryanarayana, C. Yang and J.E. Pask. Two-level chebyshev filter based complementary subspace method: Pushing the envelope of large-scale electronic structure calculations. *Journal of Chemical Theory and Computation* **14**(6), 2930–2946 (2018).
12. A.S. Banerjee, P. Suryanarayana and J.E. Pask. Periodic Pulay method for robust and efficient convergence acceleration of self-consistent field iterations. *Chemical Physics Letters* **647**, 31–35 (2016).
13. S. Baroni, S. De Gironcoli, A. Dal Corso and P. Giannozzi. Phonons and related crystal properties from density-functional perturbation theory. *Reviews of Modern Physics* **73**(2), 515 (2001).

14. S Baroni and P Giannozzi. Towards very large-scale electronic-structure calculations. *Europhysics Letters (EPL)* **17**(6), 547–552 (1992).
15. J. Barzilai and J.M. Borwein. Two-point step size gradient methods. *IMA Journal of Numerical Analysis* **8**(1), 141–148 (1988).
16. M. Benzi, P. Boito and N. Razouk. Decay properties of spectral projectors with applications to electronic structure. *SIAM Rev.* **55**, 3–64 (2013).
17. L.S. Blackford, J. Choi, A. Cleary, E. D'Azeuedo, J. Demmel, I. Dhillon, S. Hammarling, G. Henry, A. Petitet, K. Stanley, et al. *ScaLAPACK User's Guide*. Society for Industrial and Applied Mathematics, USA (1997).
18. P.E. Blöchl. Projector augmented-wave method. *Physical review B* **50**(24), 17953 (1994).
19. S.F. Boys. Electronic wavefuntions. I. A general method of calculation for the stationary states of any molecular system. *Proc. Roy. Soc. A* **200**, 542–554 (1950).
20. C. Brezinski, M. Redivo-Zaglia and Y. Saad. Shanks sequence transformations and anderson acceleration. *SIAM Review* **60**(3), 646–669 (2018).
21. C.G. Broyden. A class of methods for solving nonlinear simultaneous equations. *Math. Comp.* **19**, 577–593 (1965).
22. E. Cancès. Self-consistent field algorithms for Kohn–Sham models with fractional occupation numbers. *The Journal of Chemical Physics* **114**(24), 10616–10622 (2001).
23. E. Cancès, R. Chakir and Y. Maday. Numerical analysis of nonlinear eigenvalue problems. *Journal of Scientific Computing* **45**(1), 90–117 (2010).
24. E. Cancès, R. Chakir and Y. Maday. Numerical analysis of the planewave discretization of some orbital-free and Kohn–Sham models. *ESAIM: M2AN* **46**(2), 341–388 (2012).
25. E. Cancès, M. Defranceschi, W. Kutzelnigg, C. Le Bris and Y. Maday. *Computational quantum chemistry: a primer*. Volume X of *Handbook of Numerical Analysis*, pages 3–270. North-Holland, Amsterdam (2003).
26. E. Cancès, K.N. Kudin, G.E. Scuseria and G. Turinici. Quadratically convergent algorithm for fractional occupation numbers in density functional theory. *The Journal of Chemical Physics* **118**(12), 5364–5368 (2003).
27. E. Cancès and C. Le Bris. Can we outperform the DIIS approach for electronic structure calculations? *International Journal of Quantum Chemistry* **79**(2), 82–90 (2000).
28. E. Cancès and C. Le Bris. On the convergence of SCF algorithms for the Hartree–Fock equations. *ESAIM: Mathematical Modelling and Numerical Analysis* **34**(4), 749–774 (2000).
29. E. Cancès and M. Lewin. The dielectric permittivity of crystals in the reduced Hartree–Fock approximation. *Archive for Rational Mechanics and Analysis* **197**(1), 139–177 (2010).
30. E. Cancès and N. Mourad. Existence of a type of optimal norm-conserving pseudopotentials for Kohn–Sham models. *Communications in Mathematical Sciences* **14**(5), 1315–1352 (2015).
31. E. Cancès and K. Pernal. Projected gradient algorithms for Hartree–Fock and density matrix functional theory calculations. *The Journal of Chemical Physics* **128**(13), 134108 (2008).
32. E. Cancès, R. Chakir, L. He and Y. Maday. Two-grid methods for a class of nonlinear elliptic eigenvalue problems. *IMA Journal of Numerical Analysis* **38**(2), 605–645 (2018).
33. E. Cancès, G. Dusson, Y. Maday, B. Stamm and M. Vohralík. A perturbation-method-based a posteriori estimator for the planewave discretization of nonlinear schrödinger equations. *Comptes Rendus Mathematique* **352**(11), 941–946 (2014).
34. E. Cancès, G. Dusson, Y. Maday, B. Stamm and M. Vohralík. Guaranteed and robust a posteriori bounds for laplace eigenvalues and eigenvectors: conforming approximations. *SIAM Journal on Numerical Analysis* **55**(5), 2228–2254 (2017).
35. E. Cancès, G. Dusson, Y. Maday, B. Stamm and M. Vohralík. Guaranteed and robust a posteriori bounds for laplace eigenvalues and eigenvectors: a unified framework. *Numerische Mathematik* **140**(4), 1033–1079 (2018).
36. E. Cancès, V. Ehrlacher, D. Gontier, A. Levitt and D. Lombardi. Numerical quadrature in the Brillouin zone for periodic Schrödinger operators. *Numerische Mathematik* **144**, 479–526 (2020).

37. E. Cancès, G. Kemlin and A. Levitt. Convergence analysis of direct minimization and self-consistent iterations. *arXiv:2004.09088* (2020).
38. D.M. Ceperley and B.J. Alder. Ground state of the electron gas by a stochastic method. *Phys. Rev. Lett.* **45**, 566–569 (1980).
39. M. Challacombe and E. Schwegler. Linear scaling computation of the Fock matrix. *The Journal of Chemical Physics* **106**(13), 5526–5536 (1997).
40. J.R. Chelikowsky, N. Troullier and Y. Saad. Finite-difference-pseudopotential method: Electronic structure calculations without a basis. *Phys. Rev. Lett.* **72**, 1240–1243 (1994).
41. H. Chen, X. Gong, L. He, Z. Yang and A. Zhou. Numerical analysis of finite dimensional approximations of Kohn–Sham models. *Advances in Computational Mathematics* **38**, 225–256 (2013).
42. H.Chen and R. Schneider. Numerical analysis of augmented plane wave methods for full-potential electronic structure calculations. *ESAIM: M2AN* **49**(3), 755–785 (2015).
43. M. Chupin, M.-S. Dupuy, G. Legendre and E. Séré. Convergence analysis of adaptive diis algorithms with application to electronic ground state calculations. *arXiv preprint arXiv:2002.12850* (2020).
44. S.J. Clark, M.D. Segall, C.J. Pickard, et al. First principles methods using CASTEP. *Zeitschrift für Kristallographie - Crystalline Materials* **220**(5/6), 567–570 (2009).
45. X. Dai, J. Xu and A. Zhou. Convergence and optimal complexity of adaptive finite element eigenvalue computations. *Numerische Mathematik* **110**(3), 313–355 (2008).
46. E.R. Davidson. The iterative calculation of a few of the lowest eigenvalues and corresponding eigenvectors of large real symmetric matrices. *Journal of Computational Physics* **17**, 87–94 (1975).
47. P. Destuynder and B. Métivet. Explicit error bounds for a nonconforming finite element method. *SIAM journal on numerical analysis* **35**(5), 2099–2115 (1998).
48. P. Destuynder and B. Métivet. Explicit error bounds in a conforming finite element method. *Mathematics of Computation* **68**(228), 1379–1396 (1999).
49. M. Dolg. Effective core potentials. In: *Modern Methods and Algorithms of Quantum Chemistry, Proceedings*, John von Neumann Institute for Computing, NIC Series volume 3, edited by J. Grotendorst, pp. 507–540, Jülich, (2000).
50. R.M. Dreizler and E.K.U. Gross. *Density functional theory*. Springer (1990).
51. M.-S. Dupuy. Projector augmented-wave method: an analysis in a one-dimensional setting. *ESAIM Math. Mod. Numer. Anal. (M2AN)* **54**(1) (2020).
52. G. Dusson and Y. Maday. A posteriori analysis of a nonlinear Gross–Pitaevskii-type eigenvalue problem. *IMA Journal of Numerical Analysis* **37**(1), 94–137 (2017).
53. A. Edelman, T.A. Arias and S.T. Smith. The geometry of algorithms with orthogonality constraints. *SIAM Journal of Matrix Analysis and Applications* **92**(3), 609–625 (1953).
54. A. Edelman, T.A. Arias and S.T. Smith. The geometry of algorithms with orthonormality constraints. *J. Matrix Anal. Appl.* **20**, 303–353 (1998).
55. A. Ern and J.-L. Guermond. *Theory and practice of finite elements*, volume 159. Springer Science & Business Media (2013).
56. P. Giannozzi et al. QUANTUM ESPRESSO: a modular and open-source software project for quantum simulations of materials. *Journal of Physics: Condensed Matter* **21**(39), 395502 (2009).
57. H. Fang and Y. Saad. Two classes of multisecant methods for nonlinear acceleration. *Numerical Linear Algebra with Applications* **16**(3), 197–221 (2009).
58. J.-L. Fattebert. Finite difference schemes and block Rayleigh quotient iteration for electronic structure calculations on composite grids. *Journal of Computational Physics* **149**(1), 75–94 (1999).
59. W.M.C. Foulkes and R. Haydock. Tight-binding models and density-functional theory. *Phys. Rev. B* **39**, 12520–12536 (1989).
60. J. Francisco, J.M. Martinez and L. Martinez. Globally convergent trust-region methods for self-consistent field electronic structure calculations. *J. Chem. Phys.* **121**, 10863–10878 (2004).

61. C. Freysoldt, S. Boeck and J. Neugebauer. Direct minimization technique for metals in density functional theory. *Physical Review B* **79**(24), 241103 (2009).
62. W. Gao, C. Yang and J. Meza. Solving a class of nonlinear eigenvalue problems by Newton's method. Technical report, Lawrence Berkeley National Laboratory, Berkeley (2009).
63. L. Genovese, A. Neelov, S. Goedecker, T. Deutsch, S.A. Ghasemi, A. Willand, D. Caliste, O. Zilberberg, M. Rayson, A. Bergman and R. Schneider. Daubechies wavelets as a basis set for density functional pseudopotential calculations. *The Journal of Chemical Physics* **129**(1), 014109 (2008).
64. S. Ghosh and P. Suryanarayana. SPARC: Accurate and efficient finite-difference formulation and parallel implementation of density functional theory: Isolated clusters. *Computer Physics Communications* **212**, 189 – 204 (2017).
65. P. Giannozzi. Notes on pseudopotential generation. https://www.quantum-espresso.org/Doc/pseudo-gen.pdf (2010).
66. G. Giuliani and G. Vignale. *Quantum theory of the electron liquid.* Cambridge university press (2005).
67. S. Goedecker. Linear scaling electronic structure methods. *Reviews of Modern Physics* **71**(4), 1085–1123 (1999).
68. S. Goedecker and L. Colombo. Efficient linear scaling algorithm for tight-binding molecular dynamics. *Phys. Rev. Lett.* **73**, 122–125 (1994).
69. S. Goedecker and M. Teter. Tight-binding electronic-structure calculations and tight-binding molecular dynamics with localized orbitals. *Phys. Rev. B* **51**, 9455–9464 (1995).
70. S. Goedecker, M. Teter and J. Hutter. Separable dual-space Gaussian pseudopotentials. *Physical Review B* **54**(3), 1703 (1996).
71. X. Gonze, J.-M. Beuken, R. Caracas, F. Detraux, M. Fuchs, G.-M. Rignanese, L. Sindic, M. Verstraete, G. Zerah, F. Jollet, M. Torrent, A. Roy, M. Mikami, Ph. Ghosez, J.-Y. Raty and D.C. Allan. First-principles computation of material properties: the ABINIT software project. *Computational Materials Science* **25**(3), 478–492 (2002).
72. L. Greengard and V. Rokhlin. A fast algorithm for particle simulations. *Journal of Computational Physics* **73**(2), 325–348 (1987).
73. L. Greengard and V. Rokhlin. A new version of the Fast Multipole Method for the Laplace equation in three dimensions. *Acta Numerica* **6**, 229–269 (1997).
74. D.R. Hamann. Optimized norm-conserving Vanderbilt pseudopotentials. *Physical Review B* **88**(8), 085117 (2013).
75. J. Harris. Simplified method for calculating the energy of weakly interacting fragments. *Phys. Rev. B* **31**, 1770–1779 (1985).
76. C. Hartwigsen, S. Gœdecker and J. Hutter. Relativistic separable dual-space Gaussian pseudopotentials from H to Rn. *Physical Review B* **58**(7), 3641 (1998).
77. M.F. Herbst and A. Levitt. Black-box inhomogeneous preconditioning for self-consistent field iterations in density functional theory. *Journal of Physics: Condensed Matter* **33**(8), 085503 (2020).
78. W. Hu, L. Lin and C. Yang. Interpolative separable density fitting decomposition for accelerating hybrid density functional calculations with applications to defects in silicon. *Journal of Chemical Theory and Computation* **13**(11), 5420–5431 (2017).
79. D.D. Johnson. Modified Broyden's method for accelerating convergence in self-consistent calculations. *Phys. Rev. B* **38**, 12807–12813 (1988).
80. B. Kanungo and V. Gavini. Large-scale all-electron density functional theory calculations using an enriched finite-element basis. *Phys. Rev. B* **95**, 035112 (2017).
81. G.P. Kerker. Efficient iteration scheme for self-consistent pseudopotential calculations. *Phys. Rev. B* **23**, 3082–3084 (1981).
82. L. Kleinman and D.M. Bylander. Efficacious form for model pseudopotentials. *Physical Review Letters* **48**(20), 1425 (1982).
83. D.A. Knoll and D.E. Keyes. Jacobian-free Newton–Krylov methods: a survey of approaches and applications. *Journal of Computational Physics* **193**(2), 357–397 (2004).

84. A. Knyazev. Toward the optimal preconditioned eigensolver: Locally optimal block preconditioned conjugate gradient method. *SIAM J. Sci. Comput.* **22**(2), 517–541 (2001).
85. F.R. Krajewski and M. Parrinello. Linear scaling electronic structure monte carlo method for metals. *Phys. Rev. B* **75**, 235108 (2007).
86. G. Kresse and J. Furthmüller. Efficient iterative schemes for ab initio total-energy calculations using a plane-wave basis set. *Phys. Rev. B* **54**, 11169–11186 (1996).
87. K. Kudin, G.E. Scuseria and E. Cancès. A black-box self-consistent field convergence algorithm: one step closer. *J. Chem. Phys.* **116**, 8255–8261 (2002).
88. S. Lang. *Fundamentals of Differential Geometry*. Graduate Texts in Mathematics. Springer, New York (2001).
89. K. Lejaeghere et al. Reproducibility in density functional theory calculations of solids. *Science* **351**(6280) (2016).
90. A. Levitt. Convergence of gradient-based algorithms for the Hartree–Fock equations. *ESAIM: Mathematical Modelling and Numerical Analysis* **46**(6), 1321–1336 (2012).
91. A. Levitt. Screening in the finite-temperature reduced Hartree–Fock model. *Archive for Rational Mechanics and Analysis*, Arch. Ration. Mech. Anal. **238**, 901–927 (2020).
92. W. Liang, C. Saravanan, Y. Shao, R. Baer, A.T. Bell and M. Head-Gordon. Improved fermi operator expansion methods for fast electronic structure calculations. *The Journal of Chemical Physics* **119**(8), 4117–4125 (2003).
93. E.H. Lieb. Variational principle for many-fermion systems. *Phys. Rev. Lett.* **46**, 457–459 (1981).
94. E.H. Lieb. Density functionals for Coulomb systems. *International Journal of Quantum Chemistry* **24**(3), 243–277 (1983).
95. L. Lin, J. Lu, L. Ying and W. E. Adaptive local basis set for Kohn–Sham density functional theory in a discontinuous Galerkin framework I: Total energy calculation. *Journal of Computational Physics* **231**(4), 2140–2154 (2012).
96. L. Lin and C. Yang. Elliptic preconditioner for accelerating the self-consistent field iteration in Kohn–Sham density functional theory. *SIAM Journal on Scientific Computing* **35**(5), S277–S298 (2013).
97. L. Lin. Adaptively compressed exchange operator. *Journal of Chemical Theory and Computation* **12**(5), 2242–2249 (2016).
98. L. Lin, J. Lu, L. Ying and W. E. Pole-based approximation of the fermi-dirac function. *Chinese Annals of Mathematics, Series B* **30**(6), 729 (2009).
99. B. Liu. The simultaneous expansion method for the iterative solution of several of the lowest eigenvalues and corresponding eigenvectors of large real-symmetric matrices. Technical Report LBL-8158, Lawrence Berkeley Laboratory, University of California, Berkeley, 1978.
100. X. Liu, Z. Wen, X. Wang, M. Ulbrich and Y. Yuan. On the analysis of the discretized Kohn–Sham density functional theory. *SIAM J. Numer. Anal.* **53**(4), 1758–1785 (2015).
101. Y. Maday and G. Turinici. Error bars and quadratically convergent methods for the numerical simulation of the hartree-fock equations. *Numerische Mathematik* **94**(4), 739–770 (2003).
102. A. Marek, V. Blum, R. Johanni, V. Havu, B. Lang, T. Auckenthaler, A. Heinecke, H.-J. Bungartz and H. Lederer. The ELPA library: scalable parallel eigenvalue solutions for electronic structure theory and computational science. *J. Phys.: Condesened Matter* **26**, 213201 (2014).
103. L.D. Marks and D.R. Luke. Robust mixing for *ab initio* quantum mechanical calculations. *Phys. Rev. B* **78**, 075114 (2008).
104. N. Marzari, D. Vanderbilt and M.C. Payne. Ensemble density-functional theory for *ab initio* molecular dynamics of metals and finite-temperature insulators. *Physical review letters* **79**(7), 1337 (1997).
105. A. Masud and R. Kannan. B-splines and NURBS based finite element methods for Kohn–Sham equations. *Computer Methods in Applied Mechanics and Engineering* **241–244**, 112–127 (2012).
106. V. Michaud-Rioux, L. Zhang and H. Guo. Rescu: A real space electronic structure method. *Journal of Computational Physics* **307**, 593–613 (2016).

107. A.A. Mostofi, P.D. Haynes, C.-K. Skylaris and M.C. Payne. Preconditioned iterative minimization for linear-scaling electronic structure calculations. *The Journal of Chemical Physics* **119**(17), 8842–8848 (2003).
108. J. Nocedal and S. Wright. *Numerical optimization*. Springer Science & Business Media (2006).
109. P. Norman, K. Ruud and T. Saue. *Principles and practices of molecular properties: Theory, modeling, and simulations*. John Wiley & Sons (2018).
110. J.P. Perdew, K. Burke and M. Ernzerhof. Generalized gradient approximation made simple. *Phys. Rev. Lett.* **77**, 3865–3868 (1996).
111. J.P. Perdew and Y. Wang. Accurate and simple analytic representation of the electron-gas correlation energy. *Phys. Rev. B* **45**, 13244–13249 (1992).
112. C.J. Pickard and F. Mauri. All-electron magnetic response with pseudopotentials: Nmr chemical shifts. *Physical Review B* **63**(24), 245101 (2001).
113. W. Prager and J.L. Synge. Approximations in elasticity based on the concept of function space. *Quarterly of Applied Mathematics* **5**(3), 241–269 (1947).
114. P.P. Pratapa and P. Suryanarayana. Restarted pulay mixing for efficient and robust acceleration of fixed-point iterations. *Chemical Physics Letters* **635**, 69–74 (2015).
115. E. Prodan and W. Kohn. Nearsightedness of electronic matter. *Proceedings of the National Academy of Sciences* **102**(33), 11635–11638 (2005).
116. P. Pulay. Convergence acceleration of iterative sequences. the case of SCF iteration. *Chemical Physics Letters* **73**(2), 393–398 (1980).
117. P. Pulay. Improved SCF convergence acceleration. *Journal of Computational Chemistry* **3**(4), 556–560 (1982).
118. P. Pulay. Analytical derivatives, forces, force constants, molecular geometries, and related response properties in electronic structure theory. *WIREs Computational Molecular Science* **4**(3), 169–181 (2014).
119. D. Raczkowski, A. Canning, and L.W. Wang. Thomas-Fermi charge mixing for obtaining self-consistency in density functional calculations. *Phys. Rev. B* **64**, 121101 (2001).
120. A.M. Rappe, K.M. Rabe, E. Kaxiras and J.D. Joannopoulos. Optimized pseudopotentials. *Physical Review B* **41**(2), 1227 (1990).
121. M.C. Reed and B. Simon. *Methods of Modern Mathematical Physics. IV. Analysis of operators*. Academic Press, New York (1978).
122. M.C. Reed and B. Simon. *Methods of Modern Mathematical Physics. I. Functional Analysis*. Academic Press, New York (1980).
123. X. Ren, P. Rinke, V. Blum, J. Wieferink, A. Tkatchenko, A. Sanfilippo, K. Reuter and M. Scheffler. Resolution-of-identity approach to Hartree–Fock, hybrid density functionals, RPA, MP2 andGWwith numeric atom-centered orbital basis functions. *New Journal of Physics* **14**(5), 053020 (2012).
124. T. Rohwedder and R. Schneider. An analysis for the DIIS acceleration method used in quantum chemistry calculations. *Journal of Mathematical Chemistry* **49**(9), 1889 (2011).
125. C.C.J. Roothaan. New developments in molecular orbital theory. *Rev. Mod. Phys.* **23**, 69–89 (1951).
126. Y. Saad and M.H. Schultz. GMRES: A generalized minimal residual algorithm for solving nonsymmetric linear systems. *SIAM Journal on Scientific and Statistical Computing* **7**(3), 856–869 (1986).
127. V.R. Saunders and I.H. Hillier. A level–shifting method for converging closed shell Hartree–Fock wave functions. *International Journal of Quantum Chemistry* **7**(4), 699–705 (1973).
128. R. Schneider, T. Rohwedder, A. Neelov and J. Blauert. Direct minimization for calculating invariant subspaces in density functional computations of the electronic structure. *J. Comput. Math.* **27**, 360–387 (2009).
129. K. Schwarz and P. Blaha. Solid state calculations using WIEN2k. Proceedings of the Symposium on Software Development for Process and Materials Design. *Computational Materials Science* **28**(2), 259–273 (2003).

130. Y. Shiihara, O. Kuwazuru and N. Yoshikawa. Real-space Kerker method for self-consistent calculation using non-orthogonal basis functions. *Modelling and Simulation in Materials Science and Engineering* **16**(3), 035004 (2008).
131. C.-K. Skylaris, P.D. Haynes, A.A. Mostofi and M.C. Payne. Introducing ONETEP: Linear-scaling density functional simulations on parallel computers. *The Journal of Chemical Physics* **122**(8), 084119 (2005).
132. J. C. Slater. An augmented plane wave method for the periodic potential problem. *Phys. Rev.* **92**, 603–608 (1953).
133. J.C. Slater. Atomic shielding constants. *Phys. Rev.* **36**, 57–64 (1930).
134. J.M. Soler, E. Artacho, J.D. Gale, A. García, J. Junquera, P. Ordejón and D. Sánchez-Portal. The SIESTA method for *ab initio* order-N materials simulation. *Journal of Physics: Condensed Matter* **14**(11), 2745–2779 (2002).
135. G.P. Srivastava. Broyden's method for self-consistent field convergence acceleration. *Journal of Physics A: Mathematical and General* **17**(6), L317–L321 (1984).
136. R.M. Sternheimer. Electronic polarizabilities of ions from the Hartree–Fock wave functions. *Phys. Rev.* **96**, 951–968 (1954).
137. L. Thorgersen, J. Olsen, D. Yeager, L. Jorgensen, P. Salek and T. Helgaker. The trust-region self-consistent field method: Towards a black-box optimization in Hartree–Fock and Kohn–Sham theories. *The Journal of Chemical Physics* **121**(1), 16–27 (2004).
138. J. Tomasi, B. Mennucci and R. Cammi. Quantum mechanical continuum solvation models. *Chemical Reviews* **105**(8), 2999–3094 (2005).
139. N. Troullier and J.L. Martins. Efficient pseudopotentials for plane-wave calculations. *Physical review B* **43**(3), 1993 (1991).
140. M. Ulbrich, Z. Wen, C. Yang, D. Klöckner and Z. Lu. A proximal gradient method for ensemble density functional theory. *SIAM Journal on Scientific Computing* **37**(4), A1975–A2002 (2015).
141. S.M. Valone. A one-to-one mapping between one-particle densities and some N-particle ensembles. *The Journal of Chemical Physics* **73**(9), 4653–4655 (1980).
142. H.F. Walker and P. Ni. Anderson acceleration for fixed-point iterations. *SIAM Journal on Numerical Analysis* **49**(4), 1715–1735 (2011).
143. Z. Wen and W. Yin. A feasible method for optimization with orthogonality constraints. *Math. Program., Ser. A* **142**, 397–434 (2013).
144. N.D. Woods, M.C. Payne and P.J. Hasnip. Computing the self-consistent field in Kohn–Sham density functional theory. *Journal of Physics: Condensed Matter* **31**(45), 453001 (2019).
145. J. Xu and A. Zhou. Local and parallel finite element algorithms based on two-grid discretizations. *Mathematics of Computation* **69**(231), 881–909 (2000).
146. C. Yang, J. Meza and L. Wang. A constrained optimization algorithm for total energy minimization in electronic structure calculations. *J. Comput. Phys.* **217**, 709–721 (2006).
147. D. Zhang, L. Shen, A. Zhou and X.-G. Gong. Finite element method for solving Kohn–Sham equations based on self-adaptive tetrahedral mesh. *Physics Letters A* **372**(30), 5071–5076 (2008).
148. G. Zhang, L. Lin, W. Hu, C. Yang and J.E. Pask. Adaptive local basis set for Kohn–Sham density functional theory in a discontinuous Galerkin framework II: Force, vibration, and molecular dynamics calculations. *Journal of Computational Physics* **335**, 426–443 (2017).
149. X. Zhang, J. Zhu, Z. Wen and A. Zhou. Gradient type optimization methods for electronic structure calculations. *SISC* **36**, 265–289 (2014).
150. Z. Zhao, Z. Bai and X. Jin. A Riemannian Newton algorithm for nonlinear eigenvalue problems. *SIMAX* **36**, 752–774 (2015).
151. Y. Zhou, H. Wang, Y. Liu, X. Gao and H. Song. Applicability of kerker preconditioning scheme to the self-consistent density functional theory calculations of inhomogeneous systems. *Phys. Rev. E* **97**, 033305 (2018).

152. Y. Zhou, J.R. Chelikowsky and Y. Saad. Chebyshev-filtered subspace iteration method free of sparse diagonalization for solving the Kohn–Sham equation. *Journal of Computational Physics* **274**, 770–782 (2014).
153. Y. Zhou, Y. Saad, M.L. Tiago and J.R. Chelikowsky. Self-consistent-field calculations using Chebyshev-filtered subspace iteration. *Journal of Computational Physics* **219**(1), 172–184 (2006).

Chapter 8
Recent Progress in Evaluating the Kohn–Sham Map

Lin Lin, Jianfeng Lu, and Lexing Ying

Abstract In the framework of self-consistent field iteration for solving Kohn–Sham density functional theory (DFT), the evaluation of the Kohn–Sham map is typically the most time-consuming step. We review some recent progress in numerical methods to evaluate the Kohn–Sham map, which are suitable for large scale electronic structure calculations.

8.1 Introduction

The evaluation of the Kohn–Sham map is typically the computational bottleneck when solving Kohn–Sham equations for large and complex quantum systems (i.e. systems with thousands to tens of thousands of atoms). The first two rungs on the ladder of exchange-correlation functionals [47] comprise the local and semi-local exchange-correlation functionals. These functionals include local density approximation (LDA) and generalized gradient approximation (GGA) functionals. The effective Kohn–Sham potential $V_{\text{eff}}(\rho)$ and hence the Hamiltonian depends only on the electron density ρ. Formally, the density-to-density Kohn–Sham map can be

L. Lin (✉)
Department of Mathematics, University of California, Berkeley and Applied Mathematics and Computational Research Division, Lawrence Berkeley National Laboratory, Berkeley, CA, USA
e-mail: linlin@math.berkeley.edu

J. Lu
Departments of Mathematics, Physics, and Chemistry, Duke University, Durham, NC, USA
e-mail: jianfeng@math.duke.edu

L. Ying
Department of Mathematics and Institute for Computational and Mathematical Engineering, Stanford University, Stanford, CA, USA
e-mail: lexing@stanford.edu

E. Cancès, G. Friesecke (eds.), *Density Functional Theory*, Mathematics and Molecular Modeling, https://doi.org/10.1007/978-3-031-22340-2_8

written as

$$\mathcal{F}_{\mathrm{KS}}(\rho) := \rho_{f_\beta\left(-\frac{1}{2}\Delta+V_{\mathrm{eff}}(\rho)-\mu\right)}. \tag{8.1}$$

Here $f_\beta(\varepsilon) = (1+\mathrm{e}^{\beta\varepsilon})^{-1}$ is the Fermi–Dirac function, β is the inverse temperature, and the chemical potential μ should be chosen to satisfy the normalization condition (N is the number of electrons)

$$\mathrm{Tr}\left(f_\beta\left(-\frac{1}{2}\Delta + V_{\mathrm{eff}}(\rho) - \mu\right)\right) = N. \tag{8.2}$$

The potential to density mapping is defined accordingly as $V_{\mathrm{eff}} \mapsto \rho_{f_\beta\left(-\frac{1}{2}\Delta+V_{\mathrm{eff}}-\mu\right)}$ satisfying the constraint (8.2).

$\mathcal{F}_{\mathrm{KS}}$ maps the effective potential (external potential due to the nuclei, and the Hartree-exchange-correlation potential) to the electron density. The third rung on the ladder of exchange-correlation functionals are the meta-GGA functionals, of which the treatment is similar to that of the GGA functionals. However, meta-GGA functionals require some modification to the Laplacian operator and will not be discussed here.

The main challenge is the cost of orthogonalizing the Kohn–Sham orbitals. While such cost is often insignificant for small systems, it scales cubically with respect to the system size, and can quickly dominate the computational cost for large systems. The orthogonalization procedure is essential in all diagonalization-based methods, and leads to the "cubic scaling wall". As a result, any algorithm with improved computational complexity must not perform an eigen-decomposition of the Kohn–Sham Hamiltonian.

Kohn–Sham DFT calculations with nonlocal functionals, such as rung-4 functionals (hybrid functionals) and rung-5 functionals, can be considerably more costly than calculations with exchange-correlation functionals from the first three rungs of the ladder. More specifically, Kohn–Sham equations with local and semi-local functionals can be viewed as eigenvalue problems corresponding to differential operators. When a rung-4 functional is used, the Kohn–Sham Hamiltonian operator becomes an integro-differential operator due to the Fock exchange term. For rung-5 functionals, the self-consistency is computationally rather challenging, and most calculations are done as a post-processing step to obtain the correlation energy as a perturbation to the self-consistent solution for a semi-local functional. For rung-4 nonlocal exchange-correlation functionals, since the effective potential depends on the density matrix P, one needs to evaluate the density matrix-to-density matrix (generalized) Kohn–Sham map

$$\mathcal{F}_{\mathrm{DM}}(P) = f_\beta\left(-\frac{1}{2}\Delta + V_{\mathrm{eff}}(P) - \mu\right), \tag{8.3}$$

where V_{eff} might be a non-local effective potential, and $\mathcal{F}_{\text{DM}}$ maps the effective potential to the density matrix P. Hence the additional computational challenge of large scale Kohn–Sham DFT calculations with nonlocal functionals is the efficient evaluation of nonlocal exchange-correlation functionals.

There has been much progress in the past two decades towards addressing the challenges of evaluating the Kohn–Sham maps [2–4, 10, 16, 17, 21, 23, 29, 30, 32, 38, 39, 45, 48, 49, 55, 56]. The most works involve "linear scaling" methods. The linear scaling property relies on the near-sightedness of the electron matter, as formulated by Kohn [27]: Vaguely speaking, the dependence of electron density at $\mathbf{r}$ on the effective potential at $\mathbf{r}'$ is negligible when $|\mathbf{r}-\mathbf{r}'|$ is large. More precisely, the Jacobian matrix of the Kohn–Sham map (also called the irreducible polarizability operator in the physics literature) can be well-approximated by a sparse matrix. Therefore one may evaluate the Kohn–Sham map using divide-and-conquer and/or sparse matrix techniques. We refer the readers to [6, 16] for extensive reviews of linear scaling methods. In order to apply linear scaling methods, a necessary condition is that the system should be insulating, i.e. have a positive energy gap. On the other hand, there have been much fewer options for reducing the computational complexity of evaluating the Kohn–Sham map for both insulating and metallic systems, which is the focus of this chapter. Our discussion is heavily biased towards our own work in the past few years, and some omissions are inevitable. We also refer readers to [36] for a more detailed review of numerical methods for solving Kohn–Sham equations.

The rest of the chapter is organized as follows. Section 8.2 first introduces the filtering methods, which aim at "postponing" the cubic scaling barrier for large systems. In particular, we introduce Chebyshev filtering type techniques to reduce the preconstant of the cubic scaling component. Then Sect. 8.3 introduces the pole expansion and selected inversion (PEXSI) method, of which the cost for evaluating the Kohn–Sham map with semi-local functionals is at most quadratic with respect to the system size. In Section 8.4 we introduce a method called the interpolative separable density fitting (ISDF), which provides a compact representation of the pair product of Kohn–Sham orbitals, and is a useful tool for reducing the cost of calculations with nonlocal functionals. Section 8.5 discusses techniques for reducing the cost of hybrid functional calculations, where we also illustrate how to use ISDF to reduce the cost of evaluating hybrid functionals and a technique called the adaptive compression to mitigate the impact due to the cost of the hybrid functional in the context of self-consistent field iterations. For simplicity and without loss of generality, we consider isolated, charge neutral, spinless systems throughout this chapter.

8.2 Filtering Methods

When the matrix-vector multiplication $H\psi$ can be performed efficiently, it is common to apply iterative methods, such as conjugate gradient type methods [26, 46], to obtain low-energy eigenvectors of H. In each iteration, these methods apply H to the occupied orbitals once, followed by a Rayleigh–Ritz procedure to update the occupied orbitals. The cost of the latter step scales cubically with respect to the system size, which can dominate the computational cost for large systems.

The main idea of filtering methods is to apply H to the occupied orbitals multiple times to obtain a better approximation to the *subspace* spanned by the desired low-energy eigenvectors. Once such a subspace is identified, we still need to perform a Rayleigh–Ritz operation to diagonalize the projected Hamiltonian (i.e. the Hamiltonian operator restricted to the subspace). However, the frequency of applying cubic scaling steps is much reduced, which leads to improved algorithmic efficiency for large systems.

In order to filter out the unwanted information, we may choose a polynomial $p(z)$ which maximizes the magnitudes at the eigenvalues $\varepsilon_1, \ldots, \varepsilon_N$ and at the same time minimizes at the eigenvalues $\varepsilon_{N+1}, \ldots, \varepsilon_{N_b}$. This can be optimally done by using a properly shifted and scaled Chebyshev polynomial $P_k(z)$, where the subscript k denotes the degree of the polynomial [56]. The Chebyshev polynomials are bounded by 1 within the interval $[-1, 1]$ and grow rapidly outside this interval. Hence we can map the standard interval $[-1, 1]$ to $[\varepsilon_{N+1}, \varepsilon_{N_b}]$, and the shifted-and-scaled Chebyshev polynomial $P_k(z)$ only amplifies the occupied states of the H. At the beginning of the SCF iteration, one does not know a priori the values of ε_{N+1}, ε_{N_b}, or the order of the Chebyshev polynomial. One way to estimate these values is to perform a few steps of the Lanczos iteration [55], which also allows us to update these parameters on the fly. When applying the Chebyshev filtering algorithm, we often choose the number of states N_s to be slightly larger than N to accelerate the convergence, and the convergence is monitored with respect to the orbitals corresponding to the lowest N eigenvalues (see Algorithm 1 for a pseudo-code of the Chebyshev filtering algorithm) .

Algorithm 1: Chebyshev filtering algorithm.

Input: Hamiltonian matrix H and an orthonormal matrix $X \in \mathbb{C}^{N_b \times N_s}$
Output: Eigenvalues $\{\varepsilon_i\}_{i=1}^N$ and eigenfunctions $\{\psi_i\}_{i=1}^N$.

Estimate ε_{N+1} and ε_{N_b}.
while convergence not reached **do**
  Apply the Chebyshev polynomial $P_k(H)$ to X: $Y = P_k(H)X$.
  Orthonormalize columns of Y.
  Compute the projected Hamiltonian matrix $\widetilde{H} = Y^* HY$ and solve the eigenproblem $\widetilde{H}C = C\widetilde{D}$.
  Subspace rotation $X = YC$.
end while
Update $\{\psi_i\}_{i=1}^N$ from the first N columns of X.

We remark that although the Chebyshev filtering algorithm can be used to accurately obtain the lowest N orbitals by repeatedly applying the Chebyshev polynomials, in many implementations, we often perform a small number of iterations and do not evaluate the Kohn–Sham map accurately. In other words, we may combine the outer SCF loop with the inner diagonalization loop, and sometimes even a single inner iteration per SCF step can be sufficient. For many problems, such a combination does not increase the number of outer SCF iterations and hence may significantly reduce the overall running time.

The idea of filtering the Hamiltonian matrix also appears in the spectral slicing approach [1, 48, 49, 52, 54], where a number of matrix functions $f_i(H)$ are applied to the occupied orbitals. Here the scalar functions $f_i(\cdot)$ can be polynomials or rational functions approximately supported only on a small energy interval. One can then perform the Rayleigh–Ritz step to compute the eigenvalues restricted to each interval. At the end, the eigenvalues obtained from different intervals are merged together. The spectral splicing approach is naturally suited for massive parallelization to reduce the wall clock time.

The preconstants of the cubic scaling steps (more specifically, the Rayleigh–Ritz step in step 5 of Algorithm 1) can also be significantly reduced using the complementary subspace strategy with two levels of Chebyshev filtering (CS2CF) [3]. In the Chebyshev filtering approach, the density matrix is computed as

$$P = Y\widetilde{P}Y^*,$$

where $\widetilde{P}$ is the projected density matrix of size N_s, with its eigenvalues given by the occupation numbers $\{f_i\}_{i=1}^{N_s}$. Here we have used the finite temperature formulation of Kohn–Sham DFT. Since N_s is only slightly larger than N, most of the occupation numbers are equal to 1. States 1 through N_1 are as those with occupation numbers equal to 1. The remaining states, from N_1+1 through N_s, have occupations numbers less than 1. Let N_t be the number of these fractionally occupied states, i.e. $N_t = N_s - N_1$. The eigenvectors of the projected density matrix are $\{\widetilde{\psi}_i\}_{i=1}^{N_s}$. Denoting the identity matrix of size N_s by I_{N_s}, from the resolution of the identity we may rewrite the expression for the projected density matrix as

$$\begin{aligned}\widetilde{P} &= \sum_{i=1}^{N_s} f_i\, \widetilde{\psi}_i\, \widetilde{\psi}_i^* \\ &= \sum_{i=1}^{N_s} \widetilde{\psi}_i\, \widetilde{\psi}_i^* - \sum_{i=N_1+1}^{N_s} \widetilde{\psi}_i\, \widetilde{\psi}_i^* + \sum_{i=N_1+1}^{N_s} f_i\, \widetilde{\psi}_i\, \widetilde{\psi}_i^* \\ &= I_{N_s} - \sum_{i=N_1+1}^{N_s} (1-f_i)\, \widetilde{\psi}_i\, \widetilde{\psi}_i^* .\end{aligned} \tag{8.4}$$

Table 8.1 Wall clock times for one SCF iteration large systems using the CS2CF strategy in DGDFT (Credit: [3])

System	# atoms (# electrons)	Computational cores	CS2CF (subspace time) [s]	Diagonalization via ELPA [s]
Electrolyte3D$_{3\times3\times3}$	8586 (29,808)	3456	34 (19)	647
SiDiamond3D$_{10\times10\times10}$	8000 (32,000)	3456	40 (24)	648
Graphene2D$_{8\times8}$	11,520 (23,040)	4608	35 (27)	262
CuFCC3D$_{10\times10\times10}$	4000 (44,000)	3000	75 (46)	199
LiBCC3D$_{12\times12\times12}$	27,648 (82,944)	12,960	180 (165)	5844

Hence if the N_t top eigenvectors $\widetilde{\psi}_i$ and corresponding occupation numbers f_i are known, the projected density matrix $\widetilde{P}$ may be computed. Thus, instead of determining the full $N_s \times N_s$ set of vectors, we need to determine only an extremal block of vectors (of dimension $N_s \times N_t$), corresponding to the states $i = N_1 + 1$ to N_s.

Moreover, physical quantities such as the electron density, energy, entropy and atomic force can all be computed by knowing only the top eigenstates. These eigenstates can be efficiently evaluated via another Chebyshev polynomial filtering steps applied to the projected Hamiltonian matrix $\widetilde{H}$. Using this technique, the Rayleigh–Ritz step may be avoided altogether for insulating systems. For metallic systems, the cost for the Rayleigh–Ritz step can also be significantly reduced. Table 8.1 shows that the CS2CF strategy can be applied to insulating and metallic systems with $O(10^4)$ atoms and efficiently parallelized over $O(10^4)$ computational cores. The wall clock time to solution can be of an order of magnitude faster than parallel dense diagonalization methods such as ScaLAPACK or ELPA [43] for large systems.

8.3 Pole Expansion and Selected Inversion Method

To reduce the computational cost of Kohn–Sham DFT calculations with semi-local exchange-correlation functionals, it is important to realize that we require only the electron density in the Kohn–Sham map, but not the eigenfunctions or eigenvalues. This offers the opportunity to develop more efficient algorithms for the Kohn–Sham map. Since one does not need each individual eigenvector, it is perfectly fine to use an alternative representation for the occupied subspace, such as a density matrix or localized orbitals.

One such example is the Fermi operator expansion (FOE) method, which was originally designed to be used as a linear scaling method. Consider the density matrix at the finite temperature

$$P = f_\beta(H - \mu).$$

The right-hand side is a matrix function with respect to the Hamiltonian matrix H. Instead of diagonalizing H and evaluating the matrix function using the eigen-decomposition, the basic idea of FOE is to expand the Fermi–Dirac function $f_\beta(\cdot)$ into an m-term expansion as

$$f_\beta(\varepsilon) \approx f_{\beta,m}(\varepsilon) = \sum_{n=1}^{m} g_n(\varepsilon). \tag{8.5}$$

The corresponding matrix function approximation is

$$f_\beta(H-\mu) \approx f_{\beta,m}(H-\mu) = \sum_{n=1}^{m} g_n(H-\mu). \tag{8.6}$$

The above formulation is quite general: we only require each term $g_n(H-\mu)$ to be a simple function so that the corresponding matrix function can be evaluated directly without diagonalizing the matrix.

One example of FOE is to expand the Fermi–Dirac function into polynomials [17]

$$f_\beta(\varepsilon) \approx \sum_{n=1}^{m} c_n \varepsilon^{n-1}.$$

The corresponding matrix function version is

$$f_\beta(H-\mu) \approx \sum_{n=1}^{m} c_n (H-\mu)^{n-1}. \tag{8.7}$$

Note that each term of (8.7) is simply a matrix power $(H-\mu)^n$, which can be evaluated using only matrix-matrix multiplication recursively, without diagonalizing H. When H is a sparse matrix, this means that the polynomial approximation to $f_\beta(H-\mu)$ can be a sparse matrix as well.

Besides the polynomial expansion, another possibility is to approximate the Fermi–Dirac function using rational functions. A rational function can be decomposed into a linear combination of terms of the form $(\varepsilon - z)^{-p}$, where $z \in \mathbb{C}$ and $p \geq 1$. In particular, if all terms use $p = 1$ the resulting expansion is called a simple pole expansion, or just a pole expansion. Compared to the polynomial expansion, there are two main advantages of using the rational expansion: First, the number of terms needed for the rational expansion can be much smaller than that required for the polynomial expansion to achieve the same accuracy. This is particularly the case for small gapped systems. Second, the use of the pole expansion can yield fast algorithms with reduced complexity even for metallic systems. This is called the pole expansion and selected inversion algorithm (PEXSI). To our knowledge, PEXSI is so far *the only* algorithm allowing such reduction of complexity.

While linear scaling algorithms in principle yield fast algorithms for the evaluation of the Kohn–Sham map, their accuracy often crucially depends on the decay of orbitals or density matrices, making them usually only suitable for insulating systems with a large gap. Another drawback from the point of view of practical use is that they often require user input on the support of truncation and other tuning parameters to achieve a balance between efficiency and accuracy. The pole expansion and selected inversion method (PEXSI) [30, 32] is a reduced scaling algorithm with computational scaling at most $O(N^2)$ and smaller for lower dimensional systems. While it has a worse computational scaling than linear, the PEXSI algorithm can be applied to general systems and gives accurate results.

As an example of the FOE method, the PEXSI algorithm uses the following pole expansion to approximate the Fermi–Dirac distribution

$$P \approx \sum_{l=1}^{m} \omega_l (H - z_l)^{-1}. \tag{8.8}$$

Here we focus on the scaling with respect to the dimensionless quantity $\beta \Delta E$, where ΔE is the spectral radius of the shifted operator $H - \mu$. To reach a fixed target accuracy, although the number of terms of a straightforward construction of the pole expansion also scales as $O(\beta \Delta E)$ [4], it has subsequently been improved to $O\left((\beta \Delta E)^{\frac{1}{2}}\right)$ [45], $O\left((\beta \Delta E)^{\frac{1}{3}}\right)$ [8], and finally to $O(\log(\beta \Delta E))$ [33]. The work [44] introduces a near-optimal min-max approximation to the Fermi–Dirac function, which further reduces the number of poles needed in practice.

In order to obtain such a pole expansion, one possibility is to use the Cauchy contour integral formulation. Note that the Fermi–Dirac function $f_\beta(\varepsilon)$ is a meromorphic function in $\mathbb{C}$, and the only poles are at $\varepsilon = (2n+1)\mathrm{i}\pi/\beta, n \in \mathbb{Z}$. Furthermore, $f_\beta(\varepsilon)$ can be expanded using the following Matsubara expansion [42]

$$f_\beta(\varepsilon) = \frac{1}{2} - \frac{1}{\beta} \sum_{n \in \mathbb{Z}} \frac{1}{\varepsilon - (2n+1)\mathrm{i}\pi/\beta}. \tag{8.9}$$

Note that (8.9) only converges conditionally, and the infinite summation must be performed symmetrically with respect to the position and negative choice of n. The number of terms needed in the direct truncation of the Matsubara series naturally scales as $O(\beta \Delta E)$.

The efficiency of the pole expansion can be improved by using a contour integral formulation:

$$f_\beta(H - \mu) = \frac{1}{2\pi \mathrm{i}} \oint_{\mathscr{C}} f_\beta(z)((z+\mu)I - H)^{-1} \, \mathrm{d}z. \tag{8.10}$$

Here the contour $\mathscr{C}$ should be chosen so that it encloses all the (real) eigenvalues of $H - \mu$, but without any poles of $f_\beta(z)$, i.e., $(2n+1)\mathrm{i}\pi/\beta, n \in \mathbb{Z}$. This leads to the "dumbbell shaped" contour used in [33] (see Fig. 8.1 for an illustration). The

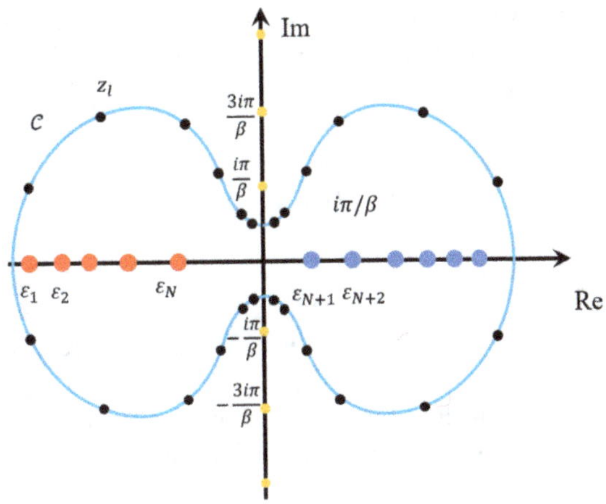

Fig. 8.1 Illustration of the contour integral representation and pole expansion for the density matrix at finite temperature β^{-1}

contour is symmetric with respect to the chemical potential μ. The discretization points are chosen to be denser around μ in order to resolve the sharp transition of the Fermi–Dirac function at μ. At finite temperature, the contour integral formulation remains well defined for gapless systems, i.e. $\varepsilon_N = \varepsilon_{N+1}$.

Each term in the pole expansion corresponds to a matrix inverse, or Green's function $(z_l - H)^{-1}$, which can be evaluated directly without diagonalizing the matrix H. Equation (8.8) converts the problem of computing P to the problem of computing m Green's functions. In order to find the Kohn–Sham map, instead of the entire density matrix P we only need the electron density, which corresponds to the diagonal matrix of P (again for simplicity we assume that a real space representation is used, e.g. the Kohn–Sham model is discretized on a grid). This amounts to the question of finding the diagonal of a Green's function. Note that even if H is a sparse matrix, the matrix inverse $(z_l - H)^{-1}$ can be a fully dense matrix. One direct method is to first evaluate each Green's function and extracts its diagonal elements. However, when H is a sparse matrix, the computation of diagonal entries, and more generally the entries of G_l corresponding to the sparsity pattern of H, can be evaluated much more efficiently by means of the selected inversion method [11, 22, 32, 35].

Although H has been a general Hermitian matrix throughout this chapter, for simplicity we assume H to be a real symmetric matrix. This makes $A = z_l - H \in \mathbb{C}^{N_b \times N_b}$ a complex symmetric, non-singular matrix. For such a matrix, the standard approach for computing A^{-1} is to first decompose A as

$$A = LDL^\top, \tag{8.11}$$

where L is a unit lower triangular matrix and D is a diagonal or a block-diagonal matrix. Equation (8.11) is often known as the $LDL^\top$ factorization of A. Given such a factorization, we can obtain $A^{-1} = (x_1, x_2, \ldots, x_{N_b})$ by solving a number of triangular systems

$$Ly = e_j, \quad Dw = y, \quad L^\top x_j = w, \tag{8.12}$$

for $j = 1, 2, \ldots, N_b$, where e_j is the j-th column of the identity matrix I. The computational cost of such algorithm is generally $O(N_b^3)$. However, when A is sparse, we can exploit the sparsity structure of L and e_j to reduce the complexity of computing selected components of A^{-1}.

The selected inversion algorithm can be heuristically understood as follows [35]. Let A be partitioned into a 2×2 matrix block form as

$$A = \begin{pmatrix} \alpha & b^\top \\ b & \widetilde{A} \end{pmatrix}. \tag{8.13}$$

The first step of an $LDL^\top$ factorization produces a decomposition of A that can be expressed by

$$A = \begin{pmatrix} 1 & \\ \ell & I \end{pmatrix} \begin{pmatrix} \alpha & \\ & \widetilde{A} - b\alpha^{-1}b^\top \end{pmatrix} \begin{pmatrix} 1 & \ell^\top \\ & I \end{pmatrix}, \tag{8.14}$$

where α is often referred to as a pivot, $\ell = b\alpha^{-1}$ and $S = \widetilde{A} - b\alpha^{-1}b^\top$ is known as the *Schur complement*. The same type of decomposition can be applied recursively to the Schur complement S until its dimension becomes 1. The product of lower triangular matrices produced from the recursive procedure, which all have the form

$$\begin{pmatrix} I & & \\ & 1 & \\ & \ell^{(i)} & I \end{pmatrix},$$

where $\ell^{(1)} = \ell = b\alpha^{-1}$, yields the final L factor. At this last step the matrix in the middle becomes diagonal, which is the D matrix.

From (8.14), A^{-1} can be expressed by

$$A^{-1} = \begin{pmatrix} \alpha^{-1} + \ell^\top S^{-1}\ell & -\ell^\top S^{-1} \\ -S^{-1}\ell & S^{-1} \end{pmatrix}. \tag{8.15}$$

This expression suggests that, once α and ℓ are known, the task of computing A^{-1} can be reduced to that of computing S^{-1}. Because a sequence of Schur complements is produced recursively in the $LDL^\top$ factorization of A, the computation of A^{-1} can be organized in a recursive fashion too. Clearly, the reciprocal of the last entry of D is the (N_b, N_b)-th entry of A^{-1}. Starting from this entry, which is also the 1×1

Schur complement produced in the $(N_b - 1)$-th step of the $LDL^\top$ factorization procedure, we can construct the inverse of the 2×2 Schur complement produced at the $(N_b - 2)$-th step of the factorization procedure, using the recipe given by (8.15). This 2×2 matrix is the trailing 2×2 block of A^{-1}. As we proceed from the lower right corner of L and D towards their upper left corner, more and more elements of A^{-1} are recovered. At the end, we recover all the diagonal entries of A^{-1} *exactly*. In fact, given the factorization $A = LDL^\top$, the selected inversion algorithm can be used to efficiently compute all entries $\{A^{-1}_{i,j} \mid (L + L^\top)_{i,j} \neq 0\}$.

In order to understand the asymptotic complexity of the selected inversion algorithm, we assume without loss of generality that the sparsity pattern of H is similar to that obtained by the second-order central difference discretization of a Laplacian operator. The computational cost associated with the $LDL^\top$ factorization, as well as the selected inversion algorithm, scales as $O(N_b)$, $O(N_b^{1.5})$ and $O(N_b^2)$ for one, two and three-dimensional systems, respectively [33]. We remark that this complexity count is robust under changes of the discretization scheme as long as local basis sets are used. Hence for quasi-1D systems (such as nanotubes) and quasi-2D systems (such as monolayer systems and surfaces), the computational cost also scales as $O(N_b)$ and $O(N_b^{1.5})$, respectively. A pseudo-code for the selected inversion algorithm is given in Algorithm 2.

Algorithm 2: Selected inversion algorithm based on $LDL^\top$ factorization.

Input: $LDL^\top$ factorization of a symmetric matrix $A \in \mathbb{C}^{N_b \times N_b}$.
Output: Selected elements of A^{-1}, i.e., $\{A^{-1}_{i,j} \mid (L + L^\top)_{i,j} \neq 0\}$.

Calculate $A^{-1}_{N_b,N_b} \leftarrow (D_{N_b,N_b})^{-1}$.
for $k = N_b - 1, \ldots, 1$ **do**
  Find the collection of indices $C = \{i \mid i > k, L_{i,k} \neq 0\}$.
  Calculate $A^{-1}_{C,k} \leftarrow -A^{-1}_{C,C} L_{C,k}$.
  Calculate $A^{-1}_{k,C} \leftarrow (A^{-1}_{C,k})^\top$.
  Calculate $A^{-1}_{k,k} \leftarrow (D_{k,k})^{-1} - A^{-1}_{k,C} L_{C,k}$.
end for

The pole expansion and selected inversion (PEXSI) method [22, 33, 35] therefore combines the pole expansion and the selected inversion, and evaluates the Kohn–Sham map without solving any eigenvalues or eigenfunctions. The selected inversion method is an exact method if exact arithmetic is used, i.e. the only error in the selected inversion method is due to round off errors. Hence the accuracy of the PEXSI method is determined by the pole expansion, which can be systematically improved by increasing the number of poles. The PEXSI method is ideally suited for massively parallel computers. The treatment of the poles can be parallelized in a straightforward fashion with communication needed only at the end to construct the density matrix. The selected inversion method itself can also be massively parallelizable to thousands of processors [22], and the total number of processors that can be efficiently used by PEXSI can be over 100,000. The PEXSI software package (available at http://www.pexsi.org, distributed under the BSD license) has now been integrated into electronic structure software packages such as BigDFT, CP2K, DFTB+, DGDFT, FHI-aims, QuantumWise ATK, SIESTA, and is part of the "Electronic Structure Infrastructure" (ELSI) package [53].

In addition to computing the charge density at a reduced computational complexity in each SCF iteration, PEXSI can be leveraged to compute energy, free energy and the atomic forces efficiently without diagonalizing the Kohn–Sham Hamiltonian, using *the same set of poles* as those used for computing the charge density [30]. Another numerical issue associated with the PEXSI technique, as well as the Fermi operator expansion techniques in general, is to determine the chemical potential, so that the condition

$$N = N_\beta(\mu) := \mathrm{Tr}[P] \tag{8.16}$$

is satisfied. Note that $N_\beta(\cdot)$ is a non-decreasing function of μ. Hence the chemical potential can be determined via a bisection strategy, or Newton's method. When Newton's method is used, the chemical potential converges rapidly near its correct value. However, the standard Newton's method may not be robust enough when the initial guess is far away from the correct chemical potential. It may give, for example, too large a correction when $N'_\beta(\mu)$ is close to zero, as when μ is near the edge or in the middle of a band gap.

One way to overcome the above difficulty is to use an approximation to the function $N_\beta(\varepsilon)$ to narrow down the region in which the correct μ must lie. This function can be seen effectively as a (temperature smeared) cumulative density of states, counting the number of eigenvalues in the interval $(-\infty, \varepsilon)$. We can evaluate such zero temperature limit, denoted by $N_\infty(\varepsilon)$, again without computing any eigenvalues of H, by using Sylvester's law of inertia [50]. It requires fewer floating point operations than the complex arithmetic direct sparse factorization used in PEXSI. During the self-consistency field iteration, the zero temperature limit $N_\infty(\varepsilon)$ evaluated from the inertia counting procedure may be used to construct upper and lower bounds of the chemical potential. This makes it possible to perform PEXSI calculations over multiple energy points *only once per SCF iteration*, without sacrificing the accuracy at convergence [24].

As an example, we apply the parallel PEXSI method to two systems DG_Graphene_2048 and DG_Graphene_8192, which are disordered graphene systems with 2048 and 8192 atoms, respectively, and compare its performance with a standard approach that requires a partial diagonalization of (H, S). The ScaLAPACK subroutine pdsyevr [51] based on the multiple relatively robust representations (MRRR) algorithm is used to perform such diagonalization. Though both H and S are sparse matrices, the MRRR algorithm treats them as dense matrices. For $H, S \in \mathbb{R}^{N\times N}$, the MRRR algorithm first performs a tridiagonalization procedure with $O(N^3)$ cost, then solves the eigenvalues and eigenvectors of the tridiagonal system with $O(N^2)$ cost, and finally constructs the eigenvectors with $O(N^3)$ cost. Figure 8.2 shows that for graphene problem with 2048 and 8192 processors, the PEXSI technique is nearly two orders of magnitude

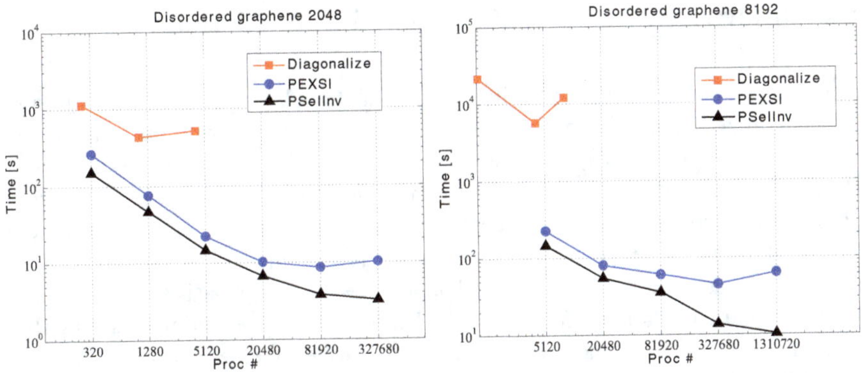

Fig. 8.2 The wall clock time versus the number of cores for a graphene system (Credit: [22])

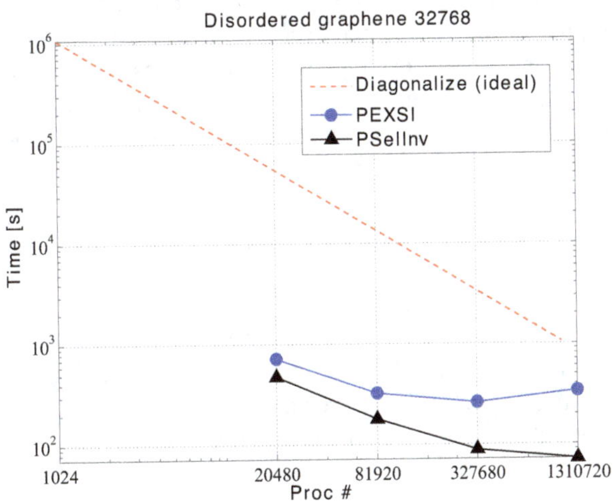

Fig. 8.3 The wall clock time versus the number of cores for a graphene system with 32,768 atoms (Credit: [22])

faster than the ScaLAPACK routine pdsyevr, and can be scalable to a much larger number of processors. The advantage of PEXSI becomes even clearer for a disordered graphene system with 32,768 atoms. For this case, the diagonalization routine is no longer feasible, while the time to solution for the PEXSI technique can be as small as 241 s (Fig. 8.3).

8.4 Interpolative Separable Density Fitting Method

Kohn–Sham DFT calculations with rung-4 functionals require the evaluation of the generalized Kohn–Sham map. They often involve the pair product of orbitals, in the form $\{\varphi_i^*(\mathbf{r})\psi_j(\mathbf{r})\}_{i,j=1}^{N}$, and the number of the pair products is $O(N^2)$. However, since these pair products are not independent from each other, it is possible to approximate pair products using only $O(N)$ functions, by a technique called interpolative separable density fitting (ISDF) [39, 40]. This can be used to reduce the cost of rung-4 functional calculations with a large basis set [9, 21], large scale phonon calculations [34], and post-Hartree–Fock calculations such as RPA correlation energy [38] and second/third order Møller–Plesset perturbation theory calculations [28].

As an intuitive explanation that a compressed representation of the pair products is possible, let us take a real space representation and note that the number of grid points N_g to represent the orbital pairs scales linearly with respect to N, but not quadratically as the number of orbital pairs. This suggests that the numerical rank of $\{\varphi_i^*\psi_j\}$, viewed as a matrix of size $N_g \times N^2$, must scale asymptotically as $O(N)$. It has been proved in [37] that for any ϵ and N, there exists a subspace B_N of dimension on the order of $\epsilon^{-\delta}N^{1+\delta}$ such that

$$\left\|\psi_i\psi_j - \Pi_{B_N}(\psi_i\psi_j)\right\|_{L^2} \leqslant \epsilon,$$

where $\{\psi_i\}$, $i = 1, \ldots, N$ are the first N eigenfunctions of an effective Hamiltonian on a compact domain or closed manifold, and δ is an arbitrary positive constant. This shows that the N^2 pair products of eigenfunctions can be approximated with nearly $O(N)$ auxiliary basis functions, regardless of the discretization level.

The ISDF method uses the following compression format

$$\varphi_i^*(\mathbf{r})\psi_j(\mathbf{r}) \approx \sum_{\mu=1}^{N_\mu} \varphi_i^*(\widehat{\mathbf{r}}_\mu)\psi_j(\widehat{\mathbf{r}}_\mu)\zeta_\mu(\mathbf{r}). \tag{8.17}$$

Here $\{\widehat{\mathbf{r}}_\mu\}_{\mu=1}^{N_\mu}$ is a subset of real space grid points $\{\mathbf{r}_i\}_{i=1}^{N_g}$ on which the orbitals are evaluated. We will refer to $\{\widehat{\mathbf{r}}_\mu\}_{\mu=1}^{N_\mu}$ as the interpolation points, and $\{\zeta_\mu(\mathbf{r})\}_{\mu=1}^{N_\mu}$ sampled on $\{\mathbf{r}_i\}_{i=1}^{N_g}$ the interpolation vectors. These terms come from the understanding of ISDF from the perspective of interpolation. Let $\{\widehat{\mathbf{r}}_\mu\}$ denote a set of grid points in the real space, and let $\zeta_\mu(\mathbf{r})$ be the Lagrange interpolation function on these grid points satisfying

$$\zeta_\mu(\widehat{\mathbf{r}}_{\mu'}) = \delta_{\mu\mu'}.$$

The ISDF decomposition becomes sufficiently accurate as one systematically refines the set $\{\widehat{\mathbf{r}}_\mu\}_{\mu=1}^{N_\mu}$. In the extreme case, all grid points need to be selected and

$N_\mu = N_g$. In practice, the interpolation often becomes sufficiently accurate when $N < N_\mu \ll N_g$, especially when both $\{\varphi_i(\mathbf{r})\}$ and $\{\psi_i(\mathbf{r})\}$ consist of functions that are sufficiently smooth. In general we may set $N_\mu = cN$ where c is a relatively small constant ($4 \sim 10$).

Let us first discuss how to find the interpolation vectors, assuming that the interpolation points $\{\widehat{\mathbf{r}}_\mu\}$ are given. Define

$$\{Z_{ij}(\mathbf{r}) := \varphi_i^*(\mathbf{r})\psi_j(\mathbf{r})\}_{1\le i\le N, 1\le j\le N}. \tag{8.18}$$

Equation (8.17) can be written as

$$Z = \Theta C, \tag{8.19}$$

where each column of Z is defined by Eq. (8.18) sampled on real space grids $\{\mathbf{r}_i\}_{i=1}^{N_g}$. $\Theta = [\zeta_1, \zeta_2, \ldots, \zeta_{N_\mu}]$ contains the interpolating vectors, and each column of C with a multi-index (i, j) is given by

$$[\varphi_i^*(\widehat{\mathbf{r}}_1)\psi_j(\widehat{\mathbf{r}}_1), \cdots, \varphi_i^*(\widehat{\mathbf{r}}_\mu)\psi_j(\widehat{\mathbf{r}}_\mu), \cdots, \varphi_i^*(\widehat{\mathbf{r}}_{N_\mu})\psi_j(\widehat{\mathbf{r}}_{N_\mu})]^\top.$$

Equation (8.19) is an over-determined linear system with respect to the interpolation vectors Θ. One possible way to solve the over-determined system is to impose the Galerkin condition

$$ZC^* = \Theta CC^*. \tag{8.20}$$

The solution to the above is given by

$$\Theta = ZC^*(CC^*)^{-1}, \tag{8.21}$$

which can also be understood as a least squares approximation to the solution of Eq. (8.17).

It may appear that the cost for matrix-matrix multiplications ZC^* and CC^* scales as $O(N_g N^3)$ and $O(N^4)$, respectively, because the size of Z is $N_g \times N^2$ and the size of C is $N_\mu \times N^2$ (here for simplicity we use that $N_\mu \sim N$). However, both multiplications can be carried out with fewer operations due to the separable structure of Z and C, and the computational complexity for computing the interpolation vectors can be reduced to $O(N_g N^2 + N^3)$ [39].

In order to optimize the set of interpolation points, a general strategy is to use the QR factorization with column pivoting (QRCP). The pseudocode of the ISDF algorithm is given in Algorithm 3.

Algorithm 3: Interpolative separable density fitting (ISDF) method

Input: Functions $\{\varphi_i(\mathbf{r})\}$ and $\{\psi_j(\mathbf{r})\}$
Output: Interpolation points $\{\widehat{\mathbf{r}}_\mu\}$, C, and Θ
1: Form matrix Z of size $N_g \times N^2$ with column ij given by $\{Z_{ij}(\mathbf{r}) := \varphi_i^*(\mathbf{r})\psi_j(\mathbf{r})\}_{1\le i\le N, 1\le j\le jN}$,
2: Perform QRCP to Z^* to obtain $Z^*\Pi = QR$. Here Q is an $N^2 \times N_g$ matrix that has orthonormal columns, R is an upper triangular matrix, and Π is a permutation matrix. Though a naive implementation takes $O(N^2 N_g^2)$ steps, randomized projection using the pair product structure of Z can reduce the complexity to $O(N_g N^2)$.
3: The locations of the nonzero entries of the first N_μ columns of Π gives $\{\widehat{\mathbf{r}}_\mu\}$ and C.
4: $\Theta = ZC^*(CC^*)^{-1}$.

The above procedure can be further accelerated. Using Eq. (8.19) directly is sub-optimal as the storage requirement for the matrix Z is $O(N_g N^2)$ and the computational cost associated with a standard QRCP procedure is $O(N_g^2 N^2)$. One possibility is to lower the cost of QRCP by using a random matrix to subsample columns of the matrix Z to form a smaller matrix $\widetilde{Z}$ of size $N_g \times \widetilde{N}_\mu$, where $\widetilde{N}_\mu$ is only slightly larger than N_μ [39, 40]. The reduced matrix size allows the computational cost of the QRCP procedure to be reduced to $O(N_g N^2)$. The implementation and parallelization of ISDF is then relatively straightforward, as QRCP algorithms are available in standard linear algebra software packages such as LAPACK and ScaLAPACK,

Another possibility for choosing interpolation points is to use a heuristic strategy. Note that an effective choice of the set of interpolation points should satisfy the following two conditions. (1) The distribution of the interpolation points should roughly follow the electron density. In particular, there should be more points in regions with higher electron density. (2) The interpolation points should be separated from each other, as otherwise the matrix formed by the interpolation vectors will be highly ill-conditioned. The QRCP procedure satisfies both (1) and (2) simultaneously. On the other hand, the conditions above can also be satisfied through a much simpler centroidal Voronoi tessellation (CVT) procedure applied to a weight vector, such as the electron density [9]. More specifically, in the centroidal Voronoi tessellation (CVT) approach, we may partition the grid points in the global domain into N_μ Voronoi cells. The interpolation points can then be simply chosen to be the centroids corresponding to each cell. The CVT procedure can be effectively implemented through a k-Means algorithm [41]. Besides reduction of the computational cost, the use of a k-Means algorithm also produces a smoother potential energy surface particularly in the context of ab initio molecular dynamics. We refer to [9] for more details of this approach.

8.5 Hybrid Functionals

For rung-4 exchange-correlation functionals, the Kohn–Sham equations take the form

$$H[P]\varphi_i = \left(-\frac{1}{2}\Delta + V_{\text{ext}} + V_{\text{Hxc}}[\rho] + V_{\text{x}}^{\text{EX}}[P]\right)\varphi_i = \varepsilon_i\varphi_i,$$
$$\int \varphi_i^*(\mathbf{r})\varphi_j(\mathbf{r})\,\mathrm{d}\mathbf{r} = \delta_{ij}, \quad P(\mathbf{r},\mathbf{r}') = \sum_{i=1}^{N}\varphi_i(\mathbf{r})\varphi_i^*(\mathbf{r}'). \tag{8.22}$$

Here V_{Hxc} is the Hartree and exchange-correlation contribution from the electron density ρ only, and $V_{\text{x}}^{\text{EX}}[P]$ is derived from the exact exchange functional $E_{\text{x}}^{\text{EX}}[P]$, with kernel

$$V_{\text{x}}^{\text{EX}}[P](\mathbf{r},\mathbf{r}') = -P(\mathbf{r},\mathbf{r}')K(\mathbf{r},\mathbf{r}'). \tag{8.23}$$

Here $K(\mathbf{r},\mathbf{r}')$ is the kernel for the electron-electron interaction. When the (bare) Hartree–Fock exchange is used, $K(\mathbf{r},\mathbf{r}') = 1/|\mathbf{r}-\mathbf{r}'|$ is the Coulomb kernel. When screened Fock exchange interactions [19] are used, K takes the form of a screened Coulomb kernel: $K(\mathbf{r},\mathbf{r}') = \operatorname{erfc}(\alpha_s|\mathbf{r}-\mathbf{r}'|)/|\mathbf{r}-\mathbf{r}'|$, where α_s is called the inverse screening length parameter. Hence the Kohn–Sham Hamiltonian is a nonlocal operator. $V_{\text{x}}^{\text{EX}}[P]$ is also called the Fock exchange operator.

The kernel of $V_{\text{x}}^{\text{EX}}[P]$ is a low rank operator, due to the Hadamard product (i.e., element-wise product) between the kernels of P and K. $V_{\text{x}}^{\text{EX}}[P]$ is also negative semi-definite. In quantum chemistry calculations with a small basis set, denoted by $\Phi = [\varphi_1,\ldots,\varphi_{N_b}]$, the matrix elements of the Fock exchange operator projected to the span of the basis set is often computed:

$$\begin{aligned}(V_{\text{x}}^{\text{EX}}[P])_{pq} &= -\int \varphi_p^*(\mathbf{r})P(\mathbf{r},\mathbf{r}')K(\mathbf{r},\mathbf{r}')\varphi_q(\mathbf{r}')\,\mathrm{d}\mathbf{r}\,\mathrm{d}\mathbf{r}' \\ &= -\sum_{i=1}^{N}\sum_{r,s} c_{r,i}c_{s,i}^*\int \varphi_p^*(\mathbf{r})\varphi_r(\mathbf{r})\varphi_s^*(\mathbf{r}')\varphi_q(\mathbf{r}')K(\mathbf{r},\mathbf{r}')\,\mathrm{d}\mathbf{r}\,\mathrm{d}\mathbf{r}'.\end{aligned} \tag{8.24}$$

Here we have expanded the occupied orbital as

$$\varphi_i(\mathbf{r}) = \sum_p \varphi_p(\mathbf{r})c_{p,i}, \quad i = 1,\ldots,N.$$

The computation of the last term in Eq. (8.24) involves the so-called two-electron integrals, and the computational cost for constructing the Fock exchange matrix $(V_{\text{x}}^{\text{EX}}[P])_{pq}$ scales as $O(N^4)$.

The formation of the Fock exchange matrix allows us to solve the Kohn–Sham equations by directly diagonalizing the Kohn–Sham Hamiltonian. When iterative methods (conjugate gradient, Chebyshev filtering etc) are used, the associated asymptotic complexity can in fact be reduced to $O(N^3)$ *without any approximation.* We only have access to the Fock exchange operator via its application to an occupied orbital φ_j as

$$\left(V_{\mathrm{x}}^{\mathrm{EX}}[P]\varphi_j\right)(\mathbf{r}) = -\sum_{i=1}^{N}\varphi_i(\mathbf{r})\int K(\mathbf{r},\mathbf{r}')\varphi_i^*(\mathbf{r}')\varphi_j(\mathbf{r}')\,\mathrm{d}\mathbf{r}', \tag{8.25}$$

for $P=\sum_{i=1}^{N}|\varphi_i\rangle\langle\varphi_i|$. Here we deliberately distinguish the orbitals in the density matrix ($\{\varphi_i\}$) and the orbitals $V_{\mathrm{x}}^{\mathrm{EX}}$ acts on ($\{\varphi_j\}$) to emphasize that they may correspond to different density matrices before self-consistency is achieved. The integral in Eq. (8.25) can be implemented by solving N Poisson-like equations, with the effective charge given by the pair product $\varphi_i^*(\mathbf{r})\varphi_j(\mathbf{r})$. Let N_g denote the grid size, and the computational cost of solving each Poisson-like equation using FFT scales as $O(N_g \log N_g)$. Therefore the cost for applying the Fock exchange operator to all occupied orbitals scales as $O(N_g \log N_g N^2) \sim O(N^3)$. However, the preconstant of this cubic scaling component can be very large. In practical Hartree–Fock calculations, the application of the Fock exchange operator can often take more than 95% of the overall computational time.

8.5.1 Adaptive Compression Method

The adaptively compressed exchange operator (ACE) method [29] accelerates hybrid functional calculations by reducing the frequency of applying the Fock exchange operator, without compromising the accuracy of the self-consistent solution. This is similar in spirit to the filtering method introduced in Sect. 8.2, which improves the efficiency by reducing the frequency of performing the expensive Rayleigh–Ritz procedure. There is a related approach called the projector based compression of the exchange operator [5, 10], which can also reduce the frequency of applying the Fock exchange operator.

Although the Fock exchange operator $V_{\mathrm{x}}^{\mathrm{EX}}$ is a full rank operator, ACE constructs a low rank surrogate operator, denoted by $\widetilde{V}_{\mathrm{x}}^{\mathrm{EX}}$, to approximate $V_{\mathrm{x}}^{\mathrm{EX}}$. Note that $V_{\mathrm{x}}^{\mathrm{EX}}$ is generally a dense, full rank operator. Hence the low rank surrogate cannot be expected to be accurate when applied to an arbitrary vector. Instead we only require $\widetilde{V}_{\mathrm{x}}^{\mathrm{EX}}$ to be accurate when applied to all occupied orbitals.

Consider a set of orbitals $\{\varphi_i\}_{i=1}^{N}$ which defines implicitly the density matrix P, which further gives $V_{\mathrm{x}}^{\mathrm{EX}}[P]$. We first apply $V_{\mathrm{x}}^{\mathrm{EX}}[P]$ to $\{\varphi_i\}_{i=1}^{N}$ as

$$W_i(\mathbf{r}) = (V_{\mathrm{x}}^{\mathrm{EX}}[P]\varphi_i)(\mathbf{r}) \quad i=1,\ldots,N. \tag{8.26}$$

The adaptively compressed exchange operator should satisfy the conditions

$$(\widetilde{V}_{\mathrm{X}}^{\mathrm{EX}}\varphi_i)(\mathbf{r}) = W_i(\mathbf{r}) \quad \text{and} \quad \widetilde{V}_{\mathrm{X}}^{\mathrm{EX}}(\mathbf{r}, \mathbf{r}') = \left(\widetilde{V}_{\mathrm{X}}^{\mathrm{EX}}(\mathbf{r}', \mathbf{r})\right)^* . \tag{8.27}$$

The choice of this surrogate operator is not unique. One possible choice satisfying the conditions (8.27) is given by

$$\widetilde{V}_{\mathrm{X}}^{\mathrm{EX}}(\mathbf{r}, \mathbf{r}') = \sum_{i,j=1}^{N} W_i(\mathbf{r}) B_{ij} W_j^*(\mathbf{r}'), \tag{8.28}$$

where $B = M^{-1}$ is a negative definite matrix, and

$$M_{kl} = \int \varphi_k^*(\mathbf{r}) W_l(\mathbf{r})\, \mathrm{d}\mathbf{r}. \tag{8.29}$$

Let us perform Cholesky factorization for $-M$, i.e., $M = -LL^*$, where L is a lower triangular matrix. Then we get $B = -L^{-*}L^{-1}$. Define the projection vector in the ACE formulation as

$$\xi_k(\mathbf{r}) = \sum_{i=1}^{N} W_i(\mathbf{r})(L^{-*})_{ik}. \tag{8.30}$$

The adaptively compressed exchange operator is then given by

$$\widetilde{V}_{\mathrm{X}}^{\mathrm{EX}}(\mathbf{r}, \mathbf{r}') = -\sum_{k=1}^{N} \xi_k(\mathbf{r}) \xi_k(\mathbf{r}'). \tag{8.31}$$

Once ACE is constructed, the cost of applying $\widetilde{V}_{\mathrm{X}}^{\mathrm{EX}}$ to any orbital φ is similar to the application of a nonlocal pseudopotential operator, thanks to its low-rank structure. ACE only needs to be constructed once per outer iteration, and we can repeatedly use the operator for all the subsequent SCF iterations, until the exchange operator is updated. The ACE formulation has been integrated in electronic structure software packages such as Quantum ESPRESSO [7, 13] and ABINIT [18].

Table 8.2 demonstrates the accuracy of the ACE formulation for Kohn–Sham DFT calculations with the HSE06 [19] hybrid functional, for silicon systems ranging from 64 to 1000 atoms. The ACE formulation can evaluate the HF energy and energy gap accurately (total energy difference is under 10^{-4} Hartree) even for large systems. Furthermore, ACE can perform hybrid functional calculations at a fraction of the cost of conventional methods [20].

The efficiency of the ACE formulation also rests on the assumption that the magnitude of the Fock exchange operator is relatively small compared to other components of the Hamiltonian. In particular, in linearized Hartree–Fock-like

Table 8.2 Comparison between the conventional hybrid DFT calculations and ACE enabled hybrid DFT calculations in terms of the HF energy E_x^{EX} (Hartree) and the energy gap E_{gap} (Hartree) for the Si_{64}, Si_{216}, Si_{512} and Si_{1000} systems. The corresponding relative errors of the HF energy are shown in parentheses (Credit: [20])

Methods	ACE HSE06		Conventional HSE06	
systems	E_x^{EX}	E_{gap}	E_x^{EX}	E_{gap}
Si_{64}	-13.541616 (10^{-6})	1.488335	-13.541629	1.488352
Si_{216}	-45.471192 (10^{-7})	1.449790	-45.471190	1.449790
Si_{512}	-107.698011 (10^{-7})	1.324901	-107.698016	1.324902
Si_{1000}	-210.300628 (10^{-6})	1.289162	-210.300524	1.289128

equations, the convergence properties of the adaptive compression formulation can be rigorously analyzed for Hamiltonians of the form $H = A+B$, where the operator norm $\|A\|_2$ is much larger than that of $\|B\|_2$. We refer readers to [31] for details.

8.5.2 *Interpolative Separable Density Fitting Method*

Since Eq. (8.25) involves solving Poisson-like equations for all pair products of occupied orbitals, we may reduce the cost by using the interpolative separable density fitting (ISDF) technique in Sect. 8.4. Using the decomposition (8.17), we can rewrite (8.25) as

$$\begin{aligned}\left(V_x^{EX}[P]\varphi_j\right)(\mathbf{r}) &\approx -\sum_{i=1}^{N}\sum_{\mu=1}^{N_\mu}\varphi_i(\mathbf{r})\left(\int K(\mathbf{r},\mathbf{r}')\zeta_\mu(\mathbf{r}')\,\mathrm{d}\mathbf{r}'\right)\varphi_i^*(\widehat{\mathbf{r}}_\mu)\varphi_j(\widehat{\mathbf{r}}_\mu)\\ &:= -\sum_{\mu=1}^{N_\mu} P(\mathbf{r},\widehat{\mathbf{r}}_\mu)V_\mu^\zeta(\mathbf{r})\varphi_j(\widehat{\mathbf{r}}_\mu).\end{aligned} \tag{8.32}$$

Here $V_\mu^\zeta(\mathbf{r}) = \int K(\mathbf{r},\mathbf{r}')\zeta_\mu(\mathbf{r}')\,\mathrm{d}\mathbf{r}'$ is the Coulomb-like potential with the interpolation vector ζ_μ being the charge-like quantity.

One may also combine ISDF and ACE together to further improve the accuracy. Note that the matrix M in Eq. (8.29) is negative semidefinite. Since the ISDF compression introduces numerical error, if we directly use Eq. (8.29) to evaluate M, the resulting matrix may be neither Hermitian nor negative semidefinite, which leads to numerical instability. One way to overcome this problem is to apply the

ISDF decomposition to both sides of the kernel K as

$$
\begin{aligned}
M_{ij} &\approx -\sum_{l=1}^{N}\sum_{\mu,\nu=1}^{N_\mu}\left(\int \zeta_\mu^*(\mathbf{r})K(\mathbf{r},\mathbf{r}')\zeta_\nu(\mathbf{r}')\,\mathrm{d}\mathbf{r}\,\mathrm{d}\mathbf{r}'\right)\varphi_l(\mathbf{r}_\mu)\varphi_l^*(\mathbf{r}_\nu)\varphi_i^*(\mathbf{r}_\mu)\varphi_j(\mathbf{r}_\nu)\\
&= -\sum_{\mu,\nu=1}^{N_\mu}\left(\int \zeta_\mu^*(\mathbf{r})K(\mathbf{r},\mathbf{r}')\zeta_\nu(\mathbf{r}')\,\mathrm{d}\mathbf{r}\,\mathrm{d}\mathbf{r}'\right)P(\mathbf{r}_\mu,\mathbf{r}_\nu)\varphi_i^*(\mathbf{r}_\mu)\varphi_j(\mathbf{r}_\nu)\\
&:= \sum_{\mu,\nu=1}^{N_\mu}\varphi_i^*(\mathbf{r}_\mu)\widetilde{M}_{\mu\nu}\varphi_j(\mathbf{r}_\nu),
\end{aligned}
\tag{8.33}
$$

Then M is Hermitian negative semidefinite, which directly follows from the fact that $\widetilde{M}$ is constructed to be Hermitian negative semidefinite. We remark that although the number of Poisson-like equations is reduced from $O(N^2)$ to $O(N)$, the cost of ISDF-based hybrid functional calculation still scales cubically due to the other matrix-matrix multiplication and inversion operations. However, the preconstant associated with such operations can be smaller than that associated with solving Poisson-like equations. Hence ISDF is effective when it is relatively costly to solve Poisson-like equations, such as in the context of planewave methods, but even more so for finite difference and finite element methods.

As an example, the accuracy of ACE-ISDF in hybrid functional calculation for the semiconducting Si_{216} system (with 216 atoms) and metallic $Al_{176}Si_{24}$ system (with 200 atoms) respectively are shown in Fig. 8.4. The accuracy is controlled by a single rank parameter c, which sets the number of interpolation points to be $N_\mu = cN_e$. The accuracy of ACE-ISDF improves systematically as the rank parameter c increases. When the rank parameter is large enough (≥ 20.0), the results are fully comparable to those obtained from the benchmark calculations. Furthermore, for a

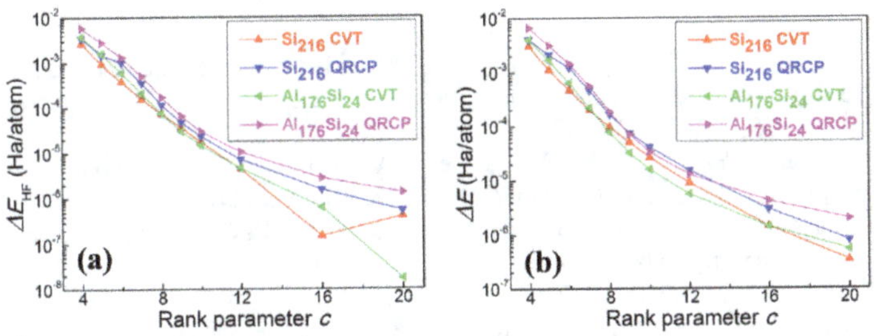

Fig. 8.4 The accuracy of ACE-ISDF-based hybrid functional calculations (HSE06) obtained by using the centroidal Voronoi tessellation (CVT) and QR decomposition with column pivoting (QRCP) procedures to select the interpolation points, with varying rank parameter c from 4 to 20 for Si_{216} and $Al_{176}Si_{24}$, including the error of (**a**) Hartree–Fock exchange energy ΔE_{HF} (Ha/atom) and (**b**) total energy ΔE (Ha/atom) (Credit: [9])

moderate choice of the rank parameter $c = 6.0$, the error of the energy per atom reaches below the chemical accuracy of 1 kcal/mol. This demonstrates that ACE-ISDF is applicable to both insulating and metallic systems.

8.6 Conclusion and Future Directions

The past decade has witnessed significant progress in accelerating the evaluation of the Kohn–Sham map and the generalized Kohn–Sham map in large scale electronic structure calculations. The methods reviewed in this paper are heavily biased towards our own work in this direction. The accuracy and the applicability range of Kohn–Sham DFT are ultimately determined by the choice of exchange-correlation functionals. In recent years, an increasing number of numerical calculations are now performed using the more accurate rung-4 and rung-5 functionals. In particular, the family of rung-5 functionals is still under development and the formulation and numerical treatment of self-consistency for such functionals is much more complicated, since these functionals involve virtual orbitals and orbital energies. Most current strategies for rung-5 functional treat them as post-DFT methods: First obtain the effective Hamiltonian and corresponding Kohn–Sham orbitals based on a semi-local or hybrid functional, and then calculate the energy of the rung-5 functional as a post-processing step (using perturbation theory). Self-consistency calculations have been performed using the optimized effective potential (OEP) framework [12, 14, 15] or more recently the generalized optimized effective potential approach [25], but they require considerable numerical effort. An active research direction is to find an efficient approach for a self-consistent treatment of rung-5 functionals for large scale systems.

References

1. H.M. Aktulga, L. Lin, C. Haine, E.G. Ng and C. Yang. Parallel eigenvalue calculation based on multiple shift-invert Lanczos and contour integral based spectral projection method. *Parallel Comput.* **40**, 195–212 (2014).
2. A.S. Banerjee, R.S. Elliott, and R.D. James. A spectral scheme for Kohn–Sham density functional theory of clusters. *J. Comput. Phys.* **287**, 226–253 (2015).
3. A.S. Banerjee, L. Lin, P. Suryanarayana, C. Yang, and J.E. Pask. Two-level Chebyshev filter based complementary subspace method for pushing the envelope of large-scale electronic structure calculations. *J. Chem. Theory Comput.* **14**, 2930 (2018).
4. S. Baroni and P. Giannozzi. Towards very large-scale electronic-structure calculations. *Europhys. Lett.* **17**(6), 547 (1992).
5. N.M. Boffi, M. Jain and A. Natan. Efficient Computation of the Hartree–Fock Exchange in Real-Space with Projection Operators. *J. Chem. Theory Comput.* **12**, 3614–3622 (2016).
6. D.R. Bowler and T. Miyazaki. O(N) methods in electronic structure calculations. *Rep. Prog. Phys.* **75**, 036503 (2012).

7. I. Carnimeo, S. Baroni, and P. Giannozzi. Fast hybrid density-functional computations using plane-wave basis sets. *Electronic Structure* (2018).
8. M. Ceriotti, T.D. Kühne and M. Parrinello. An efficient and accurate decomposition of the Fermi operator. *J. Chem. Phys.* **129**, 024707 (2008).
9. K. Dong, W. Hu and L. Lin. Interpolative separable density fitting through centroidal Voronoi tessellation with applications to hybrid functional electronic structure calculations. *J. Chem. Theory Comput.* **14**, 1311 (2018).
10. I. Duchemin and F. Gygi. A scalable and accurate algorithm for the computation of Hartree–Fock exchange. *Comput. Phys. Commun.* **181**, 855–860 (2010).
11. A. Erisman and W. Tinney. On computing certain elements of the inverse of a sparse matrix. *Comm. ACM* **18**, 177 (1975).
12. T. Fukazawa and H. Akai. Optimized effective potential method and application to static RPA correlation. *J. Phys. Condens. Matter* **27**, 115502 (2015).
13. P. Giannozzi, S. Baroni, N. Bonini, M. Calandra, R. Car, C. Cavazzoni, D. Ceresoli, G.L. Chiarotti, M. Cococcioni, I. Dabo, A. Dal Corso, S. de Gironcoli, S. Fabris, G. Fratesi, R. Gebauer, U. Gerstmann, C. Gougoussis, A. Kokalj, M. Lazzeri, L. Martin-Samos, N. Marzari, F. Mauri, R. Mazzarello, S. Paolini, A. Pasquarello, L. Paulatto, C. Sbraccia, S. Scandolo, G. Sclauzero, A.P. Seitsonen, A. Smogunov, P. Umari and R.M. Wentzcovitch. QUANTUM ESPRESSO: a modular and open-source software project for quantum simulations of materials. *J. Phys.: Condens. Matter* **21**, 395502–395520 (2009).
14. R.W. Godby, M. Schlüter and L.J. Sham. Accurate exchange-correlation potential for Silicon and its discontinuity on addition of an electron. *Phys. Rev. Lett.* **56**, 2415 (1986).
15. R.W. Godby, M. Schlüter and L.J. Sham. Self-energy operators and exchange-correlation potentials in semiconductors. *Phys. Rev. B* **37**, 10159 (1988).
16. S. Goedecker. Linear scaling electronic structure methods. *Rev. Mod. Phys.* **71**, 1085–1123 (1999).
17. S. Goedecker and L. Colombo. Efficient linear scaling algorithm for tight-binding molecular dynamics. *Phys. Rev. Lett.* **73**, 122 (1994).
18. X. Gonze, F. Jollet, F. Abreu Araujo, D. Adams, B. Amadon, T. Applencourt, C. Audouze, J.-M. Beuken, J. Bieder, A. Bokhanchuk, E. Bousquet, F. Bruneval, D. Caliste, M. CÃ'tÃ©, F. Dahm, F. Da Pieve, M. Delaveau, M. Di Gennaro, B. Dorado, C. Espejo, G. Geneste, L. Genovese, A. Gerossier, M. Giantomassi, Y. Gillet, D.R. Hamann, L. He, G. Jomard, J. Laflamme Janssen, S. Le Roux, A. Levitt, A. Lherbier, F. Liu, I. Lukačević?, A. Martin, C. Martins, M.J.T. Oliveira, S. Poncé, Y. Pouillon, T. Rangel, G.-M. Rignanese, A.H. Romero, B. Rousseau, O. Rubel, A.A. Shukri, M. Stankovski, M. Torrent, M.J. Van Setten, B. Van Troeye, M.J. Verstraete, D. Waroquiers, J. Wiktor, B. Xu, A. Zhou and J.W. Zwanziger. Recent developments in the ABINIT software package. *Comput. Phys. Commun.* **205**, 106–131 (2016).
19. J. Heyd, G.E. Scuseria and M. Ernzerhof. Hybrid functionals based on a screened coulomb potential. *J. Chem. Phys.* **118**(18), 8207–8215 (2003).
20. W. Hu, L. Lin, A. Banerjee, E. Vecharynski and C. Yang. Adaptively compressed exchange operator for large scale hybrid density functional calculations with applications to the adsorption of water on silicene. *submitted* (2017).
21. W. Hu, L. Lin and C. Yang. Interpolative separable density fitting decomposition for accelerating hybrid density functional calculations with applications to defects in silicon. *J. Chem. Theory Comput.* **13**, 5420 (2017).
22. M. Jacquelin, L. Lin and C. Yang. PSelInv–a distributed memory parallel algorithm for selected inversion: the symmetric case. *ACM Trans. Math. Software* **43**, 21 (2016).
23. M. Jacquelin, L. Lin and C. Yang. PSelInv–a distributed memory parallel algorithm for selected inversion: the non-symmetric case. *Parallel Comput.* **74**, 84 (2018).
24. W. Jia and L. Lin. Robust determination of the chemical potential in the pole expansion and selected inversion method for solving Kohn–Sham density functional theory. *J. Chem. Phys.* **147**, 144107 (2017).

25. Y. Jin, D. Zhang, Z. Chen, N.Q. Su and W. Yang. Generalized optimized effective potential for orbital functionals and self-consistent calculation of random phase approximation. *J. Phys. Chem. Lett.* **8**, 4746–4751 (2017).
26. A.V. Knyazev. Toward the optimal preconditioned eigensolver: Locally optimal block preconditioned conjugate gradient method. *SIAM J. Sci. Comp.* **23**, 517–541 (2001).
27. W. Kohn. Density functional and density matrix method scaling linearly with the number of atoms. *Phys. Rev. Lett.* **76**, 3168–3171 (1996).
28. J. Lee, L. Lin and M. Head-Gordon. Systematically improvable tensor hypercontraction: Interpolative separable density-fitting for molecules applied to exact exchange, second-and third-order møller–plesset perturbation theory. *J. Chem. Theory Comput.* **16**, 243–263 (2019).
29. L. Lin. Adaptively compressed exchange operator. *J. Chem. Theory Comput.* **12**, 2242 (2016).
30. L. Lin, M. Chen, C. Yang and L. He. Accelerating atomic orbital-based electronic structure calculation via pole expansion and selected inversion. *J. Phys.: Condens. Matter* **25**, 295501 (2013).
31. L. Lin and M. Lindsey. Convergence of adaptive compression methods for Hartree–Fock-like equations. *Commun. Pure Appl. Math.* **72**, 0451 (2019).
32. L. Lin, J. Lu, L. Ying, R. Car and W. E. Fast algorithm for extracting the diagonal of the inverse matrix with application to the electronic structure analysis of metallic systems. *Commun. Math. Sci.* **7**, 755 (2009).
33. L. Lin, J. Lu, L. Ying and W. E. Pole-based approximation of the Fermi-Dirac function. *Chin. Ann. Math.* **30B**, 729 (2009).
34. L. Lin, Z. Xu and L. Ying. Adaptively compressed polarizability operator for accelerating large scale ab initio phonon calculations. *Multiscale Model. Simul.* **15**, 29–55 (2017).
35. L. Lin, C. Yang, J. Meza, J. Lu, L. Ying and W. E. SelInv – An algorithm for selected inversion of a sparse symmetric matrix. *ACM. Trans. Math. Software* **37**, 40 (2011).
36. L. Lin, J. Lu and L. Ying. Numerical methods for Kohn–Sham density functional theory. *Acta Numer.* **28**, 405–539 (2019).
37. J. Lu, C.D. Sogge and S. Steinerberger. Approximating pointwise products of Laplacian eigenfunctions. *J. Funct. Anal.* **277**, 3271–3282 (2019).
38. J. Lu and K. Thicke. Cubic scaling algorithm for RPA correlation using interpolative separable density fitting. *J. Comput. Phys.* **351**, 187–202 (2017).
39. J. Lu and L. Ying. Compression of the electron repulsion integral tensor in tensor hypercontraction format with cubic scaling cost. *J. Comput. Phys.* **302**, 329 (2015).
40. J. Lu and L. Ying. Fast algorithm for periodic density fitting for Bloch waves. *Ann. Math. Sci. Appl.* **1**, 321–339 (2016).
41. J. MacQueen. Some methods for classification and analysis of multivariate observations. In: *Proc. of the Fifth Berkeley Symp. On Math. Stat. and Prob.*, volume 1, pp. 281–297 (1967).
42. G.D. Mahan. *Many-particle Physics*. Plenum Pub Corp (2000).
43. A. Marek, V. Blum, R. Johanni, V. Havu, B. Lang, T. Auckenthaler, A. Heinecke, H.-J. Bungartz and H. Lederer. The ELPA library: scalable parallel eigenvalue solutions for electronic structure theory and computational science. *J. Phys.: Condens. Matter* **26**, 213201 (2014).
44. J.E. Moussa. Minimax rational approximation of the Fermi–Dirac distribution. *J. Chem. Phys.* **145**, 164108 (2016).
45. T. Ozaki. Continued fraction representation of the Fermi–Dirac function for large-scale electronic structure calculations. *Phys. Rev. B* **75**, 035123 (2007).
46. M.C. Payne, M.P. Teter, D.C. Allen, T.A. Arias and J.D. Joannopoulos. Iterative minimization techniques for *ab initio* total energy calculation: molecular dynamics and conjugate gradients. *Rev. Mod. Phys.* **64**, 1045–1097 (1992).
47. J.P. Perdew and K. Schmidt. Jacob's ladder of density functional approximations for the exchange-correlation energy. In: *AIP Conference Proceedings*, pp. 1–20 (2001).
48. E. Polizzi. Density-matrix-based algorithm for solving eigenvalue problems. *Phys. Rev. B* **79**, 115112–115117 (2009).

49. G. Schofield, J.R. Chelikowsky and Y. Saad. A spectrum slicing method for the Kohn–Sham problem. *Comput. Phys. Commun.* **183**, 497–505 (2012).
50. J.J. Sylvester. A demonstration of the theorem that every homogeneous quadratic polynomial is reducible by real orthogonal substitutions to the form of a sum of positive and negative squares. *Philos. Mag.* **4**, 138–142 (1852).
51. C. Vömel. ScaLAPACK's MRRR algorithm. *ACM Trans. Math. Software* **37**, 1 (2010).
52. D.B. Williams-Young, P.G. Beckman and C. Yang. A shift selection strategy for parallel shift-invert spectrum slicing in symmetric self-consistent eigenvalue computation. *arXiv preprint arXiv:1908.06043* (2019).
53. V.W. Yu, F. Corsetti, A. García, W.P. Huhn, M. Jacquelin, W. Jia, B. Lange, L. Lin, J. Lu, W. Mi, A. Seifitokaldani, A. Vazquez-Mayagoitia, C. Yang, H. Yang and V. Blum. ELSI: A unified software interface for Kohn–Sham electronic structure solvers. *Computer Phys. Commun.* **222**, 267–285 (2018).
54. H. Zhang, B. Smith, M. Sternberg and P. Zapol. SIPs: Shift-and-invert parallel spectral transformations. *ACM Trans. Math. Software* **33**, 9–19 (2007).
55. Y. Zhou, J.R. Chelikowsky and Y. Saad. Chebyshev-filtered subspace iteration method free of sparse diagonalization for solving the Kohn–Sham equation. *J. Comput. Phys.* **274**, 770–782 (2014).
56. Y. Zhou, Y. Saad, M.L. Tiago and J.R. Chelikowsky. Self-consistent-field calculations using Chebyshev-filtered subspace iteration. *J. Comput. Phys.* **219**, 172–184 (2006).

Chapter 9
Augmented Plane Wave Methods for Full-Potential Calculations

Huajie Chen and Reinhold Schneider

Abstract In this chapter, we review the family of augmented plane wave methods, which is among the most accurate methods for full-potential/all-electron DFT calculations. Further, we view this type of approach as a nonconforming mortar method and provide some rigorous numerical analysis results. This provides an understanding of the efficiency of the methods from a mathematical point of view.

9.1 Introduction

First principles simulations using density functional theory (DFT) [33, 35] have become one of the most frequently used and computationally tractable tools in condensed matter physics. Along with the advances in computing technology that have occurred during the last decade, there have been important algorithmic improvements, particularly for plane wave-based methods. For certain classes of materials it is now feasible to simulate systems containing thousands of atoms in a unit cell. This opens the door for the direct application of these techniques in studying a substantial set of "real materials" problems. Further, it is possible to use first principles calculations to create sophisticated data sets that can parameterize model Hamiltonians, which then can be used to model even more complex materials problems. At this time, DFT practitioners are divided into two nearly disjoint communities: one employing pseudopotentials [44] and relatively simple basis sets, particularly plane waves; and the other using methods with complex but efficient basis sets, such as the augmented plane wave, the muffin-tin orbital and related methods [44]. The latter community has traditionally dominated research on transition metals and their compounds.

H. Chen
School of Mathematical Sciences, Beijing Normal University, Beijing, China
e-mail: chen.huajie@bnu.edu.cn

R. Schneider (✉)
Institut für Mathematik, Technische Universität Berlin, Berlin, Germany
e-mail: schneidr@math.tu-berlin.de

E. Cancès, G. Friesecke (eds.), *Density Functional Theory*, Mathematics and Molecular Modeling, https://doi.org/10.1007/978-3-031-22340-2_9

Plane waves with pseudopotentials are natural methods for periodic systems, which are simple to implement and give relatively accurate simulations. The pseudopotential approximations replace singular nuclear attraction potentials and complicated effects of the motion of core electrons by smooth potentials. They give satisfactory results in many cases, but sometimes fail, especially for transition metals. The mathematical analysis of the pseudopotential approximations is very rare, and we refer to [12, 13, 16, 21] for some recent works. Moreover, the core states are fixed in an atomic reference configuration. The frozen core approximation is generally reliable, but breaks down for some elements with extended core states. The core electrons also have to be considered sometimes and are responsible for some properties. Therefore, the full-potential/all-electron calculations are necessary.

For eigenvalue problems with singular potentials in full-potential calculations, plane waves are inefficient basis functions for describing the cusps and rapidly varying wavefunctions near the nuclei [28, 29, 32]. In contrast, it is observed that a significant part of the rapid oscillations can be captured by atomic orbitals such as Gaussians and Slater-type orbitals [44], which have been widely used in quantum chemistry and studied through rigorous mathematical analysis [8, 17, 38, 39]. Therefore, it would be practically efficient to approximate the wavefunction in a crystal by using combinations of plane waves and appropriate atomic orbitals, which comes to the idea of augmented plane waves (APW) [53], linearized augmented plane waves (LAPW) [51], and their extensions by including local atomic orbitals (APW+lo) [43, 49, 52]. In this chapter, we will mainly discuss this type of augmented plane wave-based approach.

A more realistic scenario is that these traditionally distinct approaches will eventually converge, and there are strong relationships between plane wave pseudopotential methods and the APW-based methods [51]. Both approaches have a common starting point, i.e. a plane wave basis set. Further, both approaches are motivated by the observation that plane waves themselves are inefficient for direct simulations of the wavefunctions in a crystal. In both cases, the basis functions are labeled as plane waves, by wave vector $\boldsymbol{G}$, and the Hamiltonian matrix elements are modified in such a way that rapid convergence with the maximum $|\boldsymbol{G}|$ (energy cutoff) can be obtained. In the plane wave pseudopotential approach, this problem is avoided by replacing the Hamiltonian near the nuclei with a smoother pseudo-Hamiltonian in such a way that the valence energy spectrum is reproduced, but the core states are removed. Although this approach may have offered significant computational advantages, much of the complexity of the method is transferred from the calculation itself to the generation of the pseudopotentials. In the APW-based methods, the plane waves are modified near the atoms rather than the Hamiltonian. This modification/augmentation is such that augmented plane waves with small $|\boldsymbol{G}|$ can reproduce the rapid variations in the wavefunctions. Thus, as in the pseudopotential method, the valence energy spectrum can be reproduced with a low energy cutoff for plane waves. The appearance and success of APW-based methods have already triggered the development of improved pseudopotentials, the so-called projector augmented wave (PAW) method [14, 37], which has been used, for example, in the VASP code [55], see also the discussions in [12, 13, 21].

The goal of this chapter is twofold. First, it provides a detailed and self-contained exposition of the APW-based methods. Secondly, it gives an understanding of this approach from a mathematical point of view. The organization of this chapter is as follows. In Sect. 9.2, we give a self-contained exposition of the APW-based methods. In Sect. 9.3, we discuss the numerical discretizations based on domain decomposition techniques, which are based on a similar idea as the APW-based methods. In Sect. 9.4, we provide some numerical analysis of the APW-based methods. In Sect. 9.5, we give some conclusions and future perspectives.

9.2 The Augmented Plane Wave Methods

The augmented plane wave (APW) method was originally proposed by Slater [53] in 1937. In spite of its demanding computational cost due to the energy dependency, it has been successfully used (see e.g., [23]). Several improvements of the basis set were tried to get rid of the energy dependency, the first really successful one was the linearization scheme introduced by Andersen [1, 2], leading to the linearized augmented plane wave (LAPW) method [34]. It was further developed by including local atomic orbitals (APW+lo) to have enough variational flexibility in the radial basis functions [43, 49, 52]. Several widely used quantum chemistry and solid-state physics software programs are based on these methods, such as *Exciting* [25], *FLEUR* [27], and *WIEN2k* [58].

The philosophy behind the family of APW methods is to simulate many-electron systems by introducing a basis set that is in some ways the "best of both worlds". The smoothly varying parts of the wave functions between the atoms are represented by plane waves, and the rapidly varying parts near the nuclei are represented by radial atomic functions time spherical harmonics. We will review relevant aspects of these methods in the following.

For simplicity, we will consider a linear Schrödinger type equation on $\mathbb{R}^3$. Let $A \in \mathbb{R}^{3\times 3}$ be a non-singular matrix and $\mathbb{L} = A\mathbb{Z}^3$ be the Bravais lattice. Let $\mathbb{L}'$ be the dual/reciprocal lattice of $\mathbb{L}$ and Ω the unit cell of $\mathbb{L}$. Then we consider the eigenvalue problem: Find $(\varepsilon_i, u_i) \in \mathbb{R} \times H^1_{\mathrm{per}}(\Omega)$ such that

$$\left(-\frac{1}{2}\Delta + V_{\mathrm{per}}(\mathbf{r})\right) u_i(\mathbf{r}) = \varepsilon_i u_i(\mathbf{r}) \qquad i = 1, 2, \cdots, \tag{9.1}$$

where $V_{\mathrm{per}}(\mathbf{r})$ is a periodic potential with respect to $\mathbb{L}$ and has singularities at nuclear positions. This linear eigenvalue problem can be viewed as a linearization of the Kohn–Sham equations in DFT [44], and has to be solved to obtain the low-lying eigenvalues in each step of the self-consistent field (SCF) iterations. We mention that the algorithms and analysis discussed in this chapter can be extended to nonlinear Kohn–Sham DFT problems [18, 51].

Throughout this chapter, we shall denote by $r = |\mathbf{r}|$ and $\hat{r} = \mathbf{r}/r$ for a vector $\mathbf{r} \in \mathbb{R}^d$. Moreover, we shall denote $\sum_{l=0}^{\infty}\sum_{m=-l}^{l}$ by $\sum_{lm}$, and $\sum_{l=0}^{L}\sum_{m=-l}^{l}$ by $\sum_{lm}^{L}$ for simplicity. We will denote by $(\cdot, \cdot)$ the inner-product $L^2_{\rm per}(\Omega) \times L^2_{\rm per}(\Omega) \to \mathbb{R}$:

$$(u, v) := (u, v)_{L^2_{\rm per}} := \int_{\Omega} u(\boldsymbol{r})v(\boldsymbol{r})\,\mathrm{d}\boldsymbol{r}.$$

Since we only consider $\mathbb{L}$-periodic functions in this chapter, the subscript $L^2_{\rm per}$ of the inner-product will be suppressed for simplicity.

9.2.1 The APW Method

In the APW method, the unit cell Ω is partitioned into two types of regions (the so-called "muffin-tin" division [44, 51], see Fig. 9.1): (1) spheres $\mathscr{C}_i$ centered at atomic sites $\mathbf{R}_i$ with radius R_i, in which the effective potential is assumed to be spherically symmetric, i.e.

$$V_{\rm per}(\mathbf{r}) = V(r_i) \qquad \text{for } r_i = |\mathbf{r} - \mathbf{R}_i| \leqslant R_i,$$

(2) the remaining interstitial region $\mathscr{D}$. In fact, the muffin-tin approximation is widely used and very good for close packed (fcc and ideal hcp) materials. It is not as good but still reasonable for bcc and related materials [24], and becomes increasingly less reliable as the site symmetry and coordination decrease.

The APW method constructs a set of basis functions that are "augmented" from plane waves. Each individual APW basis function, also labeled by the wave vector $\mathbf{G}$, consists of a single plane wave (with the same wave vector $\mathbf{G}$) in the interstitial region matched to radial functions in the atomic spheres. More precisely,

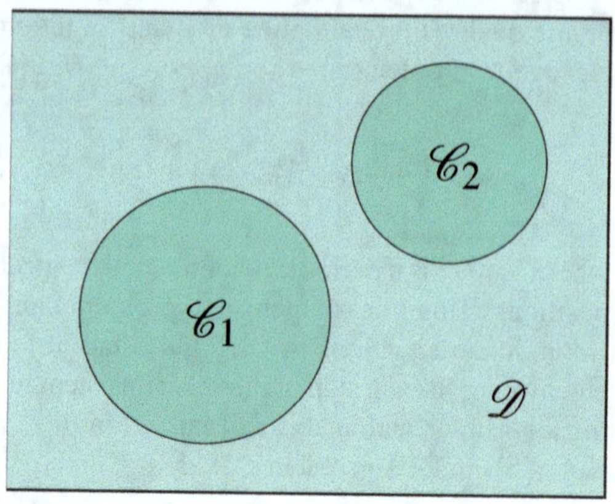

Fig. 9.1 The "muffin-tin" division of the unit cell Ω into spheres $\mathscr{C}_i$ centered at atoms and an interstitial region $\mathscr{D}$

the augmentation from plane waves to APW basis functions $\psi_{\mathbf{G}}^{\varepsilon L}(\mathbf{r})$ can be written as follows

$$|\Omega|^{-\frac{1}{2}} e^{i\mathbf{G}\cdot\mathbf{r}} \rightarrow \psi_{\mathbf{G}}^{\varepsilon L}(\mathbf{r}) := \begin{cases} |\Omega|^{-\frac{1}{2}} e^{i\mathbf{G}\cdot\mathbf{r}} & \text{in } \mathscr{D} \\ \sum_{lm}^{L} \alpha_{lm}^{\mathbf{G}} \chi_l(r_i, \varepsilon) Y_{lm}(\widehat{\mathbf{r}}_i) & \text{in } \mathscr{C}_i \end{cases} \quad \text{for } \mathbf{G} \in \mathbb{L}', \tag{9.2}$$

where L is the truncation of the angular momentum (i.e. $|l| \leqslant L$), $\mathbf{r}_i = \mathbf{r} - \mathbf{R}_i$, $r_i = |\mathbf{r}_i|$, $\widehat{\mathbf{r}}_i = \mathbf{r}_i / r_i$, $Y_{lm}(\widehat{\mathbf{r}})$ denotes the spherical harmonic functions on S^2, χ_l is the solution of the radial Schrödinger equation at energy parameter ε

$$-\frac{1}{2r^2}\frac{\mathrm{d}}{\mathrm{d}r}\left(r^2 \frac{d\chi_l(r,\varepsilon)}{dr}\right) + \left(\frac{l(l+1)}{2r^2} + V(r) - \varepsilon\right)\chi_l(r,\varepsilon) = 0, \tag{9.3}$$

and the coefficients $\alpha_{lm}^{\mathbf{G}}$ are determined such that each angular component of the basis function is matched through the spherical surface. There are two remaining issues to implement the APW method: (a) how to determine the coefficients $\alpha_{lm}^{\mathbf{G}}$, and (b) the choice of energy parameter ε for the basis functions. For simplicity of presentation, we will assume in the following that there is only one atom located at the origin and denoted $\mathscr{C}_i$ by $\mathscr{C}$. All results can be generalized to many-atom systems without any difficulty.

In the APW method, the continuity of the basis functions through the spherical surface $\Gamma(= \partial\mathscr{C})$ is guaranteed by imposing the constraint

$$b(\psi_{\mathbf{G}}^{\varepsilon L}, Y_{lm}) = 0 \quad \forall\, |l| \leqslant L, \quad \text{where } b(v, w) = \int_{\Gamma} w(v^+ - v^-)\, \mathrm{d}\Gamma \tag{9.4}$$

and $v^{\pm}$ are the traces of v taken from outside and inside the sphere, respectively. This is essentially a "weak" continuity that guarantees each Y_{lm} ($|l| \leqslant L$) component is matched at the spherical surface, but the basis functions cannot be exactly continuous due to the truncation $|l| \leqslant L$. Then by using the scattering expansion [45]

$$e^{i\mathbf{G}\cdot\mathbf{r}} = 4\pi \sum_{lm} i^l j_l(Gr) Y_{lm}^*(\widehat{\mathbf{G}}) Y_{lm}(\widehat{\mathbf{r}}), \tag{9.5}$$

one can derive the formula for the coefficients so that the constraint in (9.4) can be satisfied

$$\alpha_{lm}^{\mathbf{G}} = 4\pi i^l j_l(GR) Y_{lm}^*(\widehat{\mathbf{G}}) / \chi_l(R, \varepsilon). \tag{9.6}$$

For the choice of energy parameter ε, if it was taken as a fixed parameter, then the APW method would result in a standard linear eigenvalue problem. The solution/eigenpairs would then yield the band energies and wavefunctions, which is unfortunately not a workable scheme. The APW basis functions are solutions of the Schrödinger equation inside the spheres, but only at the energy parameter ε. They lack variational freedom to allow for changes in the wavefunction as the band energy deviates from this reference. Therefore, ε must be set equal to the band energy ε_i (eigenvalues of (9.1)) to be computed. This means that the energy bands cannot be obtained from a single diagonalization. Rather, it is necessary to solve the determinant as a function of energy parameter ε and determine its roots (or equivalent nonlinear problems [18]), which is a much more computationally demanding procedure.

Another, less serious, difficulty of the APW method is the so-called asymptote problem [51]. The eigenvalue dependent APW basis functions must be evaluated for a large number of parameters, and sometimes one might hit a parameter ε for which $\chi_l(R, \varepsilon)$ equals zero at the spherical surface. Inserting this into (9.6) will yield infinite coefficients $\alpha_{lm}^{\mathbf{G}}$ and cannot match the basis functions at the spherical surface. Therefore, we shall always assume that $\chi_l(R, \varepsilon) \geqslant \epsilon$ with some constant $\epsilon > 0$ for any $|l| \leqslant L$, which can be achieved by varying the radii of the atomic spheres.

9.2.2 *The LAPW Method*

Several modifications of the APW method were proposed with the aim of circumventing the difficulties mentioned above, in particular, to get rid of the energy dependency. The first really successful one was the linearization scheme introduced by Andersen [2]. The LAPW method uses the combination of $\chi_l(r, \varepsilon)$ and its energy derivative at fixed parameter ε as the radial basis function. The energy derivative is defined by

$$\dot{\chi}_l(r, \varepsilon) := \frac{\partial}{\partial t}\chi_l(r, t)|_{t=\varepsilon},$$

where χ_l is kept normalized to the same value in the atomic sphere. The properties of $\dot{\chi}_l(r, \varepsilon)$ can be read in [2, 44, 51]); this function satisfies the following equation

$$\left(-\frac{\mathrm{d}^2}{\mathrm{d}R^2} + \frac{l(l+1)}{r^2} + V(r) - \varepsilon\right) r\dot{\chi}_l(r, \varepsilon) = r\dot{\chi}_l(r, \varepsilon).$$

Then the LAPW basis functions $\zeta_{\mathbf{G}}^{\varepsilon L}(\mathbf{r})$ can be written as

$$\zeta_{\mathbf{G}}^{\varepsilon L}(\mathbf{r}) = \begin{cases} |\Omega|^{-\frac{1}{2}} e^{i\mathbf{G}\cdot\mathbf{r}} & \text{in } \mathscr{D} \\ \displaystyle\sum_{lm}^{L} \Big(\alpha_{lm}^{\mathbf{G}} \chi_l(r_i, \varepsilon) + \beta_{lm}^{\mathbf{G}} \dot{\chi}_l(r_i, \varepsilon)\Big) Y_{lm}(\widehat{\mathbf{r}}_i) & \text{in } \mathscr{C}_i \end{cases} \qquad \text{for } \mathbf{G} \in \mathbb{L}', \tag{9.7}$$

The coefficients $\alpha_{lm}^{\mathbf{G}}$ and $\beta_{lm}^{\mathbf{G}}$ are determined by requiring that the basis functions match both the value and slope, in the weak sense of (9.4).

The LAPW method provides flexible basis functions with fixed parameter ε to properly describe eigenfunctions that correspond to eigenvalues near ε. In contrast to the APW method, this scheme allows us to obtain eigenvalues of (9.1) by solving a linear eigenvalue problem. The LAPW method introduces errors of order $|\varepsilon - \varepsilon_i|^2$ for the eigenfunction in the atomic sphere (which can be easily observed by a Taylor expansion with respect to the energy parameter [44, 51]), and it yields errors of order $|\varepsilon - \varepsilon_i|^4$ for the eigenvalues. Although the LAPW method cannot achieve a systematically convergence due to this error, the LAPW basis functions form a good basis set over a relatively large energy region. In the few instances for which this is not possible, the energy region of interest may be divided into a few (very rarely more than 2) windows and separate calculations need to be carried out for each. This is an enormous simplification over the standard APW method, since the energy independent LAPW basis functions result in a linear eigenvalue problem and the accurate energy bands can be obtained with a single diagonalization.

9.2.3 The APW+lo Method

Singh [50] modified the LAPW approach by adding specially constructed local orbitals $\varphi_{lm}^{\varepsilon'}$, with energy parameters ε', to the basis set

$$\varphi_{lm}^{\varepsilon'}(\mathbf{r}) = \begin{cases} 0 & \text{in } \mathscr{D} \\ \sum\limits_{lm}^{L} \left(\alpha_{lm}^{\varepsilon'} \chi(r, \varepsilon') + \beta_{lm}^{\varepsilon'} \dot{\chi}(r, \varepsilon')\right) Y_{lm}(\widehat{\mathbf{r}}) & \text{in } \mathscr{C}_i \end{cases} \quad \text{for } |l| \leqslant L, \tag{9.8}$$

where the coefficients $\alpha_{lm}^{\varepsilon'}$ and $\beta_{lm}^{\varepsilon'}$ do not depend on the plane wave vector $\mathbf{G}$, but are determined by the requirement that $\varphi_{lm}^{\varepsilon'}$ is zero at the sphere boundary and normalized. Note that different parameters ε' can be chosen so that different states can be described simultaneously.

This method, known as APW+lo [51, 52], was shown to be highly effective in simulations of many systems, especially for materials with large interstitial spaces or mixtures of atoms that require high plane wave cutoffs with those requiring lower cutoffs. The added local atomic orbitals not only permit relaxation of the original basis set without an increase in the plane wave cutoff, but also suggest the possibility of avoiding the non-vanishing error of the LAPW method. Therefore by using a similar construction, the APW+lo method can enhance the accuracy with nearly no extra cost. Moreover, the APW+lo method can be used for general full-potentials, and is not restricted to muffin-tin potentials or other symmetric approximations of the charge density and potential.

9.3 Discretizations Based on Domain Decompositions

The essence and motivation of the APW-based methods have been clearly and concisely stated before: near nuclear positions, the potential and wavefunctions are strongly varying; while in the interstitial space between the nuclei, both the potential and wavefunctions are smoother; accordingly, space is divided into regions and different basis expansions are used in different regions. Similar ideas have long been explored in the community of numerical mathematics: the discretizations based on domain decompositions, which have been widely used to design efficient and parallel algorithms for solving partial differential equations. They use separate finite-dimensional discretization on non-overlapping subdomains, the basis functions on different subdomains do not match on the interface, and the "continuity" of the solution is enforced by additional techniques, for example, Lagrange multipliers, penalties and so on. The domain decomposition technique allows us to benefit from the presence of the subdomains in order to choose the discretization method best adapted to the local behavior of the solution. A further advantage is that more flexible and economical adaptive procedures such as refinements can be made on the subdomains, where needed.

In this section, we will review several widely used discretization methods based on domain decomposition techniques. Since only the mortar method is closely related to the family of APW-based methods (which essentially belong to the mortar method), we will discuss the other two techniques, discontinuous Galerkin (DG) and weak Galerkin (WG) methods, only very briefly.

9.3.1 *The Mortar Method*

The idea of the mortar method is to use different basis functions to approximate the solutions on different subdomains, and then match incompatible discretizations with a suitable variational operator, which ensure an optimal transmission of information between adjacent subdomains (see, e.g., [9–11]).

The discrete problem is constructed via the Galerkin process applied to the variational formulation of the partial differential equations. Even if the local discrete spaces, i.e., the discrete spaces on each subdomain, are included in the local variational spaces, this is usually no longer the case for the global space, since the matching conditions that are enforced on the interfaces between the subdomains are too weak to ensure the conformity. The mortar approximations involve constraints on the space, which is the so-called "weak continuity". The continuity of the solution across the subdomain interfaces is imposed in a weak sense by using the multiplier space. These constraints could be treated as Lagrange multipliers leading to a saddle point problem.

Let $\mathcal{S}(\Omega)$ be the finite-dimensional space spanned by the basis functions on all subdomains, and $\mathcal{T}(\Gamma)$ be the multiplier space on the interface. The discretization of (9.1) by mortar method reads: Find $\varepsilon_i^{\mathrm{M}} \in \mathbb{R}$ and $\varphi_i^{\mathrm{M}} \in \mathcal{S}(\Omega)$, such that $\|\varphi_i^{\mathrm{M}}\|_{L^2(\Omega)} = 1$ and

$$\begin{cases} a(\varphi_i^{\mathrm{M}}, v) = \varepsilon_i^{\mathrm{M}}(\varphi_i^{\mathrm{M}}, v) & \forall\, v \in \mathcal{S}(\Omega) \\ b(\varphi_i^{\mathrm{M}}, w) = 0 & \forall\, w \in \mathcal{T}(\Gamma) \end{cases} \qquad i = 1, 2, \cdots, \tag{9.9}$$

where the bilinear forms a and b are defined by $a(\cdot,\cdot) : \mathcal{S}(\Omega) \times \mathcal{S}(\Omega) \to \mathbb{C}$:

$$a(u, v) := \int_{\mathscr{C}} \nabla u \cdot \nabla v + \int_{\mathscr{D}} \nabla u \cdot \nabla v + \int_{\Omega} V_{\mathrm{per}} u v$$

and $b(\cdot,\cdot) : \mathcal{S}(\Omega) \times \mathcal{T}(\Gamma) \to \mathbb{C}$:

$$b(u, w) := \int_{\Gamma} (u^+ - u^-) w \, \mathrm{d}\Gamma.$$

A key argument is that the choice of integral type matching conditions on the interfaces leads to an optimal evaluation of the consistency error issued from the nonconformity of the discretization [10]. The mortar method has been successfully applied to quantum eigenvalue problems (see, e.g. [20, 31]).

We mention that the family of APW approaches actually belongs to the category of mortar methods. The convergence of APW-based methods will be derived in the framework of the mortar nonconforming methods, see Sect. 9.4.

9.3.2 The Discontinuous Galerkin Method

The DG framework has been widely used in numerical solutions of partial differential equations and investigated theoretically in a lot of works (see, e.g., [6, 7, 15, 54] and references cited therein). For second-order elliptic problems, the development of DG methods is based on the idea of introducing an interior penalty [5, 47, 57]. DG discretization methods have also recently been developed for DFT calculations [40–42, 60].

We shall briefly discuss the DG discretization for (9.1), which is related to the framework in [4]. For respectively vector-valued and scalar-valued functions $\mathbf{w}$ and u which are not continuous on the spherical surface Γ, we define the jumps by

$$[\mathbf{w}] = \mathbf{w}^+ \cdot \mathbf{n}^+ + \mathbf{w}^- \cdot \mathbf{n}^-, \qquad [u] = u^+\mathbf{n}^+ + u^-\mathbf{n}^-$$

and the averages by

$$\{\mathbf{w}\} = \frac{1}{2}(\mathbf{w}^+ + \mathbf{w}^-), \qquad \{u\} = \frac{1}{2}(u^+ + u^-),$$

where $\mathbf{w}^{\pm}$ and $u^{\pm}$ are traces of $\mathbf{w}$ and u on Γ taken from inside and outside the sphere, and $\mathbf{n}^{\pm}$ are the normal unit vectors. Let $\mathcal{S}(\Omega)$ be the finite-dimensional space spanned by the basis functions on all subdomains. We then construct the DG methods for eigenvalue problem (9.1): Find $\varepsilon_i^{\mathrm{DG}} \in \mathbb{R}$ and $\varphi_i^{\mathrm{DG}} \in \mathcal{S}(\Omega)$, such that $\|\varphi_i^{\mathrm{DG}}\|_{L^2(\Omega)} = 1$ and

$$a^{\mathrm{DG}}(\varphi_i^{\mathrm{DG}}, v) = \varepsilon_i^{\mathrm{DG}}(\varphi_i^{\mathrm{DG}}, v) \qquad \forall\, v \in \mathcal{S}(\Omega), \qquad i = 1, 2, \cdots, \tag{9.10}$$

where the bilinear form $a^{\mathrm{DG}}(\cdot,\cdot) : \mathcal{S}(\Omega) \times \mathcal{S}(\Omega) \to \mathbb{C}$ is defined by

$$\begin{aligned} a^{\mathrm{DG}}(u, v) :=& \int_{\mathscr{C}} \left(\frac{1}{2}\nabla u \cdot \nabla v + V_{\mathrm{per}} u v\right) + \int_{\mathscr{D}} \left(\frac{1}{2}\nabla u \cdot \nabla v + V_{\mathrm{per}} u v\right) \\ & - \frac{1}{2}\int_{\Gamma} \{\nabla u\} \cdot [v] \, \mathrm{d}\Gamma - \frac{1}{2}\int_{\Gamma} \{\nabla v\} \cdot [u] \, \mathrm{d}\Gamma + \int_{\Gamma} \sigma [u] \cdot [v] \, \mathrm{d}\Gamma \end{aligned}$$

with $\sigma \in \mathbb{R}^{+}$ the discontinuity-penalization parameter. Note that there are many other types of DG formulations, and (9.10) is the classical symmetric interior penalty method [4, 5, 41]. We refer to [48] for an overview of DG methods.

9.3.3 *The Weak Galerkin Method*

The WG method was first proposed in [56], and further developed in [46, 59, 61]. In the WG method, differential operators are approximated by weak forms as distributions over a set of generalized functions. It has been demonstrated that the WG method is highly flexible and robust as a numerical technique employing discontinuous piecewise polynomials on polygonal or polyhedral finite element partitions. For simplicity of presentations, we will skip the WG formulations of the eigenvalue problems (9.1) but refer to [59] for details.

9.4 Convergence Analysis of the APW-Based Methods

In this chapter, we will provide a mathematical understanding of the APW-based methods. The theory relies on two aspects: (a) the asymptotic regularity of the wavefunctions, and (b) viewing the APW family as mortar methods. Our analysis shows that, in principle, the APW and APW+lo approaches can provide super algebraic and nearly exponential convergence for full-potential calculations. This explains why the APW-based codes are frequently used for sufficiently accurate electronic structure calculations from a numerical point of view.

We shall first review the regularity of the eigenvalue problems (9.1), then construct a non-conforming mortar method based on polynomial radial basis functions and a similar construction of the APW methods, and finally discuss the convergence and error estimates of the APW methods in the framework of mortar methods.

9.4.1 Regularity

In full potential calculations, the potential V_{per} has singularities at nuclear positions, which gives rise to cusps at the singular points of the eigenfunctions. Then a first and natural question to ask is: What is the regularity of the solution of the eigenvalue problem (9.1)? By a careful analysis [28, 29, 32, 52] it has been shown that the solution to such systems is analytic away from the nuclei but satisfies some cusp condition at the nuclear positions, if the exact singular Coulomb potential is used. Due to the cusps at the nuclear positions, the plane wave approximations of (9.1) with singular Coulomb potential cannot achieve a good spectral convergence rate.

In our analysis of APW-based methods, we rely heavily on the regularity result in weighted Sobolev space for Schrödinger type eigenvalue problems developed in [26] (see also [42]). This type of analysis was introduced to investigate singularities of boundary value problems in conical domains with corners and edges (we refer to [19, 22, 30, 36] for more details). In our case the geometry is fairly simple, while the singular electrostatic potential generated by the nuclei fits this treatment perfectly. We will skip the definitions of weighted Sobolev spaces but refer to the above references for details. Instead, we state the following corollary, which can be used directly in the analysis.

Proposition 9.1 *Let φ be an eigenfunction of* (9.1)*. If $V_{\mathrm{per}} \in L^2_{\mathrm{per}}(\Omega)$ is equal to $-Z/r$ ($Z \in \mathbb{R}^+$) in the neighborhood of a nuclear position and is sufficiently smooth away from it, then $\varphi(\mathbf{r}) \in H^s([0, R] \times S^2)$ in $\mathscr{C}$ and $\varphi \in H^s_{\mathrm{per}}(\mathscr{D})$ for any $s \in \mathbb{R}^+$.*

This regularity result indicates that locally, when expressed in spherical coordinates around the nuclei, the solution is infinitely differentiable. Therefore it is actually possible to propose combined approximations which have exponential convergence with respect to the number of degrees of freedom.

9.4.2 A Nonconforming Method

We construct a nonconforming mortar method analogous to the APW methods, but using polynomials as the radial basis functions in the atomic spheres. This method has the same philosophy as the APW methods. Meanwhile, it is easier to see that it has a systematic spectral convergence rate.

We first give the basis functions on the subdomains. Define the space of functions on $\mathscr{D}$ expanded by plane waves

$$\mathcal{P}_K(\mathscr{D}) = \left\{ u \in H^1(\mathscr{D}) : u(\mathbf{r}) = \sum_{|\mathbf{G}| \leqslant K} c_{\mathbf{G}} e^{i\mathbf{G}\cdot\mathbf{r}}|_{\mathscr{D}} \right\}$$

and the space of functions on $\mathscr{C}$ expanded by polynomial times spherical harmonics

$$\mathcal{B}_{NL}(\mathscr{C}) = \left\{ u \in H^1(\mathscr{C}) : u(\mathbf{r}) = \sum_{lm}^{L} \sum_{n=0}^{N} c_{nlm} \chi_n(r) Y_{lm}(\widehat{\mathbf{r}}),\ 0 \leqslant r \leqslant R \right\},$$

where $\{\chi_n\}_{n=0}^N$ forms a basis set of the space of polynomials on $[0, R]$ with degree no more than N. Let

$$S_{NL}^K(\Omega) = \left\{ u \in L^2_{\text{per}}(\Omega) : u|_{\mathscr{C}} \in \mathcal{B}_{NL}(\mathscr{C}) \text{ and } u|_{\mathscr{D}} \in \mathcal{P}_K(\mathscr{D}) \right\} \tag{9.11}$$

and $\varrho = \min\{K, N, L\}$. We may assume in the sequel that $\varrho \geq C \max\{K, N, L\}$ with some constant $C > 0$. With the approximation space $S_{NL}^K(\Omega)$, we can follow (9.9) to write down the nonconforming mortar approximation of (9.1): Find $\varepsilon_{i,\varrho} (= \varepsilon_{i,NL}^K) \in \mathbb{R}$ and $\varphi_{i,\varrho} (= \varphi_{i,NL}^K) \in S_{NL}^K(\Omega)$ such that $\|\varphi_{i,\varrho}\|_{L^2(\Omega)} = 1$ and

$$\begin{cases} a(\varphi_{i,\varrho}, v) = \varepsilon_{i,\varrho}(\varphi_{i,\varrho}, v) & \forall\, v \in S_{NL}^K(\Omega) \\ b(\varphi_{i,\varrho}, Y_{lm}) = 0 & \forall\, |l| \leqslant L \end{cases} \quad i = 1, 2, \cdots. \tag{9.12}$$

We then design an equivalent discretization form that uses the same construction as the APW methods. Let $\{\chi_n\}_{n=0}^N$ be a basis set that spans the space of polynomials on $[0, R]$ with degree at most N and satisfies (see, Fig. 9.2)

$$\chi_n(R) = 0 \quad \text{for} \quad n = 0, \cdots, N-1, \quad \text{and} \quad \chi_N(r) = \frac{r}{R}. \tag{9.13}$$

Note that the subspace $\widetilde{\mathcal{B}}_{NL} \subset \mathcal{B}_{NL}$ defined by

$$\widetilde{\mathcal{B}}_{NL} = \left\{ u \in H_0^1(\mathscr{C}) : u(\mathbf{r}) = \sum_{lm}^{L} \sum_{n=0}^{N-1} c_{nlm} \chi_n(r) Y_{lm}(\widehat{\mathbf{r}}) \right\}$$

is a finite-dimensional subspace of $H_0^1(\mathscr{C})$.

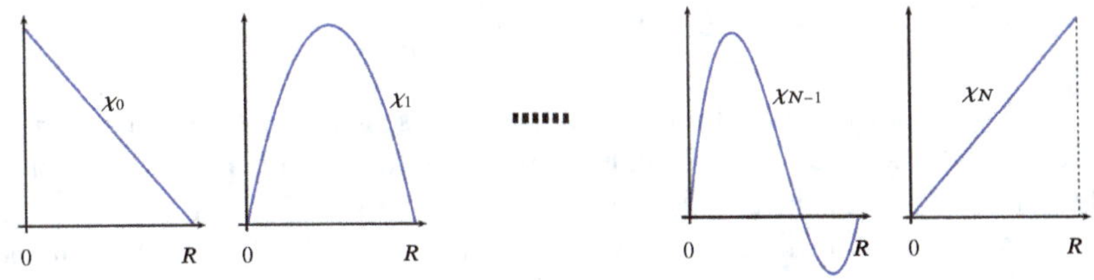

Fig. 9.2 Schematic plots of radial basis functions $\chi_n(r)$ $n = 0, \ldots, N$ satisfying (9.13)

We further augment the plane waves with the above radial basis functions as follows

$$\omega_{\mathbf{G}}(\mathbf{r}) = \begin{cases} |\Omega|^{-\frac{1}{2}} e^{i\mathbf{G}\cdot\mathbf{r}} & \text{in } \mathscr{D} \\ \sum_{lm}^{L} \beta_{lm}^{\mathbf{G}} \chi_N(r_i) Y_{lm}(\widehat{\mathbf{r}}_i) & \text{in } \mathscr{C} \end{cases} \qquad \text{for } \mathbf{G} \in \mathbb{L}', \tag{9.14}$$

where the coefficients

$$\beta_{lm}^{\mathbf{G}} = |\Omega|^{-1/2} 4\pi i^l j_l(GR) Y_{lm}^*(\widehat{\mathbf{G}}) / \chi_N(R)$$

are determined by the continuity constraint (9.4) and the scattering expansion (9.5).

Set $\widetilde{S}_{NL}^K = \widetilde{\mathscr{B}}_{NL} \oplus \text{span}\{\omega_{\mathbf{G}}(\mathbf{r}) : |\mathbf{G}| \leqslant K\}$ and define a direct Galerkin approximation of (9.1): Find $\varepsilon_{i,\varrho} \in \mathbb{R}$ and $\varphi_{i,\varrho} \in \widetilde{S}_{NL}^K(\Omega)$ such that $\|\varphi_{i,\varrho}\|_{L^2(\Omega)} = 1$ and

$$a(\varphi_{i,\varrho}, v) = \varepsilon_{i,\varrho}(\varphi_{i,\varrho}, v) \qquad \forall\, v \in \widetilde{S}_{NL}^K(\Omega) \qquad i = 1, 2, \ldots. \tag{9.15}$$

It is obvious from the definition of $\omega_{\mathbf{G}}$ and comparison of the dimensions (see [18]) that the discretizations (9.12) and (9.15) are equivalent.

With the above constructions, we can then derive the following convergence and a priori error estimate of the approximations, by using a careful numerical analysis and the regularity of the eigenfunctions. We refer to [18] for the proof of this theorem. We mention that the result is not restricted to muffin-tin potentials but is valid for general potentials.

Theorem 9.1 *Let ε_i be an eigenvalue of* (9.1) *with* $\dim\big(\mathscr{R}(\varepsilon_i)\big) = m$, *where $\mathscr{R}(\varepsilon_i)$ denotes the eigenspace of ε_i. If the assumption of Proposition 9.1 for V_{per} is satisfied and ϱ is sufficiently large, then there exist m eigenvalues $\varepsilon_{i1,\varrho}, \ldots, \varepsilon_{im,\varrho}$ of* (9.12), *or* (9.15), *such that*

$$\sup_{1 \leqslant j \leqslant m} |\varepsilon_i - \varepsilon_{ij,\varrho}| \leqslant C_s \varrho^{-(s-3/2)} \qquad \forall\, s > 3/2, \tag{9.16}$$

where C_s is a constant depending only on s, Ω and V_{per}.

9.4.3 Error Estimates of the APW Method

The APW method can be viewed as a modified scheme of the nonconforming method introduced in Sect. 9.4.2, under the assumption that the potential is spherically symmetric inside the atomic sphere $\mathscr{C}$, say $V_{\rm per}(\mathbf{r}) = V(r)$ for $r \leqslant R$.

Compared with the discretization introduced in Sect. 9.4.2, the APW method ignores the subspace $\widetilde{\mathscr{B}}_{NL}$ and replaces the basis function $\omega_{\mathbf{G}}$ in (9.14) by $\psi_{\mathbf{G}}^{\varepsilon L}$ in (9.2), where ε is the eigenvalue to be computed. This leads to a discrete nonlinear eigenvalue problem: Find $\varepsilon_{KL} \in \mathbb{R}$ and $\varphi_{KL} \in S_{KL}^{\varepsilon_{KL}}(\Omega) := \text{span}\{\psi_{\mathbf{G}}^{\varepsilon_{KL}L},\ |\mathbf{G}| \leqslant K\}$, such that

$$a(\varphi_{KL}, v) = \varepsilon_{KL}(\varphi_{KL}, v) \qquad \forall\, v \in S_{KL}^{\varepsilon_{KL}}(\Omega). \tag{9.17}$$

This is a nonlinear eigenvalue problem since the basis functions and the variational subspace depend on the eigenvalue to be computed. We have discussed its disadvantages in Sect. 9.2.

We can show that the solutions of (9.17) have limiting points as K and L go to infinity (see [18]). Let ε_∞ be any accumulation point, i.e., there exists a subsequence of eigenpairs, which we still denote by $\{(\varepsilon_{KL}, \varphi_{KL})\}_{K,L\in\mathbb{N}}$, such that

$$\lim_{K,L\to\infty} |\varepsilon_{KL} - \varepsilon_\infty| = 0.$$

We show in the following result that the limiting point is an eigenpair of (9.1) and the convergence rate for the APW approximations is optimal. We refer to [18] for the proof of this theorem.

Theorem 9.2 *If $V_{\rm per}$ is a spherically symmetric potential in $\mathscr{C}$ and the assumption of $V_{\rm per}$ in Proposition 9.1 is satisfied, then the limiting pair $(\varepsilon_\infty, \varphi_\infty)$ is an eigenpair of (9.1), that is*

$$a(\varphi_\infty, v) = \varepsilon_\infty(\varphi_\infty, v) \qquad \forall\, v \in H^1_{\rm per}(\Omega).$$

Moreover, for the sequence of eigenpairs $\big\{(\varepsilon_{K_jL_j}, \varphi_{K_jL_j})\big\}_{j\in\mathbb{N}}$ that converges to $(\varepsilon_\infty, \varphi_\infty)$, there exists a constant C_s depending only on s, Ω and $V_{\rm per}$, such that

$$|\varepsilon_\infty - \varepsilon_{K_jL_j}| \leqslant C_s\big(K_j^{-(s-1)} + L_j^{-(s-1)}\big) \qquad \forall\, s > 1.$$

When the radius R is well chosen, the assumption of a spherically symmetric potential is reasonable for many systems, as discussed in Sect. 9.2, so this result explains why the APW method has been successfully used in many computations.

9.4.4 Error Estimates of the LAPW Method

Since the accuracy of the LAPW method depends heavily on the choices of parameters, we cannot provide a systematic convergence analysis but only some error estimates.

Compared with the mortar nonconforming method introduced in Sect. 9.4.2, the LAPW method ignores the subspace $\widetilde{\mathcal{B}}_{NL}$ and replaces the basis function $\omega_{\mathbf{G}}$ in (9.14) by $\zeta_{\mathbf{G}}^{\varepsilon L}$ in (9.7), where ε is a fixed parameter. Combined with the error estimate in Theorem 9.2, we can derive for the LAPW approximations that

$$|\bar{\varepsilon} - \varepsilon_{K_j L_j}| \leqslant C_s\big(K_j^{-(s-1)} + L_j^{-(s-1)} + |\bar{\varepsilon} - \varepsilon|^4\big) \qquad \forall\, s > 1,$$

where $\bar{\varepsilon}$ is the eigenvalue to be computed.

Although the errors of the eigenvalue approximations cannot systematically converge to zero and rely on the choice of the parameter ε, the high order of $|\bar{\varepsilon} - \varepsilon|$ means that the LAPW method forms a good basis set over a relatively large eigenvalue region. In most materials, it is quite adequate to choose ε near the center of the region of eigenvalues to be computed. However, in a few instances, there is no single choice of ε that is adequate for all the eigenvalues that must be considered. Then the eigenvalue region may be divided into a few windows and separate computations with different parameters should be carried out for each.

9.4.5 Error Estimates of the APW+lo Method

We shall finally discuss the convergence of APW+lo method. This method adds local basis functions to the LAPW method in order to have enough variational flexibility for the radial basis functions in the atomic spheres.

Compared with the mortar nonconforming method introduced in Sect. 9.4.2, the APW+lo method replaces the basis function $\omega_{\mathbf{G}}$ in (9.14) by $\zeta_{\mathbf{G}}^{\varepsilon L}$ in (9.7) with ε a fixed parameter, and uses the local atomic orbitals $\varphi_{lm}^{\varepsilon'}$ in (9.8) to span the subspace $\widetilde{\mathcal{B}}_{NL}$. We observe that the APW+lo method is very similar to the method in Sect. 9.4.2, the only difference being that the polynomial radial basis functions are replaced by atomic orbitals with compact support. This method not only gets rid of the nonlinearity from energy dependence of the APW basis functions, but also systematically achieves convergence as long as the local atomic orbital sets are well chosen. By using the same arguments as for Theorem 9.1, we can easily justify the similar convergence and error estimate.

As shown by the numerical experiments in [43], the APW+lo method converges practically to the same result as the LAPW method, but allows a significantly smaller basis set (up to 50%) and thus reduces the computational cost drastically.

We finally use the package *Exciting* to perform some full-potential computations for aluminum (Al) and lithium-fluorine (LiF) crystals. *Exciting* is a full-potential

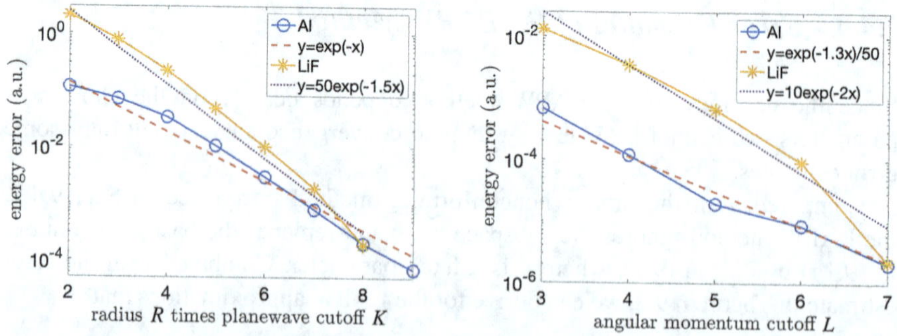

Fig. 9.3 Errors of the ground state energy approximations for Al and LiF, by using the package *Exciting*. Left: decay with respect to the wave vector cutoff. Right: decay with respect to the angular momentum cutoff

DFT package based on the APW+lo method which uses an SCF iteration to solve the nonlinear Kohn–Sham equations. The numerical errors of the ground state energy approximations with respect to the plane wave and angular momentum cutoffs are presented in Fig. 9.3, from which we clearly observe exponential convergence rates with respect to both discretization parameters K and L.

9.5 Conclusions and Perspectives

In this chapter, we have reviewed the family of APW methods that are widely used in full-potential electronic structure calculations. These methods enable highly accurate and reliable simulations for crystals, since they avoid the use of pseudopotentials that might be a source of modeling errors, and which are difficult to control.

We further introduce a nonconforming method based on the idea of the APW methods, which uses polynomial radial basis functions inside the atomic spheres. We obtain an a priori error estimate of the nonconforming approximations and generalize the results to APW-based methods. Due to the fast spectral convergence rate, the APW methods provide high precision using only approximately the same number of degrees of freedom as the plane wave methods.

As an extension, we mention at the end the muffin-tin orbital (MTO) approach [2, 3, 44], which is another type of atomic sphere method similar to the APW methods. It exploits the same idea of dividing the simulation domain and provides efficient representations for atomic-like features that are rapidly varying near each nucleus and smoothly varying between the atoms. The MTO method reformulates the multiple-scattering (MST, also called KKR since it was invented independently by Korringa, Kohn and Rostoker [44]) method, and leads to physically meaningful descriptions of the electronic states in terms of a small basis set of localized, augmented functions. Since the solutions obtained by the MTO method satisfy the equation both inside and outside the spheres, the convergence of the MTO approximations can be proven using similar arguments as those in this chapter.

References

1. O.K. Andersen. Simple approach to the band structure problem. *Solid State Commun.* **13**, 133–136 (1973).
2. O.K. Andersen. Linear methods in band theory. *Phys. Rev. B* **12**, 3060–3083 (1975).
3. O.K. Andersen and R.V. Kasowski. Electronic states as linear combinations of muffin-tin orbitals. *Phys. Rev. B* **4**, 1064–1069 (1971).
4. P.F. Antonietti, A. Buffa and I. Perugia. Discontinuous Galerkin approximation of the Laplace eigenproblem. *Comput. Methods Appl. Mech. Engrg.* **195**(25), 3483–3503 (2006).
5. D.N. Arnold. An interior penalty finite element method with discontinuous elements. *SIAM J. Numer. Anal.* **19**, 742–760 (1982).
6. D.N. Arnold, F. Brezzi, B. Cockburn and D. Marini. Discontinuous Galerkin methods for elliptic problems. In: *Discontinuous Galerkin Methods. Theory, Computation and Applications*, edited by B. Cockburn, G. Karniadakis and C.W. Shu, volume 11, pp. 89–101, Springer-Verlag (2000).
7. I. Babuška and M. Zlámal. Nonconforming elements in the finite element method with penalty. *SIAM J. Numer. Anal.* **10**(5), 863–875 (1973).
8. M. Bachmayr, H. Chen and R. Schneider. Error estimates for hermite and even-tempered Gaussian approximations in quantum chemistry. *Numer. Math.* **128**, 137–165 (2014).
9. F. Ben Belgacem and Y. Maday. The mortar element method for three dimensional finite elements. *ESIAM Math. Model. Numer. Anal.* **31**, 289–302 (1997).
10. C. Bernardi, N. Debit and Y. Maday. Coupling finite element and spectral methods: First results. *Math. Comput.* **54**, 21–39 (1990).
11. C. Bernardi, Y. Maday and A.T. Patera. Domain decomposition by the mortar element method. In: *Asymptotic and Numerical Methods for Partial Differential Equations with Critical Parameters*, edited by H.G. Kaper and M. Garbey, pp. 269–286. ASI Series C 384, Kluwer (1993).
12. X. Blanc, E. Cancès and M.S. Dupuy. Variational projector augmented-wave method. *CR Math.* **355**, 665–670 (2017).
13. X. Blanc, E. Cancès and M.S. Dupuy. Variational projector augmented-wave method: Theoretical analysis and preliminary numerical results. *Numer. Math.* **144**, 271–321 (2020).
14. P.E. Blöchl. Projector augmented-wave method. *Phys. Rev. B* **50**, 17953 (1994).
15. A. Buffa and P. Monk. Error estimates for the ultra weak variational formulation of the Helmholtz equation. *ESAIM Math. Model. Numer. Anal.* **42**(6), 925–940 (2008).
16. E. Cancès and N. Mourad. Existence of a type of optimal norm-conserving pseudopotentials for Kohn–Sham models. *Comm. Math. Sci.* **14**, 1315–1352 (2016).
17. H. Chen and R. Schneider. Error estimates of some numerical atomic orbitals in molecular simulations. *Commun. Comput. Phys.* **18**, 125–146 (2015).
18. H. Chen and R. Schneider. Numerical analysis of augmented plane waves methods for full-potential electronic structure calculations. *ESIAM: Math. Model. Numer. Anal.* **49**, 755–785 (2015).
19. M. Costabel, M. Dauge and S. Nicaise. Analytic regularity for linear elliptic systems in polygons and polyhedra. *Math. Mod. Meth. Appl. Sci.* **22**, 1250015 (2012).
20. L. Drescher. *A Mortar Finite Element Method for Full-Potential Kohn–Sham Density Functional Theory*. Master thesis of Science ETH in Physics (2014).
21. M.S. Dupuy. Projector augmented-wave method: An analysis in a one-dimensional setting. *ESAIM: M2AN* **54**, 25–58 (2020).

22. Y.V. Egorov and B.W. Schulze. *Pseudo-differential Operators, Singularities, Applications*. Birkhäuser, Basel (1997).
23. H. Ehrenreich, F. Seitz and D. Turnbull. *Solid State Physics*. New York, London (1971).
24. N. Elyashar and D.D. Koelling. Effect of non-muffin-tin terms on the electronic structure of transition metals: Niobium. *Phys. Rev. B* **13**, 5362 (1976).
25. The Exciting Code. http://exciting-code.org/.
26. H.J. Flad, R. Schneider and B.W. Schulze. Asymptotic regularity of solutions to Hartree–Fock equations with Coulomb potential. *Math. Meth. Appl. Sci.* **31**(18), 2172–2201 (2008).
27. FLEUR: The Jülich FLAPW Code Family. http://www.flapw.de/pm/.
28. S. Fournais, M. Hoffmann-Ostenhof, T. Hoffmann-Ostenhof and T. Østergaard Sørensen. The electron density is smooth away from the nuclei. *Comm. Math. Phys.* **228**, 401–415 (2002).
29. S. Fournais, M. Hoffmann-Ostenhof, T. Hoffmann-Ostenhof and T. Østergaard Sørensen. Analyticity of the density of electronic wavefunctions. *Arkiv för Matematik* **42**, 87–106 (2004).
30. P. Grisvard. *Singularities in boundary value problems*. volume 22 of Research in Applied Mathematics, Masson, Paris (1992).
31. Y. Guo, L. Jia, H. Chen, H. Li and Z. Zhang. A mortar spectral element method for full-potential electronic structure calculations. *Commun. Comput. Phys.* **29**, 1541–1569 (2021).
32. M. Hoffmann-Ostenhof, T. Hoffmann-Ostenhof and T. Østergaard Sørensen. Electron wavefunctions and densities for atoms. *Annales Henri Poincaré* **2**, 77–100 (2001).
33. P. Hohenberg and W. Kohn. Inhomogeneous electron gas. *Phys. Rev. B* **136**, 864–871 (1964).
34. D.D. Koelling and G.O. Arbman. Use of energy derivative of the radial solution in an augmented plane wave method: application to copper. *J. Phys. F: Metal Phys.* **5**, 2041–2054 (1975).
35. W. Kohn and L.J. Sham. Self-consistent equations including exchange and correlation effects. *Phys. Rev. A* **140**, 1133–1138 (1965).
36. V. Kozlov, V.G. Mazya and J. Rossmann. *Elliptic Boundary Value Problems in Domains with Point Singularities*. American Mathematical Society (1997).
37. G. Kresse and J. Joubert. From ultrasoft pseudopotentials to the projector augmented wave method. *Phys. Rev. B* **59**, 1758 (1999).
38. W. Kutzelnigg. Theory of the expansion of wave functions in a gaussian basis. *Int. J. Quantum Chem.* **51**, 447–463 (1994).
39. W. Kutzelnigg. Convergence of expansions in a gaussian basis. In: *Strategies and Applications in Quantum Chemistry Topics in Molecular Organization and Engineering*, edited by M. Defranceschi and Y. Ellinger, pp. 79–101, Springer (2002).
40. X. Li and H. Chen. A discontinuous Galerkin scheme for full-potential electronic structure calculations. *J. Comput. Phys.* **385**, 33–50 (2019).
41. L. Lin, J. Lu, L. Ying and W. E. Adaptive local basis set for Kohn–Sham density functional theory in a discontinuous Galerkin framework I: Total energy calculation. *J. Comput. Phys.* **231**, 2140–2154 (2012).
42. Y. Maday and C. Marcati. Regularity and hp discontinuous Galerkin finite element approximation of linear elliptic eigenvalue problems with singular potentials. *Math. Mod. Meth. Appl. Sci.* **29**, 1585–1617 (2019).
43. G.K.H. Madsen, P. Blaha, K. Schwarz, E. Sjöstedt and L. Nordström. Efficient linearization of the augmented plane-wave method. *Phys. Rev. B* **64**, 1951341 (2001).
44. R.M. Martin. *Electronic Structure: Basic Theory and Practical Methods*. Cambridge University Press (2004).
45. A. Messiah. *Quantum Mechanics*. Vol. I, Wiley, New York (1964).
46. L. Mu, J. Wang, Y. Wang and X. Ye. A computational study of the weak Galerkin method for second-order elliptic equations. *Numer. Algorithms* **63**, 753–777 (2013).
47. J. Nitsche. On Dirichlet problems using subspaces with nearly zero boundary conditions. In: *The Mathematical Foundations of the Finite Element Method with Applications to Partial Differential Equations*, edited by A.K. Aziz, pp. 603–627, Elsevier (1972).
48. B. Rivière. *Discontinuous Galerkin methods for solving elliptic and parabolic equations*. Society for Industrial and Applied Mathematics (2008).

49. K. Schwarz, P. Blaha and G.K.H. Madsen. Electronic structure calculations of solids using the WIEN2k package for material sciences. *Comput. Phys. Commun.* **147**, 71–76 (2002).
50. D.J. Singh. Ground-state properties of lanthanum: Treatment of extended-core states. *Phys. Rev. B* **43**, 6388 (1991).
51. D.J. Singh and L. Nordstrom. *Planewaves, Pseudopotentials, and the LAPW Method*. Springer-Berlin (2006).
52. E. Sjöstedt, L. Nordström and D.J. Singh. An alternative way of linearizing the APW method. *Solid State Commun.* **114**, 15–20 (2000).
53. J.C. Slater. Wave functions in a periodic potential. *Phys. Rev.*, **51**, 846–851 (1937).
54. E. Süli, C. Schwab and P. Houston. hp-DGFEM for partial differential equations with nonnegative characteristic form. In: *Discontinuous Galerkin Finite Element Methods. Theory, Computation and Applications*, edited by B. Cockburn, G. Karniadakis and C.W. Shu, Springer-Verlag (2000).
55. VASP. https://www.vasp.at/.
56. J. Wang and X. Ye. A weak Galerkin finite element method for second-order elliptic problems. *J. Comput. Appl. Math.* **241**, 103–115 (2013).
57. M.F. Wheeler. An elliptic collocation-finite element method with interiorppenalties. *SIAM J. Numer. Anal.* **15**, 152–161 (1978).
58. WIEN2k. http://www.wien2k.at/.
59. Q. Zhai, H. Xie, R. Zhang and Z. Zhang. The weak Galerkin method for elliptic eigenvalue problems. *Commun. Comput. Phys.* **26**, 160–191 (2019).
60. G. Zhang, L. Lin, W. Hu, C. Yang and J.E. Pask. Adaptive local basis set for Kohn-Sham density functional theory in a discontinuous Galerkin framework II: Force, vibration, and molecular dynamics calculations. *J. Comput. Phys.* **335**, 426–443 (2017).
61. R. Zhang and Q. Zhai. A weak Galerkin finite element scheme for the biharmonic equations by using polynomials of reduced order. *J. Sci. Comput.*, **64**, 559–585 (2015).

Chapter 10
Finite Element Methods for Density Functional Theory

Xiaoying Dai and Aihui Zhou

Abstract In this chapter, we introduce the finite element method, an efficient and widely used discretization tool for the solution of partial differential equations, and apply it to density functional theory-based electronic structure calculations. We focus on the mathematical theory of the finite element method, including the a priori and the a posteriori error estimations of finite element approximations, the convergent rate and complexity of adaptive finite element approximations. In addition, we also address high performance computing for large scale electronic structure calculations based on finite element discretizations.

10.1 Introduction

The physical properties of a system of M nuclei and N interacting electrons can, in principle, be obtained by solving the time-independent Schrödinger equation if the relativistic effect is not taken into account:

$$\left[\sum_{i=1}^{N}\left(-\frac{1}{2}\Delta + v_{\text{ext}}(\mathbf{r})\right) + \frac{1}{2}\sum_{\substack{i,j=1\\ i\neq j}}^{N}\frac{1}{|\mathbf{r}_i - \mathbf{r}_j|}\right]\Psi(\mathbf{r}_1, \mathbf{r}_2, \ldots, \mathbf{r}_N)$$
$$= E^{\text{el}}\Psi(\mathbf{r}_1, \mathbf{r}_2, \ldots, \mathbf{r}_N) \text{ in } \mathbb{R}^{3N}, \tag{10.1}$$

X. Dai · A. Zhou (✉)
LSEC, Institute of Computational Mathematics and Scientific/Engineering Computing, Academy of Mathematics and Systems Science, Chinese Academy of Sciences, Beijing, China
e-mail: daixy@lsec.cc.ac.cn; azhou@lsec.cc.ac.cn

E. Cancès, G. Friesecke (eds.), *Density Functional Theory*, Mathematics and Molecular Modeling, https://doi.org/10.1007/978-3-031-22340-2_10

with the external potential being the Coulomb potential

$$v_{\text{ext}}(\mathbf{r}) = -\sum_{m=1}^{M} \frac{Z_m}{|\mathbf{r} - \mathbf{R}_m|}$$

due to the M nuclei. Here, the spin state has been fixed, Z_m and $\mathbf{R}_m$ are the charges and the positions of the nuclei ($m = 1, 2, \ldots, M$), $\mathbf{r}_i$ are the coordinates of the electrons ($i = 1, 2, \ldots, N$), E^{el} is the total electronic energy of the ground state and Ψ is the associated electronic wave function. Atomic units are used throughout this chapter ($\hbar = m_e = \frac{4\pi e^2}{\epsilon_0} = 1$).

The Schrödinger equation (10.1) is a linear eigenvalue problem defined in $\mathbb{R}^{3N}$, which is of high dimension. In principle, it is an intractable model. Therefore, people have to resort to equivalent or reduced models that are computable. Among them, the density functional theory (DFT) Kohn–Sham model is the most successful one.

Instead of using the many-body wave function as the basic quantity as in the Schrödinger equation, DFT employs the electron density

$$\rho(\mathbf{r}) = N \int \ldots \int \Psi^*(\mathbf{r}, \mathbf{r}_2, \ldots, \mathbf{r}_N)\Psi(\mathbf{r}, \mathbf{r}_2, \ldots, \mathbf{r}_N)d\mathbf{r}_2 \ldots d\mathbf{r}_N$$

as the basic physical variable, and any physical quantities can be obtained from the electron density ρ [51, 60, 64].

We may date DFT back to the early works of Thomas [84] and Fermi [42] in 1927, where the kinetic energy of the system is approximated as an explicit functional of density ρ, which is originally from free electron (or homogeneous non-interacting electron) gas systems. Although their approximations are too rough for modern electronic structure calculations, the approach of using the electron density ρ as the basic physical variable inspired later researchers. In 1964, Hohenberg and Kohn [51] presented their famous work, mainly consisting of the two Hohenberg–Kohn theorems: the Hohenberg–Kohn theorem and the Hohenberg–Kohn variational theorem. The two theorems formulate DFT as an exact theory of many-body systems [60]. We mention that by using the Fundamental Theorem of Algebra, Zhou [102] gave a simple, self-contained and mathematically rigorous proof of the Hohenberg–Kohn theorem in the case of the Coulomb type potential. We refer to [57, 59, 103] and references cited therein, and Chap. 3 by Lewin, Lieb and Seiringer in this volume for more discussions on Hohenberg–Kohn theorems.

The Kohn–Sham models are basic models of DFT and were introduced in [55]. Although the Kohn–Sham equation is much easier to deal with than the original Schrödinger equation, the solution of the Kohn–Sham equation is still full of difficulties and challenges. For instance, the Kohn–Sham equation is a nonlinear eigenvalue problem, the Coulomb potential v_{ext} is singular, the number of eigenpairs required is large, many eigenvalues are degenerate, and the gaps between two eigenvalues are usually of multiscale. We see that its solution requires the design of highly accurate and efficient algorithms for simulating large scale systems [32, 73].

In addition, it is important to provide a mathematical theory for the reliability and efficiency of the numerical approximations in electronic structure calculations. We also note that the error control of the computations is very much a concern of chemists, physicists, materials scientists as well as biologists.

Based on the choice of the basis functions, we may divide the existing discretization methods for solving the Kohn–Sham models into three classes: the plane wave method, the atomic orbital basis function method (local basis set method), and the real space method. Each method has its own advantages and disadvantages. Currently, most of the mature software for first principles electronic structure calculations are based on the plane wave discretization and/or the atomic orbital basis function discretization. The real space methods include the finite difference method, the wavelet method, the finite element method, and the finite volume method. Because of the locality of the real space method, the resulting matrix is sparse, which may make it more competitive than the other discretizations in parallel computing. Indeed, the real space method becomes more and more attractive to DFT calculations. Among these real space methods, due to its simple construction and completeness of the basis functions, and the easy implementation of adaptive computation, the finite element method should have more potential than the others. In addition, the ability of adaptive computation of the finite element discretization makes it convenient to deal with both the full potential calculation and the pseudopotential calculation.

The finite element method [9, 23] is widely used in the solution of partial differential and integral equations and has achieved great success. It has been used in electronic structure calculations since the 1980s. In 1985, Levin and Shertzer [58] used the finite element method on non-uniform meshes to solve the Schrödinger equation of the He atom. In 1993, Hackel et al. [48] applied higher order finite elements to calculate the total energy of the H, H_2^+ and H_3^{2+} molecules based on non-uniform meshes. Ackermann and Roitzsch [1, 2] utilized an adaptive finite element method together with higher order elements to compute the electronic structure of the H_3^{2+} molecule to obtain highly accurate approximations. In 1995, Tsuchida and Tsukada et al. [87] calculated the H_2 molecule and Si atom by using non-uniform meshes. In 1999, Pask et al. [65] used the finite element method to compute the energy band of some solids together with a numerical analysis. Sterne [80] applied a finite element method to compute positron systems with hundreds of atoms. In 2004, Zheng et al. [100, 101] used the polar coordinate-based finite element approach to compute the Schrödinger equation of the H and Li atoms. Tsuchida [86] carried out the ab initio molecular-dynamics simulation for systems with hundreds of atoms based on the finite element discretization. Shen [75] and Zhang [95] introduced an adaptive finite element method together with a two-grid approach to solve systems with tens of atoms. Fang [36] used a hexahedral mesh-based finite element scheme to compute the electronic structure of systems with thousands of atoms using the symmetry of the systems, and has also carried out a relaxation simulation for systems with tens of atoms. In 2011, Pask et al. [67] introduced the PUFEM (Partition-of-unity finite-element method) to the solution of the Kohn–Sham equation. We refer to [14, 16, 17, 25, 37–40, 43, 66, 75, 83, 92, 95] and references therein for more applications of the finite element method in electronic structure calculations.

The numerical analysis of the Kohn–Sham models plays a central role in our understanding of the reliability and efficiency of the numerical methods of electronic structure calculations. Due to the difficulties and challenges, there is very little work on these topics and only in recent decades has some progress been made. The main difficulties involved in the numerical analysis of Kohn–Sham models is that we have to either handle global and nonconvex minimization problems with orthogonality constraints or deal with nonlinear eigenvalue problems with a large number of (possibly degenerate) eigenvalues. Under a coercivity assumption of the so-called second-order optimality condition, Cancès et al. [12] provided a numerical analysis of plane wave approximations and showed that some ground state solutions can be approximated by plane wave solutions with a certain convergence rate. Suryanarayana et al. [83] showed the convergence of ground state energy approximations based on finite element discretizations. In [18], Chen et al. presented a systematic analysis for a general finite-dimensional discretization and proved that all the limit points of finite-dimensional approximations are ground state solutions of the system and some ground state solutions can be well approximated by finite-dimensional solutions if the associated local isomorphism condition is satisfied. They provided not only convergence of the ground state energy approximations but also convergence rates of both eigenvalue and eigenfunction approximations in [18]. It should be pointed out that the local isomorphism condition used in [18] should be very mild and is indeed satisfied if the second-order optimality condition is provided. In further work, Chen et al. [15] carried out an a posteriori analysis for the finite element approximations of the Kohn–Sham equation. For instance, they proposed the residual type a posteriori error estimators and showed that the a posteriori error estimators can be used as the upper and lower bounds of the approximate errors.

We see that the eigenfunctions of (10.8) still vary rapidly around nuclei or chemical bonds even in the pseudopotential setting [7, 26, 47, 99]. It is natural and reasonable to apply adaptive computational approaches to improve the accuracy of the finite element approximations and reduce the computational cost. Indeed, Tsuchida and Tsukada combined the finite element method with the adaptive curvilinear coordinate approach for electronic structure calculations of some molecules [88, 89]. Shen and Zhang introduced some adaptive tetrahedral finite element discretizations in their theses [75, 95] and calculated several typical molecular systems efficiently [47, 76, 96, 97]. Bylaska et al. used an adaptive piecewise linear finite element method on completely unstructured simplex meshes to resolve the rapid variation electronic wave functions around atomic nuclei [10]. Dai et al. designed some parallel adaptive and localization based finite element algorithms for typical quantum chemistry and nanomaterials computations containing more than one thousand atoms using tens of hundreds of processors on computer cluster [25, 26, 29, 31]. Gavini et al. constructed a finite element mesh using an unstructured coarse-graining technique and computed materials systems [62, 83]. Yang successfully scaled their adaptive finite element simulations to over 6000 CPU cores on the Tianhe-1A supercomputer, see his thesis [92]. Bao et al. [6] proposed an adaptive method to solve the Kohn–Sham equation by optimizing the distribution of mesh points without increasing their number. Chen et al. [15] proved not only by numerical experiments but also by theory the robustness and efficiency of the adaptive finite element computations in electronic structure calculations. We refer to [15, 19, 32, 41, 43, 52, 61, 85] and references cited therein for other interesting discussions and investigations on the adaptive finite element method (AFEM).

Although there are many works on applying adaptive finite element approaches to electronic structure calculations, there was no work on carrying out the numerical analysis for adaptive finite element approximations to the Kohn–Sham models until [15]. Under some reasonable assumptions, Chen et al. [15] and Yang and Zhou [91] proved that all limit points of the adaptive finite element approximations are Kohn–Sham ground state solutions. In addition, they also proved that the Kohn–Sham ground state solutions can be well approximated by adaptive finite element approximations with a certain convergence rate, and the computational complexity is quasi-optimal. We refer to [19, 20] for more analyses of AFEM for DFT models.

Due to the needs of science and engineering and the availability of high performance computers, there is a great demand for highly efficient and scalable algorithms for electronic structure calculations. There have been many contributions toward this effort to design high performance algorithms (see, e.g., [27, 28, 36, 49, 61, 63] and references cited therein). We mention here, for instance, the symmetry-based decomposition approach [36, 40], the parallel orbital-updating approach [27, 28, 63], the local and high order finite element discretizations [25, 29], the three-scale schemes [31, 43, 44], and the hexahedron-grid-based two-scale higher order finite element scheme [21]. The first two approaches can be carried out in the two-level parallelization setting.

In this chapter, we expose some existing mathematical theory for the finite element method in DFT calculations, mainly focusing on its numerical analysis, including the a priori error estimate and the a posteriori error estimate for the finite element approximations, the convergence rate and complexity for adaptive finite element approximations. In addition, we also address the symmetry-based decomposition approach and the parallel orbital-updating approaches for high performance computing for large scale electronic structure calculations in modern supercomputers.

The rest of this chapter is organized as follows. In Sect. 10.2, we give a brief introduction to the Kohn–Sham DFT models as well as some necessary preliminaries. In Sect. 10.3, we introduce the standard finite element discretization of the Kohn–Sham equation. Then, we present the existing analysis for the a priori error estimate and the a posteriori error estimate for finite element discretizations of the Kohn–Sham equation in Sect. 10.4. In Sect. 10.5, we provide a mathematical theory of the adaptive finite element approximations for solving the Kohn–Sham equation. We also address the high performance issues for the electronic structure calculations in Sect. 10.6. Finally, we conclude with some remarks.

10.2 Preliminaries

Let $\Omega = \mathbb{R}^3$ or $\Omega \subset \mathbb{R}^3$ be a polyhedral domain. We shall use the standard notation for Sobolev spaces $W^{s,p}(\Omega)$ and their associated norms and seminorms ($s = 1, 2, \cdots, 1 \le p \le \infty$), see, e.g., [3]. For $p = 2$, we denote by $H^s(\Omega) = W^{s,2}(\Omega)$ with the norm $\|\cdot\|_{s,\Omega} = \|\cdot\|_{s,2,\Omega}$. For $s = 1$, $||\cdot||_{s,\Omega} = \left(\int_\Omega |v|^2 + |\nabla v|^2\right)^{1/2}$ and $H_0^1(\Omega) = \{v \in H^1(\Omega) : v\,|_{\partial\Omega} = 0\}$, where $v\,|_{\partial\Omega} = 0$ is understood in the sense of trace. The space $H^{-1}(\Omega)$, the dual of $H_0^1(\Omega)$, will also be used. We consider real-valued functions throughout.

For convenience, the symbol $\lesssim$ will be used throughout this paper, and $A \lesssim B$ means that $A \leqslant CB$ for some constant C that is independent of mesh parameters. To address the mathematical theory of the finite element method, we need to introduce several function spaces. Let $\mathscr{P}(p, (c_1, c_2))$ be a class of functions satisfying some growth conditions:

$$\mathscr{P}(p, (c_1, c_2)) =$$
$$\{f : \exists\, a_1, a_2 \in \mathbb{R} \text{ such that } c_1 t^p + a_1 \leqslant f(t) \leqslant c_2 t^p + a_2 \quad \forall t \ge 0\}$$

with $c_1 \in \mathbb{R}$ and $c_2, p \in [0, \infty)$. Let $\mathcal{H} = (H_0^1(\Omega))^N$ be the Hilbert space with H^1 inner product

$$(\Phi, \widetilde{\Phi})_{\mathcal{H}} = \sum_{i=1}^N \int_\Omega \nabla\varphi_i \cdot \nabla\psi_i \text{ for } \Phi = (\varphi_1, \varphi_2, \cdots, \varphi_N),$$
$$\widetilde{\Phi} = (\psi_1, \psi_2, \cdots, \psi_N) \in \mathcal{H}.$$

Let $\mathbb{Q}$ be a subspace with orthonormality constraints:

$$\mathbb{Q} = \{\Phi \in \mathcal{H} : \Phi^T \Phi = I^{N\times N}\},$$

where $\Phi^T\Psi = \left(\int_\Omega \varphi_i \psi_j\right)_{i,j=1}^N \in \mathbb{R}^{N\times N}$. For $\Phi \in \mathcal{H}$ and a subdomain $\omega \subset \Omega$, we define

$$\|\Phi\|_{s,\omega} = \left(\sum_{i=1}^N \|\varphi_i\|_{s,\omega}^2\right)^{1/2}, s = 0, 1;$$

$$\|\Phi\|_{0,p,\omega} = \left(\sum_{i=1}^N \|\varphi_i\|_{0,p,\omega}^p\right)^{1/p}, 1 \le p \le 6,$$

where $||\varphi_i||_{0,p,\omega} = \left(\int_\omega |\varphi_i|^p\right)^{1/p}$ $||\varphi_i||_{s,\omega} = ||\varphi_i||_{s,2,\omega}$ and $||\varphi_i||_{1,2,\omega} = \left(||\varphi_i||^2_{0,2,\omega} + ||\nabla\varphi_i||^2_{0,2,\omega}\right)^{1/2}$. In our discussions, we will use the following sets:

$$\mathcal{S}^{N\times N} = \{M \in \mathbb{R}^{N\times N} : M^T = M\}, \ \mathcal{A}^{N\times N} = \{M \in \mathbb{R}^{N\times N} : M^T = -M\}.$$

For any $\Phi \in \mathbb{Q}$, we may decompose $\mathcal{H}$ into a direct sum of three subspaces (see, e.g., [12, 35]):

$$\mathcal{H} = \mathcal{S}_\Phi \oplus \mathcal{A}_\Phi \oplus \mathcal{T}_\Phi,$$

where $\mathcal{S}_\Phi = \Phi\mathcal{S}^{N\times N}$, $\mathcal{A}_\Phi = \Phi\mathcal{A}^{N\times N}$, and $\mathcal{T}_\Phi = \left\{\widetilde{\Phi} \in \mathcal{H} : \widetilde{\Phi}^T\Phi = 0 \in \mathbb{R}^{N\times N}\right\}$.

10.2.1 Kohn–Sham Models

Based on the Kohn–Sham DFT, the total energy of the system is defined by the following Kohn–Sham energy functional

$$E(\Phi) = \int_{\mathbb{R}^3}\left(\frac{1}{2}\sum_{i=1}^N |\nabla\varphi_i|^2 + v_{\text{ext}}\rho_\Phi + e_{\text{xc}}(\rho_\Phi)\right) + \frac{1}{2}\int_{\mathbb{R}^3\times\mathbb{R}^3} \frac{\rho_\Phi(\mathbf{r})\rho_\Phi(\mathbf{r}')}{|\mathbf{r}-\mathbf{r}'|}\mathrm{d}\mathbf{r}\mathrm{d}\mathbf{r}',$$

where $\Phi = (\varphi_1, \varphi_2, \cdots, \varphi_N)$, $\rho_\Phi(\mathbf{r}) = \sum_{i=1}^N |\varphi_i(\mathbf{r})|^2$ is the electron density (sometimes it is convenient to abuse the notation Φ), $e_{\text{xc}}(\rho_\Phi)$ is the exchange-correlation energy, and v_{ext} is the external potential. For simplicity, we use the assumption that each Kohn–Sham orbital is occupied by one electron, and therefore the number of occupied Kohn–Sham orbitals is also N. Again for simplicity, we will not introduce a new notation but still use N to denote the number of occupied Kohn–Sham orbitals. In most cases, we consider v_{ext} to be the Coulomb potential v_{ne}, which is defined by

$$v_{\text{ne}}(\mathbf{r}) = -\sum_{m=1}^M \frac{Z_m}{|\mathbf{r}-\mathbf{R}_m|}.$$

The ground state charge density of the system is achieved by solving the minimization problem

$$\inf\{E(\Phi) : \Phi \in \mathbb{Q}\}. \tag{10.2}$$

We refer to [5, 18, 83] for a discussion on the existence of a minimizer of problem (10.2). If the energy functional is differentiable, then a minimizer $\Phi =$

$(\varphi_1, \varphi_2, \cdots, \varphi_N)$ of (10.2) should satisfy the following equation (i.e. the Euler–Lagrange equation associated with the minimization problem): Find $\varphi_i(\mathbf{r}) \in H_0^1(\mathbb{R}^3)(i = 1, 2, \ldots, N)$ such that

$$\begin{cases} (-\frac{1}{2}\Delta + v_{\text{eff}}(\rho_\Phi))\varphi_i(\mathbf{r}) = \sum_{j=1}^{N} \lambda_{ij}\varphi_j(\mathbf{r}) \text{ in } \mathbb{R}^3, \\ \int_{\mathbb{R}^3} \varphi_i(\mathbf{r})\varphi_j(\mathbf{r})d\mathbf{r} = \delta_{ij}, i, j = 1, 2, \ldots, N, \end{cases} \tag{10.3}$$

where

$$v_{\text{eff}}(\rho_\Phi) = v_{\text{ext}} + v_{\text{H}}(\rho_\Phi) + v_{\text{xc}}(\rho_\Phi),$$

$$\Lambda = (\lambda_{ij})_{i,j=1}^{N} = \left(\int_{\mathbb{R}^3} \varphi_j H_\Phi \varphi_i\right)_{i,j=1}^{N}$$

is the Lagrange multiplier, and $H_\Phi : H_0^1(\mathbb{R}^3) \to H^{-1}(\mathbb{R}^3)$ is the Kohn–Sham Hamiltonian operator defined by

$$\langle H_\Phi u, v\rangle = \frac{1}{2}(\nabla u, \nabla v) + (v_{\text{eff}}(\rho_\Phi)u, v)$$

for all $u, v \in H_0^1(\mathbb{R}^3)$. Here,

$$v_{\text{H}}(\rho)(\mathbf{r}) = \int_{\mathbb{R}^3} \frac{\rho(\mathbf{r}')}{|\mathbf{r} - \mathbf{r}'|} d\mathbf{r}' \tag{10.4}$$

is the Hartree potential, which is the electrostatic potential generated by the electronic density $\rho(\mathbf{r})$, and $v_{\text{xc}}(\rho) = e'_{\text{xc}}(\rho_\Phi)$ is the exchange-correlation potential.

Note that

$$E(\Phi) = E(\Phi U) = E\big((\sum_{j=1}^{N} u_{ji}\varphi_j)_{i=1}^{N}\big) \quad \forall U = (u_{ij})_{i,j=1}^{N} \in \mathcal{O}^{N\times N}, \tag{10.5}$$

where $\mathcal{O}^{N\times N}$ is the set of orthogonal matrices. We see from (10.5) that if Φ is a minimizer of (10.2), then ΦU is also a minimizer for any orthogonal matrix U.

For any $\Phi \in \mathcal{H}$, we define

$$[\Phi] = \{\Phi U : U \in \mathcal{O}^{N\times N}\}$$

and the set of ground state solutions

$$\mathcal{G} = \{\Phi \in \mathbb{Q} : E(\Phi) = \min_{\widetilde{\Phi} \in \mathbb{Q}} E(\widetilde{\Phi}).$$

Since the electron density ρ_Φ and operator $v_{eff}(\rho_\Phi)$ are also invariant under any unitary transform of Φ, we may diagonalize the Lagrange multiplier Λ. More precisely, there exists a unitary matrix U such that the Lagrange multiplier is diagonal for $\widetilde{\Phi} = \Phi U = (\psi_1, \psi_2, \cdots, \psi_N)$:

$$\int_{\mathbb{R}^3} \psi_j H_{\widetilde{\Phi}} \psi_i = \mu_i \delta_{ij}, i, j = 1, 2, \ldots, N.$$

Consequently, instead of (10.3), we obtain

$$\begin{cases} (-\frac{1}{2}\Delta + v_{\text{eff}}(\rho_{\widetilde{\Phi}}))\psi_i(\mathbf{r}) = \varepsilon_i \psi_i(\mathbf{r}) \text{ in } \mathbb{R}^3, i = 1, 2, \ldots, N, \\ \displaystyle\int_{\mathbb{R}^3} \psi_i(\mathbf{r})\psi_j(\mathbf{r})d\mathbf{r} = \delta_{ij}, i, j = 1, 2, \ldots, N, \end{cases}$$

which is the well-known Kohn–Sham equation. It is a nonlinear eigenvalue problem defined in $\mathbb{R}^3$.

The exact exchange-correlation energy density $e_{\text{xc}}(\rho)$ and exchange-correlation potential $v_{\text{xc}}(\rho)$ are unknown and must be approximated in computations. The commonly used approximations include local density approximation (LDA), local spin density approximation (LSDA), and generalized gradient approximation (GGA) and so on (see, e.g., [60] or Chap. 1 by Toulouse in this volume for more details). We refer to [50] for a comparison of different approximations. Since $e_{\text{xc}} : [0, \infty) \to \mathbb{R}$ does not have an analytical expression, we need to consider some approximations and may mathematically assume throughout this chapter that

$$e_{\text{xc}}(t) \in \mathscr{P}(3, (c_1, c_2)) \text{ with } c_1 \geqslant 0 \quad \text{or} \quad e_{\text{xc}}(t) \in \mathscr{P}(4/3, (c_1, c_2)), \tag{10.6}$$

which is satisfied by many LDAs used in computation.

For convenience, we focus on the finite systems in our following discussion. We mention that similar conclusions may be applied to periodic systems.

Physically, the Kohn–Sham model is set in $\mathbb{R}^3$. However, due to the exponential decay of the ground state wavefunction of the Schrödinger equation for finite systems (cf., e.g., [4, 93]), the solutions of the Kohn–Sham models for finite systems exhibit exponential decay [98, 99], too. Therefore, $\mathbb{R}^3$ is usually replaced by some polyhedral domain $\Omega \subset \mathbb{R}^3$ in practical computations. Consequently, we shall focus on the following Kohn–Sham equation

$$\begin{cases} (-\frac{1}{2}\Delta + v_{\text{eff}}(\rho_\Phi))\varphi_i(\mathbf{r}) = \sum_{j=1}^N \lambda_{ij}\varphi_j(\mathbf{r}) \text{ in } \Omega, i = 1, 2, \dots, N, \\ \int_\Omega \varphi_i(\mathbf{r})\varphi_j(\mathbf{r})d\mathbf{r} = \delta_{ij}, i, j = 1, 2, \dots, N, \\ \varphi_i(\mathbf{r}) = 0 \text{ on } \partial\Omega, i = 1, 2, \dots, N, \end{cases} \tag{10.7}$$

or

$$\begin{cases} (-\frac{1}{2}\Delta + v_{\text{eff}}(\rho))\psi_i(\mathbf{r}) = \varepsilon_i\psi_i(\mathbf{r}) \text{ in } \Omega, i = 1, 2, \dots, N, \\ \int_\Omega \psi_i(\mathbf{r})\psi_j(\mathbf{r})d\mathbf{r} = \delta_{ij}, i, j = 1, 2, \dots, N, \\ \psi_i(\mathbf{r}) = 0, \text{ on } \partial\Omega, i = 1, 2, \dots, N. \end{cases} \tag{10.8}$$

As usual, the ground state charge density can be obtained by solving the lowest N eigenpairs of (10.8) in practice [60].

Although the Kohn–Sham equation is tractable, the solution of the Kohn–Sham equation is still full of difficulties and challenges. For instance, the Kohn–Sham equation is a nonlinear eigenvalue problem, the Coulomb potential $v_{\text{ext}}(\mathbf{r})$ is singular, the number of eigenpairs required is large, some of the eigenvalues are degenerate, and the gaps between two eigenvalues are of multiscale.

10.2.1.1 Hartree Potential

The Hartree potential (10.4) represents the Coulomb interaction between electrons. Instead of using (10.4), we usually prefer to solve the following Poisson equation

$$-\Delta v_H = 4\pi\rho. \tag{10.9}$$

To get a highly accurate approximation, an appropriate boundary value condition is required. Since the Hartree potential $v_H(\rho)(\mathbf{r})$ decays to 0 with speed of $O(\frac{1}{|\mathbf{r}|})$ only, it is not a good choice to simply take 0 as the boundary value. To avoid the long range effect, we introduce a neutralizing charge $\rho_c(\mathbf{r})$ which cancels out $\rho(\mathbf{r})$ in the computational domain so that

$$\int_\Omega (\rho(\mathbf{r}) + \rho_c(\mathbf{r}))\, d\mathbf{r} = 0. \tag{10.10}$$

Denote v_c the potential generated by $\rho_c(\mathbf{r})$, the solution of (10.9) is then replaced by solving the following equivalent equation

$$\begin{cases} -\Delta\widetilde{v} = 4\pi(\rho + \rho_c) \text{ in } \Omega, \\ \widetilde{v} = 0 \text{ on } \partial\Omega \end{cases}$$

and $v_H = \widetilde{v} - v_c$. We refer to [25, 92] for some other choices of the boundary value conditions.

10.2.1.2 Pseudopotential

The Coulomb potential v_{ne} is singular, which makes the eigenfunctions oscillate rapidly near the core regions. Since the states in the core region in many cases have a negligible contribution to the electronic properties of matter, in computations, we replace the Coulomb potential in the core regions by a pseudopotential which is constructed to reproduce the atomic scattering properties: it has a Coulombic form outside the core region but is smoother inside. The remaining states, called valence states, are described by pseudo-wavefunctions which are significantly smoother without loss of accuracy [24, 68].

The pseudopotential consists of two parts: one local part v_{loc} and one nonlocal part v_{nl}. In the pseudopotential setting, the Kohn–Sham equation is still formulated as (10.8), but v_{ext} now becomes $v_{\text{loc}} + v_{\text{nl}}$, N is the number of valence electrons, and $\{\psi\}_{i=1}^N$ is the set of the pseudo-wavefunctions of the valence electrons.

10.2.1.3 Self-consistent Field Iteration

The Kohn–Sham equation is a nonlinear eigenvalue problem. It is usually solved by using a self-consistent field (SCF) iteration approach [56, 60]. Typically one starts with an initial guess for ρ, then calculates the corresponding $v_{\text{eff}}(\rho)$ and solves the Kohn–Sham equation for $\psi_i (i = 1, 2, \cdots, N)$, with which one updates the density and repeats the calculations. This procedure is then repeated until convergence is reached. The following is the general procedure for SCF iteration.

1. Give an initial input charge density ρ_{in}.
2. Compute the effective potential $v_{\text{eff}}(\rho_{\text{in}})$.
3. Solve the following linear eigenvalue problem

$$\begin{cases} (-\frac{1}{2}\Delta + v_{\text{eff}}(\rho_{in}))\psi_i(\mathbf{r}) = \varepsilon_i \psi_i(\mathbf{r}) \text{ in } \Omega, i = 1, 2, \ldots, N, \\ \int_\Omega \psi_i(\mathbf{r})\psi_j(\mathbf{r})d\mathbf{r} = \delta_{ij}, i, j = 1, 2, \ldots, N, \\ \psi_i(\mathbf{r}) = 0 \text{ on } \partial\Omega, i = 1, 2, \ldots, N. \end{cases}$$

4. Compute the new output charge density ρ_{out}.
5. Convergence check: if not converged, use some density mixing method to get a new input charge density ρ_{in}, go to step 2; else, stop.

The variation of the charge density is often used as the criterion for the convergence of the SCF iteration in a quantum chemistry calculation. For the density mixing method in step 5, if we simply use Picard iteration: take ρ_{out} as the initial density of the next iteration, the iteration sequence converges too slowly or even does not converge. Therefore, it is very important to choose a proper density mixing method. Many such density mixing methods have been proposed so far. The most widely used ones include simple mixing [53], Pulay mixing [69, 70], Broyden's mixing method [77, 79], and a modified Broyden mixing method [53, 56]. We refer to [90] for the most recent progress on designing SCF iteration schemes.

10.2.2 Weak Form and Functional Assumption

To state the numerical theory of finite element approximations to the Kohn–Sham equation, we need some notation and assumptions.

Define the trilinear form $a(\cdot;\cdot,\cdot)$ by:

$$a(\rho_\Phi;\varphi,\phi) = \langle H_{\rho_\Phi}\varphi,\phi\rangle = \frac{1}{2}(\nabla\varphi,\nabla\phi) + (v_{\text{eff}}(\rho_\Phi)\varphi,\phi) \quad \forall\varphi,\phi\in H^1(\mathbb{R}^3).$$

We then have the weak form of (10.7) as follows

$$\begin{cases} a(\rho_\Phi;\varphi_i,\phi) = (\sum_{j=1}^N \lambda_{ij}\varphi_j,\phi)\ \forall\phi\in H^1(\Omega), i=1,2,\ldots,N, \\ \int_{\mathbb{R}^3}\varphi_i(\mathbf{r})\varphi_j(\mathbf{r})d\mathbf{r} = \delta_{ij}, i,j=1,2,\ldots,N. \end{cases} \tag{10.11}$$

We call (Λ,Φ) with $\Lambda = (\lambda_{ij})$ and $\Phi = (\varphi_1,\cdots,\varphi_N)$ satisfying (10.11) a Kohn–Sham ground state. Since the Kohn–Sham energy functional is nonconvex and invariant with respect to any unitary transform, we define the set of Kohn–Sham ground states by

$$\Theta = \left\{(\Lambda,\Phi)\in\mathbb{R}^{N\times N}\times\mathbb{Q} : E(\Phi) = \min_{\widetilde{\Phi}\in\mathbb{Q}} E(\widetilde{\Phi}) \text{ and } (\Lambda,\Phi) \text{ solves } (10.11)\right\}.$$

We also define the set of states that satisfy (10.11) as:

$$\mathcal{W} = \{(\Lambda,\Phi)\in\mathbb{R}^{N\times N}\times\mathcal{H} : (\Lambda,\Phi) \text{ solves } (10.11)\}.$$

Note that $\Theta \subsetneq \mathcal{W}$. Sometimes we assume that there is a gap between the energies of corresponding to states in Θ and other states in $\mathcal{W}$:

$$\min_{\widetilde{\Phi}\in\mathbb{Q}} E(\widetilde{\Phi}) < \inf_{(M,\widetilde{\Phi})\in\mathcal{W}\setminus\Theta} E(\widetilde{\Phi}), \tag{10.12}$$

which is reasonable [15].

We see that (10.8) can be transferred to the following weak formulation: Find $(\varepsilon_i, \psi_i) \in \mathbb{R} \times H_0^1(\Omega)(i = 1, \cdots, N)$ such that

$$\begin{cases} a(\rho_{\widetilde{\Phi}}; \psi_i, \varphi) = (\varepsilon_i \psi_i, \varphi) \quad \forall \varphi \in H^1(\Omega), i = 1, 2, \dots, N, \\ \displaystyle\int_{\mathbb{R}^3} \psi_i(\mathbf{r}) \psi_j(\mathbf{r}) \mathrm{d}\mathbf{r} = \delta_{ij}, i, j = 1, 2, \dots, N, \end{cases} \tag{10.13}$$

where $\widetilde{\Phi} = (\psi_1, \cdots, \psi_N)$.

Note that any solution of (10.11) can be obtained from a unitary transform of some solution of (10.13). That is, once we get all solutions of (10.13), we then obtain all solutions of (10.11).

We define the operator $\mathcal{F} : \mathbb{R}^{N \times N} \times \mathcal{H} \to \mathcal{H}^*$ as

$$\langle \mathcal{F}(\Lambda, \Phi), \Gamma \rangle = \sum_{i=1}^{N} \Big(H_\Phi \varphi_i - \sum_{j=1}^{N} \lambda_{ji} \varphi_j, \gamma_i \Big) \quad \forall \Gamma = (\gamma_i)_{i=1}^N \in \mathcal{H}.$$

The Fréchet derivative of $\mathcal{F}$ with respect to Φ at (Λ, Φ), denoted by $\mathcal{F}'_\Phi(\Lambda, \Phi)$: $\mathcal{H} \to \mathcal{H}^*$, is as follows

$$\begin{aligned} \langle \mathcal{F}'_\Phi(\Lambda, \Phi) \widetilde{\Phi}, \Gamma \rangle &= \frac{1}{4} E''(\Phi)(\widetilde{\Phi}, \Gamma) - \sum_{i,j=1}^{N} (\lambda_{ji} \psi_j, \gamma_i) \\ &= \sum_{i=1}^{N} \Big(H_\Phi \psi_i - \sum_{j=1}^{N} \lambda_{ji} \psi_j, \gamma_i \Big) + 4 \sum_{i,j=1}^{N} \big(e''_{\mathrm{xc}}(\rho_\Phi) \varphi_i \psi_i, \varphi_j \gamma_j \big) \\ &\quad + \sum_{i,j=1}^{N} 4D(\varphi_i \psi_i, \varphi_j \gamma_j). \end{aligned}$$

To study the convergence and complexity, we use the following assumptions [18].

Assumptions 10.1

(A1) $|e'_{\mathrm{xc}}(t)| + |t e''_{\mathrm{xc}}(t)| \in \mathscr{P}(p_1, (c_1, c_2))$ for some $p_1 \in [0, 2]$.
(A2) There exists a constant $\alpha \in (0, 1]$ such that

$$|e''_{\mathrm{xc}}(t)| + |t e'''_{\mathrm{xc}}(t)| \lesssim 1 + t^{\alpha - 1} \quad \forall t > 0.$$

(A3) (Λ, Φ) is a solution of (10.3) and there exists a constant $\beta > 0$ depending on (Λ, Φ) such that

$$\inf_{\Gamma \in \mathcal{T}_\Phi} \sup_{\widetilde{\Phi} \in \mathcal{T}_\Phi} \frac{\langle \mathcal{F}'_\Phi(\Lambda, \Phi) \widetilde{\Phi}, \Gamma \rangle}{\|\widetilde{\Phi}\|_{1,\Omega} \|\Gamma\|_{1,\Omega}} \geqslant \beta. \tag{10.14}$$

We remark here that Assumption **(A2)** implies Assumption **(A1)** and the commonly used X_α and LDA exchange-correlation energy functionals satisfy Assumption **(A2)**. Assumption **(A3)** is equivalent to $\mathcal{F}'_\Phi(\Lambda, \Phi)$ being an isomorphism from $\mathcal{T}_\Phi$ to $\mathcal{T}_\Phi$. We observe that if Assumption **(A3)** is satisfied for $\Phi \in \mathbb{Q}$, then Assumption **(A3)** is satisfied for any $\widetilde{\Phi} \in [\Phi]$ with the same constant β, too. A stronger condition than (10.14), namely that

$$\langle \mathcal{F}'_\Phi(\Lambda, \Phi)\Gamma, \Gamma\rangle \geqslant \gamma \|\Gamma\|^2_{1,\Omega} \quad \forall \Gamma \in \mathcal{T}_\Phi,$$

is used in [12, 74], which is satisfied for a linear self-adjoint operator when there is a gap between the Nth lowest eigenvalue and $(N+1)$th eigenvalue [74].

10.3 Finite Element Discretization

Let d_Ω be the diameter of Ω and $\{\mathcal{T}_h\}$ be a shape regular family of nested conforming meshes over Ω with size $h \in (0, d_\Omega)$: there exists a constant γ^* such that

$$\frac{h_\tau}{\rho_\tau} \leqslant \gamma^* \quad \forall \tau \in \mathcal{T}_h, \tag{10.15}$$

where h_τ is the diameter of τ for each $\tau \in \mathcal{T}_h$, ρ_τ is the diameter of the biggest ball contained in τ, and $h = \max\{h_\tau : \tau \in \mathcal{T}_h\}$. Let $\mathcal{E}_h$ denote the set of interior faces of $\mathcal{T}_h$.

Let $S^{h,k}(\Omega)$ be a subspace of continuous functions on Ω such that

$$S^{h,k}(\Omega) = \{v \in C(\bar{\Omega}) : \; v|_\tau \in P^k_\tau \quad \forall \tau \in \mathcal{T}_h\},$$

where P^k_τ is the space of polynomials of degrees no greater than k over τ. Let $S^{h,k}_0(\Omega) = S^{h,k}(\Omega) \cap H^1_0(\Omega)$. To simplify the notation we shall denote $S^{h,k}_0(\Omega)$ by $S^h_0(\Omega)$ and let $V_h = (S^h_0(\Omega))^N$.

Then the finite element approximation of (10.2) is stated as:

$$\inf\{E(\Phi_h) : \Phi_h \in V_h \cap \mathbb{Q}\}. \tag{10.16}$$

We see from [5, 18, 83] that the minimizer of (10.16) exists under condition (10.6). Similarly, the finite element approximation of (10.11) reads

$$\begin{cases} a(\rho_{\Phi_h}; \varphi_{i,h}, v) = \Big(\sum\limits_{j=1}^{N} \lambda_{ji,h}\varphi_{j,h}, v\Big) \quad \forall v \in S^h_0(\Omega), \quad i = 1, 2, \cdots, N, \\ \displaystyle\int_\Omega \varphi_{i,h}\varphi_{j,h} = \delta_{ij}, i, j = 1, 2, \ldots, N, \end{cases} \tag{10.17}$$

with the Lagrange multiplier

$$\Lambda_h = (\lambda_{ij,h})_{i,j=1}^N = \left(\int_\Omega \varphi_{j,h} H_{\Phi_h} \varphi_{i,h}\right)_{i,j=1}^N.$$

Define the set of finite element approximations for Kohn–Sham ground states:

$$\Theta_h = \Big\{(\Lambda_h, \Phi_h) \in \mathbb{R}^{N\times N} \times (\mathbb{Q} \cap V_h) : E(\Phi_h) = \min_{\Psi \in \mathbb{Q}\cap V_h} E(\Psi), (\Lambda_h, \Phi_h) \text{ solves } (10.17)\Big\}$$

and the set of ground state finite element solutions:

$$\mathcal{G}_h = \{\Phi_h \in \mathbb{Q} \cap V_h : E(\Phi_h) = \min_{\widetilde{\Phi}_h \in \mathbb{Q}\cap V_h} E(\widetilde{\Phi}_h)\}.$$

We have from [18] that the finite element approximations are uniformly bounded, i.e., there exists a constant C such that

$$\sup_{(\Lambda_h, \Phi_h) \in \Theta_h, h \in (0, d_\Omega)} (\|\Phi_h\|_{1,\Omega} + |\Lambda_h|) < C.$$

To carry out the numerical analysis, we consider the model (10.17). To get the numerical solution of the Kohn–Sham model, we use the following discretized Kohn–Sham equation

$$\begin{cases} a(\rho_{\widetilde{\Phi}_h}; \psi_{i,h}, \varphi_h) = \varepsilon_{i,h}(\psi_{i,h}, \varphi_h) \ \forall \varphi_h \in S_0^h(\Omega), \ i = 1, 2, \cdots, N, \\ \int_\Omega \psi_{i,h}\psi_{j,h} = \delta_{ij}, \end{cases} \tag{10.18}$$

where $\widetilde{\Phi}_h = (\psi_{1,h}, \cdots, \psi_{N,h})$.

Similarly, any solution of (10.17) can be obtained from a unitary transform of a solution of (10.18). That is, once we get solutions of (10.18), we then obtain solutions of (10.17).

By using the SCF iteration addressed in Sect. 10.2.1.3, the solution of (10.18) may be obtained by the repeated solution of the following linear eigenvalue problem

$$\begin{cases} a(\rho_{in,h}; \psi_{i,h}, \varphi_h) = \varepsilon_{i,h}(\psi_{i,h}, \varphi_h) \ \forall \varphi_h \in S_0^h(\Omega), \ i = 1, 2, \cdots, N, \\ \int_\Omega \psi_{i,h}\psi_{j,h} = \delta_{ij}, \end{cases} \tag{10.19}$$

where $\rho_{in,h}$ is an input charge density approximation.

Assume $\{\varphi_{1,h}, \varphi_{2,h}, \cdots, \varphi_{i,h}, \cdots, \varphi_{N_b,h}\}$ is a basis of $S^{h,k}(\Omega)$, namely,

$$S^{h,k}(\Omega) = \text{span}\{\varphi_{1,h}, \varphi_{2,h}, \cdots, \varphi_{i,h}, \cdots, \varphi_{N_b,h}\},$$

where N_b is the dimension of space $S^{h,k}(\Omega)$. We then see that each $\psi \in S^{h,k}(\Omega)$ can be expressed as

$$\psi = \sum_{j=1}^{N_b} c_j \varphi_{j,h}, \tag{10.20}$$

where $c_j \in \mathbb{R}(j = 1, 2, \cdots, N_b)$.

Inserting (10.20) into (10.19) and choosing φ_h to be $\varphi_{l,h}(l = 1, \cdots, N_b)$, respectively, we obtain the following algebraic eigenvalue problem

$$H\mathbf{u}_{i,h} = \lambda_{i,h} B\mathbf{u}_{i,h}, \tag{10.21}$$

with H and B being the stiffness and mass matrices in the finite element bases, respectively. That is,

$$H = (H_{kl})_{k,l=1}^{N_b}, \quad B = (B_{kl})_{k,l=1}^{N_b},$$

with $H_{kl} = a(\rho_{in,h}; \varphi_{l,h}, \varphi_{k,h})$ and $B_{kl} = (\varphi_{l,h}, \varphi_{k,h})$.

We then call an existing eigensolver, e.g., the Lanczos method [11], the LOBPCG method [54], the Jacobi–Davidson method [78], to solve the algebraic eigenvalue problem (10.21). We refer to [71] for various eigensolvers. Note that the algebraic version of (10.9) under the finite element discretization becomes

$$Ax = b, \tag{10.22}$$

where A is symmetric and positive definite. We may apply a preconditioned conjugate gradient method (PCG) [94] or an algebraic multigrid method [8, 82] to solve (10.22) (cf., e.g., [72] for more details).

10.4 The A Priori and A Posteriori Analysis

The numerical analysis of Kohn–Sham models plays an important part in our understanding of the efficiency of the numerical methods used in electronic structure calculations. However, significant results in this area have only been obtained very recently. The main difficulties involved in the numerical analysis of Kohn–Sham models are that we have to either handle nonconvex minimization problems with orthogonality constraints or deal with nonlinear eigenvalue problems with a large number of (possibly degenerate) eigenvalues.

We point out that a systematic a priori analysis for a general finite-dimensional discretization is presented in [18], where the following results are obtained: all the limit points of finite-dimensional approximations are ground state solutions of the system, and some ground state solution can be well approximated by finite-dimensional solutions if the associated local isomorphism condition is satisfied. Note that not only convergence of ground state energy approximations but also convergence rates of both eigenvalue and eigenfunction approximations are provided in [18].

The a posteriori analysis for finite element discretizations of the Kohn–Sham equation was later provided in [15]. It is proved in [15] that the error of the finite element approximations for the Kohn–Sham equation is bounded from below and above by the residual type a posteriori error estimators which are also proposed in [15]. This is the first result on the a posteriori error estimate for numerical discretizations of the Kohn–Sham equation.

10.4.1 *The a priori Analysis*

In this subsection, we show the a priori error estimates for the finite element approximations, whose proofs can be found in [18]. Indeed, all the analyses in [18] are focused on finite-dimensional approximations in a class of finite-dimensional subspaces $S_n \in H_0^1(\Omega)$ satisfying the following assumption (see also (3.1) of [18])

$$\lim_{n\to\infty} \inf_{\psi\in S_n} \|\psi - \varphi\|_{1,\Omega} = 0 \ \forall \varphi \in H_0^1(\Omega). \tag{10.23}$$

Assumption (10.23) is a very mild assumption satisfied by many typical finite-dimensional spaces used in practice, including the piecewise polynomial finite element spaces [22].

The following convergence conclusions are proved in [18].

Theorem 10.2 *The following convergences hold:*

$$\lim_{h\to 0} d_{\mathcal{H}}(\Theta_h, \Theta) = 0,$$

$$\lim_{h\to 0} E_h = \min_{\Phi\in\mathcal{Q}} E(\Phi),$$

where $E_h = E(\Phi_h)$ for any $\Phi \in \mathcal{G}_h$, and the distance between sets $X, Y \subset \mathbb{R}^{N\times N} \times \mathcal{H}$ is defined by

$$d_{\mathcal{H}}(X, Y) = \sup_{(\Lambda,\Phi)\in X} \inf_{(\mu,\Psi)\in Y} (|\Lambda - \mu| + \|\Phi - \Psi\|_{1,\Omega}).$$

Theorem 10.2 says that all the finite element approximations for eigenvalues and eigenfunctions together with the total energy converge and the limit points are ground state solutions of the system.

It is also shown in [18] that the quadratic convergence rate of ground state energy approximations can be further achieved.

Theorem 10.3 *Let E_0 be the ground state energy of (10.2) and E_h be the ground state energy of (10.16), namely, $E_0 = E(\Phi)$ for all $\Phi \in \mathcal{G}$, and $E_h = E(\Phi_h)$ for all $\Phi_h \in \mathcal{G}_h$. If Assumption **(A1)** holds, then*

$$|E_0 - E_h| \lesssim \mathcal{D}^2_{\mathcal{H}}(\mathcal{G}, \mathcal{G}_h),$$

where the distance between sets $X, Y \subset \mathcal{H}$ is defined by

$$\mathcal{D}_{\mathcal{H}}(X, Y) = \sup_{\Phi \in X} \inf_{\Psi \in Y} \|\Phi - \Psi\|_{1,\Omega}.$$

We observe that Theorem 10.3 is a generalization and improvement of results in [12, 83]. Furthermore, we see that under certain assumptions every ground state solution can be approximated with some approximation rate by finite-dimensional solutions.

For any $(\Lambda, \Phi) \in \mathbb{R}^{N\times N} \times \mathcal{H}$, we define $B_\delta(\Lambda, \Phi)$ as

$$B_\delta(\Lambda, \Phi) := \{(\mu, \Psi) \in \mathbb{R}^{N\times N} \times \mathcal{H} : |\Lambda - \mu| + \|\Phi - \Psi\|_{1,\Omega} \leqslant \delta\}.$$

Let

$$X_{\Phi,h} = \mathcal{S}^{N\times N} \times (V_h \cap (\mathcal{S}_\Phi \oplus \mathcal{T}_\Phi)),$$

we then have the following conclusion (see [18] for the proof.)

Theorem 10.4 *Let $(\Lambda, \Phi) \in \Theta$. If Assumption **(A2)** is true and (Λ, Φ) satisfies Assumption **(A3)**, then there exists a $\delta > 0$ such that for sufficiently small h, (10.17) has a unique local solution $(\Lambda_h, \Phi_h) \in X_{\Phi,h} \cap B_\delta((\Lambda, \Phi))$. Moreover,*

$$\|\Phi - \Phi_h\|_{1,\Omega} \lesssim \inf_{\widetilde{\Phi} \in V_h} \|\Phi - \widetilde{\Phi}\|_{1,\Omega}, \tag{10.24}$$

$$|\Lambda_h - \Lambda| \lesssim \|\Phi_h - \Phi\|^2_{1,\Omega} + \|\Phi_h - \Phi\|_{0,\Omega}, \tag{10.25}$$

$$\|\Phi - \Phi_h\|_{0,\Omega} \lesssim r(h)\|\Phi - \Phi_h\|_{1,\Omega}, \tag{10.26}$$

with $r(h) \to 0$ as $h \to 0$.

10.4.2 The a posteriori Analysis

In this subsection, we turn to the residual type-based a posteriori analysis.

The residual type a posteriori error estimators for the finite element approximation of the Kohn–Sham equation consist of two parts: the residuals and the jumps [15]. We define the element residual $\mathcal{R}_\tau(\Phi_h)$ and the jump $J_e(\Phi_h)$ by

$$\mathcal{R}_\tau(\Phi_h) = \Big(H_{\Phi_h}\varphi_{i,h} - \sum_{j=1}^{N} \lambda_{ji,h}\varphi_{j,h}\Big)_{i=1}^{N} \quad \text{in } \tau \in \mathcal{T}_h,$$

$$J_e(\Phi_h) = \Big(j_e(\varphi_{i,h})\Big)_{i=1}^{N}, \quad j_e(\varphi_{i,h}) = \frac{1}{2}\nabla\varphi_{i,h}|_{\tau_1}\cdot\overrightarrow{n_1} + \frac{1}{2}\nabla\varphi_{i,h}|_{\tau_2}\cdot\overrightarrow{n_2},$$

where e is the common face of elements τ_1 and τ_2 with unit outward normals $\overrightarrow{n_1}$ and $\overrightarrow{n_2}$, respectively.

Let $\omega_h(e)$ be the union of elements that share the face e. For $\tau \in \mathcal{T}_h$, define the local error indicator $\eta_h(\Phi_h, \tau)$ and oscillation $\mathrm{osc}_h(\Phi_h, \tau)$ by

$$\eta_h^2(\Phi_h, \tau) = h_\tau^2\|\mathcal{R}_\tau(\Phi_h)\|_{0,\tau}^2 + \sum_{e\in\mathcal{E}_h, e\subset\partial\tau} h_e\|J_e(\Phi_h)\|_{0,e}^2, \tag{10.27}$$

$$\mathrm{osc}_h(\Phi_h, \tau) = h_\tau\|\mathcal{R}_\tau(\Phi_h) - \overline{\mathcal{R}_\tau(\Phi_h)}\|_{0,\tau}, \tag{10.28}$$

where $\overline{w}$ is the L^2-projection of $w \in L^2(\Omega)$ to polynomials of some fixed degree k on τ or e. Given a subset $\omega \subset \Omega$, define the error estimator $\eta_h(\Phi_h, \omega)$ and oscillation $\mathrm{osc}_h(\Phi_h, \omega)$ by

$$\eta_h^2(\Phi_h, \omega) = \sum_{\tau\in\mathcal{T}_h, \tau\subset\omega} \eta_h^2(\Phi_h, \tau) \quad \text{and} \quad \mathrm{osc}_h^2(\Phi_h, \omega) = \sum_{\tau\in\mathcal{T}_h, \tau\subset\omega} \mathrm{osc}_h^2(\Phi_h, \tau).$$

It is proved in [15] that the errors of the finite element approximations for the Kohn–Sham equation are bounded from below and above by the residual type a posteriori error estimators designed above. Indeed, we have (see [15] for details)

Theorem 10.5 *Suppose $h_0 \ll 1$ and $h \in (0, h_0]$. Let $(\Lambda, \Phi) \in \Theta$. If Assumptions (A2) and (A3) are satisfied, then there exist positive constants C_1, C_2 and C_3 depending on the shape regularity constant γ^* (in (10.15)), such that*

$$\|\Phi - \Phi_h\|_{1,\Omega}^2 \leqslant C_1\eta_h^2(\Phi_h, \Omega),$$

$$C_2\eta_h^2(\Phi_h, \Omega) \leq \|\Phi - \Phi_h\|_{1,\Omega}^2 + C_3\mathrm{osc}_h^2(\Phi_h, \Omega),$$

where $(\Lambda_h, \Phi_h) \in X_{\Phi,h}$ *is the finite element approximation satisfying* (10.24)–(10.26).

Further, we obtain (see, also, [15])

Theorem 10.6 *Suppose* $h_0 \ll 1$ *and* $h \in (0, h_0]$. *Let* $(\mu_h, \widetilde{\Phi}_h)$ *be a solution of (10.18). If Assumptions* **(A2)** *and* **(A3)** *are satisfied, then*

$$d^2_{\mathcal{H}}(\Theta_{\widetilde{\Phi}_h}, \Theta) \lesssim \eta^2_h(\widetilde{\Phi}_h, \Omega),$$

$$\eta^2_h(\widetilde{\Phi}_h, \Omega) \lesssim d^2_{\mathcal{H}}(\Theta_{\widetilde{\Phi}_h}, \Theta) + \mathrm{osc}^2_h(\widetilde{\Phi}_h, \Omega),$$

here

$$\Theta_{\widetilde{\Phi}_h} = \left\{(\Lambda_h, \Phi_h) \in \mathbb{R}^{N\times N} \times (\mathbb{Q} \cap V_h) : \Phi_h \in [\widetilde{\Phi}_h], \text{ and } \Lambda_h = \Phi_h^T H_{\Phi_h} \Phi_h\right\}$$

and $\Theta_{\widetilde{\Phi}_h} \subseteq \Theta_h$.

10.5 Adaptive Finite Element Approximation

Even in the pseudopotential setting, the eigenfunctions of (10.8) still vary rapidly around nuclei and chemical bonds [7, 26, 47, 99]. Hence it is quite natural and reasonable to apply adaptive finite element approaches to improve the approximation accuracy and reduce the computational cost. Although there are many works on implementing adaptive finite element methods for electronic structure calculations, as we have seen in Sect. 10.1, there was no work on numerical analysis for an adaptive finite element discretization of the Kohn–Sham models until [15] appeared. We refer to [19, 20] for the adaptive finite element analysis of orbital-free models.

We note that an adaptive mesh-refining algorithm usually consists of the following loop [13, 30]:

$$\textbf{Solve} \rightarrow \textbf{Estimate} \rightarrow \textbf{Mark} \rightarrow \textbf{Refine}.$$

Let us address each step of the loop in more detail.

Solve This step computes the piecewise polynomial finite element approximation with respect to a given mesh.

Estimate Given a partition $\mathcal{T}_h$ and the corresponding output (Λ_h, Φ_h) from the "Solve" step, "Estimate" computes the a posteriori error estimators $\{\eta_h(\Phi_h, \tau)\}_{\tau\in\mathcal{T}_h}$, e.g., the ones defined in (10.27).

Mark Based on the a posteriori error indicators $\{\eta_k(\Phi_k, \tau)\}_{\tau\in\mathcal{T}_k}$, "Mark" gives a strategy to choose a subset of elements $\mathcal{M}_k$ of $\mathcal{T}_k$ for refinement. Here and hereafter, we replace the subscript h (or h_k) by an iteration counter k whenever convenient.

One of the most widely used marking strategies to enforce error reduction is the so-called Dörfler strategy [33], which is stated as follows:

Dörfler Strategy: Given a marking parameter $0 < \theta < 1$:

1. Construct a subset $\mathcal{M}_k \subset \mathcal{T}_k$ such that

$$\sum_{\tau\in\mathcal{M}_k} \eta_k^2(\Phi_k, \tau) \geqslant \theta \sum_{\tau\in\mathcal{T}_k} \eta_k^2(\Phi_k, \tau). \tag{10.29}$$

2. Mark all the elements in $\mathcal{M}_k$.

Another commonly used marking strategy, which is called the "Maximum strategy", only requires that the set of marked elements $\mathcal{M}_k$ contains at least one element of $\mathcal{T}_k$ holding the largest value estimator [45, 46]. Namely, there exists at least one element $\tau_k^{\max} \in \mathcal{M}_k$ such that

$$\eta_k(\Phi_k, \tau_k^{\max}) = \max_{\tau\in\mathcal{T}_k} \eta_k(\Phi_k, \tau). \tag{10.30}$$

The Maximum strategy is weaker than the Dörfler strategy. We observe that the most commonly used marking strategies, e.g., the Dörfler strategy and the equidistribution strategy, satisfy (10.30).

Refine Given the partition $\mathcal{T}_k$ and the set of marked elements $\mathcal{M}_k$, "Refine" produces a new partition $\mathcal{T}_{k+1}$ by refining all elements in $\mathcal{M}_k$ at least once. Define

$$\mathcal{R}_{\mathcal{T}_k\to\mathcal{T}_{k+1}} = \mathcal{T}_k \backslash (\mathcal{T}_k \cap \mathcal{T}_{k+1})$$

as the set of refined elements, we have $\mathcal{M}_k \subset \mathcal{R}_{\mathcal{T}_k\to\mathcal{T}_{k+1}}$. Note that usually more than just the marked elements in $\mathcal{M}_k$ are refined in order to keep the mesh conforming.

Combining the basic loop of an adaptive element algorithm and the residual type a posteriori error estimators (10.27), two adaptive finite element algorithms for the Kohn–Sham equation, the adaptive finite element algorithm with the Maximum strategy and the adaptive finite element algorithm with the Dörfler strategy, are designed in [15], which are stated as follows.

10.5.1 *Convergence and Convergence Rate*

The numerical analysis for Algorithm 1 has been carried out in [15], for instance, the following theorem has been proved there.

Algorithm 1

1: Pick an initial mesh $\mathcal{T}_0$ and let $k = 0$.
2: Solve (10.18) on $\mathcal{T}_k$ to get discrete solutions $(\mu_{i,k}, \psi_{i,k})(i = 1, 2, \cdots, N)$.
3: Compute the local error indictors $\eta_k(\widetilde{\Phi}_k, \tau)$ as (10.27) for all $\tau \in \mathcal{T}_k$ with $\widetilde{\Phi}_k = (\psi_{1,k}, \cdots, \psi_{N,k})$.
4: Construct $\mathcal{M}_k \subset \mathcal{T}_k$ by the Maximum strategy.
5: Refine $\mathcal{T}_k$ to get a new conforming mesh $\mathcal{T}_{k+1}$.
6: Let $k = k + 1$ and go to 2.

Algorithm 2

Pick an initial mesh $\mathcal{T}_0$ and let $k = 0$.
Solve (10.18) on $\mathcal{T}_k$ to get discrete solutions $(\mu_{i,k}, \psi_{i,k})(i = 1, 2, \cdots, N)$.
Compute the local error indictors $\eta_k(\widetilde{\Phi}_k, \tau)$ (10.27) for all $\tau \in \mathcal{T}_k$ with $\widetilde{\Phi}_k = (\psi_{1,k}, \cdots, \psi_{N,k})$.
Construct $\mathcal{M}_k \subset \mathcal{T}_k$ by the Dörfler strategy.
Refine $\mathcal{T}_k$ to get a new conforming mesh $\mathcal{T}_{k+1}$.
Let $k = k + 1$ and go to 2.

Theorem 10.7 *Let $\{\Theta_k\}_{k\in\mathbb{N}_0}$ be the sequence generated by Algorithm 1 with a sufficiently fine initial mesh $\mathcal{T}_0$. If* (10.12) *and Assumption* **(A1)** *are satisfied, then*

$$\lim_{k\to\infty} E_k = \min_{\Phi\in\mathcal{Q}} E(\Phi),$$

$$\lim_{k\to\infty} d_{\mathcal{H}}(\Theta_k, \Theta) = 0.$$

It is shown by Theorem 10.7 that all limits of the adaptive finite element approximations for the Kohn–Sham equation are Kohn–Sham ground state solutions when the initial mesh is sufficiently fine. Indeed, the initial mesh is not necessary fine enough for the convergence, we refer to [91] for more details.

Thanks to the use of a stronger Dörfler marking strategy, not only the convergence but also the convergence rate can be obtained for Algorithm 2 [15]. Let us first do some preparation for addressing the results.

For $(\Lambda, \Phi) \in \Theta$ and $\Phi_h \in V_h$, we say the equivalence class $[\Phi_h]$ approximates the equivalence class $[\Phi]$ if

$$\mathcal{D}_{\mathcal{H}}([\Phi_h], [\Phi]) < \mathcal{D}_{\mathcal{H}}([\Phi_h], [\widetilde{\Phi}]), \quad \forall (\widetilde{\Lambda}, \widetilde{\Phi}) \in \Theta \text{ and } [\Phi] \neq [\widetilde{\Phi}].$$

A sequence $\{[\widetilde{\Phi}_k]\}_{k\in\mathbb{N}_0}$ is said to converge to an equivalence class $[\Phi]$ if there exists a sequence of unitary matrices $\{U_k\}_{k\in\mathbb{N}_0}$, such that

$$\lim_{k\to\infty} \|\widetilde{\Phi}_k U_k - \Phi\|_{1,\Omega} = 0.$$

It is proved in [15] that the error of the adaptive finite element approximations produced by Algorithm 2 is reduced. Indeed, we have (see [15] for the proof)

Theorem 10.8 *Let $\theta \in (0,1)$ and $h_0 \ll 1$. Let $\{\widetilde{\Phi}_k\}_{k\in\mathbb{N}_0}$ be a sequence of finite element approximations obtained by Algorithm 2 and $\{[\widetilde{\Phi}_{k_i}]\}_{i\in\mathbb{N}_0}$ be the subsequence that converges to some $[\Phi]$ with $(\Lambda,\Phi)\in\Theta$. If Assumption **(A2)** is true and (Λ,Φ) satisfies Assumptions **(A3)**, then*

$$\|\Phi-\Phi_{k_{i+1}}\|_{1,\Omega}^2+\gamma\eta_{k_{i+1}}^2(\Phi_{k_{i+1}},\mathcal{T}_{k_{i+1}})\leqslant\xi^2\left(\|\Phi-\Phi_{k_i}\|_{1,\Omega}^2+\gamma\eta_{k_i}^2(\Phi_{k_i},\mathcal{T}_{k_i})\right),$$

where $\Phi_{k_{i+1}}\in X_{\Phi,k_{i+1}}$ and $\Phi_{k_i}\in X_{\Phi,k_i}$ satisfy (10.24)–(10.26) with h being replaced by $h_{k_{i+1}}$ and h_{k_i}, respectively, $\gamma>0$ and $\xi\in(0,1)$ are constants depending only on the coercivity constant c_a, the shape regularity constant γ^*, and the marking parameter θ. Therefore, the k_m-th iteration solution of Algorithm 2 satisfies

$$\|\Phi-\Phi_{k_m}\|_{1,\Omega}^2+\gamma\eta_{k_m}^2(\Phi_{k_m},\mathcal{T}_{k_m})\leqslant\xi^{2m}\left(\|\Phi-\Phi_{k_0}\|_{1,\Omega}^2+\gamma\eta_{k_0}^2(\Phi_{k_0},\mathcal{T}_{k_0})\right),$$

and

$$|\Lambda-\Lambda_{k_m}|\lesssim\xi^{2m}.$$

Furthermore, there holds

$$d_{\mathcal{H}}(\Theta_{\Psi_{k_m}},\Theta)\lesssim\xi^{2m}.$$

It is shown by Theorem 10.8 that the Kohn–Sham ground state solutions can be well approximated by the adaptive finite element approximations with a certain convergence rate when the initial mesh is fine enough.

10.5.2 Complexity

The quasi-optimal complexity of adaptive finite element approximations may be derived, for which we define [15].

$$\mathcal{A}_\gamma^s=\{\Phi\in\mathcal{H}:|\Phi|_{s,\gamma}<\infty\},$$

where $\gamma>0$ is some constant and

$$|\Phi|_{s,\gamma}=\sup_{\varepsilon>0}\varepsilon\inf_{\{\mathcal{T}\subset\mathcal{T}_0:\ \inf_{\Phi_\mathcal{T}\in V_\mathcal{T}}(\|\Phi-\Phi_\mathcal{T}\|_{1,\Omega}^2+(\gamma+1)\mathrm{osc}_\mathcal{T}^2(\Phi_\mathcal{T},\mathcal{T}))^{1/2}\leqslant\varepsilon\}}(\#\mathcal{T}-\#\mathcal{T}_0)^s$$

and $\mathcal{T} \subset \mathcal{T}_0$ means $\mathcal{T}$ is a refinement of $\mathcal{T}_0$. We see that, for all $\gamma > 0$, $\mathcal{A}^s_\gamma = \mathcal{A}^s_1$. So we may use $\mathcal{A}^s$ to stand for $\mathcal{A}^s_1$ and use $|\Phi|_s$ to denote $|\Phi|_{s,\gamma}$. We note that $\mathcal{A}^s$ is the class of functions that can be approximated within a given tolerance ε by continuous piecewise polynomial functions over a partition $\mathcal{T}_k$ with the number of degrees of freedom satisfying $\#\mathcal{T}_k - \#\mathcal{T}_0 \lesssim \varepsilon^{-1/s}|\Phi|_s^{1/s}$, where $\#\mathcal{T}$ denotes the number of elements in a mesh $\mathcal{T}$.

To investigate the computational complexity of Algorithm 2, we need some assumptions as follows: Let

$$\theta_* = \frac{1}{N^2} \frac{C_2\gamma}{C_3(C_1 + (1 + 2C_*^2 C_1)\gamma)},$$

where C_* is a positive constant depending on the shape regularity constant γ^* (see [15] for details).

Assumptions 10.9

1. The marking parameter θ satisfies $\theta \in (0, \theta_*)$.
2. The marked $\mathcal{M}_{h_k}$ satisfies (10.29) with minimal cardinality.
3. The distribution of refinement edges on $\mathcal{T}_{h_0}$ satisfies condition (b) of Section 4 in [81].

With Assumptions 10.9, we then have the optimal complexity [15].

Theorem 10.10 *Suppose $\theta \in (0, 1)$ and $h_0 \ll 1$. Assume that Assumption* **(A2)** *is satisfied and* (10.11) *has m solutions in Θ (up to invariance under unitary transforms), which are denoted as $[\Phi^{(l)}](l = 1, 2, \ldots, m)$ where $m \in \mathbb{N}$ or can be chosen to be ∞. Let $\{\Psi_k\}_{k\in\mathbb{N}_0}$ be a sequence of finite element solutions produced by Algorithm 2 with Assumptions 10.9 being satisfied. Then the following quasi-optimal bound is valid*

$$\#\mathcal{T}_n - \#\mathcal{T}_0 \lesssim \sum_{l=1}^{m} \left(\|\Phi^l - \Phi^l_{k_{n_l}}\|^2_{1,\Omega} + \gamma \operatorname{osc}^2_{k_{n_l}}(\Phi^l_{k_{n_l}}, \Omega) \right)^{-1/2s}, \qquad (10.31)$$

where $\Phi^l \in [\Phi^{(l)}] \cap \mathcal{A}^s$ satisfies (10.14), *$\Phi^l_{k_{n_l}} \in X_{\Phi^l, k_{n_l}}$ satisfy* (10.24) *and* (10.26) *with h being replaced by $h_{k_{n_l}}$, and the hidden constant depends on the exact solution Φ^l and the discrepancy between θ and $\theta_*(l = 1, 2, \ldots, m)$. Here, n_l and k_{n_l} are the total number and the maximal index of iterations which approximate $[\Phi^{(l)}](l = 1, 2, \ldots, m)$ among the n iterations, respectively.*

It is shown by Theorem 10.10 that under some mild and reasonable assumptions, the computational complexity of the adaptive finite element algorithm with the Dörfler strategy is quasi-optimal when the initial mesh is fine enough.

Applying the non-polynomial behavior of the eigenfunctions, we see from [91] that the adaptive finite element method has an asymptotic linear convergence rate and an asymptotic optimal complexity from any initial mesh. More precisely, the adaptive finite element method has a linear convergence rate and an optimal complexity after finite iteration steps.

10.6 High Performance Computing

Due to the needs of science and engineering and the availability of high performance computers, there is a great demand for highly efficient and scalable algorithms for electronic structure calculations. There have been many contributions toward this effort to design high performance algorithms. We mention here, for instance, the symmetry-based decomposition approach [36, 40], the parallel orbital-updating approach [27, 28, 63], the local and high order finite element discretization [29], the three-scale schemes [31, 44], and the hexahedron-grid-based two-scale higher order finite element scheme [21]. In the following, we focus on a brief introduction to the first two approaches that involve a two-level parallelization.

10.6.1 Symmetry-Based Decomposition Approach

A symmetry-based decomposition approach is proposed to deal with linear differential eigenvalue problems as well as the Kohn–Sham equation in [40] by making use of the Abelian or non-Abelian symmetries of the systems. By this approach, for instance, the solution of the associated algebraic eigenvalue problem is then replaced by the solutions of some independent eigenvalue subproblems that are of small scale. Due to the independence, as we see, the computation can be carried out in parallel.

The group theory-based decomposition approach can be seamlessly incorporated with grid-based discretizations such as finite difference, finite element, or finite volume methods. We may place the approach into a two-level parallelization setting, which consequently reduces the CPU time notably.

To show the main idea of the group theory-based decomposition approach, we consider a model problem

$$\begin{cases} Lu = \lambda u \ \text{ in } \Omega, \\ \ \ u = 0 \ \text{ on } \partial\Omega, \end{cases} \tag{10.32}$$

where L is a Hermitian operator on a Hilbert space $V \subset L^2(\Omega)$.

A group G is said to be a symmetry group associated with eigenvalue problem (10.32) if

$$R\Omega = \Omega, \ P_R L = L P_R \quad \forall R \in G,$$

and the imposed boundary condition is also invariant under $\{P_R : R \in G\}$, which is defined as follows: for each $R \in G$ and $f \in V$, there holds

$$P_R f(Rx) = f(x) \quad \forall x \in \Omega.$$

We see that $\{P_R : R \in G\}$ forms a group isomorphic to G[34].

Associated with the model (10.32), we have [40]

Theorem 10.11 *Suppose a finite group $G = \{R\}$ is a symmetry group associated with eigenvalue problem (10.32). Denote all the inequivalent, irreducible, unitary representations of G as $\{\Gamma^{(\nu)} : \nu = 1, 2, \ldots, n_c\}$. Then the eigenvalue problem can be decomposed into $\sum_{\nu=1}^{n_c} d_\nu$ subproblems. For each $\nu \in \{1, 2, \ldots, n_c\}$, the corresponding d_ν subproblems are*

$$\begin{cases} Lu_l^{(\nu)} = \lambda^{(\nu)} u_l^{(\nu)} & \text{in } \Omega, \\ u_l^{(\nu)} = 0 & \text{on } \partial\Omega, \qquad l = 1, 2, \ldots, d_\nu, \\ u_l^{(\nu)} = \mathscr{P}_{lk}^{(\nu)} u_k^{(\nu)} & \text{in } \Omega, \end{cases}$$

where k is any chosen number in $\{1, 2, \ldots, d_\nu\}$.

Suppose the finite group $G = \{R\}$ of order n is the symmetric group associated with the problem (10.32). Denote all the inequivalent, irreducible, unitary representations of group G by $\{\Gamma^{(\nu)}, \nu = 1, 2, \ldots, n_c\}$, with d_ν being the dimension of Γ^ν. Then the original eigenvalue problem (10.32) can be decomposed into $\sum_{\nu=1}^{n} d_\nu$ subproblems. Define $n_{\text{sub}} = \sum_{\nu=1}^{n} d_\nu$. For any ν, the corresponding d_ν subproblems can be expressed as

$$\begin{cases} Lu_l^\nu = \lambda_l^\nu u_l^\nu \ \text{ in } \Omega, \\ u_l^\nu = 0 \ \text{ on } \partial\Omega, \qquad\qquad l = 1, 2, \cdots, d_\nu, \\ u_l^\nu(Rx) = \sum_{m=1}^{d_\nu} \Gamma^\nu(R)_{lm}^* u_m^\nu(x) \ \text{ in } \Omega, \forall R \in G. \end{cases} \tag{10.33}$$

For illustration, we provide an analysis of this approach for electronic structure calculations of symmetric cluster systems based on the finite element discretizations. Assume the number of degrees of freedom for discretizing (10.32) is N_g and the number of eigenpairs required is N. By the traditional approaches, the computational cost scales as $N^2 N_g$. By applying the above symmetric-based decomposition approach, the solution of (10.32) is transferred to the solutions of n_{sub} subproblems (10.33), with each subproblem being discretized by N_g/n degrees of freedom and only requiring the smallest N/n_{sub} eigenvalues and their corresponding eigenfunctions to be computed. Therefore, the total computational cost is significantly reduced. More importantly, since the subproblems (10.33) are independent of each other, they can be solved in parallel intrinsically. Thus the algorithm allows two level parallelization: one level is the solution of these N independent subproblems in parallel intrinsically, the other level is to solve each boundary problem in parallel by traditional algebraic parallel strategies or domain decomposition approaches. We refer to [36, 40] for the detailed implementation issues.

10.6.2 Parallel Orbital-Updating Approach

Now we turn to introduce the parallel orbital-updating approach, which is first proposed in [27]. The basic version is stated as follows [27].

Algorithm 3

1: Give the initial guess $(\lambda_i^{(0)}, \psi_i^{(0)})(i = 0, 1, \ldots, N)$ and initial finite-dimensional space V_0, let $k = 0$.
2: Enlarge the finite-dimensional space V_k to V_{k+1} by some strategy (e.g., adaptive refinement).
3: For $i = 1, 2, \ldots, N$, update each $\psi_i^{(k)}$ by solving some source problems in parallel, and obtain new orbitals $\widetilde{\psi}_i^{(k+1)} \in V_{k+1}$.
4: Solve (10.18) in $\widetilde{V}_{n+1}$ and obtain $(\varepsilon_i^{(k+1)}, \psi_i^{(k+1)})(i = 1, 2, \ldots, N)$.
5: Let $k = k + 1$, go to step 2.

Here, $\widetilde{V}_{k+1} = \text{span}\,\{\widetilde{\psi}_1^{(k+1)}, \widetilde{\psi}_2^{(k+1)}, \ldots, \widetilde{\psi}_N^{(k+1)}\}$.

With the parallel orbital-updating approach, the solution of the eigenvalue problem is replaced by solutions of a series of independent source problems and some small scale eigenvalue problems. Because of the independence of the source problems, these source problems can be solved in parallel essentially. For each source problem, traditional parallel strategies (for example, domain decomposition or parallelization in matrix-vector multiplication) can be used. Therefore, it allows for a two-level parallelization: one level of parallelization is obtained by partitioning these source problems into different groups of processors, another level of parallelization is obtained by assigning each source problem to several processors contained in each group. This two-level parallelization makes the parallel orbital-updating approach more competitive for large scale calculations.

The approach is then extended to solutions of the constrained minimization problem for the Kohn–Sham energy functional [28]. Some extensions to plane wave discretizations for the Kohn–Sham equation as well as modified versions are investigated in [63].

10.7 Concluding Remarks

In this chapter, we have introduced the finite element method to DFT calculations, together with its mathematical theory. We have presented the a priori and a posteriori error estimations of the finite element approximations to the Kohn–Sham equation. We have also provided the convergence rate for adaptive finite element approximations and described its complexity theory. Consequently, we have shown the efficiency and reliability of finite element approximations of the Kohn–Sham DFT models, together with a mathematical theory of the error control during the calculations. In addition, we have presented two high performance algorithms, the symmetry-based decomposition approach and the parallel orbital-updating approach, which allow for the two-level parallelization and may have the potential for exascale computing. Finally, we mention that the emphasis of this chapter has been on the work of our group, and related research.

References

1. J. Ackermann, B. Erdmann and R. Roitzsch. A self-adaptive multilevel finite element method for the stationary Schrödinger equation in three space dimensions. *J. Chem. Phys.* **101**, 7643–7650 (1994).
2. J. Ackermann and R. Roitzsch. A two-dimensional multilevel adaptive finite element method for the time-independent Schrödinger equation. *Chem. Phys. Lett.* **214**, 109–117 (1993).
3. R.A. Adams. *Sobolev Spaces*. Academic Press, New York (1975).
4. S. Agmon. *Lectures on the Exponential Decay of Solutions of Second-Order Elliptic Operators*. Princeton University Press, Princeton (1981).
5. A. Anantharaman and E. Cancès. Existence of minimizers for Kohn–Sham models in quantum chemistry. *Ann. I. H. Poincaré-AN* **26**, 2425–2455 (2009).
6. G. Bao, G. Hu and D. Liu. Numerical solution of the Kohn–Sham equation by finite element methods with an adaptive mesh redistribution technique. *J. Sci. Comput.* **55**, 372–391 (2013).
7. T.L. Beck. Real-space mesh techniques in density-function theory. *Rev. Mod. Phys.* **72**, 1041–1080 (2000).
8. A. Brandt. Multilevel adaptive solutions to boundary-value problems. *Math. Comput.* **31**, 333–390 (1977).
9. S.C. Brenner and L.R. Scott. *The Mathematical Theory of Finite Element Methods*. Springer-Verlag, New York (1994).
10. E.J. Bylaska, M. Holst and J.H. Weare. Adaptive finite element method for solving the exact Kohn–Sham equation of density functional theory. *J. Chem. Theory Comput.* **5**, 937–948 (2009).
11. D. Calvetti, L. Reichel and D.C. Sorensen. An implicitly restarted Lanczos method for large symmetric eigenvalue problems. *Elec. Trans. Numer. Anal.* **2** (1994), 1–21 (1994).
12. E. Cancès, R. Chakir and Y. Maday. Numerical analysis of the planewave discretization of some orbital-free and Kohn–Sham models. *M2AN* **46**, 341–388 (2012).
13. J.M. Cascon, C. Kreuzer, R.H. Nochetto and K.G. Siebert. Quasi-optimal convergence rate for an adaptive finite element method. *SIAM J. Numer. Anal.* **46**, 2524–2550 (2008).
14. H. Chen. *Finite Dimensional Approximations in Density Functional Theory* (in Chinese). Ph.D. Thesis, Academy of Mathematics and Systems Science, Chinese Academy of Sciences, Beijing (2010).

15. H. Chen, X. Dai, X. Gong, L. He and A. Zhou. Adaptive finite element approximations for Kohn–Sham models. *Multiscale Model. Simul.* **12**, 1828–1869 (2014).
16. H. Chen and G. Friesecke. Pair densities in density functional theory. *Multiscale Model. Simul.* **13**, 1259–1289 (2015).
17. H. Chen, G. Friesecke and C. Mendl. Numerical methods for a Kohn–Sham density functional model based on optimal transport. *J. Chem. Theory Comput.* **10**, 4360–4368 (2014).
18. H. Chen, X. Gong, L. He, Z. Yang and A. Zhou. Numerical analysis of finite dimensional approximations of Kohn–Sham equations. *Adv. Comput. Math.* **38**, 225–256 (2013).
19. H. Chen, X. Gong, L. He and A. Zhou. Adaptive finite element approximations for a class of nonlinear eigenvalue problems in quantum physics. *Adv. Appl. Math., Mech.* **3**, 493–518 (2011).
20. H. Chen, L. He and A. Zhou. Finite element approximations of nonlinear eigenvalue problems in quantum physics. *Comput. Methods Appl. Mech. Engrg.* **200**, 1846–1865 (2011).
21. H. Chen, F. Liu and A. Zhou. A two-scale higher order finite element discretization for Schrödinger equations. *J. Comput. Math.* **27**, 315–337 (2009).
22. P.G. Ciarlet. *The Finite Element Method for Elliptic Problems*. North-Holland, Amsterdam (1978).
23. P.G. Ciarlet and J.L. Lions, eds. *Finite Element Methods, Volume II of Handbook of Numerical Analysis*. North-Holland, Amsterdam (1991).
24. M.L. Cohen and V. Heine. The fitting of pseudopotentials to experimental data and their subsequent application. *Solid State Phys.* **24**, 37–248 (1970).
25. X. Dai. *Adaptive and Localization Based Finite Element Discretizations for the First-Principles Electronic Structure Calculations* (in Chinese). Ph.D. Thesis, Academy of Mathematics and Systems Science, Chinese Academy of Sciences, Beijing (2008).
26. X. Dai, X. Gong, Z. Yang, D. Zhang and A. Zhou. Finite volume discretizations for eigenvalue problems with applications to electronic structure calculations. *Multiscale Model. Simul.* **9**, 208–240 (2011).
27. X. Dai, X. Gong, A. Zhou and J. Zhu. A parallel orbital-updating approach for electronic structure calculations. *arXiv1405.0260* (2014).
28. X. Dai, Z. Liu, X. Zhang and A. Zhou. A parallel orbital-updating based optimization method for electronic structure calculations. *J. Comput. Phys.* **445**, 110622 (2021).
29. X. Dai, L. Shen and A. Zhou. A local computational scheme for higher order finite element eigenvalue approximations. *Inter. J. Numer. Anal. Model.* **5**, 570–589 (2008).
30. X. Dai, J. Xu and A. Zhou. Convergence and optimal complexity of adaptive finite element eigenvalue computations. *Numer. Math.* **110**, 313–355 (2008).
31. X. Dai and A. Zhou. Three-scale finite element discretizations for quantum eigenvalue problems. *SIAM J. Numer. Anal.* **46**, 295–324 (2008).
32. X. Dai and A. Zhou. Finite element methods for electronic structure calculations (in Chinese). *Sci. Sin. Chim.* **45**, 800–811 (2015).
33. W. Dörfler. A convergent adaptive algorithm for Poisson's equation. *SIAM J. Numer. Anal.* **33**, 1106–1124 (1996).
34. C.C. Douglas and J. Mandel. An abstract theory for the domain reduction method. *Computing* **48**, 73–96 (1992).
35. A. Edelman, T.A. Arias and S.T. Smith. The geometry of algorithms with orthogonality constraints. *SIAM J. Matrix Anal. Appl.* **20**, 303–353 (1998).
36. J. Fang. *Algorithm Study for Real-Space First-Principles Calculations* (in Chinese). Academy of Mathematics and Systems Science, Chinese Academy of Sciences, Beijing (2013).
37. J. Fang, X. Gao, X. Gong and A. Zhou. Interpolation based local postprocessing for adaptive finite element approximations in electronic structure calculations. In: *Domain Decomposition Methods in Science and Engineering XIX*, edited by Y. Huang, R. Kornhuber, O. Widlund and J. Xu, pp. 51–61, volume 78 of Lecture Notes in Computational Science and Engineering, Berlin, Heidelberg, Springer (2011).
38. J. Fang, X. Gao and A. Zhou. A Kohn–Sham equation solver based on hexahedral finite elements. *J. Comput. Phys.* **231**, 3166–3180 (2012).

39. J. Fang, X. Gao and A. Zhou. A finite element recovery approach to eigenvalue approximations with applications to electronic structure calculations. *J. Sci. Comput.* **55**, 432–454 (2013).
40. J. Fang, X. Gao and A. Zhou. A symmetry-based decomposition approach to eigenvalue problems. *J. Sci. Comput.* **57**, 638–669 (2013).
41. J.L. Fattebert, R.D. Hornung and A.M. Wissink. Finite element approach for density functional theory calculations on locally refined meshes. *J. Comput. Phys.* **223**, 759–773 (2007).
42. E. Fermi. Un metodo statistics per la determinazione di alcune proprieta dell'atomoi. *Rend. Accad. Lincei* **6**, 602–607 (1927).
43. X. Gao. *Hexahedral Finite Element Methods for the First-Principles Electronic Structure Calculations* (in Chinese). Ph.D. Thesis, Academy of Mathematics and Systems Science, Chinese Academy of Sciences, Beijing (2009).
44. X. Gao, F. Liu and A. Zhou. Three-scale finite element eigenvalue discretizations. *BIT Numer. Math.* **48**, 533–562 (2008).
45. E.M. Garau and P. Morin. Convergence and quasi-optimality of adaptive FEM for Steklov eigenvalue problems. *IMA J. Numer. Anal.* **31**, 914–946 (2011).
46. E.M. Garau, P. Morin and C. Zuppa. Convergence of adaptive finite element methods for eigenvalue problems. *M3AS* **19**, 721–747 (2009).
47. X. Gong, L. Shen, D. Zhang and A. Zhou. Finite element approximations for Schrödinger equations with applications to electronic structure computations. *J. Comput. Math.* **23**, 310–327 (2008).
48. S. Hackel, D. Heinemann, D. Kolb and B. Fricke. Calculations of the polycentric linear molecule H_3^{2+} with the finite element method. *Chem. Phys. Lett.* **206**, 91–95 (1993).
49. Y. Hasegawa, J.-I. Iwata, M. Tsuji, D. Takahashi, A. Oshiyama, K. Minami, T. Boku, F. Shoji, A. Uno, M. Kurokawa, H. Inoue, I. Miyoshi and M. Yokokawa. First-principles calculations of electron states of a silicon nanowire with 100,000 atoms on the K computer. In: *Proceedings of 2011 International Conference for High Performance Computing, Networking, Storage and Analysis (SC2011)*, pp. 1–11 (2011).
50. L. He, F. Liu, G. Hautier, M.J.T. Oliveira, M.A.L. Marques, F.D. Vila, J.J. Rehr, G.-M. Rignanese and A. Zhou. Accuracy of generalized gradient approximation functionals for density functional perturbation theory calculations. *Phys. Rev. B.* **89**, 064305–064320 (2014).
51. P. Hohenberg and W. Kohn. Inhomogeneous electron gas. *Phys. Rev.* **136**, B864–B871 (1964).
52. G. Hu, H. Xie and F. Xu. A multilevel correction adaptive finite element method for Kohn–Sham equation. *J. Comput. Phys.* **355**, 436–449 (2018).
53. D.D. Johnson. Modified Broyden's method for accelerating convergence in self-consistent calculations. *Phys. Rev. B* **38**, 12807–12813 (1988).
54. A.V. Knyazev. Toward the optimal preconditioned eigensolver: Locally optimal block preconditioned conjugate gradient method. *SIAM J. Sci. Comput.* **23**, 517–541 (2001).
55. W. Kohn and L.J. Sham. Self-consistent equations including exchange and correlation effects. *Phys. Rev. A* **140**, 1133–1138 (1965).
56. G. Kresse and J. Furthmüller. Efficient iterative schemes for *ab initio* total-energy calculations using a plane-wave basis set. *Phys. Rev. B* **54**, 11169–11186 (1996).
57. P.E. Lammert. In search of the Hohenberg–Kohn theorem. *J. Math. Phys.* **59**, 042110 (2018).
58. F.S. Levin and J. Shertzer. Finite-element solution of the Schrödinger equation for the helium ground state. *Phys. Rev. A* **32**, 3285–3290 (1985).
59. E.H. Lieb. Density functionals for Coulomb systems. *Inter. J. Quantum Chem.* **24**), 243–277 (1983.
60. R.M. Martin. *Electronic Structure: Basic Theory and Practical Methods*. Cambridge University Press, Cambridge (2004).
61. P. Motamarri, S. Das, S. Rudraraju, K. Ghosh, D. Davydov and V. Gavini. DFT-FE: A massively parallel adaptive finite-element code for large-scale density functional theory calculations. *Comput. Phys. Commun.* **246**, 106853 (2020).

62. P. Motamarri, M.R. Nowak, K. Leiter, J. Knap and V. Gavini. Higher-order adaptive finite-element methods for Kohn–Sham density functional theory. *J. Comput. Phys.* **253**, 308–343 (2013).
63. Y. Pan, X. Dai, S. de Gironcoli, X. Gong, G. Rignanese and A. Zhou. A parallel orbital-updating based plane-wave basis method for electronic structure calculations. *J. Comput. Phys.* **348**, 482–492 (2017).
64. R.G. Parr and W.T. Yang. *Density-Functional Theory of Atoms and Molecules*. Oxford University Press, New York, Clarendon Press, Oxford (1994).
65. J.E. Pask, B.M. Klein, C.Y. Fong and P.A. Sterne. Real-space local polynomial basis for solid-state electronic-structure calculations: A finite-element approach . *Phys. Rev. B* **59**, 12352–12358 (1999).
66. J.E. Pask and P.A. Sterne. Finite element methods in ab initio electronic structure calculations. *Model. Simul. Mater. Sci. Eng.* **13**, 71–96 (2005).
67. J.E. Pask, N. Sukumar, M. Guney and W. Hu. Partition-of-unity finite elements for large, accurate quantum-mechanical materials calculations. *Plenary Lecture, ECCOMAS Thematic Conference on the Extended Finite Element Method (XFEM 2011)*, Cardiff (2011).
68. J.C. Phillips. Energy-band interpolation scheme based on a pseudopotential. *Phys. Rev.* **112**, 685–695 (1958).
69. P. Pulay. Convergence acceleration of iterative sequences the case of SCF iteration. *Chem. Phys. Lett.* **73**, 393–398 (1980).
70. P. Pulay. Improved SCF convergence acceleration. *J. Comput. Chem.* **3**, 556–560 (1982).
71. Y. Saad. *Numerical Methods for Large Eigenvalue Problems*. New York: Halstead Press (1992).
72. Y. Saad. *Iterative Methods for Sparse Linear Systems*. 2nd. ed. SIAM (2003).
73. Y. Saad, J.R. Chelikowsky and S.M. Shontz. Numerical methods for electronic structure calculations of materials. *SIAM Review* **52**, 3–54 (2010).
74. R. Schneider, T. Rohwedder, A. Neelov and J. Blauert. Direct minimization for calculating invariant subspaces in density functional computations of the electronic structure. *J. Comput. Math.* **27**, 360–387 (2009).
75. L. Shen. *Parallel Adaptive Finite Element Algorithms for Electronic Structure Computing based on Density Functional Theory* (in Chinese). Ph.D. Thesis, Academy of Mathematics and Systems Science, Chinese Academy of Sciences, Beijing (2005).
76. L. Shen and A. Zhou. A defect correction scheme for finite element eigenvalues with applications to quantum chemistry. *SIAM J. Sci. Comput.* **28**, 321–338 (2006).
77. D. Singh, H. Krakauer and C.S. Wang. Accelerating the convergence of self-consistent linearized augmented-plane-wave calculations. *J. Phys. Rev. B* **34** (1986), 8391-8393.
78. G.L.G. Sleijpen and H.A. van der Vorst. A Jacobi-Davidson iteration method for linear eigenvalue problems. *SIAM J. Matrix Anal. Appl.* **17**, 401–425 (1996).
79. G.P. Srivastava. Broyden's method for self-consistent field convergence acceleration. *J. Phys. A* **17**, 317–321 (1984).
80. P.A. Sterne, J.E. Pask and B.M. Klein. Calculation of positron observables using a finite element-based approach. *Appl. Surf. Sci.* **149**, 238–243 (1999).
81. R. Stevenson. The completion of locally refined simplicial partitions created by bisection. *Math. Comput.* **77**, 227–241 (2008).
82. L. Strakhovskaya. An iterative method for evaluating the first eigenvalue of an elliptic operator. *USSR Comput. Math. Math. Phys.* **17**, 88–101 (1977).
83. P. Suryanarayana, V. Gavini, T. Blesgen, K. Bhattacharya and M. Ortiz. Non-periodic finite-element formulation of Kohn–Sham density functional theory. *J. Mech. Phys. Solids* **58**, 256–280 (2010).
84. L.H. Thomas. The calculation of atomic fields. *Proc. Cambridge Phil. Soc.* **23**, 542–548 (1927).

85. T. Torsti, T. Eirola, J. Enkovaara, T. Hakala, P. Havu, V. Havu, T. Höynälänmaa, J. Ignatius, M. Lyly, I. Makkonen, T.T. Rantala, J. Ruokolainen, K. Ruotsalainen, E. Räsänen, H. Saarikoski and M.J. Puska. Three real-space discretization techniques in electronic structure calculations. *Phys. Stat. Sol.* **B243**, 1016–1053 (2006).
86. E. Tsuchida. *Ab initio* molecular-dynamics study of liquid formamide. *J. Chem. Phys.* **121**, 4740-4746 (2004).
87. E. Tsuchida and M. Tsukada. Electronic-structure calculations based on the finite-element method. *Phys. Rev. B* **52**, 5573–5578 (1995).
88. E. Tsuchida and M. Tsukada. Adaptive finite-element method for electronic-structure calculations. *Phys. Rev. B* **54**, 7602–7605 (1996).
89. E. Tsuchida and M. Tesukada. Large-scale electronic-structure calculations based on the adaptive finite element method. *J. Phy. Soc. Jpn.* **67**, 3844–3858 (1998).
90. Q. Wang. *Gradient Flow Models and Their Computations for Electronic Structures* (in Chinese). Ph.D. Thesis, University of Chinese Academy of Sciences, Beijing (2019).
91. B. Yang and A. Zhou. Eigenfunction behaviors and adaptive finite element approximations of nonlinear eigenvalue problems in quantum physics. *ESAIM:M2AN* **55**, 209–227 (2021).
92. Z. Yang. *Finite Volume Discretization Based First-Principles Electronic Structure Calculations* (in Chinese). Ph.D. Thesis, Academy of Mathematics and Systems Science, Chinese Academy of Sciences, Beijing (2011).
93. H. Yserentant. *Regularity and Approximability of Electronic Wave Functions*. Lecture Notes in Mathematics, Springer-Verlag, Berlin (2010).
94. Y. Yuan and W. Sun. *Optimization Theory and Methods* (in Chinese). Beijing, Science Press, Beijing (1997).
95. D. Zhang. *Applications of Finite Element Methods in Electronic Structure Calculations* (in Chinese). Ph.D. Thesis, Fudan University (2007).
96. D. Zhang, L. Shen, A. Zhou and X. Gong. Finite element method for solving Kohn–Sham equations based on self-adaptive tetrahedral mesh. *Phys. Lett. A* **372**, 5071–5076 (2008).
97. D. Zhang, A. Zhou and X. Gong. Parallel mesh refinement of higher order finite elements for electronic structure calculations. *Commun. Comput. Phys.* **4**, 1086–1105 (2008).
98. X. Zhang. *Algorithm Study for Real-Space Ground and Excited States in First-Principles Calculations* (in Chinese). Ph.D. Thesis, University of Chinese Academy of Sciences, Beijing (2015).
99. X. Zhang and A. Zhou. A singularity-based eigenfunction decomposition for Kohn–Sham equations. *Sci. Sin. Math.* **59**, 1623–1634 (2016).
100. W. Zheng and L. Ying. Finite element calculations for the helium atom. *Inter. J. Quantum Chem.* **97**, 659–669(2004).
101. W. Zheng, L. Ying and P. Ding. Numerical solutions of the schrödinger equation for the ground lithium by the finite element method. *Appl. Math. Comput.* **153**, 685–695 (2004).
102. A. Zhou. Hohenberg–Kohn theorem for Coulomb type systems and its generalization. *J. Math. Chem.* **50**, 2746–2754 (2012).
103. A. Zhou. A mathematical aspect of Hohenberg–Kohn theorem. *Sci. China Math.* **62**, 63–68 (2019).

Chapter 11
Flexibilities of Wavelets as a Computational Basis Set for Large-Scale Electronic Structure Calculations

Luigi Genovese and Thierry Deutsch

Abstract The `BigDFT` project started in 2005 with the aim of testing the advantages of using a Daubechies wavelet basis set for Kohn–Sham density functional theory with pseudopotentials. This project led to the creation of the `BigDFT` code, which employs a computational approach with optimal features for flexibility, performance and precision of the results. In particular, the employed formalism has enabled the implementation of an algorithm able to tackle DFT calculations of large systems, up to many thousands of atoms, with a computational effort which scales linearly with the number of atoms. In this work we show how the localised description of the Kohn–Sham problem, emerging from the features of the basis set, are helpful in providing a simplified description of large-scale electronic structure calculations. We recall some of the features that have been made possible by the peculiar mathematical properties of Daubechies wavelets and also interpolating scaling functions.

11.1 Introduction

Since their foundation, disciplines like computational physics and quantum chemistry have had to deal with the question of the computational reliability of results. The reliability of a given approach can be defined in terms of two key concepts, namely "accuracy", i.e. the ability of the model to predict quantities which can be externally verified, e.g. through experiment, and "precision", i.e. the ability of the employed numerical approach to find the solution to a given physical model. A precise approach should therefore reduce the computational uncertainties of quantities extracted from a well-defined model, and provide reference results which can be compared to other computer codes employing the same model. For theoretical approaches wherein no analytic solution exists, reducing the computational

L. Genovese · T. Deutsch (✉)
Univ. Grenoble Alpes, CEA, IRIG-MEM, Grenoble, France
e-mail: luigi.genovese@cea.fr; thierry.deutsch@cea.fr

E. Cancès, G. Friesecke (eds.), *Density Functional Theory*, Mathematics and Molecular Modeling, https://doi.org/10.1007/978-3-031-22340-2_11

uncertainty is the only way to shed light on the predictive power of the model. The accuracy of a result with respect to experimental data may therefore only be reliably quantified if the computational uncertainty is significantly lower than the observed discrepancy.

Density functional theory (DFT) [38, 43] has had widespread success in simulating a range of materials, from molecules to solids, and has therefore become the most popular approach to electronic structure simulations. While the accuracy of DFT is dominated by the approximations made to the exchange-correlation (XC) functional, the precision of a given simulation depends on a number of factors, in particular the choice of the basis set. Thus two different DFT codes might use the same physical formalism (including the same XC functional), but differ in results due to the use of different numerical approaches. In order to compare results across DFT codes, careful attention must therefore be paid to the precision of the results, as seen for example in the DeltaCode project, in which a systematic comparison of a number of periodic DFT codes was undertaken [45].

In this context, an important distinction should be made between codes which use systematic and non-systematic basis sets. A systematic basis set allows one to calculate the exact solution of the Kohn–Sham (KS) equations with arbitrarily high precision by increasing the number of basis functions. In other terms, the numerical precision of the results is related to the number of basis functions used to expand the KS orbitals. With such a basis set it is thus possible to obtain results which are free of errors related to the choice of the basis, eliminating a source of uncertainty. As such, it is highly desirable to have at hand a computational formalism which is able to provide, at the same time:

- a set of reliable results, which can be systematically improved by the end-user, in view of increasing—when needed—the *precision* of the calculations;
- a flexible approach, in which the desired models can be explicitly implemented without having to deal with correction terms and intrinsic approximations;
- an efficient computer program, which enables the optimal use of computational resources, especially in the context of high performance computing;
- the ability to connect together different levels of theory, where various approaches might be linked within a given computational setup.

In 2005, the EU FP6-STREP-NEST `BigDFT` project funded a consortium of four European laboratories (L_Sim, CEA-Grenoble, France; Basel University, Switzerland; Louvain-la-Neuve University, Belgium; and Kiel University, Germany), with the aim of developing a novel approach for DFT calculations based on Daubechies wavelets [16]. Beyond building a DFT code from scratch, the objective of this 3-year project was to test the potential benefit of a new formalism in the context of electronic structure calculations.

This project was motivated by the fact that Daubechies wavelets exhibit a set of properties which make them ideal for a precise and optimized DFT approach. In particular, their systematicity provides a reliable basis set for high-precision results, whereas their locality (both in real and reciprocal space) is highly desirable to improve the efficiency and the flexibility of the treatment. Indeed, a localized basis set allows the optimization of the number of degrees of freedom for a required accuracy, which is highly desirable given the complexity and inhomogeneity of systems under investigation nowadays. Moreover, an approach based on localized functions makes it possible to explicitly control the nature of the boundaries of the simulation domain, allowing complex environments like mixed boundary conditions and/or systems with a net charge.

We organize this chapter as follows. We first develop shortly the KS formalism and the concepts of operators for any basis set. We then present some basic illustrations of the properties of Daubechies wavelets associated to the multi-resolution analysis, and their peculiarities in the context of computational structure electronic calculations. We briefly outline, based on the interpolating scaling functions, the main features of the Poisson Solver implemented in the code. Finally, we will then explain in more detail how the solution of the KS problem is implemented in the code and how the properties of wavelets enable the realization of a computational algorithm whose time-to-solution is linearly scaling with the number of atoms in the system.

11.1.1 Basis Set of Coordinate Representation

We start the chapter by illustrating the notation employed in the following. We represent the state associated to a particle coordinate by a label $|\mathbf{x}\rangle \equiv |\mathbf{r}\rangle \otimes |\alpha\rangle$ with $\mathbf{r}$ labelling a point in real space and the value of $\alpha = 1, 2$ corresponding to the spinorial state up or down, respectively. The coordinate representation might be used to express orbitals and operators according to the conventions

$$|\psi\rangle = \int \mathrm{d}\mathbf{x}\, \langle \mathbf{x}|\boldsymbol{\psi}\rangle\, |\mathbf{x}\rangle = \sum_{\alpha} \int \mathrm{d}\mathbf{r}\, \psi^{\alpha}(\mathbf{r})\, |\mathbf{r}\rangle \otimes |\alpha\rangle \ .$$

The completeness relation in the coordinate representation might be written as follows

$$\widehat{\mathbb{I}} = \sum_{\alpha} \int \mathrm{d}\mathbf{r}\, |\mathbf{r}\rangle \langle \mathbf{r}| \otimes |\alpha\rangle \langle \alpha| = \int \mathrm{d}\mathbf{x}\, |\mathbf{x}\rangle \langle \mathbf{x}| = \sum_{\alpha} \int \mathrm{d}\mathbf{r}\, |\mathbf{r}, \alpha\rangle \langle \mathbf{r}, \alpha| \ .$$

We may also express $|\mathbf{x}\rangle = |\mathbf{r}, \alpha\rangle$ as a shorthand.

The electronic charge density $\rho(\mathbf{r})$ might be defined from the above equations as follows:

$$\rho(\mathbf{r}) = \mathrm{tr}\left(\sum_{\alpha} \widehat{F}\, |\mathbf{r}, \alpha\rangle \langle \mathbf{r}, \alpha| \right) ,$$

where the spinorial degrees of freedom have been traced out. The Electronic Density Matrix $\widehat{F}$ is an operator that can be implicitly defined via its kernel

$$\widehat{F} = \int \mathrm{d}\mathbf{x}\mathrm{d}\mathbf{x}' \,|\mathbf{x}'\rangle\, F(\mathbf{x}, \mathbf{x}')\, \langle \mathbf{x}| \ .$$

With the above notation we imply that the kernel $F(\mathbf{x}, \mathbf{x}')$ is expressed in the basis of the position representation. Let us point out the notable relations:

$$\frac{\delta F(\mathbf{y}, \mathbf{y}')}{\delta F(\mathbf{x}, \mathbf{x}')} = \frac{\delta \langle \mathbf{y}'|\, \hat{F}\, |\mathbf{y}\rangle}{\delta \langle \mathbf{x}'|\, \hat{F}\, |\mathbf{x}\rangle} = \delta(\mathbf{r}_x - \mathbf{r}_y)\delta(\mathbf{r}'_x - \mathbf{r}'_y)\delta_{\alpha_x\alpha_y}\delta_{\beta_x\beta_y}$$

$$\frac{\delta \widehat{F}}{\delta F(\mathbf{x}, \mathbf{x}')} = \int \mathrm{d}\mathbf{y}\mathrm{d}\mathbf{y}' \,|\mathbf{y}'\rangle \langle \mathbf{y}|\, \frac{\delta F(\mathbf{y}, \mathbf{y}')}{\delta F(\mathbf{x}, \mathbf{x}')} = |\mathbf{x}'\rangle \langle \mathbf{x}|$$

$$\int \mathrm{d}\mathbf{x}\mathrm{d}\mathbf{x}' \frac{\delta \mathrm{tr}\left(\hat{F}\hat{O}\right)}{\delta F(\mathbf{x}, \mathbf{x}')} = \int \mathrm{d}\mathbf{x}\mathrm{d}\mathbf{x}' \mathrm{tr}\left(|\mathbf{x}'\rangle \langle \mathbf{x}|\, \hat{O}\right) = \int \mathrm{d}\mathbf{x}\mathrm{d}\mathbf{x}' \hat{O}(\mathbf{x}', \mathbf{x})$$

$$\frac{\delta \rho(\mathbf{r})}{\delta F(\mathbf{x}, \mathbf{x}')} = \delta(\mathbf{x}, \mathbf{x}') = \delta(\mathbf{r}_x - \mathbf{r})\delta(\mathbf{r}'_x - \mathbf{r})\delta_{\alpha_x\beta_{x'}}$$

$$\int \mathrm{d}\mathbf{x}\mathrm{d}\mathbf{x}' \frac{\delta \mathrm{tr}\left(\hat{F}\hat{O}\right)}{\delta F(\mathbf{x}, \mathbf{x}')} |\mathbf{x}\rangle \langle \mathbf{x}'| = \hat{O} \,,$$

which would hold for any one-body operator $\hat{O}$.

11.1.2 Definition of the KS Hamiltonian from the Total Energy

In the constrained search formulation we minimize the Hohenberg–Kohn functional $E[\rho]$ with respect to the internal degrees of freedom of the density operator $\widehat{F}$, by *imposing* the constraint $\rho[\widehat{F}] = \rho$. In this way, we can generalize the dependency of the functional to the density matrix, and the terms which explicitly depend on non-diagonal terms of $\hat{F}$ can be included in this way. We might therefore consider, with more generality, that the Kohn–Sham energy functional is defined in terms of the density matrix, namely $E[\rho[\widehat{F}]] = E[\widehat{F}]$.

It is interesting to consider the *linearization* of the energy functional with respect to the density operator. We define the one-body operator

$$\widehat{H}_{\mathrm{KS}}[\widehat{F}] \equiv \int \mathrm{d}\mathbf{x}\mathrm{d}\mathbf{x}' \frac{\delta E[\widehat{F}]}{\delta F(\mathbf{x}, \mathbf{x}')} |\mathbf{x}\rangle \langle \mathbf{x}'| \ ,$$

which is self-adjoint (as $F(\mathbf{x}, \mathbf{x}') \equiv \langle \mathbf{x}' | F | \mathbf{x} \rangle$). We may split the total energy into a *band structure* and a *double counting* term. The band structure term is defined as

$$E_{\text{BS}}[\widehat{F}] = \text{tr}\left(\widehat{H}_{\text{KS}}[\widehat{F}]\widehat{F}\right) ,$$

and it is therefore thought of as a functional of *the sole* density matrix $\hat{F}$.

The double counting term would then express the difference between the band structure term and the total energy, this time interpreted as a function of the charge density of the system:

$$E_{\text{DC}}[\widehat{F}] = E[\rho[\hat{F}]] - E_{\text{BS}}[\widehat{F}] . \tag{11.1}$$

The above definitions are valid for functionals which are a linear combination of each other, namely if $L[\rho] = E[\rho] + F[\rho]$, then $L_{\text{BS,DC}} = E_{\text{BS,DC}} + F_{\text{BS,DC}}$, and the same is valid for the respective KS Hamiltonians. Yet, obviously the ground-state (GS) density of the functional L will not be in relation with the GS densities of E and F.

Therefore we might formally write

$$E[\rho] = \min_{\hat{F}:\rho[\hat{F}]=\rho} \left\{E_{\text{BS}}[\widehat{F}] + E_{\text{DC}}[\widehat{F}]\right\} ,$$

where with the above notation we indicate the minimum of the functional over all the density matrices that have ρ as a charge density.

The GS energy of the system is found by minimizing the functional $E[\rho]$ over all N-representable densities, with N the number of electrons. Practically, such N-representability is achieved by representing the density in terms of the Fermi (super-) operator, defined by

$$\widehat{\widehat{\mathcal{F}}}(T, N)\hat{O} \equiv \frac{1}{1 + \exp\left(-\frac{\hat{O}-\mu(N)}{kT}\right)}$$

applied onto the KS Hamiltonian operator. The chemical potential μ is fixed by imposing that the total number of electrons corresponds to N, namely by $\text{tr}\left(\widehat{\widehat{\mathcal{F}}}(T, N)\widehat{H}_{\text{KS}}[\rho]\right) = N$. The constrained search minimization can therefore be expressed in the following way:

$$E_{GS}[N, V_{\text{ext}}; T] = \min_{\hat{F}:\hat{F}=\hat{\hat{\mathcal{F}}}(T,N)\widehat{H}_{\text{KS}}[\hat{F}]} \left\{\text{tr}\left(\widehat{H}_{\text{KS}}[\widehat{F}]\widehat{F}\right) + E_{\text{DC}}[\widehat{F}]\right\} ,$$

where the GS energy functional is, as pointed out by Hohenberg and Kohn in their seminal work, written in terms of the total number of electron N and of the external potential V_{ext}.

11.2 Operators of Generalized Kohn–Sham Formalism

As we discussed before, each of the terms contributing to the total energy creates a contribution to the KS Hamiltonian, to the double counting term and their respective derivatives. We list here such contributions:

Kinetic Term
The single particle term for the kinetic energy depends only on the density matrix:

$$T_s[\hat{F}] = -\frac{1}{2}\int \mathbf{dx}\,\mathrm{tr}\left(|\mathbf{x}\rangle\,\nabla^2\,\langle\mathbf{x}|\,\hat{F}\right).$$

This term gives rise to the kinetic operator of $\hat{H}_{\mathrm{KS}}$:

$$\hat{T}_s \equiv -\frac{1}{2}\int \mathbf{dx}\,|\mathbf{x}\rangle\,\nabla^2\,\langle\mathbf{x}|\,,$$

and as, for this term $E_{\mathrm{BS}} = T_s$, we have $E_{\mathrm{DC}} = 0$. Once again, we have employed this notation to highlight that the Laplacian operator is associated to the kinetic operator in the position representation.

External Potential
Another important term for the energy comes from the external potential. Let us assume a generic non-local form for a one-body potential:

$$V[\hat{F}] = \mathrm{tr}\left(\hat{V}_{\mathrm{ext}}\,\hat{F}\right) = \int \mathbf{dxdx}'\;V_{\mathrm{ext}}(\mathbf{x},\mathbf{x}')F(\mathbf{x}',\mathbf{x})$$

where the external potential term is

$$\hat{V}_{\mathrm{ext}} = \int \mathbf{dxdx}'\;|\mathbf{x}'\rangle\,V_{\mathrm{ext}}(\mathbf{x},\mathbf{x}')\,\langle\mathbf{x}|\,.$$

The double-counting term is also zero in this case.

Exchange and Correlation Energy
In the traditional semi-local approximations to the XC terms, the functional $E_{\mathrm{xc}}[\rho]$ is considered as a function of the *spin* density $\rho^{\alpha\beta}(\mathbf{r}) \equiv \langle\mathbf{r},\beta|\,\hat{F}\,|\mathbf{r},\alpha\rangle$. The one-body exchange and correlation potential is the functional derivative

$$\hat{V}_{\mathrm{xc}} = \sum_{\alpha,\beta}\hat{V}_{\mathrm{xc}}^{\alpha\beta}\,,\qquad \hat{V}_{\mathrm{xc}}^{\alpha\beta} = \int \mathbf{dr}\,|\mathbf{r},\alpha\rangle\,\frac{\delta E_{\mathrm{xc}}[\rho[\hat{F}]]}{\delta\rho^{\alpha\beta}(\mathbf{r})}\,\langle\mathbf{r},\beta|\,,$$

evaluated for the density ρ. This potential will appear in the $\hat{H}_{\mathrm{KS}}$ operator, and the corresponding band structure energy will be $E_{\mathrm{BS}} = \mathrm{tr}\left(\hat{V}_{\mathrm{xc}}\hat{F}\right)$, which will be expressed as a function of $\rho^{\alpha\beta}(\mathbf{r})$. The double counting term will thus be a functional of the spin density alone:

$$E_{\mathrm{DC}}[\rho] = E_{\mathrm{xc}}[\rho] - \sum_{\alpha\beta} \int d\mathbf{r} V_{\mathrm{xc}}^{\alpha\beta}(\mathbf{r})\rho^{\alpha\beta}(\mathbf{r}) \, .$$

Hartree Potential from the Electrostatic Energy
Another important term is related to the electrostatic energy. To generalize the treatment let us suppose that such term is provided by

$$E_{\mathrm{H}}[\epsilon, \rho] \equiv \frac{1}{2} \int d\mathbf{r} \rho(\mathbf{r}) \mathcal{V}_{\mathrm{H}}[\epsilon](\mathbf{r}, \mathbf{r}')\rho(\mathbf{r}') \, ,$$

where $\mathcal{V}_{\mathrm{H}}[\epsilon]$ denotes the Green's function of the Poisson operator $\mathcal{P}[\epsilon]$ with a generic dielectric function ϵ. We assume that this function can also be dependent on the charge density ρ. Such operators are implicitly defined by the following relations:

$$\int d\mathbf{r}' \mathcal{P}[\epsilon](\mathbf{r}, \mathbf{r}')\varphi[\epsilon, \rho](\mathbf{r}') \equiv -\frac{1}{4\pi} \nabla \cdot (\epsilon(\mathbf{r}) \nabla \varphi[\epsilon, \rho](\mathbf{r})) = \rho(\mathbf{r}) \, . \tag{11.2}$$

The Hartree term that has to be included in the KS Hamiltonian is therefore

$$\hat{V}_{\mathrm{H}}[\rho] = \sum_{\alpha} \int d\mathbf{r} \, |\mathbf{r}, \alpha\rangle \, V_{\mathrm{H}}[\epsilon, \rho](\mathbf{r}) \, \langle \mathbf{r}, \alpha|$$

with the Hartree potential being

$$V_{\mathrm{H}}[\epsilon, \rho](\mathbf{r}) \equiv \frac{\delta E_{\mathrm{H}}[\epsilon, \rho]}{\delta \rho(\mathbf{r})} = \varphi[\epsilon, \rho](\mathbf{r}) \, .$$

The band structure term in this case is $E_{\mathrm{BS}} = 2E_{\mathrm{H}}$ and the double counting term is thus $E_{\mathrm{DC}} = -E_{\mathrm{H}}$.

11.2.1 KS-DFT Formalism: Nearsightedness and Support Functions

An equivalent expression of this functional can be provided by giving the expression of the electronic density in term of the Kohn–Sham (KS) orbitals $|\Psi_i\rangle$. In our approach the KS orbitals are expressed as a linear combination of *intermediate*, possibly minimal, basis functions $|\varphi_\alpha\rangle$, also referred to as support functions (SFs):

$$|\Psi_i\rangle = \sum_{\alpha} c_i^{\alpha} \, |\varphi_\alpha\rangle \, . \tag{11.3}$$

In other terms, we assume that the density matrix of the system $\widehat{F}$ can be defined from a set of localized SFs as follows:

$$\widehat{F} \equiv \sum_i f(\epsilon_i) |\Psi_i\rangle \langle\Psi_i| = \sum_{\alpha,\beta} |\varphi_\alpha\rangle K^{\alpha\beta} \langle\varphi_\beta| , \tag{11.4}$$

with an SF overlap matrix $S_{\alpha\beta} = \langle\varphi_\alpha|\varphi_\beta\rangle$, which can be chosen to have a unit diagonal and where $K^{\alpha\beta}$ is the so-called density kernel. Here the f_i denote the occupation numbers associated with the KS orbitals $|\Psi_i\rangle$, which determine the density $\rho(\mathbf{x}) = F(\mathbf{x}, \mathbf{x})$. Such occupation numbers are dependent on the KS eigenvalues ϵ_i, $f_i = f(\epsilon_i)$, as by definition the KS orbitals satisfy the eigenvalue problem of H_{KS}.

This kernel is related to the density matrix formulation of Hernández and Gillan [37], and has to be thought of as functionally dependent on the KS Hamiltonian, namely $\mathbf{K} = \mathbf{K}[H_{\mathrm{KS}}]$. The density matrix $F(\mathbf{r}, \mathbf{r}')$ decays exponentially with respect to the distance $|\mathbf{r} - \mathbf{r}'|$ for systems with a finite gap or for metals at finite temperature [3, 11, 12, 26, 36, 39, 42]; for metals at zero temperature it decays algebraically [47]. Therefore in these cases it can be represented by strictly localized basis functions. A natural and exact choice for these would be the maximally localized Wannier functions (MLWFs) which have the same exponential decay [48]. In our case, the localized functions are constructed in situ during the self-consistency cycle in terms of a underlying wavelet basis set.

The energy of the system in the KS formalism can then be defined by employing the external potential V_{ext}, which contains local and non-local pseudopotential (PSP) terms, and depends on a set of electron-independent parameters λ which label the system, for instance by indicating the positions of atoms. We also consider the core charge density $\rho_c[\lambda]$, which depends on such a label, but it is assumed to be independent of the KS orbitals. We also include the exact exchange term $E_X[F]$ and the associated Fock operator $\widehat{D}_X$ already defined above, multiplied by a parameter α_X which quantifies the fraction of exact exchange introduced in the formalism. We then obtain:

$$\begin{aligned} E[\lambda, \rho_c, \mathbf{K}, \{\varphi_\alpha\}] &= -\frac{1}{2}\int \mathbf{dr}\, \mathrm{tr}\left(\nabla^2 |\mathbf{r}\rangle \langle\mathbf{r}|\, F\right) \\ &+ \mathrm{tr}\left(F V_{\mathrm{ext}}[\lambda]\right) + E_{\mathrm{xc}}[\rho + \rho_c[\lambda]] + E_{\mathrm{H}}[\rho] + \alpha_X E_X[F] \\ &= \sum_{\alpha\beta} H_{\mathrm{KS}\alpha\beta} K^{\alpha\beta} \\ &- E_{\mathrm{H}}[\rho] - \alpha_X E_X[F] + E_{\mathrm{xc}}[\rho + \rho_c] - \int \mathbf{dr}\rho(\mathbf{r}) V_{\mathrm{xc}}[\rho + \rho_c](\mathbf{r}) , \end{aligned} \tag{11.5}$$

The KS Hamiltonian is defined as

$$H_{\rm KS}[\lambda, \rho_c, \mathbf{K}, \{\varphi_\alpha\}] \equiv -\frac{1}{2}\nabla^2 + V_{\rm H}[\rho] + V_{\rm xc}[\rho+\rho_c] + \alpha_X \widehat{D}_X + V_{\rm ext}[\lambda]. \quad (11.6)$$

BigDFT efficiently treats *Gaussian pseudopotentials* of the Goedecker–Teter–Hutter (GTH) [29] and Hartwigsen–Goedecker–Hutter (HGH) [35] types (see also [44, 64]), since the intrinsic separability of both the basis set and Gaussian pseudopotentials allows for the simplification of several 3D operations into a sum of 1D products. The approximation of the all-electron KS quantities induced by the PSP terms has been shown to be much less severe than the exchange and correlation terms. Such PSP terms have proven to yield all-electron precision for most of the quantities of interest in ground-state DFT calculations, as can be seen in the DeltaTest initiative [45], where an accuracy of 0.1 meV/atom—the best among the PSP calculations—can be obtained for the set of atoms belonging to the first three rows of the periodic table, or in Willand et al. [64], where we show that the accuracy of the G2-1 and S22 test sets is comparable with all-electron calculations made by highly precise Gaussian basis sets.

11.2.2 Atomic Forces

The atomic forces are, by definition, the opposite of the derivative of the total energy with respect to the atom position. In this notation we should thus calculate

$$\frac{\mathrm{d}E}{\mathrm{d}\lambda} = \sum_i f_i \langle \Psi_i | \frac{\mathrm{d}V_{\rm ext}}{\mathrm{d}\lambda} | \Psi_i \rangle + \sum_\sigma \int \mathrm{d}\mathbf{r} \frac{\mathrm{d}\rho_c^\sigma(\mathbf{r})}{\mathrm{d}\lambda} V_{\rm xc}^\sigma[\rho + \rho_c](\mathbf{r}) \,. \quad (11.7)$$

Clearly, numerically, the set of $|\Psi_i\rangle$ is expressed in a finite basis set. This means that the action of $\hat{H}_{\rm KS}$ can in principle lie *outside* the span of the $|\Psi_i\rangle$. We can define therefore a residual function

$$|\chi_i\rangle = \hat{H}_{\rm KS} |\Psi_i\rangle - \epsilon_i |\Psi_i\rangle \,, \quad (11.8)$$

which represents the deviation of the *numerical* KS orbital from being the *exact* KS orbital. By definition $\langle \Psi_j | \chi_i \rangle = 0 \, \forall i, j$. The norm of this vector, *once projected* onto the basis set used to express $|\Psi_i\rangle$, is often used as a convergence criterion for the ground state energy. However, even though the basis set is finite, the orthogonality of KS orbitals holds exactly. It is thus easy to show that the *numerical* atomic forces are defined as follows:

$$-\frac{dE_{\mathrm{BS}}}{d\mathbf{R}_a} = -\sum_i f_i \langle\Psi_i| \frac{\partial \hat{H}_{\mathrm{KS}}}{\partial \mathbf{R}_a} |\Psi_i\rangle - 2\sum_i \mathrm{Re}\left(\left\langle\chi_i|\frac{\partial \Psi_i}{\partial \mathbf{R}_a}\right\rangle\right) , \qquad (11.9)$$

where the first term of the right-hand side of the above equation is the Hellman–Feynman contribution to the forces. The second term should not exist in a exact representation and only originates from the fact that the residual vector is non-zero.

Let us now suppose that the KS Hamiltonian and orbitals are expressed in a basis set which is complete enough to describing the orbitals *and their derivatives* within a targeted error. In this case the norm of $|\chi_i\rangle$ can be reduced within the *same* basis set so as to meet this targeted precision. Therefore the projection of $\left|\frac{\partial \Psi_i}{\partial \mathbf{R}_a}\right\rangle$ onto the basis set used for the calculation can be safely neglected as it is associated to the same numerical precision. When the basis set is complete enough to express also $\left|\frac{\partial \Psi_i}{\partial \mathbf{R}_a}\right\rangle$, then the atomic forces can be evaluated by the Hellmann–Feynman term only, as the remaining part is proportional to the desired precision. We can then neglect the second term of Eq. (11.9); this is an example of how having a systematic basis set, where we have *estimators* of the error for *generic* quantities, may help to drastically simplify the calculation of GS-derived quantities.

11.3 Wavelets as a Computational Basis Set

In the first two decades of the development of density functional theory, wavelet basis sets have rarely been used for electronic structure calculations, with most efforts having been devoted to their use in all-electron calculations, e.g. in MRChem (see [40, 53]) and most applications of MADNESS [32]. Since such a basis is therefore rather uncommon, we explain its use in the context of KS-DFT calculations. While referring the reader to Goedecker [30] for an exhaustive presentation of how wavelet basis sets can be used for numerical simulations, here we summarize the main properties of Daubechies wavelets, with a special focus on the representation of the objects (wavefunctions and operators) involved in the KS-DFT formalism. We will start by illustrating the principles of one-dimensional Daubechies wavelets basis.

Wavelet theory [15, 17] comes from multi-resolution analysis (MRA), which has the goal of properly defining how we can enlarge a given basis set for a given accuracy. MRA defines nested sets and their complements with the idea to have, for a given function, a suite of different approximated functions in order to control the accuracy. Before formally defining MRA, we prefer to first define the scaling function.

11.3.1 Introduction to Wavelet Theory

The key concept of wavelet theory is the scaling relation which links two levels of resolution. A mother scaling function $\Phi(x)$ is defined by the scaling relation

$$\Phi(x) = \sum_{j=-m}^{m} h_j \Phi(2x - j) \tag{11.10}$$

where the $\Phi(x-j)$ functions are centred on an equidistant grid. The integer m gives the length $2m+1$ of the compact support where the coefficients are non-zero. This scaling relation with the filter $\{h_j\}_{j=-m,m}$ fully defines the mother scaling function as a sum of the same function but at a twice finer resolution. To apply any operation, we need to manipulate this filter $\{h_j\}$ by means of convolutions.

The simplest scaling function, called the Haar function, is a simple rectangle function which is a sum of two smaller rectangle functions. Graphically we can express the scaling relation for the Haar functions as

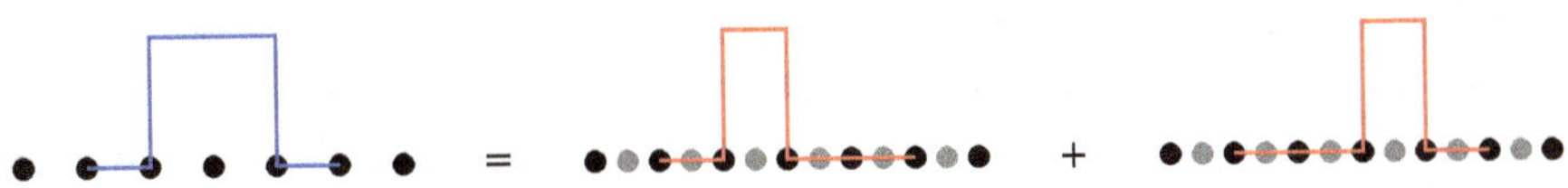

For the inverse operation, which consists in going from a fine resolution to a coarse one, we need to introduce a new function, called the wavelet function $\Psi(x)$, which gives the details allowing us to avoid a loss of information. For the Haar scaling function, we have

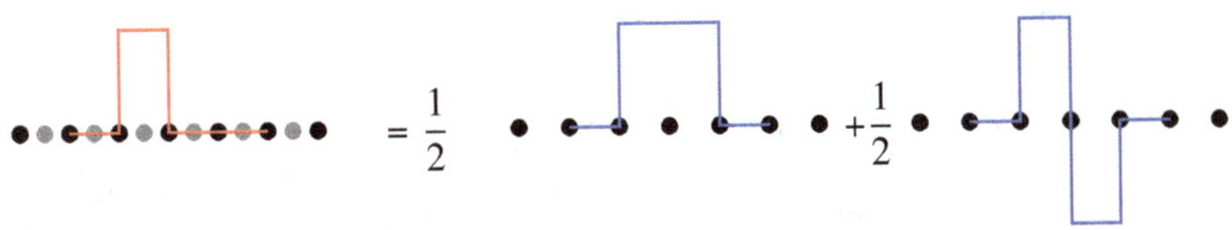

In this figure, we point out that a fine resolution (the grid with black and grey points) can be expressed using a coarse resolution (the grid with only black points) with a **double grid step** and **two functions per node**, the scaling function and the wavelet function. In the case of the `BigDFT` code, the mesh is uniform on the simulation domain but the number of basis functions is different per node depending on the resolution we want.

Like the scaling relation (11.10) for the mother scaling function, the wavelet function $\Psi(x)$ can also be expressed as a sum of scaling functions at a finer resolution

$$\Psi(x) = \sum_{j=-m}^{m} g_j \Phi(2x - j)\,, \tag{11.11}$$

which we sketch for the Haar wavelet function as

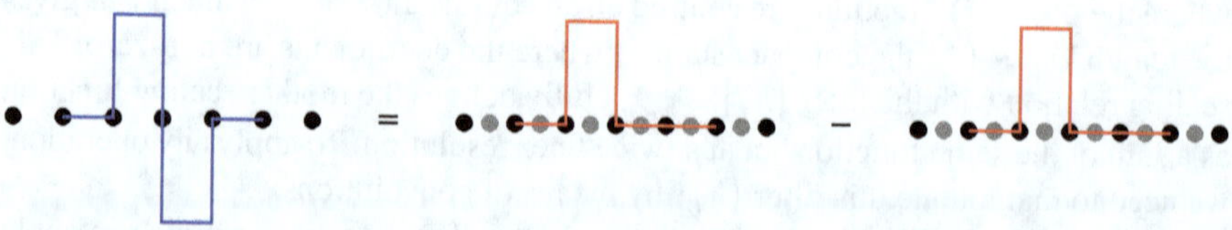

The scaling functions $\Phi(2x - j)$ come from the mother scaling function $\Phi(2x)$ centred on the points of abscissa $j/2$ which are on a finer mesh. The wavelet families that we use have a compact support, they have non-zero values only in a given interval $[-m, m]$, but non-compact wavelet families also exist [46]. This is particularly interesting because all the operations can be applied only in this interval. This is the reason why we call this basis set a localized basis set in real space.

Using orthogonal wavelets, as we did, means that we only manipulate the two filters $\{h_j\}$, $\{g_j\}$ which are linked by the scaling relations. These filters can decompose any function into scaling and wavelet functions at different levels of resolution.

The Daubechies wavelets [15, 17] are an important family of orthogonal wavelets. To develop any function $f(x)$ into Daubechies scaling functions, we apply the scalar product

$$\int \Phi_i(x)\Phi_j(x)\mathrm{d}x = \delta_{ij}\,, \quad f_0(x) = \sum_i f_i \Phi_i(x) \quad \text{where} \quad f_i = \int f(x)\Phi_i(x)\mathrm{d}x \tag{11.12}$$

to obtain an approximation $f_0(x)$ of the given function.

This property of orthogonality is especially interesting when expressing the electronic orbitals for each electron state avoiding the use of an overlap matrix. In this case we don't have to solve a generalized eigenvalue problem $HX = ESX$ to find the Kohn–Sham orbitals.

The number of continuous derivatives of the Daubechies functions, i.e. their smoothness, which is well described by the Sobolev spaces, depends strongly on the wavelet order. The Haar function is a Daubechies function of order 1 which is piecewise continuous.

BigDFT uses the least asymmetric Daubechies wavelet family of order $m = 8$ (see Fig. 11.1) for the electronic orbitals with two levels of resolution.

A homogeneous grid is created with a constant step grid h. For each atom type, two radii are defined: one, r_c, gives the extension of the mesh around each atom of the same nature, the second one, r_f, gives the extension where the electronic orbitals have one coefficient for the scaling function per point but also other ones for the corresponding wavelet functions.

To express the electronic density and the potentials, we prefer to use biorthogonal scaling functions (see Sect. 11.4.1). After some clues about wavelets, we come back to a formal definition of multi-resolution analysis following the notation from the book of S. Goedecker [31].

11.3.2 Definition of Multi-Resolution Analysis

We can introduce many levels of resolution using the full possibility of multi-resolution analysis (MRA). We here sketch some notions of MRA. A k ($k \in \mathbb{N}$) factor is used to specify the resolution level and i ($i \in \mathbb{Z}$) corresponds to the centre of the Daubechies basis function. We define

$$\Phi_i^{(k)}(x) \equiv 2^{k/2}\Phi(2^k x - i)\,, \qquad \Psi_i^{(k)}(x) \equiv 2^{k/2}\Psi(2^k x - i)\,. \tag{11.13}$$

Using the bra and ket notation, we can write the orthogonality conditions as

$$\langle\Phi_i^{(k)}|\Phi_j^{(k)}\rangle = \delta_{ij}; \qquad \langle\Phi_i^{(k)}|\Psi_j^{(k')}\rangle = 0; \qquad \langle\Psi_i^{(k)}|\Psi_j^{(k')}\rangle = \delta_{ij}\delta_{kk'}\,, \tag{11.14}$$

We then introduce the vector spaces spanned by $\Phi^{(k)}$ and $\Psi^{(k)}$, respectively:

$$\mathcal{V}_k \equiv \text{span}\left\{|\Phi_i^{(k)}\rangle\right\}, \qquad \mathcal{W}_k \equiv \text{span}\left\{|\Psi_i^{(k)}\rangle\right\}, \tag{11.15}$$

with $\mathcal{V}_k \cap \mathcal{W}_k = \{0\}$, from the orthogonality conditions. The scaling relations are valid for any level k

$$|\Phi_i^{(k-1)}\rangle = \sum_j h_j\,|\Phi_{2i+j}^{(k)}\rangle\,, \qquad |\Psi_i^{(k-1)}\rangle = \sum_j g_j\,|\Phi_{2i+j}^{(k)}\rangle\,, \tag{11.16}$$

which means that $\mathcal{V}_k = \mathcal{V}_{k-1} \oplus \mathcal{W}_{k-1}$, or stated otherwise, that the wavelet spaces $\mathcal{W}_k$ are the orthogonal complement of $\mathcal{V}_k$ in the space $\mathcal{V}_{k+1}$. Thus we define a sequence of nested sets

$$\mathcal{V}_0 \subset \mathcal{V}_1 \subset \cdots \mathcal{V}_k \subset \mathcal{V}_{k+1} \cdots \subset L^2(\mathbb{R})\,. \tag{11.17}$$

For all $k \geqslant 0$, the space of square integrable functions $L^2(\mathbb{R})$ can be decomposed into an infinite sum of spaces:

$$L^2(\mathbb{R}) = \text{closure of} \left(\mathcal{V}_k \bigoplus_{q=k}^{\infty} \mathcal{W}_q \right) . \tag{11.18}$$

This can be generalized into biorthogonal wavelet families with two nested sets, one for the direct space and another one for the dual space.

The relation (11.18) shows that any square integrable function can be uniquely decomposed as

$$f(x) = \sum_i c_i^k \Phi_i^{(k)}(x) + \delta f(x), \qquad \delta f(x) = \sum_{q=k}^{\infty} \sum_i d_i^k \Psi_i^{(k)}(x) . \tag{11.19}$$

In contrast to `BigDFT` which uses only two levels of resolutions, two groups have developed multi-resolution basis sets based on multiwavelets [33, 34, 40]. They can achieve very accurate results which are very useful for reference calculations to decouple the different approximations coming from the basis sets, the choice of algorithms, or which are inherited from the used methods such as the DFT functional. The high level of resolution is only used near the nuclei to describe the oscillations and the nuclei cusp of each electronic orbital.

11.3.3 Daubechies Wavelets

In the `BigDFT` code, the least asymmetric Daubechies wavelet family of order $m = 8$ is the basis set used to describe the KS orbitals as shown in Fig. 11.1. These functions feature a compact support $[-m+1, m]$ and are also localized in Fourier space. The use of Daubechies wavelets families is guided by different criteria. Daubechies wavelets represent the best compromise between compact support, continuity and orthogonality for a wavelet family. We chose the family of order 16 as it is the most compact one which has a degree of continuity of at least two, thereby enabling a unbiased evaluation of the kinetic operator. Nonetheless, such a family exhibits polynomial exactness of degree 8, which means that it is able to represent exactly the Taylor expansion of a Kohn Sham orbital up to the eighth order. This observation, combined with the Magic Filter method (see [23]), enables an accurate and efficient approach for the evaluation of the potential energy in KS-DFT calculations.

The coefficients of the filters h_j and $g_j = (-1)^j h_{-j}$ of the scaling relation (or "refinement relations") for the Daubechies families are called *low-* and *high-pass* filters, respectively. A wavelet family is therefore completely defined by its low-pass filter. In the case of Daubechies-$2m$ wavelets, $j \in [1-m, m]$.

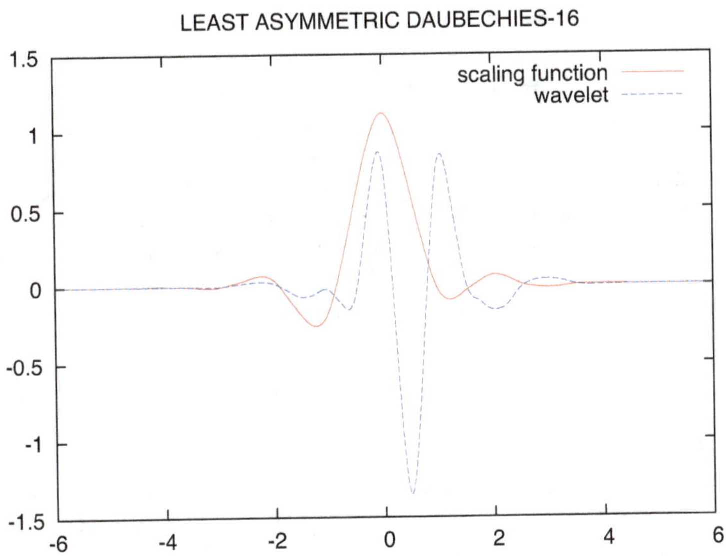

Fig. 11.1 Least asymmetric Daubechies wavelet family of order $m = 8$. Note that both the scaling function $\Phi(x)$ and the wavelet $\Psi(x)$ are different from zero only within the interval $[1 - m, m]$

11.3.4 One-Dimensional Operators with Daubechies Wavelets

The multiresolution property also plays a fundamental role in the wavelet representation of differential operators. For example, it can be shown that the *exact* matrix elements of the kinetic operator can be written in the form of a circulant matrix, namely:

$$T_{ij} = T_{i-j} \equiv -\frac{1}{2} \int \mathrm{d}x \, \Phi_i^{(0)}(x) \partial^2 \Phi_j^{(0)}(x) \,, \tag{11.20}$$

and are equal to the entries of an eigenvector of a matrix which solely depends on the low-pass filter (see e.g. [30]).

Daubechies-$2m$ wavelets exhibit m vanishing moments, thus any polynomial of degree less than m can be represented exactly by an expansion over the sole scaling functions of order m. For higher order polynomials the error is $O(h^m)$, i.e. vanishingly small as soon as the grid is sufficiently fine. Hence, the difference between the discretized representation and the exact function f decreases as h^m. The discretization error due to Daubechies-$2m$ wavelets is therefore controlled by the grid spacing. Among all the orthogonal wavelet families, Daubechies wavelets feature the minimum support length for a given number of vanishing moments.

Given a potential V known numerically on the points $\{x_k\}$ of a uniform grid, it is possible to identify an effective approximation for the potential matrix elements $V_{ij} \equiv \langle \Phi_j | V | \Phi_i \rangle$. It has been shown [23, 55] that a quadrature filter $\{\omega_k\}$ can be defined such that the matrix elements given by

$$V_{ij} \equiv \langle \Phi_j | \, V \, | \Phi_i \rangle = \sum_k \omega_{k-i} \, V(x_k) \, \omega_{k-j} \tag{11.21}$$

yield excellent accuracy with the optimal convergence rate $O(h^{2m})$ for the potential energy. The same quadrature filter can be used to express the grid point values of a (wave)function given its expansion coefficients in terms of scaling functions:

$$f(x_k) = \sum_i c_i \, \omega_{k-i} + O(h^m) \, ;$$

$$c_i = \sum_k f(x_k) \, \omega_{k-i} + O(h^m) \, .$$

As a result, the potential energy can equivalently be computed either in real space or in the wavelet space, i.e. $\langle f | \, V \, | f \rangle \simeq \sum_k f(x_k) V(x_k) f(x_k) \simeq \sum_{ij} c_i V_{ij} c_j$. The quadrature filter elements can therefore be considered as the most reliable transformation between grid point values $f(x_k)$ and scaling function coefficients c_i, as they provide exact results for polynomials of order up to $m - 1$ and do not alter the convergence properties of the basis set discretization. The filter $\{\omega_k\}$ is of length $2m$ and is defined unambiguously by the moments of the scaling functions (which in turn depend only on the low-pass filter) [30].

Using the above formulae, the (so far one-dimensional) Hamiltonian matrix $H_{ij} = T_{ij} + V_{ij}$ can be constructed. Note that, in contrast to other discretization schemes (finite differences, plane waves etc.), in the wavelet basis set *neither* the potential *nor* the kinetic terms have diagonal representations. Instead, $\widehat{H}$ is represented by a *band matrix* of width $2m$. We will discuss the mathematical considerations beyond these arguments in the forthcoming section. First, we will introduce the discretization we will employ for the three-dimensional domain.

11.3.5 Three-Dimensional Wavelet Basis

A support function $\varphi_\alpha(\mathbf{r})$ can thus be expanded in the wavelet basis as follows:

$$\varphi_\alpha(\mathbf{r}) = \sum_{i_1,i_2,i_3} s_{i_1,i_2,i_3} \Phi^{(0)}_{i_1,i_2,i_3}(\mathbf{r}) + \sum_{j_1,j_2,j_3} \sum_{l=1}^{7} d^{(\ell)}_{j_1,j_2,j_3} \Psi^{(\ell)}_{i_1,i_2,i_3}(\mathbf{r}). \tag{11.22}$$

Here $\Phi^{(0)}_{i_1,i_2,i_3}(\mathbf{r}) = \Phi(x - i_1)\Phi(y - i_2)\Phi(z - i_3)$ denotes the tensor product of three one-dimensional scaling functions, whereas $\Psi^{(\ell)}_{j_1,j_2,j_3}(\mathbf{r})$ are the seven tensor products containing at least one one-dimensional wavelet. The sums over i_1, i_2, i_3 (j_1, j_2, j_3) run over all grid points where the scaling functions (wavelets) are centered.

In other words, the three-dimensional basis functions are a tensor product of one-dimensional basis functions, a mixed basis set of scaling functions augmented by a set of 7 wavelets. These points are associated with regions of low and high resolution levels, respectively. Note that we are using a cubic grid, where the grid spacing is the same in all directions, but the following description can be straightforwardly applied to general orthorhombic and non-orthorhombic grids.

In a simulation domain, there are therefore three regions: those which are closest to the atoms ("fine region") carry one (three-dimensional) scaling function and seven (three-dimensional) wavelets; those which are further from the atoms ("coarse region") carry only one scaling function, corresponding to a resolution which is half that of the fine region; and those which are even further away ("empty region") carry neither scaling functions nor wavelets. To determine these regions of different resolution, we construct two spheres around each atom a; a small one with radius $R_a^f = \lambda^f \cdot r_a^f$ and a large one with radius $R_a^c = \lambda^c \cdot r_a^c$ ($R_a^c > R_a^f$). The values of r_a^f and r_a^c are fixed for each atom type, whereas λ^f and λ^c can be specified by the user in order to control the accuracy of the calculation. The fine (coarse) region is then given by the union of all the small (large) spheres, as shown in Fig. 11.2. Hence in `BigDFT` the basis set is controlled by three user specified parameters; systematic

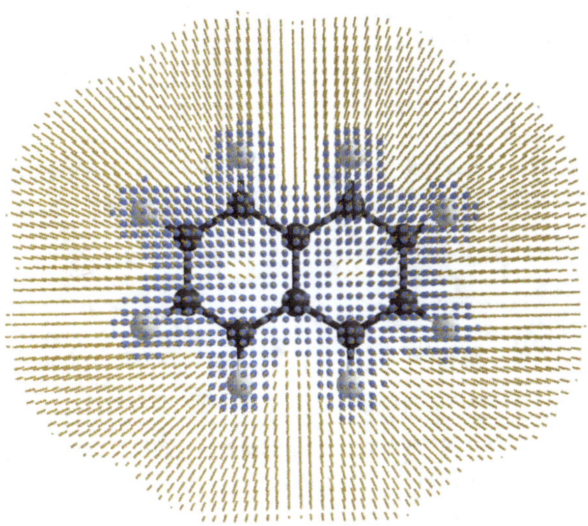

Fig. 11.2 Example simulation grid for a molecule with coarse (fine) grid points depicted in (gold) blue

convergence of the total energy is achieved by increasing the values of λ^c and λ^f while reducing the value of h.

We have all the tools to express any 3D functions as Kohn–Sham orbitals in a 3D wavelet basis set. Our goal is to use the properties of the compactness of the wavelets to build a linear-scaling structure electronic code. To do this, we need to introduce the concept of localization regions.

11.3.6 *Localization Regions*

Thanks to the nearsightedness principle, it is possible to define an approach in which the computational cost is linear scaling (LS) with respect to the number of atoms, N, rather than the cubic scaling (CS) which arises when extended KS orbitals are used. Such approaches allow one to go beyond the treatment of a few hundred atoms as is typically seen with $O(N^3)$ DFT approaches and instead treat systems containing several thousands of atoms. This has the benefit of also opening up the treatment of new types of materials and simulations using a pure quantum mechanical approach, as discussed, for example, in Ratcliff et al. [60] and Zaccaria et al. [65]. The $O(N)$ formalism implemented in the `BigDFT` code exploits the possibility that, for systems with suitable electronic structure, the support functions φ can be optimized while preserving their strict locality, namely so that their support is within a pre-defined localization region. A similar approach is for example used in the ONETEP [61] and Conquest [9] codes.

For large systems where the nearsightedness principle guarantees that a local description of the orbitals is possible, the large number of degrees of freedom offered by the wavelet basis is a waste. It is therefore advantageous to build a minimal basis formed of localized (e.g. atom centered) functions. Of course, these functions will also be expanded in terms of the underlying wavelet basis, but to strictly impose locality, they will be expressed only in a subset of this global basis set. To do so we set to zero all scaling function and wavelet coefficients if they lie outside of a sphere with radius R_{loc} around the point $\mathbf{R}_\alpha$ on which the function is centered:

$$\begin{aligned} s_{i_1,i_2,i_3} &= 0 \quad &\Leftarrow \quad &|\mathbf{R}_{(i_1,i_2,i_3)} - \mathbf{R}_\alpha| > R_{loc} \\ d_{j_1,j_2,j_3} &= 0 \quad &\Leftarrow \quad &|\mathbf{R}_{(j_1,j_2,j_3)} - \mathbf{R}_\alpha| > R_{loc}. \end{aligned} \tag{11.23}$$

Here $\mathbf{R}_{(i_1,i_2,i_3)}$ is the position of the grid point (i_1, i_2, i_3) and $\mathbf{R}_\alpha$ that of the atom on which the minimal basis function $\varphi_\alpha(\mathbf{r})$ is centered. These localization regions can still contain various resolution levels, as they are constructed on top of the global simulation domain. The index α is instead used in the following formulae to label kets that are associated to SFs.

In other terms, instead of working directly with the function $|\varphi_\alpha\rangle$, we work with the *localized* function $\mathcal{L}^{(\alpha)}|\varphi_\alpha\rangle$, where the definition of the localization projector operator in the Daubechies basis space is, as described:

$$\mathcal{L}^{(\alpha)}_{i_1,i_2,i_3;j_1,j_2,j_3} = \delta_{i_1 j_1}\delta_{i_2 j_2}\delta_{i_3 j_3}\theta(R_{loc} - |\mathbf{R}_{(i_1,i_2,i_3)} - \mathbf{R}_\alpha|)\,, \tag{11.24}$$

from which it becomes apparent that such a projection operator $\mathcal{L}^{(\alpha)}$ explicitly depends on the localization radius R_{loc} and the localization region center $\mathbf{R}_\alpha$. Clearly, if $|\varphi_\alpha\rangle$ is localized around $\mathbf{R}_\alpha$ and R_{loc} is large enough, $\mathcal{L}^{(\alpha)}$ leaves $|\varphi_\alpha\rangle$ unchanged and no approximation is introduced to the KS equations.

It is important to emphasize that, since the Daubechies basis set is independent of $\mathbf{R}_\alpha$, $|\varphi_\alpha\rangle$ depends on the center of the localization region by the introduction of the projector $\mathcal{L}^{(\alpha)}$:

$$|\varphi_\alpha\rangle = \mathcal{L}^{(\alpha)}|\varphi_\alpha\rangle\,. \tag{11.25}$$

By taking the derivative of this equation with respect to $\mathbf{R}_\beta$ it is easy to find

$$\left(1 - \mathcal{L}^{(\alpha)}\right)|\frac{\partial\varphi_\alpha}{\partial\mathbf{R}_\beta}\rangle = \delta_{\alpha\beta}\frac{\partial\mathcal{L}^{(\alpha)}}{\partial\mathbf{R}_\alpha}|\varphi_\alpha\rangle\,. \tag{11.26}$$

Let us now employ this result in the calculation of the atomic forces. When the KS orbitals are expressed in terms of the SFs, the non-Hellmann–Feynman term can be written as follows:

$$\mathbf{F}_a - \mathbf{F}_a^{(HF)} = -2\sum_{\alpha\beta}\mathrm{Re}\left(K^{\alpha\beta}\right)\left\langle\chi_\beta|\frac{\partial\varphi_\alpha}{\partial\mathbf{R}_a}\right\rangle \tag{11.27}$$

where the SF residue is

$$|\chi_\alpha\rangle = \hat{H}_{\mathrm{KS}}\,|\varphi_\alpha\rangle - \sum_{j\rho\sigma} c_j^\rho\epsilon_j c_j^\sigma\, S_{\sigma\alpha}\,|\varphi_\rho\rangle\,. \tag{11.28}$$

This result would be completely identical to Eq. (11.9) when no localization projectors are applied on the SF. Therefore the only term of the forces which cannot be captured by the localization regions is the part which is projected outside the localization regions (but still inside the computational domain of the CS approach). The extra Pulay term due to the localization constraint is therefore

$$\mathbf{F}^{(P)} = -2\sum_{\alpha\beta}\mathrm{Re}\left(K^{\alpha\beta}\right)\left\langle\chi_\beta|(1 - \mathcal{L}^{(\alpha)})|\frac{\partial\varphi_\alpha}{\partial\mathbf{R}_a}\right\rangle\,. \tag{11.29}$$

From Eq. (11.26) we obtain

$$\mathbf{F}^{(P)} = -2 \sum_{\alpha\beta} \mathrm{Re}\left(K^{\alpha\beta}\right) \left\langle \chi_\beta | \frac{\partial \mathcal{L}^{(\alpha)}}{\partial \mathbf{R}_a} | \varphi_\alpha \right\rangle \tag{11.30}$$

and from Eq. (11.24) we derive:

$$\begin{aligned} \frac{\partial \mathcal{L}^{(\alpha)}}{\partial \mathbf{R}_\beta}_{i_1,i_2,i_3;j_1,j_2,j_3} &= \delta_{\alpha\beta}\delta_{i_1 j_1}\delta_{i_2 j_2}\delta_{i_3 j_3} \frac{\mathbf{R}_{(i_1,i_2,i_3)} - \mathbf{R}_\alpha}{R_{loc}} \\ &\times \delta(R_{loc} - |\mathbf{R}_{(i_1,i_2,i_3)} - \mathbf{R}_\alpha|) \,. \end{aligned} \tag{11.31}$$

Therefore if the support functions are zero at the border of the localization region, there is no Pulay term in the atomic forces.

The Hellman-Feynman force, given by the expression

$$\begin{aligned} \mathbf{F}_a^{HF} &= -\sum_i f_i \left\langle \Psi_i | \frac{\partial \hat{H}_{\mathrm{KS}}}{\partial \mathbf{R}_a} | \Psi_i \right\rangle \\ &= -\sum_{\alpha,\beta} K^{\alpha\beta} \left\langle \varphi_\alpha | \frac{\partial \hat{H}_{\mathrm{KS}}}{\partial \mathbf{R}_a} | \varphi_\beta \right\rangle, \end{aligned} \tag{11.32}$$

involves only the functional derivative of the Hamiltonian operator, which is independent of the localization regions. The CS and the LS implementations of the atomic forces are therefore identical.

Daubechies wavelets are used to describe all terms related to KS orbitals. For the electronic density and potential, the orthogonality property is not the key parameter, but efficiently discretizing a function in a uniform real-space mesh is very important. The next section is devoted to this problem and develops tools related to the interpolating scaling function used in `BigDFT`.

11.4 Collocation Problems for Discretized Functions

Discretizing an analytic function on a uniform real-space grid is often done via a straightforward collocation method. This is ubiquitous in all areas of computational physics and quantum chemistry. An example in DFT is given by the external potential describing the interaction between ions and electrons. Notable examples are also given by the analytic functions defining compensation charges for range-separated electrostatic treatments.

A real space approach is mandatory in the solution of complex partial differential equation problems, as well as for the treatment of complex environments and non-trivial boundary conditions. The solution of the Poisson equation in vacuum and in the presence of continuum solvents is a notable example. In this framework, the collocation method is a straightforward procedure that is used to discretize a *known* function, to express its values in the real-space domain.

The accuracy of the collocation method used is therefore very important for the reliability of subsequent treatments like self-consistent field solutions of electronic structure problems. When the real-space grid is too coarse, the collocation method introduces numerical artifacts, spoiling the numerical stability of the description.

We present in this section a new quadrature scheme that is able to exactly preserve the multipoles of the original function for a wide range of grid spacings, in the spirit of the so-called "Magic-Filter" method, which has been used to identify a passage matrix between Daubechies wavelets and real-space grid meshes used to express the electronic density and potential.

For discretization on uniform grid spacings, the collocation method is well-justified when the original function can be reasonably approximated by an *interpolation* of its values on the grid mesh points. Let us consider a one-dimensional function f. Suppose we want to discretize this function on a uniform grid of spacing h and coordinates $x_k = hk$. Given a family of interpolating functions $\{L_k(t)\}$, if the approximation

$$f(x) \simeq f_L(x) \equiv \sum_k L_k(x) f(x_k) \tag{11.33}$$

is reasonably accurate, the collocation method can be applied. This fact stems from the interpolating property of the family $\{L_k(t)\}$. Indeed, an interpolating family comprises a set of functions L_k, each one associated to a point k of the grid, such that $L_k(j) = \delta_{kj}$. Given Eq. (11.33), then $f_L(x_k) = f(x_k)$ and the continuous representation of $f(x)$ may be given by $f_L(x)$.

Given the interpolating property, it is also said that an interpolating function family is *dual* to the Dirac deltas. In other terms, denoting the above function by the bra-ket notations, we have

$$|f\rangle \simeq |f_L\rangle = \sum_k |L_k\rangle\langle\delta_k|f\rangle\,,$$

where $|\delta_k\rangle$ represents the Dirac distribution centered at point x_k, i.e. $\langle\delta_k|f\rangle = f(x_k)$. The above defined interpolating property implies that the duality relation $\delta_{k\ell} = \langle\delta_\ell|L_k\rangle$ holds.

11.4.1 Bi-orthogonal Wavelets

Examples of interpolating functions are the interpolating scaling functions used in `BigDFT` to represent the electronic density and potential and solve the Poisson equation in order to calculate the Hartree potential.

The interpolating scaling functions, which are an important biorthogonal wavelet family, have the delta function as dual functions, which considerably simplifies the expansion of a given function $f(x)$:

$$f(x) = \sum_j f_j \varphi_j(x) \quad \text{where} \quad f_j = f(j)\,. \tag{11.34}$$

There are different families of interpolating wavelets. In BigDFT, we use the Deslauriers–Dubuc scaling functions, which are the auto-correlation scaling functions of the Daubechies wavelet [4]:

$$\varphi(x) = \int \Phi(s)\Phi(s-x)\mathrm{d}s\,. \tag{11.35}$$

In Fig. 11.3 we show the interpolating scaling Deslauriers–Dubuc function and its associated wavelet. Unlike the Daubechies wavelets, these functions are symmetric.

We use these interpolating scaling functions in order to express the electronic density, the different potentials and to solve the Poisson equation. Because `BigDFT` uses the pseudopotential approximation, two levels of resolution are enough and `BigDFT` uses only interpolating scaling functions expressed in the same mesh as the one for the electronic orbitals but with the fine resolution in the whole simulation domain.

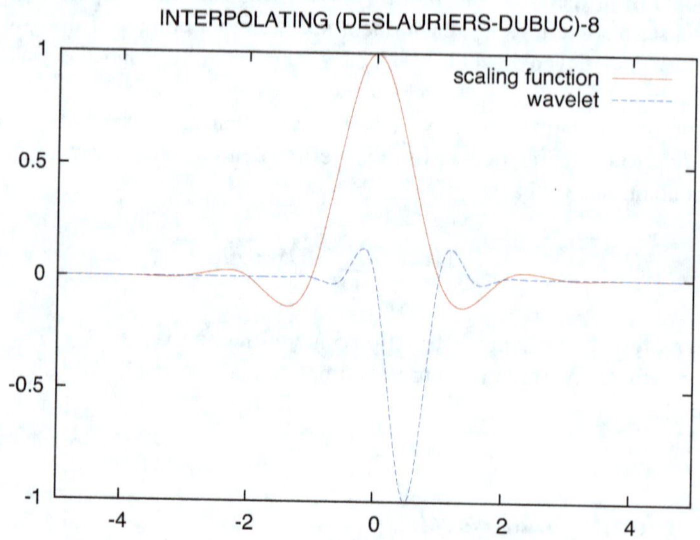

Fig. 11.3 Eighth order scaling and wavelet interpolating functions

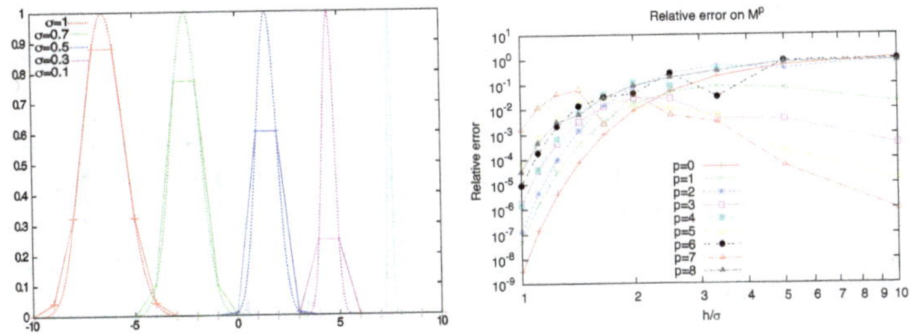

Fig. 11.4 Collocation of Gaussians of different standard deviation σ on a grid of spacing $h = 1$. All Gaussians are centered in points which lie between two grid spacings. In the left panel, it is easy to see that the collocated values are no longer reliable when the ratio h/σ grows above 1. This fact can also be confirmed by the discrete multipoles of the collocated function, whose error is presented in the right panel

11.4.2 Polynomial Exactness and Discrete Multipoles

The collocation method is therefore meaningful for the functions for which the action of the projector operator $\sum_k |L_k\rangle\langle\delta_k|$ approaches the action of the identity operator. It is easy to understand that this condition is valid only when the grid spacing size h is *considerably* smaller than the typical oscillations of the function $|f\rangle$ we want to represent. As soon as this is not the case, the function $|f_L\rangle$ becomes so different from $|f\rangle$ that the numerical accuracy of the approximation is severely affected. To have an idea of how rapidly the accuracy of this approximation is spoiled, in Fig. 11.4 we consider the collocation of a Gaussian function centered in x_0 and with standard deviation σ. When the ratio h/σ becomes bigger than one, the interpolated function f_L given by Eq. (11.33) becomes too crude an approximation. This is particularly visible for a localized function centered between two grid points. As the function becomes too sharp, the collocated values are nearly zero and the function f_L is not representative of the original f. This loosening of the accuracy can be easily quantified by having a look at the multipoles of the discretized functions.

This situation seems unavoidable: as the expansion coefficients of the function $|f_L\rangle$ are given in terms of the scalar products $\langle\delta_j|f\rangle$, the grid has to provide a reasonable sampling of the function f.

11.4.3 Polynomial Exactness

The accuracy of the approximation (11.33) is of great importance for a reliable computational treatment. Clearly, such accuracy is intimately related to the family of interpolating functions chosen.

Interpolating function families are normally constructed using families of polynomial functions. An interpolating family $\{L_k(x)\}$ is said to be of order m if any monomial function x^p, with $0 < p < m$, is exactly expressed by the interpolated function f_L of Eq. (11.33). This is the concept of *polynomial exactness*. This concept is important in determining the accuracy of the interpolation: a smooth function can reasonably be approximated by its Taylor polynomial around a given point. The higher the order of the polynomial exactness of the functions L_k, the better the Taylor expansion of the original function would be approximated by the function $f_L(x)$, therefore the difference

$$R_L(x) \equiv |f(x) - f_L(x)|$$

will be reduced. Let us now demonstrate that polynomial exactness of order m implies the moment preserving property. Polynomial exactness implies that, for a monomial function x^p, indicated by $|p\rangle$,

$$\sum_j x_j^p |L_j\rangle = \sum_j |L_j\rangle\langle\delta_j|p\rangle = |p\rangle \quad \forall p < m \,.$$

Therefore for all positive integers p, q such that $p + q < m$ we have

$$\langle q|p\rangle = \langle 0|p+q\rangle = \sum_j \langle q|L_j\rangle x_j^p = \sum_j \langle 0|L_j\rangle x_j^{p+q} \,,$$

and this is valid only if

$$\langle q|L_j\rangle = x_j^q \,, \quad 0 \leqslant q < m \,. \tag{11.36}$$

The above quantity is the q-moment (or multipole) of the interpolating function $L_j(x)$. Polynomial exactness therefore implies that the first m multipoles of the interpolated function $|f_L\rangle$ can actually be calculated exactly with their discrete multipoles: Indeed, by Eq. (11.36), we can express the moments of the function f_L in terms of discrete moments of the collocated function:

$$M_p[f_L] = \langle p|f_L\rangle = \sum_j \langle p|\varphi_j\rangle f(x_j) = \sum_j x_j^p f(x_j) \,, 0 \leqslant p < m \,.$$

As $|f\rangle$ is an analytic function, the accuracy of the multipoles $M_p[f_L]$ provides a quantitative evaluation of the accuracy of f_L. In Fig. 11.4 we see that the collocation method becomes unstable as soon as the grid spacing is bigger than the typical oscillations of the original function.

Let us now see if it is possible to define an alternative set of dual functions, such that the multipoles of the original functions are preserved. In other terms, we search for a family of dual functions such that

$$M_p[f_L] = \sum_j \langle p|L_j\rangle\langle\widetilde{L}_j|f\rangle = \sum_j x_j^p \langle\widetilde{L}_j|f\rangle = \langle p|f\rangle = M_p[f]\,, 0 \leqslant p < m\,.$$

It is easy to see that this condition can be obtained by imposing the polynomial exactness of the *dual* set $\widetilde{L}_k$. In other terms, if the set of $\widetilde{L}_k$ is such that

$$|p\rangle = \sum_j |\widetilde{L}_j\rangle\langle L_j|p\rangle = \sum_j |\widetilde{L}_j\rangle x_j^p\,, 0 \leqslant p < m$$

then the multipole preserving property is guaranteed.

A particular family of interpolating functions is given by the Interpolating Scaling Function (ISF):

$$\int \mathrm{d}x\,(x-x_j)^\ell \varphi(x/h) = h x_j^\ell\,, \forall \ell < m\,.$$

Indeed the ISF basis $\{\varphi_j\}_{j\in\mathbb{Z}}$ contains precisely the polynomials up to order $m-1$. This means that this basis is, *in principle*, able to express the Taylor expansion of the original function f. However, this does not happen when extracting the coefficients with the collocation method, as can be seen from Fig. 11.4. Therefore, there should exists a set of coefficients f_j such that

$$|f_\varphi\rangle \equiv \sum_j f_j|\varphi_j\rangle = |f\rangle + O(h^m)\,.$$

In the above discussion we have proceeded as if we have defined the basis of $|\varphi_j\rangle$ as a self-dual basis, modulo an error of $O(h^m)$. This is indeed a general property of all interpolating functions. In the following we will demonstrate that *any* interpolating function $L(t)$ which has a Taylor expansion is orthogonal to the ISF family up to $O(h^m)$. To demonstrate this point, let us define the Taylor expansion of $L(t) = \sum_k \alpha_k t^k$. We have

$$\begin{aligned}\int \mathrm{d}s L(s)\varphi(s-j) &= \sum_k \alpha_k \int_{\chi(\varphi)} \varphi(s-j)t^k \mathrm{d}s = \sum_{k=0}^m \alpha_k j^k + O(h^m)\\ &= L(j) + O(h^m) = \delta_j + O(h^m)\end{aligned}$$

This results stems from three factors: the interpolating property of $L(t)$, the fact that L can be defined (at least up to $O(h^m)$) by its Taylor expansion on its support $\chi(\varphi)$ and the vanishing moment property of φ. Therefore, if L can be exactly defined in terms of polynomials of order less than m, the duality relation is preserved exactly.

Under these hypotheses, let us now consider what would happen to the evaluation of the discrete moments $M_p[f]$ when using such basis $\{L_k\}$ as a dual basis. In order to do that, let us redefine the polynomial expansion of the function as $L_j(x) = \sum_{p=0}^{m} \beta_{j,p} x^p$ on its support. This can easily be done by considering

$$\sum_{k=0}^{m} \alpha_k (x-j)^k = \sum_{k=0}^{m} \alpha_k \sum_{p=0}^{k} \binom{k}{p} (-j)^{k-p} x^p = \sum_{p=0}^{m} \underbrace{\sum_{k=p}^{m} \alpha_k \binom{k}{p} (-j)^{k-p}}_{\beta_{j,p}} x^p .$$

In this way we obtain

$$\begin{aligned} f_j = \langle L_j | f \rangle &= \sum_{p=0}^{m} \beta_{j,p} \int_{\chi(L_j)} x^p f(x) \mathrm{d}x = \sum_{p=0}^{m} \beta_{j,p} \sum_{i=0}^{2m-1} \int_{x_{i+j-m}}^{x_{i+j-m+1}} x^p f(x) \mathrm{d}x \\ &= \sum_{p=0}^{m} \beta_{j,p} \sum_{i=j-m}^{j+m} \int_{x_i}^{x_{i+1}} x^p f(x) \mathrm{d}x , \end{aligned}$$

where we have split the integral over the support of L_j in chunks of size of one grid spacing. With this relation we have

$$\begin{aligned} M_p[f] = \sum_j x_j^p f_j &= \sum_j x_j^p \sum_{q=0}^{m} \beta_{j,q} \sum_{i=j-m}^{j+m} \int_{x_i}^{x_{i+1}} x^q f(x) \mathrm{d}x \\ &= \sum_i \sum_{q=0}^{m} \int_{x_i}^{x_{i+1}} x^q f(x) \mathrm{d}x \sum_{j=i-m}^{i+m} x_j^p \beta_{j,q} . \end{aligned}$$

If the condition

$$\sum_{j=i-m}^{i+m} x_j^p \beta_{j,q} = \delta_{p,q} \tag{11.37}$$

held for all i, then we would have

$$M_p[f] = \sum_j \int_{x_j}^{x_{j+1}} x^p f(x) \mathrm{d}x = \int x^p f(x) ,$$

and the discrete and the continuous moments would coincide.

11.5 Multipole-Preserving Collocations for Daubechies Wavelets: The Magic Filter

Let us now apply the concepts of the above section to functions which are discretized in this basis set. Given a set of $2m$-family Daubechies scaling functions Φ centered in a uniform mesh of spacing h, the expansion coefficients of a given function $f(x)$ in this set are defined as

$$c_i = \frac{1}{\sqrt{h}} \int \mathrm{d}x \, \Phi(\frac{x}{h} - i) f(x) = \sqrt{h} \int \mathrm{d}x \, \Phi(x - i) f(hx) \,. \tag{11.38}$$

Here we have defined the normalisation factor of the basis set in order to preserve the equivalence $||f||_2 = \sum_i c_i^2$, if the function f is exactly represented. Given the momentum-preserving property of Daubechies scaling functions, the discretization of the function f in this basis set has an algebraic h^m convergence rate. In other terms

$$f(x) - \sum_i c_i \Phi(\frac{x}{h} - i) = O(h^m) \,.$$

A wavelet quadrature in this context is based on the idea of approximating the expression above by a collocation formula. In other terms, we should define some coefficients w_i, where $i = 1 - m, \ldots, m$ such that

$$c_i = \sqrt{h} \sum_j w_{j-i} f(hj) + \sqrt{h} O(h^{2m}) \,. \tag{11.39}$$

This will be possible only if the above formula gives the exact result when $f(x) = x^p$, $p = 0, \ldots, 2m-1$. By comparing (11.38) and (11.39) in the case of polynomials we thus find the equation defining the Magic Filter:

$$\sum_j w_j (j+i)^p = \int \mathrm{d}x \, \Phi(x)(x+i)^p \qquad \forall i, p = 0, \cdots, 2m-1 \,,$$

which is solved for all i if and only if

$$\sum_j w_j j^p = \int \mathrm{d}x \, \Phi(x) x^p \equiv M_p \qquad \forall p = 0, \cdots, 2m-1 \,, \tag{11.40}$$

which can be written in matricial form $\mathbf{A} \cdot \mathbf{w} = \mathbf{M}$, where the matrix $A_{k,j} = j^k$ is the Vandermonde matrix, which has an inverse written in terms of Lagrange polynomial coefficients in the basis of monomials:

$$L_k^{2m-1}(x) = \prod_{i \neq p} \frac{x-i}{k-i} = \sum_{j=0}^{2m-1} \bar{A}_{k,j} x^j \,,$$

therefore, given $\sum_p \bar{A}_{k,j} A_{k,i} = \delta_{i,j}$, the Magic Filters are given by Neelov and Goedecker [55, Eq. (10)]:

$$w_k = \sum_j \bar{A}_{k,j} M_j = \int \mathrm{d}x\, \Phi(x) L_k^{2m-1}(x) \,. \tag{11.41}$$

This equation shows that the Magic Filter can be viewed as the *expansion coefficients of Daubechies scaling functions in the basis of Lagrange polynomials.* Johnson et al. [41] can be used as a reference in this regard.

11.5.1 From Magic Filters to a Passage Matrix

We proceed as if we had found a recipe to *interpolate* the values of the scaling function $\Phi(j) \simeq \widetilde{\Phi}(j) = w_j$ such that the collocation formula

$$c_i = \sqrt{h} \sum_j \widetilde{\varphi}(j-i) f(hj)$$

would apply *exactly* for functions f which can be expressed by polynomials of order less than $2m$. This is a very interesting feature to exploit when it is desirable to calculate the collocation values of a functional of f, for instance the charge density, or the exchange and correlation potential.

Let us explore this concept in more detail and start with the following definition of equivalence class. Two functions f and g are considered as equivalent (denoted $f \Leftrightarrow g$) if they have the same expansion coefficients in the given Daubechies scaling function basis:

$$f(x) :\Leftrightarrow g(x) \text{ iff } \int \mathrm{d}x\, \Phi\left(\frac{x}{h} - i\right) (f(x) - g(x)) = 0 \quad \forall i \,.$$

It follows from the above properties that $f - g = O(h^m)$. These functions therefore have the same representation in the scaling function basis set, provided by the coefficients $\{c_i\}$ defined in (11.38). The function

$$C(x) = \sum_i c_i \Phi\left(\frac{x}{h} - i\right) ,$$

is by definition equivalent to f. Let us now consider the *moments* of the function C *up to order* $2m-1$, and suppose we would like to find a set of quadrature coefficients C_j so that the following holds:

$$M_p^C = \int \mathrm{d}x x^p C(x) = h^{p+1} \sum_j C_j j^p \quad p = 0, \ldots, 2m-1 \, .$$

The values of the coefficients C_j can be easily related to the Magic Filters. Indeed

$$\frac{1}{h^{p+1}} \int \mathrm{d}x x^p C(x) = \sum_j c_j \int \mathrm{d}x (x+j)^p \Phi(x) = \sum_j c_j \sum_{q=0}^{p} \binom{p}{q} j^q M_{p-q}$$

and by using Eq. (11.40) we have

$$\frac{1}{h^{p+1}} \int \mathrm{d}x x^p C(x) = \sum_j c_j \sum_{q=0}^{p} \binom{p}{q} j^q \sum_{k=1-m}^{m} w_k k^{p-q} = \sum_j \sum_{j'} c_{j'} w_{j-j'} j^p \, .$$

Therefore we have found $C_j = \sum_{j'} c_{j'} w_{j-j'}$. The same reasoning could have been applied if the moment had been centered at another grid point.

Let us now consider the polynomial function which has C_j coefficient as collocation values. This function is (locally) expressed as a linear combination of Lagrange polynomials, which have the well-known interpolating property.

$$D(x) = \sum_j C_j L_j^{2m-1}(x) \, .$$

By definition, the moments of this function written in collocation coincide with M_p^C. Let us now consider the expansion coefficients of $D(x)$ in the scaling function basis set. For simplicity we omit the grid spacing, i.e.

$$d_i = \int \mathrm{d}x \Phi(x-i) D(x) = \sum_j C_j \int \mathrm{d}x \Phi(x-i) L_j^{2m-1}(x) \, .$$

Moreover, by using the result of Eq. (11.40), together with the orthogonality of Lagrange polynomials, the coefficients d_i can be calculated from C_i by applying the following convolution:

$$d_j = \sum_{j'} C_{j'} w_{j'-j} \, .$$

It can be shown, see Johnson et al. [41] and Neelov and Goedecker [55, Appendix A], that for Daubechies scaling functions, $\sum_j' w_{i-j'} w_{j'-j} = \delta_{i,j}$. Therefore in this case we have proved that $d_j = c_j$.

This not only means that $D(x) :\Leftrightarrow C(x)$, but also that we have defined a prescription that is *more precise* than the collocation of C, as the difference between D and C is of order h^{2m}. It is as if the Magic Filter prescription chooses *among the functions belonging to the same equivalency class* as $C(x)$, the collocation values of $D(x)$, which is the *smoothest possible function* preserving the moments of C.

The convolution with the Magic Filters, direct or transposed, acts therefore as a Passage Matrix between the Daubechies scaling function basis and the "collocation basis" provided by the Lagrange interpolating polynomials. The interpolating property of the Deslauriers–Dubuc interpolating scaling function (which is the autocorrelation of two Daubechies scaling functions) plays a key role in this property.

11.6 Representing the Charge Density on Real Space: An ISF Poisson Solver

The Magic Filter method makes it possible to deal with real-space quantities with a very high precision and without the need to calculate complicated numerical quantities. The multipole-preserving property of this method makes it ideal for the treatment of electrostatic problems where the preservation of moments (e.g. multipoles) are important.

We illustrate the basis of our approach for the calculation of terms related to the Poisson equation. In Eq. (11.6), the Hartree potential $V_{\mathrm{H}}[\rho]$ depends on the charge density from the Poisson equation, which in atomic units, in vacuum, reads $\nabla^2 V = -4\pi\rho$. Having efficient algorithms to solve the Poisson equation is therefore essential. The large variety of situations in which this equation can be found requires us to face this problem with different choices of boundary conditions (BCs) in mind. The long-range behaviour of the inverse Laplacian operator makes this problem strongly dependent on the BCs of the system.

In Genovese et al. [24, 25] and Cerioni et al. [10], a novel method for solving the screened and unscreened Poisson equation in vacuum with free, fully periodic, surface-like and wire-like BCs was presented, including non-orthorhombic cells. Such a method is direct (rather than iterative) in that the solution along the isolated directions is found in its integral form using the Green's function method. For instance, in the case of a fully isolated (or "cluster-like") system,

$$V(\mathbf{r}) = 4\pi \int \mathrm{d}\mathbf{r}'\, G(\mu_0; |\mathbf{r} - \mathbf{r}'|)\rho(\mathbf{r}'), \tag{11.42}$$

with the Green's function being $G(\mu_0; r) \equiv \frac{e^{-\mu_0 r}}{4\pi r}$. Homogeneous Dirichlet BCs ($V = 0$ at $|\mathbf{r}| \to \infty$) along the isolated directions are explicitly enforced by the selection of the Green's function.

The method has been in use for a few years in a number of ab initio codes (see the references cited in Cerioni et al. [10]) and has proven to be highly efficient and accurate in every application attempted to date. It is based on a representation of ρ and V in terms of interpolating scaling functions (ISFs), which allows any sort of periodicity to be modelled in the most natural, clean and mathematically rigorous way. ISFs—arising in wavelet theory [30]—enjoy several properties which make them superior to other basis sets. For instance, the representation in terms of m-th order ISFs make the first m moments of the continuous and discrete charge distributions coincide [23]. As a consequence the representation is definitely faithful (more than just convenient), since the different moments of the charge distribution capture the major features of the potential. Moreover, ISFs are genuinely localized due to their compact support (the length of which is equal to $2m$) and endowed with the refinement relations which easily allow for switching from a representation on a grid with spacing h to a doubly refined grid with spacing $h/2$.

The inclusion of such functionalities is motivated by the strong theoretical, experimental and technological interest in the characterization of nanostructured materials, since solving the Poisson equation is only one of the many steps involved in state-of-the-art computer simulations and is repeated several times. Moreover, in the context of KS-DFT and extensions thereof, there are quantities which are computed via convolution integrals very similar to that in Eq. (11.42): for instance, the exact exchange term arising within those generalizations of KS-DFT employing orbital-dependent or hybrid functionals (see Ratcliff et al. [58] and references therein), or the coupling-matrix in time-dependent DFT (TDDFT) [54]. In this respect, the electrostatic problem of concern here provides the paradigm for many other computations, even well beyond the scope of electrostatics.

11.6.1 The Soft-Sphere Implicit Solvation Model

This high-degree of flexibility makes the BigDFT Poisson solver library optimal for calculations of polarized systems or systems with non-Born–von Karman boundary conditions, such as material surfaces and isolated molecules. The computational study of matter in various environments is a continuously growing field in solid state physics and chemistry. Systems of interest are, for instance, molecules, clusters or surfaces in contact with solvents [13]. An alternative to the explicit inclusion of a wet environment is to give an implicit description, while still treating the other parts of the system explicitly on an atomic quantum level [2]. Such an explicit/implicit treatment requires three main ingredients:

1. a dielectric cavity represented by a proper function $\epsilon(\mathbf{r})$ mimicking the surrounding solvent of a solute as a continuum dielectric;

2. a solver for the generalized Poisson equation [21]

$$\nabla \cdot \epsilon(\mathbf{r})\nabla\varphi(\mathbf{r}) = -4\pi\rho(\mathbf{r}), \tag{11.43}$$

where $\varphi(\mathbf{r})$ is the potential generated by a given charge density $\rho(\mathbf{r})$;
3. a model for the non-electrostatic terms to the total free energy of solvation.

The dielectric function $\epsilon(\mathbf{r})$ has to take the value of $\epsilon_0 = 1$ where the solute is placed to solve a vacuum-like quantum problem, and the bulk dielectric constant ϵ outside.

The "soft-sphere" model developed by Fisicaro et al. [20] and implemented in BigDFT improves upon previous solvation approaches (see e.g. Tomasi's method [63]). Model features are: accurate forces and a numerical cost comparable to standard vacuum calculations; feasible extensive potential energy surface (PES) explorations; a small number of model parameters; exact treatment of molecular or slab-like geometries; and the ability to treat neutral and charged molecules simultaneously in order to tackle complex interfaces (e.g. a double layer).

The interface between the quantum-mechanical solute and the surrounding environment is described by a fully continuous permittivity built up with atomic-centered "soft" spheres. This approach combines many of the advantages of the self-consistent continuum solvation model [1] in handling solutes and surfaces in contact with complex dielectric environments or electrolytes in electronic-structure calculations. In addition, it is able to describe accurately both neutral and charged systems.

We developed, tested and implemented within the BigDFT suite a solver for the generalized Poisson (Eq. 11.43) and the Poisson–Boltzmann equations to treat neutral and ionic solutions, respectively [21]. The solver for the solution of the generalized Poisson equation and the linear regime of the Poisson–Boltzmann is based on a preconditioned conjugate gradient scheme. It allows for the iterative solution of the minimization problem with some ten iterations of the ordinary Poisson equation solver. In addition, a self-consistent procedure solves the non-linear Poisson–Boltzmann problem. Both solvers exhibit very high accuracy and parallel efficiency and allow for the treatment of free, slab and wire-like boundary conditions.

The continuous function, describing the variation of the permittivity, allows for the analytic computation of the non-electrostatic contributions to the solvation free energy that are described in terms of the quantum surface. The capability of treating arbitrary molecular or slab-like geometries as well as charged molecules is key to tackling electrolytes within mixed explicit/implicit frameworks. Within the soft-sphere model two parameters are sufficient to give a mean absolute error of only 1.12 kcal/mol with respect to the experimental aqueous solvation energies for a set of 274 neutral solutes. For charged systems, the same set of parameters provides solvation energies for a set of 60 anions and 52 cations with an error of 2.96 and 2.13 kcal/mol, respectively, improving upon previous values in the literature.

The soft-sphere model has already been applied to the study of molecular doping of silicon [57], the interface of fluorite terminations with water [22] and the investigation of wet environment effects for ethanol and water adsorption on anatase TiO_2 (101) surfaces [19].

11.6.2 (Exact) Exchange and Correlation Terms

For a collinear-spin formalism, the calculation of the exact exchange energy E_X requires a double summation over all the N occupied orbitals

$$E_X[\widehat{F}] = -\frac{1}{2}\sum_{\sigma}\int \mathbf{dr}\,\mathbf{dr}'\frac{F_\sigma(\mathbf{r},\mathbf{r}')\,F_\sigma(\mathbf{r}',\mathbf{r})}{|\mathbf{r}-\mathbf{r}'|}$$
$$= -\frac{1}{2}\sum_{i,j,\sigma} f_{i,\sigma} f_{j,\sigma}\int \mathbf{dr}\,\mathbf{dr}'\,\frac{\rho^\sigma_{ij}(\mathbf{r})\,\rho^\sigma_{ji}(\mathbf{r}')}{|\mathbf{r}-\mathbf{r}'|}\,, \tag{11.44}$$

where we have defined $\rho^\sigma_{ij}(\mathbf{r}) = \psi^*_{j,\sigma}(\mathbf{r})\,\psi_{i,\sigma}(\mathbf{r})$. The diagonal $(i = j)$ contribution to E_X exactly cancels out the Hartree electrostatic energy $E_{\mathrm{H}}[\rho]$. The action of the Fock operator $\widehat{D}_X$ to be added to the KS Hamiltonian directly stems from the E_X definition:

$$\widehat{D}_X\,|\psi_{i,\sigma}\rangle = \int \mathbf{drdr}'\,\frac{\delta E_X[\widehat{F}]}{\delta F_\sigma(\mathbf{r},\mathbf{r}')}\psi_{i,\sigma}(\mathbf{r}')\,|\mathbf{r}\rangle = -\sum_j\int \mathbf{dr} f_{j,\sigma} V^\sigma_{ij}(\mathbf{r})\psi_{j,\sigma}(\mathbf{r})\,|\mathbf{r}\rangle\,, \tag{11.45}$$

where we have defined

$$V^\sigma_{ij}(\mathbf{r}) = \int \mathbf{dr}'\,\frac{\rho^\sigma_{ji}(\mathbf{r}')}{|\mathbf{r}-\mathbf{r}'|}\,, \tag{11.46}$$

which is the solution of the Poisson equation $\nabla^2 V^\sigma_{ij} = -4\pi\rho^\sigma_{ij}$. In a KS-DFT code which searches for the ground state orbitals, during the SCF procedure, for a given set of $\psi_{i,\sigma}(\mathbf{r})$, one has to repeatedly evaluate the value of E_X as well as the action of the corresponding Fock operator $\widehat{D}_X$ on the entire set of occupied orbitals.

11.7 The BigDFT Code Approach for Ground-State

The `BigDFT` code may therefore express the solution of the KS problem in two ways. The traditional approach, which has a computational overhead that scales cubically with the number of atoms in the system and therefore called the cubic

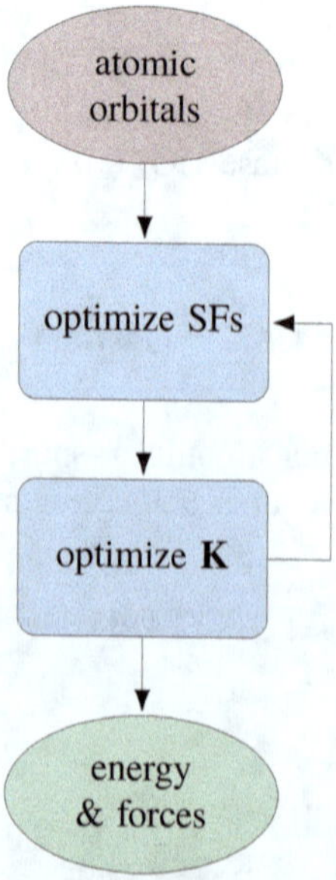

Fig. 11.5 Flowchart summarizing the high-level algorithm used in `LS-BigDFT`

scaling algorithm, expresses the KS orbitals Ψ_i directly in a wavelet basis. In this case only the KS optimization loop is needed, and no localization projection operator $\mathcal{L}_\alpha$ is considered. The orbitals are directly labelled by their index i.

The linear scaling (LS) approach in `BigDFT` instead consists of two optimization loops, as depicted by the flowchart in Fig. 11.5. The SF and kernel optimization loops are independent of each other, with the number of iterations, convergence criteria etc. specified independently.

Although additional approximations are introduced in the LS approach compared to the extended KS orbitals used in the CS approach, excellent agreement between total energies and forces calculated with the LS and CS approaches has been demonstrated for a range of materials and system sizes [51, 52].

Furthermore, systematic convergence remains possible—as the value of R_{loc} is increased, both the total energy and forces converge towards the CS result. For example for a fullerene molecule it was shown that for SF radii of 7.4 Å the total energies agree to within 0.1 meV, while the forces show better than 1 meV/Å agreement [51]. Thus, in addition to the wavelet basis parameters, the user should take care to ensure the localization radii (which may be varied independently for different atomic species) are large enough for the required accuracy. In some cases it may also be desirable to increase the number of SFs per atom, although in the majority of cases a minimal basis is sufficient, e.g. 4 SFs per C/N/O atom, 1 SF per H atom.

Aside from the basic SF parameters, a number of additional options are available in `LS-BigDFT` to allow for additional flexibility, such as whether or not to impose orthogonality on the SFs. Given the number of additional parameter choices compared to the CS approach, it is worth asking to what extent the choice of computational parameters depends on the system in question. Where a very high accuracy or optimal parallel performance are required it is important to carefully converge with respect to parameters such as the SF radii. Nonetheless we have demonstrated that a common set of parameters may be used to achieve consistent accuracy and robust performance across a wide range of systems [52]. Such parameters might easily be accessed by using the appropriate input profile. These profiles have been defined to give a suitable set of parameters for a number of common use cases.

11.7.1 The Direct Minimization Approach

We have seen that the support functions *or* the KS orbitals are represented in a Daubechies wavelet basis and are therefore susceptible to be optimized in the LS and CS algorithm respectively. We present in the following the main algorithm which is employed for their optimization.

Let us now impose that the wavefunctions have to be orthonormal with respect to a certain Hermitian metric operator $\widehat{\mathbb{S}}[\lambda]$ (not to be confused with the SF overlap matrix), that is $\langle\psi_i|\widehat{\mathbb{S}}[\lambda]|\psi_j\rangle = \delta_{ij}$. For norm-conserving PSPs, frozen core and/or all-electron calculations $\widehat{\mathbb{S}}[\lambda]$ is the identity operator, whereas it is a non-trivial quantity for projector augmented wave (PAW) [8] and ultrasoft PSPs. As in the previous sections, we denote by λ a multi-index labelling the system (indicating for instance the coordinates of the atomic positions).

Orthogonality is imposed *via* the following Lagrangian

$$L[\{\psi_i\}, \Lambda] = E - \sum_{ij} \Lambda_{ij} \left(\langle\psi_i|\widehat{\mathbb{S}}[\lambda]|\psi_j\rangle - \delta_{ij}\right) .$$

As always for Lagrange multiplier techniques, criticality with respect to Λ leads to the orthogonality constraint. The coefficients Λ_{ij} form a Hermitian matrix, due to the Hermiticity of $\widehat{\mathbb{S}}$. Criticality of this Lagrangian under variation of the KS orbitals $\langle\psi_i|$ leads to the equation:

$$0 = \frac{\delta L}{\delta\langle\psi_i|} = f_i\,|H_{\mathrm{KS}}\psi_j\rangle - \sum_j \Lambda_{ij}\widehat{\mathbb{S}}[\lambda]|\psi_j\rangle . \tag{11.47}$$

Multiplying Eq. (11.47) by $\langle\psi_k|$ leads to the condition on the Hermitian part of $\Lambda_{ik} = \frac{1}{2}(f_i + f_k)\left(\langle\psi_k|\hat{H}_{\mathrm{KS}}\psi_i\rangle + \langle\psi_i|H_{\mathrm{KS}}\psi_k\rangle\right)$. Convergence is achieved when the mean of the norms of the Lagrangian derivatives is below a user-defined numerical tolerance.

11.7.2 Support Function Optimization

We illustrate the main guidelines for the optimization of the support functions. In principle, the optimization of KS orbitals in the basis of the support functions should minimize the total energy. This is in fact equivalent to minimizing the band structure energy, assuming a fixed density matrix and Hamiltonian, i.e. the following functional

$$\min_{\varphi_\alpha} \sum_{\alpha,\beta} K^{\alpha\beta} \langle \varphi_\alpha | \widehat{H}_{\text{KS}} | \varphi_\beta \rangle \,. \tag{11.48}$$

As discussed, the SFs used in `LS-BigDFT` are strictly localized (numerical) functions which are expressed in a Daubechies wavelet basis. Starting from an atomic orbital (AO) input guess, they are optimized by minimizing the target function $\Omega = \text{tr}\left(\widehat{F}\widehat{H}_c\right)$, subject to the orthonormality condition of the KS orbitals. The operator $\widehat{H}_c = \widehat{H}_{\text{KS}}[\rho] + \widehat{V}_c$ is the sum of the density-dependent KS Hamiltonian plus a confining operator $\widehat{V}_c$ such that

$$\langle \varphi_\alpha | \, \widehat{V}_c \, | \varphi_\beta \rangle = \delta_{\alpha\beta} \, \langle \varphi_\alpha | \, \widehat{V}_c^\alpha \, | \varphi_\alpha \rangle \,, \tag{11.49}$$

$$V_c^\alpha(\mathbf{r}) = c_\alpha |\mathbf{r} - \mathbf{R}_\alpha|^4 \,. \tag{11.50}$$

We therefore have to minimize the following functional:

$$\Omega = \text{tr}\left(\widehat{F}\widehat{H}_c\right) - \sum_{\alpha,\beta,i,j} \Lambda_{ij} \left(c_i^\alpha c_j^\beta \langle \varphi_\alpha | \varphi_\beta \rangle - \delta_{ij} \right) \tag{11.51}$$

where the coefficients Λ_{ij} are determined by the relation:

$$\sum_{i,j} c_i^\alpha c_j^\beta \Lambda_{ij} = \sum_{\rho\sigma} K^{\alpha\rho} \langle \varphi_\sigma | \widehat{H}_{\text{KS}} | \varphi_\rho \rangle \, S^{-1}_{\sigma\beta} \,. \tag{11.52}$$

The KS Hamiltonian $\hat{H}_{\text{KS}}$ does not commute with the $\mathcal{L}^{(\alpha)}$ operators. Therefore when calculating $\mathcal{L}^{(\alpha)} | \hat{H}_{\text{KS}} \varphi_\beta \rangle$ the localization constraint has to be relaxed before applying the Hamiltonian operator. Practically, this is done as follows. When applying the Hamiltonian, the value of the cutoff radius R_{loc} must be increased by half of the convolution filter length times the grid spacing, corresponding to a buffer region of eight grid points around the localization region. These buffers are initialized to zero, but the convolution will result in non-zero values in those regions. When the scalar product with another basis function is evaluated, it is therefore important to keep this buffer zone. Therefore, given a set of truncation radii R_{loc}, the KS Hamiltonian can be explicitly evaluated within the applied truncation scheme, preserving the variationality of the result.

We impose therefore the localization condition $|\varphi_\alpha\rangle = \mathcal{L}^{(\alpha)}|\varphi_\alpha\rangle$ on the SFs. The functional to be minimized then becomes

$$\Omega - \sum_\alpha \left\langle \varphi_\alpha | 1 - \mathcal{L}^{(\alpha)} | \ell_\alpha \right\rangle , \tag{11.53}$$

where the components of the vector $|\ell_\alpha\rangle = \left(1 - \mathcal{L}^{(\alpha)}\right)|\ell_\alpha\rangle$ are the Lagrange multipliers of the constraints. The stationary condition on the functional $0 = \left|\frac{\delta\Omega}{\delta\langle\varphi_\alpha|}\right\rangle$ provides the following gradient:

$$\begin{aligned} |g_\alpha\rangle &= \sum_\beta K^{\alpha\beta}\mathcal{L}^{(\alpha)}\mathcal{H}_{\rm KS}|\varphi_\beta\rangle - \sum_{\beta\rho\sigma} K^{\alpha\rho}\langle\varphi_\sigma|\mathcal{H}_{\rm KS}|\varphi_\rho\rangle S^{-1}_{\sigma\beta}\mathcal{L}^{(\alpha)}|\varphi_\beta\rangle \\ &= \sum_{\beta\rho} K^{\alpha\rho} S^{1/2}_{\rho\beta}\left[\mathcal{L}^{(\alpha)}\mathcal{H}_{\rm KS}|\widetilde{\varphi}_\beta\rangle - \sum_\sigma \langle\widetilde{\varphi}_\sigma|\mathcal{H}_{\rm KS}|\widetilde{\varphi}_\rho\rangle\mathcal{L}^{(\alpha)}|\widetilde{\varphi}_\sigma\rangle\right] , \end{aligned} \tag{11.54}$$

which is explicitly localized ($|g_\alpha\rangle = \mathcal{L}^{(\alpha)}|g_\alpha\rangle$). Here the gradient is expressed in terms of the *orthogonalized* support functions $|\widetilde{\varphi}_\alpha\rangle = S^{-1/2}_{\alpha\beta}|\varphi_\beta\rangle$. The localization condition can therefore be imposed more easily by applying the constraint on (quasi-) *orthogonal* support functions, i.e. $S_{\alpha\beta} = \delta_{\alpha\beta}$. This further simplifies the evaluation of the gradient. To ensure a good compromise between locality and flexibility, in general the orthogonality is not ensured strictly for the support functions, but it is inserted into the gradient to provide a search direction which optimizes the diagonality of the overlap matrix.

Such a minimization proceeds by applying the same guidelines as the direct minimization approach of Sect. 11.7.1, assuming unit values for the occupation numbers. The coefficient c_α is dynamically adjusted during the basis set optimization procedure. This approach has the effect of keeping the SFs confined in their localization regions, centered on the position $\mathbf{R}_\alpha$, while reducing the KS band structure energy. Usually the position $\mathbf{R}_\alpha$ of the support function α coincides with the position $\mathbf{R}_a$ of the atom a where φ_α is initially centered at the beginning of the SCF optimization procedure. To some extent this enables one to associate φ_α to a particular atom a.

As illustrated in Fig. 11.6, this procedure results in a set of SFs which have adapted to their local chemical environments. For a molecular calculation, we therefore obtain a *minimal* set of molecular orbitals that, by construction, exactly represent the occupied KS orbitals. The SF basis also has a non-zero projection into the unoccupied orbitals subspace, although in general the unoccupied KS orbitals are not expected to be well represented. Although the SFs resulting from `LS-BigDFT` are entirely numerical and are therefore not constrained to any particular form, even in extended systems they generally retain some resemblance to AOs, particularly by preserving the orbital quantum number, and are thus referred to as s, or p-like SFs.

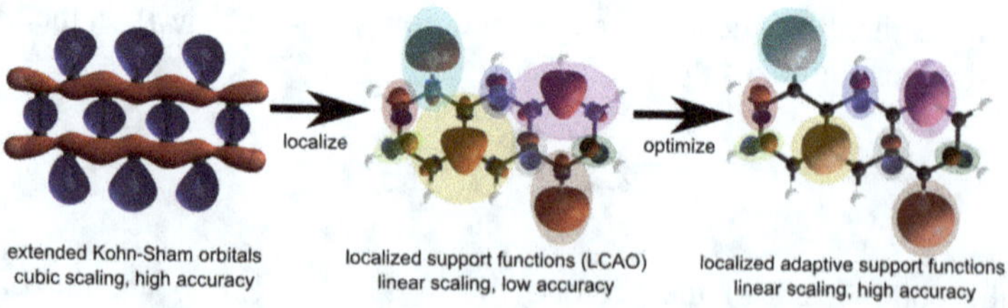

Fig. 11.6 Illustration of the different approaches in `BigDFT` for the example of anthracene. Shown are an extended KS orbital from the CS approach (left), selected unoptimized AOs used as a starting guess in the LS approach (centre) and selected resulting optimized SFs which constitute an accurate minimal basis in the LS approach. Note that the colours are used only to highlight different SFs, while the localization regions are spherical

11.7.3 Density Kernel Optimization

Given a recipe for optimizing the SFs, the question remains of how to find the density kernel for a given set of SFs. Three options are available in `BigDFT`: diagonalization, direct minimization and the Fermi Operator Expansion (FOE). The first two approaches retain explicit reference to the KS wavefunctions, while FOE works directly with the density kernel. As such, FOE is the preferred approach when strict linear scaling behaviour is required.

The first approach is straightforward and uses standard linear algebra routines in LAPACK, or optionally SCALAPACK, to solve the generalized eigenproblem defined by the SF Hamiltonian and overlap matrices. Although such an approach is of course not linear scaling, it can be useful as a benchmark approach, while the minimal size of the SF basis means that the computational cost is low compared to diagonalizing in the full wavelet basis.

The direct minimization approach works directly with the KS wavefunctions, but avoids explicit diagonalization by instead minimizing the band structure energy, subject to appropriate orthogonality constraints, as described in Mohr et al. [51]. The direct minimization approach does not scale as well with respect to the system size (see [51]), however it may nonetheless be preferred to FOE in certain cases. Notably, a few unoccupied states may be straightforwardly included in both the SF and kernel optimization steps, and so direct minimization is typically used for cases where it is important to have an SF basis which is capable of accurately representing the lowest unoccupied molecular orbital (LUMO).

Finally, in the FOE approach [27, 28] the density kernel is expressed as a function of the SF Hamiltonian matrix, i.e. $\mathbf{K} = f(\mathbf{H})$, where f is the Fermi function. The Fermi function is written as an expansion of Chebyshev polynomials, in such a way as to allow the $\mathbf{K}$ to be constructed using only matrix vector multiplications. This is combined with sparse matrix algebra, as implemented in the CheSS library [50], resulting in LS behaviour, while it can also be used to treat metallic systems at a (small) finite temperature [49]. The use of a finite temperature can also be used to ensure robust convergence even when the gap of a system closes due to a poor initial guess, bond breaking, or when computing charged systems.

The LS behaviour of `BigDFT` when using the FOE approach has been demonstrated for a number of materials, for systems containing up to tens of thousands of atoms [49, 51, 52, 59].

11.7.4 Suitability of the Linear Scaling Approach

The SF basis of `BigDFT` offers numerous benefits for linear algebra-based code bottlenecks. The in situ optimized approach allows for the accuracy of a large basis, while keeping the number of basis functions similar to the size of a minimal basis, leading to small matrices even for large systems. The use of strictly localized, quasi-orthogonal basis functions further ensures that the matrices used are sparse and well conditioned. In Table 11.1, we report the matrix dimensions and sparsities for four different systems: a 1CRN protein [62] in gas phase, a pentacene cluster, a 1L2Y protein [56] in solution, and a cluster of water molecules.

The benefits of this basis set are further reflected in the spectral quantities of these matrices, as shown in Table 11.2. We see that the spectral width of the overlap matrix is quite low, reflecting how well conditioned the basis is. The ratio of the band gap to the spectral width of the Hamiltonian is also relatively high, which leads to huge efficiency gains for diagonalization-free methods. The lower this ratio, the more polynomials that are required in order to approximate the Fermi function of the Hamiltonian. `BigDFT`'s matrices require few polynomials, similar to what would be needed for minimal basis calculations with Gaussians or tight binding calculations,

Table 11.1 Matrix properties of four example systems. NNZ refers to the percentage of non-zero elements in the Hamiltonian (**H**), overlap (**S**), and density kernel (**K**)

System	N_{atoms}	N_{SFs}	**H** NNZ	**S** NNZ	**K** NNZ
1CRN	642	1623	22.09	9.40	37.20
Pentacene	6876	19,482	2.89	1.04	5.70
1L2Y	1942	4045	5.90	2.13	11.57
Water	1719	3438	9.43	3.43	18.30

Table 11.2 Spectral properties of four example systems. The "Gap To Width" is the ratio of the band gap to the spectral width of the Hamiltonian

System	S width	H width (eV)	Gap (eV)	Gap to width
1CRN	0.9557	47.0783	1.9977	0.0424
Pentacene	0.9852	42.2971	1.0323	0.0244
1L2Y	0.8936	47.9860	1.3682	0.0285
Water	0.4496	40.8961	7.7297	0.1890

as was shown in a head-to-head comparison when using density matrix purification techniques [18].

Sparse matrices are stored in a custom Segment Storage Format, which groups together consecutive non-zero values in a matrix row. This format not only reduces the storage overhead of a matrix, but also can improve the performance of matrix-vector multiplication by using calls to dense operations. The Hamiltonian matrix is replicated across processes and columns of the density matrix are distributed. This data distribution allows each column to be computed independently to improve parallel performance.

As mentioned in Sect. 11.7.3, the diagonalization free method of choice in `BigDFT` is the Fermi operator expansion based on matrix vector multiplication. This approach is usually far more expensive than those based on recursive polynomial expansions such as density matrix purification [14]. However, the high sparsity, relatively low dimension, and good conditioning of BigDFT's matrices enables a more tailored choice of algorithm and parallelization scheme. This novel approach is made possible by the unique properties of the Daubechies wavelet basis set employed in `BigDFT`.

11.8 Perspective

Daubechies wavelets have a number of favourable properties which make them an ideal basis set for electronic structure calculations. In this work we have outlined the use of such a basis set for density functional theory calculations, as implemented in the open source `BigDFT` code. Through examples presented here and referenced within the text, we have shown how the combination of a wavelet basis set with an implementation designed for massively parallel machines allows for efficient and accurate calculations of hundreds of atoms, even within a traditional cubic scaling approach. Such a treatment also allows for the simulation of relatively large systems using hybrid functionals, particularly where GPUs are available, while the availability of different boundary conditions allows for the straightforward treatment of molecules, surfaces and solids. A number of functionalities are available in `BigDFT`, including dynamics, explicit charges and electric fields and implicit environments, while, for example, there are ongoing developments in the treatment of excited states.

Going beyond the cubic scaling approach, the localized nature of wavelets is also highly suitable for a linear scaling approach, wherein the nearsightedness of matter is exploited by imposing localization on the system *via* the use of a minimal set of localized support functions. Such an approach further expands the applicability of `BigDFT` to systems containing several thousand atoms, and has been shown to converge reliably for a range of materials. The localized support function-based approach may also be further exploited to define a fragment approach, in which the computational cost may be significantly reduced by exploiting repetition in molecular or periodic systems.

The treatment of such large systems brings new types of problems within the reach of first principles simulations. However, such simulations also bring new challenges, for example the increased size of the configuration space associated with complex materials containing large numbers of atoms. Furthermore, as well as treating large lengthscales it is desirable to treat also long timescales, which remains unfeasible within a purely quantum mechanical approach, so that QM/MM approaches are required. To this end, the support function approach allows not only the treatment of large enough systems to test and validate QM/MM approaches, but can also be used to analyze and fragment a system without requiring any a priori knowledge of the system. This offers a route to reduce the complexity of QM calculations of large systems and thereby inform the setup of multiscale simulations.

In the context of large scale electronic structure simulations, a wavelet-based approach therefore offers another significant advantage in that it facilitates the implementation of a range of approaches designed to treat different system sizes, as illustrated in Fig. 11.7. The existence of such a comprehensive framework with a single underlying formalism means that each successive approximation is applied in a systematic and controlled manner. Therefore, one can easily test and validate the approximations between different levels of theory. For example, QM/MM simulations may be benchmarked with respect to fragment calculations, which may be compared with full linear scaling calculations and so on. Furthermore, we have also introduced a number of indicators which can be used to predict whether or not a particular approximation is appropriate for a given system.

The development and improvement of a comprehensive multi-scale framework is an ongoing priority for `BigDFT`. As with other developments in `BigDFT`, a key aspect of this is a module-based approach which aims towards sustainable software development. In tandem with the implementation of new approaches, we also aim to prioritize both accessibility and reproducibility of new functionalities, by

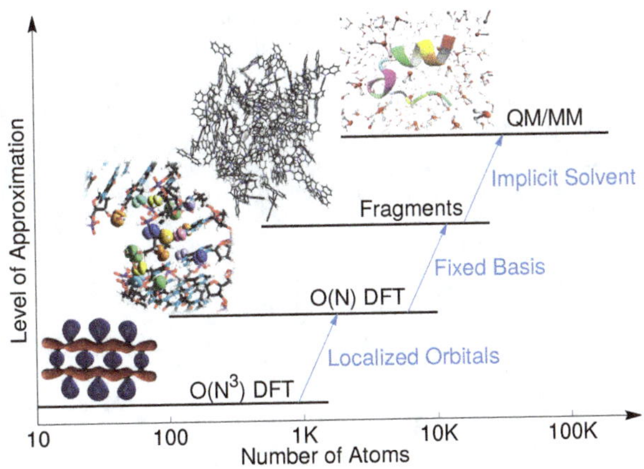

Fig. 11.7 Illustration of the different approaches available in `BigDFT`, the approximations introduced and approximate applicable lengthscales

providing Jupyter-notebooks demonstrating how such functionalities may be used *via* the PyBigDFT interface. Further information on `BigDFT` may be found on the website [6] and in the documentation [5], while the code may be downloaded from GitLab [7].

References

1. O. Andreussi, I. Dabo and N. Marzari. Revised self-consistent continuum solvation in electronic-structure calculations. *J. Chem. Phys.* **136**(6), 064102 (2012). URL https://doi.org/10.1063/1.3676407.
2. O. Andreussi and G. Fisicaro. Continuum embeddings in condensed-matter simulations. *Int. J. Quantum Chem.* **119**, e25725 (2019). URL https://onlinelibrary.wiley.com/doi/abs/10.1002/qua.25725.
3. R. Baer and M. Head-Gordon. Sparsity of the Density Matrix in Kohn–Sham Density Functional Theory and an Assessment of Linear System-Size Scaling Methods. *Phys. Rev. Lett.* **79**, 3962–3965 (1997).
4. C. Bernard. *Wavelets and ill posed problems: optic flow and scattered data interpolation*. PhD thesis, École Polytechnique (1999).
5. BigDFT documentation. URL http://bigdft-suite.readthedocs.io/en/latest.
6. BigDFT website (2020). URL http://www.bigdft.org.
7. BigDFT GitLab URL http://gitlab.com/l_sim/bigdft-suite.
8. P.E. Blöchl. Projector augmented-wave method. *Phys. Rev. B* **50**, 17953–17979 (1994). URL http://link.aps.org/doi/10.1103/PhysRevB.50.17953.
9. D.R. Bowler and T. Miyazaki. Calculations for millions of atoms with density functional theory: linear scaling shows its potential. *J. Phys.: Condens. Matter* **22**(7), 074207 (2010).
10. A. Cerioni, L. Genovese, A. Mirone and V. Armando Sole. Efficient and accurate solver of the three-dimensional screened and unscreened Poisson's equation with generic boundary conditions. *J. Chem. Phys.* **137**(13), 134108 (2012).
11. J. Des Cloizeaux. Analytical Properties of *n*-Dimensional Energy Bands and Wannier Functions. *Phys. Rev. A* **135**, 698–707 (1964).
12. J. Des Cloizeaux. Energy Bands and Projection Operators in a Crystal: Analytic and Asymptotic Properties. *Phys. Rev. A* **135**, 685–697 (1964).
13. I. Dabo, Y. Li, N. Bonnet and N. Marzari. *Ab Initio* Electrochemical Properties of Electrode Surfaces. In: *Fuel Cell Science: THeory, Fundamentals, and Biocatalysis*, edited by A. Wieckowski and J.K. Nørskov, pp. 415–431, John Wiley & Sons, Inc. (2010). ISBN 9780470630693. URL https://doi.org/10.1002/9780470630693.ch13.
14. A.D. Daniels and G.E. Scuseria. What is the best alternative to diagonalization of the Hamiltonian in large scale semiempirical calculations? *J. Chem. Phys.* **110**(3), 1321–1328 (1999).
15. I. Daubechies. Orthonormal bases of compactly supported wavelets. *Communications on Pure and Applied Mathematics* **41**(7), 909–996 (1987).
16. I. Daubechies. *Ten Lectures on Wavelets*. SIAM (1992).
17. I. Daubechies. *Ten lectures on wavelets*, volume 61 of *CBMS-NSF Regional Conference Series in Applied Mathematics*. SIAM, Philadelphia (1992).
18. W. Dawson and T. Nakajima. Massively parallel sparse matrix function calculations with NTPoly. *Comput. Phys. Commun.* **225**, 154–165 (2018).
19. G. Fisicaro, S. Filice, S. Scalese, G. Compagnini, R. Reitano, L. Genovese, S. Goedecker, I. Deretzis and A. La Magna. Wet environment effects for ethanol and water adsorption on anatase tio_2 (101) surfaces. *J. Phys. Chem. C* **124**(4), 2406–2419 (2020). URL https://doi.org/10.1021/acs.jpcc.9b05400.

20. G. Fisicaro, L. Genovese, O. Andreussi, S. Mandal, N.N. Nair, N. Marzari and S. Goedecker. Soft-sphere continuum solvation in electronic-structure calculations. *J. Chem. Theory Comput.* **13**(8), 3829–3845 (2017). URL https://doi.org/10.1021/acs.jctc.7b00375.
21. G. Fisicaro, L. Genovese, O. Andreussi, N. Marzari and S. Goedecker. A generalized poisson and Poisson-Boltzmann solver for electrostatic environments. *J. Chem. Phys.* **144**(1), 014103 (2016). URL https://doi.org/10.1063/1.4939125.
22. G. Fisicaro, M. Sicher, M. Amsler, S. Saha, L. Genovese and S. Goedecker. Surface reconstruction of fluorites in vacuum and aqueous environment. *Phys. Rev. Materials* **1**, 033609 (2017). URL https://link.aps.org/doi/10.1103/PhysRevMaterials.1.033609.
23. L. Genovese and T. Deutsch. Multipole-preserving quadratures for the discretization of functions in real-space electronic structure calculations. *Phys. Chem. Chem. Phys.* **17**, 31582–31591 (2015). URL https://doi.org/10.1039/C5CP01236H.
24. L. Genovese, T. Deutsch and S. Goedecker. Efficient and accurate three-dimensional poisson solver for surface problems. *J. Chem. Phys.* **127**(5), 054704 (2017).
25. L. Genovese, T. Deutsch, A. Neelov, S. Goedecker and G. Beylkin. Efficient solution of Poisson's equation with free boundary conditions. *J. Chem. Phys.* **125**(7), 074105 (2006).
26. S. Goedecker. Decay properties of the finite-temperature density matrix in metals. *Phys. Rev. B* **58**, 3501–3502 (1998).
27. S. Goedecker and L. Colombo. Efficient Linear Scaling Algorithm for Tight-Binding Molecular Dynamics. *Phys. Rev. Lett.* 73, 122–125 (1994).
28. S. Goedecker and M. Teter. Tight-binding electronic-structure calculations and tight-binding molecular dynamics with localized orbitals. *Phys. Rev. B* **51**, 9455–9464 (1995).
29. S. Goedecker, M. Teter and J. Hutter. Separable dual-space gaussian pseudopotentials. *Phys. Rev. B* **54**, 1703–1710 (1996). URL https://link.aps.org/doi/10.1103/PhysRevB.54.1703.
30. S. Goedecker. *Wavelets and Their Application: For the Solution of Partial Differential Equations in Physics*. Presses polytechniques et universitaires romandes (1998).
31. S. Goedecker. *Wavelets and their application for the solution of partial differential equations in physics*. Presses polytechniques et universitaires romandes, Lausanne (2009).
32. R.J. Harrison, G. Beylkin, F.A. Bischoff, J.A. Calvin, G.I. Fann, J. Fosso-Tande, D. Galindo, J.R. Hammond, R. Hartman-Baker, J.C. Hill, J. Jia, J.S. Kottmann, M-J. Yvonne Ou, J. Pei, L.E. Ratcliff, M.G. Reuter, A.C. Richie-Halford, N.A. Romero, H. Sekino, W.A. Shelton, B.E. Sundahl, W. Scott Thornton, E.F. Valeev, Á. Vázquez-Mayagoitia, N. Vence, T. Yanai and Y. Yokoi. Madness: A multiresolution, adaptive numerical environment for scientific simulation. *SIAM J. Sci. Comput.* **38**(5), S123–S142 (2016).
33. R.J. Harrison, G.I. Fann, T. Yanai, Z. Gan and G. Beylkin. Multiresolution quantum chemistry: Basic theory and initial applications. *J. Chem. Physics* **121**(23), 11587–11598 (2004).
34. R. Harrison, G.I. Fann, T. Yanai and G. Beylkin. Multiresolution quantum chemistry in multiwavelet bases. In: *Computational Science-ICCS 2003*, edited by P M A Sloot et al., pp. 103–110, volume 2660 of Lecture Notes in Computer Science, Springer-Verlag (2003).
35. C. Hartwigsen, S. Goedecker and J. Hutter. Relativistic separable dual-space Gaussian pseudopotentials from H to Rn. *Phys. Rev. B* **58**, 3641–3662 (1998).
36. L. He and D. Vanderbilt. Exponential decay properties of Wannier functions and related quantities. *Phys. Rev. Lett.* **86**, 5341–5344 (2001).
37. E. Hernández and M.J. Gillan. Self-consistent first-principles technique with linear scaling. *Phys. Rev. B* **51**(15), 10157–10160 (1995).
38. P. Hohenberg and W. Kohn. Inhomogeneous electron gas. *Phys. Rev. B* **136**(3), 864–871 (1964).
39. S. Ismail-Beigi and T.A. Arias. Locality of the Density Matrix in Metals, Semiconductors and Insulators. *Phys. Rev. Lett.* **82**, 2127–2130 (1999).
40. S.R. Jensen, T. Fla, D. Jonsson, R.S. Monstad, K. Ruud and L. Frediani. Magnetic properties with multiwavelets and DFT: the complete basis set limit achieved. *Phys. Chem. Chem. Phys.* **18**, 21145–21161 (2016). URL https://doi.org/10.1039/C6CP01294A.
41. B.R. Johnson, J.P. Modisette, P.J. Norlander and J.K. Kinsey. Quadrature integration for orthogonal wavelet systems, *J. Chem. Phys.* **110**(17), 8309–8317 (1999).

42. W. Kohn. Analytic Properties of Bloch Waves and Wannier Functions. *Phys. Rev.* **115**, 809–821 (1959).
43. W. Kohn and L.J. Sham. Self-consistent equations including exchange and correlation effects. *Phys. Rev. A* **140**(4), 1133–1138 (1965).
44. M. Krack. Pseudopotentials for H to Kr optimized for gradient-corrected exchange-correlation functionals. *Theor. Chem. Acc.* **114**, 145–152 (2005).
45. K. Lejaeghere, G. Bihlmayer, T. Björkman, P. Blaha, S. Blügel, V. Blum, D. Caliste, I.E. Castelli, S.J. Clark, A. Dal Corso, S. de Gironcoli, T. Deutsch, J.K. Dewhurst, I. Di Marco, C. Draxl, M. Dułak, O. Eriksson, J.A. Flores-Livas, K.F. Garrity, L. Genovese, P. Giannozzi, M. Giantomassi, S. Goedecker, X. Gonze, O. Grånäs, E.K.U. Gross, A. Gulans, F. Gygi, D.R. Hamann, P.J. Hasnip, N.A.W. Holzwarth, D. Iuc san, D.B. Jochym, F. Jollet, D. Jones, G. Kresse, K. Koepernik, E. Küc cükbenli, Y.O. Kvashnin, I.L.M. Locht, S. Lubeck, M. Marsman, N. Marzari, U. Nitzsche, L. Nordström, T. Ozaki, L. Paulatto, C.J. Pickard, W. Poelmans, M.I.J. Probert, K. Refson, M. Richter, G.-M. Rignanese, S. Saha, M. Scheffler, M. Schlipf, K. Schwarz, S. Sharma, F. Tavazza, P. Thunström, A. Tkatchenko, M. Torrent, D. Vanderbilt, M.J. van Setten, V. Van Speybroeck, J.M. Wills, J.R. Yates, G.-X. Zhang and S. Cottenier. Reproducibility in density functional theory calculations of solids. *Science* **351**(6280), (2016). ISSN 0036-8075.
46. S.G. Mallat. A theory for multiresolution signal decomposition: the wavelet representation. *IEEE Transactions on Pattern Analysis and Machine Intelligence* **11**(7), 674–693 (1989).
47. N.H. March, W.H. Young and S. Sampanthar. *The Many-body Problem in Quantum Mechanics*. Dover books on physics. Dover Publications, Incorporated (1967). ISBN 9780486687544.
48. N. Marzari and D. Vanderbilt. Maximally localized generalized Wannier functions for composite energy bands. *Phys. Rev. B* **56**, 12847–12865 (1997).
49. S. Mohr, M. Eixarch, M. Amsler, M.J. Mantsinen and L. Genovese. Linear scaling DFT calculations for large tungsten systems using an optimized local basis. *J. Nucl. Mater.* **15**, 64–70 (2018). ISSN 2352-1791.
50. S. Mohr, W. Dawson, M. Wagner, D. Caliste, T. Nakajima and L. Genovese. Efficient computation of sparse matrix functions for large-scale electronic structure calculations: The chess library. *J. Chem. Theory Comput.* **13**(10), 4684–4698 (2017).
51. S. Mohr, L.E. Ratcliff, P. Boulanger, L. Genovese, D. Caliste, T. Deutsch and S. Goedecker. Daubechies wavelets for linear scaling density functional theory. *J. Chem. Phys.* **140**(20), 204110 (2014).
52. S. Mohr, L.E. Ratcliff, L. Genovese, D. Caliste, P. Boulanger, S. Goedecker and T. Deutsch. Accurate and efficient linear scaling DFT calculations with universal applicability. *Phys. Chem. Chem. Phys.* **17**, 31360–31370 (2015).
53. 2020. MultiResolution Chemistry (MRChem) program package, https://mrchem.readthedocs.io/en/latest/.
54. B. Natarajan, L. Genovese, M.E. Casida, T. Deutsch, O.N. Burchak, C. Philouze and M.Y. Balakirev. Wavelet-based linear-response time-dependent density-functional theory. *Chem. Phys.* **402**, 29–40 (2012). ISSN 0301-0104. URL http://www.sciencedirect.com/science/article/pii/S0301010412001437.
55. A.I. Neelov and S. Goedecker. An efficient numerical quadrature for the calculation of the potential energy of wavefunctions expressed in the Daubechies wavelet basis. *J. Comp. Phys.* **217**(2), 312–339 (2006).
56. J.W. Neidigh, R.M. Fesinmeyer and N.H. Andersen. Designing a 20-residue protein. *Nat. Struct. Mol. Biol.* **9**(6), 425 (2002).
57. R.A. Puglisi, S. Caccamo, C. Bongiorno, G. Fisicaro, L. Genovese, S. Goedecker, G. Mannino and A. La Magna. Direct observation of single organic molecules grafted on the surface of a silicon nanowire. *Sci. Rep.* **9**, 5647 (2019). URL https://doi.org/10.1038/s41598-019-42073-5.
58. L.E. Ratcliff, A. Degomme, J.A. Flores-Livas, S. Goedecker and L. Genovese. Affordable and accurate large-scale hybrid-functional calculations on GPU-accelerated supercomputers. *J. Phys.: Condens. Matter* **30**(9), 095901 (2018).

59. L.E. Ratcliff and L. Genovese. Pseudo-fragment approach for extended systems derived from linear-scaling DFT. *J. Phys.: Condens. Matter* **31**(28), 285901 (2019).
60. L.E. Ratcliff, S. Mohr, G. Huhs, T. Deutsch, M. Masella and L. Genovese. Challenges in large scale quantum mechanical calculations. *WIREs Comput. Mol. Sci.* **7**(1), e1290 (2017).
61. C.-K. Skylaris, P.D. Haynes, A.A. Mostofi and M.C. Payne. Introducing ONETEP: linear-scaling density functional simulations on parallel computers. *J. Chem. Phys.* **122**(8), 84119 (2005).
62. M.M. Teeter. Water structure of a hydrophobic protein at atomic resolution: Pentagon rings of water molecules in crystals of crambin. *Proc. Natl. Acad. Sci. U.S.A.* **81**(19), 6014 (1984).
63. J. Tomasi, B. Mennucci and R. Cammi. Quantum mechanical continuum solvation models. *Chem. Rev.* **105**(8), 2999–3094 (2005). URL https://doi.org/10.1021/cr9904009.
64. A. Willand, Y.O. Kvashnin, L. Genovese, Á. Vázquez-Mayagoitia, A.K. Deb, A. Sadeghi, T. Deutsch and S, Goedecker. Norm-conserving pseudopotentials with chemical accuracy compared to all-electron calculations. *J. Chem. Phys.* **138**(10), 104109 (2013).
65. M. Zaccaria, W. Dawson, V. Cristiglio, M. Reverberi, L.E. Ratcliff, T. Nakajima, L. Genovese and B. Momeni. Designing a bioremediator: mechanistic models guide cellular and molecular specialization. *Curr. Opin. Biotech.* **62**, 98–105 (2020).

Chapter 12
Accurate Approximations of Density Functional Theory for Large Systems with Applications to Defects in Crystalline Solids

Kaushik Bhattacharya, Vikram Gavini, Michael Ortiz, Mauricio Ponga, and Phanish Suryanarayana

Abstract This chapter presents controlled approximations of Kohn–Sham density functional theory (DFT) that enable very large scale simulations. The work is motivated by the study of defects in crystalline solids, though the ideas can be used in other applications. The key idea is to formulate DFT as a minimization problem over the density operator, and to cast spatial and spectral discretization as systematically convergent approximations. This enables efficient and adaptive algorithms that solve the equations of DFT with no additional modeling, and up to desired accuracy, for very large systems, with linear and sublinear scaling. Various approaches based on such approximations are presented, and their numerical performance is demonstrated through selected examples. These examples also provide important insights into the mechanics and physics of defects in crystalline solids.

K. Bhattacharya (✉) · M. Ortiz
California Institute of Technology, Pasadena, CA, USA
e-mail: bhatta@caltech.edu; ortiz@aero.caltech.edu

V. Gavini
University of Michigan, Ann Arbor, MI, USA
e-mail: vikramg@umich.edu

M. Ponga
University of British Columbia, Vancouver, BC, Canada
e-mail: mponga@mech.ubc.ca

P. Suryanarayana
Georgia Institute of Technology, Atlanta, GA, USA
e-mail: phanish.suryanarayana@ce.gatech.edu

E. Cancès, G. Friesecke (eds.), *Density Functional Theory*, Mathematics and Molecular Modeling, https://doi.org/10.1007/978-3-031-22340-2_12

12.1 Introduction

Defects are common in crystalline solids [33, 65]. These include point defects (vacancies with missing atoms, substitutional elements where an atom of an impurity (solute) replaces an atom of the actual material, or interstitial atoms where an extra atom is inserted into the solid), cluster defects (vacancy cluster or prismatic dislocation loop with a missing disc of atoms), line defects (dislocations where an extra plane of atoms terminates along a line) or planar defects (twin or grain boundaries across which the crystal orientation changes or phase boundaries across which the crystal structure changes).

Defects play a critical role in determining important properties of materials. Vacancies mediate creep, solutes strengthen solids, vacancy clusters lead to void nucleation and dislocations mediate plasticity. Remarkably, they do so at extremely dilute concentrations. Vacancies affect creep at parts per million, and dislocations densities are of the order of one in a million amongst atomic columns during plastic deformation.

The reason why defects can have such a profound effect on properties at dilute concentrations is because they trigger physics at multiple length and time scales [65, 80]. In this review, we are interested in the equilibrium structure, and therefore focus only on the length scales. The defects cause an imbalance of forces on the neighboring atoms which in turn lead to deformations. Even though electronic interactions decay quickly, displacement of the atoms from their periodic equilibrium positions lead to imbalanced forces on their neighbors, and so on, leading to extremely slow decay of the displacement field. The complex quantum mechanical or chemical interactions at the defect core lead to a complex atomistic and electronic structure that need an electronic structure theory for its description. As we move away, the displacements from the periodic structure are less complex and may be understood through atomistic interactions. Even farther away, the displacements are smaller and may be described by continuum elasticity theory.

Crucially, these scales interact intimately and one scale does not dominate over others. We consider two examples. The first example is a vacancy, where a single atom is missing from an otherwise perfect lattice. In the far field, elasticity theory tells us that the displacement decays as r^{-2} so that the stress and the strain decay as r^{-3} [58]. This is a relatively fast decay and the energy is summable in the far field. The divergence at the origin is regularized by atomistic and electronic interactions. One can estimate the energy due to the chemistry of the core by the cohesive energy of a solid (the energy difference between an isolated atom and an atom in the crystal), and this is typically a few electron volts in metals. One can also estimate the energy due to the elastic field far away, and this is of the order of a few tenths of electron volts for a typical metal. While one is smaller than the other, it is not negligibly so; therefore these fields do interact with each other even in this simple defect. Further, the elastic fields generated by the vacancy are large enough to interact with macroscopic stress due to boundary conditions resulting in a stress-induced driving force on the vacancy. The second example is that of a dislocation.

The stress and strain decay as r^{-1} away from the line, which means that the elastic energy density is logarithmic, and thus divergent at both the origin and the far field [33, 58]. In other words, dislocations provide a direct link between continuum scale boundary conditions and electronic scale interactions at the core. In short, defects connect the far field to the electronic scale and it is this ability to bridge scales that result in defects having a profound effect on macroscopic properties. This also makes defects extremely difficult to study.

Kohn–Sham DFT [34, 43] has emerged as the method of choice for the study of electronic structure in condensed matter [11]. It converts the many particle Schrödinger equation to a single particle problem with an effective single-electron potential. While one can show the existence of such a theory, the functional that gives rise to the potential (or even the locality or lack thereof) remains unknown, and is modeled. Thus, while DFT is nominally ab initio in that it is agnostic about the material (other than the atomic number), it does require a constitutive model of the universal exchange-correlation functional. It has proved to be an extremely useful compromise between practical application of quantum mechanics and fidelity.

Given the importance of electronic structure in determining the structure and energetics of defects, it is desirable to study defects using DFT. Such studies require large domains including large numbers of electrons due to the slow decay and low concentrations that are typical in materials; unfortunately such large domains are beyond the capability of brute force (full-resolution) DFT calculations using existing widely-used methods with reasonable computational resources. However, the complexity and details of the electronic structure are important near the core, and less so in the large regions of slow decay. This has motivated multiscale modeling of materials, where one builds a cascade of models (DFT, atomistic, continuum) to study the phenomena at different scales [1, 6, 91, 97]. However, the interaction between the scales means that one has to link them, which requires *further modeling*. Much of this modeling is empirical, taking us away from the ab initio point of view. Moreover, such a cascade of models linked empirically do not have an inherent or quantifiable notion of error. A notion of error is important since the study of defects requires comparing the energies of different configurations, and therefore evaluating a small difference between two large numbers. Therefore an estimate of the accuracy and an ability to control the accuracy is important.

This chapter presents a line of work that seeks to *solve the equations of DFT, and only the equations of DFT, with no further modeling* on large domains relevant to defects by introducing approximations where the error can be controlled. The idea is to formulate DFT as a well-defined minimum problem over spaces of operators, and then introduce systematically convergent approximations. Specifically, there are two approximations: (1) Spatial discretization, resulting in finite-dimensional approximate problems obtained by constrained minimization of the DFT energy functional; and (2) Spectral discretization, based on approximations of the DFT energy functional itself, or *variational crimes* in the parlance of approximation theory.

A related issue is the fact that widely used DFT methods for condensed matter are limited to periodic systems. This is motivated by crystals that involve a periodic arrangement of atoms. The periodicity enables one to work in Fourier space using a plane wave basis, and this has proven to be extremely efficient on moderate computational resources. One can extend this approach to defects using "super-cells", i.e., studying a periodic arrangement of defects and the resulting unit cell. However, since defects interact over long distances, these can lead to artifacts. Further, dislocations are topological defects and therefore not amenable to periodic arrangement unless one studies defect complexes (dislocation dipoles or quadrapoles) with zero topological content, further leading to potential artifacts. Therefore, it is desirable to move away from periodic arrangements. Finally, defects are interesting because they interact with far-field stimuli. Therefore, it is desirable to study defects under arbitrary boundary conditions. These motivate the need to solve DFT in real space.

The systematically convergent approximations lead to various algorithms that enable the solution of DFT with controlled error on large systems in real space. This chapter describes three in some detail. The first involves variable spatial discretization by exploiting adaptive higher finite elements. The second involves spectral discretization using quadratures. The final method combines spectral and spatial discretization.

The chapter is organized as follows. Section 12.2 provides a variational formulation of DFT. An important result is the reformulation (12.4) which presents DFT as a nested variational problem. We then use spectral theory to rewrite the inner variational problem. This reformulation enables us to introduce spatial discretization and spectral quadratures as convergent (Rayleigh–Ritz) approximations. Section 12.3 introduces three ideas that are useful for the efficient practical implementation of the methods. Section 12.4 introduces spatial discretization using (higher order) finite-elements, and describes how this can be used for spatial adaptivity. This section also presents a series of examples that describe the efficacy of such an approach in studying defects, and the overall performance of the method. We turn to spectral quadratures in Sect. 12.5. We discuss the relation of this method to other approaches, including the recursion method (widely used in tight binding), Padé approximations and Fermi-operator expansion. We discuss convergence and demonstrate the performance of the method using various examples. We turn to combined spatial and spectral coarse-graining in Sect. 12.6. We describe a sub-linear scaling method for the study of defects, and its application to study vacancy clusters and dislocation cores in magnesium.

12.2 Variational Formulation of Density Functional Theory

In this section, we proceed formally following the notation and presentation of Refs. [3, 90] to formulate Kohn–Sham DFT as a well-defined minimum problem over spaces of operators, and the approximation schemes—spatial and spectral—that it suggests.

12.2.1 Kohn–Sham Density Functional Theory as a Minimum Problem

We consider a closed shell spin-unpolarized system in an insulated, bounded, open and Lipschitz domain $\Omega \subset \mathbb{R}^3$ for simplicity. The presentation may be extended to spin-polarized systems and unbounded domains [3]. Let $\mathcal{V} = H_0^1(\Omega)$, $\mathcal{H} = L^2(\Omega; \mathbb{C})$, $\mathcal{D}(h) = H_0^1(\Omega; \mathbb{C})$, and $\mathcal{X} = \{\gamma \in \mathcal{S}(\mathcal{H}),\ \mathcal{R}(\gamma) \subset \mathcal{D}(h)\}$, where h denotes the single-particle Hamiltonian, γ represents the single-particle density matrix, and $\mathcal{S}(\mathcal{H})$ denotes the vector space of bounded self-adjoint operators on $\mathcal{H}$. Let

$$\mathcal{K} = \{\gamma \in \mathcal{X} : 0 \leqslant \gamma \leqslant 1,\ 2\mathrm{tr}(\gamma) = N\}$$

be a constraint set defining the admissible density operators. Define the Kohn–Sham energy functional $F : \mathcal{X} \to \bar{\mathbb{R}}$ as

$$F(\gamma) = \begin{cases} 2\mathrm{tr}\left(-\frac{1}{2}\Delta\gamma\right) + G(\rho), & \text{if } \gamma \in \mathcal{K}, \\ +\infty, & \text{otherwise,} \end{cases}$$

which can be written as

$$F(\gamma) = I_{\mathcal{K}} + 2\mathrm{tr}\left(-\frac{1}{2}\Delta\gamma\right) + G(\rho),$$

where

$$\rho(\mathbf{r}) = 2\gamma(\mathbf{r}, \mathbf{r}')$$

is the electron density. In addition,

$$G(\rho) = \int_\Omega \rho v \, \mathrm{d}\mathbf{r} + J(\rho) + E_{\mathrm{xc}}(\rho),$$

where v is an external potential, and

$$J(\rho) = \sup_{\phi \in \mathcal{V}} \left\{ \int_\Omega \rho\phi \, \mathrm{d}\mathbf{r} - \frac{1}{4\pi} \int_\Omega \frac{1}{2} |\nabla\phi|^2 \, \mathrm{d}\mathbf{r} \right\} \tag{12.1}$$

is the classical electrostatic energy [37]. A formal connection between (12.1) and the oft-used equivalent expression (up to an inessential constant) based on the Coulombic interaction formula $\frac{1}{2} \int_\Omega \int_\Omega \frac{\rho(\mathbf{r})\rho(\mathbf{r}')}{|\mathbf{r}-\mathbf{r}'|} \mathrm{d}\mathbf{r}\mathrm{d}\mathbf{r}'$ can be established simply by writing out the Euler–Lagrange equation of (12.1) and solving for ϕ using the Green's function for the Laplacian. The expression (12.1) simply recognizes the fact that $J(\rho)$ is the dual of the Dirichlet functional. Representation (12.1) is advantageous over the Coulombic representation from the standpoint of approximation, which only requires local conforming interpolation of the electrostatic field ϕ. Finally, $E_{\mathrm{xc}}(\rho)$ is the exchange-correlation energy functional. It must necessarily be modeled. Here, for simplicity, we choose the local density approximation (LDA) [60]. The *Kohn–Sham DFT problem* is to find the ground state energy

$$E_{\mathrm{KS}} = \inf_{\gamma \in \mathcal{X}} F(\gamma), \tag{12.2}$$

and attendant energy-minimizing states. For subsequent purposes, we use duality to reformulate the energy in trace form. To this end, assume for definiteness that the exchange correlation $E_{\mathrm{xc}} : \mathcal{V} \to \bar{\mathbb{R}}$ is convex with dual $E^*_{\mathrm{xc}} : \mathcal{V} \to \bar{\mathbb{R}}$, so that [90]

$$E_{\mathrm{xc}}(\rho) = \sup_{v_{\mathrm{xc}} \in \mathcal{V}} \{\rho u - E^*_{\mathrm{xc}}(v_{\mathrm{xc}})\}.$$

Now, define the Hamiltonian

$$h(\varphi, v_{\mathrm{xc}}) = -\frac{1}{2}\Delta + \Phi + V_{\mathrm{xc}} + V,$$

where the electrostatic potential operator Φ, the exchange-correlation potential operator V_{xc}, and the external potential operator V are bounded self-adjoint operators over $\mathcal{H}$ defined by the properties

$$2\mathrm{tr}(\Phi\gamma) = \int_\Omega \rho\phi \, d\mathbf{r}, \quad 2\mathrm{tr}(V_{\mathrm{xc}}\gamma) = \int_\Omega \rho v_{\mathrm{xc}} \, d\mathbf{r}, \quad 2\mathrm{tr}(V\gamma) = \int_\Omega \rho v \, d\mathbf{r}.$$

Then,

$$F(\gamma) = I_{\mathcal{K}}(\gamma) + \sup_{\phi \in \mathcal{V}} \sup_{v_{\mathrm{xc}} \in \mathcal{V}} \left(2\mathrm{tr}(h(\phi, v_{\mathrm{xc}})\gamma) - \frac{1}{4\pi} \int_\Omega \frac{1}{2} |\nabla\phi|^2 \, d\mathbf{r} - E^*_{\mathrm{xc}}(v_{\mathrm{xc}}) \right)$$

and the Kohn–Sham DFT problem (12.2) becomes

$$E_{\mathrm{KS}} = \inf_{\gamma \in \mathcal{X}} \sup_{\phi \in \mathcal{V}} \sup_{v_{\mathrm{xc}} \in \mathcal{V}} \left(I_{\mathcal{K}}(\gamma) + \left(2\mathrm{tr}(h(\phi, v_{\mathrm{xc}})\gamma) - \frac{1}{4\pi} \int_\Omega \frac{1}{2} |\nabla\phi|^2 \, d\mathbf{r} - E^*_{\mathrm{xc}}(v_{\mathrm{xc}}) \right) \right). \tag{12.3}$$

It is possible to exchange the order of the inf and sup operations in the above equation [90] to arrive at the *reformulated Kohn–Sham DFT problem*,

$$E_{\mathrm{KS}} = \sup_{\phi \in \mathcal{V}} \sup_{v_{\mathrm{xc}} \in \mathcal{V}} \left[\inf_{\gamma \in \mathcal{X}} \left(I_{\mathcal{K}}(\gamma) + 2\mathrm{tr}(h(\phi, v_{\mathrm{xc}})\gamma) \right) - \frac{1}{4\pi} \int_\Omega \frac{1}{2} |\nabla\phi|^2 \, d\mathbf{r} - E^*_{\mathrm{xc}}(v_{\mathrm{xc}}) \right]. \tag{12.4}$$

This reformulation offers various advantages and serves as the basis for the approximations to follow. First, in the same spirit as in (12.1), the representation (12.4) only involves local operators and requires local or conforming interpolation of ϕ and v_{xc}. Second, the functional is expressed in terms of linear operators acting on γ only, thus paving the way for a spectral treatment of the problem, as we do presently.

We now focus on the inner inf operation that yields the energy-minimizing density matrix for fixed (ϕ, u):

$$U = \inf_{\gamma \in X} \big(I_{\mathcal{K}}(\gamma) + 2\mathrm{tr}(h\gamma) \big), \tag{12.5}$$

where, here and subsequently, we omit the dependence of h on the fixed fields (ϕ, v_{xc}) for simplicity of notation. Note that the quantity U is commonly referred to as the *band structure energy* in the physics literature.

It follows from spectral theory (cf., e.g., [70]) that the minimizing density matrix operator γ of (12.5) shares the same spectral measure as the Hamiltonian h, i.e., we may write

$$h = \int_{\mathbb{R}} \varepsilon \, d\mathcal{E}(\varepsilon) = h^{\mathrm{T}}, \quad \gamma = \int_{\mathbb{R}} f(\varepsilon) d\mathcal{E}(\varepsilon) = \gamma^{\mathrm{T}}, \tag{12.6}$$

for $0 \leq f \leq 1$, where $\mathcal{E}$ is a resolution of the identity over the Borel sets of the real line. In addition, γ and h have the same spectral measure if and only if they commute, i.e.,

$$\gamma h = h\gamma.$$

Finally, we can show that there is a minimizer $f \in \{0, 1\}$ so that $\gamma^2 = \gamma$. Therefore, the minimum problem is

$$\text{minimize}: \; U(\gamma) := 2\mathrm{tr}(h\gamma), \tag{12.7a}$$

$$\text{subject to}: \; \gamma^{\mathrm{T}} = \gamma, \; \gamma h = h\gamma, \gamma^2 = \gamma, \; 2\mathrm{tr}(\gamma) = N. \tag{12.7b}$$

The variational problem (12.4) is often solved by a fixed point iteration of *self-consistent field* (SCF) iteration where (12.7) is solved for γ with a fixed ϕ, v_{xc} and the outer sup problem in (12.4) is used to update ϕ and v_{xc} for fixed γ.

12.2.2 Approximations Resulting from Spatial Discretization

We proceed to discretize problem (12.4) *à la* Rayleigh–Ritz, i.e., by restriction to finite-dimensional spaces. To this end, let $\mathcal{V}_h$ be a nested sequence of finite-dimensional subspaces of $\mathcal{V}$ spanned by orthonormal bases $\{e_1, \ldots, e_{N_g}\}$, e.g., corresponding to a finite element discretization.[1] Let $\mathcal{H}_h = \mathcal{V}_h$ be the corresponding sequence of subspaces of $\mathcal{H}$. Then, the discrete wave function, electrostatic field and exchange-correlation potential field are of the form

$$\varphi_h(\mathbf{r}) = \sum_{a=1}^{N_g} \varphi_a e_a(\mathbf{r}), \quad \phi_h(\mathbf{r}) = \sum_{a=1}^{N_g} \phi_a e_a(\mathbf{r}), \quad v_{xc,h}(\mathbf{r}) = \sum_{a=1}^{N_g} v_{xc,a} e_a(\mathbf{r}).$$

Likewise, the discrete Hamiltonian is

$$h_h(\mathbf{r}, \mathbf{r}') = \sum_{a_1=1}^{N_g} \sum_{a_2=1}^{N_g} H_{a_1 a_2} e_{a_1}(\mathbf{r}) e_{a_2}(\mathbf{r}'),$$

with

$$H = A_h + \Phi_h + V_{xc,h} + V_h,$$

$$A_{h;a_1a_2} = \int_\Omega \frac{1}{2} \nabla e_{a_1}(\mathbf{r}) \cdot \nabla e_{a_2}(\mathbf{r})\, d\mathbf{r} = B_{h;a_1a_2},$$

$$\Phi_{h;a_1a_2} = \int_\Omega \left(\sum_{a=1}^{N_g} \phi_a e_a(\mathbf{r}) \right) e_{a_1}(\mathbf{r}) e_{a_2}(\mathbf{r})\, d\mathbf{r},$$

$$V_{xc,h;a_1a_2} = \int_\Omega \left(\sum_{a=1}^{N_g} v_{xc,a} e_a(\mathbf{r}) \right) e_{a_1}(\mathbf{r}) e_{a_2}(\mathbf{r})\, d\mathbf{r},$$

and

$$V_{h;a_1a_2} = \int_\Omega v(\mathbf{r}) e_{a_1}(\mathbf{r}) e_{a_2}(\mathbf{r})\, d\mathbf{r}.$$

We note the additional linear structure

$$\Phi_h = T_h \phi_h, \quad V_{xc,h} = T_h u_h,$$

where

$$T_{a_1a_2a_3} = \int_\Omega e_{a_1}(\mathbf{r}) e_{a_2}(\mathbf{r}) e_{a_3}(\mathbf{r})\, d\mathbf{r}.$$

---

[1] Note that we use the subscript h to index the nested spaces following the typical notation in computational science, and not to signify a relationship with the Hamiltonian.

Finally, the discrete density matrix is of the form

$$\gamma_h(\mathbf{r}, \mathbf{r}') = \sum_{a_1=1}^{N_g} \sum_{a_2=1}^{N_g} P_{a_1 a_2} e_{a_1}(\mathbf{r}) e_{a_2}(\mathbf{r}'),$$

and the discrete electron density follows as

$$\rho_h(\mathbf{r}) = 2\gamma_h(\mathbf{r}, \mathbf{r}).$$

This sequence $\{\mathcal{V}_k\}$ of finite-dimensional subspaces of $\mathcal{V}$ defines a nested sequence of subspaces of $\mathcal{X}$ of *density matrices* $\mathcal{X}_h = \{\gamma_h \in \mathcal{S}(\mathcal{H}_h)\}$, where $\mathcal{S}(\mathcal{H}_h)$ denotes the vector space of symmetric linear operators on $\mathcal{H}_h$. This in turn defines a sequence of discrete constraint sets $\mathcal{K}_h = \{\gamma_h \in \mathcal{X}_h \, : \, 0 \leqslant \gamma_h \leqslant 1, 2\text{tr}(\gamma_h) = N\}$, where $0 \leqslant \gamma_h \leqslant 1$ expresses the requirement that $0 \leqslant (\varphi_h|\gamma_h|\varphi_h) \leqslant 1$ for all $\varphi_h \in \mathcal{H}_h$. We note that, if the spaces $\mathcal{H}_h$ are nested, then $\mathcal{K}_h$ defines a decreasing sequence of sets in $\mathcal{X}$ and that $\mathcal{K} \subset \mathcal{K}_h$. Then, the corresponding sequence of discrete energies $F_h : \mathcal{X}_h \to \bar{\mathbb{R}}$ follows as

$$F_h(\gamma_h) = I_{\mathcal{K}_h}(\gamma_h) + \sup_{\phi_h \in \mathcal{V}_h} \sup_{v_{xc,h} \in \mathcal{V}_h} \left(2\text{tr}(h_h \gamma_h) - \frac{1}{8\pi}(\phi_h|B_h|\phi_h) - E^*_{\text{xc}}(v_{xc,h}) \right), \tag{12.8}$$

where $I_{\mathcal{K}_h}$ is the indicator function of $\mathcal{K}_h$, and $B_h = D^*_h D_h$ with D_h the discrete gradient operator. The discrete Kohn–Sham DFT problem becomes

$$\begin{aligned} E_{\text{KS},h} &= \inf_{\gamma_h \in \mathcal{X}_h} F_h(\gamma_h) \\ &= \sup_{\phi_h \in \mathcal{V}_h} \sup_{v_{xc,h} \in \mathcal{V}_h} \left[\inf_{\gamma_h \in \mathcal{X}_h} \left(I_{\mathcal{K}_h}(\gamma_h) + 2\text{tr}(h_h \gamma_h) \right) - \frac{1}{8\pi}(\phi_h|B_h|\phi_h) - E^*_{\text{xc}}(v_{xc,h}) \right], \end{aligned} \tag{12.9}$$

where we have again exchanged the order of the inf and sup operations [90].

As before, we may rewrite the inner inf problem as

$$\text{minimize}: \; U_h(\gamma_h) := 2\text{tr}(h_h \gamma_h), \tag{12.10a}$$

$$\text{subject to}: \; \gamma_h^{\text{T}} = \gamma_h, \; \gamma_h h_h = h_h \gamma_h, \gamma_h^2 = \gamma_h, \; 2\text{tr}(\gamma_h) = N. \tag{12.10b}$$

12.2.3 Spectral Reformulation of the Discrete Kohn–Sham Problem

By the spectral decomposition theorem (cf., e.g., [70]), we can write

$$H = \int_{\mathbb{R}} \varepsilon \, \mathrm{d}\mathcal{E}_h(\varepsilon), \quad P = \int_{\mathbb{R}} f_h(\varepsilon) \, \mathrm{d}\mathcal{E}_h(\varepsilon),$$

where $\mathcal{E}_h$ is an operator-valued measure. In this representation, we have

$$\mathrm{tr}(HP) = \int_{\mathbb{R}} \varepsilon f_h(\varepsilon) \, \mathrm{d}\mathcal{M}_h(\varepsilon) \equiv F_h(f_h),$$

and

$$\mathrm{tr}(P) = \int_{\mathbb{R}} f_h(\varepsilon) \, \mathrm{d}\mathcal{M}_h(\varepsilon) \equiv M_h(f_h),$$

where

$$\mathcal{M}_h = \mathrm{tr}(\mathcal{E}_h)$$

is a *spectral measure* with:

$$\mathrm{d}\mathcal{M}_h = \sum_{i=1}^{N_g} \delta_{\varepsilon_i},$$

where δ is the Dirac delta function. Given the spectral measure $\mathcal{M}_h$, the calculation of the energy-minimizing discrete density matrix γ_h at fixed (ϕ_h, u_h) reduces to the scalar problem

$$\inf_{f_h \in X_h} \{F_h(f_h), \ 0 \leqslant f \leqslant 1, \ 2M_h(f_h) = N\}, \tag{12.11}$$

where X_h denotes the space of bounded real-valued Borel functions over the real line.

12.2.4 Approximation by Numerical Quadrature

We proceed to reduce problem (12.11) by recourse to numerical quadrature. Let

$$\int_{\mathbb{R}} g(\varepsilon) \, \mathrm{d}\mathcal{M}_h(\varepsilon) \approx \sum_{j=0}^{k} A_j g(\varepsilon_j)$$

be a sequence of quadrature rules, parameterized by $k \in \mathbb{N}$, with weights A_j and nodes ε_j. Here,

$$A_j = \int_{\mathbb{R}} l_j(\varepsilon) \, \mathrm{d}\mathcal{M}_h(\varepsilon), \tag{12.12}$$

where

$$l_j(\varepsilon) = \prod_{\substack{i=0 \\ i \neq j}}^{k} \frac{\varepsilon - \varepsilon_i}{\varepsilon_j - \varepsilon_i},$$

for $j = 0, \ldots, k$ are the Lagrange polynomials.

Define the sequence of approximate energies

$$F_k(f_h) = \sum_{j=0}^{k} A_j \varepsilon f_h(\varepsilon_j),$$

and the sequence of approximate masses

$$M_k(f_h) = \sum_{j=0}^{k} A_j f_h(\varepsilon_j).$$

Then, we have a corresponding sequence of *discretized* problems

$$\inf_{f_h \in X_h} \{F_k(f_h), \; 0 \leqslant f_h \leqslant 1, \; 2M_k(f_h) = N\}. \tag{12.13}$$

The solution of these approximate problems then follows from the algorithm:

(i) Set $f_0(\varepsilon) = 0$, $i = 0, \ldots, k$, $I_0 = \{0, \ldots, k\}$, $N_0 = 0$, $n = 1$.
(ii) Let $i_n \in \operatorname{argmin}\{\varepsilon, \; i \in I_{n-1}\}$, $N_n = N_{n-1} + A_{i_n}$.
(iii) If $N_n < N$, set $f_n(\varepsilon_{i_n}) = 1$, $I_n = I_{n-1} \backslash \{i_n\}$, $n \leftarrow n + 1$, go to (ii).
(iv) Otherwise, set $f_n(\varepsilon_{i_n}) = (N - N_{n-1})/A_{i_n}$, $f_h = f_n$, exit.

12.2.5 Rayleigh–Ritz Interpretation

The numerical-quadrature reduction can again be given an appealing Rayleigh–Ritz interpretation. Begin by noting the identity

$$\int_{\mathbb{R}} l_i(\varepsilon)\varepsilon \, \mathrm{d}\mathcal{M}_h(\varepsilon) = \sum_{j=0}^{k} A_j l_i(\varepsilon_j)\varepsilon_j = A_i \varepsilon_i = \left(\int_{\mathbb{R}} l_i(\varepsilon) \, \mathrm{d}\mathcal{M}_h(\varepsilon) \right) \varepsilon_i.$$

From this identity and (12.12) we have

$$F_k(f_h) = \sum_{i=0}^{k} \left(\int_{\mathbb{R}} l_i(\varepsilon)\, \mathrm{d}\mathcal{M}_h(\varepsilon) \right) \varepsilon_i f(\varepsilon_i) = \int_{\mathbb{R}} \left(\sum_{i=0}^{k} l_i(\varepsilon)\varepsilon_i f(\varepsilon_i) \right) \mathrm{d}\mathcal{M}_h(x)$$

$$= \int_{\mathbb{R}} \left(\sum_{i=0}^{k} l_i(\varepsilon) f(\varepsilon_i) \right) \varepsilon\, \mathrm{d}\mathcal{M}_h(\varepsilon) = F_h(f_k),$$

where

$$f_k = \sum_{i=0}^{k} l_i(\varepsilon) f_h(\varepsilon_i).$$

Likewise,

$$M_k(f_h) = \sum_{i=0}^{k} \left(\int_{\mathbb{R}} l_i(\varepsilon)\, \mathrm{d}\mathcal{M}_h(\varepsilon) \right) f_h(\varepsilon_i)$$

$$= \int_{\mathbb{R}} \left(\sum_{i=0}^{k} l_i(\varepsilon) f_h(\varepsilon_i) \right) \mathrm{d}\mathcal{M}_h(\varepsilon) = M_h(f_k).$$

Define now the sequence of spaces

$$\mathcal{X}_k = \mathrm{span}\{l_i(H),\ i = 0, \dots, k\} = \{P = \sum_{i=0}^{k} f_i l_i(H),\ f_i \in \mathbb{R},\ i = 0, \dots, k\},$$

where $\{l_i,\ i = 0, \dots, k\}$ are the Lagrange polynomials defined by the roots of the orthogonal polynomial p_{k+1} generated by H. Define, in addition, the sequence of relaxed constraint sets

$$\mathcal{K}_k = \{P = \sum_{i=0}^{k} f_i l_i(H),\ 0 \leqslant f_i \leqslant 1,\ i = 0, \dots, k\}.$$

Then, the reduced problem (12.13) is equivalent to solving

$$\inf_{\gamma_h \in \mathcal{X}_k} \left(I_{\mathcal{K}_k}(\gamma_h) + 2\mathrm{tr}(h_h \gamma_h) \right),$$

which corresponds to a Rayleigh–Ritz reduction of problem (12.5) to the subspaces of density matrices $\mathcal{X}_k$ generated by numerical quadrature, and to the corresponding relaxed constraint sets $\mathcal{K}_k$.

12.2.6 Convexification and Thermalization

The DFT problem (12.4) and the inner minimization problem (12.7) are not convex due to the constraint that $f(\varepsilon)$ takes values in $\{0, 1\}$. We convexify the problems by allowing the function to take values in the entire interval $[0, 1]$, the resulting function henceforth referred to as $f_\beta(\varepsilon)$. We expect the minimizers to take extreme values only and thus the convexified and original problems to yield the same minimizers and the same minimum energy. We can now enforce the constraint $0 \leqslant f_\beta(\varepsilon) \leqslant 1$ by entropic penalization. We present it for the infinite-dimensional version (12.4), though it can readily be extended to the versions with spatial and spectral discretization. Introduce the entropy

$$S(\gamma) = 2\mathrm{tr}[\gamma \log(\gamma) + (\mathcal{I} - \gamma)\log(\mathcal{I} - \gamma)],$$

and the thermalized problem

$$\begin{aligned} \text{minimize}: \; U_\beta(\gamma) &= U(\gamma) + \frac{1}{\beta} S(\gamma) \\ &= 2\mathrm{tr}(h\gamma) + \frac{2}{\beta}\mathrm{tr}[\gamma \log(\gamma) + (\mathcal{I} - \gamma)\log(\mathcal{I} - \gamma)], \end{aligned} \tag{12.14a}$$

$$\text{subject to}: \; \gamma^{\mathrm{T}} = \gamma, \; \gamma h = h\gamma, \; 2\mathrm{tr}(\gamma) = N, \tag{12.14b}$$

where β is an inverse temperature. The minimizer of $U_\beta(\gamma)$ is

$$\gamma_\beta = f_\beta(h) = (\mathcal{I} + \mathrm{e}^{\beta(h - \mu\mathcal{I})})^{-1}, \tag{12.15}$$

where μ is a chemical potential introduced to enforce the number constraint. μ and f_β are commonly referred to as the *Fermi level* and *Fermi–Dirac distribution*, respectively, with $\beta \to \infty$ representing the zero-temperature limit.

The thermalized total energy

$$E_{\mathrm{KS},\beta} = \sup_{\phi \in \mathcal{V}} \sup_{v_{\mathrm{xc}} \in \mathcal{V}} \left[U_\beta(\gamma_\beta) - \frac{1}{4\pi} \int_\Omega \frac{1}{2} |\nabla \phi|^2 \, \mathrm{d}\mathbf{r} - E^*_{\mathrm{xc}}(v_{\mathrm{xc}}) \right]$$

Finally, we may estimate the zero temperature ground state energy as

$$E_{\mathrm{KS},0} \approx \sup_{\varphi \in \mathcal{V}} \sup_{v_{\mathrm{xc}} \in \mathcal{V}} \left[U_\beta(\gamma_\beta) - \frac{1}{2\beta} S(\gamma_\beta) - \frac{1}{4\pi} \int_\Omega \frac{1}{2} |\nabla \varphi|^2 \, \mathrm{d}\mathbf{r} - E^*_{\mathrm{xc}}(v_{\mathrm{xc}}) \right]. \tag{12.16}$$

12.2.7 Spatial Densities

For later use, we note that the quantities in (12.4) and (12.16) have associated spatial densities and can be rewritten in terms of volume integrals. Following (12.6), and explicitly introducing the spatial variables

$$h(\mathbf{r},\mathbf{r}') = \int_{\mathbb{R}} \varepsilon \, \mathrm{d}\mathcal{E}_{\mathbf{r},\mathbf{r}'}(\varepsilon) = h^{\mathrm{T}}, \quad \gamma_\beta(\mathbf{r},\mathbf{r}') = \int_{\mathbb{R}} f_\beta(\varepsilon)\mathrm{d}\mathcal{E}_{\mathbf{r},\mathbf{r}'}(\varepsilon) = \gamma_\beta^{\mathrm{T}}. \tag{12.17}$$

Therefore, the number of electrons, the band structure energy and the entropy may be written as

$$\begin{aligned}
N &= 2\mathrm{tr}(\gamma_\beta) = \int_\Omega \left(2\int_{\mathbb{R}} f_\beta(\varepsilon)d\mathcal{E}_{\mathbf{r},\mathbf{r}}(\varepsilon)\right)\mathbf{dr} = \int_\Omega \rho(\mathbf{r})\mathbf{dr},\\
U_\beta &= 2\mathrm{tr}(h\gamma_\beta) = \int_\Omega \left(2\int_{\mathbb{R}} \varepsilon f_\beta(\varepsilon)d\mathcal{E}_{\mathbf{r},\mathbf{r}}(\varepsilon)\right)\mathbf{dr} = \int_\Omega u(\mathbf{r})\mathbf{dr},\\
S &= 2\mathrm{tr}(\gamma_\beta \log\gamma_\beta + (I-\gamma_\beta)\log(I-\gamma_\beta))\\
&= \int_\Omega \left(2\int_{\mathbb{R}} [f_\beta(\varepsilon)\log f_\beta(\varepsilon) + (1-f_\beta(\varepsilon))\log(1-f_\beta(\varepsilon))]\mathrm{d}\mathcal{E}_{\mathbf{r},\mathbf{r}}(\varepsilon)\right)\mathbf{dr}\\
&= \int_\Omega s(\mathbf{r})\mathbf{dr}
\end{aligned}$$

in terms of the *charge or number density* ρ, *band structure energy density* u and *entropic density* s. Indeed, recall that $\rho(\mathbf{r}) = 2\gamma_\beta(\mathbf{r},\mathbf{r})$. These densities play a key role in later sections.

12.2.8 Eigenvalue Problem

We close our formulation by connecting it to the way DFT is usually presented as an eigenvalue problem. The direct solution of problem (12.9) entails the computation of $N/2$ eigenvalues and eigenvectors. To see this, consider the inner inf operation in (12.9). Write

$$P = Q_h Q_h^{\mathrm{T}},$$

$Q_h \in \mathcal{L}(\mathcal{H}_h, \mathbb{R}^{N/2})$, with

$$Q_h^{\mathrm{T}} Q_h = I_{N/2},$$

where $I_{N/2}$ denotes the identity in $\mathcal{L}(\mathbb{R}^{N/2})$. Here and subsequently, $\mathcal{L}(\mathcal{A}, \mathcal{B})$ denotes the space of linear transformations between two linear spaces $\mathcal{A}$ and $\mathcal{B}$, and $\mathcal{L}(\mathcal{A})$ the space of linear transformations from a linear space $\mathcal{A}$ to itself. Then, $P^{\mathrm{T}} = P$, $0 \leqslant P \leqslant 1$ and $\mathrm{tr}(P) = N/2$, hence $\gamma_h \in \mathcal{K}_h$. The problem under consideration thus becomes

$$Q_h \in \operatorname{argmin}\{2\mathrm{tr}(Q_h^{\mathrm{T}} H Q_h),\ Q_h^{\mathrm{T}} Q_h = I_{N/2}\},$$

The Euler–Lagrange equations of this problem are

$$H Q_h = Q_h \Lambda_h,$$

where $\Lambda_h \in \mathcal{L}(\mathbb{R}^{N/2})$, $\Lambda_h^{\mathrm{T}} = \Lambda_h$, is a Lagrange multiplier. Clearly, these Euler–Lagrange equations are solved if the columns of Q_h consist of eigenvectors of H and Λ_h stores the corresponding eigenvalues in its diagonal. In addition, if $\{\varepsilon_1, \ldots, \varepsilon_{N_g}\}$ are the ordered eigenvalues of H is ascending order and $\{\varphi_1, \ldots, \varphi_{N_g}\}$ are the corresponding eigenvectors, then the minimum problem is solved by $Q_h = \{\varphi_1, \ldots, \varphi_{N/2}\}$ and $\Lambda_h = \mathrm{diag}\{\varepsilon_1, \ldots, \varepsilon_{N/2}\}$. Finally, the energy follows as

$$2\mathrm{tr}(Q_h^{\mathrm{T}} H Q_h) = 2\mathrm{tr}(Q_h^{\mathrm{T}} Q_h \Lambda_h) = 2\mathrm{tr}(\Lambda_h).$$

Clearly, this computation becomes intractable for large material samples containing a large number electrons N. Therefore, computational tractability of large samples requires an additional reduction (beyond spatial discretization) that we refer to as *spectral reduction* above.

12.3 Filtering, Spectrum Splitting and Pseudopotentials

This section introduces three ideas that enable faster calculations. The first two, filtering and spectrum splitting, are convergent approaches and take advantage of the spectral formulation. The third, pseudopotentials, involves modeling.

12.3.1 Filtering

The discrete DFT problem (12.9) is posed as a problem in an N_g-dimensional subspace $\mathcal{V}_h$ of $\mathcal{V} = H_0^1(\Omega)$. In practice, the accurate solution of the equations requires that $N_g \gg N$. However, the solution to our problem, the density matrix γ_h, has rank N (in the thermalized problem, the thermalized density matrix $\gamma_{\beta,h}$ has rank larger than but close to N). Therefore, one can obtain significant savings in computational effort by identifying a priori a sub-space $\mathcal{V}_h^f$ such that $\text{range}(\gamma_h) \subset \mathcal{V}_h^f \subset\subset \mathcal{V}_h$, and restricting the problem (12.10) and specifically the Hamiltonian h_h to the sub-subspace $\mathcal{V}_h^f$. This can be achieved using filtering. While many approaches have been proposed based on filtering such as purification (cf. e.g. [31, 48, 59, 76]) and approximations to the Fermi–Dirac functions (cf. e.g. [4, 24, 49]), the Chebyshev filtering technique [98, 99] is adopted in many recent DFT codes [21, 22, 51, 53]. The main idea in Chebyshev filtering is to approximate the subspace $\mathcal{V}_h^f$ as

$$\mathcal{V}_h^f \approx T_m(g(h_h))X_h,$$

where $X_h \subset \mathcal{V}_h$ with $\dim(X_h) = \dim(\mathcal{V}_h^f)$, T_m is a Chebyshev polynomial of order m, and

$$g(x) = \frac{2}{b-a}\left(x - \frac{b+a}{2}\right), \quad b > a$$

with $b = \max \sigma(h_h) >> 1$ (σ denoting the spectrum) and $a = \max \sigma(\gamma_h h_h) + O(1)$. In particular, $a \approx \mu + O(1)$ is a reasonable choice. We note that g transforms the spectrum of h_h such that $\sigma(g(h_h)) < 1$ and $\sigma(g(\gamma_h h_h))) \in (-\infty, -1)$. Thus, as $T_m(x) > 1$ for $x \in (-\infty, -1)$, $T_m(g(h_h))X_h$ provides a good approximation to $\mathcal{V}_h^f$. We note that a suitable choice of m depends on the value $(b-a)$, a larger m being needed for larger values of $(b-a)$. For instance, based on numerical studies, if $(b-a) \sim O(10^2)$, values of $m \sim 10 - 30$ are sufficient to construct a good approximation to $\mathcal{V}_h^f$ [53, 99]. However, if $(b-a) \sim O(10^6)$, values of $m \sim 1000$ are needed [71].

If $\mathcal{P}_f : \mathcal{V}_h \to \mathcal{V}_h^f$ denotes the projection operator onto the filtered subspace, then the solution to the DFT problem can be obtained by replacing h_h in (12.10) with $\mathcal{P}_f h_h \mathcal{P}_f$. As the spectral width $\Sigma(\mathcal{P}_f h_h \mathcal{P}_f) \ll \Sigma(h_h)$, it enables faster numerical solution of the DFT problem, and has been the basis for subspace projection methods (cf. e.g. [16, 54]).

12.3.2 Spectrum Splitting

The next idea combines filtering with a feature of the solution of a typical problem. Here, we assume that the DFT problem has already been projected onto $\mathcal{V}_h^f$, and denote $\mathcal{P}_f h_h \mathcal{P}_f$ by h_f. We denote by $\sigma_h = \sigma(h_f)$ the spectrum of h_f, and assume in the following that $\max \sigma_h < 0$ (i.e., h_f is appropriately shifted such that this condition is satisfied). It has long been recognized that the spectrum of h_f has a gap that separates the so-called *core*, or deeply bound states at the lower end, from the rest. In other words, the spectrum $\sigma_h = \sigma_h^c \cup \sigma_h^r$ with $\varepsilon' + E_g \leqslant \varepsilon'' \; \forall \varepsilon' \in \sigma_h^c, \varepsilon'' \in \sigma_h^r$

for a gap $E_g > 0$. We can therefore split the Hamiltonian h_f and the density operator γ_h (corresponding to h_f) into

$$h_f = h_f^c + h_f^r\,, \quad \gamma_h = \gamma_h^c + \gamma_h^r\,, \tag{12.18}$$

where the spectrum of h_f^c is $\sigma_h^c \cup \{0\}$. It follows that we can divide $\mathcal{V}_h^f$ into two orthogonal subspaces,

$$\mathcal{V}_h^f = \mathcal{V}_h^c \oplus \mathcal{V}_h^r$$

where $\mathcal{V}_h^{c,r}$ is the range of $h_f^{c,r}$. Further, since σ_h^c is the lower end of the spectrum, it follows that

$$\gamma_h^c = \mathcal{P}_h^c \tag{12.19}$$

is the projection operator from $\mathcal{V}_h^f$ to $\mathcal{V}_h^c$.

Now, in light of the spectral gap, we can again use filtering on h_f, and then readily identify $\mathcal{V}_h^c$ as the range of h_f^c. Therefore, we can use (12.19) to easily compute γ_h^c. Further, using the orthogonality of the subspaces,

$$h_f^r = (\mathcal{I} - \mathcal{P}_h^c) h_f (\mathcal{I} - \mathcal{P}_h^c)\,.$$

Since the spectra σ_h^c and σ_h^r are disjoint, it follows that

$$\mathrm{tr}(h_f \gamma_h) = \mathrm{tr}(h_f^c \gamma_h^c) + \mathrm{tr}(h_f^r \gamma_h^r)\,. \tag{12.20}$$

We may now reduce (12.10) as

$$\text{minimize}: \ U_h^r(\gamma_h^r) := 2\mathrm{tr}(h_f^r \gamma_h^r), \tag{12.21a}$$

$$\text{subject to}: \ (\gamma_h^r)^{\mathrm{T}} = \gamma_h^r,\ \gamma_h^r h_f^r = h_f^r \gamma_h^r, (\gamma_h^r)^2 = \gamma_h^r,\ 2\mathrm{tr}(\gamma_h^r) = N - N^c\,, \tag{12.21b}$$

where N_c denotes the number of core electrons. This approach of spectrum splitting provides a number of advantages. First, the computation of γ_h^c, the core part of the density matrix, is relatively simple as described above. Second, in practice, the width of the spectrum of h_f^r ($\Sigma(h_f^r)$) is significantly smaller than that of h_f ($\Sigma(h_f)$), and this allows for a more efficient numerical solution. Finally, the core subspace $\mathcal{V}_h^c$ consists of functions which have a compact support close to the nuclei. In other words, this is the subspace spanned by the orbitals of the core electrons. This can be further exploited to gain numerical efficiency. Further, its complement, $\mathcal{V}_h^r$, which contains so-called valance and conduction electrons, consists of functions that vary smoothly outside a core region around the nucleus. Therefore, we can use a spatially adaptive resolution to discretize it.

We may proceed similarly in the thermalized problem to find that (12.19) and (12.20) still hold, and

$$\gamma^r_{h,\beta} = f_\beta(h^r_f) \,. \tag{12.22}$$

It is common to compute this by expanding it in a polynomial basis (Fermi operator expansion [24, 25]), which we shall show later in Sect. 12.5 is related to the spectral quadratures. Therefore, the advantages of spectrum splitting carry over to the thermalized setting.

The accuracy and efficacy of this approach for large-scale all-electron DFT calculations has been demonstrated in [56]. Here, we present some representative results on Si and Au nanoclusters. Figure 12.1 shows the results from ground-state energies computed using two approaches: (1) SubPJ-FE: A subspace projection approach via filtering (Sect. 12.3.1) implemented in a finite-element basis, where $\gamma_{h,\beta} = f_\beta(h_f)$ is computed via Fermi-operator expansion using Chebyshev polynomials for various orders; (2) Spectrum-splitting method: In addition to the subspace projection via filtering, spectrum splitting is used, where $\gamma_{h,\beta} = \gamma^c_h + f_\beta(h^r_f)$ and $f_\beta(h^r_f)$ is evaluated via Fermi-operator expansion using Chebyshev polynomials for various orders. The results for Si_{95} are provided for two values of β corresponding to $T = 500$ and 1000 K, and results for the Au_6 cluster are shown for $T = 500$ K. As is evident, spectrum splitting not only provides computationally efficiency—due to a substantial reduction in the polynomial order required in Fermi operator expansion—it is indispensable to obtain the desired accuracy for systems with large atomic numbers, like Au.

We conclude this subsection by noting that spectrum splitting is also closely related to the so-called *enrichment* methods. Note that the identity (12.19) means that we can use any basis set to represent $\mathcal{V}^c_h$. Therefore, picking $\sim N_c/2$ functions that are computationally convenient and approximate the span of $\mathcal{V}^c_h$ provides a

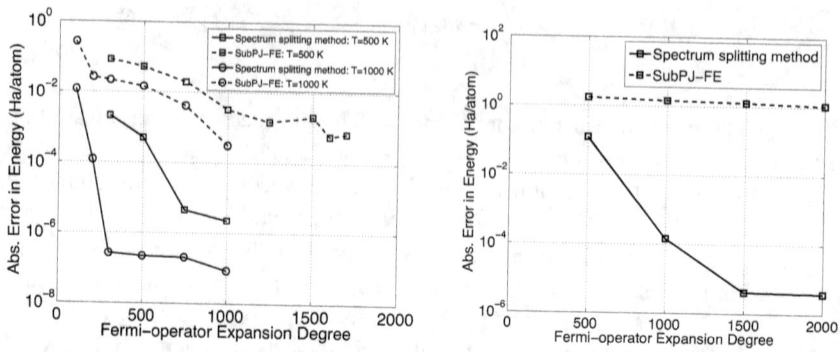

Fig. 12.1 Accuracy and computational efficiency afforded by spectrum splitting in all-electron calculations via Fermi operator expansion. (**Left**) Si_{95} nanocluster; (**Right**) Au_6 nanocluster. Adapted from [56]

good starting point. Subsequently, choosing a spatial discretization sufficient to span $\mathcal{V}_h^f$ provides the desired accuracy. This is computationally effective since the spatial discretization does not have to be so fine as to represent the core electrons. The basis set approximately spanning $\mathcal{V}_h^c$ can be iteratively updated as the calculation proceeds. These ideas lead to augmented plane wave (APW), linearized augmented plane wave (LAPW) [73], and enriched finite basis [40, 71, 95]. We refer the reader to Chap. 9 by Chen and Schneider in this volume for a detailed discussion of these methods.

12.3.3 *Frozen Core Approximation and Pseudopotentials*

The formulations discussed till now have considered all electrons in the system. However, it is a long-held observation in the field that core electrons play a minimal role in the bonding between atoms. Specifically, it is observed that the γ_h^c is relatively independent of the external potentials v that arise in molecules and crystals. This motivates the desire to exclude these electrons from the calculations, and to focus on the valance and conduction electrons.

One approach to doing so is the so-called *frozen core approximation*. Here, a high resolution all-electron calculation for a single atom is conducted to obtain the core density matrix, $\gamma_{h,Z}^c$ for a single atom (the subscript Z here refers to the single atom of atomic number Z with the nucleus located at the origin). Subsequently, this is used as an ansatz for the core electrons for any given problem. Specifically, for a problem with N_a atoms with atomic numbers $\{Z_i\}$ located at $\{\mathbf{r}_i\}$,

$$\bar{\gamma}_h^c(\mathbf{r}, \mathbf{r}') = \sum_{i=1}^{N_a} \gamma_{h,Z_i}^c(\mathbf{r} - \mathbf{r}_i, \mathbf{r}' - \mathbf{r}_i)$$

is used as an ansatz for

$$\gamma_h = \bar{\gamma}_h^c + \gamma_h^r \tag{12.23}$$

in (12.10) to solve for γ_h^r. Note that the computational complexity of the problem is now reduced from N electrons to $N - N^c$ electrons. Further, as noted above, the range of γ_h^r is spanned by relatively smooth functions outside the core, and therefore one can use a spatially adaptive discretization to represent this problem.

Note that this is an uncontrolled approximation since it is based on an ansatz. Table 12.1 from Ref. [52] shows the errors from the frozen core approximation for a range of systems. In particular, the two metrics used to measure the approximation are: (1) the relative error in the core electron density at the ground-state $||\rho_0^c - \bar{\rho}^c||_{L^2}/||\rho_0^c||_{L^2}$, where ρ_0^c is the core electron density at the ground-state from the all-electron calculation and $\bar{\rho}^c = \bar{\gamma}_h^c(\mathbf{r}, \mathbf{r})$; (2) the relative error in the total electron density at the ground-state $||\rho_0 - \bar{\rho}_0||_{L^2}/||\rho_0||_{L^2}$, where ρ_0 is the

Table 12.1 Error incurred from the frozen core approximation for various systems [52]

System	$\|\|\rho_0^c - \bar{\rho}^c\|\|_{L^2}/\|\|\rho_0^c\|\|_{L^2}$	$\|\|\rho_0 - \bar{\rho}_0\|\|_{L^2}/\|\|\rho_0\|\|_{L^2}$
Li_2	0.00703	0.00787
O_2	0.00102	0.00128
CO	0.00181	0.00129
Si_{18}	0.01272	0.0130
Si_{31}	0.01273	0.0134

total electron density at the ground-state from the all-electron calculation, and $\bar{\rho}_0$ is the total electron density at the ground-state from the frozen core approximation. As evident, while the approximation is good for some systems, it can incur larger errors for others (such as Si nanoclusters).

A closely related idea is that of a *pseudopotential*. Here, the objective is to fully exclude the core states by using a fictitious potentials, namely pseudopotentials, thus replacing h_h with h_{PS}. The pseudopotentials are generated such that $\gamma_{PS}^r = f_\beta(h_{PS})$ closely approximates γ_h^r outside a core radius around each atom, but the range of γ_{PS}^r is smooth all through the simulation domain. Thus, this alleviates the need for a spatially refined basis to resolve the core states. Various pseudopotentials have been proposed and are widely used (cf. e.g. [8, 29, 88]). Despite the errors and the uncontrolled nature of these approximations, it is often the only practical route to proceed in large systems of interest.

12.4 Spatial Coarse-Graining: Finite-Element Discretization

Spatial discretization (cf. Sect. 12.2.2) plays a central role in the practical aspects of computing the solution to the Kohn–Sham DFT problem in an efficient manner. Many discretization schemes have been adopted by the scientific community in solving the Kohn–Sham problem, and besides the algorithms employed, the discretization schemes have been the main differentiator for the various DFT codes and their performance based on computational efficiency and scalability. The widely used discretization methods include the plane-wave basis (cf. e.g. [23, 27, 44]) and atomic orbital type basis functions (cf. e.g. [9, 32, 36, 86]). While the plane-wave basis offers spectral convergence, it is primarily efficient for periodic problems owing to the lack of spatial adaptivity, and is constrained by limited parallel scalability of numerical implementations. The atomic orbital type basis functions present a reduced order basis, but in practice may not guarantee a robust and systematically convergent solution, especially for metallic systems. Also, they suffer from limited parallel scalability owing to the global nature of the basis functions. Although the finite-element and finite difference discretization schemes were explored over two decades ago [45, 62, 63, 83, 84], they have only recently started to gain traction as efficient and scalable approaches for solving the Kohn–Sham problem [21, 22, 53, 57].

The finite-element discretization in particular offers many attractive features, including the following: (1) Systematic convergence. Piecewise polynomials of a fixed degree p are dense in $H^1(\Omega)$ as the finite-element mesh-size h becomes small. Further, polynomials of increasing p are dense for a fixed h. (2) Flexibility. It has the ability to easily handle complex geometries and mixed boundary conditions, which is especially important when treating defects where periodicity may not be appropriate. (3) Spatial adaptivity. The discretization can be exploited to provide desired basis resolution in regions of interest and coarse-graining elsewhere. (4) Parallel scalability. The locality of the FE basis provides for efficient parallel scalability of numerical implementation. We also refer the reader to Chap. 10 by Dai and Zhou in the present volume for a broad discussion of the application of finite element discretization to DFT.

12.4.1 *Higher-Order Spectral Finite-Elements*

Despite the aforementioned advantages of the finite-element basis, and many prior efforts that explored the use of a finite-element basis for electronic structure calculations, they have not been competitive with widely used plane-wave and atomic orbital basis sets until recently. The two main issues limiting the performance of a finite-element basis in Kohn–Sham DFT had been: (1) the significant degree of freedom disadvantage of commonly used linear finite-elements in comparison to a plane-wave basis that affects the computational efficiency in practical DFT calculations; (2) the non-orthogonality of the finite-element basis that either limits the available solution schemes or requires an additional evaluation of the inverse of the overlap matrix.

Figure 12.2 provides insights into the lack of computational efficiency of linear finite-elements observed in prior studies. The figure shows the error in the ground-state energy for various finite element discretizations of different finite-element orders for two materials systems The higher order finite-elements employed in the study are hexahedral finite-elements, where the finite-element basis functions are constructed as a tensor product of basis functions in each dimension. The hexahedral finite-element basis functions in the isoparametric formulation are constructed from polynomial basis functions in the reference domain $[-1, 1]^3$ as

$$P_{i,j,k}(\xi, \eta, \kappa) = l_i(\xi)l_j(\eta)l_k(\kappa), \quad l_i(\xi) := \prod_{\substack{0 \leqslant m < p \\ m \neq i}} \frac{\xi - \xi_m}{\xi_i - \xi_m}, \quad i, j, k = 0, 1, \dots p \tag{12.24}$$

where $l_i(\xi)$ is a Lagrange polynomial of degree p constructed based on the $p + 1$ nodes of the finite-element. Conventionally, the finite-element nodes are chosen to be equidistant, however the conditioning of basis functions is known to deteriorate with increasing order [10]. Instead, spectral finite-elements, where the finite-element

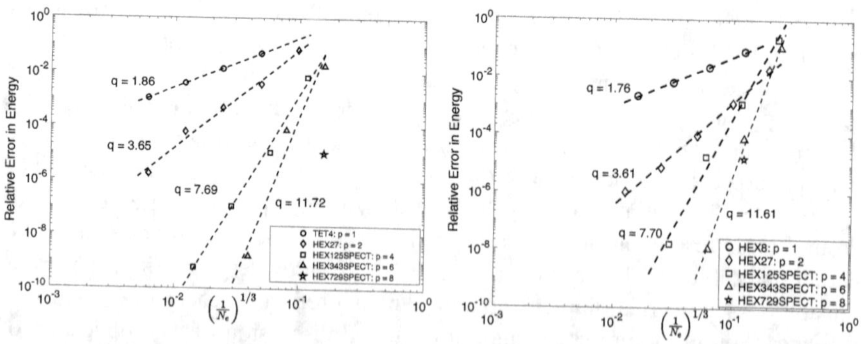

Fig. 12.2 Convergence of finite-element discretization for various finite-element orders. N_e denotes the number of elements with $(\frac{1}{N_e})^{1/3}$ providing a measure of the finite-element mesh size h; TET4 denotes a linear tetrahedral element; HEX27 denotes a quadratic hexahedral element (p = 2); HEX125SPECT denotes a quartic spectral hexahedral element (p = 4); HEX343SPECT denotes a sixth order spectral hexahedral element (p = 6); HEX729SPECT denotes an eighth order spectral hexahedral element (p = 8). The benchmark systems comprise of (**Left**) Barium cluster, non-periodic system; (**Right**) Face-centered cubic Calcium unit cell, periodic system. Adapted from [57]

nodes are chosen to be the roots of the Chebyshev polynomial, or the roots of the derivative of the Legendre polynomial, are known to provide a better conditioned basis for higher-order discretizations. From the results in Fig. 12.2, we note that for all orders of finite-element discretizations, the relative error in ground-state energy $|\frac{E_{KS,h}-E_{KS}}{E_{KS}}| \sim Ch^q$, where $E_{KS,h}$ is the discrete ground-state energy, E_{KS} is the converged ground-state energy, and h is a measure of the finite-element mesh size, chosen to be $(\frac{1}{N_e})^{1/3}$ where N_e is the number of elements. The results show that q is close to $2p$ with p denoting the finite-element order (degree of the Lagrange polynomial l_i). These results also show that the faster convergence of higher-order finite element approximations also provide a substantial reduction in the number of finite-elements required to achieve chemical accuracy ($\sim 10^{-5}$ relative errors in energy). This suggests the use of higher-order finite-element discretization as a potential path to bridging the significant degree of freedom disadvantage with a plane-wave basis.

Figure 12.3 (left) shows the degrees of freedom needed to solve two benchmark systems—a copper nanocluster with 55 atoms (non-periodic systems) and Mo supercell with a monovacancy containing 53 atoms—to chemical accuracy (0.1 mHa/atom in energy and 0.1 mHa/Bohr in force) with various orders of hexahedral spectral finite elements. It is evident that by using a fourth order finite element in comparison to a linear finite element, the basis function requirement can be reduced by $\sim 1000\times$. This subsequently translates into a $\sim 1000\times$ improvement in computational efficiency, as shown in Fig. 12.3 (right), which provides the corresponding computational times in CPU-Hrs. While the gap between the number of basis functions required to achieve chemical accuracy is substantially reduced between plane-wave and higher-order finite-element discretization, the number

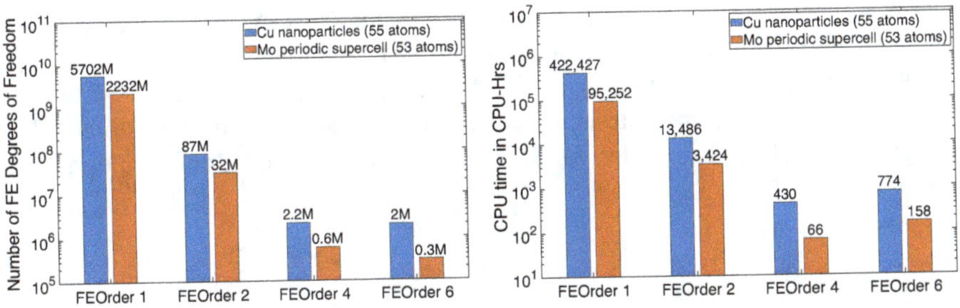

Fig. 12.3 **(Left)** The number of finite element degrees of freedom required to achieve chemical accuracy for the various orders of finite elements for two benchmark systems involving a Cu nanoparticle (non-periodic system) and a Mo supercell with a mono-vacancy (periodic system). **(Right)** The corresponding computational efficiency afforded by the various orders of finite elements. Results obtained using DFT-FE code [53]

of basis functions using finite-element discretization is still ~5-fold larger than plane-waves. However, computational cost per basis function is typically lower compared to plane-waves, and, given the better parallel scalability, finite-element discretization is emerging as an alternative to plane-waves for systematically convergent, fast and scalable DFT calculations.

The use of spectral higher-order finite-elements, while improving the conditioning of the basis, provides a path to addressing the non-orthogonality of the finite element basis. In particular, the Löwdin orthonormalized finite element basis ($\{e_1, \ldots, e_{N_b}\}$) can be constructed from the Lagrange finite element basis ($\{q_1, \ldots, q_{N_b}\}$) as

$$e_\alpha(\mathbf{r}) = \sum_{\beta=1}^{N_b} (M^{-1/2})_{\alpha\beta}\, q_\beta(\mathbf{r})\,, \qquad M_{\alpha\beta} = \int_\Omega q_\alpha(\mathbf{r}) q_\beta(\mathbf{r}) \mathrm{d}\mathbf{r}\,.$$

However, such a transformation requires the computation of $M^{-1/2}$, which can be prohibitively expensive for large N_b. We note that by using spectral finite elements with the nodes located at the derivative of the Legendre polynomial (in addition to nodes at the end points) in conjunction with Gauss–Lobatto–Legendre (GLL) quadrature rules, M is rendered diagonal and the transformation is trivial. In particular,

$$\begin{aligned} M_{\alpha\beta} = \int_\Omega q_\alpha(\mathbf{r}) q_\beta(\mathbf{r}) \mathrm{d}\mathbf{r} &= \sum_{el=1}^{N_e} \int_{\Omega_{el}} q_\alpha(\mathbf{r}) q_\beta(\mathbf{r}) \mathrm{d}\mathbf{r} \\ &= \sum_{el=1}^{N_e} \int_{[-1,1]^3} P_\alpha(\xi,\eta,\kappa) P_\beta(\xi,\eta,\kappa) J_{\Omega_{el}} \mathrm{d}\xi \mathrm{d}\eta \mathrm{d}\kappa\,, \end{aligned}$$

where Ω_{el}, $el = 1, 2, \ldots, N_e$ denote the domains corresponding to each finite-element, and $J_{\Omega_{el}}$ is the Jacobian of the transformation from Ω_{el} to $[-1, 1]^3$. $P_\alpha(\xi, \eta, \kappa)$ is the Lagrange polynomial defined on $[-1, 1]^3$ (Eq. 12.24) with $\alpha = (i, j, k)$ denoting a composite index corresponding to a node in the element. The integral in the above expression for $M_{\alpha\beta}$ is evaluated using quadrature rules as

$$\int_{[-1,1]^3} P_\alpha(\xi, \eta, \kappa) P_\beta(\xi, \eta, \kappa) J_{\Omega_{el}} \mathrm{d}\xi \mathrm{d}\eta \mathrm{d}\kappa$$

$$= \sum_{q=1}^{N_q} w_q P_\alpha(\bar{\xi}_q, \bar{\eta}_q, \bar{\kappa}_q) P_\beta(\bar{\xi}_q, \bar{\eta}_q, \bar{\kappa}_q) J_{\Omega_{el}}(\bar{\xi}_q, \bar{\eta}_q, \bar{\kappa}_q),$$

where N_q denotes the number of quadrature points and w_q are the weights associated with the quadrature points $(\bar{\xi}_q, \bar{\eta}_q, \bar{\kappa}_q)$ for $q = 1, 2, \ldots, N_q$. In particular, while using spectral finite elements (Legendre) in conjunction with the GLL quadrature rule, the quadrature points are coincident with the nodes, i.e., $(\bar{\xi}_q, \bar{\eta}_q, \bar{\kappa}_q) = (\xi_i, \eta_j, \kappa_k)$ with $q = (i, j, k)$ denoting a composite index, $i, j, k = 0, 1, \ldots, p$. Further, noting the Kronecker delta property of Lagrange polynomials, $P_{(i,j,k)}(\xi_{i'}, \eta_{j'}, \kappa_{k'}) = \delta_{ii'}\delta_{jj'}\delta_{kk'}$, it is easy to infer $P_\alpha P_\beta = \delta_{\alpha\beta}$. Thus, for spectral finite elements (Legendre) with the GLL quadrature rule,

$$M_{\alpha\beta} = C\delta_{\alpha\beta}.$$

Thus, the evaluation of $M^{-1/2}$, and subsequently the construction of Löwdin orthonormalized finite element basis, is rendered trivial. We note that numerical results show that the use of a reduced order quadrature rule for the evaluation of $M^{-1/2}$ does not affect the convergence rates or limit the accuracy of the calculation [57]. This can be rationalized by noting that the quadrature error for the GLL quadrature rule is $O(h^{2p})$, which is also the order of discretization error. Further, the GLL quadrature is needed for the aforementioned simplification only in the evaluation of M, whereas all other integrals are evaluated using Gauss quadrature.

Thus, by addressing the two main limitations of the finite element discretization—the degree of freedom disadvantage when using higher-order spectral finite element discretizations and the nonorthogonality of the basis when using spectral finite elements in conjunction with GLL quadrature—the finite-element discretization has emerged as a competing basis to plane-waves in practical DFT calculations (cf. Sect. 12.4.3), especially owing to the benefits derived from it being a real-space basis, the locality of the basis functions, and its potential for excellent parallel scalability.

12.4.2 Spatial Adaptivity

Spatial adaptivity can naturally be realized in finite-element discretization by using a spatially refined mesh in regions of interest and coarsening elsewhere. Figure 12.4 shows a spatially adaptive mesh for a Cu nanoparticle with spatial refinement around the Cu atoms and coarse-graining away from the atoms. In addition to higher-order finite-elements, spatial adaptivity can be leveraged to further reduce the dimensionality of the finite-element subspace to achieve the desired accuracy. In particular, spatial adaptivity can significantly aid computational efficiency of all-electron DFT calculations where the solution to the Kohn–Sham problem can be sharply varying. Pseudopotential calculations involving transition metals, where electrons in the penultimate shell are also treated as valence electrons, can also benefit from spatial adaptivity of finite-element discretization. Further, spatial adaptivity can provide a substantial benefit in reducing the number of basis functions for non-periodic problems such as clusters of atoms as evidenced by the results in Table 12.2—8× reduction in the basis functions, in comparison to a uniform mesh—which, in turn, translates to improved computational efficiency. The spatial adaptivity is realized via a priori and a posteriori mesh adaption strategies based on error estimates obtained from numerical analysis of the finite-element discretization of the Kohn–Sham problem. We refer to Chap. 10 by Dai and Zhou in this volume for a detailed discussion on the finite-element error estimates for the Kohn–Sham DFT problem, and refer to [12, 14, 53, 57] for the mesh adaption strategies proposed in the context of the Kohn–Sham problem.

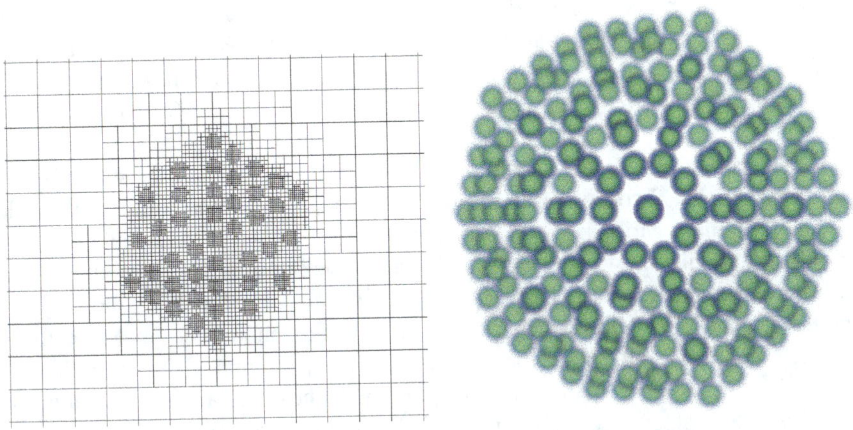

Fig. 12.4 (**Left**) Spatially adaptive mesh on the mid-plane of a Cu icosahedral nanoparticle (309 atoms, 5871 e-); (**Right**) Electron-density contours for the nanoparticle. Results obtained using DFT-FE code [53]

Table 12.2 Comparison of uniform and spatially adaptive finite-element discretizations for Cu icosahedral nanoparticle (309 atoms, 5871 e-). The discretizations are chosen such that basis discretization errors in ground-state energy and forces are under 0.1 mHa/atom and 0.1 mHa/Bohr, respectively. The simulations are performed on NERSC-Cori. Results obtained using DFT-FE code [53]

FE mesh	# Basis functions	Energy (Ha/atom)	CPU-time
Uniform (FE order = 6)	81,182,737	−1.82590939e+02	16.33 node-hrs
Adaptive (FE order = 6)	9,804,717	−1.82590932e+02	1.94 node-hrs

12.4.3 DFT-FE: A Massively Parallel Code for Real-Space Finite-Element DFT Calculations

In addition to systematic convergence and being amenable to spatial adaptivity, the finite-element basis also has potential for excellent parallel scalability owing to the locality of the basis. Further, the data structures inherent to the finite-element basis make it amenable to GPU acceleration to take advantage of the hybrid CPU-GPU computing architectures. The recent development of DFT-FE [53], a massively parallel open-source code for Kohn–Sham DFT calculations using adaptive higher order finite-element discretization, is an effort in the direction of enabling fast and accurate large-scale DFT calculations. The ionic forces and stresses in DFT-FE are computed via configurational forces corresponding to inner variations of the Kohn–Sham variational problem [55]. Recent benchmark studies [53] have shown that DFT-FE outperforms state-of-the-art plane-wave codes in computational efficiency for systems containing a few thousand electrons, and beyond. Further, the parallel scalability of DFT-FE and the GPU acceleration [13] have enabled fast DFT calculations with wall-times of a few seconds per self-consistent field (SCF) iteration—the eigenvalue problem corresponding to the inner minimization problem in Eq. (12.9) (cf. Sect. 12.2.8)—on systems containing ~30,000 electrons. Figure 12.5a shows the comparison of minimum wall-times for an SCF iteration achieved[2] using DFT-FE and Quantum Espresso (QE)—a widely used state-of-the-art plane-wave DFT code—on the NERSC Cori supercomputer, for a benchmark system containing Mo supercells with a monovacancy (periodic calculation). In addition, the minimum wall-times for DFT-FE on the Summit supercomputer using GPUs are also provided. These benchmark results suggest that, by exploiting the parallel scalability and the GPU acceleration, DFT-FE can provide a ~100× boost over QE. Figure 12.5b shows the electron density contours of the pyramidal II dislocation in Mg computed using DFT-FE, with the calculation representing a fully resolved defect core containing ~6000 atoms (~60,000 electrons). These recent developments have provided the capability to conduct fast and accurate fully resolved DFT calculations containing 10,000s of electrons that enables an efficient and accurate treatment of the defect core.

[2] Minimum wall-times computed using a metric of 40% parallel efficiency.

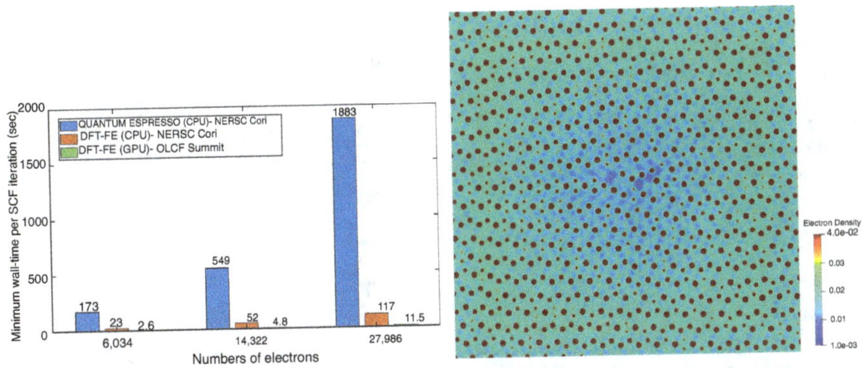

Fig. 12.5 (**Left**) Wall-time comparison of DFT-FE and Quantum Espresso on NERSC-Cori and Summit supercomputers for benchmark systems comprising of a mono-vacancy in Molybdenum supercells with varying system sizes. The wall-time for DFT-FE on Summit is using GPUs. All benchmark calculations have been conducted using ONCV pseudopotentials with discretization errors commensurate with chemical accuracy, 0.1 mHa/atom in ground-state energy and 0.1 mHa/Bohr in forces. (**Right**) Electron density contour of pyramidal II screw dislocation system in Mg, with the fully resolved defect core containing 6164 Mg atoms (61,640 electrons) [13]

The spatial adaptivity of the finite-element basis in DFT-FE has enabled systematically convergent pseudopotential and all-electron calculations in the same framework. While pseudopotential calculations have been the workhorse of DFT calculations owing to their computational efficiency, there are many scenarios where all-electron calculations are indispensable—such as material properties under extreme environments, prediction of ionization potentials, magnetizability and spectroscopic properties. In particular, a systematically convergent approach for computing the spin Hamiltonian parameters that are crucial to understanding the properties of spin defects in semiconductors—promising quantum bits for quantum computing—was lacking, until recently. The systematic convergence of the finite-element basis for all-electron calculations in DFT-FE has filled this gap [18]. Further, as pseudopotential and all-electron calculations are treated using the same framework, this has opened the possibility of mixed all-electron and pseudopotential calculations, where only a subset of atoms are treated using all-electron accuracy, while other atoms are treated using a pseudopotential approximation. These mixed calculations have enabled the calculation of spin Hamiltonian parameters of spin defects with all-electron accuracy using simulation domains that provide cell-size converged properties [19].

12.4.4 *Enriched Finite-Element Bases*

The finite-element basis with spatial adaptivity provides a systematically convergent approach for conducting all-electron calculations. However, numerical studies

Table 12.3 Comparison of classical and enriched finite element (FE) bases: Energy per atom (E in Ha), degrees of freedom per atom (DoF), and the computational CPU time (in node-hours) for various silicon nanoclusters [40]

	Classical FE	Enriched FE
Si $1 \times 1 \times 1$ (252 e-)		
E	−288.320035	−288.319450
DoF	402,112	14,728
CPU Hrs	66.63	1.03
Si $2 \times 2 \times 2$ (1,330 e-)		
E	−288.359459	−288.359266
DoF	360,467	10,642
CPU Hrs	3164	23.1

Table 12.4 Comparison of the ground-state energy per atom (E) in Ha and computational CPU time (C) in node-hours of various NV-diamond supercells, using enriched FE (EFE) and LAPW+lo bases. All reported energies are evaluated at Γ-point [71]

Supercell	Atoms (Electrons)	EFE (E)	LAPW+lo (E)	EFE (C)	LAPW+lo (C)
$2 \times 2 \times 2$	63 (379)	−38.0520	−38.0522	0.19	0.32
$3 \times 3 \times 3$	215 (1291)	−37.8716	−37.8720	1.6	15.1
$4 \times 4 \times 4$	511 (3067)	−37.8276	–	16.1	–

have shown that, despite using higher order adaptive finite-elements, they require a substantially larger number of basis functions than atomic orbital type basis functions or approaches such as APW, LAPW and LAPW+lo (see Chap. 9 of this volume, by Chen and Schneider). This limitation of the finite-element basis can be mitigated by using augmentation techniques in the finite element basis (similar to augmentation in the plane-wave basis), where the finite element basis is enriched with compactly supported atomic orbitals [40, 71], or via the partition of unity finite element method [2, 61]. We note that these augmentation techniques are in the spirit of coarse-graining presented in this chapter, where enrichment functions capturing the known oscillatory behavior of wavefunctions near the atom are used to numerically coarsegrain the higher-order adaptive finite element basis. Table 12.3 shows the comparison of the (classical) finite-element basis with an enriched finite-element basis in terms of basis functions required to achieve chemical accuracy, and the respective computational times, for all-electron calculations of Si nanoclusters. As is evident, there is $\sim$30× reduction in the finite-element basis functions using enrichments, and this translates to a staggering $\sim$100× improvement in computational efficiency. Table 12.4 shows the comparison with Elk code—a state-of-the-art LAPW code—on the accuracy and computational efficiency afforded by an enriched finite-element basis for all-electron periodic calculations on supercells of NV Diamond. For a more comprehensive discussion on the convergence properties of enriched finite-element bases, accuracy, computational efficiency and scalability of parallel implementation, we refer to recent works [40, 71] where benchmark all-electron calculations on systems containing up to $\sim$10,000 electrons are reported.

12.5 Spectral Coarse-Graining: The Spectral Quadrature Method

In the previous section, we exploited spatial coarse-graining for numerical efficiency and consideration of large-scale materials systems. In this section, we discuss another aspect of coarse-graining that concerns the eigenspectrum to enable even larger systems. In particular, the quantities of interest in the Kohn–Sham problem can be directly evaluated without having to calculate all the occupied eigenvalues and corresponding orbitals of the Hamiltonian, a strategy that we refer to as spectral coarse-graining. One such technique is the recently proposed Spectral Quadrature (SQ) method [68, 75, 78, 79], which is the focus here. Notably, the SQ method allows the development of the infinite-cell approach [75, 78], which enables non-traditional boundary conditions [20], an important aspect for the study of crystal defects.

For simplicity, let us suppose that the Hamiltonian operator is discretized using an orthogonal basis that satisfies the Kronecker-delta property. Though we make this assumption, the discussion that follows can be easily generalized. In particular, we describe the calculation of the diagonal components of the density matrix, band structure energy, and electronic entropy—quantities that need to be determined from the inner variational problem (12.14) or linear eigenvalue problem arising in each SCF iteration—using the Gauss and Clenshaw–Curtis variants of the SQ method. Indeed, the electronic ground state energy can be determined using the knowledge of these quantities. The off-diagonal components of the density matrix are also available, and these are needed to calculate the Hellmann–Feynman atomic forces [68, 79] and stress tensor [72].

12.5.1 Spectral Integrals and Quadrature

We start by rewriting the expression for the density matrix:

$$P = f_\beta(H, \mu) = f_{\widehat{\beta}}(\widehat{H}, \widehat{\mu}) \,,$$

where the shifted and scaled quantities are as follows:

$$\widehat{H} = (H - \chi I_{N_g})/\xi \,, \quad \widehat{\mu} = (\mu - \chi)/\xi \,, \quad \widehat{\beta} = \beta\xi \,.$$

Above, I denotes the identity matrix of size indicated by the subscript, and the shift and scale factors are:

$$\chi = (\varepsilon_{N_g} + \varepsilon_1)/2 \,, \quad \xi = (\varepsilon_{N_g} - \varepsilon_1)/2 \,.$$

Next, analogous to their continuous versions in Sect. 12.2.7, the constraint on the number of electrons, electron density, band structure energy, and electronic entropy can be written in the discrete setting as [26, 75, 78]:

$$N = 2\sum_{n=1}^{N_g} \rho_n \,, \tag{12.25}$$

$$\rho_n = 2e_n^{\mathrm{T}} P e_n = 2\int_{-1}^{1} f_{\widehat{\beta}}(\widehat{\varepsilon}, \widehat{\mu}) \mathrm{d}\mathcal{E}_n(\widehat{\varepsilon}) \,, \tag{12.26}$$

$$U = 2\sum_{n=1}^{N_g} e_n^{\mathrm{T}} \widehat{H} P e_n = 2\sum_{n=1}^{N_g} \int_{-1}^{1} (\xi\widehat{\varepsilon} + \chi) f_{\widehat{\beta}}(\widehat{\varepsilon}, \widehat{\mu}) \mathrm{d}\mathcal{E}_n(\widehat{\varepsilon}) \,, \tag{12.27}$$

$$\begin{aligned} S &= 2\sum_{n=1}^{N_g} e_n^{\mathrm{T}} \left[P \log P + (I_{N_g} - P)\log(I_{N_g} - P)\right] e_n \\ &= 2\sum_{n=1}^{N_g} \int_{-1}^{1} [f_{\widehat{\beta}}(\widehat{\varepsilon}, \widehat{\mu}) \log f_{\widehat{\beta}}(\widehat{\varepsilon}, \widehat{\mu}) + (1 - f_{\widehat{\beta}}(\widehat{\varepsilon}, \widehat{\mu})) \log(1 - f_{\widehat{\beta}}(\widehat{\varepsilon}, \widehat{\mu}))] \mathrm{d}\mathcal{E}_n(\widehat{\varepsilon}) \,. \end{aligned} \tag{12.28}$$

where ρ_n denotes the nth component of ρ and e_n denotes the standard basis vector. Note that it is from (12.25) that the scaled chemical potential $\widehat{\mu}$ is determined, which can then be used for the calculation of the electron density (12.26), band structure energy (12.27), and electronic entropy energy (12.28). Also, note that we have dropped the subscript β in the band structure energy U, for simplicity of notation.

The key idea underlying the SQ method is the approximation of the integrals arising in the definition of the above quantities using a quadrature rule:

$$\int_{-1}^{1} g(\widehat{\varepsilon}, \widehat{\mu}) \mathrm{d}\mathcal{E}_n(\widehat{\varepsilon}) \approx \sum_{j=1}^{k} \widehat{w}_j^n g(\widehat{\varepsilon}_j^n, \widehat{\mu}) \,, \tag{12.29}$$

where g is any one of the functions arising in the integrals presented in (12.25–12.28), and $\{\widehat{\varepsilon}_j^n\}_{j=1}^k$ and $\{\widehat{w}_j^n\}_{j=1}^k$ are the nodes and weights of the quadrature rule, respectively. Among the various quadrature schemes possible, Gauss and Clenshaw–Curtis quadrature present themselves as attractive choices [81, 82], whose evaluation in the current spectral setting is described in Sects. 12.5.3 and 12.5.4, respectively. In order to evaluate these quadrature rules efficiently and make them more amenable to spatial coarse graining, it is common to employ spatial localization, as described in Sect. 12.5.2.

Remark The SQ method does not require computation of the eigenvalues and eigenvectors of the Hamiltonian $\widehat{H}$ and uses (12.29) instead, for which we note the connection. The measure $\mathcal{E}_n$ may be written as:

$$\mathcal{E}_n(\widehat{\varepsilon}) = \begin{cases} 0\,, & \text{if } \widehat{\varepsilon} < \widehat{\varepsilon}_1 = -1 \\ \sum_{i=1}^{m} \varphi_{i,n}^2\,, & \text{if } \widehat{\varepsilon}_m \leqslant \widehat{\varepsilon} < \widehat{\varepsilon}_{m+1}\,, \\ \sum_{i=1}^{N_g} \varphi_{i,n}^2\,, & \text{if } 1 = \widehat{\varepsilon}_{N_g} < \widehat{\varepsilon} \end{cases} \tag{12.30}$$

where $\varphi_{i,n}$ denotes the nth component of φ_i. In using (12.29), the SQ method avoids the calculation of the eigenvalues and eigenvectors of the Hamiltonian, thereby circumventing the bottleneck encountered in traditional diagonalization-based Kohn–Sham DFT calculations.

12.5.2 *Spectral Integrals and Quadrature with Spatial Localization*

To significantly reduce the computational cost as well as make the quantities amenable to coarse-graining, we now introduce spatial localization by taking advantage of the nearsightedness of electronic correlations, i.e., exponential decay of the density matrix for metals at nonzero smearing values as well as insulators [5, 69, 77]. To do so, we introduce the 'nodal' density matrices [68, 79]

$$P^n = f_\beta(H^n, \mu) = f_{\widehat{\beta}^n}(\widehat{H}^n, \widehat{\mu})\,, \quad n = 1, \ldots, N_g\,,$$

where

$$\widehat{H}^n = (H^n - \chi^n I_{N_g^n})/\xi^n\,, \quad \widehat{\mu} = (\mu - \chi^n)/\xi^n\,, \quad \widehat{\beta}^n = \beta\xi^n\,,$$

with

$$\chi^n = (\varepsilon_{N_g^n}^n + \varepsilon_1^n)/2\,, \quad \xi^n = (\varepsilon_{N_g^n}^n - \varepsilon_1^n)/2\,.$$

Above, H^n is the submatrix of the Hamiltonian H formed by spatially localizing it around the point of interest, i.e., a matrix formed by the N_g^n rows and columns of H that are 'near' the nth row and column. In addition, ε_m^n, $m = 1, \ldots, N_g^n$, are the eigenvalues of H^n. Thereafter, we approximate the constraint on the number of electrons, electron density, band structure energy, and electronic entropy given in Eqs. 12.25–12.28 as:

$$N = 2\sum_{n=1}^{N_g} \rho_n\,, \tag{12.31}$$

$$\rho_n \approx 2e_s^T P^n e_s = 2\int_{-1}^{1} f_{\widehat{\beta}^n}(\widehat{\varepsilon},\widehat{\mu})\mathrm{d}\mathcal{E}_s^n(\widehat{\varepsilon}), \tag{12.32}$$

$$U \approx 2\sum_{n=1}^{N_g} e_s^{\mathrm{T}} \widehat{H}^n P^n e_s = 2\sum_{n=1}^{N_g}\int_{-1}^{1} (\xi^n\widehat{\varepsilon}+\chi^n) f_{\widehat{\beta}^n}(\widehat{\varepsilon},\widehat{\mu})\mathrm{d}\mathcal{E}_s^n(\widehat{\varepsilon}), \tag{12.33}$$

$$\begin{aligned} S &\approx 2\sum_{n=1}^{N_g} e_s^{\mathrm{T}} \Big[P^n\log P^n + (I_{N_g^n} - P^n)\log(I_{N_g^n} - P^n)\Big] e_s \\ &= 2\sum_{i=1}^{N_g}\int_{-1}^{1} [f_{\widehat{\beta}^n}(\widehat{\varepsilon},\widehat{\mu})\log f_{\widehat{\beta}^n}(\widehat{\varepsilon},\widehat{\mu}) + (1-f_{\widehat{\beta}^n}(\widehat{\varepsilon},\widehat{\mu}))\log(1-f_{\widehat{\beta}^n}(\widehat{\varepsilon},\widehat{\mu}))]\mathrm{d}\mathcal{E}_s^n(\widehat{\varepsilon}), \end{aligned} \tag{12.34}$$

where e_s denotes the standard basis vector corresponding to the node of interest in the truncated Hamiltonian, i.e., the row and column corresponding to the node around which spatial truncation has been performed.

We now proceed to approximate the integrals arising in the definition of the above quantities using a quadrature rule:

$$\int_{-1}^{1} g(\widehat{\varepsilon},\widehat{\mu})\mathrm{d}\mathcal{E}_s^n(\widehat{\varepsilon}) \approx \sum_{j=1}^{k} \widehat{w}_j^n g(\widehat{\varepsilon}_j^n,\widehat{\mu}), \tag{12.35}$$

where g is any one of the functions arising in the integrals presented in (12.32–12.34), and $\{\widehat{\varepsilon}_j^n\}_{j=1}^k$ and $\{\widehat{w}_j^n\}_{j=1}^k$ are the nodes and weights of the quadrature rule (dropped index s, for simplicity of notation). Specifically, we describe the evaluation of the Gauss and Clenshaw–Curtis spectral quadrature rules in Sects. 12.5.3 and 12.5.4, respectively.

Remark The measure $\mathcal{E}_s^n$ can be written as:

$$\mathcal{E}_s^n(\widehat{\varepsilon}) = \begin{cases} 0, & \text{if } \ \widehat{\varepsilon} < \widehat{\varepsilon}_1^n = -1 \\ \sum_{i=1}^m (\varphi_{i,s}^n)^2, & \text{if } \ \widehat{\varepsilon}_m^n \leqslant \widehat{\varepsilon} < \widehat{\varepsilon}_{m+1}^n \\ \sum_{i=1}^{N_g^n} (\varphi_{i,s}^n)^2, & \text{if } \ 1 = \widehat{\varepsilon}_{N_g^n}^n < \widehat{\varepsilon} \end{cases},$$

where φ_i^n denote the eigenvectors of the truncated Hamiltonian H^n. As stated previously, the SQ method does not require the calculation of the measure $\mathcal{E}_s^n$ explicitly, thereby avoiding the need to calculate the eigenvalues and eigenvectors of the truncated Hamiltonians H^n, $n = 1,\ldots,N_g$, resulting in significant computational savings.

12.5.3 Gauss Spectral Quadrature

To generate the Gauss SQ rule for the integral in (12.35), we use the Lanczos type iteration [26, 75, 78]

$$b^n_{j+1} v^n_{j+1} = \widehat{H}^n v^n_j - a^n_{j+1} v^n_j - b^n_j v^n_{j-1}\,, \quad j = 0, \ldots, k-1\,,$$
$$v^n_{-1} = 0\,, \quad v^n_0 = e_s\,, \quad b^n_0 = 1\,, \qquad (12.36)$$

where

$$a^n_{j+1} = (v^n_j)^{\mathrm{T}} \widehat{H}^n v^n_j\,, \quad j = 0, \ldots, k-1\,,$$

and b^n_j is computed such that $(v^n_j)^{\mathrm{T}} v^n_j = 1\,,\ j = 0, \ldots, k-1$. Subsequently, we form the symmetric tridiagonal Jacobi matrix:

$$J^n_k = \begin{pmatrix} a^n_1 & b^n_1 & & & \\ b^n_1 & a^n_2 & b^n_2 & & \\ & \ddots & \ddots & \ddots & \\ & & b^n_{k-2} & a^n_{k-1} & b^n_{k-1} \\ & & & b^n_{k-1} & a^n_k \end{pmatrix}, \qquad (12.37)$$

whose eigenvalues and squares of the first elements of the normalized eigenvectors are the nodes $\{\widehat{\varepsilon}^n_j\}_{j=1}^k$ and weights $\{\widehat{w}^n_j\}_{j=1}^k$ of the quadrature rule, respectively. To show this result, the above procedure can be viewed as first performing the following decomposition of the nodal Hamiltonian:

$$\widehat{H}^n \approx V^n_k J^n_k (V^n_k)^{\mathrm{T}}\,,$$

where V^n_k is a matrix with the $j+1$ column being the vector v^n_j generated during the Lanczos iteration in (12.36). Thereafter,

$$\begin{aligned} e_s^{\mathrm{T}} g(\widehat{H}^n, \widehat{\mu}) e_s &\approx (e_s^{\mathrm{T}} V^n_k) g(J^n_k, \widehat{\mu}) (V^{n^{\mathrm{T}}}_k e_s) \\ &= e_1^{\mathrm{T}} g(J^n_k, \widehat{\mu}) e_1 \\ &= \sum_{j=1}^{k} \widehat{w}^n_j g(\widehat{\varepsilon}^n_j, \widehat{\mu})\,, \end{aligned}$$

where $\{\widehat{\varepsilon}^n_j\}_{j=1}^k$ and $\{\widehat{w}^n_j\}_{j=1}^k$ are the eigenvalues and squares of the first elements of the normalized eigenvectors of J^n_k, respectively. Note that the nodes and weights are independent of the function being integrated within the above scheme.

In Gauss SQ, the constraint on the number of electrons, electron density, band structure energy, and electronic entropy can then be written as:

$$N = 2\sum_{n=1}^{N_g}\sum_{j=1}^{k} \widehat{w}_j^n f_{\widehat{\beta}^n}(\widehat{\varepsilon}_j^n, \widehat{\mu}),$$

$$\rho_n = 2\sum_{j=1}^{k} \widehat{w}_j^n f_{\widehat{\beta}^n}(\widehat{\varepsilon}_j^n, \widehat{\mu}),$$

$$U = 2\sum_{n=1}^{N_g}\sum_{j=1}^{k} \widehat{w}_j^n (\xi^n \widehat{\varepsilon}_j^n + \chi^n) f_{\widehat{\beta}^n}(\widehat{\varepsilon}_j^n, \widehat{\mu}),$$

$$S = 2\sum_{n=1}^{N_g}\sum_{j=1}^{k} \widehat{w}_j^n [f_{\widehat{\beta}^n}(\widehat{\varepsilon}_j^n, \widehat{\mu}) \log f_{\widehat{\beta}^n}(\widehat{\varepsilon}_j^n, \widehat{\mu}) + (1 - f_{\widehat{\beta}^n}(\widehat{\varepsilon}_j^n, \widehat{\mu})) \log(1 - f_{\widehat{\beta}^n}(\widehat{\varepsilon}_j^n, \widehat{\mu}))].$$

Since the nodes and weights are independent of the Fermi level, they do not need to be recomputed for the different quantities above, nor do they need to be recomputed for the different guesses for the Fermi level in solving for the constraint on the number of electrons.

In cases where the off-diagonal components of the density matrix are required, e.g., the computation of Hellmann–Feynman atomic forces and stress tensor, the nth column of the density matrix can be obtained using the relation:

$$P^n e_s \approx V_k^n g(J_k^n, \widehat{\mu}) e_1 .$$

Indeed, all these quantities are already computed as part of the above procedure, and so do not introduce any additional cost.

Relation to the Recursion Method and Padé Approximation

The spectral Gauss SQ method bears a resemblance to the recursion method [30] that had been developed in the context of the tight binding method. To see this, we note the relation [28]:

$$g(\widehat{H}^n, \widehat{\mu}) = \frac{1}{2\pi i}\oint_C g(z, \widehat{\mu})(z I_{N_g^n} - \widehat{H}^n)^{-1} \mathrm{d}z ,$$

where $i = \sqrt{-1}$, and $\oint_C$ represents a contour that encloses the spectrum of $\widehat{H}^n$ in the complex plane, from which it follows:

$$e_s^{\mathrm{T}} g(\widehat{H}^n, \widehat{\mu}) e_s = \frac{1}{2\pi i} \oint_C g(z, \widehat{\mu}) e_s^T (z I_{N_g^n} - \widehat{H}^n)^{-1} e_s \mathrm{d}z \,. \tag{12.38}$$

In the current framework, the recursion method involves using the following approximation:

$$e_s^{\mathrm{T}} (z I_{N_g^n} - \widehat{H}^n)^{-1} e_s \approx \cfrac{1}{z - a_1^n - \cfrac{(b_1^n)^2}{z - a_2^n - \cdots - \cfrac{(b_{k-1}^n)^2}{z - a_k}}} = \frac{q_k^n(z)}{p_k^n(z)} \,.$$

In particular, the continued fraction above is used within the integral of (12.38) to evaluate the quantity of interest. Since the rational function has zeros of $p_k^n(z)$, a number of techniques to smoothen it have been developed [30]. It can however be shown that [78]:

$$\frac{q_k^n(z)}{p_k^n(z)} = e_1^{\mathrm{T}} (z I_k - J_k^n)^{-1} e_1 \,,$$

which when substituted into (12.38) along with the spectral theorem recovers the Gauss SQ quadrature rule:

$$\begin{aligned} e_s^{\mathrm{T}} g(\widehat{H}^n, \widehat{\mu}) e_s = \int_{-1}^{1} g(\widehat{\varepsilon}, \widehat{\mu}) \mathrm{d}\mathcal{E}^n(\widehat{\varepsilon}) &\approx e_1^{\mathrm{T}} \left[\frac{1}{2\pi i} \oint_C g(z, \widehat{\mu}) (z I_k - J_k^n)^{-1} dz \right] e_1 \\ &= e_1^{\mathrm{T}} g(J_k^n, \widehat{\mu}) e_1 \\ &= \sum_{j=1}^{k} \widehat{w}_j^n g(\widehat{\varepsilon}_j^n, \widehat{\mu}) \,. \end{aligned}$$

Note that the rational functions satisfy the following best approximation property [74]:

$$e_s^{\mathrm{T}} (z I_{N_g^n} - \widehat{H}^n)^{-1} e_s - \frac{q_k^n(z)}{p_k^n(z)} = O\left(\frac{1}{z^{2k+1}} \right) ,$$

which make them the Padé approximants. Indeed, it can be shown from the above equation—multiplying both sides by a polynomial of degree $2k - 1$ and integrating along a contour encircling the real line [87]—that polynomials of degree $2k - 1$ are integrated exactly using the above quadrature rule, as is the property of Gauss quadrature.

12.5.4 Clenshaw–Curtis Spectral Quadrature

In Clenshaw–Curtis SQ [68, 75, 79], rather than determining quadrature weights corresponding to the quadrature nodes (zeros of the Chebyshev polynomials), it is advantageous to perform the following expansion in terms of Chebyshev polynomials:

$$\int_{-1}^{1} g(\widehat{\varepsilon}, \widehat{\mu}) \mathrm{d}\mathcal{E}_s^n(\widehat{\varepsilon}) \approx \sum_{j=0}^{k}{}' c_j(\widehat{\mu}) \int_{-1}^{1} T_j(\widehat{\varepsilon}) \mathrm{d}\mathcal{E}_s^n(\widehat{\varepsilon}) \,,$$

where the summation with a prime indicates that the first term is halved, and the Chebyshev coefficients are given by

$$c_j(\widehat{\mu}) = \frac{2}{\pi} \int_{-1}^{1} \frac{g(\widehat{\varepsilon}, \widehat{\mu}) T_m(\widehat{\varepsilon})}{\sqrt{1-\widehat{\varepsilon}^2}} \mathrm{d}\widehat{\varepsilon} \,, \quad j = 0, \ldots, k \,. \tag{12.39}$$

We can then write

$$\int_{-1}^{1} T_j(\widehat{\varepsilon}) \mathrm{d}\mathcal{E}_s^n(\widehat{\varepsilon}) = e_s^T t_j^n = t_{j,s}^n \,,$$

where t_j^i are evaluated from the three-term recurrence relation:

$$\begin{aligned} t_{j+1}^n &= 2\widehat{H}^n t_j^n - t_{j-1}^n \,, \quad j = 1, \ldots k-1 \,, \\ t_1^n &= \widehat{H}^n e_s \,, \; t_0^n = e_s \,. \end{aligned} \tag{12.40}$$

In Clenshaw–Curtis SQ, the constraints on the number of electrons, electron density, band structure energy, and electronic entropy take the form:

$$N = 2 \sum_{n=1}^{N_g} \sum_{j=0}^{k}{}' c_j^{\rho}(\widehat{\mu}) t_{j,s}^n \,, \quad c_j^{\rho}(\widehat{\mu}) = \frac{2}{\pi} \int_{-1}^{1} \frac{f_{\widehat{\beta}^n}(\widehat{\varepsilon}, \widehat{\mu}) T_m(\widehat{\varepsilon})}{\sqrt{1-\widehat{\varepsilon}^2}} \mathrm{d}\widehat{\varepsilon} \,,$$

$$\rho_n = 2 \sum_{j=0}^{k}{}' c_j^{\rho} t_{j,s}^n \,, \quad c_j^{\rho} = \frac{2}{\pi} \int_{-1}^{1} \frac{f_{\widehat{\beta}^n}(\widehat{\varepsilon}, \widehat{\mu}) T_m(\widehat{\varepsilon})}{\sqrt{1-\widehat{\varepsilon}^2}} \mathrm{d}\widehat{\varepsilon} \,,$$

$$U = 2 \sum_{n=1}^{N_g} \sum_{j=0}^{k}{}' (\xi^n c_j^U + \chi^n c_j^{\rho}) t_{j,s}^n \,, \quad c_j^U = \frac{2}{\pi} \int_{-1}^{1} \frac{\widehat{\varepsilon} f_{\widehat{\beta}^n}(\widehat{\varepsilon}, \widehat{\mu}) T_m(\widehat{\varepsilon})}{\sqrt{1-\widehat{\varepsilon}^2}} \mathrm{d}\widehat{\varepsilon} \,,$$

$S = 2 \sum_{n=1}^{N_g} \sum_{j=0}^{k}{}' c_j^S t_{j,s}^n$, where c_j^S is equal to

$$\frac{2}{\pi} \int_{-1}^{1} \frac{[f_{\widehat{\beta}^n}(\widehat{\varepsilon}, \widehat{\mu}) \log f_{\widehat{\beta}^n}(\widehat{\varepsilon}, \widehat{\mu}) + (1 - f_{\widehat{\beta}^n}(\widehat{\varepsilon}, \widehat{\mu})) \log(1 - f_{\widehat{\beta}^n}(\widehat{\varepsilon}, \widehat{\mu}))] T_m(\widehat{\varepsilon})}{\sqrt{1-\widehat{\varepsilon}^2}} \mathrm{d}\widehat{\varepsilon} \,.$$

Note that in cases where the off-diagonal components of the density matrix are required, the nth column of the density matrix can be obtained using the relation:

$$P^n e_s \approx \sum_{j=0}^{k}{}' c_j^\rho t_j^n .$$

Indeed, all these quantities are already computed as part of the above procedure, and so do not incur any additional cost.

Relation to Fermi Operator Expansion (FOE)

The Clenshaw–Curtis quadrature bears a resemblance to the classical Fermi Operator Expansion (FOE) [24, 25]. In particular, the FOE method employs the following expansion of the density matrix in terms of Chebyshev polynomials:

$$P = \sum_{j=0}^{k} c_j^\rho(\widehat{\mu}) T_j(\widehat{H}) ,$$

where the matrices $T_j(\widehat{H})$ are evaluated using the three-term recurrence relation:

$$T_{j+1}(\widehat{H}) = 2\widehat{H} T_j(\widehat{H}) - T_{j-1}(\widehat{H}) , \quad j = 1, \ldots k-1 .$$

In order to achieve linear scaling with system size, truncation is introduced into the matrix-matrix multiplication routines. In spite of the similarity of this approach with Clenshaw–Curtis SQ, there are a number of key differences. First, compared to the sparse matrix-vector routines in Clenshaw–Curtis SQ, the operations involved in FOE are sparse matrix-matrix routines, which are challenging to write, particularly for efficient scaling to a large number of processors. Second, the effect of truncation is not automatically incorporated into FOE, as it is done in Clenshaw–Curtis SQ. Third and finally, since the Chebshev matrices cannot be generally stored, an outer loop on the Fermi level is required, which makes the FOE significantly more costly as well.

12.5.5 Convergence Rates

In the SQ method, the error with respect to the quadrature order decays as [75]:

$$\left| \int_{-1}^{1} g(\widehat{\varepsilon}, \widehat{\mu}) \mathrm{d}\mathcal{E}_s^n(\widehat{\varepsilon}) - \sum_{j=1}^{k} \widehat{w}_j^n g(\widehat{\varepsilon}_j^n, \widehat{\mu}) \right| \sim O(\mathrm{e}^{-\alpha k}) ,$$

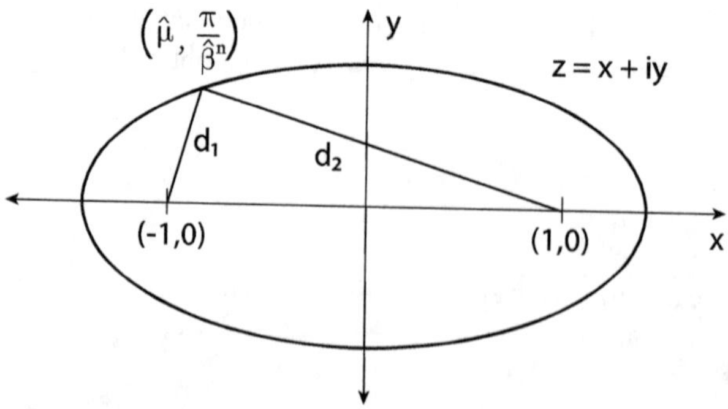

Fig. 12.6 Largest ellipse in the complex plane where the Fermi–Dirac function $f_{\widehat{\beta}^n}$ is analytic

where

$$\alpha = n_q \log r \tag{12.41}$$

is the rate of convergence. Here, $n_q = 1$ and $n_q = 2$ for the Clenshaw–Curtis and Gauss SQ methods, respectively. In addition, r is the sum of the semi-major and semi-minor axes for the largest ellipse in the complex plane where the function g is analytic. In the current context, the closest singularity of the Fermi–Dirac function $f_{\widehat{\beta}^n}$ to the interval $[-1, 1]$ is at

$$z = \widehat{\mu} \pm i \frac{\pi}{\widehat{\beta}^n} .$$

The corresponding ellipse is as shown in Fig. 12.6, for which we have:

$$\begin{aligned} r &= a + \sqrt{a^2 - 1} , \\ a &= \frac{1}{2}(d_1 + d_2) , \\ d_1 &= \sqrt{(1 + \widehat{\mu})^2 + \left(\frac{\pi}{\widehat{\beta}^n}\right)^2} , \\ d_2 &= \sqrt{(1 - \widehat{\mu})^2 + \left(\frac{\pi}{\widehat{\beta}^n}\right)^2} . \end{aligned} \tag{12.42}$$

Performing a first-order Taylor series expansion for the convergence rate α about $1/\widehat{\beta}^n \to 0$, we obtain

$$\alpha \approx \frac{n_q \pi}{\widehat{\beta}^n \sqrt{1 - \widehat{\mu}^2}} . \tag{12.43}$$

This expression represents a very good approximation for practical DFT calculations, since the spectral width of the Hamiltonian (2ξ) is generally large and the smearing ($1/\widehat{\beta}^n$) used for ambient conditions is typically small.

Although the above error estimates are also valid for insulating systems, the bounds are not expected to be tight, especially as the smearing becomes smaller. In fact, it is common to not use any smearing for insulators, i.e., $1/\widehat{\beta}^n \to 0$. It has been predicted that for an insulating system with band-gap E_g and smearing $1/\widehat{\beta}^n \to 0$ [75]:

$$\alpha \approx \frac{n_q \widehat{E}_g}{2\sqrt{1-\widehat{\mu}^2}}. \tag{12.44}$$

Above, the Fermi level has been assumed to be in the middle of the band-gap and $\widehat{E}_g = E_g/\xi$.

We now compare the predicted convergence rate with that obtained numerically within a DFT calculation. Specifically, we consider a 107-atom system consisting of a vacancy in face-centered cubic (FCC) aluminum. We choose a smearing of 1 eV, commensurate with that adopted for metallic systems in practical Kohn–Sham calculations. In Fig. 12.7, we plot the convergence in electron density with quadrature order for a specific point in space, while choosing a large enough truncation radius, so as to put associated errors well below the quadrature errors of interest. All simulations are performed using the real-space Kohn–Sham DFT code SPARC [21, 22, 94], in which the SQ method has been recently implemented.

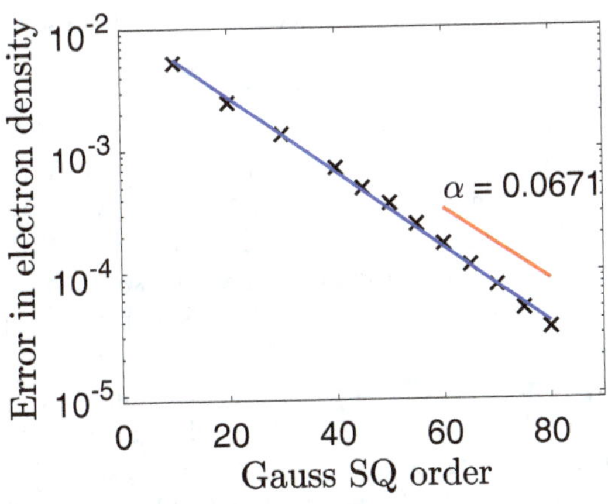

Fig. 12.7 Convergence in electron density with Gauss SQ order at a spatial point. The thick red line represents the theoretically predicted convergence rate. The system under consideration is a 107-atom system consisting of a vacancy in FCC aluminum, with smearing of 1 eV

12.5.6 Scaling Estimates

The computational cost of the Gauss and Clenshaw–Curtis SQ methods is dictated by the cost of the matrix-vector products appearing in the iteration described by Eqs. 12.36 and 12.40, respectively. Given the sparse nature of $\widehat{H}^n$, the cost of each matrix-vector product scales as $O(N_g^n)$. Since there are k such matrix-vector products in the iteration and n ranges from 1 to N_g, the total computational cost scales as $O(N_g^n k N_g)$. As can be seen from the theoretical results presented above, the quadrature order k required for a certain accuracy is independent of the number of grid points N_g. Moreover, for large enough system sizes, N_g^n is also independent of N_g. Therefore, the scaling of the SQ method is $O(N_g)$, which makes it $O(N)$ with the number of electrons in the system, i.e., linear scaling with system size. Therefore, the cubic scaling bottleneck inherent to traditional diagonalization approaches can be overcome using the SQ method, enabling the study of large system sizes that were previously intractable. Note that the unlike orbital-based diagonalization and linear scaling approaches, the cost of the SQ method decreases with increasing temperature [68, 77], making it ideal for the study of materials under extreme conditions [7, 92, 96].

12.5.7 Numerical Results

We now study the accuracy and efficiency of the aforedescribed Gauss and Clenshaw–Curtis SQ methods. As a representative example, we choose an unrelaxed vacancy in FCC aluminum, which is modeled by removing a single atom within a supercell of FCC aluminum.

In Fig. 12.8, considering a 107-atom system, we plot the convergence of the ground state energy, Hellmann–Feynman atomic forces, and Hellman–Feynman stress tensor with quadrature order and truncation radius, which are the two new parameters introduced within the SQ method. Note that we employ Gauss SQ for the calculation of the electron density and energy in each SCF iteration, and Clenshaw–Curtis SQ for the atomic forces and stress tensor. It is clear that there is systematic geometric convergence in all quantities, demonstrating the accuracy of the SQ method.

In Fig. 12.9, we plot the strong and weak parallel scaling of the SQ method, as implemented in the SPARC code [21, 22, 94]. All parameters, including mesh-size, quadrature order and truncation radius have been chosen so that the error in energy and force are within 0.001 Ha/atom and 0.001 Ha/Bohr, numbers that are representative of the accuracy targeted in typical DFT simulations. For the strong scaling, we use a 107-atom system, while increasing the number of processors from 24 to 960. For the weak scaling, we increase the system size from 107 to 10,975, while proportionally increasing the processors from 27 to 2744. It is clear that the SQ method demonstrates excellent strong and weak scaling, enabling the study of large systems needed in the study of crystal defects.

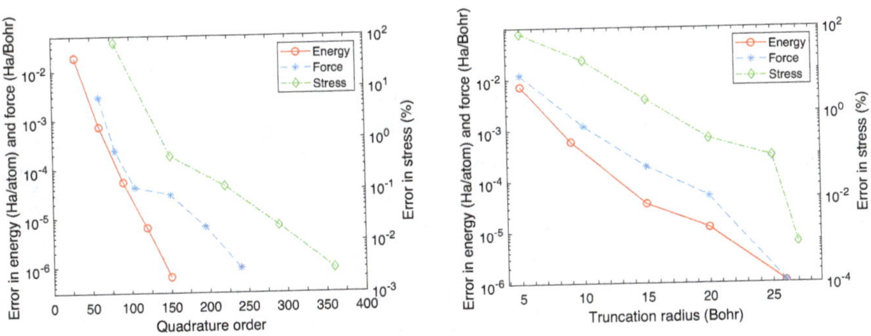

(a) (b)

Fig. 12.8 Convergence of the energy, atomic forces, and stress tensor with quadrature order and truncation radius for the SQ method as implemented in the SPARC code. The system under consideration is a 107-atom system containing a vacancy with smearing of 1 eV. Gauss SQ has been used for the energy, and Clenshaw–Curtis SQ is used for the force and stress. The error in force and stress correspond to the maximum difference in any component. (**a**) Convergence with quadrature order. (**b**) Convergence with truncation radius

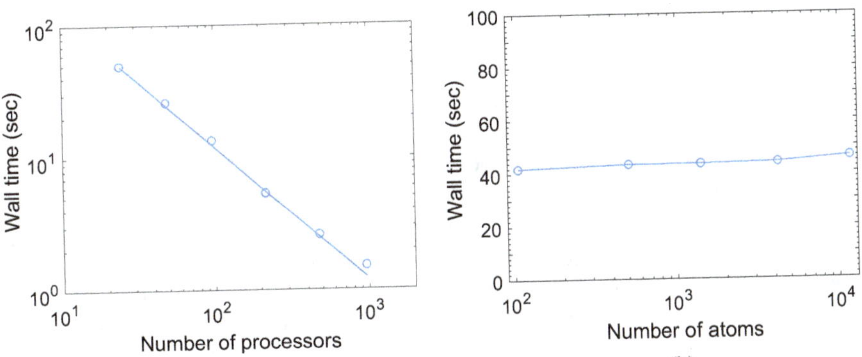

(a) (b)

Fig. 12.9 Strong and weak scaling of the SQ method, with timings reported for a single SCF iteration. The system under consideration for strong scaling is a 107-atom system representing a vacancy in aluminum. The systems for weak scaling are larger and larger supercells, each with a vacancy. (**a**) Strong scaling. (**b**) Weak scaling

12.6 Spatial and Spectral Coarse-Graining

In this section, we combine spatial and spectral coarse-graining to enable very large scale studies of defects in crystalline materials. It exploits the nature of defects where the electronic and atomistic fields decay away from the defect to those associated with a periodic crystal to construct a controlled approximation to DFT.

12.6.1 *Periodic Systems*

The presentation in Sect. 12.2 can be adapted to the periodic setting. The complete basis consists not only of periodic functions but Bloch–Floquet waves. Consequently the orbitals, the operator γ and the partition of unity $\mathcal{E}$ are not periodic (i.e., $(r, r') \mapsto \gamma(r, r')$ is not periodic). This leads to plane waves and k-point sampling. However, the measure $\mathcal{M} = \operatorname{tr} \mathcal{E}$ is periodic.

However, and this is the key observation, the densities ρ, u, s are in fact periodic since they depend on the trace of $\mathcal{E}$. It also follows that the dual variables, the electrostatic potential ϕ and exchange correlation potential V_{xc} are also periodic.

12.6.2 *Coarse-Grained Representation*

We consider a Bravais lattice first, and then describe the extension to other lattices.

12.6.2.1 Atoms

Consider a crystalline solid whose crystal structure is given by a Bravais lattice. Introduce a defect at the origin (e.g., a vacancy cluster by removing a cluster of atoms at the origin) and consider the restriction of the lattice (with a defect) to a simply connected domain $\mathcal{D}$. Let $\{x_m\}_{m=1}^M$ denote the positions of the atoms and pick these to be the reference configuration. There are unbalanced forces on the atoms near the core and they deform. We are interested in finding the deformed positions $\{r_m\}_{m=1}^M$ of these atoms. We can find a smooth deformation $y : \mathcal{D} \to \mathbb{R}^3$ such that $r_m = y(x_m),\ m = 1, \ldots, M$. We expect the displacements $y(x) - x$ to be large and oscillate on a fine scale (that of the lattice) near the core (origin), but vary smoothly on the scale of the lattice and decay as we go away from the defect. Thus, we need a fine discretization near the defect core, but can coarsen as we move away.

Therefore, we use a quasi-continuum approximation [17, 41, 66, 67, 78, 80] to represent the positions of the atoms. We consider a subset of atoms $\mathcal{P}_a$ we call the *representative atoms*, and introduce a Lagrangian triangulation $\mathcal{T}_a$ with the representative atoms as nodes. We track the position of the representative atoms $\{r_a\}_{a=1}^A$ and represent the positions of the remaining atoms using the interpolation Γ^a_{an} induced by the triangulation $\mathcal{T}_a$:

$$\bar{y}_m = \sum_{a=1}^{A} \Gamma^a_{am} \bar{y}_a, \quad m = 1, \ldots M. \tag{12.45}$$

We pick $\mathcal{P}_a$ to be dense near the core and gradually coarsen away from it.

12.6.2.2 Electronic Fields

We now turn to the electronic fields—electron density, electrostatic potential—for the specimen of a crystalline solid with a defect at its center. Consider a region distant from the defect where the deformation is smooth and the deformation gradient F is uniform on a scale large compared to the lattice: i.e., $F = O(1)$ and $\nabla F = O(a/L)$ where a is a typical lattice spacing and L is the radius of the computational domain. The atomic positions are periodic to a good approximation, and we expect the electronic fields to be periodic to a good approximation in that region due to the short-sightedness of electronic matter [42]. In other words, for an electronic field Q of interest, we expect

$$Q(r) \approx \widetilde{Q}\left(\frac{a}{L}r, F^{-1}(r)r\right) \quad \text{for } r \gg a,$$

where $\widetilde{Q}(\cdot, z)$ is periodic with the periodicity of the reference unit cell. In other words, we expect

$$Q(r) \approx Q^p(r) + Q^c(r),$$

where $Q^p(r) = \widetilde{Q}\left(\frac{a}{L}r, F^{-1}(r)\right)$ and $Q^c(r)$ decays smoothly for large r. The idea then is to represent Q^p (the projection onto continuous functions of piecewise periodic functions) and Q^c on a grid that is fine near the defect core and coarsens away from it. We call Q^p the *predictor* and Q^c the *corrector*.

We achieve this representation using two spatial meshes. The first is the *fine electronic mesh* $\mathcal{P}_f$, which is a uniform finite difference mesh. We use this to represent the Hamiltonian and in our Lanczos algorithm. The second is the *coarse electronic mesh* $\mathcal{P}_c$, which is a subset of the fine electronic mesh $\mathcal{P}_f$. We compute the electronic quantities on this mesh and therefore call the elements of $\mathcal{P}_c$ the *electronic sampling points* (ESPs). As with the atomistic grid, the coarse grid $\mathcal{P}_c$ is fine (includes all points in $\mathcal{P}_c$) close to the defect but gradually coarsens away.

We represent an electronic field Q as follows. First, we define the predictor. Recall that the deformation (12.45) is affine in each element of the Lagrangian atomistic triangulation $\mathcal{T}_a$, and that it convects the reference lattice to a deformed periodic lattice. We perform a unit cell calculation based on this deformed periodic lattice in each element Ω_e of the $\mathcal{T}_a$ to obtain the electron density $Q_e(y)$ on the image $\bar{y}(\Omega_e)$ of the element, and define the *predictor* as the $L^2 \to H^1$ projection of this piecewise periodic function

$$Q^p_f = P_{L^2 \to H^1}\left(\chi_{\bar{y}(\Omega_e}Q_e(y_f)\right),$$

where y_f is the position of the fth node of $\mathcal{P}_f$ and $\chi_{\mathcal{A}}$ is the characteristic function of a set $\mathcal{A}$.

We now turn our attention to the corrector. Let Q_c the quantity of interest at an ESP labelled c. We define the corrector at the ESP as the difference between the computed electron density and predictor:

$$Q_c^c = Q_c - Q_c^p.$$

We then extend the definition of the *corrector* to the fine grid $\mathcal{P}_f$ through interpolation:

$$Q_f^c = \sum_{c=1}^{C} \Gamma_{cf}^c Q_c^c,$$

where Γ^c is the interpolation associated with the triangulation induced by $\mathcal{P}_c$. In summary, we represent the electron density as

$$Q_f = Q_f^p + \sum_{c=1}^{C} \Gamma_{cf}^c (Q_c - Q_c^p). \tag{12.46}$$

While we have the representation on the fine grid, we do not need to evaluate the quantities on the fine grid. Since we seek to perform the Lanczos procedure only at the ESPs, we need the Hamiltonian in a sufficiently large neighborhood of each ESP. Therefore, we create clusters of fine grid points around each ESP, and collect these points into the set $\mathcal{P}_f^{\text{eval}}$. We evaluate the electronic quantities only on $\mathcal{P}_f^{\text{eval}}$. Note that $\mathcal{P}_f^{\text{eval}}$ is fully dense near the core, but becomes sparse as we go away.

Finally, to compute global quantities like energy, we have to compute sums like

$$\sum_{f \in \mathcal{P}_f} Q_f.$$

We do so following the cluster summation approach of Knap and Ortiz [41] using $\mathcal{P}_c$ and $\mathcal{P}_f^{\text{eval}}$ (see [66] for details).

The overall approach is summarized in Algorithm 1.

12.6.2.3 Crystals

In the case of a crystal where one has more than one atom per unit cell, we limit the representative atoms to belong to the skeletal lattice as we coarsen, and use the periodic calculation within each element of $\mathcal{T}_a$ to determine the positions of the other atoms in the unit cell.

Algorithm 1: Spatial and spectral coarse-grained approach

Given an initial configuration of atoms,
while *representative atoms are not in equilibrium* **do**
  perform a periodic DFT calculation in each element of the triangulation $\mathcal{T}_a$;
  find the predictor on $\mathcal{P}_f^{\text{eval}}$;
  initial guess of the corrector on $\mathcal{P}_f^{\text{eval}}$;
  while *electronic fields have not converged* **do**
    form the Hamiltonian on $\mathcal{P}_f^{\text{eval}}$;
    use the Gauss SQ to find the electronic quantities at the ESPs $\mathcal{P}_c$;
    find the correctors at the ESPs $\mathcal{P}_c$;
    update the corrector on $\mathcal{P}_f^{\text{eval}}$;
    check convergence
  end
  compute the forces on the atoms;
  check equilibrium
end

12.6.3 Selected Results

We now demonstrate the approach using a few selected examples from magnesium, which forms a hexagonal close-packed (HCP) crystal structure. Magnesium and its alloys have received recent interest due to their high strength to weight ratio (with a density of 1.8 g/cm^3 and yield strength exceeding 100 MPa), and have been explored for automotive, biomedical and other engineering applications. However, these alloys often have limited ductility and suffer sudden, almost brittle, failure. We refer the reader to recent reviews [39, 46, 47, 93]. Therefore the study of defects in magnesium and its alloys have been a topic of much recent interest.

These examples are drawn from [66, 67]. We take the exchange-correlation function to be the parametrized form of Perdew and Wang [64], and a local pseudopotential proposed by Huang and Carter [35]. We take $\sigma = 0.8$ eV corresponding to a temperature of 10,000 K. We use a sixth-order finite difference stencil adopted to hexagonal symmetry that combines a triangular stencil on the basal plane with a normal stencil normal to it [15]. The energy and force convergence thresholds are 10^{-5} and 10^{-3} eV Å^{-1} respectively.

Figure 12.10, adapted from [66], shows the capabilities of the proposed approach using a vacancy. Figure 12.10a shows the computed total energy of a series of calculations with various amounts of coarse-graining. The computational domain in each of these calculations consists of 93,312 atoms discretized with 8.1×10^7 nodes. The six calculations have a progressively larger number of electronic sampling points: we see that the total energy converges at about 1.8×10^5 electronic sampling points. In other words, a calculations with 1.8×10^5 degrees of freedom is able to correctly reproduce the energy of a calculation with 8.1×10^7 degrees of

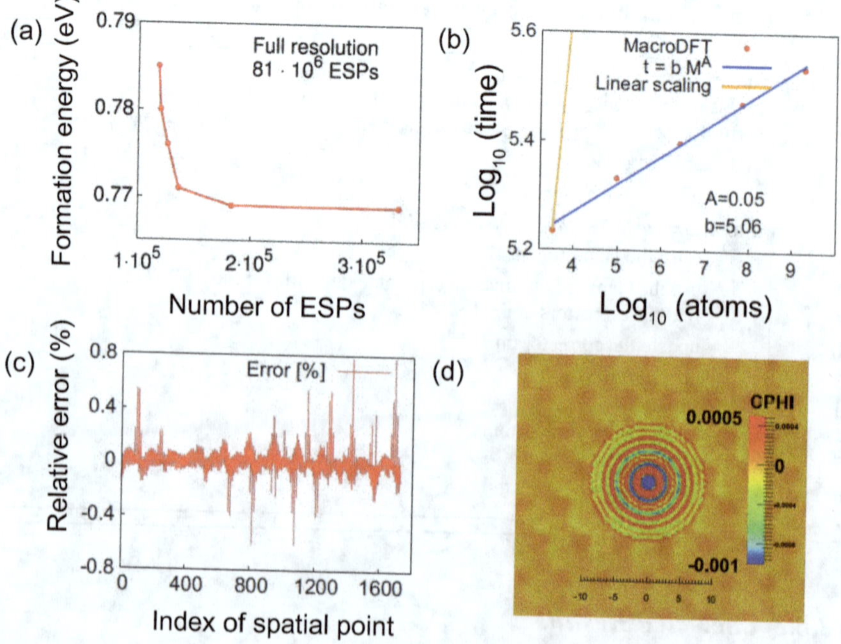

Fig. 12.10 Computational results from the study of a vacancy in HCP magnesium. (**a**) The computed total energy of a specimen with 93,312 atoms discretized with 8.1×10^7 nodes computed with various levels of coarse-graining. (**b**) Computational time vs domain size shows dramatic sublinear scaling. (**c**) Relative error between the electron density computed with and without coarse-graining at 1600 grid points. (**d**) The corrector electron density on the basal plane shows fine oscillations close to the vacancy. Reprinted from [66] with permission from Elsevier

freedom, a saving factor of 440. Remarkably, this factor increases as the size of the computational domain increases since larger domains have larger regions of coarser discretization.

Consequently one obtains *dramatic sub-linear performance* as shown in Fig. 12.10b. In this example, also with a vacancy, we see that the computational time t scales as a power law of the number of atoms M with an exponent 0.05 ($t = bM^a$, $a = 0.05$) up to a billion atoms. Of course simplicity of the example where the defect is confined to a small area contributes to the remarkable sublinearity, but we expect at least square-root scaling in all examples of defects.

Importantly, this saving in computational cost does not come at the cost of accuracy. This is demonstrated in Fig. 12.10c. This shows the relative error $(\rho_n^{\mathrm{CG}} - \rho_n^{\mathrm{full}})/\rho_n^{\mathrm{full}}$ at the nth grid point where ρ_n^{CG} is the electron density computed by the coarse-grained approximation (by recourse or (12.46)) and ρ_n^{full} is the electron density computed without any coarse-graining over about 1600 grid points. We observe that the relative error is less than 0.8% in any of these grid points. In fact the average and root-mean-square errors are 10^{-5} times the mean density.

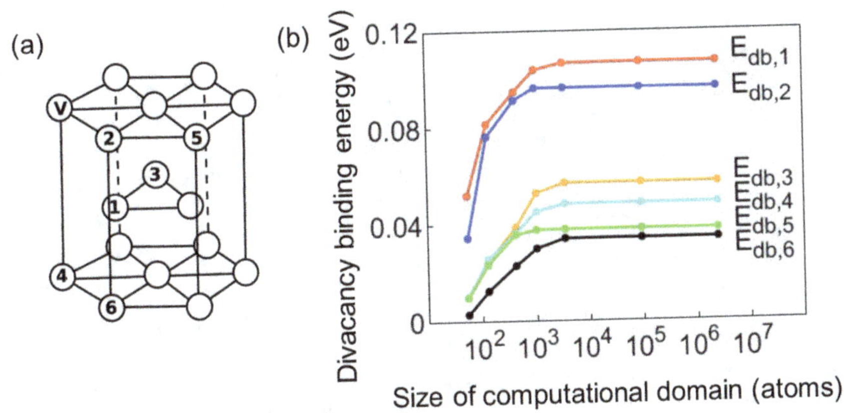

Fig. 12.11 Computational results from the study of divacancies in HCP magnesium. (**a**) Various divacancy complexes: one vacancy is located at the site labeled V while the others are at sites labeled with numbers. (**b**) Computed divacancy binding energy of various divacancy complexes for computational domains of varying sizes. Reprinted from [66] with permission from Elsevier

This efficacy of the coarse-graining method shows that subgrid sampling can be effective away from the defects. However, the details are complex and important near the core and require full resolution. Figure 12.10d shows the corrector electron density on the basal plane in the vicinity of the vacancy. We see oscillations on a scale finer than the atomic spacing—these are the analogs of the Friedel oscillations on interfaces and contribute to the electronic character of the defects. Therefore, it is important to resolve these carefully. Further, they interact with the far field stresses, and one reason why the decay length of defects tend to be high and why defects require large computational cells.

We now turn to the importance of sufficiently large computational unit cells in accurately calculating the binding energy of a divacancy. The binding energy is the energy difference between two isolated vacancies and a divacancy complex. This is illustrated in Fig. 12.11, adapted from [66]: it shows the divacancy binding energy of various divacancy complexes computed with computational domains of varying sizes. We see that we need a sufficiently large computational domain with $>10^3$ atoms to accurately predict the divacancy binding energy. Importantly, the result leads to qualitative differences: calculations with small computational domains incorrectly predict that some vacancies barely bind, while the large computational domains predict strong binding consistent with experimental observations [38, 50, 85, 89].

The final example is adapted from Ref. [67] and concerns the study of dislocations. Recall that the elastic energy of a dislocation scales logarithmically with the size of the domain. Figure 12.12 shows the computed excess energy—the difference in total energy between a domain with a dislocation and a domain without for two types of screw dislocations for domains of various sizes. It shows that our coarse-

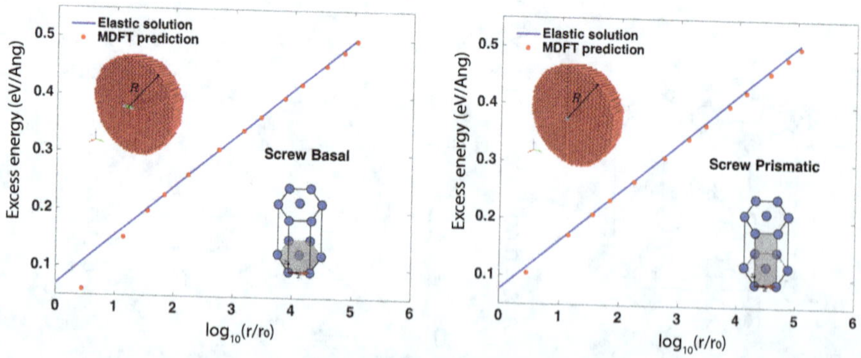

Fig. 12.12 Computational results from the study of screw dislocations in HCP magnesium. Scaling of the excess energy computed with various domain sizes for (**a**) basal $\frac{a_0}{3}[11\bar{2}0]\{0001\}$ screw and (**b**) $\frac{a_0}{3}[11\bar{2}0]\{10\bar{1}0\}$ prismatic dislocations. Reprinted from [67] with permission from Elsevier

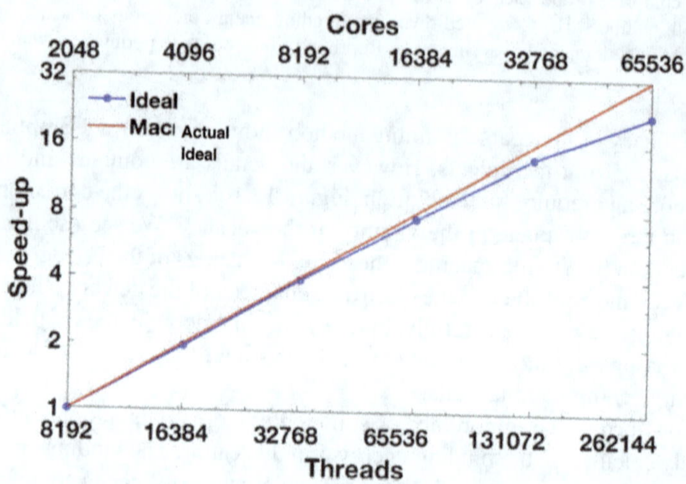

Fig. 12.13 Parallel performance in a benchmark problem of a seven vacancy cluster. Reprinted from [67] with permission from Elsevier

grained DFT approach correctly predicts this elastic scaling. The details (see [67]) describe the core structure, and the intercept at $r = r_0$ provides the "core energy".

We end by noting the excellent numerical performance with respect to parallelization in Fig. 12.13 in a benchmark problem of a seven vacancy cluster obtained on MIRA, an IBM BG/Q 1.6 GHz PowerPC A2 supercomputer at Argonne National Laboratory.

Acknowledgments We are grateful to Phani Motamarri for sharing the unpublished results shown in Table 12.1. We acknowledge the help of Arpit Bhardwaj, Sambit Das and Xin Jing in running some of the DFT-FE and SQ simulations, and generating the corresponding figures. KB, MO and MP acknowledge the support of the Army Research Laboratory under Cooperative Agreement Number W911NF-12-2-0022. VG acknowledges the support of the U.S. Department of Energy, Office of Science through grants DE-SC0008637 and DE-SC0017380. VG also gratefully acknowledges the support of the Army Research Office through the DURIP grant W911NF1810242. PS acknowledges support of the U.S. Department of Energy, Office of Science through grant DE-SC0019410. The computations presented here were conducted on the Resnick High Performance Cluster at Caltech, the Great Lakes High Performance Cluster at the University of Michigan, the Oak Ridge Leadership Computing Facility, a DOE Office of Science User Facility operated by the Oak Ridge National Laboratory under contract DE-AC05-00OR22725, and the National Energy Research Scientific Computing Center, a DOE Office of Science User Facility supported by the Office of Science of the U.S. Department of Energy under Contract No. DE-AC02-05CH11231. The views and conclusions contained in this document are those of the authors and should not be interpreted as representing the official policies, either expressed or implied, of the Army Research Laboratory, Department of Energy, or the U.S. Government. The U.S. Government is authorized to reproduce and distribute reprints for Government purposes notwithstanding any copyright notation herein.

Appendix: Crystalline solids and the Cauchy–Born Rule

A *Bravais lattice* is a lattice with a single atom in its unit cell:

$$\mathcal{L}_B(a_i, o) = \{r \in \mathbb{R}^3 : r = \sum_{i=1}^{3} \nu^i a_i,\ \ \nu_i \text{ integers}\}$$

where a set of linearly independent vectors or *lattice vectors* $\{a_i\}_{i=1}^3$ describes the unit cell, or translational symmetry, and o signifies the presence of an atom at the origin. A *crystal* (also called lattice with a basis) is a periodic arrangement of atoms (points) in $\mathbb{R}^3$ with a finite number M of atoms in the unit cell. It may be regarded as a union of P congruent Bravais lattices which are displaced from each other:

$$\mathcal{L}(a_i, p_\alpha) = \cup_{\alpha=1}^{M} \mathcal{L}_B(a_i, p_\alpha)$$

where $\{a_i\}_{i=1}^3$ are the lattice vectors and the shift vectors $p_\alpha, \alpha = 1, \ldots, M$ describe the relative positions of the atoms within the unit cell. It is conventional to take $p_1 = o$, but this is not necessary. The underlying Bravais lattice is often referred to as the skeletal lattice.

A *crystalline solid* is a restriction of a lattice to a domain $\Omega \subset \mathbb{R}^3$. Let $\{r_a\}_{a=1}^A$ denote the positions of the atoms in a crystalline solid $\mathcal{L}(a_i^0, p_\alpha^0) \cap \Omega$ in the reference domain $\Omega \subset \mathbb{R}^3$. As the solid deforms, the current position of the atoms are given by $\{y_a\}_{a=1}^A \subset \mathbb{R}^3$. Let $y : \Omega \to \mathbb{R}^3$ denote a smooth deformation that maps the positions of the underlying skeletal lattice, i.e., $y_a = y(r_a)\ \forall\ r_a \in \mathcal{L}_B(a_i^0, p_1^0)$. We call y the *macroscopic deformation*. Now, if the scale of the lattice is small compared to the size of the domain, and if the deformation y varies slowly on the scale of the lattice, i.e., it may be approximated by an affine map of a scale large compared to that of a_i, then at any $r_0 \in \Omega$, the current positions of the atoms in the neighborhood of $y(r_0)$ are arranged in a lattice $\mathcal{L}(a_i, q_\alpha)$ where

$$a_i = \nabla y(r_0) a_i^0.$$

In other words, for moderate macroscopic deformations, the deformation gradient convects the lattice vectors. This is known as the *Cauchy–Born rule*. Note that the macroscopic deformation only constrains the skeletal Bravais lattice and the atoms are free to “shuffle” within the unit cell.

References

1. F.F. Abraham, J.Q. Broughton, N. Bernstein and E. Kaxiras. Spanning the length scales in dynamic simulation. *Comput. Phys.* **12**, 538–546 (1998).
2. C. Albrecht, C. Klaar, J.E. Pask, M.A. Schweitzer, N. Sukumar and A. Ziegenhagel. Orbital-enriched flat-top partition of unity method for the Schrödinger eigenproblem. *Comput. Methods Appl. Mech. Eng.* **342**, 224–239 (2018).
3. A. Anantharaman and E. Cancès. Existence of minimizers for Kohn–Sham models in quantum chemistry. *Ann. Inst. Henri Poincaré (C) Anal. Non-Lineaire* **26**(6), 2425–2455 (2009).
4. R. Baer and M. Head-Gordon. Chebyshev expansion methods for electronic structure calculations on large molecular systems. *J. Chem. Phys.* **107** (1997).
5. M. Benzi, P. Boito and N. Razouk. Decay properties of spectral projectors with applications to electronic structure. *SIAM Rev.* **55**(1), 3–64 (2013).
6. N. Bernstein, J.R. Kermode and G. Csanyi. Hybrid atomistic simulation methods for materials systems. *Rep. Prog. Phys.* **72**(2), 026501 Jan. (2009).
7. M. Bethkenhagen, A. Sharma, P. Suryanarayana, J.E. Pask, B. Sadigh and S. Hamel. Properties of carbon up to 10 million kelvin from Kohn-Sham density functional theory molecular dynamics. *Phys. Rev. E* **107**(1), 015306 (2023). https://doi.org/10.1103/PhysRevE.107.015306.
8. P.E. Blöchl. Projector augmented-wave method. *Phys. Rev. B* **50**, 17953–17979 Dec. (1994).
9. V. Blum, R. Gehrke, F. Hanke, P. Havu, V. Havu, X. Ren, K. Reuter and M. Scheffler. *Ab initio* molecular simulations with numeric atom-centered orbitals. *Comput. Phys. Commun.* **180**(11), 2175–2196 (2009).
10. J.P. Boyd. *Chebyshev and Fourier spectral methods*. Dover (2001).
11. K. Burke. Perspective on density functional theory. *J. Chem. Phys.* 136, 150901, Apr. (2012).
12. H. Chen, X. Dai, X. Gong, L. He and A. Zhou. Adaptive Finite Element Approximations for Kohn–Sham Models. *Multiscale Model. Simul.* **12**(4), 1828–1869 (2014).
13. S. Das, P. Motamarri, V. Gavini, B. Turcksin, Y.W. Li and B. Leback. Fast, scalable and accurate finite-element based ab initio calculations using mixed precision computing: 46 PFLOPS simulation of a metallic dislocation system. In *Proceedings of the 6th European Conference on Computer Systems*, number 2, pp. 1–11 (2019).
14. D. Davydov, T.D. Young and P. Steinmann. On the adaptive finite element analysis of the Kohn–Sham equations: methods, algorithms, and implementation. *Int. J. Numer. Methods Eng.* **106**(11), 863–888 (2016).
15. B. Fornberg. *A practical guide to pseudospectral methods*. Number 1. Cambridge Univ. Press (1998).

16. C. J. García-Cervera, J. Lu, Y. Xuan and W. E. Linear-scaling subspace-iteration algorithm with optimally localized nonorthogonal wave functions for Kohn–Sham density functional theory. *Phys. Rev. B* **79**, 115110 Mar. (2009).
17. V. Gavini, K. Bhattacharya and M. Ortiz. Quasi-continuum orbital-free density-functional theory: A route to multi-million atom non-periodic DFT calculation. *J. Mech. Phys. Solids* **55**(4), 697–718 (2007).
18. K. Ghosh, H. Ma, V. Gavini and G. Galli. All-electron density functional calculations for electron and nuclear spin interactions in molecules and solids. *Phys. Rev. Mater.* **3**, 043801 Apr. (2019).
19. K. Ghosh, H. Ma, M. Onizhuk, V. Gavini and G. Galli. Spin-spin interactions in defects in solids from mixed all-electron and pseudopotential first-principles calculations. *npj Computational Materials* **7**, 2021.
20. S. Ghosh and K. Bhattacharya. Spectral quadrature for the first principles study of crystal defects: Application to magnesium. *J. Comput. Phys.* **456**, 111035 (2022).
21. S. Ghosh and P. Suryanarayana. SPARC: Accurate and efficient finite-difference formulation and parallel implementation of Density Functional Theory: Extended systems. *Comput. Phys. Commun.* **216**, 109–125 (2017).
22. S. Ghosh and P. Suryanarayana. SPARC: Accurate and efficient finite-difference formulation and parallel implementation of Density Functional Theory: Isolated clusters. *Comput. Phys. Commun.* **212**, 189–204 (2017).
23. P. Giannozzi, S. Baroni, N. Bonini, M. Calandra, R. Car, C. Cavazzoni, D. Ceresoli, G. L. Chiarotti, M. Cococcioni, I. Dabo, A. Dal Corso, S. de Gironcoli, S. Fabris, G. Fratesi, R. Gebauer, U. Gerstmann, C. Gougoussis, A. Kokalj, M. Lazzeri, L. Martin-Samos, N. Marzari, F. Mauri, R. Mazzarello, S. Paolini, A. Pasquarello, L. Paulatto, C. Sbraccia, S. Scandolo, G. Sclauzero, A. P. Seitsonen, A. Smogunov, P. Umari and R. M. Wentzcovitch. QUANTUM ESPRESSO: a modular and open-source software project for quantum simulations of materials. *J. Phys. Condens. Matter* **21**(39), 395502 (2009).
24. S. Goedecker and L. Colombo. Efficient linear scaling algorithm for tight-binding molecular dynamics. *Phys. Rev. Lett.* **73**(1), 122 (1994).
25. S. Goedecker and M. Teter. Tight-binding electronic-structure calculations and tight-binding molecular dynamics with localized orbitals. *Phys. Rev. B* **51**(15), 9455 (1995).
26. G.H. Golub and G. Meurant. *Matrices, moments and quadrature with applications*. Princeton Univ. Press (2009).
27. X. Gonze, J.-M. Beuken, R. Caracas, F. Detraux, M. Fuchs, G.-M. Rignanese, L. Sindic, M. Verstraete, G. Zerah, F. Jollet, M. Torrent, A. Roy, M. Mikami, P. Ghosez, J.-Y. Raty and D. Allan. First-principles computation of material properties: the {ABINIT} software project. *Comput. Mater. Sci.* **25**(3), 478–492 (2002).
28. N. Hale, N.J. Higham and L.N. Trefethen. Computing A^α,\log(A), and related matrix functions by contour integrals. *SIAM J. Numer. Anal.* **46**(5), 2505–2523 (2008).
29. D.R. Hamann. Optimized norm-conserving Vanderbilt pseudopotentials. *Phys. Rev. B* **88**, 085117 Aug. (2013).
30. R. Haydock. *Solid State Phys.*, volume 35. Academic Press (1980).
31. P.D. Haynes, C.-K. Skylaris, A.A. Mostofi and M.C. Payne. ONETEP: linear-scaling density-functional theory with local orbitals and plane waves. *Phys. Status Solidi B* **243**(11), 2489–2499 (2006).
32. W.J. Hehre, R.F. Stewart and J.A. Pople. Self-Consistent molecular-orbital methods. I. Use of Gaussian expansions of Slater-type atomic orbitals. *J. Chem. Phys.* **51**(6), 2657–2664 (1969).
33. J.P. Hirth and J. Lothe. *Theory of dislocations*. Wiley, New York, 2nd edition (1982).
34. P. Hohenberg and W. Kohn. Inhomogeneous electron gas. *Phys. Rev. B* **136**, 864–871 (1964).
35. C. Huang and E.A. Carter. Transferable local pseudopotentials for magnesium, aluminum and silicon. *Phys. Chem. Chem. Phys.* **10**(47), 7109–7120 (2008).
36. J. Hutter, M. Iannuzzi, F. Schiffmann and J. VandeVondele. CP2K: Atomistic simulations of condensed matter systems. *Wiley Interdiscip. Rev.: Comput. Mol. Sci.* **4** (2014).

37. S. Ismail-Beigi and T. Arias. New Algebraic Formulation of Density Functional Calculation. *Comput. Phys. Commun.* **128**, 1–45 (2000).
38. C. Janot, D. Malléjac and B. George. Vacancy-formation energy and entropy in magnesium single crystals. *Phys. Rev. B* **2**(8), 3088 (1970).
39. W.J. Joost and P.E. Krajewski. Towards magnesium alloys for high-volume automotive applications. *Scr. Mater.* **128**, 107–112 (2017).
40. B. Kanungo and V. Gavini. Large-scale all-electron density functional theory calculations using an enriched finite-element basis. *Phys. Rev. B* **95**, 035112 Jan. (2017).
41. J. Knap and M. Ortiz. An analysis of the quasicontinuum method. *J. Mech. Phys. Solids* **49**(9), 1899–1923 (2001).
42. W. Kohn. Density functional and density matrix method scaling linearly with the number of atoms. *Phys. Rev. Lett.* **76**(17), 3168 (1996).
43. W. Kohn and L.J. Sham. Self-Consistent Equations Including Exchange and Correlation Effects. *Phys. Rev. A* **140**, 1133–1138 (1965).
44. G. Kresse and J. Furthmüller. Efficient iterative schemes for ab initio total-energy calculations using a plane-wave basis set. *Phys. Rev. B* **54**(16), 11169–11186 (1996).
45. L. Kronik, A. Makmal, M. L. Tiago, M. M. G. Alemany, M. Jain, X. Huang, Y. Saad and J. R. Chelikowsky. PARSEC — the pseudopotential algorithm for real-space electronic structure calculations: recent advances and novel applications to nano-structures. *Phys. Status Solidi B* **243**(5), 1063–1079 (2006).
46. M.K. Kulekci. Magnesium and its alloys applications in automotive industry. *Int. J. Adv. Manuf. Technol.* **39**, 851–865 (2008).
47. K. Kuśnierczyk and M. Basista. Recent advances in research on magnesium alloys and magnesium-calcium phosphate composites as biodegradable implant materials. *J Biomater Appl.* **31**(6), 878–900 (2017).
48. X.-P. Li, R.W. Nunes and D. Vanderbilt. Density-matrix electronic-structure method with linear system-size scaling. *Phys. Rev. B* 47, 10891–10894 Apr. (1993).
49. L. Lin, M. Chen, C. Yang and Y. He. Accelerating atomic orbital-based electronic structure calculation via pole expansion and selected inversion. *J. Phys. Condens. Matter* **25**(29) (2013).
50. C. Mairy, J. Hillairet and D. Schumacher. Energie de formation et concentration d'équilibre des lacunes dans le magnésium. *Acta Metall.* **15**(7), 1258–1261 (1967).
51. V. Michaud-Rioux, L. Zhang and H. Guo. RESCU: A real space electronic structure method. *J. Comput. Phys.* **307**, 593–613 (2016).
52. P. Motamarri, Personal communication with the authors (2021).
53. P. Motamarri, S. Das, S. Rudraraju, K. Ghosh, D. Davydov and V. Gavini. DFT-FE - A massively parallel adaptive finite-element code for large-scale density functional theory calculations. *Comput. Phys. Commun.* **246**, 106853 (2020).
54. P. Motamarri and V. Gavini. Subquadratic-scaling subspace projection method for large-scale Kohn–Sham density functional theory calculations using spectral finite-element discretization. *Phys. Rev. B* **90**, 115127 Sep. (2014).
55. P. Motamarri and V. Gavini. Configurational forces in electronic structure calculations using Kohn–Sham density functional theory. *Phys. Rev. B* **97**, 165132 Apr. (2018).
56. P. Motamarri, V. Gavini, K. Bhattacharya and M. Ortiz. Spectrum-splitting approach for Fermi-operator expansion in all-electron Kohn–Sham DFT calculations. *Phys. Rev. B* **95**, 035111 Jan. (2017).
57. P. Motamarri, M. Nowak, K. Leiter, J. Knap and V. Gavini. Higher-order adaptive finite-element methods for Kohn–Sham density functional theory. *J. Comput. Phys.* **253**, 308–343 (2013).
58. T. Mura. *Micromechanics of Defects in Solids*. Nijhoff Publishers (1987).
59. A.M.N. Niklasson. Implicit purification for temperature-dependent density matrices. *Phys. Rev. B* **68**, 233104 Dec. (2003).
60. R. Parr and W. Yang. *Density-Functional Theory of Atoms and Molecules*. Oxford Univ. Press (1994).

61. J. Pask and N. Sukumar. Partition of unity finite element method for quantum mechanical materials calculations. *Extreme Mech. Lett.* **11**, 8–17 (2017).
62. J.E. Pask, B.M. Klein, C.Y. Fong and P.A. Sterne. Real-space local polynomial basis for solid-state electronic-structure calculations: A finite-element approach. *Phys. Rev. B* **59**, 12352–12358 (1999).
63. J.E. Pask and P.A. Sterne. Finite element methods in ab initio electronic structure calculations. *Modell. Simul. Mater. Sci. Eng.* **13**(3), R71 (2005).
64. J.P. Perdew and Y. Wang. Accurate and simple analytic representation of the electron-gas correlation energy. *Phys. Rev. B* **45**, 13244 (1992).
65. R. Phillips. *Crystals, defects and microstructures: Modeling across scales*. Cambridge Univ. Press (2001).
66. M. Ponga, K. Bhattacharya and M. Ortiz. A sublinear-scaling approach to density-functional-theory analysis of crystal defects. *J. Mech. Phys. Solids* **95**, 530–556 Oct. (2016).
67. M. Ponga, K. Bhattacharya and M. Ortiz. Large scale ab-initio simulations of dislocations. *J. Comput. Phys.*, 1–25 (2020).
68. P.P. Pratapa, P. Suryanarayana and J.E. Pask. Spectral Quadrature method for accurate $O(N)$ electronic structure calculations of metals and insulators. *Comput. Phys. Commun.* **200**, 96–107 (2016).
69. E. Prodan and W. Kohn. Nearsightedness of electronic matter. *PNAS* **102**(33), 11635–11638 (2005).
70. W. Rudin. *Functional Analysis*. McGraw-Hill, Boston (1991).
71. N.D. Rufus, B. Kanungo and V. Gavini. Fast and robust all-electron density functional theory calculations in solids using orthogonalized enriched finite elements. *Phys. Rev. B* **104**, 085112 Aug. (2021).
72. A. Sharma, S. Hamel, M. Bethkenhagen, J.E. Pask and P. Suryanarayana. Real-space formulation of the stress tensor for $O(N)$ density functional theory: Application to high temperature calculations. *J. Chem. Phys.* **153**(3), 034112 (2020).
73. D. Singh and L. Nordström. *Plane waves, pseudopotentials and the LAPW method*. Springer Verlag (2006).
74. S. P. Suetin. Padé approximants and efficient analytic continuation of a power series. *Russian Math. Surv.* **57**(1), 43 (2002).
75. P. Suryanarayana. On spectral quadrature for linear-scaling Density Functional Theory. *Chem. Phys. Lett.* **584**, 182–187 (2013).
76. P. Suryanarayana. Optimized purification for density matrix calculation. *Chem. Phys. Lett.* **555**, 291–295 (2013).
77. P. Suryanarayana. On nearsightedness in metallic systems for $O(N)$ Density Functional Theory calculations: A case study on aluminum. *Chem. Phys. Lett.* **679**, 146–151 (2017).
78. P. Suryanarayana, K. Bhattacharya and M. Ortiz. Coarse-graining Kohn–Sham Density Functional Theory. *J. Mech. Phys. Solids* **61**(1), 38–60 (2013).
79. P. Suryanarayana, P.P. Pratapa, A. Sharma and J.E. Pask. SQDFT: Spectral Quadrature method for large-scale parallel $O(N)$ Kohn–Sham calculations at high temperature. *Comput. Phys. Commun.* **224**, 288–298 (2018).
80. E.B. Tadmor, M. Ortiz and R. Phillips. Quasicontinuum analysis of defects in solids. *Philos. Mag. A* **73**(6), 1529–1563 (1996).
81. L.N. Trefethen. Is Gauss quadrature better than Clenshaw–Curtis? *SIAM Rev.* **50**(1), 67–87 (2008).
82. L.N. Trefethen. *Approximation Theory and Approximation Practice, Extended Edition*. SIAM (2019).
83. E. Tsuchida and M. Tsukada. Electronic-structure calculations based on the finite-element method. *Phys. Rev. B* **52**, 5573–5578 (1995).
84. E. Tsuchida and M. Tsukada. Adaptive finite-element method for electronic-structure calculations. *Phys. Rev. B* **54**, 7602–7605, Sep. (1996).
85. P. Tzanetakis, J. Hillairet and G. Revel. The formation energy of vacancies in aluminium and magnesium. *Phys. Status Solidi B* **75**(2), 433–439 (1976).

86. M. Valiev, E. Bylaska, N. Govind, K. Kowalski, T. Straatsma, H.V. Dam, D. Wang, J. Nieplocha, E. Apra, T. Windus and W. de Jong. NWChem: A comprehensive and scalable open-source solution for large scale molecular simulations. *Comput. Phys. Commun.* **181**(9), 1477–1489 (2010).
87. W. Van Assche. Padé and Hermite-Padé approximation and orthogonality. *Surv. in Approx. Theory* **2**, 61–91 (2006).
88. D. Vanderbilt. Soft self-consistent pseudopotentials in a generalized eigenvalue formalism. *Phys. Rev. B* **41**, 7892–7895, Apr. (1990).
89. A. Vehanen and K. Rytsölä. *Proceedings of the International School of Physics "Enrico Fermi," Course LXXXIII*, Varenna 1981, p. 659, North Holland, Amsterdam (1983)
90. X.-C. Wang, T. Blesgen, K. Bhattacharya and M. Ortiz. A Variational Framework for Spectral Approximations of Kohn–Sham Density Functional Theory. *Arch. Ration. Mech. Anal.* **221**(2), 1035–1075 (2016).
91. C. Woodward. First-principles simulations of dislocation cores. *Mater. Sci. Eng. A* **400–401**, 59–67 (2005).
92. C.J. Wu, P.C. Myint, J.E. Pask, C.J. Prisbrey, A.A. Correa, P. Suryanarayana and J. B. Varley. Development of a multiphase beryllium equation of state and physics-based variations. *J. Phys. Chem. A* **125**(7), 1610–1636 (2021).
93. C. Xianhua, G. Yuxiao and P. Fusheng. Research progress in magnesium alloys as functional materials. *Rare Metal Mater. Eng.* **45**, 2269–2274 (2016).
94. Q. Xu, A. Sharma and P. Suryanarayana. M-SPARC: Matlab-simulation package for ab-initio real-space calculations. *SoftwareX* **11**, 100423 (2020).
95. S. Yamakawa and S.-a. Hyodo. Gaussian finite-element mixed-basis method for electronic structure calculations. *Phys. Rev. B* **71**, 035113 Jan. (2005).
96. S. Zhang, A. Lazicki, B. Militzer, L.H. Yang, K. Caspersen, J.A. Gaffney, M.W. Däne, J.E. Pask, W.R. Johnson, A. Sharma, et al. Equation of state of boron nitride combining computation, modeling, and experiment. *Phys. Rev. B* **99**(16), 165103 (2019).
97. X. Zhang, G. Lu and W. Curtin. Multiscale quantum/atomistic coupling using constrained density functional theory. *Phys. Rev. B* **87**, 054113 (2013).
98. Y. Zhou, Y. Saad, M.L. Tiago and J.R. Chelikowsky. Parallel self-consistent-field calculations via Chebyshev-filtered subspace acceleration. *Phys. Rev. E* **74**, 066704 Dec. (2006).
99. Y. Zhou, Y. Saad, M.L. Tiago and J.R. Chelikowsky. Self-consistent-field calculations using Chebyshev-filtered subspace iteration. *J. Comput. Phys.* **219**(1), 172–184 Nov. (2006).

Glossary

DFT Density functional theory. Has its origin in the work of Thomas, Fermi, and Dirac in the 1920s. In its modern form, it was introduced by Hohenberg, Kohn and Sham in 1964/1965. Encompasses both *exact DFT*, which re-formulates the quantum mechanical ground state problem for many-electron systems in terms of the electron density, and *Kohn–Sham DFT*, which is concerned with computationally tractable approximations of the ground state problem in terms of the electron density.

Exchange-correlation functional Functional giving the exchange-correlation energy or an approximation thereof in terms of the single-particle density. Often denoted E_{xc}.

Exchange energy functional Exchange contribution to the exchange-correlation functional. Often denoted E_{x}.

Correlation functional Correlation contribution to the exchange-correlation functional. Often denoted E_{c}.

Exchange-correlation potential Functional derivative of the exchange-correlation functional with respect to the density. Often denoted $v_{\mathrm{xc}}[\rho]$. Together with the Hartree potential and the external potential, constitutes the effective one-body potential in the Kohn–Sham equations.

Exchange potential Exchange contribution to the exchange-correlation potential. Often denoted v_{x}.

Correlation potential Correlation contribution to the exchange-correlation potential. Often denoted v_{c}.

Hartree functional Classical Coulomb self-repulsion energy of the electron density as a functional of the latter. Important ingredient in most Kohn–Sham DFT models. Often denoted E_{H}.

Hartree potential Electrostatic potential of the electron density. Also arises as the functional derivative of the Hartree functional with respect to the density. Often denoted v_{H}.

E. Cancès, G. Friesecke (eds.), *Density Functional Theory*, Mathematics and Molecular Modeling, https://doi.org/10.1007/978-3-031-22340-2

Hartree-exchange-correlation functional Sum of the Hartree and exchange-correlation functional. Often denoted E_{Hxc}.

Hartree-exchange-correlation potential Sum of the Hartree and exchange-correlation potential. Together with the external potential, constitutes the effective one-body potential in the Kohn–Sham equations. Often denoted v_{Hxc}.

LDA Local density approximation to the exchange-correlation functional and potential, obtained by approximating the exchange-correlation energy density of an inhomogeneous system at a point by the exchange-correlation energy density of the uniform electron gas of the same density. Often denoted, respectively, $E_{\mathrm{xc}}^{\mathrm{LDA}}$ and $v_{\mathrm{xc}}^{\mathrm{LDA}}$.

SCE functional Strictly correlated electrons functional. Minimum Coulomb interaction energy of a many-electron system as a functional of its density. Corresponds to the exact Hartree-exchange-correlation functional in the strong-interaction limit.

www.ingramcontent.com/pod-product-compliance
Lightning Source LLC
Chambersburg PA
CBHW051651290626
47387CB00003B/290
* 9 7 8 3 0 3 1 2 2 3 3 9 6 *